Theoretical and Analytical Service-Focused Systems Design and Development

Dickson K.W. Chiu
Dickson Computer Systems, Hong Kong

Managing Director: Lindsay Johnston
Senior Editorial Director: Heather A. Probst
Book Production Manager: Sean Woznicki
Development Manager: Joel Gamon
Acquisitions Editor: Erika Gallagher
Typesetter: Russell A. Spangler
Cover Design: Nick Newcomer, Lisandro Gonzalez

Published in the United States of America by
Information Science Reference (an imprint of IGI Global)
701 E. Chocolate Avenue
Hershey PA 17033
Tel: 717-533-8845
Fax: 717-533-8661
E-mail: cust@igi-global.com
Web site: http://www.igi-global.com

Library of Congress Cataloging-in-Publication Data

Theoretical and analytical service-focused systems design and development / Dickson K.W. Chiu, editor.
p. cm.
Includes bibliographical references and index.
Summary: "This book provides solutions to these challenges, practices and understanding of contemporary theories and empirical analysis for systems engineering in a way that achieves service excellence"--Provided by publisher.
ISBN 978-1-4666-1767-4 (hardcover) -- ISBN 978-1-4666-1768-1 (ebook) -- ISBN 978-1-4666-1769-8 (print & perpetual access) 1. Systems engineering. I. Chiu, Dickson K. W., 1966-
TA168.T48 2012
620'.0042--dc23

2012002480

British Cataloguing in Publication Data
A Cataloguing in Publication record for this book is available from the British Library.

International Advisory Board

Associate Editors

Table of Contents

Preface xx

Acknowledgment xxxii

Section 1
Emerging Technologies and Killer Applications

Chapter 1
Engineering e-Collaboration Services with a Multi-Agent System Approach 1
Dickson K.W. Chiu, Dickson Computer Systems, Hong Kong
S.C. Cheung, The Hong Kong University of Science and Technology, Hong Kong
Ho-fung Leung, The Chinese University of Hong Kong, Hong Kong
Patrick C.K. Hung, University of Ontario Institute of Technology, Canada
Eleanna Kafeza, Athens University of Economics and Business, Greece
Hua Hu, Hangzhou Dianzi University, China
Minhong Wang, The University of Hong Kong, Hong Kong
Haiyang Hu, Zhejiang Gongshang University, China
Yi Zhuang, Zhejiang Gongshang University, China

Chapter 2
A Survey of Web Services Provision 24
An Liu, University of Science & Technology of China, CityU-USTC Advanced Research Institute, and City University of Hong Kong, China
Hai Liu, University of Science & Technology of China, CityU-USTC Advanced Research Institute, and City University of Hong Kong, China
Baoping Lin, University of Science & Technology of China, CityU-USTC Advanced Research Institute, and City University of Hong Kong, China
Liusheng Huang, University of Science & Technology of China, and CityU-USTC Advanced Research Institute, China
Naijie Gu, University of Science & Technology of China, and CityU-USTC Advanced Research Institute, China
Qing Li, CityU-USTC Advanced Research Institute, and City University of Hong Kong, China

Detailed Table of Contents

Dickson K.W. Chiu, Dickson Computer Systems, Hong Kong
S.C. Cheung, The Hong Kong University of Science and Technology, Hong Kong
Ho-fung Leung, The Chinese University of Hong Kong, Hong Kong
Patrick C.K. Hung, University of Ontario Institute of Technology, Canada
Eleanna Kafeza, Athens University of Economics and Business, Greece
Hua Hu, Hangzhou Dianzi University, China
Minhong Wang, The University of Hong Kong, Hong Kong
Haiyang Hu, Zhejiang Gongshang University, China
Yi Zhuang, Zhejiang Gongshang University, China

With recent advances in mobile technologies and e-commerce infrastructures, there have been increasing demands for the expansion of collaboration services within and across systems. In particular, human collaboration requirements should be considered together with those for systems and their components. Agent technologies have been deployed in order to model and implement e-commerce activities as multi-agent systems (MAS). Agents are able to provide assistance on behalf of their users or systems in collaboration services. As such, we advocate the engineering of e-collaboration support by means of MAS in the following three key dimensions: (i) across multiple platforms, (ii) across organization boundaries, and (iii) agent-based intelligent support. To archive this, we present a MAS infrastructure to facilitate systems and human collaboration (or e-collaboration) activities based on the belief-desire-intension (BDI) agent architecture, constraint technology, and contemporary Web Services. Further, the MAS infrastructure also provides users with different options of agent support on different platforms. Motivated by the requirements of mobile professional workforces in large enterprises, the authors present their development and adaptation methodology for e-collaboration services with a case study of constraint-based collaboration protocol from a three-tier implementation architecture aspect. They evaluate our approach from the perspective of three main stakeholders of e-collaboration, which include users, management, and systems developers.

An Liu, University of Science & Technology of China, CityU-USTC Advanced Research Institute, and City University of Hong Kong, China
Hai Liu, University of Science & Technology of China, CityU-USTC Advanced Research Institute, and City University of Hong Kong, China
Baoping Lin, University of Science & Technology of China, CityU-USTC Advanced Research Institute, and City University of Hong Kong, China
Liusheng Huang, University of Science & Technology of China, and CityU-USTC Advanced Research Institute, China
Naijie Gu, University of Science & Technology of China, and CityU-USTC Advanced Research Institute, China
Qing Li, CityU-USTC Advanced Research Institute, and City University of Hong Kong, China

Web services technologies promise to create new business applications by composing existing services and to publish these applications as services for further composition. The business logic of applications is described by abstract processes consisting of tasks which specify the required functionality. Web services provision refers to assigning concrete Web services to perform the constituent tasks of abstract processes. It describes a promising scenario where Web services are dynamically chosen and invoked according to their up-to-date functional and non-functional capabilities. It introduces many challenging problems and has therefore received much attention. In this article, the authors provide a comprehensive overview of current research efforts. The authors divide the lifecycle of Web services provision into three steps: service discovery, service selection, and service contracting. They also distinguish three types of Web services provision according to the functional relationship between services and tasks: independent provision, cooperative provision and multiple provision. Following this taxonomy, we investigate existing works in Web services provision, discuss open problems, and shed some light on potential research directions.

V. Pouli, National Technical University of Athens, Greece
C. Marinos, National Technical University of Athens, Greece
M. Grammatikou, National Technical University of Athens, Greece
S. Papavassiliou, National Technical University of Athens, Greece
V. Maglaris, National Technical University of Athens, Greece

Traditionally, network Service Providers specify Service Level Agreements (SLAs) to guarantee service availability and performance to their customers. However, these SLAs are rather static and span a single provider domain. Thus, they are not applicable to a multi–domain environment. In this paper, the authors present a framework for automatic creation and management of SLAs in a multi-domain environment. The framework is based on Service Oriented Computing (SOC) and contains a collection of web service calls and modules that allow for the automatic creation, configuration, and delivery of an end-to-end SLA, created from the merging of the per-domain SLAs. This paper also presents a monitoring procedure to monitor the QoS guarantees stipulated in the SLA. The SLA establishment and monitoring procedures are tested through a Grid application scenario targeted to perform remote control and monitoring of instrument elements distributed across the Grid.

Chapter 4
Execution Management for Mobile Service-Oriented Environments 62
Kleopatra Konstanteli, National Technical University of Athens, Greece
Tom Kirkham, Nottingham University, UK
Julian Gallop, STFC Rutherford Appleton Laboratory, UK
Brian Matthews, STFC Rutherford Appleton Laboratory, UK
Ian Johnson, STFC Rutherford Appleton Laboratory, UK
Magdalini Kardara, National Technical University of Athens, Greece
Theodora Varvarigou, National Technical University of Athens, Greece

This paper presents an Execution Management System (EMS) for Grid services that builds on the Open Grid Services Architecture (OGSA) while achieving "mobile awareness" by establishing a WS-Notification mechanism with mobile network session middleware. It builds heavily on the Session Initiation Protocol (SIP), used for managing sessions with mobile terminals (such as laptops and PDAs) where the services are running. Although the management of mobile services is different to that of ubiquitous services, the enhanced EMS manages both of them in a seamless fashion and incorporates all resources into one Mobile Dynamic Virtual Organization (MDVO). The described EMS has been implemented within the framework of the Akogrimo EU IST project and has been used to support mission critical application scenarios in public demonstrations, including composite and distributed applications made of both ubiquitous and mobile services within multiple domains.

Chapter 5
Developing Local Association Network Based IoT Solutions for Body Parts
Tagging and Tracking 83
Zongwei Luo, The University of Hong Kong, China
Martin Lai, The University of Hong Kong, China
Mary Cheung, The University of Hong Kong, China
ShuiHua Han, Xiamen University, China
Tianle Zhang, BUPT, China
Zhongjun Luo, FDA, USA
James W. M. Ting, The University of Hong Kong, China
Patrick W. L. Wong, The University of Hong Kong, China
Sam K. Y. Chan, The University of Hong Kong, China
Kwok Fai So, The University of Hong Kong, China
George L. Tipoe, The University of Hong Kong, China

Traditional Internet is commonly wired with machine to machine persistent connections. Evolving towards mobile and wireless pervasive networks, Internet has to entertain dynamic, transient, and changing interconnections. The vision of the Internet of Things furthers technology development by creating an interactive environment where smart objects are connected and can sense and react to the environment. Adopting such an innovative technology often requires extensive intelligence research. A major value indicator is how the potentials of RFID can translate into actions to improve business operational efficiency (Luo et al., 2008). In this paper, the authors will introduce a local association network with a coordinated P2P message delivery mechanism to develop Internet of Things based solutions body parts tagging and tracking. On site testing and performance evaluation validate the proposed approach. User feedback strengthens the belief that the proposed approach would help facilitate the technology adoption in body parts tagging and tracking.

Section 2
Intelligence Computing for Systems and Services

Chapter 6
Secure Key Generation for Static Visual Watermarking by Machine Learning in
Intelligent Systems and Services 106
Kensuke Naoe, Keio University, Japan
Hideyasu Sasaki, Ritsumeikan University, Japan
Yoshiyasu Takefuji, Keio University, Japan

The Service-Oriented Architecture (SOA) demands supportive technologies and new requirements for mobile collaboration across multiple platforms. One of its representative solutions is intelligent information security of enterprise resources for collaboration systems and services. Digital watermarking became a key technology for protecting copyrights. In this article, the authors propose a method of key generation scheme for static visual digital watermarking by using machine learning technology, neural network as its exemplary approach for machine learning method. The proposed method is to provide intelligent mobile collaboration with secure data transactions using machine learning approaches, herein neural network approach as an exemplary technology. First, the proposed method of key generation is to extract certain type of bit patterns in the forms of visual features out of visual objects or data as training data set for machine learning of digital watermark. Second, the proposed method of watermark extraction is processed by presenting visual features of the target visual image into extraction key or herein is a classifier generated in advance by the training approach of machine learning technology. Third, the training approach is to generate the extraction key, which is conditioned to generate watermark signal patterns, only if proper visual features are presented to the classifier. In the proposed method, this classifier which is generated by the machine learning process is used as watermark extraction key. The proposed method is to contribute to secure visual information hiding without losing any detailed data of visual objects or any additional resources of hiding visual objects as molds to embed hidden visual objects. In the experiments, they have shown that our proposed method is robust to high pass filtering and JPEG compression. The proposed method is limited in its applications on the positions of the feature sub-blocks, especially on geometric attacks like shrinking or rotation of the image.

Chapter 7
Modelling Self-Led Trust Value Management in Grid and Service
Oriented Infrastructures: A Graph Theoretic Social Network Mediated Approach........................... 122
Antony Brown, University of Bedfordshire, UK
Paul Sant, University of Bedfordshire, UK
Nik Bessis, University of Bedfordshire, UK
Tim French, University of Reading and University of Bedfordshire, UK
Carsten Maple, University of Bedfordshire, UK

Current developments in grid and service oriented technologies involve fluid and dynamic, ad hoc based interactions between delegates, which in turn, serves to challenge conventional centralised structured trust and security assurance approaches. Delegates ranging from individuals to large-scale VO (Virtual Organisations) require the establishment of trust across all parties as a prerequisite for trusted and meaningful e-collaboration. In this paper, a notable obstacle, namely how such delegates (modelled as nodes) operating within complex collaborative environment spaces can best evaluate

in context to optimally and dynamically select the most trustworthy ad hoc based resource/service for e-consumption. A number of aggregated service case scenarios are herein employed in order to consider the manner in which virtual consumers and provider ad hoc based communities converge. In this paper, the authors take the view that the use of graph-theoretic modelling naturally leads to a self-led trust management decision based approach in which delegates are continuously informed of relevant up-to-date trust levels. This will lead to an increased confidence level, which trustful service delegation can occur. The key notion is of a self-led trust model that is suited to an inherently low latency, decentralised trust security paradigm.

Chapter 8
Autonomic Business-Driven Dynamic Adaptation of Service-Oriented Systems and the WS-Policy4MASC Support for Such Adaptation 140
Vladimir Tosic, NICTA, The University of New South Wales, Australia and The University of Western Ontario, Canada

When a need for dynamic adaptation of an information technology (IT) system arises, often several alternative approaches can be taken. Maximization of technical quality of service (QoS) metrics (e.g., throughput, availability) need not maximize business value metrics (e.g., profit, customer satisfaction). The goal of autonomic business-driven IT system management (BDIM) is to ensure that operation and adaptation of IT systems maximizes business value metrics, with minimal human intervention. The author presents how his WS-Policy4MASC language for specification of management policies for service-oriented systems supports autonomic BDIM. WS-Policy4MASC extends WS-Policy with new types of policy assertions: goal, action, probability, utility, and meta-policy assertions. Its main distinctive characteristics are description of diverse business value metrics and specification of policy conflict resolution strategies for business value maximization according to various business strategies. The author's decision making algorithms use this additional WS-Policy4MASC information to choose the adaptation approach best from the business viewpoint.

Chapter 9
ODACE SLA: Ontology Driven Approach for Automatic Establishment of Service Level Agreements 157
Kaouthar Fakhfakh, LAAS-CNRS, Université de Toulouse, France and ReDCAD, University of Sfax, Tunisia
Tarak Chaari, ReDCAD and University of Sfax, Tunisia
Said Tazi, LAAS-CNRS and Université de Toulouse, France
Mohamed Jmaiel, ReDCAD and University of Sfax, Tunisia
Khalil Drira, LAAS-CNRS and Université de Toulouse, France

The establishment of Service Level Agreements between service providers and clients remains a complex task regarding the uninterrupted growth of the IT market. In fact, it is important to ensure a clear and fair establishment of these SLAs especially when providers and clients do not share the same technical knowledge. To address this problem, the authors started modeling client intentions and provider offers using ontologies. These models helped them in establishing and implementing a complete semantic matching approach containing four main steps. The first step consists of generating correspondences between the client and the provider terms by assigning certainties for their equivalence. The second step corrects and refines these certainties. In the third step, the authors evaluate the matching results using inference rules, and in the fourth step, a draft version of a Service Level Agreement is automatically generated in case of compatibility.

Chapter 10
Using Description Logics for the Provision of Context-Driven Content
Adaptation Services 176
Stephen J.H. Yang, National Central University, Taiwan
Jia Zhang, Northern Illinois University, USA
Jeff J.S. Huang, National Central University, Taiwan
Jeffrey J.P. Tsai, The University of Illinois at Chicago, USA

This article presents our design and development of a description logics-based planner for providing context-driven content adaptation services. This approach dynamically transforms requested Web content into a proper format conforming to receiving contexts (e.g., access condition, network connection, and receiving device). Aiming to establish a semantic foundation for content adaptation, we apply description logics to formally define context profiles and requirements. We also propose a formal Object Structure Model as the basis of content adaptation management for higher reusability and adaptability. To automate content adaptation decision, our content adaptation planner is driven by a stepwise procedure equipped with algorithms and techniques to enable rule-based context-driven content adaptation over the mobile Internet. Experimental results prove the effectiveness and efficiency of our content adaptation planner on saving transmission bandwidth, when users are using handheld devices. By reducing the size of adapted content, we moderately decrease the computational overhead caused by content adaptation.

Section 3
Knowledge Engineering for Systems and Services

Chapter 11
Two Heads Are Better Than One: Leveraging Web 2.0 for Business Intelligence 211
Ravi S. Sharma, Nanyang Technological University, Singapore
Dwight Tan, Nanyang Technological University, Singapore
Winston Cheng, Nanyang Technological University, Singapore

This paper examines how Web 2.0 may be used in organizations to support business intelligence activities. Five leading professional services firms in the Energy, IT, software and health industries were used as the field research sites and action research performed on their Web 2.0 tools and environment. Business intelligence was the most significant driver of service value to their clients. From the data, five key findings were observed on the strategic use of Web 2.0 in the leading services firms. Firstly, the firm is aware that social networking tools can improve employees' performance. Secondly, there are more tools for tacit-to-tacit and tacit-to-explicit knowledge transfer than explicit-to-explicit and explicit-to-tacit. Thirdly, the firm has a higher number of tools where knowledge flows within itself and almost none for external knowledge flows. Fourthly, social network is part of normal work responsibilities. Finally, among KM tools that were most recognized as assisting social network use were of the Web 2.0 genre such as wikis, RSS feeds and instant messaging and blogging. The authors show that using Web 2.0 improves social networking and may be linked to a service professional's individual performance.

Chapter 12
Knowledge Capture in E-Services Development: A Prosperous Marriage? 236
Eva Söderström, University of Skövde, Sweden
Lena Aggestam, University of Skövde, Sweden
Jesper Holgersson, University of Skövde, Sweden

In this paper, the authors examine whether the union of Knowledge Management with e-services development would be successful in performing as a collaborative functioning unit. The focus of this research is examining the potential for using Knowledge Management as a means for improving research and practice in e-services development. The authors analyze a real-life case against the Knowledge Capture model and its associated knowledge loss. The results show that KM theory has definite potential to elevate e-services research and practice, for example, by adding analysis and decision points concerning what knowledge to use and how to collect it. This is particularly relevant when collecting requirements, information, and desires from potential users of an e-service at the start of a development project.

Chapter 13
Knowledge Extraction from a Computational Consumer Model Based on Questionnaire Data Observed in Retail Service.. 252
Tsukasa Ishigaki, National Institute of Advanced Industrial Science and Technology, Japan
Yoichi Motomura, National Institute of Advanced Industrial Science and Technology, Japan
Masako Dohi, Otsuma Women's University, Japan
Makiko Kouchi, National Institute of Advanced Industrial Science and Technology, Japan
Masaaki Mochimaru, National Institute of Advanced Industrial Science and Technology, Japan

In service industries, matching the level of demand of the consumer and the level of service of the provider is important because it requires the service provider to have knowledge of consumer-related factors. Therefore, an intelligent model of the consumer is needed to estimate such factors because they cannot be observed directly by the service provider. This paper describes a method for computational modeling of the consumer by understanding his or her behavior based on datasets observed in real services. The proposed method constructs a probabilistic structure model by integrating questionnaire data and a Bayesian network, which incorporates nonlinear and non-Gaussian variables as conditional probabilities. The proposed method is applied to an analysis of the requested function from customers regarding the continued use of an item of interest. The authors obtained useful knowledge for function design and marketing from the constructed model by a simulation and sensitivity analysis.

Chapter 14
Non-Monotonic Modeling for Personalized Services Retrieval and Selection 267
Raymond Y. K. Lau, City University of Hong Kong, China
Wenping Zhang, City University of Hong Kong, China

With growing interest in Semantic Web services and emerging standards, such as OWL, WSMO, and SWSL in particular, the importance of applying logic-based models to develop core elements of the intelligent Semantic Web has been more closely examined. However, little research has been conducted in Semantic Web services on issues of non-mono-tonicity and uncertainty of Web services retrieval and selection. In this paper, the authors propose a non-monotonic modeling and uncertainty reasoning framework to address problems related to adaptive and personalized services retrieval and selection in the context of micro-payment processing of electronic commerce. As intelligent payment service agents are faced with uncertain and incomplete service information available on the Internet, non-monotonic modeling and reasoning provides a robust and powerful framework to enable agents to make service-related decisions quickly and effectively with reference to an electronic payment processing cycle.

Chapter 15
Adaptable Services for Novelty Mining 280
Flora S. Tsai, Nanyang Technological University, Singapore
Agus T. Kwee, Nanyang Technological University, Singapore
Wenyin Tang, Nanyang Technological University, Singapore
Kap Luk Chan, Nanyang Technological University, Singapore

Novelty mining is the process of mining relevant information on a given topic. However, designing adaptable services for real-world novelty mining faces several challenges like real-time processing of incoming documents, computational efficiency, multi-user working environment, diverse system requirements, and integration of domain knowledge from different users. In this paper, the authors bridge the gap between generic data mining methodologies and domain-specific constraints by providing adaptable services for intelligent novelty mining that model user preferences by synthesizing the parameters of novelty scoring, threshold setting, performance monitoring, and contextual information access. The resulting novelty mining system has been tested in a variety of performance situations and user settings. By considering the special issues based on domain knowledge, the authors' adaptable novelty mining services can be used to support a real-life enterprise.

Chapter 16
Engineering Financial Enterprise Content Management Services:
Integration and Control 298
Dickson K. W. Chiu, Dickson Computer Systems, China
Patrick C. K. Hung, University of Ontario Institute of Technology, Canada
Kevin H. S. Kwok, The Chinese University of Hong Kong, China

The demand is increasing to replace the current cost ineffective and bad time-to-market hardcopy publishing and delivery of content in the financial world. Financial Enterprise Content Management Services (FECMS) have been deployed in intra-enterprises and over the Internet to network with customers. This paper presents Web service technologies that enable a unified scalable FECMS framework for intra-enterprise content flow and inter-enterprise interactions, combining existing sub-systems and disparate business functions. Additionally, the authors demonstrate the key privacy and access control policies for internal content flow management (such as content editing, approval, and usage) as well as external access control for the Web portal and institutional programmatic users. Through the modular design of an integrated FECMS, this research illustrates how to systematically specify privacy and access control policies in each part of the system with Enterprise Privacy Authorization Language (EPAL). Finally, a case study in an international banking enterprise demonstrates how both integration and control can be achieved.

Section 4
Systems Engineering and Service-Oriented Engineering

Chapter 17
RT-Llama: Providing Middleware Support for Real-Time SOA 328
Mark Panahi, University of California, USA
Weiran Nie, University of California, USA
Kwei-Jay Lin, University of California, USA

Service-oriented architectures (SOA) are being adopted in a variety of industries. Some of them must support real-time activities. In this paper, the authors present RT-Llama, a novel architecture for real-time SOA to support predictability in business processes. Upon receiving a user-requested process and deadline, our proposed architecture can reserve resources in advance for each service in the process to ensure it meets its end-to-end deadline. The architecture contains global resource management and business process composition components. They also create a real-time enterprise middleware that manages utilization of local resources by using efficient data structures and handles service requests via reserved CPU bandwidth. They demonstrate that RT-Llama's reservation components are both effective and adaptable to dynamic real-time environments.

Chapter 18
SLIM: Service Location and Invocation Middleware for Mobile Wireless Sensor and Actuator Networks 346
Gianpaolo Cugola, Politecnico di Milano, Italy
Alessandro Margara, Politecnico di Milano, Italy

One of the main obstacles to the adoption of Wireless Sensor Networks (WSNs) outside the research community is the lack of high level mechanisms to easily program them. This problem affects distributed applications in general, and it has been replied by the Software Engineering community, which recently embraced Service Oriented Programming (SOP) as a powerful abstraction to ease development of distributed applications in traditional networking scenarios. In this paper, the authors move from these two observations to propose SLIM: a middleware to support service oriented programming in mobile Wireless Sensor and Actuator Networks (WSANs). The presence of actuators into the network, and the capability of SLIM to support efficient multicast invocation within an advanced protocol explicitly tailored to mobile scenarios, make it a good candidate to ease development of complex monitoring and controlling applications. In the paper the authors describe SLIM in detail and show how its performance easily exceeds traditional approaches to service invocation in mobile ad-hoc networks.

Chapter 19
A Method to Support Fault Tolerance Design in Service Oriented Computing Systems 362
Domenico Cotroneo, Università degli Studi di Napoli Federico II and Complesso Universitario Monte Sant'Angelo, Italy
Antonio Pecchia, Università degli Studi di Napoli Federico II, Italy
Roberto Pietrantuono, Università degli Studi di Napoli Federico II, Italy
Stefano Russo, Università degli Studi di Napoli Federico II and Complesso Universitario Monte Sant'Angelo, Italy

Service Oriented Computing relies on the integration of heterogeneous software technologies and infrastructures that provide developers with a common ground for composing services and producing applications flexibly. However, this approach eases software development but makes dependability a big challenge. Integrating such diverse software items raise issues that traditional testing is not able to exhaustively cope with. In this context, tolerating faults, rather than attempt to detect them solely by testing, is a more suitable solution. This paper proposes a method to support a tailored design of fault tolerance actions for the system being developed. This paper describes system failure behavior through an extensive fault injection campaign to figure out its criticalities and adopt the most appropriate countermeasures to tolerate operational faults. The proposed method is applied to two distinct SOC-enabling technologies. Results show how the achieved findings allow designers to understand the system failure behavior and plan fault tolerance.

Chapter 20
Demand-Driven Development of Service Compositions in Organizational Networks 377
Ralph Feenstra, Delft University of Technology, The Netherlands
Marijn Janssen, Delft University of Technology, The Netherlands
Sietse Overbeek, Delft University of Technology, The Netherlands

Organizational collaborate more and more in organizational networks to remain competitive. New systems can be created by assembling a set of elementary services provided by various organizations. Several composition methods are available, yet these methods are not adopted in practice as they are primarily supply-driven and cannot deal the complex characteristics of organizational networks. In this paper, the authors present a service composition development method and a quasi-experiment to evaluate this method by comparing it with existing ones. The development method is able to deal with incomplete information, to take the demand as a starting point, to deal with news services that do not exist yet, to include and to evaluate non-functional requirements, to show various stakeholder views, and to help to create a shared vision. Visualization and evaluation of alternative compositions and negotiation about the desired results are important functions of any composition method in organizational networks.

Chapter 21
Capacity-Driven Web Services: Concepts, Definitions, Issues, and Solutions.................................. 392
Samir Tata, TELECOM SudParis and CNRS UMR Samovar, France
Zakaria Maamar, Zayed University, UAE
Djamel Belaïd, TELECOM SudParis and CNRS UMR Samovar, France
Khouloud Boukadi, SFAX, Tunisia

This paper presents the concepts, definitions, issues, and solutions that revolve around the adoption of capacity-driven Web services. Because of the intrinsic characteristics of these Web services compared to regular, mono-capacity Web services, they are examined in a different way and across four steps denoted by description, discovery, composition, and enactment. Implemented as operations to execute at run-time, the capacities that empower a Web service are selected with respect to requirements put on this Web service such as data quality and network bandwidth. In addition, this paper reports on first the experiments that were conducted to demonstrate the feasibility of capacity-driven Web services, and also the research opportunities that will be pursued in the future.

Chapter 22
Investigating the Role of Service Encounter in Enhancing Customer Satisfaction 416
Irene Y. L. Chen, National Changhua University of Education, Taiwan

Recently, it is found that several pure e-tailers set up a customer service center where on-line shoppers can access a real person over the phone to answer their questions. However, there has been little systematic research examining how service encounter help to enhance customer satisfaction when a pure e-retail company set up a call center to provide additional services. This study conducted a questionnaire survey and collected data from persons who shopped on-line and had experiences in requesting help from customer service centers. 116 responses were collected and the data were then analyzed to examine the four relationships posited in the research model.

The proposed research model suggests that service encounter significantly influences service quality and information quality, which can jointly predict customer satisfaction. Findings of this study help to advance the understanding of the role that service encounters play in enhancing customer satisfaction.

Preface

ABSTRACT

Service-focused systems design and development (SFSDD) has been widely adopted in modern organizations to tackle the complexity and dynamism of the current globalized computing environment. To achieve this aim, the industry and academia has been shifting the focus to a cross-disciplinary approach from pure Service-Oriented Computing (SOC). This preface highlights some resultant challenges and opportunities for SFSDD, which can be classified into four key areas: (i) emerging technologies and killer applications; (ii) intelligence computing; (iii) knowledge engineering; and (iv) system and service-oriented engineering. The contents of this book are also classified into these four categories and introduced accordingly in this preface.

INTRODUCTION

The global economy and organizations are evolving to become service-oriented. There have recently been more and more research works on services provision. Beyond the Services Oriented Computing (SOC), intelligence in computing and knowledge from various disciplines (such as marketing, economics, management science, operations research, and psychology) is required to achieve service excellence for the ever complicating requirements in the rapidly evolving global environment. Therefore, although SOC (Zhang, 2004) were originated mainly from a computer science perspective, the industry and academia has been shifting the focus to a cross-disciplinary approach. Service-focused systems design and development (SFSDD) now also becomes a mainstream approach to tackle the complexity of contemporary business requirements.

On one hand, from the perspective on SOC, the engineering of services aims to compose high-quality interoperable and collaborative systems, which is essential to fulfill general business requirements. On the other hand, the perspective of SFSDD used in this book is the realization of successful systems that fulfils the goal of service excellence. Therefore, the emphasis of Systems Engineering (Kasser, 2007) is on the interdisciplinary, holistic approaches and means to deal with complex and intelligent systems is in line with this perspective of SFSDD. In fact, the perspective of systems engineering and SFSDD can be roughly compared with top-down and bottom-up approaches: they often offer complimentary perspectives and solutions to complex problems and requirements.

This preface highlights some challenges and opportunities for SFSDD. Such research areas include four major areas including: (i) the impact of related emerging technologies and killer applications; (ii) the application of intelligence computing; (iii) knowledge engineering for systems and services; and (iv) system engineering methodologies to service-oriented engineering. The preface introduces the contents of this book, which are also classified into these four categories. Then it introduces some recent professional activities in promoting SFSDD before concluding this preface.

Emerging Technologies and Killer Applications

As Zhang (2004) has pointed out, killer applications are required to drive Web Services research. Since the publication of *International Journal of Web Services Research* (Zhang, 2004), basic researches for services have been steadily progressing. However, the big challenge of the design and development of killer applications from systems perspective is still emerging based on the accumulating experiences of services deployment within and across organizations.

The current basis of services and most systems is the Web, which is ever evolving, towards Web 2.0, Web 3.0 (Lassila and Hendler, 2007), etc., so called Web x.0. Web 2.0 refers to a second generation of the Web, facilitating communication, information sharing, interoperability, and collaboration based on user-centered design. Virtual systems and virtual communities based on autonomous and peer-to-peer systems are therefore among the hottest research topics in line with the approach, relating systems and services.

Further, the widespread applications of mobile technologies have further resulted in an increasing demand for the support of mobile services across multiple platforms anytime and anywhere (Chiu, Cheung, et al., 2004). Examples include supply-chain logistics, group calendars, and dynamic human resources planning. As mobile devices become more powerful, the adoption of mobile computing is imminent. However, the realization of mobile services is not merely porting the software with an alternative user interface, but rather involves a wide range of new requirements, constraints, and technical challenges.

On the other hand, to scale up service provision, the Grid is a high-potential technology for the solution (Foster and Kesselman, 2004). Based on the Grid, concepts like software as a service (SaaS), Infrastructure as a Service (IaaS), Platform as a Service (PaaS), Communications as a Service (CaaS), utility computing, meta-services, and recently Cloud Computing have emerged (Hayes, 2008). Cloud computing emphasizes the realization of computing as a large collection of services rather than a single product, whereby shared resources, software, and information are provided to organizations and individual users anywhere anytime. Thus, many mature technologies are used as components in Cloud Computing, but still there are many unresolved and open problems. In particular, how traditional information systems can be (re-) engineered and migrated to new cloud platforms is a key issue of its adoption, especially intelligent ones.

Such emerging system architectures and computing paradigms bring new power to the engineering of systems and services, and provide opportunities in new application domains (such as aviation services). However, they also bring ever increasing complexities that calls for innovations and standardization. In addition, social and legal issues of such emerging technologies and systems must not be ignored (Chiu, Kafeza, and Hung, 2011).

Section 1 Introduction

To illustrate some theoretical and analytical advancement in this perspective, the first section of this book contains five chapters and centers on killer applications and emerging technologies, such as collaborative systems, grid environment, mobile computing, and Internet of Things (IoT).

First, the keynote chapter, by Chiu, Cheung, et al. (2010), illustrates the importance of intelligence and flexibility in the engineering of e-collaboration services to provide for service excellence. The authors advocate the engineering of e-collaboration support by means of multi-agent systems (MAS) in three key dimensions, namely, across multiple platforms, across organization boundaries, and agent-based intelligent support. Motivated by the requirements of mobile professional workforces in large enterprises, the authors present their development and adaptation methodology for e-collaboration services with a case study of constraint-based collaboration protocol from a three-tier implementation architecture aspect. They evaluate their approach from the perspective of three main stakeholders of e-collaboration, which include users, management, and systems developers. This chapter also reflects the vision and experiences of some of the key members of the editorial board in the application of emerging technologies to the killer application of collaboration systems.

As for review, Liu et al. (2010) analyze the three main steps of Web services provision lifecycle, namely, service discovery, service selection, and service contracting, and relate the lifecycle with three types of Web services provision including independent, cooperative, and multiple provision. They also distinguish three types of Web services provision according to the functional relationship between services and tasks: independent provision, cooperative provision and multiple provision. These are important aspects of contemporary Web service research. The chapter also discusses some potential research directions of Web service provision.

For Grid environment, Pouli et al. (2010) examine Service Level Agreements (SLAs) in Grid computing systems that combine computing resources from multiple administrative domains. The authors provide a set of Web services for automatic creation of end-to-end SLAs by merging per-domain SLAs, as well as for management (particularly monitoring) of these SLAs. The authors' solution has been successfully deployed in the European Research & Education Networks (NRENs) Advanced Multi-domain Provisioning System (AMPS).

Konstanteli et al. (2010) address the problem of execution management for mobile service-oriented Grid environments, which is mainly due to the significantly higher rate of various changes (e.g., related to connectivity and underlying network quality of service) experienced under mobile environments than under non-mobile environments. Their Execution Management System (EMS) supports additional capabilities for reliability and dependability of composite applications. The EMS has been implemented upon industry standards to seamlessly manage mobile and other services across multiple domains to create a mobile dynamic virtual organization (MDVO), and tested within the European Akogrimo project.

To conclude this section, Luo et al. (2010) introduce a local association network with a coordinated P2P message delivery mechanism to develop IoT based solutions for body parts tagging and tracking. The IoT vision extends technology development by creating an interactive environment for mobile and wireless pervasive networks, where smart objects can sense, connect, and react towards dynamic, transient, and changing interconnections. On-site testing, performance evaluation, and user feedback further strengthens the positive aspects of adopting the proposed approach.

Intelligence Computing for Systems and Services

Recently, although there is still much debate on the scope of Web 3.0, widely accepted key components of Web 3.0 include semantics and intelligence. Ontology is a formal semantic representation of concepts from a particular domain, as well as their relationships. Ontologies are frequently used for resolving terminological incompatibilities. Ontologies provides users and systems semantics for better understanding of their requirements, as well as related trust, reputation, security, forensic, and privacy issues for a better foundation of intelligent system behaviors. Example application areas include but not limit to service management, service marketing, relationship management, negotiations, auctions, and electronic marketplaces (Chiu et al., 2005).

On the other hand, agent based technologies are one of the most promising solution for the integration of systems and services in an intelligent context (He at al., 2003; Chiu, Cheung, et al., 2010). Intelligent agents are considered as autonomous entities with abilities to execute tasks independently. Various technologies from artificial intelligence can be applied at services, agents, and systems level, including computational intelligence, soft computing, game theory, genetic algorithms, evolutionary computing, logics, machine learning, cybernetics, planning, optimization, and so on. Such intelligence can be further enhanced when application semantics are available from ontologies, and is vital for effective service matchmaking, recommendation, personalization, operation, and monitoring (Chiu, Yueh, et al., 2009), which are the basis of quality services.

Example application areas include but not limit to service management, service marketing, relationship management, negotiations, auctions, and electronic marketplaces. With further increasing popularity of mobile and ubiquitous computing, location and pervasive intelligence for services (Hong et al., 2007; Chiu, Yueh, et al., 2009) is also an active research area with a wide horizon of possible advancement.

Section 2 Introduction

To illustrate some theoretical and analytical advancement in this perspective, the second section of this book centers on intelligent systems and services and contains five chapters, which can be roughly divided into two categories: intelligent system support and intelligent process management.

For practical intelligent system support, Naoe et al. (2010) address the provision of intelligent security support in collaborative systems and services through the generation of secure keys for static visual watermarking of document exchange by using machine learning technology. The proposed services contribute to secure visual information hiding without losing any detailed data of visual objects or any embedded hidden resource objects. In the experiments, they have shown that their proposed method is robust to high pass filtering and JPEG compression.

Brown et al. (2010) model self-led trust value management in grid and SOC infrastructures with a graph theoretic social network mediated decision based approach, in which delegates (ranging from individuals to large-scale virtual organizations) are continuously informed of relevant up-to-date trust levels. This fulfills the requirements of fluid, dynamic, ad hoc based interactions between delegates for intelligent applications in complex collaborative environment, which cannot be adequately provided by centralized approaches. It can therefore solve an important problem in both the SOC and MAS context of evaluating in context to optimally and dynamically select the most trustworthy ad hoc service for consumption. The key advantage of this approach is suited to an inherently low latency, decentralized trust security paradigm, especially when services provision scales up.

As for intelligent process management, Tosic (2010) proposes a novel application of autonomic business-driven system management to facilitate the maximization of business value metrics for the operation and adaptation of systems and services with minimal human intervention. He proposes the WS-Policy4MASC language for specifying such SOC management policies, characterized by the description of diverse business value metrics and specification of policy conflict resolution strategies for business value maximization. The author's decision making algorithms use this additional WS-Policy4MASC information to choose the adaptation approach intelligently from business viewpoint.

Fakhfakh et al. (2010) propose an ontology driven approach for automatic establishment of SLAs, which are contracts between service providers and clients, describing quality of service (such as expected response time and availability) and pricing aspects. This intelligently addresses frequent problems of the syntactic and semantic incompatibility between descriptions of client intentions and provider offers during auto- and semi-auto establishment of SLA, via a novel four-step approach to automatically generate a draft version of a SLA compatible both with client intentions and provider offers.

To conclude this section, Yang et al. (2010) address the issue of providing context-driven content adaptation services over the mobile Internet based on semantics with a description logic based planner approach. This approach dynamically transforms requested Web content into a proper format conforming to receiving contexts (e.g., access condition, network connection, and receiving device). The content adaptation planner is intelligently driven by a stepwise procedure equipped with algorithms and rule-based techniques. Experimental results prove the effectiveness and efficiency of the content adaptation planner on saving transmission bandwidth of handheld devices. By reducing the size of adapted content, they also achieve some reduction in the computational overhead caused by content adaptation.

Knowledge Engineering for Systems and Services

To empower systems and services with intelligence, knowledge acquired beforehand and during system execution is the key ingredient. Therefore, knowledge engineering is a key to excellence in systems and services, which includes knowledge modeling, architectures, acquisition, discovery, integration, and applications (Chiu, Lau, et al., 2010). Typical knowledge engineering services include content, multimedia, and metadata management, design management, engineering management, electronic education, and so on. As for knowledge application, they can be deployed for decision support and strategic information systems, integration of research and practice, and the management of service personnel and workforce, et cetera.

To address the growing complexity of the service sector at the age of Information and Communication Technology (ICT), traditional KM research and practices have to be extended and adapted, particularly aiming at effective learning, application, and management of new knowledge. On the other hand, the SOC provides new opportunities and challenges for knowledge and learning processes, such as integration and outsourcing (Hung, Chiu, et al., 2007).

One key source of knowledge comes from Business intelligence (BI), while knowledge provides the basis of further analysis for intelligence. BI is evolving: from the traditional data-and-analysis exercise for supporting management decisions to an integral part of business processes, providing analytics for a wide range of users across organizations. In this end, KM, BI, SOC, and many other related technologies are required to address the current challenges in the provision of high-quality services to meet the ever-increasing complexity of the current globalized economy.

Section 3 Introduction

To illustrate some theoretical and analytical advancement in knowledge engineering for systems and services, the third section of this book has six chapters on knowledge engineering that can be roughly divided into three categories: (i) survey of fundamental issues of this emerging field, (ii) modeling for KM and BI, and (iii) novel practical systems.

As for survey, Sharma, Tan, & Cheng (2010) examines how Web 2.0 may be strategically used in organizations to support business intelligence and KM, especially for knowledge creation, through field research on service firms in energy, information technology, software, and health industries. They summarize their survey into five significant findings: (i) firms are aware that social networking tools can improve employees' performance and thus use large numbers of such tools; (ii) there are more tools for tacit-to-tacit and tacit-to-explicit knowledge transfer than explicit-to-explicit and explicit-to-tacit; (iii) firms have a higher number of tools where knowledge flows within itself and almost none for external knowledge flows; (iv) social network is part of normal work responsibilities, particularly for knowledge intensive services where tacit collaboration is critical; and (v) among KM tools that were most recognized as assisting social network use were of the Web 2.0 genre such as wikis, RSS feeds, instant messaging, and blogging.

Next, Söderström, Aggestam, & Holgersson (2010) analyze the real-life case against the Knowledge Capture model and its associated knowledge loss and demonstrate the potential of using KM research as a means for improving research and practice in e-services development, especially for requirements engineering at the start of a project.

As for modeling, Ishigaki et al. (2010) propose an intelligent computational consumer model based on questionnaire data observed in retail services using Bayesian network techniques. The proposed method has been successfully applied to an analysis of the requested function from customers regarding the continued use of an item of interest to obtain useful knowledge for function design and marketing from the constructed model by a simulation and sensitivity analysis.

Lau & Zhang (2010) propose a novel non-monotonic model for the retrieval and selection of adaptive personalized services in the context of micro-payment processing of electronic commerce, based on semantics and uncertainty reasoning. Their payment service agents can *proactively* monitor the latest information in a service market and *autonomously* select the best payment service option on behalf of their users. To make prompt decisions based on uncertain and incomplete service information, non-monotonic modeling and reasoning provides a robust and powerful framework. This research work also opens the door to the application of non-classical reasoning framework to streamline Semantic Web services.

As for practical systems, Tsai et al. (2010) propose the use of adaptable and composable services for the mining of novelty for knowledge management in order to bridge the gap between generic data mining methodologies and domain-specific constraints. They model user preferences by synthesizing the parameters of novelty scoring, threshold setting, performance monitoring, and contextual information access. Their system has been tested in a variety of performance situations and user settings, and its adaptable novelty mining services can effectively help discover novelty in real-life enterprises.

To conclude this section, Chiu, Hung, & Kwok (2010) report their design of a financial enterprise content management system for an international bank based on the service-oriented architecture with a focus on integration and control. Such systems provide a high value to customer relations through value-added information and knowledge services as well as to the internal knowledge management of the enterprise. They show how Web service technologies enable a unified scalable framework for intra-enterprise content flow and inter-enterprise interactions, thus integrating sub-systems and disparate business functions. At the same time, they demonstrate the implementation of key privacy and access control policies with the Enterprise Privacy Authorization Language (EPAL) for internal content flow management (such as content editing, approval, and usage) as well as external access control for the Web portal and institutional programmatic users.

Systems Engineering and Service-Oriented Engineering

To provide a holistic solution, systems engineering provides the theoretical foundation, engineering processes, and methodology for the emerging field of service-oriented engineering. In addition, systems management and service management is essential for achieving service excellence.

Some emphasis on system quality include: communications, control, integration among systems, human, organizations, and virtual communities, requirement elicitation and analysis, design science, reliability and quality engineering, exception handling, uncertainty, and risk management. Topics to facilitate the understanding of complex systems include infrastructure systems and services, self-organizing systems, mobile, ubiquitous, and pervasive systems, and system of systems. Although individual topics may have already been well-studied on its own, the design and development of complex systems often requires advanced combination of knowledge from multiple disciplines, which has not been adequately studied before. Further challenges arise from emerging and ever-evolving computing environment, under which completely new problem-solving strategies and tactics need to be developed.

A prominent approach to provide coherent and quality services of ever increasing complexity and diversity is through *middleware*, which is placed between the network operating system and the application layer (Göschka et al., 2010). Middleware simplifies the task of programming and managing distributed applications by providing programming abstractions and runtime tools. Under the SOC, middleware plays an additional role to help abstract software constituents as services through dynamically discovery, composition, and reuse to provide greater flexibility and implementation independence through abstract programming models, quality-of-service provisions, transactions, monitoring, management, and so on.

In addition, middleware enables traditional business-to-business (B2B) and enterprise application integration (EAI) applications to migrate smoothly into the emerging Cloud Computing paradigm, as software on top of SOC middleware need not worry about the underlying implementation. At the same time, middleware provides accounting, service selection, orchestration of data and process flows that move across cloud and legacy environments, and other necessary additional support to upper-layer applications for providing quality services. On the other hand, middleware can also reduce the complexity in designing and developing embedded systems, wireless sensor networks, and complex context-sensitive ubiquitous systems (Aitenbichler et al., 2007), through middleware support of communication overhead balancing, power consumption management, uniformly addressable and reachable service end-points, heterogeneous devices support, context detection and management, etc.

Further, evaluation of systems and services provide important feedback mechanisms for their improvement. Besides modeling and simulation, testing and performance evaluation of systems and services in a practical way, if possible, provides more realistic and accurate evaluations. On the other hand, technology assessment, adoption, and diffusion studies provide assessment on how well systems and services are received by users. These studies will remain to be the fundamental pillars even though technologies and computing environments evolve.

Section 4 Introduction

To illustrate some theoretical and analytical advancement in this perspective, the fourth section of this book centers on system and service-oriented engineering and contains six chapters, which can be roughly divided into three categories: middleware design, service composition design, and service evaluation.

For middleware design, Panahi et al. (2010) address the issue of real-time support and predictability in SOC with a novel middleware architecture for resource management called RT-Llama, in order to face the challenge of adaptability to dynamic real-time environments. The proposed architecture can reserve resources in advance for each service in the process to ensure it meets user-specified deadline through global resource management and appropriate business process composition, together with real-time management of local resources and CPU bandwidth reservation.

Cugola and Margara (2010) study how to make programming of wireless sensor and actuator networks (WSANs) easier under the SOC paradigm with their Service Location and Invocation Middleware (SLIM), as mobility significantly complicates the management of WSANs. SLIM adopts an advanced routing protocol designed for reliable discovery and invocation of services in mobile scenarios. It also supports efficient multicast invocations, invoking at once all services satisfying a given query. Their simulations have shown that the performance achieved by using SLIM is much better compared to alternative approaches.

As for service composition design, Cotroneo et al. (2010) address the challenge in gaining knowledge about system behavior when various faults occur in SOC systems. Various faults can occur during operation of SOC systems, but not all of these faults can be discovered during testing. The authors overcome this problem with a novel method that supports design of fault tolerance in SOC systems by determining where to place fault tolerance code and what functionality this code should provide. The authors have applied their method to the Apache Web server and the TAO Open Data Distribution Service (DDS).

Feenstra et al. (2010) present a demand-driven service composition development method in organizational networks and a quasi-experiment to evaluate this method by comparing it with existing ones. Their approach can deal with incomplete information, take the demand as a starting point, deal with news services not yet exist, include and evaluate non-functional requirements, show various stakeholder views, and help create a shared vision of the composition. The key contribution is to provide the important functions of visualization and evaluation of alternative compositions and negotiation about the desired results in organizational networks.

Tata et al. (2010) present the concepts, definitions, issues, and solutions of capacity-driven Web services. Implemented as operations to execute at run-time, the capacities that empower a Web service are selected with respect to requirements put on this Web service such as data quality and network bandwidth. Experiments results demonstrate the feasibility of this novel approach, and related future research opportunities are discussed.

Last but not least, for service evaluation, Chen (2010) examines how service encounter helps enhance customer satisfaction when a pure e-retail company set up a call center to provide additional services, which is a current trend of business strategy for e-retailers. This study conducted a questionnaire survey and collected data from persons who shopped on-line and had experiences in requesting help from customer service centers. The result supports a practically significant research hypothesis that service encounter significantly influences service quality and information quality, which can jointly predict customer satisfaction.

SUMMARY

SOC has become an integral part of ICT infrastructure of modern organizations, governments, and even individual computing needs to facilitate the modeling, analysis, design, integration, development, and deployment of information systems and services. The tools and methodologies based on SOC generally provide cross-platform compatibility, agility, flexibility, and cost-efficiency (Chiu, Cheung, et al., 2010, Göschka et al., 2010). This in turn facilitates the achievement of intelligent and quality services in a wide range of application context.

The creation, operation, and evolution of the research and practice in SFSDD raise concerns that range from high-level requirements and policy modeling through to the deployment of specific implementation technologies and paradigms, as well as involve a wide (and ever growing) range of methods, tools, and technologies. They also cover a broad spectrum of vertical domains, industry segments, and even government sectors.

Researchers are continuously seeking collaborations and carrying out various scholarly activities, including workshops, conference special tracks, and journal special issues on this topic. Efforts can be dated back to the first International Workshop on Service Intelligence and Service Science (SISS) held in Hong Kong, October 2006 and later in Beijing, China, September 2008. Following up is the book entitled "Service Intelligence and Service Science: Evolutionary Technologies and Challenges" (Leung, Chiu, & Hung, 2010) and a special issue on "Service Intelligence" in the *International Journal of Organizational and Collective Intelligence* (IJOCI) (Chiu, Hung, & Leung, 2010). Recently, the First International Symposium on Web Intelligent Systems & Services (WISS 2010), held in Hong Kong, December 2010, has accepted about 30 papers from various countries all over the world (Chiu, Bellatreche, et al. 2011). Then, further were two special issues in the *International Journal of Systems and Service-Oriented Engineering* (IJSSOE) on "Intelligent Services" and "Intelligent Systems," respectively (Chiu & Sasaki, 2011).

The contributors of this book also promote SFSSD in more specific technological contexts such as mobile computing and cloud computing. For mobile computing, they founded the workshop on "Mobile Business Collaboration" in 2009, Brisbane, Australia, and continued it in 2010, Hong Kong, and 2012, Kun Ming, China. They have also founded our recent workshop on "Cloud Information System Engineering" (CISE) in 2010, Hong Kong (Chiu, Bellatreche, et al. 2011).

The authors have discussed a wide range of challenges and opportunities in this emerging field. These chapters illustrate some of the current research areas pertinent to service-focused systems design and development (SFSDD); while, in many ways, also amplifying the many challenges, which remain to be addressed. It is expected that new topics will emerge while existing research will shift concentration into these areas in the coming years.

Therefore, there has been tight collaboration with IJOCI and IJSSOE to provide an open, formal publication for high-quality research work developed by theoreticians, educators, developers, researchers, and practitioners across disciplines to advance the practice and understanding of contemporary theories and empirical analysis in the perspective of our emerging field towards the goal of achieving service excellence under the current globalized service-oriented economy.

Dickson K.W. Chiu
Dickson Computer Systems, Hong Kong

REFERENCES

Aitenbichler, E., Kangashraju, J., & Mühlhäuser, M. (2007). MundoCore: A light-weight infrastructure for pervasive computing. *Pervasive and Mobile Computing*, *3*(4), 332–361. doi:10.1016/j.pmcj.2007.04.002

Brown, A., Sant, P., Bessis, N., French, T., & Maple, C. (2010). Modelling self-led trust value management in grid and service oriented infrastructures: A graph theoretic social network mediated approach. *International Journal of Systems and Service-Oriented Engineering, 1*(4), 1–18. doi:10.4018/jssoe.2010100101

Chen, I. (2010). Investigating the role of service encounter in enhancing customer satisfaction. *International Journal of Systems and Service-Oriented Engineering, 1*(4), 19–26. doi:10.4018/jssoe.2010100102

Chiu, D. K. W., Bellatreche, L., et al. (Eds.). (2011) *Web information systems engineering: WISE 2010 International Symposium WISS, and International Workshops CISE, MBC LNCS 6724.* Hong Kong, China, December 12-14, 2010. Revised Selected Papers. Springer.

Chiu, D. K. W., Cheung, S. C., Kafeza, E., & Leung, H.-F. (2004). A three-tier view methodology for adapting m-services. *IEEE Transactions on Systems, Man and Cybernetics . Part A, 33*(6), 725–741.

Chiu, D. K. W., Cheung, S.-C., Leung, H.-F., Hung, P. C. K., Kafeza, E., & Hu, H. (2010). Engineering e-collaboration services with a multi-agent system approach. *International Journal of Systems and Service-Oriented Engineering, 1*(1), 1–25. doi:10.4018/jssoe.2010092101

Chiu, D. K. W., Hung, P. C. K., & Kwok, K. H. S. (2010). Engineering financial enterprise content management services: Integration and control. *International Journal of Systems and Service-Oriented Engineering, 1*(2), 86–113. doi:10.4018/jssoe.2010040106

Chiu, D. K. W., Hung, P. C. K., & Leung, H.-F. (2010). Guest editorial preface: Challenges and opportunities for service intelligence: The next wave of service computing. *International Journal of Organizational and Collective Intelligence, 1*(2), i–iv.

Chiu, D. K. W., Kafeza, E., & Hung, P. C. K. (2011). ISF special issue on emerging social and legal aspects of information systems with Web 2.0. *Information Systems Frontiers, 13*(2), 153–155. doi:10.1007/s10796-009-9168-x

Chiu, D. K. W., Lau, R. Y. K., Cheung, W. K., & Karduck, A. (2010). Editorial preface: Service-oriented knowledge management and business intelligence. *International Journal of Systems and Service-Oriented Engineering, 1*(2), i–iii.

Chiu, D. K. W., Poon, J. K. M., Lam, W. C., Tse, C. Y., Siu, W. H. T., & Poon, W. S. (2005). *How ontologies can help in an e-marketplace*? ECIS (European Conference on Information Systems) 2005. Retrieved from http://aisel.aisnet.org/ecis2005/70/

Chiu, D. K. W., & Sasaki, H. (2011). Editorial preface: Special issue on intelligent web systems / web services. *International Journal of Systems and Service-Oriented Engineering, 2*(1), i–iii.

Chiu, D. K. W., Yueh, Y. T. F., Leung, H.-F., & Hung, P. C. K. (2009). Towards ubiquitous tourist service coordination and process integration: A collaborative travel agent system with Semantic Web services. *Information Systems Frontiers, 11*(3), 241–256. doi:10.1007/s10796-008-9087-2

Cotroneo, D., Pecchia, A., Pietrantuono, R., & Russo, S. (2010). A method to support fault tolerance design in service oriented computing systems. *International Journal of Systems and Service-Oriented Engineering, 1*(3), 75–89. doi:10.4018/jssoe.2010070105

Cugola, G., & Margara, A. (2010). SLIM: Service location and invocation middleware for mobile wireless sensor and actuator networks. *International Journal of Systems and Service-Oriented Engineering, 1*(3), 60–74. doi:10.4018/jssoe.2010070104

Fakhfakh, K., Chaari, T., Tazi, S., Jmaiel, M., & Drira, K. (2010). ODACE SLA: Ontology driven approach for automatic establishment of service level agreements. *International Journal of Systems and Service-Oriented Engineering, 1*(3), 1–20. doi:10.4018/jssoe.2010070101

Feenstra, R., Janssen, M., & Overbeek, S. (2010). Demand-driven development of service compositions in organizational networks. *International Journal of Systems and Service-Oriented Engineering, 1*(4), 27–41. doi:10.4018/jssoe.2010100103

Foster, I., & Kesselman, C. (Eds.). (2004). *The Grid 2: Blueprint for a new computing infrastructure* (2nd ed.). Morgan Kaufmann.

Göschka, K. M., Paik, H.-Y., & Tosic, V. (2010). Engineering middleware for service-oriented computing: Editorial note. *International Journal of Systems and Service-Oriented Engineering, 1*(3), i–v.

Hayes, B. (2008). Cloud computing. *Communications of the ACM, 51*(7), 9–11. doi:10.1145/1364782.1364786

He, M., Jennings, N. R., & Leung, H.-F. (2003). On agent-mediated electronic commerce. *IEEE Transactions on Knowledge and Data Engineering, 15*(4), 985–1003. doi:10.1109/TKDE.2003.1209014

Hong, D., Chiu, D. K. W., Cheung, S. C., Shen, V. Y., & Kafeza, E. (2007). Ubiquitous enterprise service adaptations based on contextual user behavior. *Information Systems Frontiers, 9*(4), 343–358. doi:10.1007/s10796-007-9039-2

Hung, P. C. K., Chiu, D. K. W., Fung, W. W., Cheung, W., Wong, R., & Choi, S. P. M. (2007). End-to-end privacy control in service outsourcing of human intensive processes: A multi-layered web service integration approach. *Information Systems Frontiers, 9*(1), 85–101. doi:10.1007/s10796-006-9019-y

Ishigaki, T., Motomura, Y., Dohi, M., Kouchi, M., & Mochimaru, M. (2010). Knowledge extraction from a computational consumer model based on questionnaire data observed in retail service. *International Journal of Systems and Service-Oriented Engineering, 1*(2), 40–54. doi:10.4018/jssoe.2010040103

Kasser, J. E. (2007). *A framework for understanding systems engineering*. BookSurge Publishing.

Konstanteli, K., Kirkham, T., Gallop, J. R., Matthews, N., Johnson, I. J., Kardara, M., & Varvarigou, T. A. (2010). Execution management for mobile service-oriented environments. *International Journal of Systems and Service-Oriented Engineering, 1*(3), 39–59. doi:10.4018/jssoe.2010070103

Lassila, O., & Hendler, J. (2007). Embracing "Web 3.0". *IEEE Internet Computing, 11*(3), 90–93. doi:10.1109/MIC.2007.52

Lau, R. Y. K., & Zhang, W. (2010). Non-monotonic modeling for personalized services retrieval and selection. *International Journal of Systems and Service-Oriented Engineering, 1*(2), 55–63. doi:10.4018/jssoe.2010040104

Leung, H.-F., Chiu, D. K. W., & Hung, P. C. K. (Eds.). (2010). *Service intelligence and service science: Evolutionary technologies and challenges*. Hershey, PA: IGI Global. doi:10.4018/978-1-61520-819-7

Liu, A., Liu, H., Lin, B., Huang, L., Gu, N., & Li, Q. (2010). A survey of web services provision. *International Journal of Systems and Service-Oriented Engineering, 1*(1), 26–45. doi:10.4018/jssoe.2010092102

Luo, Z., & Lai, M. (2010). Developing local association network based IoT solutions for body parts tagging and tracking. *International Journal of Systems and Service-Oriented Engineering, 1*(4), 42–64. doi:10.4018/jssoe.2010100104

Naoe, K., Sasaki, H., & Takefuji, Y. (2010). Secure key generation for static visual watermarking by machine learning in intelligent systems and services. *International Journal of Systems and Service-Oriented Engineering, 1*(1), 46–61.

Panahi, M., Nie, W., & Lin, K.-J. (2010). (20100. RT-Llama: Providing middleware support for real-time SOA. *International Journal of Systems and Service-Oriented Engineering, 1*(1), 62–78.

Pouli, V., Marinos, C., Grammatikou, M., Papavassiliou, S., & Maglaris, V. (2010). A service oriented SLA management framework for grid environments. *International Journal of Systems and Service-Oriented Engineering, 1*(3), 21–38. doi:10.4018/jssoe.2010070102

Sharma, R. S., Tan, D., & Cheng, W. (2010). Two heads are better than one: Leveraging Web 2.0 for business intelligence. *International Journal of Systems and Service-Oriented Engineering, 1*(2), 1–24. doi:10.4018/jssoe.2010040101

Söderström, E., Aggestam, L., & Holgersson, J. (2010). Knowledge capture in e-services development: A prosperous marriage? *International Journal of Systems and Service-Oriented Engineering, 1*(2), 25–39. doi:10.4018/jssoe.2010040102

Tata, S., Maamar, Z., Belaïd, D., & Boukadi, K. (2010). Capacity-driven Web services: Concepts, definitions, issues, and solutions. *International Journal of Systems and Service-Oriented Engineering, 1*(4), 65–88. doi:10.4018/jssoe.2010100105

Tosic, V. (2010). Autonomic business-driven dynamic adaptation of service-oriented systems and the WS-Policy4MASC support for such adaptation. *International Journal of Systems and Service-Oriented Engineering, 1*(1), 79–95. doi:10.4018/jssoe.2010092105

Tsai, F. S., Kwee, A. T., Tang, W., & Chan, K. L. (2010). Adaptable services for novelty mining. *International Journal of Systems and Service-Oriented Engineering, 1*(2), 69–85. doi:10.4018/jssoe.2010040105

Yang, S. J. H., Zhang, J., Huang, J. J. S., & Tsai, J. J. P. (2010). Using description logics for the provision of context-driven content adaptation services. *International Journal of Systems and Service-Oriented Engineering, 1*(1), 96–129. doi:10.4018/jssoe.2010092106

Zhang, L.-J. (2004). Challenges and opportunities for web services research, editorial preface. *International Journal of Web Services Research, 1*(1), vii–xii.

Acknowledgment

I thank IGI Global and other publishers for supporting our publications, authors for their innovative contributions, conference organizers and research grants for supporting our meetings, as well as editorial board and program committee members for evaluating our manuscripts.

Dickson K.W. Chiu
Dickson Computer Systems, Hong Kong

Section 1
Emerging Technologies and Killer Applications

Chapter 1
Engineering e-Collaboration Services with a Multi-Agent System Approach

Dickson K.W. Chiu
Dickson Computer Systems, Hong Kong

S.C. Cheung
The Hong Kong University of Science and Technology, Hong Kong

Ho-fung Leung
The Chinese University of Hong Kong, Hong Kong

Patrick C.K. Hung
University of Ontario Institute of Technology, Canada

Eleanna Kafeza
Athens University of Economics and Business, Greece

Hua Hu
Hangzhou Dianzi University, China

Minhong Wang
The University of Hong Kong, Hong Kong

Haiyang Hu
Zhejiang Gongshang University, China

Yi Zhuang
Zhejiang Gongshang University, China

ABSTRACT

With recent advances in mobile technologies and e-commerce infrastructures, there have been increasing demands for the expansion of collaboration services within and across systems. In particular, human collaboration requirements should be considered together with those for systems and their components. Agent technologies have been deployed in order to model and implement e-commerce activities as multi-agent systems (MAS). Agents are able to provide assistance on behalf of their users or systems in collaboration services. As such, we advocate the engineering of e-collaboration support by means of MAS in the following three key dimensions: (i) across multiple platforms, (ii) across organization boundaries, and (iii) agent-based intelligent support. To archive this, we present a MAS infrastructure to facilitate systems and human collaboration (or e-collaboration) activities based on the belief-desire-intension (BDI) agent architecture, constraint technology, and contemporary Web Services. Further, the MAS infrastructure also provides users with different options of agent support on different platforms. Motivated by the requirements of mobile professional workforces in large enterprises, the authors

DOI: 10.4018/978-1-4666-1767-4.ch001

present their development and adaptation methodology for e-collaboration services with a case study of constraint-based collaboration protocol from a three-tier implementation architecture aspect. They evaluate our approach from the perspective of three main stakeholders of e-collaboration, which include users, management, and systems developers.

INTRODUCTION

The Internet has become a common platform where organizations and individuals communicate amongst each other to carry out various e-commerce activities (Chiu et al., 2004). As such, systems and human collaboration has recently become a focus of systems engineering and design. In particular, as the objective of all systems has the fundamental mission of serving humans, human collaboration requirements should never bc ignored along with those for systems and their components. We refer to such integrated systems and human collaboration as e-collaboration. However, the recent turbulence of the globalized economy together with fast-evolving information and communication technologies (ICT) has caused great impact on how businesses are being carried out currently and in the future. Such ever-growing complexity and requirement evolution demand a critical rethinking on the methodologies for the engineering of e-collaboration services, which also demands for much flexibility and intelligence.

Recent advances in information and communications technologies (ICT) have created a plethora of mobile devices (Lin & Chlamtac, 2000) with ever increasing range of communication, computing, and storage capabilities. New mobile applications running on such devices provide users with easy access to remote services regardless of differing locations (Hong et al., 2004). Interesting mobile applications taking advantage of the ubiquity of wireless networking emerge to create new virtual worlds through expanded e-collaboration services (Hong et al. 2007; Chiu et al., 2009). An example application is the management of mobile workforce in this knowledge and service based economy (Chiu et al., 2006), which is applicable to a wide range of domains such as supply-chain logistics (Wang et al., 2009), dynamic human resources planning, negotiation support (Chiu et al., 2004b), and tourism (Chiu et al., 2009). At any time and regardless of location, the management as well as professional workforces needs to collaborate with colleagues and communicates with information systems, not only within their own organization but also among other broader systems not directly synchronized with their own.

Mobile devices nowadays are generally equipped with adequate processing power to support reasonably complex computations (Chiu et al., 2004). Sophisticated intelligence for collaboration and decision support, usually in the form of agents, is now possible on these platforms. This functionality is able to relieve people from repetitive tasks like their personal assistants (He et al., 2003; Chiu et al., 2004; 2006). Thus, there are increasing demands to expand e-collaboration services in the following three dimensions: (i) across multiple platforms, (ii) across organization boundaries, and (iii) agent-based intelligent support.

However, to the best of our knowledge, no previous research effort has been made to systematically study the engineering of e-collaboration services to these three dimensions specifically. In particular, there is no unified implementation framework yet to incorporate all these three types of expansion, integrating all the enabling technologies together coherently. New challenges arise from such new computing and communications environment, such as the handling of mobility, handsets with reduced screens and varying bandwidth to support users' ubiquitous access, privacy and security of individual users and across organizational boundaries, and so on. Therefore, it is necessary to address these new challenges with a platform independent solution based on a coherent technological framework. To achieve this, we consolidate and extend our previous research in negotiation support system (NSS) (Chiu et al., 2004; 2004b; 2005), cross-organizational process management (Chiu et al., 2001), and mobile service (m-service) adaptation methodology (Chiu et al. 2004; Hong et al., 2007).

In this paper, we summarize our experience in engineering e-collaboration services with the use of agent technologies, and particularly a *Multi-Agent System* (MAS). A MAS consists of multiple intelligent software agents that exchange messages and collaborate on behalf of their human users. With MAS support, an end-user (known as the delegator) can delegate an agent to process or search for information; or to provide a set of value-added services for some business purposes. The MAS architecture also enables multi-platform support, based on belief-desire-intension (BDI) agent architecture (Rao & Georgeff, 1995), constraint technology (Tsang, 1993), mobile technologies (Lin & Chlamtac, 2000), and contemporary Web Services (Weerawarana et al., 2005).

The rest of our paper is organized as follows. We first introduce the theoretical background including the MAS conceptual model, BDI agents, and constraint based collaboration, comparing related work. Then, we present our adaptation based methodology for engineering e-collaboration services with standard three-tier implementation architecture, addressing flexible agent intelligence and the heterogeneity of mobile platforms. Next, we evaluate the applicability of our methodology in response to different stakeholders' perspective of an e-collaboration services. Finally, we conclude this paper with our outlook and plans for further research.

THEORETICAL BACKGROUND

A MAS framework provides an infrastructure under which multiple agents and users can collaborate under a pre-defined protocol. Agents in the MAS are distributed and autonomous; each carrying out actions based on their own strategy. In this section, we explain our MAS conceptual model; in which the computational model of an agent can be described using a BDI framework. Acceptable collaboration arrangements are solved by mapping the constraints generated in e-Negotiation according to the well-known Constraint Satisfaction Problem (CSP) (Tsang, 1993), for which efficient solvers are widely available.

Multi-Agent Information System Conceptual Model

Figure 1 summarizes our layered framework. Conventionally, e-collaboration is driven by human representatives. It could be a tedious, repetitive, and error-prone process, especially when the professional workforces have to commute frequently. Furthermore, agents facilitate the protection of privacy and security (as explained later in Section 4). The provision of computerized personal assistance to individual users across organizations by means of agents is a sensible choice. These agents, acting on behalf of their delegates, collaborate through both wired and wireless Internet, forming a dynamic MAS whenever required. Such repeatable processes can be adequately supported and the cost of developing the infrastructure is well justified.

The Believe-Desire-Intention (BDI) framework is a well-established computational model for deliberative intelligent agents (Wegner et al., 1996). A BDI agent constantly monitors the changes in the environment and updates its information accordingly. Possible goals are then generated, from which intentions to be pursued are identified and a sequence of actions will be performed to achieve the intentions. BDI agents are proactive. They do

Figure 1. A layered framework for e-collaboration services

e-Collaboration Services (e.g., e-Negotiation)	
Multi-agent System (MAS)	
BDI Agents	
	Collaboration Protocol
Three-tier Implementation Architecture (Interface Tier / Application Tier/ Data Tier)	

this by taking initiatives to achieve their goals, yet adaptive by reacting to the changes in the environment in a timely manner. They are also able to accumulate experience from previous interactions with the environment and other agents.

Internet applications are generally developed with a three-tier architecture comprising front, application and data tiers. This is known as three-tier architecture. Figure 2 depicts the architecture of a three-tier MAS system. Though the use of three-tier architecture in the agent community is new, it is a well-accepted pattern to provide flexibility in each tier. This is absolutely required in the expansion of e-collaboration services. Such flexibility is particularly important to the front tier, which often involves the support of different solutions on multiple platforms. In our architecture, users may either interact manually with other collaborators or delegate an agent to make decision on behalf of himself. Thus, users without agent support can still participate through flexible user interface for multiple platforms, while enterprise business processes can interact with the e-collaboration Web Services.

A Web Service (Weerawarana, 2005) is defined as an autonomous unit of application logic that provides either some business functionality or information to other applications through an Internet connection. Web Services are based on a set of eXtended Markup Langauge (XML) standards such as Simple Object Access Protocol (SOAP), Universal Description, Discovery and Integration (UDDI) and Web Services Description Language (WSDL). The benefits of adopting Web Services over traditional business-to-business applications include faster time to production, convergence of disparate business functionalities, a significant reduction in total cost of development, and easy to deploy business applications for trading partners.

Figure 3 describes the meta-model of a MAS system in a class diagram of the Unified Model-

Figure 2. Architecture of a three-tier MAS system with multi-platform support

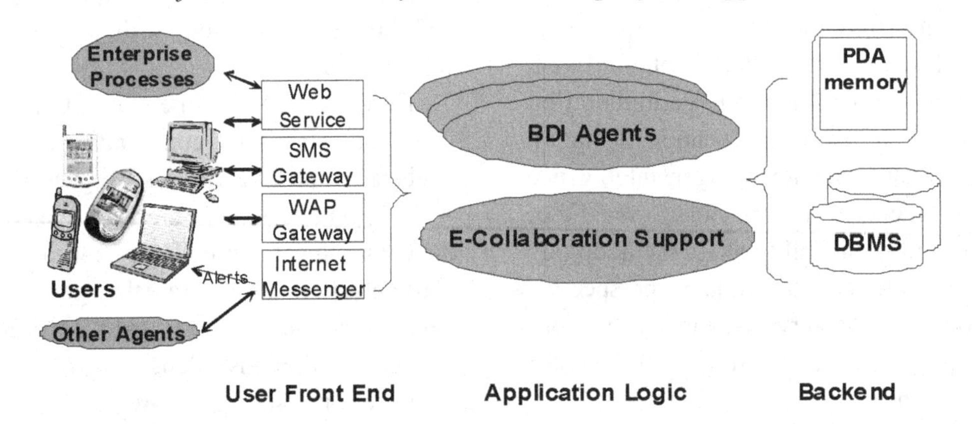

Figure 3. Meta-model of a MAS system in UML class diagram

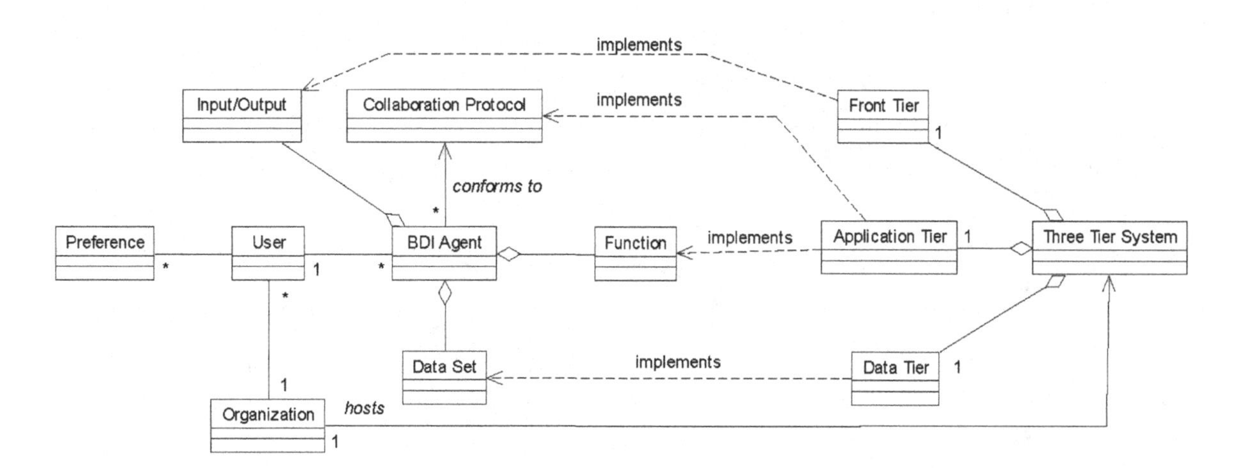

ing Language (UML) (Carlson, 2001). This language is widely used for visualizing, specifying, constructing, and documenting the artifacts of a software-intensive system. It summarizes our mapping between the components of a BDI agent to individual tiers of a three-tier system hosted by an organization. A BDI agent is made up of three major components: input/output, functions and data sets. Detailed discussion of the mapping is presented in Section 3. A BDI agent acts on behalf of a user in an organization in order to interact with other agents according to a predefined collaboration protocol. The agent receives inputs and generates outputs through the front tier. The agent's functions and the protocol logic can be implemented at the application tier. The data tier can be used to implement the various data sets of an agent.

BDI Agent Model

As depicted in Figure 4, a BDI computational model is composed of three (3) main data sets, these include: belief, desire, and intention. Information and/or data are passed from one data set to another through the application of various functions. Once a *stimulus* is sensed as input, the *belief revision function* (*brf*) converts it to a belief. The *desired* set is updated by generating some options based on the data in belief set. Options in the desired set are then filtered to new intentions of the agent, and a corresponding plan of actions is then generated. The BDI agent simulates an assistant for decision on behalf of a human user for collaboration.

Even though the agent can receive signals from the environment (e.g., user location), the *stimulus* inputs are mainly incoming requests and responses from the other agents and users. These inputs are usually associated with a set of constraints and/or options (solution) to a proposal. As a result, the *belief* set contains several sets of constraints representing the requirements of a proposal. All solutions or even future options should satisfy these sets of constraints. Figure 4 illustrates an example of the *brf* function. The *brf* function scans the received stimulus to examine the elements and values of the requirements. In Figure 2, there are 4 elements - *Date, Time, Location,* and *Duration*. The *brf* function then converts them into a constraint format. In the above example, the belief generated is: *(Date=6/1/2002 and Time>9:00 and Time<12:00 and Location='CUHK' and Duration=30minutes).*

When the belief is updated, the next process is to generate some options for the counter pro-

Figure 4. Computational model and the internal structure of a belief-desire-intention agent in UML class diagram with example

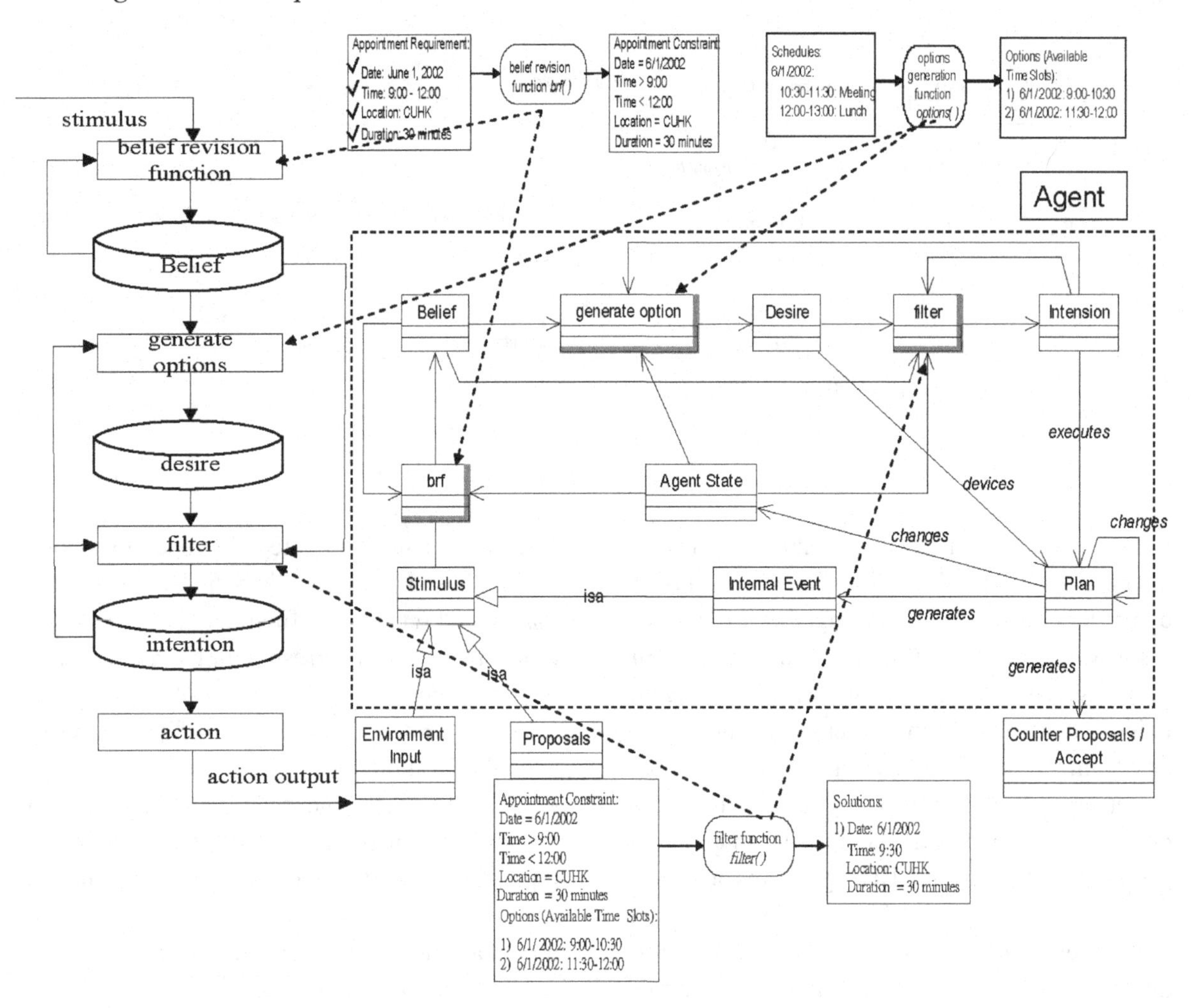

posals. The agent looks up the user's schedule in their Personal Digital Assistant (PDA) device in order to see which time slots are free for a new appointment. The time periods available are converted to constraints again and then stored in a desire set. Figure 4 illustrate an example of the *options generation function*. The *options generation function* accesses the schedule database to get all appointments within the specified date, say '*6/1/2002*'. In the above example, two appointments are retrieved. By retrieving the existing appointments, the agent knows whether the user is available or not. The *option generation function* then generates a list of valid options (in constraints format). In the above example, two constrains are generated: *(Date=6/1/2002 and Time>9:00 and Time<10:30) and (Date=6/1/2002 and Time >11:30 and Time<12:00).*

Although both the *belief* and *desire* sets contain constraints, their nature is different. Constraints in the *belief* set represent the requirements and preferences of users and agents. These constraints do not reflect any information about their availability. However, on the other hand, constraints in the *desire* set are created by looking up the users' schedules and reflect their actual availability. The agent can determine solutions from these options in later step.

After generating some options, the agent determines the intentions by filtering. The filter function takes the *belief* and *desire* sets as input and then returns intentions as output. In a meeting scheduling multi-agent system, this can be accomplished by *tree searching* (Kumar, 1998), which systematically tests different combinations of values of date (or weekday), time, location, duration, and attendees against the constraints. If a combination of values for these variables satisfies all the constraints, and the agent likes such a combination of values, then the combination is returned as output of the *filter* function, e.g., *(Date=6/1/2002, Time=9:00, Location='CUHK')*. This triggers an update to the intention. Figure 4 also includes an example to illustrate the task of *filter* function.

Constraint Satisfaction

Constraint satisfaction (Kumar, 1998; Tsang, 1993) is one of the major research areas within Artificial Intelligence. Many real life problems, which cannot be easily modeled in specific mathematical formulae, or be solved by conventional Operations Research methods, may be expressed as constraint satisfaction problems (CSPs). One relevant classical example is the "car-sequencing" problem (Dincbas et al., 1998). In its basic form, a constraint satisfaction problem consists of a finite number of variables, each ranging over its own finite domain of discrete values, and a finite number of constraints. Each constraint is a relation over a subset of the variables, restricting the combinations of values these variables that they can take. Formally, a *Constraint Satisfaction Problem* is a tuple $<X, D, C>$, where X is a finite set of variables, D a function mapping a variable to its domain, and C a finite set of constraints. For any variable x in X, we require that $D(x)$ is a finite set of discrete constants. A constraint $c(y_1,y_2,\ldots,y_n)$ in C is a relation over a finite subset $\{y_1,y_2,\ldots,y_n\}$ of n variables in X. A solution to a constraint satisfaction problem is an assignment of values to variables so that all constraints are satisfied.

To illustrate this collaboration problem we propose a hypothetical dilemma. Suppose that the two main issues to be determined are the time and place for a meeting. Usually, a user has a number of possible options for the time and place of an appointment. Each of these options is usually associated with a degree of preference so that the feasibility of these options can be interrelated. For example, a user (Franklin) may wish to make a 30-minute appointment at CUHK, the Hong Kong University of Science and Technology (HKUST), or the University of Hong Kong (HKU). He can be there any time from 9:00 to 11:00 if the appointment is either CUHK or HKUST, but only after 14:00 if the appointment is HKU. In addition, the possibility of appointment at HKUST is excluded if it is on a Wednesday. On Sundays and Saturdays, Franklin is not available. Figure 5[REMOVED REF FIELD] shows the constraint satisfaction problem that models Franklin's requirements. As such, a user's requirements can naturally be formulated as a constraint satisfaction problem for collaboration activities.

RELATED WORK

E-collaboration (Bafoutsou & Mentzas, 2001) supports communication, coordination, and cooperation for a set of geographically dispersed users. Thus, e-collaboration requires a framework based on the strategy, organization, processes, and information technology (Gerst, 2003). Further, Rutkowski et al. (2002) address the importance of structuring activities for balancing electronic communication during e-collaboration. This is used to prevent and solve conflicts.

Intelligent agents are considered autonomous entities with abilities to execute tasks independently. He et al. (2003) present a comprehensive survey on agent-mediated e-commerce. An agent should be proactive and subject to personalization, with a high degree of autonomy. In particular; due to the different limitations on different

Figure 5. Some sample constraints for meeting scheduling

```
Variables:
  Day, Time, Place
Domains:
  D(Day) = {Monday, Tuesday, Wednesday, Thursday, Friday, Saturday, Sunday}
  D(Time) = {9:00, 9:15, 9:30, ..., 14:00}
  D(Place) = {CUHK, HKUST, HKU}
  D(Duration) = {15min, 30min, 60min}
Constraints:
  SuitableTime(Time)
  SuitableTime(Day)
  SuitablePlace(Place)
  Day-Place-Requirement(Day, Place)
  Time-Place-Requirement(Time, Place)
  Duration-Requirement(Duration)
  Plus: other personal constraints on Time, Day and Place
Extents of Constraints:
  SuitableTime = {9:00, ..., 11:00, 14:00}
  SuitableDay = {Monday, Tuesday, Wednesday, Thursday, Friday}
  SuitablePlace = {CUHK, HKUST, HKU}
  Day-Place-Requirement =
          {<Monday, CUHK>, <Monday, HKUST>, <Monday, HKU>,
          <Tuesday, CUHK >, <Tuesday, HKUST>, <Tuesday, HKU>,
          <Wednesday, CUHK >, <Wednesday, HKUST>,
          <Thursday, CUHK >, <Thursday, HKUST>, <Thursday, HKU>,
          <Friday, CUHK >, <Friday, HKUST>, <Friday, HKU>,
          <Saturday, CUHK >, < Saturday, HKUST>, < Saturday, HKU>,
          <Sunday, CUHK >, < Sunday, HKUST>, < Sunday, HKU>}
  Time-Place-Requirement = {<9:00, CUHK>, ..., <11:00, CUHK>,
<9:00, HKUST>, ..., <11:00, HKUST>, <14:00, HKU>}
  Duration-Requirement = {30min}
```

platforms, users may need varying options in agent delegation. Prior research studies usually focus on the technical issues in a domain-specific application. For example, MIT Media Lab's Kasbah (Chavez & Maes, 1996) is an online, multi-agent consumer-to-consumer transaction system. Users create autonomous agents to buy and sell goods on their behalf, and also specify parameters to guide and constrain an agent's overall behavior. Similarly, Further, Teich et al. (1999) present heuristic algorithms for multiple issues electronic markets and auctions that were based on dovetailing buyers' and sellers' interests and preferences. In addition, Lo & Kersten (1999) present an integrated negotiation environment by using software agent technologies for supporting negotiators. However, all of them did not support their model on different platforms.

The emergence of MAS dates back to Sycara & Zeng (1996), who discussed the issues in coordination of multiple intelligent software agents. In general, a MAS provides a platform that brings the multiple types of expertise for any decision making (Luo et al., 2001). For examples, Lin et al. (1998) present a MAS composed of four main components: agents, tasks, organizations, and information infrastructure for modeling the order fulfillment process in a supply chain network. Further, Lin & Pai (2000) discuss the implementation of MAS based on a multi-agent simulation platform called Swarm. Next, Shakshuki et al. (2000) present a MAS architecture, in which each agent is autonomous, cooperative, coordinated, intelligent, rational, and able to communicate with other agents to fulfill the users' needs. Choy et al. (2003) propose the use of mobile agents to aid in meeting the critical requirement of universal access in an efficient manner. Chiu et al. (2004; 2006) also propose the use of a three-tier view-based methodology for adapting human-agent collaborative systems for multiple mobile platforms. In order to ensure interoperability of MAS, standardization between different levels is required (Gerst, 2003). Thus, based on all these prior works, our proposed MAS framework adapts agents with wireless as well as Web Services standardized technologies in order to expand e-collaboration support in multiple dimensions.

For logic based collaboration, Bui (1987) describes various protocols for multi-criteria group decision support in an organization. Bui et al. (1998) further propose a formal language based on first order-logic to support and document argumentation, claims, decision, negotiation, and coordination in network-based organizations. In this context, a constrained-based collaboration can be modeled as a specific example of the ARBAS language.

Wegner et al. (1997) present a multi-agent collaboration algorithm using the concepts of belief, desire and intention. In addition, Fraile et al. (1999) provides a negotiation, collaboration, and cooperation model for supporting a team of distributed agents to achieve the goals of assembly tasks. However, this paper is mainly focused on supporting negotiation activities in the context of e-collaboration environment.

As meeting scheduling is a well-known and non-trivial collaboration problem, we utilize this as a case study to illustrate our framework and methodology in this paper. In practice, scheduling is a time-consuming and tedious task. It involves intensive communications among multiple persons, taking into account many factors or constraints. In our daily life, meeting scheduling is often performed personally or by our secretaries via telephone or e-mail. Most of the time, each attendee has some uncertainties and incomplete knowledge about the preferences and the intentions of other attendees. Thus, a meeting scheduler is a useful tool for group collaborations. Historically, meeting scheduling was emerged as a classic problem in artificial intelligence and multi-agent systems. There are some commercial products but they are just calendars or simple diaries with special features, such as availability checkers and meeting reminders (Garrido et al., 1996). Shitani et al. (2000) highlights a negotiation approach among agents for a distributed meeting scheduler based on the multi-attribute utility theory. Van Lamsweerde et al. (1995) discuss goal-directed elaboration of requirements for a meeting scheduler, but do not discuss any implementation frameworks. Sandip (1997) summarizes an agent-based system for automated distribution meeting scheduler, but is not based on BDI agent architecture. However, all these systems cannot adequately support human interactions in the decision process or address any mobile support issues. Therefore, we use this application as a relevant example for our methodology.

In summary, none of the existing or prior works considered multi-dimensional expansion of e-collaboration in a unified framework. There is neither any work describing a concrete implementation and adaptation methodology by means of a portfolio of contemporary enabling technologies.

SYSTEM ARCHITECTURE AND IMPLEMENTATION

In this section, our purpose is to discuss our methodology and implementation framework for expanding e-collaboration support with a case study of meeting scheduling.

Methodology

Figure 6 summarizes the adaptation on each tier required to address the additional requirements and technological limitations of differing plat-

Figure 6. Adaptation for users on various platforms

Platforms / Views	Enterprise	PDA	WAP	SMS
User Front-end	Web Service interface for programmed interactions	Simplified screen layout Low resolution graphics Panning and zooming	WML translation Highly reduced screens	SMS message presentation
Application Logic	Cross-organizational process collaboration	Simplified process steps and procedures		SMS dialog presentation
Backend	Mutually agreed schema and semantics for interoperation	Omit some fields Summarized information May need access to PDA memory	Mandatory fields only Highly summarized information	Highly summarized and mandatory information as message content

forms. Manual collaborators should be supported on different platforms, such as web browsers on PDAs and Short Message Service (SMS) or Wireless Application Protocol (WAP) on mobile phones (Lin & Chlamtac, 2000). For example, a traveling collaborator may be alerted regarding immediate decision through mobile devices, such as his/her mobile phone via SMS. Then the collaborator may connect via the use of a mobile phone or simply send back a reply with SMS. Alternatively, if more information is required, the collaborator may access the NSS or other applications for more information before making the decision with a PDA on a wireless network or a PC in a net cafe. Thus, such extensions to our previous approach by alerting users just with ICQ (Weverka, 2001) or emails can further reduce decision delay when collaborators are away from their home or office. This approach also separates alerts from sessions to improve the flexibility. If a collaborator is not online or does not reply within a pre-defined period, the application server will send the alert again by email. Whatever the alert channel has been, the collaborator need not connect to application server on the same device, or even on the same platform.

The following summarize the procedures to adapt a conventional e-collaboration application to support multiple mobile platforms and cross organizational interactions. The adaptation assumes a MAS infrastructure where agents act on behalf of their owning organizations. As discussed in Section 3.1, a BDI agent can be implemented on a three-tier system.

1. Gather the requirements of the target frequently repeated collaboration process. Identify overall data requirement to support the process in the backend databases (e.g., represent them with class diagrams in UML).
2. Identify the issues to be determined, and the information to be exchanged among collaborators and formulate their representation as constraints for an efficient exchange.
3. Design the basic collaboration protocol to be implemented its logic at the application tier (e.g., represent the protocol with activity diagram in UML).
4. Design the required Web Service access points to support the execution of the protocol.
5. Identify the (mobile) platforms to be supported. See if any adaptations of the basic collaboration protocol are required.
6. Design and adapt the user interface required for users to input their preferences. Customize displays to individual users and platforms.

7. Determine how user preferences are mapped into constraints.
8. Formulate these constraints as Web Services policies (WS-Policy) in order to exchange them in a standardized format.
9. Consider automated decision support with agents: identify the stimulus, collaboration parameters, and output actions to be performed by a BDI agent.
10. Partition the collaboration parameters into three data sets including: belief, desire and intention. Formulate a data sub-schema for each of these data sets. Implement the schema at the data tier.
11. Derive transformations amongst the three data sets. Implement these transformations at the application tier.

e-Collaboration Protocol Design and Adaptations

Figure 7 depicts the activity diagram in UML of an agent-based collaboration protocol. The user who initiates a proposal (or proposer) enters his/her requirements and suggestions (or options) and delegates an agent to initiate the process. The delegated agent then contacts the other collaborators or their delegated agent. Each of the collaborators or agents then determines if any options are good. If so, the decision can simply be passed back to the proposer. Otherwise, depending on the platform such as *Personal Digital Assistant* (PDA), *Wireless Application Protocol* (WAP) and *Short Message Services* (SMS) (see Figure 8), a collaborator might reply in the following three respond modes: (i) *Passive mode* – the collaborator (or his agent) just reply that all the proposed schedules are bad, without any counter proposals; (ii) *Counter-offer mode* – the collaborator (or his agent) suggests other counter proposals; and (iii) *Constraint mode* – the collaborator (or his agent) gives some of the constraints representing the collaborator's preference and availability. After gathering the feedback, the proposing agent evaluates it for a common feasible solution for all collaborators. If this is successful, all collaborators will be informed of the result. Otherwise, the proposing agent will attempt another round of proposals, with the consideration of the options already rejected by the collaborators and their constraints (if any). However, if the maximum round is exceeded, the proposing agent will consider the scheduling failed and inform all collaborators.

In particular for manual collaborators, we have to impose process restrictions on different platforms. After receiving an alert via ICQ or emails a PDA collaborator (or other users on a web browser) or WAP collaborator logs on the server to review the details of the service proposal. The collaborator then determines if any options are viable. If so, the decision can be passed back to the service provider through the Web or WAP interface respectively. Otherwise, a PDA collaborator might reply in any of these three respond mode described in the previous section. However, if the *constraint mode* is too complicated to be supported through a WAP interface it is therefore not provided to WAP users. As for a SMS collaborator, the only practical option is to reply with a SMS message, as other ways of responding is far too complicated.

In order to capture the system requirements for users of a mobile platform, a commonly adopted approach from object-oriented design is to carry out requirement analysis. We should concentrate on the differences in the decision and operation requirements between various mobile devices and standard desktop PCs. These differences are compared with the standard process to formulate the restrictions. We should identify similar or identical tasks to maximize reuse, and in particular, consider the possibility of customizing them instead of rewriting them.

Usually, a complete detailed decision process is too complicated for the mobile environment. Therefore, typical solutions are simplification of the process, reordering the work steps, delegation of tasks (work steps) to other personal, etc. For

Figure 7. UML activity diagram of agent-based collaboration protocol

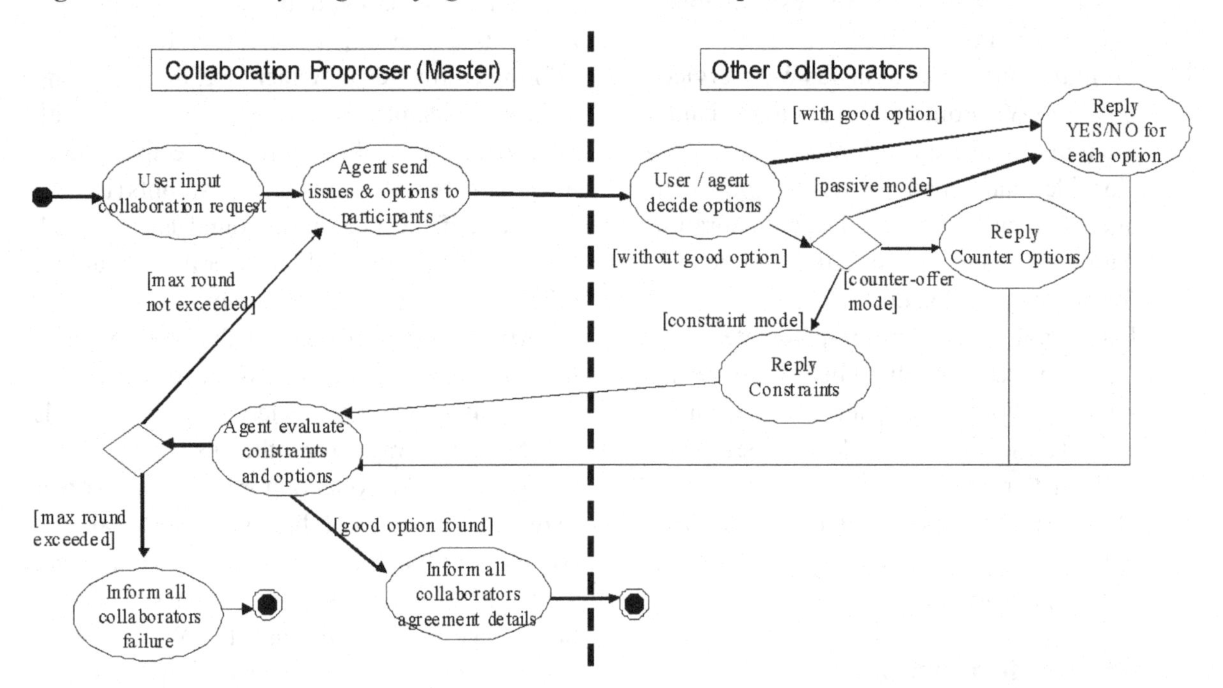

example, a user with only SMS support can make only very simple decisions or feedback. As this kind of operating environment is often error prone and may have security problems, we suggest not allowing critical options in these process views. On the other hand, it is often difficult to tell if a collaborator can tolerate a complex process because of the operating environment or even due to the collaborator's mood, which cannot be determined merely with observation facts, such as the mobile platform and physical location. Therefore, it is always a good idea to allow collaborators to choose their desired view options whenever feasible.

Another problem is the recognition of a collaborator's preferences. In our implementation, user preferences are recognized by associating a solution evaluation function defined according to the user's preferences, and enhance the tree search to a branch-and-bound search strategy

Figure 8. Different features for collaborators on different platforms

Features \ Collaborators	Enterprise Software	PDA / Desktop User	WAP User	SMS User
Service Request	Web Service	Browser interface or programmed action	WAP interface	SMS
Automation	Business Process on Server	Agent run on PDA	Agent run on server (and on some new handsets)	
Alerts	SOAP Message	ICQ, email (or SMS if user also accessible)	SMS	
User response mode	Passive, Counter-offer, Constraint	Passive, Counter-offer, Constraint	Passive, Counter-offer	Passive

Figure 9. Mapping between the data schema and the BDI architecture in Figure 4

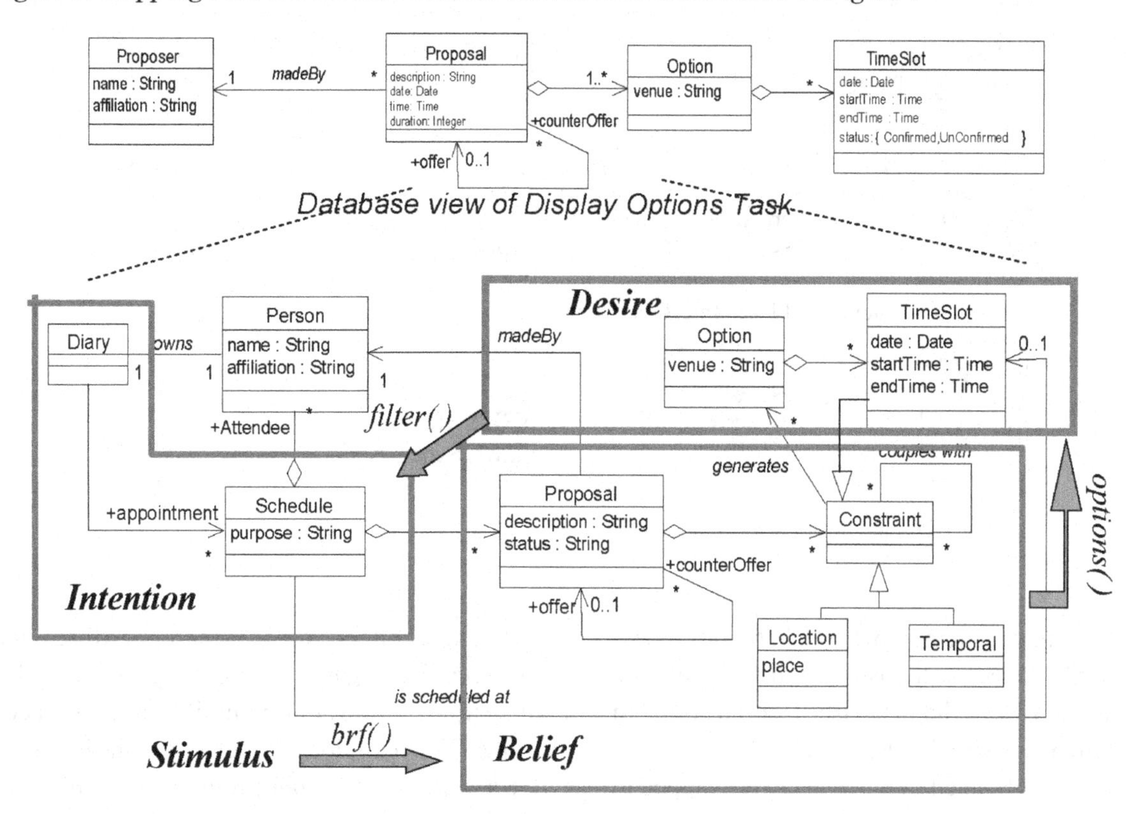

(Kumar, 1998). In adaptation towards mobility, the agents might need to be aware of the current locations of all collaborators, such as the case in the meeting scheduling example. The information is used to ensure that the collaborators can arrive at the venue on time.

Data Schema and View Design

Figure 9 summarizes the implementation of the complete data schema using BDI agent architecture, together with the schema of a database view in UML Class Diagram for the *Display options* task, based on a meeting scheduler. Please note that a class or an association in the diagram maps to a data table with columns representing the class attributes (Elmasri & Navathe, 2006). The main table in this data view is the *Proposal* table, which consists of a number of proposals made by the requestor. A proposal can be a counter-offer of a previous proposal. Each proposal contains a description and the time when the proposal was made. A proposal consists of constraints concerning the service options. Multiple alternatives can be suggested in an option.

The data view for the *Display options* task is a projection and selection of the entire data set for the NSS. As shown in the Figure 9, the data schema is a selected subset of data objects as well as a projection of some data object fields. Thick arrows indicate update operations. Here, we have not explicitly elaborated on the *plan* in the mapping because it is outside the scope of this paper. For instance, the plan to attend a meeting at 9:30am in the Chinese University of Hong Kong (CUHK) may consist of taking a bus for Kowloon at 8:00am and then change for a train to CUHK. In general, a plan of an agent involves proper resource allocation so as to realize its intentions.

Figure 10. Implementation framework for generating different interfaces

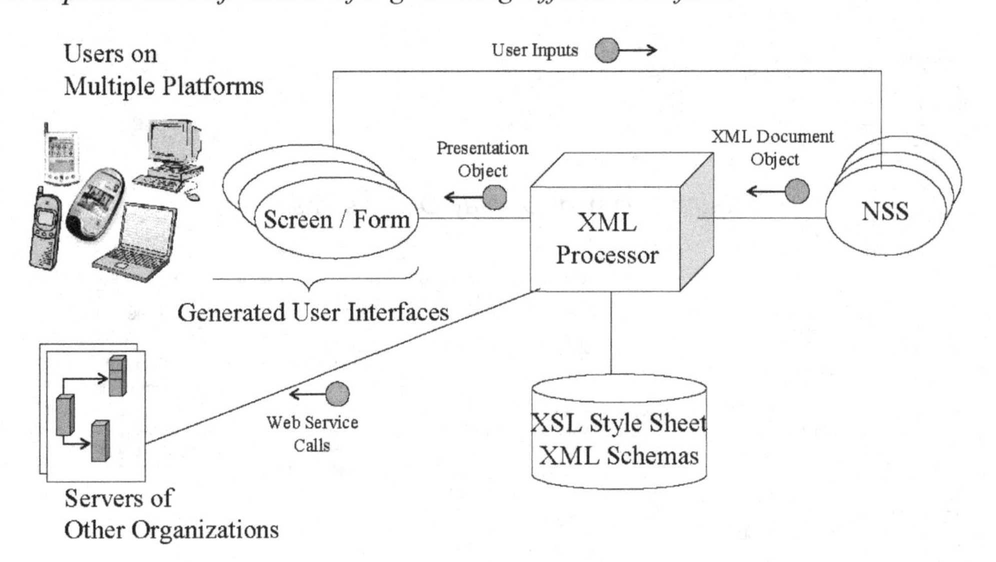

The data requirement of a collaboration process is mainly based on the issues and their relation to the data from the database in some form. In particular, we should identify mandatory fields, optional fields, and fields that are to be skipped in the view, in order to cope with the simplification required for mobile users. However, additional fields those have to be computed for summarizing information and knowledge may be required. In addition, security requirements should also be considered, for example, sensitive information may have to be restricted to collaborators within the office or to those of pre-approved locations. Database views may also be employed to hide less important data field or to show alternate summary columns.

Multi-Platform Front Tier Design

We have to provide two types of interfaces (i) for users to input their preferences and (ii) for agents to exchange information. (See Figure 10)

Different user interfaces can be facilitated by contemporary XML technologies augmented with XML Stylesheet Language (XSL) (Lin and Chlamtac, 2000) in elegantly. In most situations, collaborators on PDAs and PCs may also prefer different user interfaces to cater for the difference in screen size. Figure 11 presents a possible implementation framework for multiple user interfaces using XSL technology. For example, the interface for a web user to input preferences is different from that for a WAP user, but can be rendered by two different XSL style sheets. On the other hand, examples for interfaces definitions for business-to-business (B2B) interactions in Web Services Definition Language (WSDL) can be found in our earlier work (Chiu et al., 2004).

To accommodate user interfaces in mobile devices, we usually need to remove graphics or reduce the resolution, provide panning and zooming, shorten fields or provide summarized ones instead, break one web page into several screens, etc. For user input, we should consider the difficulties in entering data (especially typing) on mobile devices, and provide menu selections as far as possible. For PDA interface, the main problem is just a smaller screen, some of which may be black-and-white. If the original full-function user interface is too complicated (e.g., too many unnecessary features or high resolution screen layout), another simplified user interface is probably required. Pictures and documents may require to be shown in lower resolution and

Figure 11. Constraints of Figure 5 in WS-Policy

```
(01) <wsp:Policy wsu:Id="MeetingSchedule"
       xmlns:wsu="http://schemas.xmlsoap.org/ws/2002/07/utility"
(02)     xmlns:wsp="http://schemas.xmlsoap.org/ws/2002/12/policy"
(03)     xmlns:arbas=
           "http://schemas.xmlwsmp.org/ARBAS/2004/03/constraints">
(04)   <wsp:All wsp:Usage="wsp:Required" wsp:Preference="100">
(05)     <arbas:Variables>
(06)       <Day/> <Time/> <Place/>
(07)     </arbas:Variables>
(08)     <arbas:Domains>
(09)       </abras:D Id="Day">
(10) <Moday/><Tuesday/><Wednesday/><Thursday/><Friday/><Saturaday/><Sunday/>
(11)       </abras:D>
(12)       </abras:D Id="Time">
(13)         <0900/><0915/><0930/>…<1400/>
(14)       </abras:D>
(15)       </abras:D Id="Place">
(16)         <CUHK/><HKUST/><HKU/>
(17)       </abras:D>
(18)       </abras:D Id="Duration">
(19)         <900/><1800/><2700/>
(20)       </abras:D>
(21)     </arbas:Domains>
(22)     <arbas:Constraints>
(23)       <arbas:SuitableTime>
(24)         <Time/>
(25)       </arbas:SuitableTime>
(26)       <arbas:SuitableDay>
(27)         <Day/>
(28)       </arbas:SuitableDay>
(29)       <arbas:SuitablePlace>
(30)         <Place/>
(31)       </arbas:SuitablePlace>
(32)       <arbas:Day-Place-Requirement>
(33)         <Day/><Place/>
(34)       </arbas: Day-Place-Requirement >
(35)       <arbas:Time-Place-Requirement>
(36)         <Time/><Place/>
(37)       </arbas: Time-Place-Requirement >
(38)       <arbas:Duration-Requirement>
(39)         <Duration/>
(40)       </arbas: Duration-Requirement >
(41)       <!--- Plus: other personal constraints on Time, Day and Place --->
(42)     </arbas:Constraints>
(43)     <arbas:ExtentsOfConstraints>
(44)       <arbas:Extent Id="SuitableTime">
(45)         <0900/>…<1100/><1400/>
(46)       </arbas:Extent>
(47)       <arbas:Extent Id="SuitableDay">
(48)         <Monday/><Tuesday/><Wednesday/><Thursday/><Friday/>
(49)       </arbas:Extent>
(50)       <arbas:Extent Id="SuitablePlace">
(51)         <CUHK/><HKUST/><HKU/>
(52)       </arbas:Extent>
(53)       <arbas:Extent Id="Day-Place-Requirement">
(54)         <Extent-tuple Id="DPR-1"><Monday/><CUHK/></Extent-tuple>
(55)         <Extent-tuple Id="DPR-2"><Monday/><HKUST/></Extent-
tuple>
(56)         <Extent-tuple Id="DPR-3"><Monday/><HKU/></Extent-tuple>
(57)         <Extent-tuple Id="DPR-4"><Tuesday/><CUHK/></Extent-
tuple>
(58)         <Extent-tuple Id="DPR-5"><Tuesday/><HKUST/></Extent-
tuple>
(59)         <Extent-tuple Id="DPR-6"><Tuesday/><HKU/></Extent-tuple>
(60)         <Extent-tuple Id="DPR-7"><Wednesday/><CUHK/></Extent-
tuple>
(61)         <Extent-tuple Id="DPR-8"><Wednesday/><HKUST/></Extent-
tuple>
(62)         <Extent-tuple Id="DPR-9"><Thursday/><CUHK/></Extent-
tuple>
(63)         <Extent-tuple Id="DPR-10"><Thursday/><HKUST/></Extent-
tuple>
(64)         <Extent-tuple Id="DPR-11"><Thursday/><HKU/></Extent-
tuple>
(65)         <Extent-tuple Id="DPR-12"><Friday/><CUHK/></Extent-tuple>
(66)         <Extent-tuple Id="DPR-13"><Friday/><HKUST/></Extent-
tuple>
(67)         <Extent-tuple Id="DPR-14"><Friday/><HKU/></Extent-tuple>
(68)         <Extent-tuple Id="DPR-15"><Saturaday/><CUHK/></Extent-
tuple>
(69)         <Extent-tuple Id="DPR-16"><Saturaday/><HKUST/></Extent-
tuple>
(70)         <Extent-tuple Id="DPR-17"><Saturaday/><HKU/></Extent-
tuple>
(71)         <Extent-tuple Id="DPR-18><Sunday/><CUHK/></Extent-tuple>
(72)         <Extent-tuple Id="DPR-19><Sunday/><HKUST/></Extent-
tuple>
(73)         <Extent-tuple Id="DPR-20"><Sunday/><HKU/></Extent-tuple>
(74)       </arbas:Extent>
(75)       <arbas:Extents Id="Time-Place-Requirement">
(76)         <Extent-tuple Id="TPR-1"><0900/><CUHK/></Extent-tuple>
(77)           …
(78)         <Extent-tuple Id="TPR-6"><1100/><CUHK/></Extent-tuple>
(79)         <Extent-tuple Id="TPR-7"><0900/><HKUST/></Extent-tuple>
(80)           …
(81)         <Extent-tuple Id="TPR-15"><1100/><HKUST/></Extent-tuple>
(82)         <Extent-tuple Id="TPR-16"><1400/><HKU/></Extent-tuple>
(83)       </arbas:Extent>
(84)       <arbas:Extent Id="Duration-Requirement">
(85)         <900/>
(86)       </arbas:Extent>
(87)     <arbas:ExtentsOfConstraints>
(88)   </wsp:All>
(89) </wsp:Policy>
```

documents may be outlined and level-structured. Panning and zooming (supported by most browsers) should also help. For collaborators on a WAP interface connecting to the application server via a WAP gateway, the screen is extremely small and it is mandatory to translate the content into WML for display.

As for interfaces among agents, we propose an extension of constraint assertions in the newly developed WS-Policy (Weerawarana et al., 2005). WS-Policy is an XML representation that provides a grammar for expressing Web Services policies, to allow service locators to have a common interpretation of different business requirements. Thus, WS-Policy is suitable for expressing constraint policies. Referring to Figure 11, besides those standard Web Services utility (wsu) and policy (wsp) namespaces defined at http://schemas.xmlsoap.org/ws/2002/07/utility and http://schemas.xmlsoap.org/ws/2002/12/policy, we assume that we have set up a schema for specifying constraints (arbas) located at http://schemas.xmlwsmp.org/ARBAS/2004/03/constraints. Then, the <wsp:Policy> element (lines 1, 2, 3, 4, 88, and 89) is

Figure 12. Concerns of different stakeholders of expanding e-collaboration support

	General Concerns	Multi-platform Expansion	Cross-Organizational Expansion	Agent Automation
User's Perspective	**Assist their work** Ease of Use System Performance, especially response time Privacy	Convenience and connectivity – anytime anywhere	Interoperability	Reduce tedious work Reliability – retries, search alternatives
Management's Perspective	**Cost vs. Benefit** Improve productivity Scalability Security Reduction in total development cost	Locate, control, and communicate with staff Communications cost	Increase business opportunities Improve relationships Convergence of disparate business functionalities	Knowledge management
System Developer's Perspective	**Development / Maintenance effort & cost** Reusability Scalability Uncertainty in the use of new technologies Overall System Complexity Requirement Elicitation	Frequent appearance of new devices	Too many Web standards Faster time to production	Difficulties in programming captured knowledge

the top-level container for a policy expression enlisting the concerned namespaces. In addition, the wsu:id attribute is used to indicate a fragment ID in arbitrary containing elements.

The primary operator in this policy expression is <wsp:All> (lines 4 and 88), which indicates that all of its child assertions are combined to form a policy statement. The <wsp:All> element indicates that all of the contained assertions must be met. The wsp:Usage attribute is used to qualify the semantics of the leaf elements as applied to a policy subject such as the collaborator(s). In this example, the value wsp:Required means that the assertion must be applied to the collaborator(s). The wsp:preference attribute is to specify the preference of this policy statement. The higher is the value of the preference, and the greater is the weighting of the expressed preference. The remaining lines describe the set of constraints as specified in Figure 5.

DISCUSSIONS AND EVALUATIONS

In this section, we intend to evaluate the applicability of our implementation framework and methodology with respect to the major stakeholders, including users, management, and system developers (as summarized in Figure 12). As mentioned, the proposed methodology expands traditional e-collaboration in three dimensions: multiple platforms, cross-organizational support, and agent intelligence. The goal is to compare each of the three dimensions with respect to the stakeholders' general concerns. The issues considered are based on the research framework on nomadic computing proposed by Lyytinen & Yoo (2002). Our expansion to multiple (mobile) platforms and that across organizations addresses the individual level and inter-organizational level, respectively, of the research framework. Further, agent intelligence facilitates the individual, team, and organization level improvements in the research framework, while the study of inter-organizational agents is our prospective future work.

User's Perspective

Users require e-collaboration to assist their work. Expanding e-collaboration services to multiple platforms increases connectivity. The provision of anytime and anywhere connections is essential due to the fact that workforce tends to become mobile, especially for professionals such as physicians, service engineers, and sales executives as well as other staff who need to travel. In particular, the flexibility of supporting multiple front-end devices increases users' choice of hardware and therefore their means of connectivity. The adaptation of e-collaboration protocol design for different operating environment improves the ease of use. This helps overcome the impact of expanding system functionalities and operation environment.

Next, expanding e-collaboration support across organizations increase chances to interoperate with e-collaboration systems of other organizations and therefore improves the e-collaboration support with members of other organizations. The need for interoperability drives the use of the contemporary Web Services technology in our proposed e-collaboration service implementation.

As for agent intelligence, this proposition helps reduce tedious collaboration tasks that are often repeated, including meeting scheduling as well as negotiations with standardized parameter such as sales negotiation (Chiu et al., 2002). Agents help improve reliability and robustness of e-collaboration by retrying upon unsuccessful attempts, searching for alternatives, and so on.

We continue to explain why our proposed way of applying constraint technologies helps a balance of performance and privacy. Once all users' preferences are represented as constraints, conventional constraint solving method, such as systematic search, possibly enhanced by constraint propagation (Kumar, 1998), can be applied to find a feasible solution. However, since all these conventional solvers are centralized, it is inappropriate to employ any of them in a mobile setting. This is because employing a centralized solver implies having to require the collaborators to send all or part of their private information to a designated agent, which is supposed to find out a feasible solution to the appointment. Obviously, this is inappropriate as all collaborators are supposed to enjoy privacy protection and autonomy.

The conventional protocol described in the previous paragraph, which commonly known as *open-calendar protocol* is computational efficient yet provides no privacy protection. Another problem is that too much unnecessary data is sent. Therefore this approach wastes bandwidth and is not suitable for mobile users or agents. On the other extreme, the most privacy-protected protocol is to require, say, to make a sequence of specific suggestions to the proposer's agent. Each of these suggestions consists of a specific day, a specific time on that day, and a specific place. The proposer's agent then considers each of them and decides whether or not to accept the suggestion. This protocol, which we call the *passive mode*, is a simple, inefficient protocol (may be causing too many exchanges of short messages) that provides high degree of privacy protection. We can see that there is a spectrum of protocols in between these two extremes, which require the agents to exchange their private information to a certain degree. For example, in the *counter-offer* mode implemented, the proposer's agent can send several options to other collaborators, who are expected to indicate the feasibility of each of these individual options. The collaborators may also counter-propose by replying with a set of individualized constraints. As such, a balance among privacy protection, message exchange costs, and computational efficiency can be achieved.

Management's Perspective

A major concern of the management is the cost against opposed to the benefits of the expanded e-collaboration services. In particular, if any of the improvements to the organization's members (as discussed in the previous sub-section) can significantly help improve their productivities, the cost is justified. E-collaboration services provide tangible benefit for organizations by allowing information sharing between partners and participants. In addition, mobile extensions of e-collaboration systems usually imply the ability to locating mobile staff members and therefore improve staff communications. Though this may not be in the interest of every staff member, the mobile system infrastructure helps the management to control and manage them, such as for location dependent job allocation.

Next, incorporation of cross organizational e-collaboration services increases business opportunities. It also helps to improve relationships because of improved communications. In particular, through standardized technologies such as Web Services, the challenges lie in converging and interfacing different businesses across alternative organizations can be tackled in a proper approach. Web services provide a standard means to support interoperable service-to-service interaction over a network in conjunction with other Web-related standards. The disparity of heterogeneous organizational applications has created inflexible boundaries for communicating and sharing information and services among different organizations. Therefore, Web services technologies provide a standardized way to share the information and services among various heterogeneous applications, and also the standardized Web services interfaces among business operations can be established to take the convergent benefits of all the organizational applications.

Further benefits can be derived from the employing of intelligent agents. Agents help improve the quality and consistency of decision results through pre-programmed intelligence. The BDI agent architecture mimics the human practical deliberation process by clearly differentiate among the mental modalities of beliefs, desires, and intentions. Flexibility and adaptation are achieved by the agent's means- and ends-revising capabilities. As such, costs to program into the agent the operation and even management knowledge elicited are minimized. Also, expertise to handle practical problems can be incorporated into the options function to generate desires and the filter function that determines intentions.

As for cost factors, our approach are suitable for adaptation of *existing* systems as illustrated in the previous section. Through software reuse, a reduction in not only the total development cost but also training and support cost can be achieved. For security, as explained in the previous sub-section, the main benefit of using constraints is to reduce the need of revealing unnecessary information to collaborations. Thus, this improves the security when e-collaboration with other organizations is required.

System Developer's Perspective

System developers are often concerned about system development cost and subsequent maintenance efforts. These concerns can be addressed by systematic fine-grained requirement elicitation and the adaptation of existing systems using the widely-adopted three-tier architecture. Thus, with loosely coupled and tightly coherent software modules at different tiers, system complexity can be managed. These software modules are highly reusable and can be maintained with relative ease and minimal associated costs. Further, it should be noted that the use of XSL technologies and databases views as the main mechanism for adaptation in the front and backend tiers facilitates program maintenance at the application tier. This can significantly shorten the system development time, meeting management expectation to exceed the competitive edges.

Recent advances in technologies have resulted in fast evolving device models and Web standards. E-collaboration systems require much greater extents of adaptations to keep up with the blooming mobile technologies. Selected modules can be adapted to cope with new technologies and to allow for more comprehensive unit testing. Therefore, our adaptation-based approach further helps reduce uncertainties through adequate testing and experimentations of new technologies.

The proposed case study has been implemented using entry level PDAs (HP iPAQs); each unit is equipped with a 200MHz StrongArm processor and 32MB SDRAM. The implementation aimed at exploring the feasibility of supporting agents on

PDA platforms. The BDI agent model and the associated constraint solver were written in Microsoft embedded Visual C++ and executed under Windows CE. We found that an agent could comfortably solve 100 constraints with 200-300 variables in a single second. This problem size is typical for daily applications. As such, there is no need to rely on a powerful computational server to solve these constraints. In fact, distributed solutions of agents favor not only privacy but also scalability. It further eases the programming of captured knowledge as explained in the previous sub-section.

Another interesting issue related to methodology described earlier is the adaptation of displays to multiple platforms. Although it is possible to convert automatically XML into different output markup languages (such as HTML or WAP), the solution is often not satisfactory. However, further customization of information displays, including summarization, on different platforms needs prior knowledge of the information formats and contents. In fact, it was found that most of the development costs in our case study lie in this category of adaptation. We discovered problems similar to those identified in our previous work on contract-template based negotiation (Chiu et al., 2005). In daily life, the policies and requirements of recurring negotiations are similar. As such, contract templates with parameters are formulated. Common issues and options for these negotiations are identified based on their typical usages. These templates and sample tradeoff views are then stored in a repository and available for reuse and user adaptation. In particular, customization for the issues and options to be displayed can be stored at the templates.

To overcome the limitation of the small screen size of the available mobile devices, the decomposition of a complex collaboration problem into smaller sets of related issues with our developed methodology further helps. For example, a collaboration process for organization a function can be decomposed into three sets of issues like ({meeting place, time}, {activities, cost}, {cost-sharing}) and then resolved by a single set. As such, this collaboration process can be broken down into three rounds of collaboration tasks and executed with the protocol in this paper. As each round has a smaller number of tradeoff issues, the screens are thus simpler and more convenient for mobile users, overcoming both the device and recognition limitations.

As constraints can be used to express general planning problems, including those involving higher order logic (Tsang, 1993), our protocol though looks simple is adequate. It can be applied in different domains for solving different collaboration processes. For formal validation of the protocol for any particular collaboration process, Chiu et al. (2004) have also developed a methodology for consistency check based on process algebra and automata theory.

CONCLUSIONS AND OUTLOOK

In conclusion, this article has presented a pragmatic methodology of expanding e-collaboration support into a MAS environment, to support multiple platforms, in particular wireless mobile interfaces, together with accommodating for cross-organizational e-collaboration through enterprise Web Services. We have discussed practical implementation issues and adaptation required for different platforms based on a three-tier architecture and constraint technologies. We have demonstrated the support for customizable degree of agent delegation and users without agent support. We have also illustrated the feasibility of our methodology by utilizing a case study of meeting scheduling. We further evaluate our implementation framework and methodology with respect to the main stakeholders, including users, management, and system developers. Compared with other research frameworks on this topic, our approach employs an improved environment through standard state-of-the-art technologies, which can adapt to changing requirements of mobile users and devices, with extensive support for reuse.

As for outlook, primarily, we have observed that Web services with the Service-Oriented Architecture (SOA) is gradually being adopted for cross-organizational collaboration in a wide range of businesses (such as Lee et al., 2007) and industries (such as Chung et al., 2007), as well as involving governments (such as Wong et al., 2007). We have recently investigated in details on how Web services can be employed for the implementation of our MAS framework (Chiu et al., 2009). Among a variety of issues, security and privacy is one of the key emerging issues for research and practice. We advocate a judicious of Web services as a solution approach because Web services can provide a single-boarder check to information access management (e.g., facilitating logging and monitoring) together with other emerging Web service standards for security and privacy protection (Hung et al., 2007).

At the same time, mobile extensions to existing services as well as emerging applications are also getting popular. The increasing demand of anytime and anywhere access; together with the rapidly increasing number of available information and services has further led to recent developments in ubiquitous computing. We expect ubiquitous support of e-collaboration to be one important direction of relevant evolution and continuing research (Chiu et al., 2005). Towards this, we have shown that our three-tier implementation framework extended with view-based adaptation model based on context is a promising approach (Hong et al., 2007).

We perceive that our approach is applicable to other group collaboration systems and decision support systems. In fact, apart from using this for negotiation support systems (Chiu et al., 2004b), we have extended this approach to a comprehensive system architecture for mobile workforce management with multi-agent clusters (Chiu et al., 2006). Employing multi-agent clusters is one potential way to address the scalability issue. The scalability issue is one of the main reasons that agent technologies are gradually being adopted. We also apply this approach, further extended with the use of ontologies from Semantic Web technologies, for a next generation multi-platform tourist assistance system (Chiu et al., 2009). We perceive the use of semantics for collaboration, especially for automated e-collaborations, is a key research direction to enable stakeholders to understand one another so that they all can communicate and work together symbiotically. In particular, we have shown how semantics can help users to understand their requirements and therefore facilitate matchmaking, recommendation, and negotiation (Chiu et al., 2005).

Last but not least, we are expecting a gradual adoption of a variety of new technologies, which could potentially revolutionized the engineering of e-collaboration services in the future. We advocate a phase approach to such an adoption emphasizing the potential of software reuse, which we have demonstrated in the architecture of this paper. We are working towards a more detailed methodology. Together in combination with other practical evaluations we hope to summarize our current and ongoing experiences in service-oriented systems engineering in the near future.

ACKNOWLEDGMENT

This paper is partially supported by the National Natural Science Foundation of China under Grant No. 60873022 and 60736015, the Natural Science Foundation of Zhejiang Province of China under Grant No. Y1080148, and the Open Project of Zhejiang Provincial Key Laboratory of Information Network Technology of China.

REFERENCES

Bafoutsou, G., & Mentzas, G. (2001). A Comparative Analysis of Web-based Collaborative Systems. In *Proceeding 12th International. Workshop Database and Expert Systems Applications* (pp. 496-500).

Bui, T. X. (1987). *Co-oP: a Group Decision Support System for Cooperative Multiple Criteria Group Decision Making*, Springer, LNCS 290.

Bui, T. X., Bodart, F., & Ma, P.-C. (1998). ARBAS: A Formal Language to Support Argumentation in Network-Based Organization. *Journal of Management Information Systems*, *14*(3), 223–240.

Carlson, D. (2001). *Modeling XML Applications with UML*. Addison-Wesley.

Chavez, A., & Maes, P. (1996). Kasbah: An Agent Marketplace for Buying and Selling Goods. In *Proceedings of the 1st International Conference on the Practical Application of Intelligent Agents and Multi-Agent Technology*, 75-90.

Chiu, D. K. W., Cheung, S. C., & Hung, P. C. K. (2002). A Contract Template Driven Approach to e-Negotiation Processes. In *Proceedings of the 6th Pacific Asia Conf. on Information Systems*, Tokyo, Japan, CDROM.

Chiu, D. K. W., Cheung, S. C., Hung, P. C. K., Chiu, S. Y. Y., & Chung, K. K. (2005). Developing e-Negotiation Process Support with a Metamodeling Approach in a Web Services Environment. *Decision Support Systems*, *40*(1), 51–69. doi:10.1016/j.dss.2004.04.004

Chiu, D. K. W., Cheung, S. C., Hung, P. C. K., & Leung, H.-F. (2004). Constraint-based Negotiation in a Multi-Agent Information System with Multiple Platform Support. In *Proceedings of the 37th Hawaii International Conference on System Sciences* (HICSS37), Big Island, Hawaii, CDROM, IEEE Computer Society Press.

Chiu, D. K. W., Cheung, S. C., Kafeza, E., & Leung, H.-F. (2004). A Three-Tier View Methodology for adapting M-services. *IEEE Transactions on Systems, Man and Cybernetics. Part A*, *33*(6), 725–741.

Chiu, D. K. W., Cheung, S. C., & Leung, H.-F. (2006). Mobile Workforce Management in a Service-Oriented Enterprise: Capturing Concepts and Requirements in a Multi-Agent Infrastructure. In R. Qiu (Ed.), *Enterprise Service Computing: From Concept to Deployment. Enterprise Service Computing: From Concept to Deployment*, Idea Group Publishing.

Chiu, D. K. W., Karlapalem, K., Li, Q., & Kafeza, E. (2002). Workflow Views Based E-Contracts in a Cross-Organization E-Service Environment. *Distributed and Parallel Databases*, *12*(2-3), 193–216. doi:10.1023/A:1016503218569

Chiu, D. K. W., Yueh, Y. T. F., Leung, H.-f., & Hung, P. C. K. (2009). Towards Ubiquitous Tourist Service Coordination and Process Integration: a Collaborative Travel Agent System with Semantic Web Services. *Information Systems Frontiers*, *11*(3), 241–256. doi:10.1007/s10796-008-9087-2

Choy, M. C., Srinivasan, D., & Cheu, R. L. (2003). Cooperative, Hybrid Agent Architecture for Real-time Traffic Signal Control. *IEEE Transactions on Systems, Man and Cybernetics. Part A*, *33*(5), 597–607.

Chung, J. C. S., Chiu, D. K. W., & Kafeza, E. (2007). An Alert Management System for Concrete Batching Plant. In *Proceedings of the 12th International Conference on Emerging Technologies and Factory Automation* (pp. 591-598), Patras, Greece.

Dincbas, M., Simons, H., & van Hentenryck, H. (1998). Solving the Car-Sequencing Problem in Constraint Logic Programming. In *Proceeding of ECAI-88* (pp. 290-295).

Elmasri, R. A., & Navathe, S. B. (2006) *Fundamentals of Database Systems* (5th ed.). Addison-Wesley.

Fraile, J.-C., Paredis, C. J. J., Wang, C.-H., & Khosla, P. K. (1999). Agent-based Planning & Control of a Multi-manipulator Assembly System. In. *Proceedings of the IEEE International Conference on Robotics & Automation, 2*, 1219–1225.

Garrido, L., Brena, R., & Sycara, K. (1996). Cognitive Modeling & Group Adaptation in Intelligent Multi-Agent Meeting Scheduling. In *Proceedings of the 1st Iberoamerican Workshop on Distributed Artificial Intelligence & Multi-Agent Systems* (pp. 55–72).

Gerst, M. H. (2003). The Role of Standardisation in the Context of E-collaboration: ASNAP shot. In *Proceedings of the 3rd Conferernce on Standardization & Innovation in Information Technology* (pp. 113-119).

He, M., Jennings, N. R., & Leung, H.-F. (2003). On agent-mediated electronic commerce. *IEEE Transactions on Knowledge and Data Engineering, 15*(4), 985–1003. doi:10.1109/TKDE.2003.1209014

Hong, D., Chiu, D. K. W., Cheung, S. C., Shen, V. Y., & Kafeza, E. (2007). Ubiquitous Enterprise Service Adaptations Based on Contextual User Behavior. *Information Systems Frontiers, 9*(4), 343–358. doi:10.1007/s10796-007-9039-2

Hung, P. C. K., Chiu, D. K. W., Fung, W. W., Cheung, W., Wong, R., & Choi, S. P. M. (2007). End-to-End Privacy Control in Service Outsourcing of Human Intensive Processes: A Multi-layered Web Service Integration Approach. *Information Systems Frontiers, 9*(1), 85–101. doi:10.1007/s10796-006-9019-y

Kumar, V. (1998). Algorithms for Constraint-Satisfaction Problems: A Survey. *AI Magazine, 13*(1), 32–44.

Lee, R. C. M., Mark, K. P., & Chiu, D. K. W. (2007). Enhancing Workflow Automation in Insurance Underwriting Processes with Web Services and Alerts. In *Proceedings of the. 40th Hawaii International Conference on System Sciences*, Big Island, Hawaii, CDROM, IEEE Computer Society Press.

Lin, F.-R., & Pai, Y.-H. (2000). Using Multi-agent Simulation & Learning to Design New Business Processes. *IEEE Transactions on Systems, Man and Cybernetics. Part A, 30*(3), 380–384.

Lin, F.-R., Tan, G. W., & Shaw, M. J. (1998). Modeling Supply-chain Networks by a Multi-agent System. In *Proceedings of the 31st Hawaii International Conference on System Sciences* (HICSS31), *5*, 105-114.

Lin, Y.-B., & Chlamtac, I. (2000). *Wireless & Mobile Network Architectures*. John Wiley & Sons.

Liu, L., Song, H., & Liu, Y. (2001). HDBIS Supporting E-collaboration in E-business. In *Proceedings of the 6th International Conference on Computer Supported Cooperative Work in Design* (pp. 157-160).

Lo, G., & Kersten, G. K. (1999). Negotiation in Electronic Commerce: Integrating Negotiation Support & Software Agent Technologies. In *Proceedings of the 29th Atlantic Schools of Business Conference*.

Luo, Y., Liu, K., & Davis, D. N. (2002). A Multi-agent Decision Support System for Stock Trading. *IEEE Network, 16*(1), 20–27. doi:10.1109/65.980541

Lyytinen, K., & Yoo, Y. (2002). Research Commentary: The Next Wave of Nomadic Computing. *Information Systems Research, 13*(4), 377–388. doi:10.1287/isre.13.4.377.75

Naiburg, E. J., & Maksimchuk, R. A. (2001) *UML for Database Design*. Addison-Wesley.

Rao, A. S., & Georgeff, M. P. (1995). BDI Agents: from theory to practice. In *Proceedings 1st International Conference on Multiagent Systems* (pp. 312-319).

Rutkowski, A. F., Vogel, D. R., van Genuchten, M., Bemelmans, T. M. A., & Favier, M. (2002). E-collaboration: the Reality of Virtuality. *IEEE Transactions on Professional Communication, 45*(4), 219–230. doi:10.1109/TPC.2002.805147

Sandip, S. (1997). Developing an Automated Distributed Meeting Scheduler. *IEEE Expert, 12*(4), 41–45. doi:10.1109/64.608189

Shakshuki, E., Ghenniwa, H., & Kamel, M. (2000). A Multi-agent System Architecture for Information Gathering. In *Proceedings of the 11th International Workshop on Database & Expert Systems Applications* (pp. 732-736), Los Alamitos, CA, USA.

Shitani, T., Ito, T., & Sycara, K. (2000) Multiple Negotiations among Agents for a Distributed Meeting Scheduler. In *Proceedings of the 4th International Conference on MultiAgent Systems* (pp. 435-436).

Stroulia, E., & Hatch, M. P. (2003). An Intelligent-agent Architecture for Flexible Service Integration on the Web. *IEEE Transactions on Systems, Man and Cybernetics. Part C, 33*(4), 468–479.

Sycara, K., & Zeng, D. (1996). Coordination of Multiple Intelligent Software Agents. *International Journal of Cooperative Information Systems, 5*(2&3), 181–212. doi:10.1142/S0218843096000087

Teich, J., Wallenius, H., & Wallenius, J. (1999). Multiple-Issue Auction and Market Algorithms for the World Wide Web. *Decision Support Systems, 26*(1), 49–66. doi:10.1016/S0167-9236(99)00016-0

Tsang, E. (1993). *Foundations of Constraint Satisfaction*. Academic Press.

van Lamsweerde, A., Darimont, R., & Massonet, P. (1995). Goal-Directed Elaboration of Requirements for a Meeting Scheduler: Problems & Lessons Learnt. In *Proceedings of the 2nd IEEE International Symposium on Requirements Engineering* (pp. 194-203).

Wang, M., Wang, H., Vogel, D., Kumar, K., & Chiu, D. K. W. (2009). Agent-Based Negotiation and Decision Making for Dynamic Supply Chain Formation. *Engineering Applications of Artificial Intelligence, 4*(2), 36–56.

Weerawarana, S., Curbera, F., Leymann, F., Storey, T., & Ferguson, D. F. (2005). *Web Services Platform Architecture: SOAP, WSDL, WS-Policy, WS-Addressing, WS-BPEL, WS-Reliable Messaging, and More*, Prentice Hall.

Wegner, L., Paul, M., Thamm, J., & Thelemann, S. (1996). A Visual Interface for Synchronous Collaboration and Negotiated Transactions. In *Proceeding of Workshop on Advanced Visual Interfaces* (pp. 156-165).

Weverka, P. (2001). *Mastering ICQ: The Official Guide*. Hungry Minds.

Wong, J. Y. Y., Chiu, D. K. W., & Mark, K. P. (2007). Effective e-Government Process Monitoring and Interoperation: A Case Study on the Removal of Unauthorized Building Works in Hong Kong. In *Proceedings of the 40th Hawaii International Conference on System Sciences*, Big Island, Hawaii, CDROM, IEEE Press.

Yueh, Y. T. F., Chiu, D. K. W., Leung, H. F., & Hung, P. C. K. (2007). A Virtual Travel Agent System for M-Tourism with a Web Service Based Design and Implementation. In *Proceedings of the IEEE 21st International Conference on Advanced Information Networking and Applications*, Niagara Falls, Canada.

This work was previously published in International Journal of Systems and Service-Oriented Engineering, Volume 1, Issue 1, edited by Dickson K.W. Chiu, pp. 1-25, copyright 2010 by IGI Publishing (an imprint of IGI Global).

Chapter 2
A Survey of Web Services Provision

An Liu
University of Science & Technology of China, CityU-USTC Advanced Research Institute, and City University of Hong Kong, China

Hai Liu
University of Science & Technology of China, CityU-USTC Advanced Research Institute, and City University of Hong Kong, China

Baoping Lin
University of Science & Technology of China, CityU-USTC Advanced Research Institute, and City University of Hong Kong, China

Liusheng Huang
University of Science & Technology of China, and CityU-USTC Advanced Research Institute, China

Naijie Gu
University of Science & Technology of China, and CityU-USTC Advanced Research Institute, China

Qing Li
CityU-USTC Advanced Research Institute, and City University of Hong Kong, China

ABSTRACT

Web services technologies promise to create new business applications by composing existing services and to publish these applications as services for further composition. The business logic of applications is described by abstract processes consisting of tasks which specify the required functionality. Web services provision refers to assigning concrete Web services to perform the constituent tasks of abstract processes. It describes a promising scenario where Web services are dynamically chosen and invoked according to their up-to-date functional and non-functional capabilities. It introduces many challenging problems and has therefore received much attention. In this article, the authors provide a comprehensive overview of current research efforts. The authors divide the lifecycle of Web services provision into three steps: service discovery, service selection, and service contracting. They also distinguish three types of Web services provision according to the functional relationship between services and tasks: independent provision, cooperative provision and multiple provision. Following this taxonomy, we investigate existing works in Web services provision, discuss open problems, and shed some light on potential research directions.

DOI: 10.4018/978-1-4666-1767-4.ch002

INTRODUCTION

Web services are rapidly emerging as a new paradigm for developing and deploying business processes within and across enterprises. The reason for this great success achieved in Web services arena can be boiled down to its declarative lookup and invocation modes. In particular, Web services highly rely on some descriptive XML-based artifacts to accomplish communication and interaction, and henceforth some de-facto standards, such as SOAP, WSDL, and UDDI (Curbera et al., 2002), are speedily shaping to accelerate the development of Web services agenda. Currently, enterprises encapsulate their internal business processes as Web services and publish them into public directories such as UDDI so that other enterprises can invoke these business processes through well defined service interface in their business processes. Generally, a business process contains a number of tasks describing the required business functionalities. At runtime, appropriate services need to be chosen to perform these tasks.

Web services provision refers to assigning concrete Web services to perform the constituent tasks of abstract processes. In general, the lifecycle of service provision consists of three phases. First, it is necessary to discover services that can perform the functionality defined by a task. Considering the fact that many services provided by different organizations have the similar functionality, a large number of services may be obtained after the discovery phase. However, these services can be further distinguished according to their non-functional properties, that is, quality of service (QoS). Therefore, in the second phase, the best service is selected for each task from the candidates according to their QoS. Finally, the composite service makes a contract with the selected services regarding usage requirements both at high business level and low interface level.

In the real applications, however, the above three basic steps may become more complicated due to service capability. In particular, there are three types of relations between services and tasks from a functional point of view: 1) a task can be fulfilled by one service and this service can only perform this task; 2) a task cannot be fulfilled by one service but can be carried out by a set of services; 3) a service can accomplish multiple tasks in a business process. According to these relations, we distinguish three types of provision: independent provision, cooperative provision and multiple provision.

Even in the same phase of Web services provision, for example, the service discovery phase, different provision types will introduce different research issues. Up to now, a large variety of research works and industrial progress have been made in these aforementioned arenas. In this article, we provide a survey to review these works, including their contributions and potential improvements.

The rest of the article is organized as follows. Section 2 introduces a motivating example for Web services provision. Section 3 provides a survey on Web services discovery, including simple services matchmaking, services planning and process matchmaking. Section 4 presents the recent advances in QoS-aware service selection. Section 5 introduces semantic Web services contracting and heterogeneity mediation. Finally, Section 6 discusses some open problems and concludes the article.

MOTIVATING EXAMPLE

To illustrate Web services provision and provide application requirements for the research issues that we focus on in this survey, we describe here a classic scenario – travel agency – in which a customer makes a plan for his trip through a travel agent.

Suppose Bob wants to plan a trip with his family to celebrate his birthday. At first, he and his family come to an agreement that the scenic spot should be close to mountains and not be too hot, and they can go there directly by plane. Then, Bob sends a travel agent these requirements based on which candidate scenic spots can be searched. After finding these scenic spots, the agent contacts a flight company to see whether enough tickets are available. If not, it notifies Bob the unfortunate result; otherwise, it tells Bob to select and confirm one scenic spot and corresponding flight tickets. As soon as Bob acknowledges the initial plan, the agent starts to confirm the expected flight. After that, it reserves hotel rooms nearby and tickets for the scenic spot. Finally, Bob is informed of the plan details including flight tickets, scenic spot tickets, and hotel rooms.

In this context, the agent is developed by the emerging Web services technologies. In particular, it is implemented as a Web service described by WSDL and published in UDDI so that Bob can easily find and communicate with it via a personal agent service located, for example, in his laptop or mobile phone. Besides, the agent's internal business process that is used to fulfill its advertised functionality is specified by WS-BPEL (Jordan and Evdemon, 2007) that actually assembles operations supplied by different Web services. Figure 1 illustrates a graphical WS-BPEL process of the travel agent, which is designed by the ActiveBPEL Designer[1] software. In this process, each invoke activity directs a Web service to perform a one-way or request-response operation. Note that these activities only specify partner links and operations to invoke, and do not state which services are used to perform these operations. Web services provision, however, aims at assigning services to operations so that the WS-BPEL process can be executed.

Through this example, we can distinguish three types of service provision as follows:

- One service for one operation (independent provision). This is the simplest type of service provision in which an operation can be fulfilled by one service. In particular, the process sends a request to a service and then waits for the response from the service. For example, a service named ABC is found and invoked by the travel agent to perform the 'InvokeHotelRes' operation in Figure 1.
- Multiple services for one operation (cooperative provision). The above type is fundamental to service provision and can be considered as a basic assumption for a large number of research issues. In the real service applications, however, the functionality of an operation may be so complex that it cannot be fulfilled by a single service. For example, the 'InvokeScenicSpotSearch' operation in Figure 1 needs to recommend several candidate scenic spots based on user requirements such as the scenic spot should be close to mountains and not be too hot. To judge whether a scenic spot could be a candidate, at least two types of queries should be processed: map query and weather query. Unfortunately, not a single service can provide both query functions. It is therefore necessary to find a set of services that can cooperate to achieve the expected functionality. In this case, multiple services constitute a virtual composite service to fulfill an invoke operation.
- One service for multiple operations (multiple provision). Another type of service provision is a service can be used to fulfill multiple operations in a process. For example, a flight booking service XYZ can be assigned to two operations 'InvokeFlightRequest' and 'InvokeFlightRes' in the travel agent example. It should be noted that the operations performed by service XYZ actually have an implicit order in the agent process.

In the next three sections, we will investigate current works in the three provision phases. In each section, we will consider different research issues introduced by different provision types, and their corresponding solutions.

Figure 1. WS-BPEL process of travel agent

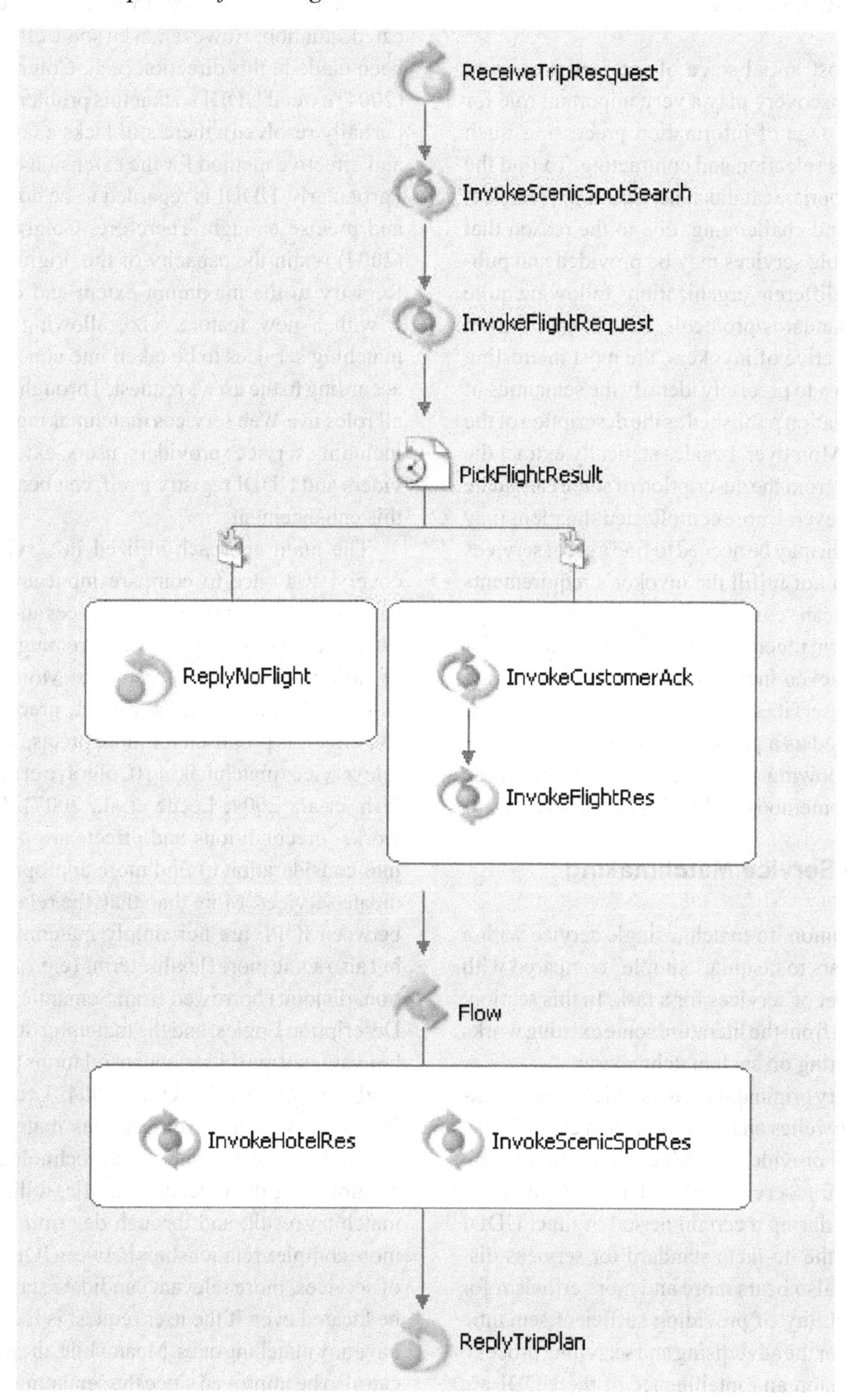

SERVICE DISCOVERY

As the most initial stage of services provision, services discovery play a very important role for the latter stage of information processing, such as services selection and contracting. To find the most appropriate candidate services is nevertheless difficult and challenging, due to the reason that the available services may be provided and published by different organizations following quite distinct standards/protocols. In this regard, from the perspective of invokers, the most interesting issue is how to precisely identify the semantics of the information published as the description of the services. Moreover, besides statically extract the semantics from the description of some candidate services, several more complicated situations may arise, e.g., it may be needed to find a set of services which can not fulfill the invoker's requirements solely, but can "cover" these requirements through putting them together. A more intellectual way for discovery even includes the dependency among candidate services, so that these services can be orchestrated as a process to satisfy the invokers. In the following subsections, we discuss these discovery methods in detail.

Simple Service Matchmaking

In our opinion, to match a single service with a task appears to be quite "simple" compared with to find a set of services for a task. In this section, we review from the literature some existing works concentrating on such matchmaking.

The very original stage for services matchmaking mainly relies on UDDI (Clement et al., 2004), which can provide both advertising and locating methods for services providers and invokers. Although during a certain period of time, UDDI serves as the de-facto standard for services discovery, it also bears more and more criticism for its incapability of providing sufficient semantic supports for the advertising and searching process. The precision and intelligence of the UDDI are considered to be enhanced to handle more complicated situation. However, even some efforts have been made in this direction (e.g., Colgrave et al. (2004) extend UDDI so that this problems can be partially resolved), there still lacks a consensual and effective method for the extension of UDDI. Particularly, UDDI is regarded to be not flexible and precise enough. Therefore, Colgrave et al. (2004) retain the capacity of the original UDDI Registry to the maximum extent and enhanced it with a new feature, viz., allowing external matching services to be taken into consideration according to the user's request. Through this way, all roles in a Web services matchmaking process, including services providers, users, external providers and UDDI registry itself, can benefit from this enhancement.

The main approach utilized in services discovery was once to compare input and output parameters between provided services and request. This method is useful but also not enough in front of various semantic requirements. More specifically, IOPE model (input, output, precondition, and effect) is proposed for more precise and flexible services matchmaking (Colgrave et al., 2004; Sirin et al., 2004; Lecue et al., 2007). In these works, preconditions and effects are also taken into consideration to find more appropriate candidate services. More than that, the relationships between IOPE are not simply matching or not, but also some more flexible terms (e.g., subsumption, disjoint) borrowed from Semantic Web and Description Logics, and the matching degree will be evaluated via those concept relations (Paolucci et al., 2002; Colgrave et al., 2004; Lecue et al., 2007). Conducting Web services matchmaking on the basis of semantic Web technology aims at improving the precision and flexibility of the matching result, and through determining these more complex relationships between IOPE modes of services, more relevant candidate services can be located even if the user request is too strict to have any matching ones. Meanwhile, the precision can also be improved since the semantics existing in services description can be somehow exploited even not so complete and comprehensive.

However, the aforesaid works suffer from a common problem, viz., their underlying logics are incapable of representing any numerical constraint, not to mention process modeling and thereby the semantic gap among different partner services can be bridged automatically. This defect can be vital for Web service contracting and composition since the agreement about the QoS and partner services' behaviors is the focus of a contracting or composition process.

Complex Services Discovery: Covering and Planning

In the previous subsection, some approaches for "simple" services matchmaking are surveyed. However, since a number of Web services are provided by different organizations and there are still no enough agreements or standards for all aspects of Web services designing and provision, a particular situation is usually be considered: what if no any single Web service is located for a user request? In the context of current Web technology and development, we should be aware of the fact that it is likely to occur, but the methods to resolve it are still premature. Although some works are dedicated to this aspect, this problem may still stay in academic stage but does not enter into practice. In this subsection, we give a survey on these methods.

The most straightforward way on this aspect should be to put a finite set of services together such that any service can not fulfill the user request solely, but they can provide an assembled functionality which is sufficient for the user. Even the idea behind the approach seems to be quite simple, the problem itself is indeed more complex (NP-hard) (Benatallah et al., 2005). The technical difficulties about this problem also include how to extract unambiguous semantics from services description (input and output parameters), and how to design a sound yet efficient algorithm to resolve the problem. In the work done by Benatallah et al. (2005), the former difficulty is overcome through adopting a variant of description logics, called ALN, to underpin the services description. Meanwhile, with the intention to verify whether or not the assembly of a set of candidate services is a best cover, Benatallah et al. (2005) introduce difference operator, which is proposed by Teege (1994), to calculate the "missing part" of a query. The latter difficulty is tackled by using hypergraph theory and greedy algorithm, which sits on the basis of a Description Logic ALN and compute a set of service descriptions so as to semantically "cover" the original request.

Although the aforementioned works provide a possible method to aggregate a finite set of Web services automatically so as to accomplish a single request, it did not take execution order into consideration and the according consequence is these services have to be invoked concurrently. To overcome this limitation, Brogi et al. (2008) propose a new algorithm, namely, SAM (Services Aggregation Matchmaking). Their method is also compatible with OWL-S standards, and will sequentialize the returned result services so that the client can invoke these services one by one. Moreover, if there is no suitable aggregation, this method will provide further recommendation and suggestion to enhance its flexibility.

However, in previous methods, no dependency and interaction among Web services are taken into account. In the following, we introduce some other works focusing on that aspect:

In the work done by Aversano et al. (2004), the authors consider a very complex situation: when a perfectly matching service can not be discovered, they will automatically compose a set of Web services to fulfill the function. Although this algorithm is claimed that it can automatically compose suitable set of services to fulfill user's goal, there is still some potential problems which are not clarified, such as the termination of the algorithm, and the complexity of the heuristic method.

It should be aware of that a number of works concentrating on automatic Web services composition, which takes planning techniques into consideration. Although these works does not directly impact on Web services discovery, they still contribute to the planning-based automatic Web services discovery. Ponnekanti and Fox (2002) make use of the input and output parameters of a set of candidate services, and using back-tracing technique to provide a planning result for a composition requirements. Agarwal et al. (2005) distinguish two types of composition: logical composition and physical composition, and develop a tool to fulfill them. In the work done by Sirin et al. (2004), Description Logic-based services model are created and the planning process mainly relies on the semantics of services description.

However, the previous works share a same defect, viz., they did not take into consideration the semantics existing in Web services descriptive document. Meanwhile, another problem deserving more attention is the efficiency of composition planning process, since it has been identified to be a particularly time-consuming process. In that spirit, Kona et al. (2007) provides a comparatively well-formed semantic Web service model, which takes both dataflow parameters (input and output) and execution effect into account. Moreover, they also use multi-step narrowing algorithm and constraint logic programming technique to improve the performance of their approach, so as to resolve the efficiency problem of automatic Web services composition.

Process-Aware Service Matchmaking

As illustrated by our motivating example, there can be three types of service provision. The former two types, independent provision and cooperative provision, are more traditional and have attracted a lot of attentions in Web services area. While the third type, multiple provision, is still an undeveloped and difficult issue. One important reason is that multiple provision involves process matching, which can be very complex. This subsection reviews some significant efforts on process-aware service matchmaking.

Taking process information into the service matchmaking can increase the matching preciseness between user-requested behavior and service-provided behavior. Currently, most of the research and industry adopt BPEL as the process description language. Therefore, the typical scenario of process matching is the comparison between two BPEL process models defined by the user and service. Konig et al. (2008) discuss the issue of whether an executable BPEL process is compatible to an abstract one. The authors propose a novel profile that extends the existing Abstract Process Profile for Observable Behavior (APPOB) by defining a behavioral relationship. This effort can facilitate the direct comparison between BPEL processes. However, due to the inherent weakness of direct BPEL comparison, more work focused on the formal model based process matching.

One of the main formal models used to describe process is Petri nets. By using Petri nets, huge theoretical investigations with a wide range of efficient algorithms are directly applicable to BPEL processes. Martens (2003) define the compatibility of Web services based on Petri nets modeling. They decide the compatibility of Web services based on whether the composed system of the two Web services is usable. Martens et al. (2006) present a business process analysis framework that integrates Petri nets into IBM's business integration tools. The framework first transforms BPEL into Petri nets, and then analyzes the result Petri nets. Aalst et al. (2008) also translates BPEL into Petri nets, and has developed Petri net-based conformance checking techniques for service behaviors. The proposed conformance checking techniques can be utilized for process matching.

The other main formal model is finite-state machine (FSM). FSM constitutes the core of state-charts, which are one of the main components of UML and are becoming a widely used formalism for specifying the dynamic behavior of entities. Martens (2005) transform a BPEL process into a communication graph, which is a finite deterministic automaton. The service broker then compares the requested behavior and published behavior based on communication graphs. The equivalence of processes is defined by using the simulation of their communication graphs. Wombacher et al. (2004) give a formal semantics to business process matchmaking based on finite state automata extended by logical expressions associated to states. Lin et al. (2007) also uses FSM to model the external behavior of Web services–Choreography. To derive the choreography compatibility, the state-level compatibility is first defined. Then a compatibility checking algorithm is designed based on the definitions. Bordeaux et al. (2004) survey, propose and compare a number of formal definitions of the compatibility notion based on a Labeled Transition System, the article defines two Web services are compatible if they have opposite behaviors.

Besides the above two formal models, Web services researchers also use Pi calculus to model processes. For example, Brogi et al. (2004) adopt a subclass of Pi calculus – CCS to capture the semantics of Web service choreographies, and discuss the compatibility of them based on the CCS models. Agarwal and Studer (2006) also use Pi calculus to model Web service processes and study the process matchmaking problem by using the formal semantics of Pi calculus.

SERVICE SELECTION

Due to the increasing number of available services, each task in a composite service can be performed by a large amount of services that offer the same functionality. We have discussed how to identify the candidate services for each task by service discovery, and we now move on to service selection, that is, how to choose for each task the most appropriate service from a set of candidates. Though the candidates for a task have the same functionality, they can be differentiated and selected by their non-functional (i.e., QoS) properties, which is known as QoS-aware service selection. Unlike service discovery, current works on QoS-aware service selection focus on independent provision. Therefore, in this section, we review the works on dependent provision from three aspects. The first aspect is QoS modeling that provides means to define, collect, and store service QoS. The second aspect is QoS aggregation, that is, to predict the QoS of a composite service according to the QoS of its component services. The last aspect is efficient selection algorithms since QoS-aware service selection is actually a NP-Hard problem (Bonatti and Festa, 2005). Next, we discuss these parts in details.

QoS Modeling

QoS is a broad concept that includes a number of non-functional properties. As the foundation of QoS-aware service selection, it is necessary to identify important and meaningful non-functional properties of Web services. By examining existing commercial services in daily life, Sullivan et al. (2002) summarize eight QoS attributes for electronic (Web) services, including temporal and spatial availability, channels, charging styles, settlement models, settlement contracts, payment, service quality, security, trust and ownership. While this piece of work investigates the non-functional properties of Web services at the business level, a lot of other works consider QoS attributes of Web services from the traditional distributed systems viewpoint since Web services are actually software applications that can be accessed via the Internet. Menasce (2002) presents availability, security, response time, and throughput as Web services QoS attributes. Ran (2003) divides QoS attributes into four categories: runtime related, transaction support related, configuration management and cost related, as well as security related. Zeng et al. (2003) consider five QoS attributes for both elementary services and composite services, including execution price, execution duration, reputation, reliability, and availability. The above works mainly discuss generic QoS attributes. Meanwhile, some domain specific QoS attributes (Aggarwal et al., 2004) are also mentioned in some works. Liu et al. (2004) study the usability of Web services in the phone service provision domain from three aspects: transaction, compensation rate, and the penalty rate. Canfora et al. (2006) consider cost, color depth, and resolution in the image manipulation domain. Though these works do not (and clearly cannot) enumerate all QoS attributes in all domains, they reveal an important fact that particular attentions must be paid to QoS-aware service selection in concrete domains since the aggregation of domain-specific QoS attributes is much different from that of generic ones. We will discuss this issue further in the next subsection.

As Web services are provided by different organizations, a common QoS attribute usually have different explanations. For example, the term availability, which is occurred frequently in the Web services QoS vocabulary, has different definitions (Shao et al., 2008). Zeng et al. (2003) compute its value by the following formula: T_a/θ T_a / θ, where T_a is the total amount of time in which a service is available during the last θ seconds. Mikic-Rakic et al. (2005) quantify availability as the ratio of the number of successfully completed interactions to the total number of attempted interactions over a period of time. Rosenberg et al. (2006) compute availability by the following formula: 1−*downtime/uptime*, where downtime and uptime are measured in minutes. From this example, it is clear that standard definitions of QoS attributes are necessary. Otherwise, it is hard to come up with correct QoS contracts between service providers and service consumers. Therefore, some researchers propose to describe service QoS by using ontology. Maximilien and Singh (2004) distinguish three ontologies for QoS: upper ontology that captures the most generic quality concepts and defines the basic concepts associated with a quality, such as quality measurement and relationships, middle ontology that incorporates several QoS attributes encountered in distributed systems such as availability and performance, as well as lower ontology. Though the proposed ontology is quite complete, the metrics concept is absent. Zhou et al. (2005) overcome this shortcoming by introducing an ontology that consists of three layers: QoS profile layer that is used for QoS matchmaking, QoS property definition layer that constrains the property's domain and range information, and QoS metrics layer that defines proper QoS metrics and precise meaning for service measurement.

Another issue in Web services QoS modeling is QoS measurement, that is, to determine service QoS values. Liu et al. (2004) distinguish two types of QoS attributes: deterministic and non-deterministic. For deterministic attributes, for example, the execution price of a service, their values are decided by service providers. For non-deterministic attributes, for example, the execution duration of a service, their values are decided either by active monitoring or by user feedbacks. According to the source of QoS attributes measures, Gibelin and Makpangou (2005) distinguish three categories of QoS attributes: service access attributes whose values are given by service providers, service delivery attributes whose values are provider by Web services proxies, and feedback metrics whose values are specified by service consumers. Serhani et al. (2005) devise a QoS broker that can automatically generate a number of test cases to measure the exact QoS values. Compared with the test case approach, user feedback has the advantage of up-to-date and objective QoS values that are collected from actual consumption of services. However, it is hard to guarantee the accuracy of QoS values due to the following two reasons. Firstly, the Internet is an open environment and therefore there may be some malicious users. Secondly, users usually have different preference and service usage context such as network bandwidth, so they likely return different scores even for the same service. To solve the first problem, Jurca et al. (2007) propose an award mechanism to encourage users to submit objective feedbacks. To solve the second problem, Shao et al. (2007) define the concept of user feedback similarity and use it to improve the accuracy of user feedbacks.

Once the service QoS values are determined, we need to consider where and how to store and query these values. Currently, WSDL and UDDI specifications do not support QoS description and query, but they provide extension mechanisms that can be used to advertise QoS. Consider the service access attributes as an example. Service providers can attach execution price to the operation elements in the WSDL to specify how much the service consumer should pay to the service provider when invoking these operations. In addition, UDDI can be extended to support publish and query QoS (Zhou et al., 2004). However, these approaches need to modify WSDL and UDDI. To follow the design strategy, separation of concern, which is widely accepted in the service oriented architecture, some other approaches are proposed recently. For example, Web services agreement specification (Andrieux et al., 2007) is a piece of work that allows service providers to advertize QoS separately. In addition, some works (e.g., Gibelin and Makpangou, 2005; Li et al., 2007) propose to store QoS values in a peer to peer network to solve the scalability and performance issues that are usually severe problems in centralized QoS repositories such as extended UDDI.

QoS Aggregation

Generally, there are two strategies for QoS-aware service selection: local selection and global selection. The former is to select a service from a set of candidates for a task without taking into account the other tasks. The latter computes the aggregated QoS values of all possible service combinations and selects the service combination that maximizes the aggregated QoS values while satisfying global QoS constraints. Obviously, global selection is more attractive than local selection as users are often mainly concerned with overall performance of composite services. In order to enable global selection, we need to evaluate the QoS values of a composite service based on that of its component services. In this subsection, we review current works on QoS aggregation.

Cardoso et al. (2004) identify six distinct building blocks such as sequential system, parallel system, and loop system, and give each a set of reduction rules to compute the overall QoS values according to that of component services involved in the building block. By iteratively using these reduction rules, the overall QoS of a composite service can be calculated. Having a similar idea, Jaeger et al. (2004) propose a set of composition patterns that are combinations of basic workflow patterns (Aalst et al., 2003) as the building blocks of composite services, and give each composition pattern a set of reduction rules. These methods are suitable for particular generic QoS attributes such as execution duration. Based on these works, Berbner et al. (2006) distinguish three types of QoS (i.e., additive, multiplicative, and min-operator) and give a computation method for each type. It is clear that many QoS attributes do not belong to these three types, so we need a way to dynamically add new computation methods. Canfora et al. (2006) propose a language for defining QoS aggregation formulae for domain-dependent QoS attributes. Their work aims at the framework level and leaves the deduction of concrete formulae to domain experts. An example can be found in the area of Web services transaction, that is, to evaluate the transactional property of a composite service based on that of its component services (Li et al., 2007; Haddad et al., 2008). Another example is about economic benefit in service oriented business applications (Liu et al., 2008). The authors study the relation between profit of service composition and various QoS attributes.

The above approaches for QoS aggregation have several problems. First, composite services are assumed to be fully structured, that is, they can be constructed by nesting several basic building blocks. However, this is not the case for WS-BPEL which is now a standard language to describe composite services. More specifically, WS-BPEL allows links to cross the boundaries of structured activities, which makes it impossible to apply the above works directly to WS-BPEL composite services. Second, QoS values are assumed to be static (e.g., the average or the worst case QoS values), which does not fit reality well. To address the first problem, Mukherjee et al. (2008) elaborately analyze the constructs of WS-BPEL and propose an approach to predict the QoS of WS-BPEL composite services. To solve the second problem, probabilistic QoS has been introduced (Hwang et al., 2004; Hwang et al., 2007). The authors explain how to compute an exact formula for the aggregated QoS in an efficient manner, but their approaches only apply to restricted forms of composition, for example, fully structured composite services. Unlike the work done by Hwang, Rosario et al. (2008) propose a Monte-Carlo simulation based approach to evaluate the aggregated QoS according to the probabilistic QoS of component services.

Selection Algorithm

As mentioned earlier, there are two strategies for QoS-aware service selection. If local selection is used, the candidates for a task are sorted according an aggregation of the values of multiple QoS attributes, which is usually done by multiple criteria decision making techniques, particularly, the simple additive weighting (SAW) method (Yoon and Hwang, 1995). Then, the selection can be done with a linear time complexity. Though this strategy performs excellent in terms of computation cost, it cannot ensure the global end-to-end QoS constraints. As a result, global selection has got more and more attention recently.

Zeng et al. (2004) consider the service selection as an integer programming (IP) problem in which the objective function is defined as a linear composition of multiple QoS attributes, including time, cost, availability, reliability, and reputation. Users have to set exact weights for these attributes and global QoS constraints according to their preferences and requirements. Ardagna and Pernici (2007) also adopt IP technique but they have a different method to eliminate loop constructs in the process of composite services. Specifically, Zeng et al. (2004) use loop unfolding technique, that is, each loop is annotated with an estimated maximum number of iterations k, and then the partial process appearing between the beginning and end of a loop are cloned k times, whereas Ardagna and Pernici (2007) adopt loop peeling technique, that is, loops are represented as a sequence of branches.

Though IP can find the optimal solution, its time complexity is very high, so it cannot be used to solve larger problems. To overcome this shortcoming, Berbner et al. (2006) first relax the IP formulation by allowing the variable representing whether a service is selected to take any real values between 0 and 1 (in standard IP formulation, the value of the variable is either 0 or 1), then consider the real values as the possibility that a service is selected in the optimal solution, and finally use backtracking algorithm to construct a near-optimal service selection. In addition, Ye and Mounla (2008) point out that the existing solutions can be used to speed up the selection process. Specifically, if a service request is totally new, they use IP technique to get the optimal solution; otherwise, they adopt case-based reasoning (CBR) to select one from existing solutions.

Besides high time complexity, IP cannot be applied when the objective function is not a linear function. Therefore, some common optimization techniques in natural computing (Kari and Rozenberg, 2008) are introduced into QoS-aware service selection. Canfora et al. (2005) adopt genetic algorithm (GA) for service selection.

They use one dimension coding schema, that is, a chromosome is represented as a one-dimension array. Their experiments showed that GA has a better performance than IP especially when a composite service has a small number of tasks, each of which has a large number of candidate services. This is because the computation cost of GA mainly depends on the length of chromosome which equals to the number of tasks in a composite service under. Though one dimension coding schema is simple and straightforward, it cannot effectively represent loops and alternative paths in composite services. Ma and Zhang (2008) propose a relation matrix coding schema as well as an enhanced initial population policy and an evolution policy to improve the performance of GA. Recently, Xu and Reiff-Marganiec (2008) apply immune algorithm into QoS-aware service selection. They propose some heuristic rules to accelerate the convergence. However, they do not compare their approach with existing ones. Their experiments show that their approach is better than pure IP, but they do not compare their approach with other methods such as GA.

Yu et al. (2007) propose two models for QoS-aware service selection: a combinatorial model and a graph model. The combinatorial model considers service selection as a multi-dimension multi-choice 0-1 knapsack problem. The graph model considers service selection as a multi-constraint optimal path problem. They propose a polynomial heuristic algorithm for the combinatorial model and an exponential heuristic algorithm for the graph model. Li et al. (2007) make use of the correlations among component services to prune the search space and therefore improve the selection efficiency. Alrifai and Risse (2009) propose a novel approach for QoS-aware service selection. The key idea is to decompose global QoS constraints into an optimal set of local QoS constraints by using IP technique. The satisfaction of local constraints guarantees the satisfaction of global constraints. By decomposition of global constraints, it is only necessary to conduct several local selections simultaneously, which significantly improves the performance of selection process.

SERVICES CONTRACTING

As opposed to traditional Web services composition, Web services contracting distinguish itself with four phases: first, a consensual agreement involving a wide spectrum of Web services characteristics has to be predefined, such that interaction and communication problems (Betin-Can et al., 2005) are not the central concern in this phase; second, the clauses and rules in a contract are cross-organizational, and will be more complicated in practice, which makes the verification process more challenging; third, contracting a set of Web services can be time-consuming, but this process can bring in the guarantee and enhance the reliability of Web services cooperation; lastly, monitoring and negotiation will play an important role in Web services contracting, which are somewhat ignored by the traditional Web services composition.

Contracting a set of Web services is however a non-trivial and even quite difficult task in the context of the current Web techniques. The technical difficulties mainly lie on the following two aspects: first, how to establish a formal language to describe the services contract, while still retaining decidable and efficient reasoning capability; second, how to actually implement and fulfill all the phases such as negotiation and monitoring of services contracting. In this section, we focus on these aspects to review the existing works.

Semantic Models

A semantic model for Web services contracting process is a necessity, since contracting process should be on the basis of some kind of logic (Davulcu et al., 2004). However, to define an appropriate semantic model is not so straightforward, due to the fact that the computational complexity will also arise with the increase of expressive power. Indeed, the more expressive a language is, the higher complexity the reasoning procedure will incur. In sequel, a balance needs to be achieved between expressive power and reasoning complexity.

Concurrent Transaction Logics (CTR) has been considered as a candidate formal language for semantic Web services contracting (Davulcu et al., 2004; Roman and Kifer, 2007). In these papers, CTR is a conservative extension for classical predicate logics, so that all classical predicate assertions, conjunction, disjunction and negation connectives are all legal expressions in CTR. Moreover, two additional connectives: serial conjunction, and concurrent conjunction, and a modal operator: isolated execution, are also allowed in CTR formulas. In parallel, the knowledge formulated in CTR appears in the form of Horn rules, which is very intuitive and convenient. In particular, Davulcu et al. (2004) and Roman and Kifer (2007) model Web services contracting and choreography using Horn goal and constraints, and the validation process is to determine whether or not a state path exists for the goal formulas.

In the area of Electronic Contract, Governatori et al. (2006) propose a semantic model in which deontic logic is adopted as the underpinning. Deontic logic introduces three types of modal operators: obligation, permission, and prohibition, representing respectively what each party has to do, is allowed to do, and is forbidden to do. They further establishes a formal contract language (FCL) developed to analyze and verify e-contracts.

Although these initiatives do take a step further about Web services contracting, they suffer from an inherent defect, viz., they are not designed for Web services from the very beginning. Thereby, their expressive powers are not sufficient for Web services contracting, and ineluctably, some very important features of Web services are neglected by them, e.g., the QoS attributes, the semantic matchmaking capability for various partner services, etc.

The verification issue of Web services composition has also gained a lot of attention recently, and the most representative works should be Bultan et al. (2003) and Deutsch et al. (2006). These works present various methodologies to verify message-level correctness for peer-services composition. However, these works only focus on the interfaces compatibility of a composition process, but are not capable of representing and verifying high-level and more sophisticated constraints.

Heterogeneity Mediation

Heterogeneity problems are inherent in the distributed Web environment. While possible conflicts in service-level contracting may need more negotiation, mismatches in the interface-level can be mediated in a (semi-) automatic manner. In the Web services provision, mismatches can occur in exchanged data as well as the process.

Since Web services interact with each other by exchanging messages, the understanding of those messages is fundamental to the correct interaction. If the message definitions of interacted Web services are not agreed, data mismatches may occur. This will be normal cases in Web services provision, as the provided services are defined by independent venders. Therefore, how to mediate mismatched data become critical to Web services provision. Fortunately, data mediation is a mature research topic that can be reshaped and re-explored in the Web services context.

Some recent efforts have been focus on this issue for Web services. Mocan and Cimpian (2005) provide solutions for ontology-to-ontology mediation, as a starting-point for the more complex problem of mediation in the context of Semantic Web Services. Nagarajan (2007) brings schema mappings to Web service descriptions and presents a data mediation architecture. The paper first articulates the types of message-level heterogeneities, and then proposes data transformation techniques using SAWSDL which is based on WSDL-S. In Cimpian et al. (2006), data mediators being able to create and execute mapping rules are devised for Web services discovery, invocation, and composition. Wu et al. (2007) present a planning-based approach to solve both the data heterogeneity and process heterogeneity problems. For the data mediation, a context-based ranking algorithm is designed for the conversion between messages. Hau et al. (2003) use ontology to annotate Web services and reasons on the annotated data to create conversion functions between mismatched data. Zeng et al. (2006) first discuss the disadvantages of keywords-based mediation methods and then propose a correlation-based mediation method. Shen et al. (2004) present a classification of data mismatches in Web services, and devise a corresponding mediation rule for each class of data mismatches.

While the data mismatches may occur in all three types of Web services provision, process mismatches only occurs in the third type of Web services provision, multiple provision. In multiple provision, services with defined processes are provided for fulfillment of complex tasks comprising multiple ordering operations. If the process definitions of the tasks and services are not compatible, the interaction based on the provision will not succeed. Since in Web services provision, process mismatches may be common cases as argued before, the need of process mediation becomes realistic.

Currently, processes mediation is still a poorly explored research field. The existing work represents only visions of mediator systems able to resolve the processes heterogeneity problems in a (semi-) automatic manner, without presenting sufficient details about their architectural elements. Altenhofen et al. (2005) define a high-level model to mathematically capture the behavioral interface of abstract virtual providers, their refinements and their composition into rich mediator structures. In Bracciali et al. (2005), the authors present a π-calculus based methodology for mediating components with mismatching interaction behavior. Cimpian and Mocan (2005) propose a solution for defining process mediators based on the Web Service Modeling Ontology (WSMO) choreographies. In Dumas et al. (2006), the authors present a declarative approach to service interface mediation based on an algebra over behavioral interfaces and a visual language that allows pairs of provided-required interfaces to be linked through algebraic expressions. Nezhad et al. (2007) present novel techniques and a tool that provides semi-automated support for identifying and resolving of mismatches between service interfaces and protocols, and for generating adapter specifications. In Yellin and Strom (1997), the authors examine the augmentation of application interfaces with protocols, and propose techniques for automatic generation of component adapters. Benatallah et al. (2005) propose a systematic methodology to Web service process mediation based on a classification of process mismatches. For each mismatch class, the paper devises a corresponding mediation template which accepts the mismatch as parameters and performs a prescribed conversion logic. Vaculín and Sycara (2007) use OWL-S processes to specify Web service processes and propose an approach to finding out possible mappings or mismatches between Web service processes.

CONCLUSION AND OPEN PROBLEMS

Web services provision aims at filling user requests in an on-demand manner, which will be common in the near future. Up to now, we have reviewed recent achievements in the area of Web services provision. In order to give a clear sketch on these works, we divide the lifecycle of Web services provision into three phases, that is, services discovery, selection and contracting. In addition, we distinguish three types of Web services provision: independent provision, cooperative provision, and multiple provision. While independent provision is the most basic pattern in Web services provision, the other two patterns exhibit more power in filling user requests. To support cooperative provision and multiple provision, we have specially studied service covering and process-aware service matchmaking. On the other hand, efforts on QoS-based service selection have been conducted to facilitate non-functional requirements in Web services provision. In order to ensure correct interaction of provided Web services, semantic models and heterogeneity mediation are also important, and have been discussed in this article as well.

Based on the existing efforts, we identify several open problems for the future research of Web services provision as follows:

- Bridging the semantic gap among different partner services. Currently, there exist various types of semantics among different business units and organizations, which brings in tremendous difficulties during Web services interaction and cooperation. In particular, many QoS parameters may be attached to different services, which are crucial for their performance evaluations and selection criteria, but these parameters may be defined by different equations and matrix (Oldham et al., 2006). This reveals a fact that the semantic gap among Web services involves not only the conceptual aspect, but also their numerical aspect. On that account, more research works should be conducted to take a step further towards this direction.
- User preference modeling. Currently, users have to manually assign weights to QoS attributes to express their individual preferences. It is however difficult and inconvenient for users to assign appropriate weights in practice since the weights are only simple real values and have no explicit semantics. For example, a user may have the following QoS preference: if execution price is larger than $100, reliability is more important than execution duration; otherwise, reliability is less important than execution duration. Obviously, simple real values cannot model this kind of preference. We believe that some advanced approaches for user preference modeling, for example, CP-nets (Wang et al., 2008), are expected in user centric Web services provision.
- QoS-Aware service selection in cooperative provision and multiple provision. As seen in the discussion of service discovery and service contracting, different provisions introduce a number of different research issues. Due to the complexity of cooperative provision and multiple provision, QoS-Aware service selection also becomes more challenging. For example, a service can accomplish multiple tasks in multiple provision, which actually adds a set of Web services dependency constraints in the optimization models of traditional service selection (Ardagna and Pernici, 2007). Therefore, some new and efficient selection algorithms are expected in the context of cooperative provision and multiple provision.
- Composition Compatibility and Composition Mediation. Once a target process needs to be matched by a composition of several source processes, the compatibility and the possible mediation of the target process and the composition of the source processes arise. Different from compatibility and mediation of single processes, the issues of composition compatibility and composition mediation involve a single process and a composition of processes. A possible approach to composition compatibility and composition mediation is to decompose the single process and map the decomposed sub processes to the processes in the composition. Through the decom-

position and mapping, the two issues can be converted to compatibility and mediation of single processes which are easy to tackle. However, methods and techniques to support the decomposition and mapping become critical issues and need to be well studied.

ACKNOWLEDGEMENT

The research described here is supported, in part, by a grant from City University of Hong Kong (7002212), a grant from the Research Grants Council of the HKSAR, China [CityU 117608], and the National Basic Research Fund of China ("973" Program) under Grant No.2003CB317006.

REFERENCES

Aalst, W. M. P., Dumas, M., Ouyang, C., Rozinat, A., & Verbeek, E. (2008). Conformance checking of service behavior. *ACM Transactions on Internet Technology, 8*(3). doi:10.1145/1361186.1361189

Aalst, W. M. P., Hofstede, A. H. M., Kiepuszewshi, B., & Barros, A. P. (2003). Workflow patterns. *Distributed and Parallel Databases, 14*(3), 5–51. doi:10.1023/A:1022883727209

Agarwal, S., & Studer, R. (2006). Automatic matchmaking of web services. In *Proceedings of IEEE International Conference on Web Services (ICWS)* (pp. 45-54).

Agarwal, V., Dasgupta, K., Karnik, N. M., Kumar, A., Kundu, A., Mittal, S., & Srivastava, B. (2005). A service creation environment based on end to end composition of web services. In *Proceedings of International Conference on World Wide Web (WWW)*, (pp. 128-137).

Aggarwal, R., Verma, K., Miller, J. A., & Milnor, W. (2004). Constraint driven web service composition in METEOR-S. In *Proceedings of IEEE International Conference on Service Computing (SCC)* (pp. 23-30).

Akkiraju, R., Farrell, J., Miller, J. A., Nagarajan, M., Schmidt, M.-T., Sheth, A., & Verma, K. (2005). *Web service semantics - WSDL-S*. Retrieved May 31, 2009, from http://www.w3.org/Submission/WSDL-S/.

Alrifai, M., & Risse, T. (2009). Combining global optimization with local selection for efficient QoS-aware service composition. In *Proceedings of International Conference on World Wide Web (WWW)* (pp. 881-890).

Altenhofen, M., Borger, E., & Lemcke, J. (2005). An abstract model for process mediation. In *Proceedings of International Conference on Formal Engineering Methods (ICFEM)* (pp. 81-95).

Andrieux, A., Czajkowski, K., Dan, A., Keahey, K., Ludwig, H., Nakata, T., et al. (2007). *Web services agreement specification (WS-Agreement)*. Retrieved May 31, 2009, from http://www.ogf.org/documents/GFD.107.pdf.

Ardagna, D., & Pernici, B. (2007). Adaptive service composition in flexible processes. *IEEE Transactions on Software Engineering, 33*(6), 369–384. doi:10.1109/TSE.2007.1011

Aversano, L., Canfora, G., & Ciampi, A. (2004). An algorithm for web service discovery through their composition. In *Proceedings of IEEE International Conference on Web Services (ICWS)* (pp. 332-339).

Baader, F., Calvanese, D., McGuinness, D. L., Nardi, D., & Patel-Schneider, P. F. (2003). *The description logic handbook: theory, implementation, and applications*. Cambridge University Press.

Benatallah, B., Casati, F., Grigori, D., Nezhad, H. R., & Toumani, F. (2005). Developing adapters for web services integration. In *Proceedings of International Conference on Advanced Information System Engineering (CAiSE)* (pp. 415–429).

Benatallah, B., Hacid, M. S., Leger, A., Rey, C., & Toumani, F. (2005). On automating web services discovery. *The VLDB Journal*, *14*(1), 84–96. doi:10.1007/s00778-003-0117-x

Berbner, R., Spahn, M., Repp, N., Heckmann, O., & Steinmetz, R. (2006). Heuristics for QoS-aware web service composition. In *Proceedings of International Conference on Web Services (ICWS)* (pp. 72-82).

Betin-Can, A., Bultan, T., & Fu, X. (2005). Design for verification for asynchronously communicating web services. In *Proceedings of International Conference on World Wide Web (WWW)* (pp. 750-759).

Bonatti, P. A., & Festa, P. (2005). On optimal service selection. In *Proceedings of International Conference on World Wide Web (WWW)* (pp. 530-538).

Bordeaux, L., Salaun, G., Berardi, D., & Mecella, M. (2004). When are two web services compatible. In *Proceedings of International Workshop on Technologies for E-Services (TES)* (pp. 15-28).

Bracciali, A., Brogi, A., & Canal, C. (2005). A formal approach to component adaptation. *Journal of Systems and Software*, *74*(1), 45–54. doi:10.1016/j.jss.2003.05.007

Brogi, A., Canal, C., Pimentel, E., & Vallecillo, A. (2004). Formalizing web service choreographies. *Electronic Notes in Theoretical Computer Science*, *105*, 73–94. doi:10.1016/j.entcs.2004.05.007

Brogi, A., Corfini, S., & Popescu, R. (2008). Semantics-based composition-oriented discovery of web services. *ACM Transactions on Internet Technology*, *8*(4). doi:10.1145/1391949.1391953

Bultan, T., Fu, X., Hull, R., & Su, J. (2003). Conversation specification: a new approach to design and analysis of e-service composition. In *Proceedings of International Conference on World Wide Web (WWW)* (pp. 403–410).

Canfora, G., Penta, M. D., Esposito, R., Perfetto, F., & Villani, M. L. (2006). Service composition (re)binding driven by application-specific QoS. In *Proceedings of International Conference on Service Oriented Computing (ICSOC)* (pp. 141-152).

Canfora, G., Penta, M. D., Esposito, R., & Villani, M. L. (2006). An approach for QoS-aware service composition based on genetic algorithms. In *Proceedings of International Conference on Genetic and Evolutionary Computation (GECCO)* (pp. 1069-1075).

Cardoso, J., Sheth, A., Miller, J., Arnold, J., & Kochut, K. (2004). Quality of service for workflows and web service processes. *Journal of Web Semantics*, *1*(3), 281–308. doi:10.1016/j.websem.2004.03.001

Cimpian, E., & Mocan, A. (2005). WSMX process mediation based on choreographies. In *Proceedings of Business Process Management Workshops* (pp. 130-143).

Cimpian, E., Mocan, A., & Stollberg, M. (2006) Mediation enabled semantic web services usage. In *Proceedings of Asian Semantic Web Conference (ASWC)* (pp. 459-473).

Clement, L., Hately, A., Riegen, C., & Rogers, T. (2004). *UDDI Version 3.0.2.* Retrieved May 31, 2009, from http://www.uddi.org/pubs/uddi_v3.htm.

Colgrave, J., Akkiraju, R., & Goodwin, R. (2004). External matching in UDDI. In *Proceedings of IEEE International Conference on Web Services (ICWS)* (pp. 226-233).

Curbera, F., Duftler, M. J., Khalaf, R., Nagy, W., Mukhi, N., & Weerawarana, S. (2002). Unraveling the web services web: an introduction to SOAP, WSDL, and UDDI. *IEEE Internet Computing*, *6*(2), 86–93. doi:10.1109/4236.991449

Davulcu, H., Kifer, M., & Ramakrishnan, I. V. (2004). CTR-S: a logic for specifying contracts in semantic web services. In *Proceedings of International Conference on World Wide Web on Alternate Track Papers & Posters (WWW)* (pp. 144-153).

Deutsch, A., Sui, L., Vianu, V., & Zhou, D. (2006). Verification of communicating data-driven web services. In *Proceedings of ACM SIGACT-SIGMOD-SIGART Symposium on Principles of Database Systems (PODS)* (pp. 90-99).

Dumas, M., Sport, M., & Wang, K. (2006). Adapt or perish: algebra and visual notation for service interface adaptation. In *Proceedings of International Conference on Business Process Management (BPM)* (pp. 65-80).

Farrell, J., & Lausen, H. (2007). *Semantic annotations for WSDL and XML schema*. Retrieved May 31, 2009, from http://www.w3.org/TR/sawsdl/.

Gibelin, N., & Makpangou, M. (2005). Efficient and transparent web-services selection. In *Proceedings of International Conference on Service Oriented Computing (ICSOC)* (pp. 527-532).

Governatori, G., Milosevic, Z., & Sadiq, S. W. (2006). Compliance checking between business process and business contracts. In *Proceedings of IEEE International Enterprise Distributed Object Computing Conference (EDOC)* (pp. 221-232).

Haddad, J. E., Manouvrier, M., Ramirez, G., & Rukoz, M. (2008). QoS-driven selection of web services for transactional composition. In *Proceedings of International Conference on Web Services (ICWS)* (pp. 653-660).

Hau, J., Lee, W., & Newhouse, S. (2003). Autonomic service adaptation in ICENI using ontological annotation. In *Proceedings of International Workshop on Grid Computing (GRID)* (pp. 10-17).

Hwang, S.-Y., Wang, H., Srivastava, J., & Paul, R. A. (2004). A probabilistic QoS model and computation framework for web services-based workflows. In *Proceedings of International Conference on Conceptual Modeling (ER)* (pp. 596-609).

Hwang, S.-Y., Wang, H., Tang, J., & Srivastava, J. (2007). A probabilistic approach to modeling and estimating the QoS of web-services-based workflows. *Information Sciences*, *177*(23), 5484–5503. doi:10.1016/j.ins.2007.07.011

Jaeger, M. C., Rojec-Goldmann, G., & Muhl, G. (2004). QoS aggregation for service composition using workflow patterns. In *Proceedings of International Conference on Enterprise Distributed Object Computing (EDOC)* (pp. 149-159).

Jordan, D., & Evdemon, J. (2007). *Web services business process execution language version 2.0*. Retrieved May 31, 2009 from http://docs.oasis-open.org/wsbpel/2.0/OS/wsbpel-v2.0-OS.html.

Jurca, R., Faltings, B., & Binder, W. (2007). Reliable QoS monitoring based on client feedback. In *Proceedings of International Conference on World Wide Web (WWW)* (pp. 1003-1012).

Kari, L., & Rozenberg, G. (2008). The many facets of natural computing. *Communications of the ACM*, *51*(10), 72–83. doi:10.1145/1400181.1400200

Kona, S., Bansal, A., & Gupta, G. (2007). Automatic composition of semantic web services. In *Proceedings of International Conference on Web Services (ICWS)* (pp. 150-158).

Konig, D., Lohmann, N., Moser, S., Stahl, C., & Wolf, K. (2008). Extending the compatibility notion for abstract WS-BPEL processes. In *Proceedings of International Conference on World Wide Web (WWW)* (pp. 785-794).

Lecue, F., Delteil, A., & Leger, A. (2007). Applying abduction in semantic web services composition. In *Proceedings of IEEE International Conference on Web Services (ICWS)* (pp. 94-101).

Li, F., Yang, F., Shuang, K., & Su, S. (2007). Q-Peer: A decentralized QoS registry architecture for web services. In *Proceedings of International Conference on Service Oriented Computing (ICSOC)* (pp. 145-156).

Li, L., Liu, C., & Wang, J. (2007). Deriving transactional properties of composite web services. In *Proceedings of IEEE International Conference on Web Services (ICWS)* (pp. 631-638).

Li, L., Wei, J., & Huang, T. (2007). High performance approach for Multi-QoS constrained web services selection. In *Proceedings of International Conference on Service Oriented Computing (ICSOC)* (pp. 283-294).

Lin, B., Li, Q., & Gu, N. (2007). A semantic specification framework for analyzing functional composability of autonomous web services. In *Proceedings of IEEE International Conference on Web Services (ICWS)* (pp. 695-702).

Liu, A., Li, Q., Huang, L., & Liu, H. (2008). Building profit-aware service-oriented business applications. In *Proceedings of IEEE International Conference on Web Services (ICWS)* (pp. 489-496).

Liu, Y. H. A., Ngu, H., & Zeng, L. (2004). QoS computation and polcing in dynamic web service selection. In *Proceedings of International Conference on World Wide Web (WWW)* (pp. 66-73).

Ma, Y., & Zhang, C. (2008). Quick convergence of genetic algorithm for QoS-Driven web service selection. *Computer Networks, 52*(5), 1093–1104. doi:10.1016/j.comnet.2007.12.003

Martens, A. (2003). On compatibility of web services. *Petri Net Newsletter, 65*, 12–20.

Martens, A. (2005). Process oriented discovery of business partners. In *Proceedings of International Conference on Enterprise Information Systems (ICEIS)* (pp. 57-64).

Martens, A., Moser, S., Gerhardt, A., & Funk, K. (2006). Analyzing compatibility of BPEL processes. In *Proceedings of Advanced International Conference on Telecommunications and International Conference on Internet and Web Applications and Services (AICT/ICIW)*.

Maximilien, E. M., & Singh, M. P. (2004). A framework and ontology for dynamic web services selection. *IEEE Internet Computing, 8*(5), 84–93. doi:10.1109/MIC.2004.27

McIlraith, S. A., Son, T. C., & Zeng, H. (2001). Semantic web services. *IEEE Intelligent Systems, 16*(2), 46–53. doi:10.1109/5254.920599

Menasce, D. A. (2002). QoS issues in web services. *IEEE Internet Computing, 6*(6), 72–75. doi:10.1109/MIC.2002.1067740

Mikic-Rakic, M., Malek, S., & Medvidovic, N. (2005). Improving availability in large, distributed component-based systems via redeployment. In *Proceedings of International Working Conference on Component Deployment (CD)* (pp. 83-98).

Mocan, A., & Cimpian, E. *WSMX data mediation*. Retrieved May 31, 2009, from http://www.wsmo.org/TR/d13/d13.3/v0.2/20051011/d13.3v0.2_20051011.pdf.

Mukherjee, D., Jalote, P., & Nanda, M. G. (2008). Determining QoS of WS-BPEL compositions. In *Proceedings of International Conference on Service Oriented Computing(ICSOC)* (pp. 378-393).

Nagarajan, M., Verma, K., Sheth, A., & Miller, J. (2007). Ontology driven data mediation in web services. *International Journal of Web Services Research, 4*(4), 104–126.

Nezhad, H., Benatallah, B., Martens, A., Curbera, F., & Casati, F. (2007). Semi-automated adaptation of service interactions. In *Proceedings of International Conference on World Wide Web (WWW)* (pp. 993-1002).

O'Sullivan, J., Edmond, D., & Hofstede, A. T. (2002). What's in a service? Towards accurate description of non-functional service properties. *Distributed and Parallel Databases, 12*, 117–133. doi:10.1023/A:1016547000822

Paolucci, M., Kawamura, T., Payne, T. R., & Sycara, K. P. (2002). Importing the semantic web in UDDI. In *Proceedings of International Workshop on Web Services, E-Business, and the Semantic Web (WES)* (pp. 225-236).

Paolucci, M., Kawamura, T., Payne, T. R., & Sycara, K. P. (2002). Semantic matching of web services capabilities. In *Proceedings of International Semantic Web Conference (ISWC)* (pp. 333-347).

Ponnekanti, S. R., & Fox, A. (2002). SWORD: A developer toolkit for web service composition. In *Proceedings of International Conference on World Wide Web on Alternate Track Papers (WWW)*.

Ran, S. (2003). A model for web services discovery with QoS. *ACM SIGecom Exchanges, 4*(1), 1–10. doi:10.1145/844357.844360

Roman, D., & Kifer, M. (2007). Reasoning about the behavior of semantic web services with concurrent transaction logic. In *Proceedings of International Conference on Very Large Data Bases (VLDB)* (pp. 627-638).

Rosario, S., Benveniste, A., Haar, S., & Jard, C. (2008). Probabilistic QoS and soft contracts for transaction-based web services orchestrations. *IEEE Transactions on Service Computing, 1*(4), 187–200. doi:10.1109/TSC.2008.17

Rosenberg, F., Platzer, C., & Dustdar, S. (2006). Bootstrapping performance and dependability attributes of web services. In *Proceedings of International Conference on Web Services (ICWS)* (pp. 205-212).

Serhani, M. A., Dssouli, R., Hafid, A., & Sahraoui, H. A. (2005). A QoS broker based architecture for efficient web services selection. In *Proceedings of International Conference on Web Services (ICWS)* (pp. 113-120).

Shao, L., Zhang, J., Wei, Y., Zhao, J., Xie, B., & Mei, H. (2007). Personalized QoS prediction for web services via collaborative filtering. In *Proceedings of International Conference on Web Services (ICWS)* (pp. 439-446).

Shao, L., Zhang, L., Xie, T., Zhao, J., Xie, B., & Mei, H. (2008). Dynamic availability estimation for service selection based on status identification. In *Proceedings of IEEE International Conference on Web Services (ICWS)* (pp. 645-652).

Shen, D., Yu, G., Yin, N., & Nie, T. (2004). Heterogeneity resolution based on ontology in web services composition. In *Proceedings of IEEE International Conference on E-Commerce Technology for Dynamic E-Business* (pp. 274-277).

Sirin, E., Parsia, B., & Hendler, J. A. (2004). Filtering and selecting semantic web services with interactive composition techniques. *IEEE Intelligent Systems, 19*(4), 42–49. doi:10.1109/MIS.2004.27

Sirin, E., Parsia, B., Wu, D., Hendler, J. A., & Nau, D. S. (2004). HTN planning for web service composition using SHOP2. *Journal of Web Semantics, 1*(4), 377–396. doi:10.1016/j.websem.2004.06.005

Teege, G. (1994). Making the difference: a subtraction operation for description logic. In *Proceedings of International Conference on Principles of Knowledge Representation and Reasoning (KR)* (pp. 540-550).

Vaculín, R., & Sycara, K. (2007). Towards automatic mediation of OWL-S process models. In *Proceedings of IEEE International Conference on Web Services (ICWS)* (pp. 1032-1039).

Wang, H., Xu, J., Li, P., & Hung, P. (2008). Incomplete preference-driven web services selection. In *Proceedings of IEEE International Conference on Service Computing (SCC)* (pp. 75-82).

Wombacher, A., Fankhauser, P., Neuhold, E., & Mahleko, B. (2004). Matchmaking for business processes based on choreographies. In *Proceedings of IEEE International Conference one-Technology, e-Commerce, and e-Science (EEE)* (pp. 359-368).

Wu, Z., Ranabahu, A., Gomadam, K., Sheth, A., & Miller, J. (2007). *Automatic composition of semantic web services using process and data mediation*. Technical Report, LSDIS Lab, University of Georgia.

Xu, J., & Reiff-Marganiec, S. (2008). Towards heuristic web services composition using immune algorithm. In *Proceedings of IEEE International Conference on Web Services (ICWS)* (pp. 238-245).

Ye, X., & Mounla, R. (2008). A hybrid approach to QoS-aware service composition. In *Proceedings of IEEE International Conference on Web Services (ICWS)* (pp. 62-69).

Yellin, D., & Strom, R. (1997). Protocol specifications and component adaptors. *ACM Transactions on Programming Languages and Systems*, *19*(2), 292–333. doi:10.1145/244795.244801

Yoon, K. P., & Hwang, C.-L. (1995). *Multiple attribute decision making: an introduction*. Sage University Paper series on Quantitative Applications in the Social Sciences, 07-104, Thousand Oaks, CA: Sage.

Yu, T., Zhang, Y., & Lin, K.-J. (2007). Efficient algorithms for web services selection with end-to-end QoS constraints. *ACM Transactions on Web*, *1*(1), 1–26. doi:10.1145/1232722.1232723

Zeng, L., Benatallah, B., Dumas, M., Kalagnanam, J., & Sheng, Q. Z. (2003). Quality driven web services composition. In *Proceedings of International Conference on World Wide Web (WWW)* (pp. 411-421).

Zeng, L., Benatallah, B., Ngu, A. H. H., Dumas, M., Kalagnanam, J., & Chang, H. (2004). QoS-Aware middleware for web services composition. *IEEE Transactions on Software Engineering*, *30*(5), 311–327. doi:10.1109/TSE.2004.11

Zeng, L., Benatallah, B., Xie, G., & Lei, H. (2006). Semantic service mediation. In *Proceedings of International Conference on Service Oriented Computing (ICSOC)* (pp. 490-495).

Zhou, C., Chia, L.-T., & Lee, B.-S. (2004). QoS-aware and federated enhancement for UDDI. *International Journal of Web Services Research*, *1*(2), 58–85.

Zhou, C., Chia, L.-T., & Lee, B.-S. (2005). Web services discovery with DAML-QoS ontology. *International Journal of Web Services Research*, *2*(2), 43–66.

ENDNOTE

[1] http://www.active-endpoints.com/

This work was previously published in International Journal of Systems and Service-Oriented Engineering, Volume 1, Issue 1, edited by Dickson K.W. Chiu, pp. 26-45, copyright 2010 by IGI Publishing (an imprint of IGI Global).

Chapter 3
A Service Oriented SLA Management Framework for Grid Environments

V. Pouli
National Technical University of Athens, Greece

M. Grammatikou
National Technical University of Athens, Greece

C. Marinos
National Technical University of Athens, Greece

S. Papavassiliou
National Technical University of Athens, Greece

V. Maglaris
National Technical University of Athens, Greece

ABSTRACT

Traditionally, network Service Providers specify Service Level Agreements (SLAs) to guarantee service availability and performance to their customers. However, these SLAs are rather static and span a single provider domain. Thus, they are not applicable to a multi–domain environment. In this paper, the authors present a framework for automatic creation and management of SLAs in a multi-domain environment. The framework is based on Service Oriented Computing (SOC) and contains a collection of web service calls and modules that allow for the automatic creation, configuration, and delivery of an end-to-end SLA, created from the merging of the per-domain SLAs. This paper also presents a monitoring procedure to monitor the QoS guarantees stipulated in the SLA. The SLA establishment and monitoring procedures are tested through a Grid application scenario targeted to perform remote control and monitoring of instrument elements distributed across the Grid.

DOI: 10.4018/978-1-4666-1767-4.ch003

INTRODUCTION

A Service Level Agreement (SLA) is a contract between a network provider and a customer defining all aspects of the offered services. It may specify the levels of availability, serviceability, performance and operation conditions (Bouras, Campanella, & Sevasti, 2002; Goderis, 1979). It may also define the procedures and the reports that must be provided to track and ensure compliance with the SLA or describe other attributes of the service, such as billing (Zhang, Tan, & Dey, 2009) and penalties in case of SLA violations (Gulliver, 1979). Service Level Agreements with the network providers enable users to use services for their applications in a more effective and reliable way (Clark, Gilmore, & Tribastone, 2009). The benefit of using an SLA for an application is that a contract can be provided to the application users for the level of Quality of Service (QoS) that they have requested and the procedures that will be followed in order to enforce this level of QoS. An SLA can also provide a means of describing the network domains' level of QoS in a unified way, irrespective of the type of technologies that are used to provide QoS across the network.

In Grid environments (Foster & Kesselman, 1998; Opitz et al., 2008) where applications or users require resources from heterogeneous domains offering various performance guarantees, the issue of acquiring a specified level of QoS is of great importance. The lack of control in traditional IP networks created the need for efficient QoS provisioning mechanisms. Among the dominant ones is the DiffServ framework (Nichols & Carpenter, 2000; Young-Tak, 2005; Dong-Jin & Young-Tak, 2003) where marking and policing of flows is implemented at the edge routers according to per–domain SLAs (Courcoubetis & Siris, 1999). This way a variety of advanced network end–to–end (e2e) services is offered across multiple, separately administrated domains (Van der Mei & Meeuwissen, 2006). Based on the DiffServ architecture, the Premium IP (PIP) service was implemented in National Research & Education Networks (NRENs). Using priorities, traffic can be classified as high priority PIP, Best Effort (BE) for standard IP applications and Less than Best Effort (LBE) for non–time critical bulk data transfer. The PIP service provides a virtual leased line service as PIP data packets are not expected to experience congestion in the network regardless of the load of the other traffic classes. PIP is suitable for real–time applications, such as Voice over IP (VoIP), IPTV, video conferencing as well as time critical applications such as remote control of instruments and devices (Andreeva et al., 2008).

In this paper we present a framework based on SOC (Dustdar & Kramer, 2008) that supports the provisioning of advanced e2e IP network services via the dynamic management of SLAs, in a real multi-domain environment that crosses the NRENs. The framework uses Web Service (WS) calls to allow for the automatic creation, configuration and delivery of an e2e SLA created by merging the SLAs provided by each domain. In this way a business-centric architectural approach is achieved that integrates all the SLA functions together between the different domains. We also present a monitoring procedure to monitor the QoS guarantees stipulated in the SLA. Our framework has been developed for the needs of the EGEE project (Enabling Grids for E-sciencE http://www.eu-egee.org) and has been deployed in the GÉANT Advanced Multi-domain Provisioning System – AMPS (http://www.geant2.net/server/show/nav.00d008009002) to offer automatic e2e SLAs for network path reservations.

The rest of the paper is organized as follows. The next section presents some related work, and then are the challenges and requirements related to multi domain SLAs, followed by the description of the SLA Framework along with the SLA establishment and monitoring procedures. Afterwards, the evaluation and testing of the SLA framework is performed through a Grid application scenario targeted to perform remote control and monitoring of instrument elements distributed across the Grid. Finally, in the last section we provide some concluding remarks.

RELATED WORK

To date several frameworks have been proposed to support the creation and management of services based on SLAs (Rong-Xiao, Jing-bo, Shu-fu, & Kai, 2009; Parkin, Kuo, & Brooke, 2006; Papazoglou & Van den Heuvel, 2005). For instance, in (Hoffner et al., 2001) a Contract Framework is presented and in (Hasselmeyer et al., 2007) a generic framework for negotiating SLAs is also introduced, where specific components automate large parts of the negotiation process with the ability for the user to maintain control. In a related work, Garg et al. (2002) presented an SLA-based framework for QoS provisioning and dynamic capacity allocation. The proposed SLA framework allows users to buy a long term capacity at a pre-specified price with a three tier pricing model with penalties. Moreover, Hudert, Ludwig, and Wirtz (2009) present a generic negotiation framework with some extensions to the WS-Agreement specification (Andrieux et al., 2008; Web Services Agreement Specification, 2007) to offer a variety of bilateral and multilateral negotiation protocols. On the vendors' side, DiffServ (Courcoubetis & Siris, 1999) is primarily indicated as the preferred technology to achieve SLA management, while delay, jitter, packet–loss, throughput and availability are considered as the most vital SLA technical parameters for an IP service. Despite the fact that the previous works present comparably competent results and solutions related to the modelling and development of SLA frameworks, however these frameworks are not applicable in multi-domain environments.

For multi domain environments significant work has been done including, but not limited to, the one of (Campanella et al., 2003) who present an approach to providing e2e QoS services across Europe. Another one is of (Van der Mei & Meeuwissen, 2006) who model a framework and provide sets of SLAs negotiated by Service Providers to realize the desired e2e QoS for transaction based services. Also Bagchi et al. (2009) have proposed a framework for multi domain SLA management. They define the relationships among service providers, suppliers and customers and the related SLAs needed to plan, fulfil and assure e2e services. Moreover, Clark, Gilmore, and Tribastone (2009) have proposed a novel Markovian process calculus to allow for accurate SLAs taking into account the uncertainty existing in service-oriented environments. All the above presented works provide remarkable solutions for managing e2e QoS, but none of them has been implemented via service oriented architecture.

With respect to the above proposals, our framework differentiates from the others in the sense that it uses service oriented techniques to provide an automated mechanism for the generation, configuration and delivery of an e2e SLA applied in a multi-domain environment. This e2e SLA is created by merging the per-domain SLAs according to specific merging rules, described in the next section. Our SLA framework offers QoS in the various applications that require resources spanning multiple domains. Along with the SLA establishment procedure we provide a monitoring procedure to monitor the QoS guarantees stipulated in the SLA. The SLA framework has already been deployed on top of the AMPS, a multi-domain reservation tool and tested in real QoS–enabled networks, such as the NRENs and GÉANT to offer e2e SLAs for the network path reservations.

MULTI-DOMAIN SLA CHALLENGES

In this section, we discuss several main research challenges related to the SLA-driven inter-domain management. When dealing with multi domain environments, the SLAs given from the various domains can differ, sometimes making it impossible to merge the per-domain SLAs into an e2e SLA. For example, the problems associated with the transformation of each provider needs to a per-domain SLA are of different kind in nature. One further issue is the propagation of the SLA requirements from end to end, something that has an impact on the negotiated e2e SLA. Therefore it will be necessary to define negotiation protocols between providers, based on standardised SLA templates. With this framework, a solution to solving these problems is offered by providing a common template and automating the otherwise manual procedure of finding and contacting each administrative domain across the path. Thus it makes the provision and delivery of the per-domain SLAs and consequently of the e2e SLA in a common way, transparently to the user. The end-user manages only one SLA, the e2e one. All necessary information for the SLA management is propagated in the network through the framework from one end to another.

On the assurance side, a challenging issue is to identify efficient monitoring procedures to rapidly instrument and manage the networks, in order to have a reliable and efficient monitoring of the SLA fulfillment and also provide accurate accounting. With our framework we propose a monitoring procedure to fulfill this purpose and honour the SLA guarantees.

SLA DESCRIPTION

An SLA is a set of technical and non-technical parameters agreed between the consumer and the provider. An SLA contract represents a description of a required service and can be formed by two different parts: (i) the Administrative Legal Object (ALO) that contains administrative information and does not depend on the particular service (e.g., user authentication, availability information, privacy and security aspects), and (ii) the Service Level Specification (SLS) part, which is a set of parameters and their corresponding values that define the technical parameters of the service provisioning for a specific flow instance.

The administrative details contained in the ALO part remain the same in each domain for each service request while the information contained in the SLS part is specific for every service instance. An SLS can contain technical information such as the Service Instance Scope, the Flow Description, the End-To-End Performance Guarantees, the Monitoring Infrastructure and the Reliability as described in Table 1.

Due to the fact that we are dealing with a multi-domain environment where QoS provisioning is each domain's responsibility, the e2e SLA will be derived by the merging of each domain's SLA, as shown in Figure 1. The merging rules applied to each SLS parameter of the per-domain SLAs in order to finally provide the e2e SLA are shown in Table 1.

RESEARCH ASSUMPTIONS AND ARCHITECTURAL REQUIREMENTS

Regarding the research assumptions of this work we assume that: a) the underlying grid middleware has to communicate via a common API with the SLA framework for the establishment of SLA contracts and b) the participating domains must agree to cooperate with each other, authorize and configure the framework's communication.

Concerning the architectural requirements, in principle, the SLA framework provides an end-to-end (e2e) SLA that has to contain all necessary parameters to provide end-to-end control for a service instance specification between two endpoints. The e2e-SLA establishment should be transparent to the end-users, regardless of the underlying network technologies and intermediate domains.

For the generation of an e2e SLA by the SLA framework several issues have to be examined:

Table 1. e2e Service Level Specification (SLS)

Parameters	Description
Service Instance Scope	This field contains a list with the technical information of the ingress interface and the egress interface of the domains involved in the path.
Flow Description	This field should contain a list of the DSCP values, the IP source-destination address pair along with the protocol and the ports that the application is targeted to.
End-To-End Performance Guarantees	The SLA performance guarantees field of the end-to-end SLA is proposed to adopt the guarantees derived from the individual SLAs involved in the service provision across the path. More specifically: - BandWidth (BW) (bps): will be less or equal to the minimum of the bandwidths in each SLA instance: $BW_{e2e} <= \min\{BW_i\}$ - One-Way-Delay (OWD) (ms): is an additive metric and therefore the corresponding field in the e2e SLA will be produced as follows: $OWD_{e2e} \geq \sum\{OWD_i\}$ - $Jitter_{e2e}$ (ms): we treat Jitter as an RMS value and use its square value as an additive parameter. - Packet loss (PL) (%): will be the sum of all the packet losses across the individual domains: $PL_{e2e} \geq \sum\{PL_i\}$
Monitoring Infrastructure	This field should contain information on the monitoring capabilities of each domain in the path in terms of which parameters are monitored, the points where measurements are possible, the availability of measurements.
Reliability	Reliability should define allowed mean downtime per unit of time for the service provision and maximum allowed Time-To-Repair (TTR) in case of breakdown for the provision of the service described by the SLA. Usually it is evaluated as the product of the reliability of the involved domains.

- Different ISPs must agree to cooperate with each other, authorize and configure the framework's communication between other cooperating ISPs and clearly specify what actions are taken and by whom.
- Depending on the service request type (unidirectional, bi-directional), a uni-directional or bi-directional end-to-end SLA should be delivered by the SLA Framework. This distinction is necessary since the downstream (closer to the destination) domain may not wish to participate in the e2e marking and policing setup requested by the upstream (closer to the source) domain.
- In order to honor the guarantees stipulated in the e2e-SLA, continuous monitoring, for the whole service duration, of all the SLA parameters in the contract between the user and the service providers is needed. The e2e monitoring SLA will be based on the investigation and monitoring of all the per-domain SLAs along the path and must not have negative impact on the network performance.

Figure 1. End-to-end SLA merging from per-domain SLAs

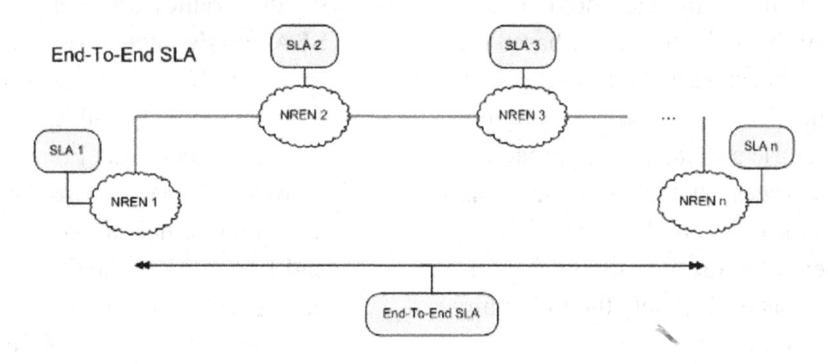

Figure 2. SLA framework architecture

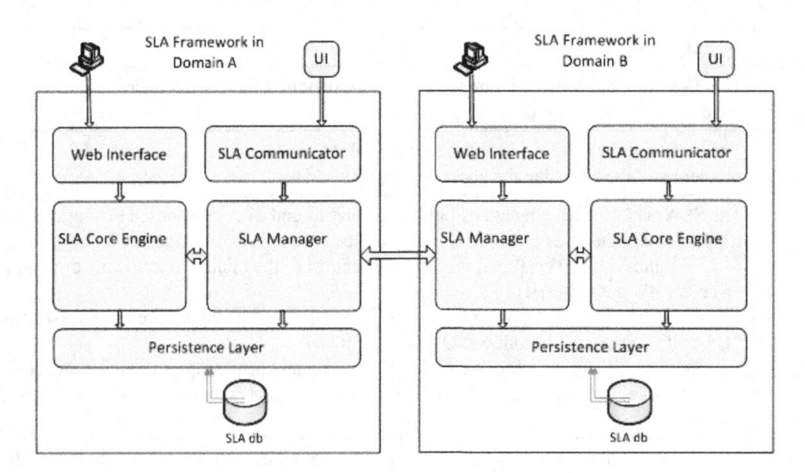

SLA FRAMEWORK DESCRIPTION

The SLA framework comprises a service-oriented middleware that provides automated end-to-end (e2e) SLA management for grid computing. It can be deployed in network reservation systems in order to provide end-to-end SLAs for network reservations. For the establishment and monitoring procedures to be feasible, the SLA framework should be installed in each domain across the end-to-end path in order to acquire from each domain its respective SLA and provide finally to the end-user the e2e SLA created by the merging of the per-domain SLAs, according to the merging rules defined in Table 1 above. Well defined APIs will carry out the WS communication between the SLA Frameworks of the involved domains, including all required intra-domain entities and the administrator who will be responsible for the SLA management within each domain. All the components of the architecture are based on service oriented architecture and realize the inter domain communication via web services. In this way a business-centric architectural approach is achieved that integrates all the SLA functions together between the different domains.

The components for the inter-domain communication can be categorized into the following levels as shown in Figure 2:

- **Web Interface**: through this web interface the domain administrator can manage and configure the SLA Framework.
- **SLA Communicator**: this component is responsible for the intra-communication between the different entities of each framework. By calling the appropriate web services, each SLA communicator of the SLA framework can collect required data and interchange messages.
- **SLA Manager**: through this entity, the SLA Frameworks in the different domains inter-communicate with each other and transfer the appropriate data via Web Services.
- **SLA Core Engine**: this component collects the information from the Web Interface and the SLA Manager and generates either the per-domain SLAs or the e2e SLA if it is the first framework in the path.
- **SLA Persistence Layer**: this layer is responsible for the communication of the core components with the SLA database.
- **SLA DB**: each SLA Framework maintains its own database where local SLAs of the domain, internal data of the framework and end-to-end SLAs are stored. The data can be viewed and edited via the administrative functions of the Web Interface.

Figure 3. User web interface

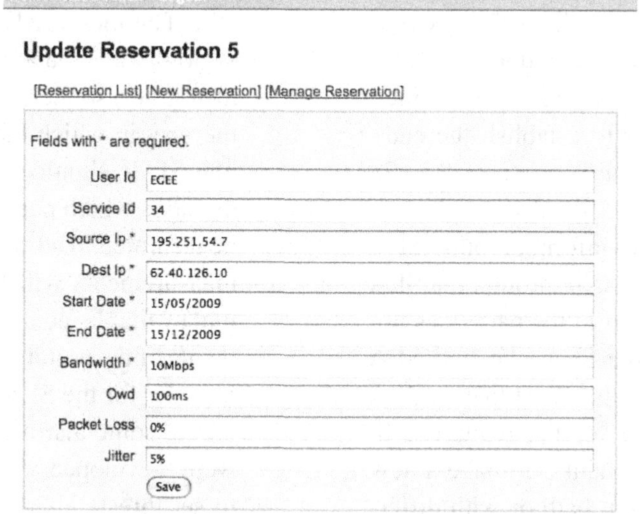

Home Reservation Logout

Update Reservation 5

[Reservation List] [New Reservation] [Manage Reservation]

Fields with * are required.

User Id	EGEE
Service Id	34
Source Ip *	195.251.54.7
Dest Ip *	62.40.126.10
Start Date *	15/05/2009
End Date *	15/12/2009
Bandwidth *	10Mbps
Owd	100ms
Packet Loss	0%
Jitter	5%

Save

Figure 2 shows the SLA Framework Architecture and its components.

For enabling the user to have access to the SLA framework and request an SLA, a User Interface (UI), as shown in Figure 3, is provided. Through this interface the user, after having been authenticated, can fill in his request with the desired level of QoS. Afterwards the UI will communicate via Web Services with the SLA Communicator (Figure 2) to forward the user request to the main components of the SLA framework in order to proceed with the generation of the corresponding SLA.

SLA ESTABLISHMENT PHASES

In the following section we will describe the SLA generation life–cycle. It involves three phases: Creation, Publication and Delivery.

- **Creation Phase:** The end user triggers the end-to-end SLA provisioning, when he requests a service. More specifically, the user requests, through the UI, an e2e SLA for a network reservation spanning multiple domains. He gives the source and destination IPs for his reservation and then the framework has a path-finding module that selects the shortest available path between the source and destination IPs.The service requests are received through the SLA Communicator, which forwards the request to the SLA Manager to communicate it a) to the next domains and also b) to the local domain's SLA Core Engine component in order to provide the per-domain SLAs. Each domain evaluates the network reservation request, produces the local SLA by filling in the corresponding Administrative Part and Service Level Specification of Table 1 and finally, stores the respective SLA to its local database (SLA db in Figure 2)
- **Publication Phase:** After the creation phase, the newly defined per–domain SLAs need to be published in order to be available for the merging of the e2e SLA. Publication consists of the transfer of the per-domain SLAs to the first SLA Framework in the reservation path, which is made through the SLA Manager component.

- **Delivery Phase:** Once the per–domain SLAs have been defined and published, the first SLA Framework in the reservation path merges all the per–domain SLAs, according to the merging rules defined in Table 1, in order to establish the end–to–end SLA. Afterwards, it stores the e2e SLA to the local SLA database and communicates via WS to inform all other SLA Frameworks in the chain through the SLA Manager. If a domain in the path does not deliver its respective SLA, the first SLA Framework informs the client that an SLA cannot be delivered for the requested IP service, so that the client can make a new request for a different path or with different level of QoS.

Monitoring Procedure

In an inter–domain environment, mechanisms for monitoring the performance parameters for a specific service instance must be available. Such information is critical for both the end user and the service provider, since the former is the entity that requests the service and the later is the entity that delivers the service. Evaluating the monitoring results, a user can ask for SLA re–negotiation if the negotiated service level is not the appropriate one.

For the SLA monitoring, several monitoring tools will be used. The entity that will be responsible for the SLA monitoring, apart from the end user, should be the one that provides the SLA framework administrator.

The SLA monitoring and troubleshooting procedures that are followed are described below:

- In order for the administrator to monitor an SLA it should access the monitoring framework and provide all the required parameters (e.g., time period over which the network monitoring for the SLA applies, source and destination IP addresses, metrics that need to be measured, such as One-Way-Delay (OWD), achievable bandwidth etc). The measured data can be viewed in various ways (data table, matrix, time plot, histogram) so the administrator can choose the way in which the resultant data is displayed. It should be mentioned that the domains which need to be measured, must be informed in advance about what kind of measurements will be taken, in order for the available configurations and deployments of the monitoring tools to take place.
- In order for the SLA guarantees to be honoured, some alarm triggering mechanisms were developed that provide the ability to set thresholds on the measured metrics and accordingly generate alarms in case of threshold violations.
- In case an SLA metric is violated due to a problem (e.g., bandwidth falls below the specified value in the SLA template) the monitoring framework tries to identify in which domain the problem occurred by taking intermediate measurements if needed and accordingly notify the Technical Contact(s) (provided in the Administrative Part of the e2e SLA) of one or more of the domains along the path in order to solve the problem. Those involved in a reservation, such as users/Virtual Organizations/Resource Centres, can install and/or run their own application oriented tests in order to verify the performance of their service and if it is degraded. In case of faults, perceived by them, the problem must then be reported and a trouble ticket should be opened. It should be mentioned that the administrator must refer to the specific SLA instance of the reservation path for which the problem exists when reporting a problem to a domain, so that the reservation details, as registered in the corresponding SLA instance, are taken into consideration for troubleshooting.

Figure 4. Monitoring procedure

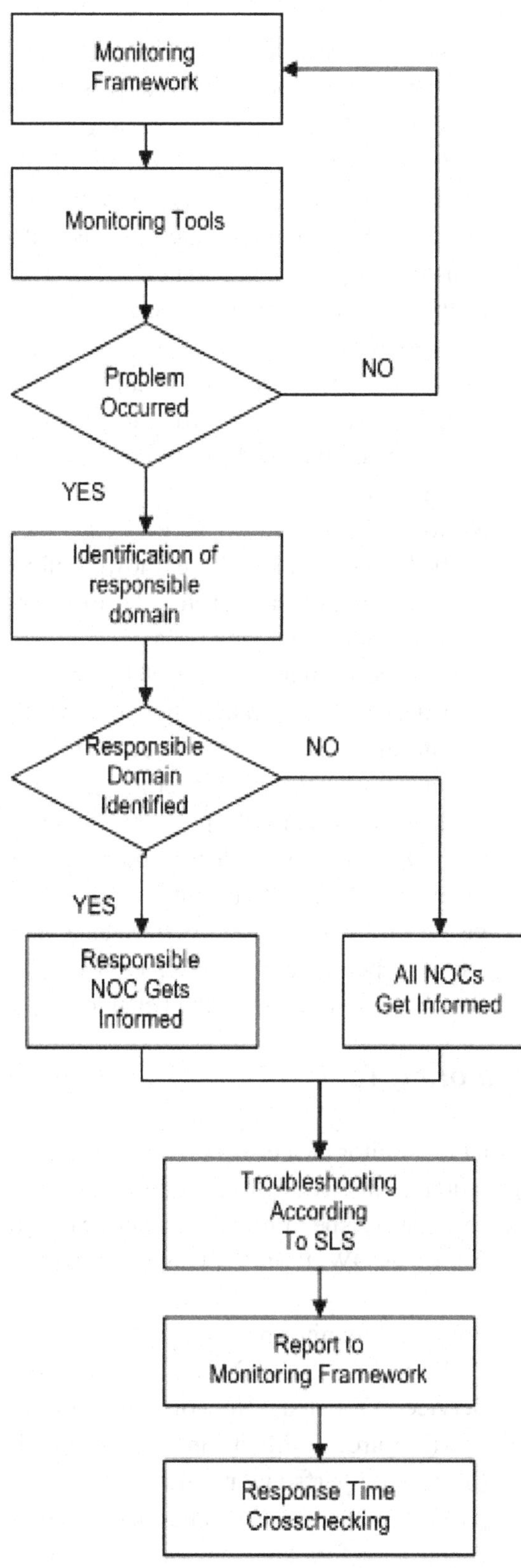

- In each domain, the fault-handling and trouble ticket procedures specified in the corresponding SLA must be followed. Normally, the domain(s) in which the problem occurs will be able to troubleshoot it.
- After troubleshooting they should report back to the administrator what were the causes of the problem and if normal operation has been restored.
- Finally, actual failure indication response times (as declared by a domain in the corresponding SLA) must be crosschecked by the administrator against the promised values.

The flowchart for the above monitoring procedure is shown in Figure 4:

In order for the monitoring procedure to be feasible, some monitoring requirements must be available:

- Network performance data should be available for every SLA that corresponds to an e2e reservation.
- The SLA should be monitored on an e2e basis, with the point of measurements as close to the end point as possible.
- For every SLA monitored, the following metrics should be collected: OWD, Round Trip Time (RTT), Jitter, Packet Loss (PL), Bandwidth.

There should also be possible to make on demand measurements frequently in order to provide an up–to–date view of the network performance but in any case they should not greatly impact the network performance.

GRID EXPERIMENT

In this section we describe the Grid experiment that was performed in order to test and evaluate the SLA establishment and monitoring procedures that were described in the previous sections. The application scenario was taken from the GRIDCC (http://www.gridcc.org) project and considers a remote control application to monitor the instruments distributed across the Grid.

The GRIDCC project produced middleware is based on a web services architecture and not only extends the middleware being developed by other Grid projects but also develops completely new middleware where needed. An innovative idea of the GRIDCC project is the Instrument Element (IE), which provides an abstraction of the instruments in order to access and control them via Web Services in a unified and standardized way from the Virtual Control Room (VCR). The VCR is the User Interface (UI) that allows users to build complex workflows and submit them to the execution services or even connect to the resources directly in order to control the instruments in real time (Montagnat et al., 2008).

GRIDCC provided a good candidate application willing to adopt the SLAs that promised to offer the stringent network QoS guarantees needed for the real-time remote control of instrumentation over the Grid. The testing gave us a real example of a network application and the ability to study the particular requirements of this application and try and identify what the network can offer it.

Description

The Grid experiment was based on Web Service invocations in order to remotely control, through the IE, the instruments distributed across the Grid environment and then transfer the data produced by these instruments to Grid Resource Centres for further processing.

In order to introduce a real network and preserve the real time behaviour needed for the remote control and data transfer of produced results from the experiment, we need to have a network which can guarantee that its network characteristics (bandwidth, delay etc) remain within a given range when a client located in one site needs to access a server in another site.

Thus, we performed the following test:

- A client in site A logs in to the VCR, described above, and makes Web Service calls (with the same SOAP message payload) to the server which hosts the IE in site B in order to perform remote control of instruments and then initiate probable data transfers of the data produced.
- Both the client and server are time synchronized.
- The test is performed with and without SLA reserved path in order to evaluate the effect of the SLA reservation on the performance.
- Network performance metrics are described in the SLA and are known before the beginning of the test in order to determine an end-to-end time behaviour.
- For the performance testing frequent e2e measurements are taken by appropriate monitoring tools in order to evaluate and monitor the level of QoS over the network.
- Results of the network performance metrics are gathered in collective diagrams, in order to have a better control and supervision of the application performance and are also permanently stored to have the possibility to track back the behaviour.

Monitoring Tools

In order to monitor the network used for the Grid application testing we deployed some monitoring tools to measure specific network performance metrics such as OWD, Jitter, PL and Bandwidth.

- For the monitoring, we installed the PerfSONAR (Performance focused Service Oriented Network monitoring Architecture) monitoring framework (http://wiki.perfsonar.net/jra1-wiki/index.php/) which forms an interoperable, distributed performance measurement mid-

Figure 5. Grid experiment test-bed: networked real time environment for accessing a remote instrument element (IE)

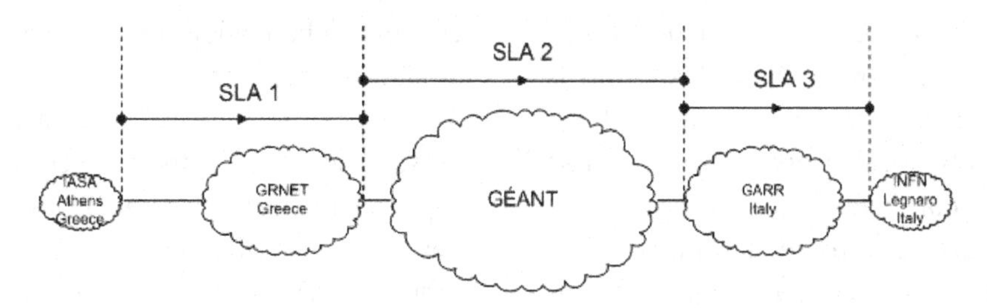

dleware framework. The release that we installed contains, amongst others, the following tool that was used to monitor the performance metrics needed for the application: Bandwidth Controller (BWCTL) Measurement Point (MP) for measuring the achievable bandwidth.

- One-Way Ping (OWAMP) application for measuring one-way latencies such as OWD, Jitter and PL. One way measurements help to better isolate a problem because the direction of the congestion becomes immediately apparent. Since traffic could be asymmetric in the two directions between hosts, one way measurements allow the user to better isolate the effects of congestion for specific parts of the network thus allowing for more informative measurements and in some cases help providers decrease the areas of congestion whenever possible by better allocating resources in their networks.
- RRD (Round Robin Database) Measurement Archive (MA) for collecting the required metrics and providing graphical displays of historical views of the data whenever required.

Deployment Procedure

For the testing of the Grid application scenario, the required software was installed on two nodes (server-client). More specifically, a VCR was installed on the client side and provided a UI for the user to access the instruments and an IE was installed on the server side to provide an interface for the remote control of various instruments. The server node was located in Legnaro, Italy within the INFN (National Institute of Nuclear Physics: http://www.infn.it/indexen.php) resource centre, while the client node used for calling the IE and storing the data produced by the remote control was located in the HELLASGRID (http://www.hellasgrid.gr/) cluster in Athens, Greece at IASA (Institute of Accelerating Systems & Applications http://www.iasa.gr/). The end-to-end path from IASA to INFN traversed the Greek NREN GRNET, the Pan-European Interconnection backbone GÉANT and the Italian NREN GARR (http://www.garr.it). As shown in Figure 5, we managed to establish three per-domain SLAs across this path, namely from IASA to GRNET, GRNET-GÉANT-GARR, GARR-INFN.

All the monitoring tools described in the previous section were deployed on both nodes in order to run the Grid application between the two nodes and test its performance by measuring specific network metrics. In order to realize the automatic SLA establishment procedure through the SLA framework and evaluate the performance of the reservation mechanism, we conducted two experiments in the same network route; the first one was under Best-Effort (no reservation to the network) while the second one was under a reserved path through the SLA framework which provided guaranteed service, PIP enabled. For the two tests we followed the steps described below in each case.

Best Effort Service

The first test was realized under a best-effort service in a specific network path. Best effort is the default service level in GÉANT and NRENs. In a best effort network all users obtain best effort service, meaning that there are no specific network performance guarantees provided by the NRENs. Thus the users and applications obtain unspecified bandwidth and delivery time, depending on current traffic load. Thus there was no need to make a service request to reserve a network path neither to offer an SLA because the Best Effort service does not provide any guarantees that data is delivered or that a user is given a guaranteed quality of service level or a certain priority. In this situation we conducted our experiments in the GÉANT/NRENs network and took the measurements without making a path reservation.

In order to monitor the test for the whole SLA duration we installed in both the previous cases the monitoring tools described in previous section and took measurements of the performance metrics every 10 minutes across the network path as shown in the following section. The experimental results are represented graphically below through the RRD diagrams.

SLA Guaranteed Service

The second test was made under a reserved network path with guaranteed QoS characteristics and an e2e SLA was offered to the end user automatically through the SLA framework. More specifically, the SLA request was requested through the site from where the user filled in his request, which was then forwarded to the AMPS in order to make the path reservation. The AMPS is an advanced reservation system that accepts PIP service requests, authenticates them according to the policy module that it has and reserves network resources in each of the domains across the full path. When requesting an SLA the required network metrics, Start and End Times of the reservation, and one IP Source and Destination Address Pair should be specified. In our application scenario we are interested to have guaranteed bandwidth for the transfer of the data produced by the instruments towards the Grid resource for further processing and also the minimum possible delay across the real network is needed to approach the near real time nature of the application. In our case the requested bandwidth was 16 Mbps, the OWD 50ms, PL 0% and Jitter 1ms. The SLA period was one week and the network path had as source IP: 195.251.54.7 and destination IP: 62.40.126.10. Each domain in the path could provide the requested values thus the merging of the per-domain SLAs into an e2e one, according to the merging rules of Table 1 was trivial. The default MTU in all domains was 1500bytes.

MEASUREMENTS

In the following section we will describe the performance testing procedure followed by the measurements that we took.

For the monitoring procedure we performed frequent (i.e., every 10 minutes) end-to-end measurements with the BWCTL and OWAMP applications, measuring the network performance metrics between the two nodes bi-directionally. The monitoring mechanisms were installed in AMPS servers at the edges of the end domains, i.e., IASA-GRNET and INFN-GARR.

More specifically, the OWAMP is used to provide measurements for the OWD, Jitter and PL, while the BWCTL is used to measure the achievable bandwidth by sending small packets in order not to affect the network performance of the application. Retrieved data are collected via the RRDtool, stored in an RRD database, and illustrated graphically in diagrams, for the two cases, as shown below.

In the sections below, the first diagrams refer to the Best-Effort experiment and following are the ones concerning the SLA reserved path experiment.

Figure 6. a) Bandwidth (Mbps) representation from IASA to INFN and b) backwards. c) Packet loss (%) representation from IASA to INFN and backwards. d) Jitter (ms) representation from IASA to INFN and backwards. e) One-Way-Delay (ms) representation from IASA to INFN and f) backwards.

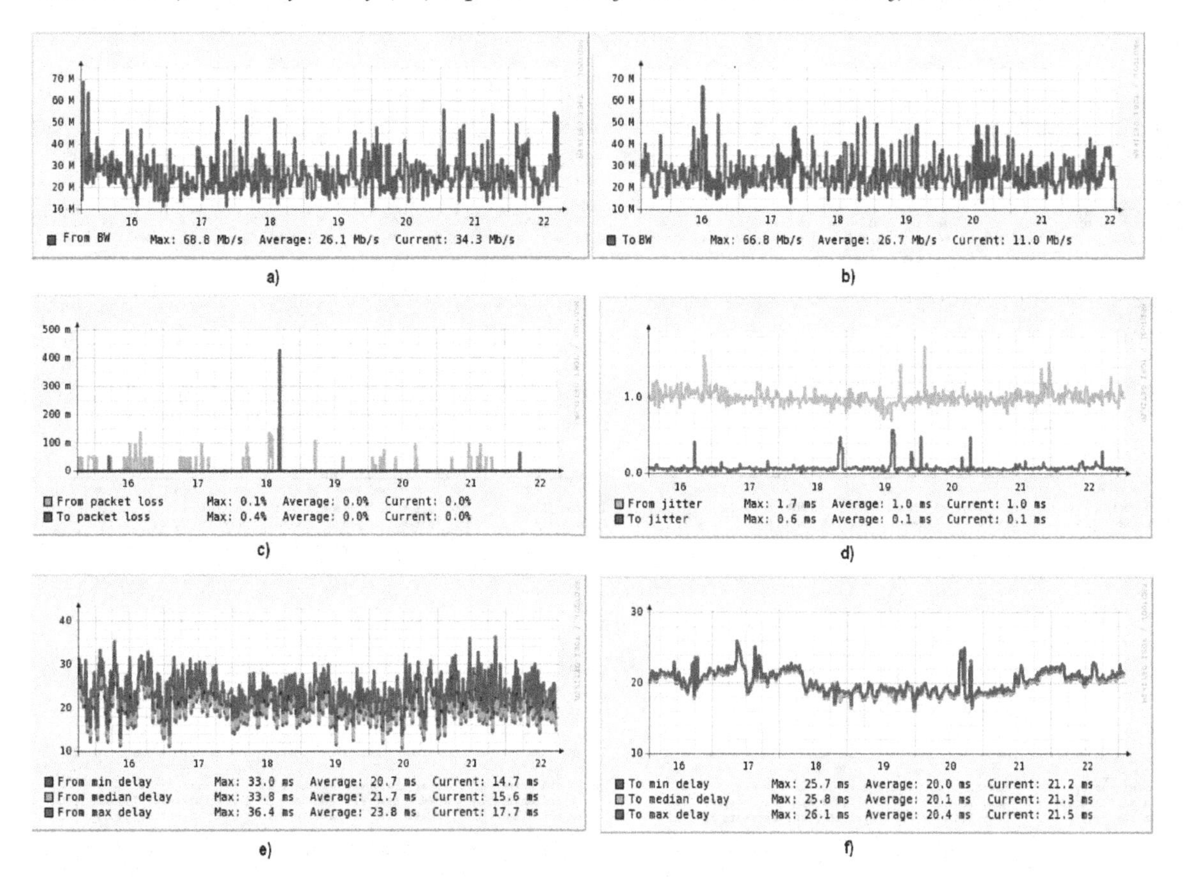

Best-Effort Diagrams

Below are the daily diagrams for the Best-Effort case. The x-axis declares the days during a week while the y-axis declares the Bandwidth, Packet-loss, OWD and Jitter in each case.

Best-Effort, as mentioned before, is the default service in the GÉANT and NRENs networks meaning that all users obtain Best-Effort service. This service does not provide guarantees for bandwidth and delivery time so the measurements that we will see in the figures below will be measurements showing the current traffic load.

In Figure 6 a) we can see the Bandwidth (Mbps) from IASA to INFN and in b) backwards. Its value as we can observe has a lot of variations ranging from 10Mbps to 70Mbps.

In Figure 6 c) we see that the Packet Loss during a week is almost 0%, but in many cases presents some peaks in both directions. The difference in these values is due to the asymmetric nature of the two directions between the two nodes in IASA and INFN.

The Jitter (ms) representation is shown in Figure 6 d). In the IASA- INFN direction we have almost no jitter with some peaks in some cases, while in the opposite direction we have an average jitter of 1ms.

Finally, in Figure 6 e), we can see the OWD (ms) from IASA to INFN and in f) backwards. Its average value is about 20ms.

Figure 7. a) Bandwidth (Mbps) representation from IASA to INFN and b) backwards. c) Packet loss (%) from IASA to INFN and backwards. d) Jitter (ms) from IASA to INFN and backwards. e) OWD (ms) from IASA to INFN and f) backwards.

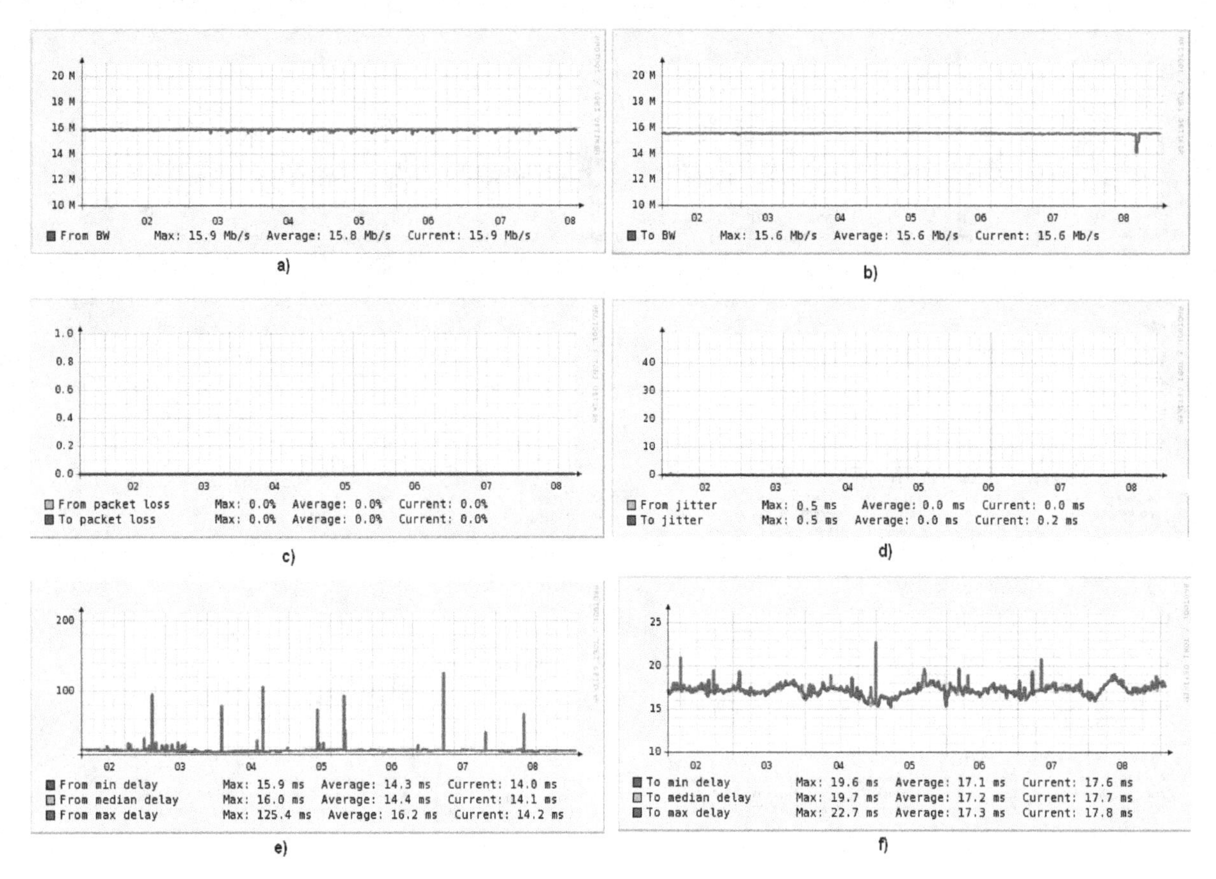

SLA Guaranteed Diagrams

Following are the weekly diagrams for the SLA reserved path. The x-axis declares the days during the week while the y-axis declares the Bandwidth, Packet-loss, OWD and Jitter in each case.

In Figure 7 a) we can see the bandwidth measurements from IASA to INFN and in b) the backwards direction respectively. The bandwidth value is almost stable at 16Mbps in each diagram respectively. Compared to the Best-Effort case, we observe a greater stability in the bandwidth measurements reaching the guaranteed bandwidth that we requested.

In Figure 7 c) we can observe the packet loss for the two directions (IASA -> INFN, INFN -> IASA). As in the Best-Effort case, packet loss is equal to zero.

The Jitter (%) representation is shown in Figure 7 d) below. The jitter is 0%. Again we can see an improvement of the jitter compared to the jitter results shown in the Best-Effort diagrams.

Finally, the OWD from IASA to INFN and backwards are shown in Figure e) and f) respectively. In the first case (IASA -> INFN) we can see that the average OWD is about 14ms while in the second case (INFN -> IASA) the OWD is about 17ms. Comparing with the Best-Effort case, the OWD in the SLA guaranteed service is a little bit better than the OWD in the Best-Effort. Some picks in the diagrams may be due to false alarms of the monitoring tools.

Comparison between the Best-Effort and SLA Reserved Path Diagrams

From the experimental results we can see that we have an improvement in performance in the reserved path through the SLA framework where the bandwidth and delay variations present smaller variations compared to the best-effort service. That is due to the fact that the service offered through the SLA framework in a PIP reserved path is the best offered service over an IP network that uses the available high priority scheduling techniques to emulate a dedicated circuit. The aim of a guaranteed service is to provide guarantees on bandwidth, OWD, Jitter and negligible percentage of PL for IP traffic across one or more interconnected domains. This is of great importance for applications and the primary beneficiaries of this service are the users of applications that demand strict QoS guarantees over the network, e.g., video conferencing and remote control of the Grid instrumentation, due to their real time nature.

CONCLUSION

In this paper we propose a framework for automated Grid SLA management, i.e., the creation, negotiation, publication and monitoring in a multi–domain distributed high performance computing & networking environment. The automated management and monitoring procedures have been presented as they have been designed for a multi-domain system, including Grid resources and high speed interconnecting communication networks. Key to our framework is the provision of per-domain SLSs in a common template as proposed in our paper.

Our framework provides an e2e automatic SLA service to a Grid end-user and describes SLS specifications (e.g., networking PIP characteristics, monitoring infrastructures and mechanisms, availability guarantees) for each administrative domain involved in the e2e path upon making the reservation. It automates the otherwise manual procedure of finding and contacting each administrative network domain across paths between Grid resources to produce and deliver per-domain SLAs and consequently the e2e SLA.

Real time and interactive Grid applications can benefit from our SLA framework as it offers an automated way to provide and monitor a contract to the end-user specifying that the network performance guarantees and the troubleshooting procedures stipulated in the SLA will be provided and enforced. As an example, a user by signing an SLA can enforce time critical constraints imposed by his experiment.

The proposed framework has been adopted within the EGEE and GÉANT operations and is deployed in the AMPS provisioning system. The integration of our e2e SLA Framework in AMPS gave us the opportunity to evaluate and measure its performance in real QoS–enabled networks such as the NRENs and their interconnecting backbone GÉANT.

ACKNOWLEDGMENT

This paper has been partially supported by the 6th Framework Programme for Research & Technological Development of the European Union – EC DG INFSO projects EGEEII, EGEEIII, GN2 (GÉANT) and GRIDCC.

The authors wish to particularly thank Xavier Jeannin (CNRS and EGEE-SA2 Activity Leader), Afrodite Sevasti (GRNET and GN2-SA3 Activity Leader), Paris Sphicas (University of Athens, IASA, CERN and GRIDCC WP6 Leader), Marios Chatziangelou (IASA and HellasGrid administrator), Anand Patil (DANTE), Waldemar Zurowski (DANTE), Claudia Battista (GARR), Mauro Campanella (GARR), Francesco Lelli (INFN) and Gaetano Maron (INFN, CERN and GRIDCC Coordinator).

REFERENCES

Andreeva, J., Campana, S., Fanzago, F., & Herrala, J. (2008). High-Energy Physics on the Grid: the ATLAS and CMS Experience. *Journal of Grid Computing, 6*(1), 3–13. doi:10.1007/s10723-007-9087-3

Andrieux, A., Czajkowski, K., Dan, A., Keahey, K., Ludwig, H., & Pruyne, J. (2008). *Web services agreement negotiation specification (WSAgreementNegotiation) (draft) (Tech. Rep.)*. Open Grid Forum.

Bagchi, A., Caruso, F., Mayer, A., Roman, R., Kumar, P., & Kowtha, S. (2009). *Framework to achieve multi-domain service management*. Paper presented at the IFIP/IEEE International Symposium on Integrated Network Management (IM '09) (pp 287-290).

Bouras, C., Campanella, M., & Sevasti, A. (2002). *SLA definition for the provision of an EF-based service*. Paper presented at the 16th International Workshop on Communications Quality & Reliability (CQR 2002), Okinawa, Japan (pp. 17-21).

Campanella, M., Przybylski, M., Roth, R., Sevasti, A., & Simar, N. (2003). *Multidomain End to End IP QoS and SLA* (LNCS 2601, pp. 171-184). New York: Springer.

Courcoubetis, C., & Siris, V. (1999). *Managing and Pricing Service Level Agreements for Differentiated Services. Paper presented in the IEEE/IFIP (IWQoS'99)*. London: UCL.

DiffServ Provisioning. In *Proceedings of the International Conference on Software Engineering Research, Management & Applications (SERA 2005)*, Mt. Pleasant, USA (pp. 325-330).

Dong-Jin, S., & Young-Tak, K. (2003). Design and Implementation of Performance Management for the DiffServ-aware-MPLS Network. In *Proceedings of the Conference on APNOMS 2003*, Fukuoka, Japan.

Dustdar, S., & Kramer, B. J. (2008): Introduction to Special Issue on Service-Oriented Computing (SOC). *ACM Transactions on the Web, 2*(2).

Foster, I., & Kesselman, C. (1998). *The Grid: Blueprint for a new computing infrastructure*. San Francisco, CA: Morgan Kaufmann Publishers Inc.

Garg, M., Saran, R., Randhawa, H., & Singh, R. S. (2002). *A SLA framework for QoS provisioning and dynamic capacity allocation*. Paper presented at the 10th IEEE International Workshop on Quality of Service.

Goderis, D., et al. (2002, January). *Service Level Specification Semantics and Parameters* (Internet Draft, <draft-tequila-sls-02.txt>).

Gulliver, P. H. (1979). *Disputes and Negotiation: A Cross-Cultural Perspective*. New York: Academic.

Hasselmeyer, P., Mersch, H., Koller, B., Quyen, H. N., Schubert, L., & Wieder, P. (2007). *Implementing an SLA Negotiation Framework*. Paper presented at eChallenges.

Hoffner, Y., Field, S., Grefen, P., & Ludwig, H. (2001). Contract driven creation and operation of virtual enterprises. *Computer Networks, 37*, 111–136. doi:10.1016/S1389-1286(01)00210-9

Hudert, S., Ludwig, H., & Wirtz, G. (2009). Negotiating SLAs-An Approach for a Generic Negotiation Framework for WS-Agreement. *Journal of Grid Computing*, 7(2), 225–246. doi:10.1007/s10723-009-9118-3

Montagnat, J., Glatard, T., Plasencia, I. C., Castejón, F., Pennec, X., & Taffoni, G. (2008). Workflow-Based Data Parallel Applications on the EGEE Production Grid Infrastructure. *Journal of Grid Computing, 6*(4), 369–383. doi:10.1007/s10723-008-9108-x

Nichols, K., & Carpenter, B. (2000). *Definition of Differentiated Services Behavior Aggregates and Rules for their Specification* (draft-ietf-diffserv-ba-def-00.txt).

Opitz, A., König, H., & Szamlewska, S. (2008). What Does Grid Computing Cost? *Journal of Grid Computing, 6*(4), 385–397. doi:10.1007/s10723-008-9098-8

Papazoglou, M. P., & Van den Heuvel, W. J. (2005). Web services management: a survey. *IEEE Internet Computing, 9*, 58–64. doi:10.1109/MIC.2005.137

Parkin, M., Kuo, D., & Brooke, J. (2006). A framework & negotiation protocol for service contracts. In Proceedings of the IEEE SCC (pp. 253-256). Washington, DC: IEEE Computer Society.

Rong-xiao, G., Jing-bo, X., Shu-fu, D., & Kai, W. (2009). Research on Multi-domain Policy-Based SLA Management Model. In. *Proceedings of the International Conference on Networking and Digital Society, 1*, 209–212.

Van der Mei, R. D., & Meeuwissen, H. B. (2006). *Modelling End-to-end Quality-of-Service for Transaction-Based Services in Multi-Domain Environments.* Paper presented at the IEEE International Conference on Web Services (ICWS'06), Hyatt Regency at 0' Hare Airport, Chicago.

Web Services Agreement Specification. (2007, March). *The Open Grid Forum.* Retrieved from http://www.ogf.org/documents/GFD.107.pdf

Young-Tak, K. (2005). Inter-AS Session & Connection Management for QoS-guaranteed

Zhang, Z., Tan, Y., & Dey, D. (2009). Price competition with service level guarantee in web services. *Decision Support Systems, 47*(2), 93–104. doi:10.1016/j.dss.2009.01.004

This work was previously published in International Journal of Systems and Service-Oriented Engineering, Volume 1, Issue 3, edited by Dickson K.W. Chiu, pp.21-38, copyright 2010 by IGI Publishing (an imprint of IGI Global).

Chapter 4
Execution Management for Mobile Service-Oriented Environments

Kleopatra Konstanteli
National Technical University of Athens, Greece

Tom Kirkham
Nottingham University, UK

Julian Gallop
STFC Rutherford Appleton Laboratory, UK

Brian Matthews
STFC Rutherford Appleton Laboratory, UK

Ian Johnson
STFC Rutherford Appleton Laboratory, UK

Magdalini Kardara
National Technical University of Athens, Greece

Theodora Varvarigou
National Technical University of Athens, Greece

ABSTRACT

This paper presents an Execution Management System (EMS) for Grid services that builds on the Open Grid Services Architecture (OGSA) while achieving "mobile awareness" by establishing a WS-Notification mechanism with mobile network session middleware. It builds heavily on the Session Initiation Protocol (SIP), used for managing sessions with mobile terminals (such as laptops and PDAs) where the services are running. Although the management of mobile services is different to that of ubiquitous services, the enhanced EMS manages both of them in a seamless fashion and incorporates all resources into one Mobile Dynamic Virtual Organization (MDVO). The described EMS has been implemented within the framework of the Akogrimo EU IST project and has been used to support mission critical application scenarios in public demonstrations, including composite and distributed applications made of both ubiquitous and mobile services within multiple domains.

DOI: 10.4018/978-1-4666-1767-4.ch004

INTRODUCTION

The development of more robust mobile communication networks and more powerful portable computing devices has encouraged the emergence of a wide range of mobile distributed computing applications. These applications use services that reside on mobile devices presenting users with information derived from a variety of sources and localities. The applications that use these emerging services are the subject of much research and business models (Waldburger & Stiller, 2005). However bringing about the enhanced functionality associated with these advances presents a challenge (Forman & Zahorjan, 1994).

Supporting mobile devices in a service-oriented computing environment (Weissman, Kim, & England, 2005) requires specialist management capabilities to ensure the reliability and dependability of a single composite application. Services in these environments have not only to present predictable behaviour in terms of connectivity but collectively a framework to aid automated decision making based on a variety of information linked to location data. For example, as mobile devices are subject to physical connection changes as they move between networks and areas of coverage, the application may adapt its characteristics in order to accommodate such events. In the case of movement of a device onto another country's network, the data the device shares may be restricted by international legislation or in the case of lower bandwidth the device again may adapt not transmit certain types of data.

In addition, the management of fluctuating connection quality as devices roam across networks may create the need for live service reselection or replacement, as an application may reselect services using more efficient networks as they become available. All these changes require the application to make real-time automated decisions and this challenge is why critical service-oriented applications that use and support multiple mobile services are something of a rarity. To ensure that applications in critical environments are dependable, issues of cross-domain co-operation and dynamic service management have to be solved. The European IST project Akogrimo (2007) addressed these issues by integrating a Grid approach into web service management and creating a framework to support mobile network-aware services, users and service providers.

This paper focuses on some results of the Akogrimo project which are relevant to this purpose. It describes how the Execution Management Service (EMS), enhanced in Akogrimo, supports distributed Grid applications that are mobile and dynamic and does this by supporting Mobile Dynamic Virtual Organizations (MDVOs). It establishes a notification to keep track of the changes that may occur in the location and availability of Grid services that reside in mobile resources. By doing so, the OGSA-compliant EMS achieves increased awareness of the changes that take place in the mobile network layer and adjusts them on the fly so that they remain transparent to the user.

RELATED WORK

The use of mobile resources within distributed computing frameworks relies on strong communication, management and understanding between the network and service middleware (Tosic et al., 2005). A common approach to this has been the management of network derived mobile context. This context management has been a focus for work in the Akogrimo project using SIP (Rosenberg et al., 2002) to link the mobile network with the grid middleware. SIP was chosen due to the familiarity of the protocol with the Akogrimo developers and its wider user community than other similar standards. Other work in this area has investigated the use of other protocols such as IP Multimedia Subsystem (IMS) (Camarillo & Garcia-Martin, 2004) which was developed around the standardization of mobile devices in UMTS networks.

Management of mobile context in service based applications has been further enhanced by standards such as the Web Service Offerings Language (WSOL) (Tosic, Kruti, & Pagurek, 2002). This standard works with Web Service Description Language (WSDL) (Christensen et al., 2001) to better categorize and explain web service interfaces. Approaches to web service orchestration and specifically choreography enhance the use of WSOL and increase the amount of service level collaboration via the use of standards such as Web Services Choreography Description Language (WS-CDL) (Kavantzas, 2004) and Web Service Choreography Interface (WSCI) (W3C, 2002).

In Akogrimo the development of the execution environment to suit Grids along with the use of specific Business Processes Execution Language (BPEL) (BPEL, 2003) templates has presented a service communication model that is hierarchical in nature. The model implemented is based around the intelligence in the workflow and the work at enhancing the web service interfaces to better handle context and choreography at this stage was not a primary focus of work. Akogrimo follows a service-oriented approach that became established in Grids with the publication of OGSA (latest version is OGSA, 2006) and allowed the application of service-oriented application design in the domain of Grid computing. Here, Grid services are aggregated in large-scale, composite applications controlled by a central set of Grid middleware services.

In these applications the service-oriented computing model is vital in terms of service discovery and execution, as services in these models are selected dynamically and are often tied to a Service Level Agreement (SLA) (Ludwig, Gimpel, Dan, & Kearney, 2005). The Grid service concept reflects a more complex model of service lifetime, which is incorporated in the adopted standards, such as WS-Resource (Czajkowski et al., 2004).

The Grid concept also encompasses the idea of the virtual organization (VO), which is essentially a collection of services that provide the environment in which distributed and composite applications can be executed and managed (Foster, Kesselman & Tuecke, 2001). A VO provides an environment for resource provision and management bringing these resources together from multiple conventional organizations.

Within the Open Grid Service Architecture (OGSA), VO functionality has been applied using service–oriented methodology (OGSA, 2006). In Akogrimo, the VO idea is extended to be mobile and dynamic i.e., a Mobile Dynamic Virtual Organization (MDVO). The OGSA architecture defines how Grid applications can be formed from service-based architectures and has identified a list of core services, which includes Execution Management Services (EMS) and Information Services among others. These services have to be implemented using clearly defined standards in order for them to be interoperable. The Web Services Resource Framework (WSRF) has combined WS-standards with Grid resources to create implementation environments in which OGSA .services can be created (Czajkowski et al., 2005). The implementation of WSRF used in Akogrimo has been instantiated by the Globus alliance in the form of the Globus Toolkit version 4 (GT4) (Foster, 2005).

Execution management according to OGSA is central to the dependable delivery of applications in service-oriented Grids. Although existing Grid toolkits have some form of execution management, Akogrimo goes beyond these in that its EMS manages WSRF services and furthermore, the main challenge in this paper, it manages services within a Mobile Dynamic Virtual Organization.

To ensure that the application is dependable in this context, a responsive service management framework is established around the EMS set of services. In Akogrimo, as in other service-oriented projects, the development of the supporting services defines the business models the Grid presents for both service providers (SPs) and users of the applications. Research areas in Grid service trust (Dimitrakos, Golby, & Kearney, 2004) dynamic workflows (Cao et al., 2003) etc, support dynamic service management. Akogrimo, therefore in this research context, is applying this management challenge in a mobile service environment (Wesner, Dimitrakos, & Jeffery, 2004).

CHALLENGES BEHIND THE EXECUTION MANAGEMENT OF MOBILE DYNAMIC VIRTUAL ORGANIZATIONS

A mobile Grid is formed from a group of mobile devices, which can be dedicated to offer computational, memory and networking resources, and to support a composite, multiple application composed of services from multiple Service Providers. The mobile Grid is a more dynamic system than the traditional static Grid, since it is subjected not only to changes, such as system or network failures and system performance degradation that the static Grid suffers, but also removal and movement of machines, variations in the cost of resources, and so on. This dynamicity has a direct impact on various system functionalities, which need to be managed through the EMS.

Supporting MDVOs

Within the Akogrimo framework, the MDVO infrastructure is based around a permanent, static VO working with dynamic VOs. The static VO is called the BaseVO and supports the core Akogrimo infrastructure services that are permanent and ever present. For example, these include the BaseVO Manager, Policy Manager, Workflow Manager and EMS. These core services are the starting point for any application request made on the Akogrimo framework by either a service or a user.

The dynamic application specific VOs are called Operative VOs (OpVOs) and are formed when required in the BaseVO using the WS-Resource scheme to create instances of an OpVO manager in the BaseVO. In addition to this the BaseVO can create instances of core services for use in specific OpVOs, such as the EMS. The use of WS-Resource also allows the BaseVO the power to destroy instances of the OpVO or other service instances in the case of application completion or a breach of security or SLA. Workflow composition is integral to the VO management of the application. The workflow description language used to describe applications that consist of various Grid services running in the Akogrimo platform is BPEL. The abstract workflow description is populated with the endpoint references (EPRs) of the selected infrastructure and application Grid services, during the OpVO set-up phase, and is orchestrated during execution by a BPEL enactment engine.

Managing Mobile Terminals

More specifically, the Virtual Organization that the Akogrimo Grid system supports is not only composed of traditional Grid services running on fixed terminals under static IP/IPv6 address, but also includes resources hosted in mobile terminals under MIPv6, which may move across different networks as well as within the same network by migrating to different mobile terminals. During the actual execution of a service that resides in a mobile terminal, this freedom of movement results in the mobile service being unavailable, followed by a possible change of IP address and consequently of the service's URI. This is translated as a loss of connectivity between the mobile service and the EMS that manages its execution. Especially, in the case when the mobile service is part of a workflow of services whose execution is dependable from each other, this may result in a chain of failures in the execution of the involved services and as a consequence the entire workflow application may be aborted (Figure 1).

Even if this loss of connectivity is temporary due to a failure of the wireless network that may occur for various reasons and the mobile service is reconnected to the network after a small period, the

Figure 1. Mobile User A interacts with Mobile User B using a mobile Grid infrastructure. After moving from Network A to Network B, Mobile Terminal A is assigned a different IP address. Using the old IP address, the EMS fails to reestablish communication with Mobile Service A, and eventually the entire workflow application is aborted.

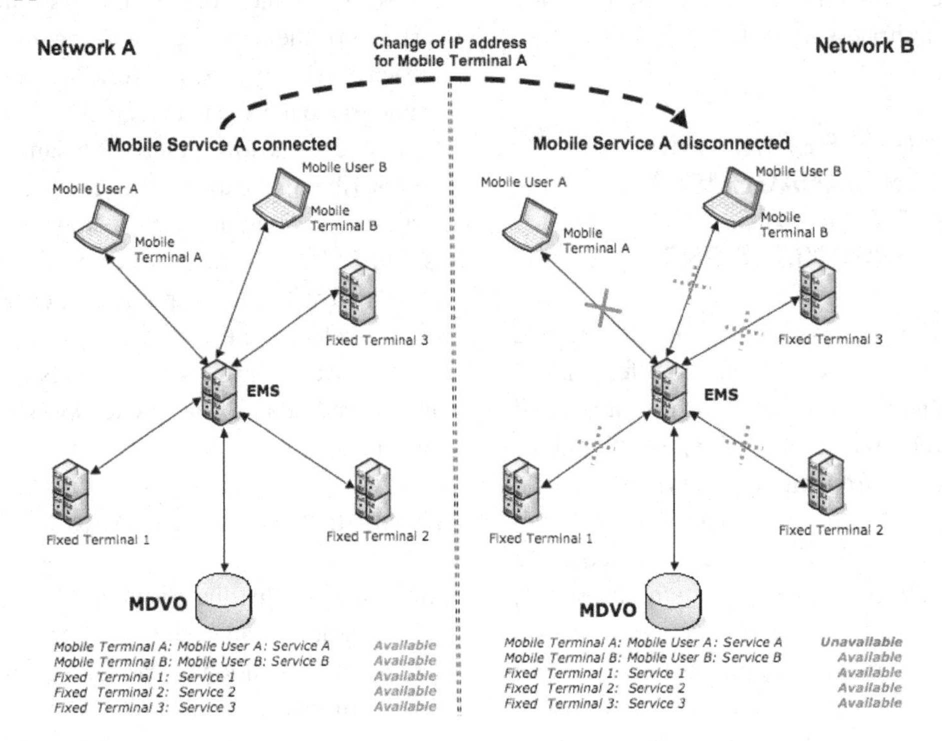

EMS in charge is forced to take corrective actions to rescue the execution of the service/workflow. Although common fault tolerance techniques such as the *retry technique*, in which the EMS performs multiple tries to establish connection with the service after a failure, could be used to deal with temporary loss of connection with the mobile services, after which the mobile services emerge under the same IP address, this is not applicable in the cases where the temporary loss of connection is due to a movement across different networks which results in a change of the IP address assigned to the mobile terminal running the service. Without a mechanism in place to communicate this information from the network to the Grid Infrastructure, the latter remains unaware that the mobile user is actually available but in a changed network, its resources are falsely tagged as unavailable and are therefore no longer part of the MDVO.

One more crucial aspect of the mobile Grid is that services that reside on mobile terminals, and therefore the users behind them, are either the actors responsible for initiating the execution of a workflow application in the Grid or the end users or both, whereas the non-mobile services are responsible for processing intermediate data. This implies that other fault tolerance techniques such as migration of the failing services to other available locations, which is applicable in non-mobile services, may not always be meaningful in the case of mobile services, given that the results are targeted towards specific mobile users/consumers and not just mobile services. The fact that the mobile service is tightly connected to its mobile user and can be considered as one entity is another key issue when dealing with mobile Grids, which calls for increased network awareness not only on network QoS level but also on possible movements.

Communication between EMS and Mobile Networks

This level of service and execution management of services in Akogrimo is a complex challenge that has not been addressed in previous research, which has largely focused on single service or end-user mobility within Grid systems. However, Akogrimo is the first project to consider mobility from the point of view of a Grid service. Instead of being one to many, Akogrimo can handle a relationship between multiple mobile and multiple static services (static services being non mobile) (Park, Ko, & Jai-Hoon, 2003). In this scenario, the VO that supports the services and users has to be flexible and the approach chosen in Akogrimo is the idea of the Operative VO, which exists for the lifetime of the application (Waldburger, 2007). This approach presents composite service-oriented applications that can act automatically and incorporate concepts of self-healing and management. These applications can be linked to specific EMS functions and can operate separately without impact on other applications and VOs.

It becomes evident that in order to effectively manage such MDVOs, the Grid layer needs to communicate with the mobile network layer in order to be informed of changes that may occur in the location and/or the availability of the mobile services. Therefore, the mobile network is needed to source data to the VO management infrastructure, such as service location and network quality of service (QoS). This sharing of data between the network and the VO is achieved through the Akogrimo EMS that joins both layers. Although the management of mobile Grid services is different to that of ubiquitous ones due to their inherent differences, EMS is able to manage both types of service in a seamless fashion and incorporate all resources into one MDVO. The EMS achieves increased awareness of the possible movements of the Grid services that reside in mobile resources such as mobile terminals by establishing communication with the Akogrimo SIP-based network layer (Olmedo, Villagrá, Konstanteli, Burgos, & Berrocal, 2009). This allows the changes that occur in the location and availability of the mobile Grid services that are registered to the SIP-based network to be propagated to the Grid layer via a WS-Notification mechanism, while they remain transparent to the clients. This service-oriented solution including the WSRF interfaces and the WS-Notification mechanism can be easily adapted by any execution management system seeking to manage mobile services running on top of an IP-based signaling network protocols.

Service Providers and Service Agents: Populating an MDVO

In order for a SP to join and offer services to Akogrimo, they must integrate on a contractual and technical basis. This linkage at the SP level allows the business process to execute multiple services from the same domain. The invocation of the services is achieved using specific service agents. These are effectively service interfaces specially created in the BaseVO to link to specifically selected endpoint references (EPRs) of target application specific services and then added to the OpVO. The management of service behaviour in a mobile Grid is a challenge as services change location that can direct affect service properties such as performance and location. It should also be noted that Akogrimo envisages the network SP as another entity that becomes part of the Akogrimo infrastructure. Agreements with the network SP are needed to ensure that the vital network data relating to mobility is passed to the VO infrastructure. Thus, within the application that the VO hosts, the interested parties are: the user, the application SP, the Grid SP, the network SP and trusted third parties. The Grid SP establishes SLAs with the network and application SPs to ensure QoS. The challenge of managing these changes is centered both in the VO and the mobile network layer (Figure 2).

Figure 2. VO relationships

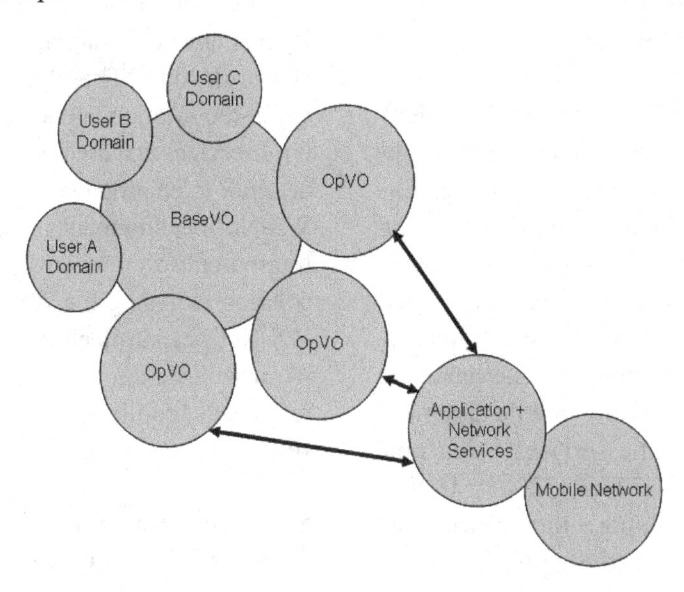

EXECUTION MANAGEMENT FOR MDVOS

The Akogrimo EMS builds heavily on the OGSA Framework specifications and can guarantee QoS enforcement by exploiting the functionalities offered by the rest of the services in the Akogrimo Framework, while at the same time it addresses the main goal of the Akogrimo project which is the management of an MDVO. It comprises the central controller of the execution of the WSRF-based Grid services that reside in the underlying OpVO and is also responsible for discovering, negotiating and reserving resources in the underlying mobile Grid (Litke et al., 2008). These services can be used to form workflow applications using the BPEL workflow description language. The reservation process includes the creation of WS-resources of each WSRF service that is present in the workflow description. These instances of the services are the ones that are being executed at runtime, allowing for multiple concurrent executions of the same service under different workflows and users. After a successful reservation of the workflow services, the abstract BPEL workflow description document that is used in the negotiation process is being populated (becomes concrete) with the endpoint references of the WS-resources that are created for each service in the workflow. At execution time, the EMS is triggered by the BPEL enactment engine in order to manage the execution of the services in the workflow in an orchestrated manner.

Although from the service client's point of view, the EMS appears to be a single service it is composed of various Grid services following the SOA paradigm where each service is responsible for different subtasks such as negotiation, reservation, discovery and execution as depicted in the Figure 3. The Core service acts as a gateway service offering a single-point of access to EMS while at the same time "hiding" its details and complexity from the clients. These Grid services build heavily on the capabilities offered by the WSRF, WS-Notifications and WS-Security and have been implemented using the GT4 platform, but are also accessible to WSRF.NET clients and services (University of Virginia, 2005). In addition, GT4's MDS4 high-level services (Globus, 2006) have been leveraged into EMS. The MDS4 Index service that is deployed in the same standalone container as the EMS, acts as the central MDVO-

Figure 3. EMS architecture

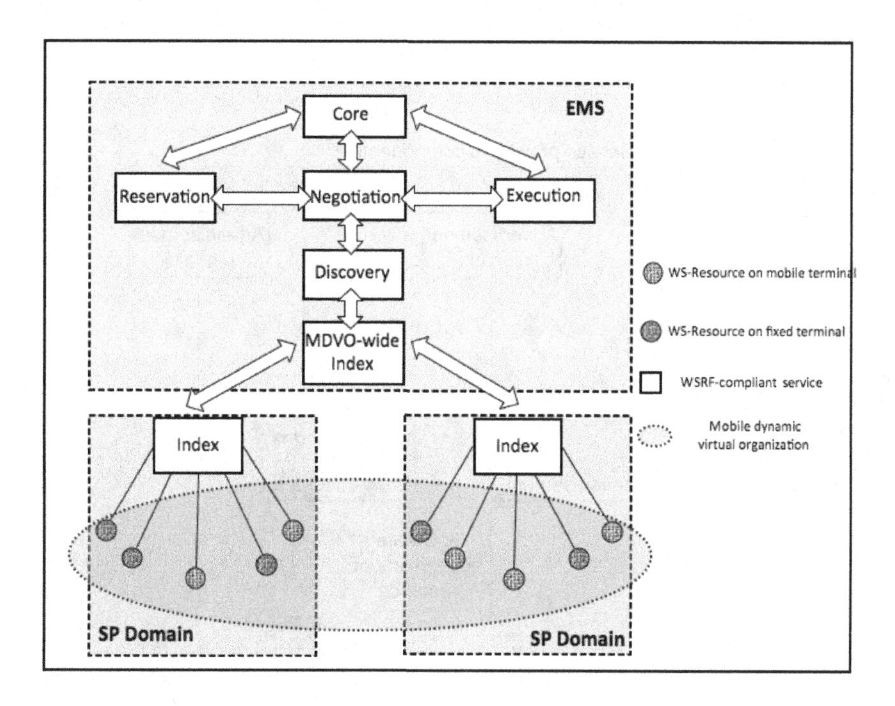

index to which all the services in the BaseVO scattered throughout the underlying Grid, whether on fixed or mobile resources and whether in the same or in different SP domains, are registered.

Registration of Ubiquitous Services to MDVO

In order to maintain an index of non-mobile services that will form the non-mobile part of the MDVO, the EMS depends upon a hierarchical structure of MDS4 Indexes that lead up to its own Index, which serves as the MDVO-wide index. Information about the services in the form of WS-resources that are registered locally in the machines hosting the services is being "pulled" by the central Index automatically after a specific period of time using inherent functionality of the MDS4 Index service. In this way, the MDVO maintains an index of available resources on ubiquitous services for the entire underlying Grid that spans across multiple SP domains as depicted in Figure 3.

The SP is responsible for creating an "advertisement" of his services with QoS properties to the EMS by specifying a unique within his domain name for the service (*serviceID*), the URI of the service (*serviceURI*) as well as the values of low level parameters related to its performance, such as such as memory, CPU, bandwidth and disk capacity. Figure 4 shows the detailed sequence diagram for this phase:

1. The SP runs an *Advertise client* application on his local machine, specifying a *serviceID*, a *serviceURI* and values for the low-level performance parameters.
2. The *Advertise client* filters given values to check if they have accepted format and are within a logical range and then invokes its local Index service, running in the SP's host under a GT4 container. In particular, at this point the Advertise client invokes the *Register* operation, requesting that the advertisement be included in the list of the advertised services by this SP.

Figure 4. Sequence diagram for the registration of non-mobile resources to the MDVO

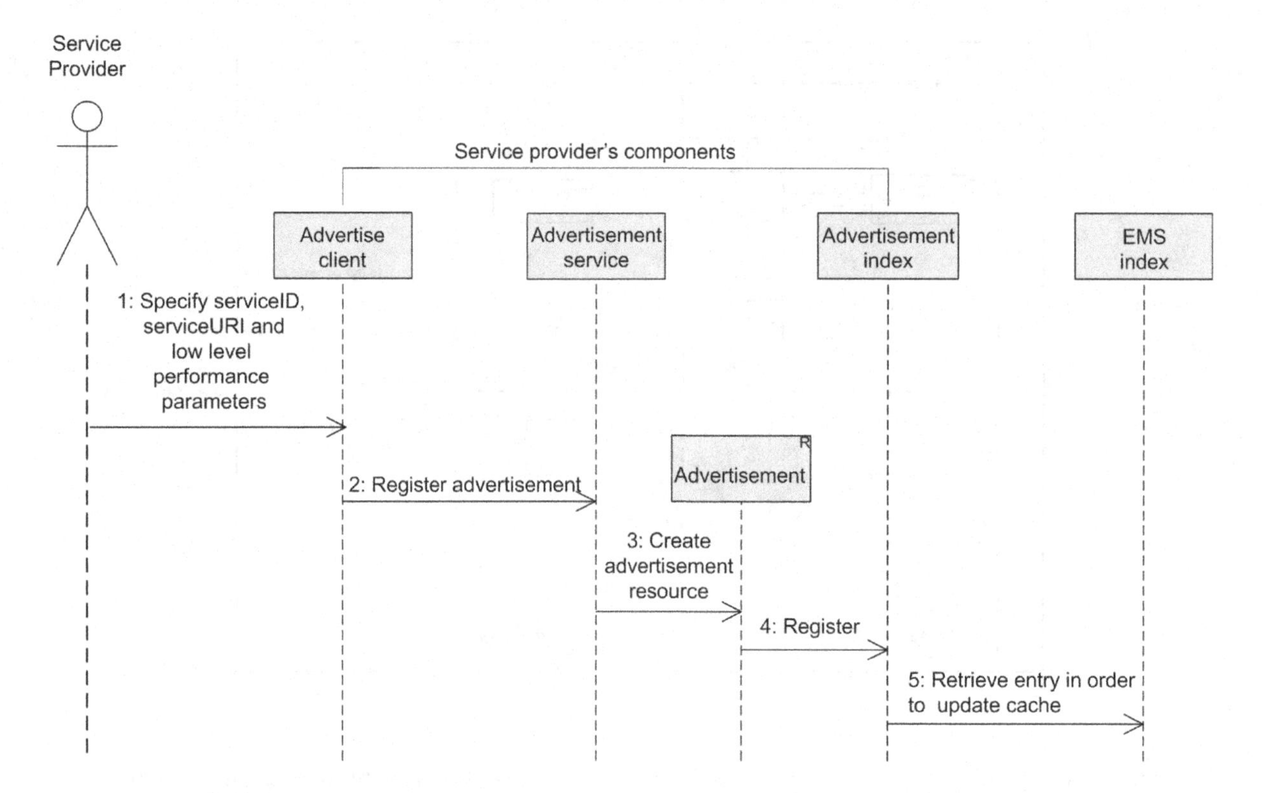

3. The Register operation of the *Advertisement service* triggers the creation of a WS-Resource called *Advertisement*[1]. Each service for sale in the MDVO is represented by an Advertisement resource that exists locally in the SP's container.
4. Upon creation, the Advertisement resource registers with the Index service running locally on the SP's container, which gathers information on the advertisements in the SP's machine.
5. After some time, which is configurable, the local registrations will propagate to the EMS's Index service, which will also cache the data stored in the SP's local Advertisement index service.

Using the same functionality, the SP can perform updates on the information included in the advertisement of the service, with which the old content in MDVO will be automatically refreshed once this information is "pulled" from the central EMS Index service. It should be noted that the registrations of the advertisements have limited lifetime. Upon failing to refresh the contents of a registration and/or once its lifetime has expired, it is automatically removed from the Index.

Registration of Mobile Services to MDVO

As previously mentioned the EMS draws information about ubiquitous services periodically with automated procedures by making use of MDS4 mechanisms. In the case of fixed terminals, services are bound to given machines and IP addresses and changes in the network are infrequent. Furthermore, in the case the registration of the service is not up to date, for example a service becomes unavailable in the short window of time between its last update and the request for execution, upon failure to establish connection, the EMS is able to redirect the execution to other alternative services in the Grid. Therefore, maintaining a dynamic registry that is periodically updated with new information after a short period of time is sufficient given that the probability of failure is low. However, in the case of mobile services this functionality is not enough since changes are frequent, and the location of the service is not bound to a single, static device, but to the device that the user is employing to access the system at a given time, and therefore fault tolerance mechanisms such as retry, migration or redundancy cannot be applied without knowledge of the location (IP address) of the mobile service at any given time.

In order for the EMS to become instantly aware of any changes in the locations of mobile resources and be able to deal with them in a seamless and completely transparent to the client of the Grid manner, the EMS establishes a WS-Notification mechanism with a network broker carrying information about the location, i.e., the IP addresses, named SIP Broker. The SIP Broker offers SIP-based services (Rosenberg & Schulzrinne, 2002) such as session establishments and transfers of Grid services running on mobile devices, and exposes this information through a WSRF-compliant interface. The SIP Broker tracks each user in order to keep the information about its IP address up to date. When the mobile user gets connected to the network, he/she issues a registration message to let the SIP Broker to communicate the IP address of the device being used and every mobile user is given a globally unique identifier called SIP URI. The mobile user's registration is updated every time the location of a service changes to allow for keeping track of the IP address of the service. It is assumed that the mobile terminals in which the mobile services are running are properly SIP-registered to the Akogrimo SIP infrastructure. More information about the Akogrimo SIP infrastructure can be found in (Akogrimo Deliverable D4.1.4, 2007)

The interactions that take place between the EMS and SIP Broker are depicted in the form of a sequence diagram in Figure 5, which also includes the required steps for setting up the WS-Notification mechanism and initialization of the mobile service's registrations (in the form of WS resources) in the MDS4 Index service. We distinguish between two different phases:

1. *Setup phase:* At first step, the EMS subscribes to the SIP Broker service in order to receive notifications about the availability and the location of the mobile users (steps 1 and 2). Then, it requests the connection details of each SIP-registered mobile user by invoking the appropriate method on the SIP Broker (step 3). The SIP Broker responds with the connection details of the mobile user (step 4) and the EMS uses this information to initialize the Index service with information about the mobile users. This implies the creation of a new "advertisement" WS-resource in case of new users (steps 5 to 10).
2. *Runtime phase:* After the establishment of the WS-Notification mechanism, whenever there is a change in the status of the mobile Grid services, a notification message is sent to the EMS (steps 1 and 2). This notification message includes information about the availability or unavailability of the specific mobile Grid service (step 3). To aid discovery, the notification message also includes the new location of the mobile Grid service in the form of a URI as well as the list of services that are co-hosted in the same mobile resource. Whenever such a notification is received, the EMS updates the appropriate registration in the MDS4 Index service to reflect the changes in the status and/or the locations of the mobile services (steps 4 to 8).

Figure 5. Establishment of WS-Notification mechanism between the EMS and the underlying SIP infrastructure

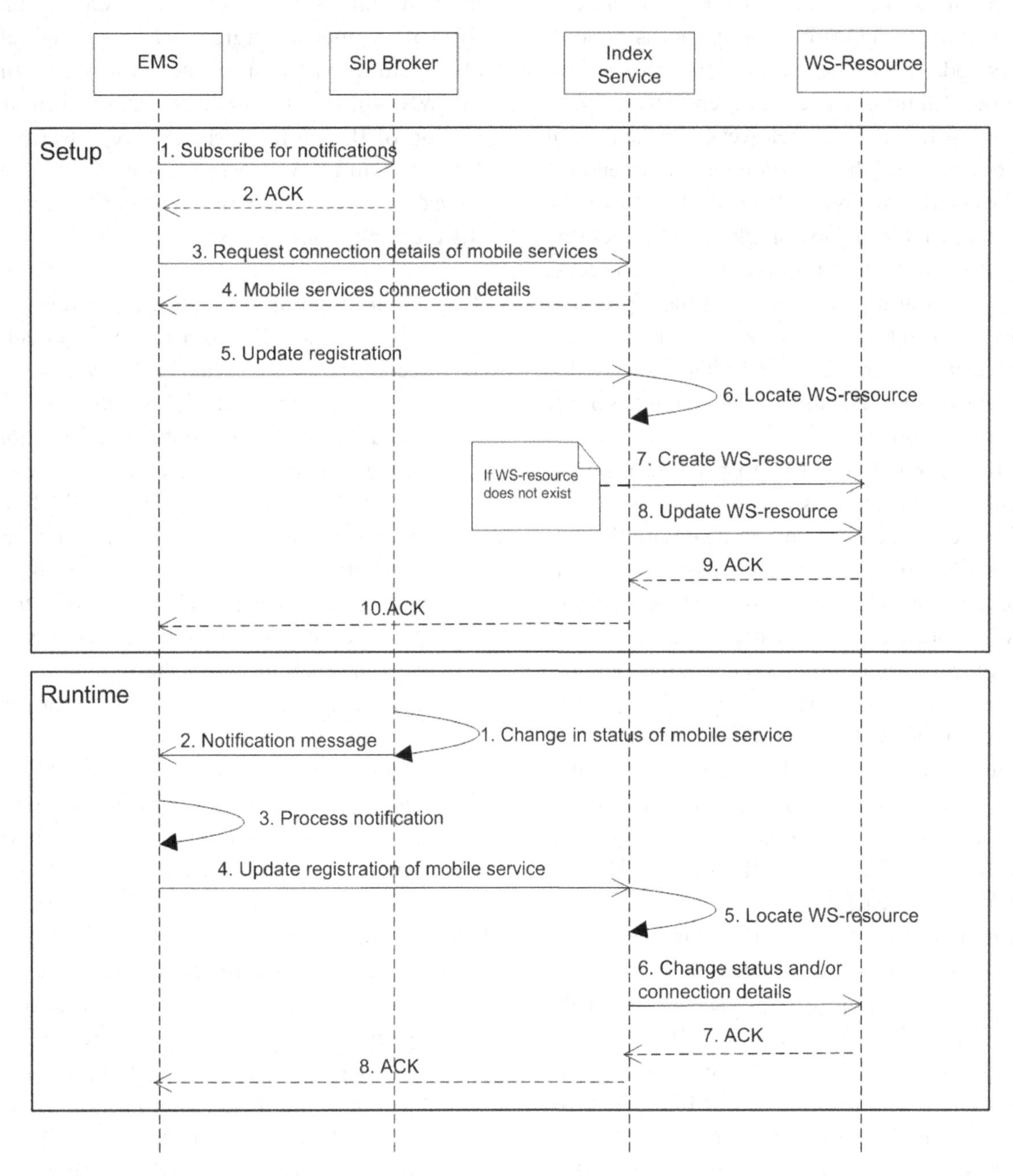

Usage of Mobile Information

After the WS-Notification between the EMS and the network infrastructure has been established, the EMS constantly "listens" for notifications about the movements of the mobile services in the Grid, i.e., possible disconnections and reconnections of the mobile services in other locations. Although, the EMS internally handles mobile and ubiquitous Grid services differently from each other, this distinction is kept transparent to the client and the services. The various tasks that the Akogrimo EMS covers are encapsulated into two distinct phases: (i) Discovery, Negotiation Reservation phase, (ii) Execution and Monitoring phase.

(i) Discovery, Negotiation and Reservation

During the Discovery, Negotiation and Reservation phase the EMS queries its central MDVO-wide MDS4 Index service in order to locate available mobile or ubiquitous services (step 4). As stated earlier, all resources throughout the MDVO are registered to this index, which is constantly updated by the EMS whenever there is change in their status and/or location.

In order to locate the most suitable, available service within the MDVO, the EMS makes use of low-level performance parameters such as CPU, memory, network bandwidth and disk capacity as well as the availability of the service and the price that the client is willing to pay. If the discovery process returns more than one candidate services, then the EMS finds the one that is closest to the client's requirements. In case there are services with identical attributes, the EMS selects the cheapest one by default. The WS-Notification mechanism in place makes sure that the information about the IP addresses of the mobile users is up to date. In the case where the client has requested for a specific mobile user that is marked as unavailable at the time of negotiation, the EMS waits for a fixed time interval, which is optional and can be set to zero by the client of the application, and then checks again to see if its availability status has changed (steps 5 and 6). If it is still unavailable, the EMS terminates negotiation and reports unavailability to the client (step 7.1). In the case where the client has not requested for a specific mobile user but has identified only a specific mobile service, the EMS proceeds with the negotiation searching for alternative mobile users running the requested service.

After a successful discovery and negotiation process, the EMS proceeds to the reservation of the requested services which involves the creation of instances (WS-resources) of the winning services, and manages their lifecycle by using the start time and end time defined by the client (steps 8 and 9 - for the sake of brevity these details are not shown in Figure 6 but can be found in (Litke et al., 2008)). In order for EMS to perform lifecycle management, each Grid service implements the WS-ResourceLifetime specification (IBM, 2004).

After a successful reservation for each service in the requested list the EMS creates a Reservation resource (RR), which is a persistent WS-resource (information is stored inside a file in order to survive container restarts). All information related to the negotiation request and the reserved resources is stored inside this RR (step 10). The complete list of RR EPRs is returned to the OpVO client (step 12.2 and 13) and is used to populate the BPEL workflow description. It should be noted that the latter is not populated by the EPRs of the actual reserved WS-resources of the services but the EPRs of the RRs are used instead. This increases the degree of security within the Akogrimo Grid Infrastructure since only the EMS is aware of the actual locations (URIs) of the services that are running within the mobile Grid and this information is extracted from the RR at execution time, given its EPR. Although the information published in the EMS Index service is constantly updated, there is always a very small window in time between the discovery of the service and the its actual reservation where this information may not reflect the current status of the selected fixed service and/or the status of the mobile service may change to unavailable. Therefore, in the event of failure during the creation of the service resource, a new discovery process is triggered (step 12.1) and the sequence is repeated with the service that generated the failure being left out of the list of candidate services.

(ii) Execution and Monitoring

The Execution and Monitoring phase begins with a client initiating the execution of a negotiated service resource by providing the EPR to the RR

Figure 6. Sequence diagram for the discovery, negotiation and reservation phase

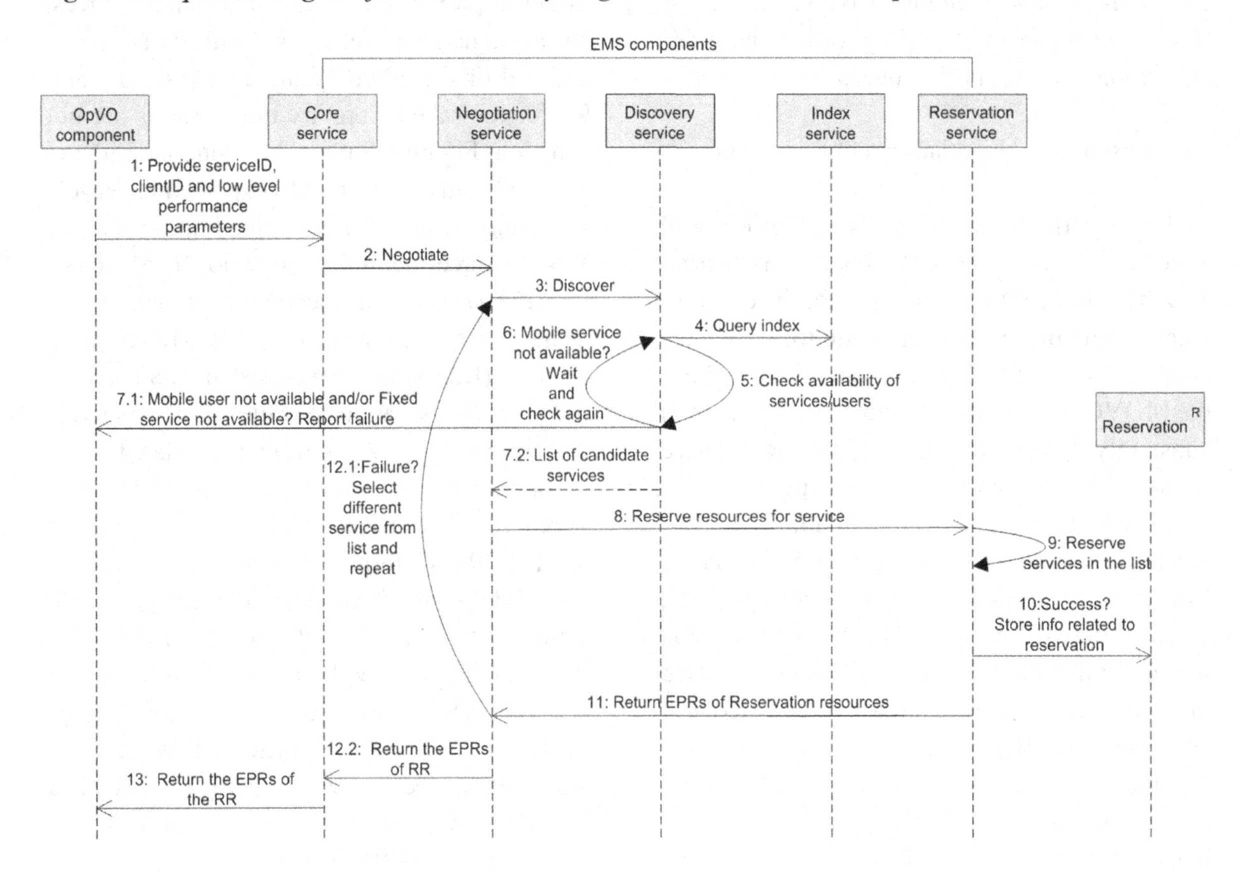

obtained at the end of the Discovery, Negotiation and Reservation phase and specifying the input needed for the execution of the specific service. The EMS uses the EPR of the RR to locate it and retrieve all information related to the reservation. Table 1 presents isolated parts of EMS's Execution class that provide insight into the recursive algorithm that is used (encapsulated inside *execute* method) and the implementation of the execution and monitoring process. Before triggering the execution of the reserved service, a check is performed on its current availability status (line 11). If the service is not available (line 13), the EMS waits for a fixed time interval (line 15), which is optional and can be set to zero, and then checks again to see if its availability status has changed (line 16). If it is still unavailable (line 18), a new negotiation is triggered, and the execution parameters are replaced with the newly discovered ones (line 19). If there is no alternative service, the process is terminated (line 20) otherwise the execution of the service is triggered (lines 23-24) during which the EMS is constantly listening for notifications (line 25). Upon receiving such a notification, the EMS aborts execution (line 26), and the process ends with a recursive call to the execute method itself (line 27).

In order for the EMS to able to handle concurrent executions of services in an asynchronous manner, class Execution builds heavily upon the *ExecutorService* Java class in cooperation with the *Future* interface, both part of Java package *java.util.concurrent*. In more detail, the *Future* interface has been used to represent the result of a Grid service execution as an asynchronous computation. This interface contains method *isDone* which is used to check if the execution of the Grid service is complete or still in progress and the result of the execution is retrieved using method get when the computation has completed. In case the service becomes unavailable during execution, cancellation is performed by the cancel method. The unavailability of the service is signaled by the change of the value of the global parameter availability to "unavailable" (line 49). This change takes place inside the deliver method (line 40), which is called when a notification is delivered from the SIP server. This method is part of the *NotifyCallback* interface of GT4's Java API that is required to be implemented by services that act as notification consumers.

The functionality described above was incorporated in a prototype EMS implementation and was used to support two diverse mission critical application scenarios that are described in the following section.

Table 1. EMS: part of class execution

```
public class Execution() implements NotifyCallback {
...
ExecutorService executor = Executors.newFixedThreadPool(100);
Long time_interval = 6000;
String availability = null;
...
/* The method below is called when a request for execution arrives */
public execute(PerformExecution params) {
/* Create response object */
Result result = null;
availability =checkAvailability(params);
if (availability = = "unavailable") {
Thread.sleep(time_interval);
availability =checkAvailability(params);
if (availability = = "unavailable") {
params = negotiation(params);
if (params = = null) return null;
}
....
Callable<Result> execute = new CallableImpl(params);
Future<Result> execution = executor.submit(execute);
while (!execution.isDone)
if (availability == "unavailable") {
execution.cancel();
this(params);
}
...
try {
result = execution.get();
} catch (Exception ex) {
result = null;
}
...
return result;
}//end of execute method
/* The method below is called when a notification is delivered */
public void deliver(List topicPath, ..., Object message) {
....
String notification = null;
...
//Extract notification from message
notification=message.getResourcePropertyValueChangeNotification();
...
if (notification!= null) {
/*A notification message has been delivered.*/
availability = notification.getAttribute("availability");
...
}
}//end of deliver method
...
}//end of Execution class
```

AKOGRIMO APPLICATIONS

The Akogrimo architecture was applied to demonstrators in two mission-critical domains. These were eHealth, in which a heart monitoring and emergency scenario was developed, and Integrated Emergency Management. In both scenarios: mobile resources are fundamental; multiple organizations need to pool resources and create teams to use those pooled resources – hence virtual organizations; and some of the services need to meet time constraints, in some cases critical ones, in order to support the activities of those VOs – hence the need for an effective execution management system.

eHealth Scenario

Within the eHealth scenario in the project framework, the EMS has been used to support a suite of e-Health services that constitute an e–Health application, offering to its clients a Heart Monitoring and Emergency Response system.

The background to the demonstrator scenario consists of a university hospital, regional hospitals, medical specialists, general practitioners, emergency medical services and an emergency dispatch center establishing a regional health network.

Figure 7. eHealth scenario

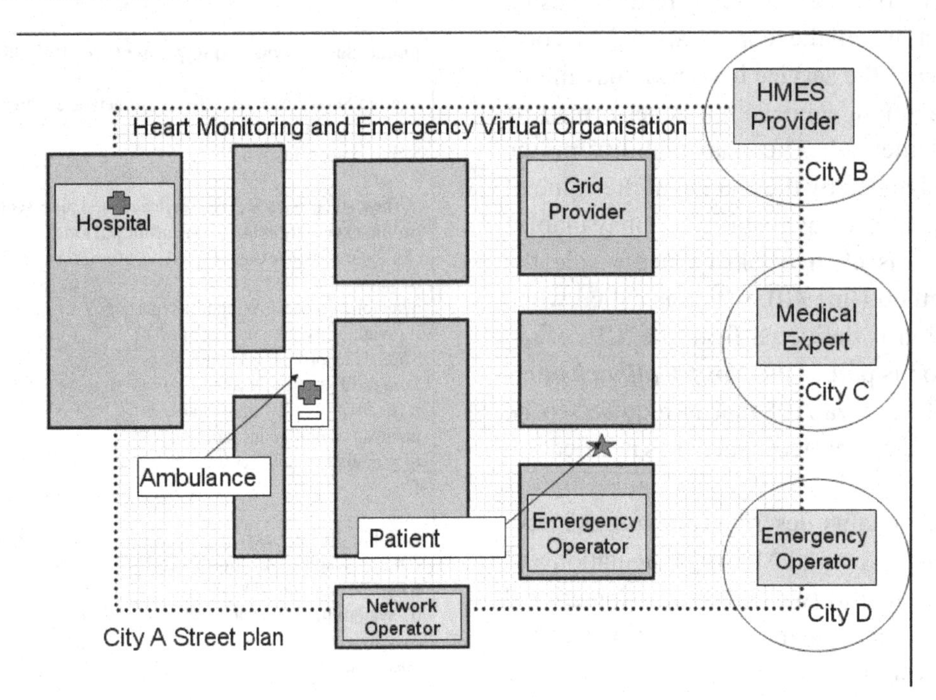

The health network is headed by the university hospital and provides telemedicine services to the partners and the patients attended by partners of the network (Figure 7). The scenario itself involves a heart monitoring and emergency response service which: monitors patients' heart indications wherever they are – and so implying mobility; assesses the overall condition of the heart of each patient based on current and past data; provides for an emergency to be triggered either by notification conventionally by voice or automatically based on the assessment calculations; continues to monitor the patient's condition while being transported in an ambulance; identifies available and appropriate specialists who might not be physically on the hospital premises and sets up the conditions for conference calls to be made; and makes relevant data available to all the specialists involved in the call through visualization.

In this scenario, the organizations initially agree to set up a Base VO, which incorporates all the Grid and web services that are required in the infrastructure and in the application. This includes services, which are located within a mobile hosting environment and also services, which are responsible for the establishment of SIP calls. When, because of an already known medical condition, a patient is identified as requiring the heart monitoring provision, a new specific Operative VO is established. Workflows associated with the monitoring and emergency system are instantiated, ready for use. By means of the EMS and the other systems it invokes, services are identified and instantiated, allowing the construction of specific workflows to be completed.

The main users and service providers for the scenario can be seen in Figure 7. Within the Operative VO the mobile data of the patient and if required and available an on-scene medical responder is linked to a specific network provider as the VO is formed. As part of the workflow the parties such as Emergency Operator and Medical Expert will be included in the VO and the ambulance directed to the patient. In this scenario the main job of the EMS will be to ensure that the best services to execute the workflow for the patient are both available and used. The best services in the eHealth scenario will be selected from context derived from the situation. This could be identifying the doctor with the appropriate specialism (such as stroke response) and a near enough location and the Ambulance that has the quickest route to the patient based on various mobile traffic monitoring nodes. Among the services to be identified, invoked and managed through EMS are ones for setting up and closing SIP calls, which will make their own demands for execution monitoring.

IEM Scenario

The other main Akogrimo scenario is concerned with Integrated Emergency Management (IEM). IEM, which adopts the concept of a flexible, managed response to emergencies and which aims to integrate the key workers of multiple agencies, has become a part of the government process at local, national and wider regional levels. The term used varies, but IEM is the phrase used in UK government planning (UK Cabinet Office, 2004) and post-Katrina in some agencies in the US (Tierney, 2007). In this demonstrator the EMS is used in the previous scenario is extended.

Although the fully integrated approach includes preparation, response and recovery phases, the Akogrimo testbed concentrated on the response phase to provide a demanding test of the mobile Grid infrastructure. In the previous eHealth scenario the mobile services were limited to numbers of two or three, in the IEM scenario there were many mobile services. The project whitepaper on the IEM scenario (Briscombe et al., 2006) contains an early description of the proposed scenario. The characteristics of such an environment are also summarized in the IEM whitepaper, in which the challenges are described as being:

- Uncertainty of knowledge of location of resources, requiring effective service discovery – where the services may be mobile.
- Drawing resources from multiple organizations, making them available within the VO but in a way that is secure, punctual and auditable.
- Optimizing the brokering of expert and specialist resources – human and machine-based.
- Detecting and managing context changes – often subtle and risk driven in Integrated Emergency Management (IEM).
- Dealing with rapid shifts in priority and activity modes.

Figure 8 (taken from the IEM whitepaper) shows the storyboard used. The specific scenario took the example of a dirty bomb emergency in a major UK city. A silver command control centre is set up, but it has to be complemented by local subsidiary contact points and in this instance one such is set up in a local school. The scenario contains several key actors: some act in a coordinating role (call centre and the command centre) and others are mobile, directly involved in the emergency as it unfolds. Lines with arrows show the flow of information between actors. The command centre maintains a Common Operational Picture (COP), which provides the necessary integration. It provides the basis for communicating information to the mobile actors and is refreshed by information from them, including multimedia information (for instance, from mobile phone cameras).

The COP and the various factors involved are unified by a Grid system and a number of application services are established, some being located in a mobile hosting environment, such as the portable devices assigned to mobile workers such as police officers, fire fighters and paramedics. In this scenario the EMS and VO infrastructure was successfully used to manage this large number of mobile users and data providers in the Akogrimo framework. One of the key points of this scenario with regard to the mobility of the involved users was the establishment of a multi-conference, which involved the fixed terminal of

Figure 8. Akogrimo demonstrator storyboard– an example of an IEM application in the response phase using mobile grids

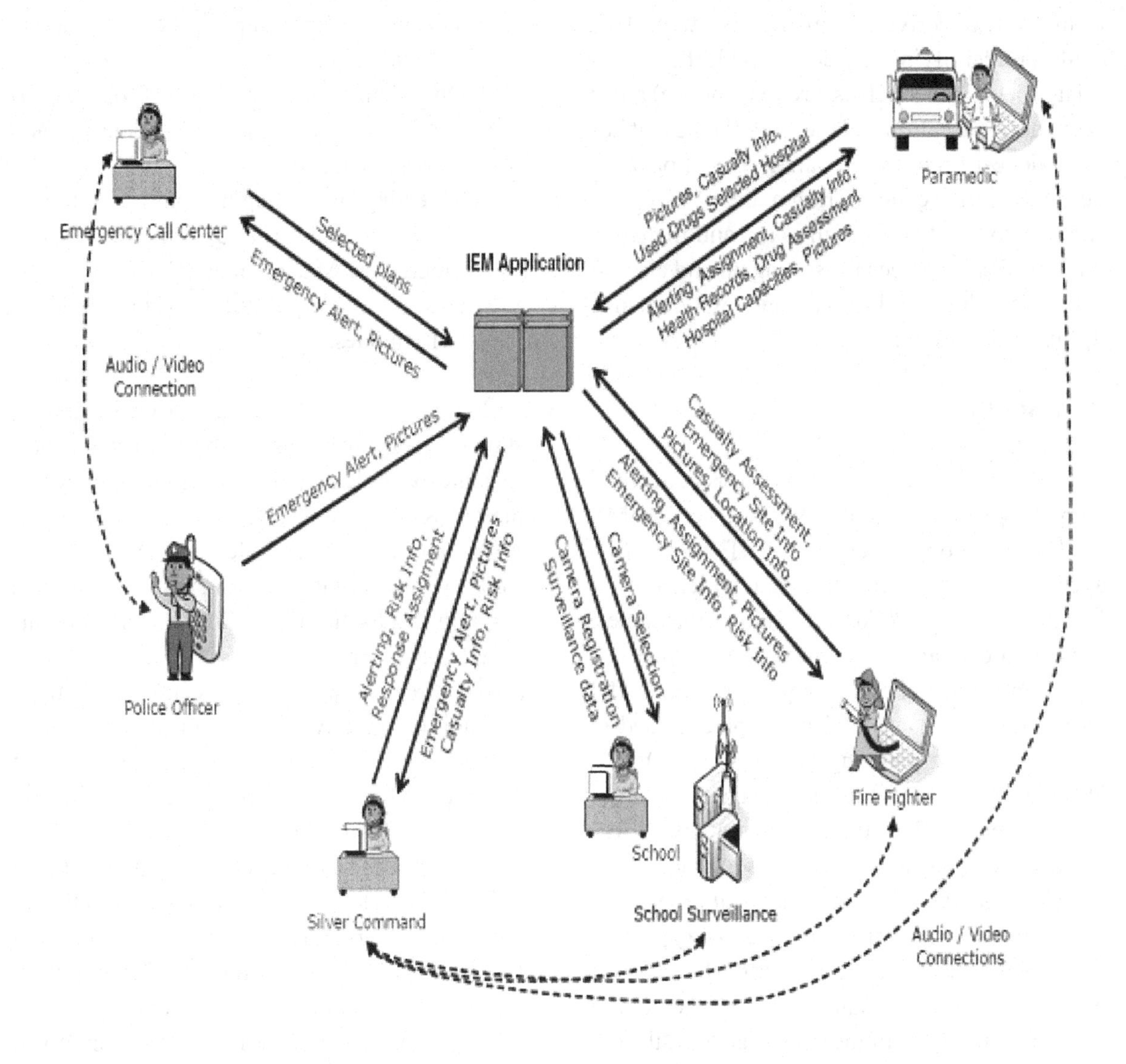

the paramedics located originally at the hospital, and the mobile devices of the fire fighters that had already reached the location where the bomb had exploded. After making a preliminary assessment, the paramedics switched from their static terminal to their mobile one, and moved to the location of the disaster while holding the multi-conference.

In conventional Grids, setting up a VO is a time consuming and often mainly manual process. Here, in both scenarios, the EMS takes a significant role in the set up of a new Operative VO and manages the initiation of the service instances and it has been found that the whole setting up of the Operative VO, based on an existing template managed by the Base VO, is reduced to a matter of a few minutes.

Compared with the eHealth scenario, this showed that the EMS could manage the execution of a more substantial workload, in terms of application services, users and terminals. The scenario was demonstrated over a distance with the Grid infrastructure in Stuttgart and the mobile users in Madrid.

FUTURE WORK

The management of mobile services in an application environment such as Akogrimo is an area of research that has not been the focus of many projects. The basic communications of mobility data and in particular application specific context data in these scenarios proved a particular challenge. Thus in future work there is plenty of scope to devise a mechanism to standardize the expression of service context change in mobile environments. The development of this vocabulary will be of great benefit to services that have to make decisions in on mobile data in often critical scenarios such as the EMS.

For the VO the development of dynamic virtual organizations is continuing in the Grid community but the further development in the mobile computing community is limited to projects. However, desktop-based Grids such as XtreemOS have the potential to revitalize this area of research. The application area of Grid computing infrastructure applied to personal computers and mobile devices is an emerging area of research. Future work in the area of execution management will be focused on extending the advance reservation mechanism to offer stronger availability while increasing the utilization of the resources. Furthermore, we will investigate techniques for improving its scalability and extending its current functionality to support differentiated networks and other types of signaling protocols, such as QSIG, and H.225.0.

CONCLUSION

Service execution management is a key element of OGSA and central to the delivery of applications that depend on composite services. In order to achieve this, various methods of service management have been proposed and are often linked to the application environment of the Grid, but not yet standardized. The Akogrimo approach to service execution management is an innovation in this field as the EMS capability presented, is the first of its kind to manage service execution in a mobile and ubiquitous service-oriented environment.

The EMS presented in this paper was based around the creation of dynamic VOs that include both ubiquitous and mobile services. The mobility of these services placed the need for the EMS to be aware of any changes in the availability and the location of the mobile services in a dependable manner. The establishment of communication between the Grid and the SIP-based network layer has been a key part in making a mobile service-oriented application dependable, as the Akogrimo VO through its EMS uses the SIP capabilities to the fullest in a seamless manner by establishing/managing sessions and processing notifications to keep track of the changes that may occur in the location/availability of the mobile services.

The execution of services in a mobile, service-oriented Grid is a delicate and responsive process that adds an extra dimension to the service-oriented lifecycle. This extra dimension that the challenge of mobility adds is best illustrated in Grid management applications and the Akogrimo testbed provided an ideal environment where it could be frequently tested and refined. Specifically, the EMS has been used to support two diverse real-life scenarios, with heavy requirements with encouraging results that prove its efficiency.

ACKNOWLEDGMENT

This work was supported in part by the Akogrimo Integrated Project (FP6-2003-IST-004293) which was part funded by the EU Framework 6 Programme. The authors wish to thank the members of the Akogrimo consortium.

REFERENCES

W3C. (2002). *Web Service Choreography Interface (WSCI) 1.0*. Retrieved from www.w3.org/TR/wsci/

Access to Knowledge through the Grid in a Mobile World (Akogrimo). (2007). *Web Site home page.* Retrieved from www.akogrimo.org

Akogrimo Deliverable D4. 1.4. (2007). *Consolidated Network Service Provisioning Concept*. Retrieved from http://www.akogrimo.org/modules02d2.pdf?name=UpDownload&req=getit&lid=124

Allcock, B. (2002). Data Management and Transfer in High Performance Computational Grid Environments. *Parallel Computing Journal, 28*(5), 749–771. doi:.doi:10.1016/S0167-8191(02)00094-7

Briscombe, N., et al. (2006). *Enabling Integrated Emergency Management: Reaping the Akogrimo Benefits.* Retrieved from http://www.akogrimo.org/download/White_Papers_and_Publications/Akogrimo_whitePaper_Disas terCrisisMgmt_v1-1.pdf

Camarillo, G., & Garcia-Martin, M. *The 3G IP Multimedia Subsystem.* New York: John Wiley and Sons.

Cao, J., et al. (2003, May 12-15). GridFlow: Workflow Management for Grid Computing. In *Proceedings of the 3rd International Symposium on Cluster Computing and the Grid (CCGrid),* Tokyo, Japan. Los Alamitos, CA: IEEECS Press.

Christensen, E., Curbera, F., Meredith, G., & Weerawarna, S. (2001). *Web Services Description Language (WSDL) 1.1., W3C, Note 15.* Retrieved from http://www.w3.org/TR/wsdl

Czajkowski, K., et al. (2004, March 5). *The WS-Resource Framework Version 1.0.* Retrieved from. http://www.106.ibm.com/developerworks/library/

Czajkowski, K., et al. (2005). *From Open Grid Services Infrastructure to WS-Resource Framework Refactoring & Evolution.* Retrieved from http://www.globus.org/wsrf/specs/ogsi_to_wsrf_1.0.pdf

Dimitrakos, T., Golby, D., & Kearney, P. (2004, October). Towards a Trust and Contract Management Framework for Dynamic Virtual Organisations. In *Proceedings of the eChallenges Conference (eChallenges 2004),* Vienna, Austria.

Forman, G. H., & Zahorjan, J. (1994). The challenges of mobile computing. *IEEE Computer, 27*(4), 38–47.

Foster, I. (2005). Globus Toolkit Version 4: Software for Service-Oriented Systems. In *Proceedings of the* IFIP *International Conference on Network and Parallel Computing* (LNCS 3779, pp. 2-13). New York: Springer Verlag.

Foster, I., Kesselman, C., & Tuecke, S. (2001). The anatomy of the grid: Enabling scalable virtual organizations. *International Journal of High Performance Computing Applications, 15*(3), 200–222. doi:.doi:10.1177/109434200101500302

Globus. (2006). *Information Services (MDS): Key Concepts.* Retrieved from http://www.globus.org/toolkit/docs/4.0/info/key-index.html

IBM. (2003, May). *Business Process Execution Language for Web Services (BPEL), Version 1.1.* Retrieved from http://www.ibm.com/developerworks/library/ws-bpel

IBM. (2004). *WS-ResourceLifetime specification.* Retrieved from http://www.ibm.com/developerworks/library/ws-resource/ws-resourcelifetime.pdf

Kavantzas, N., Burdett, D., Ritzinger, G., Fletcher, T., & Lafon, Y. (2004, December). Web Services Choreography Description Language Version 1.0. In *Proceedings of the World Wide Web Consortium.*

Litke, A., Konstanteli, K., Andronikou, V., Chatzis, S., & Varvarigou, T. (2008). Managing service level agreement contracts in OGSA-based Grids. *Future Generation Computer Systems, 24*(4), 245–258. doi:.doi:10.1016/j.future.2007.06.004

Ludwig, H., Gimpel, H., & Kearney, R. D. (2005). Template-Based Automated Service Provisioning - Supporting the Agreement-Driven Service Life-Cycle. In *Proceedings of the International Conf. Service Oriented Computing (ICSOC'05)* (pp. 283-295).

OASIS. (2005). *Web Services Base Notification 1.3 (WS-BaseNotification).* Retrieved from http://docs.oasis-open.org/wsn/wsn-ws_base_notification-1.3 -spec-os.pdf

Olmedo, V., Villagrá, V., Konstanteli, K., Burgos, J., & Berrocal, J. (2009). Network mobility support for Web Service-based Grids through the Session Initiation Protocol. *Future Generation Computer Systems, 25*(7), 758–767. doi:.doi:10.1016/j.future.2008.11.007

Park, S., Ko, Y., & Jai-Hoon. (2003). Disconnected Operation Service in Mobile Grid Computing. In *Proceedings of ICSOC 2003.* Retrieved from www.unitn.it/convegni/download/icsoc03/papers/Jai.pdf

Rosenberg, J., Schulzrinne, H., et al. (2002). *SIP: Session Initiation Protocol" IETF RFC 3261.* Retrieved from http://www.ietf.org/rfc/rfc3261.txt"http://www.ietf.org/rfc/rfc3261.txt

Rosenberg, J., Schulzrinne, H., Camarillo, G., Johnston, R., Peterson, J., Sparks, R., Handley, M., & Schooler, E. (2002, June). SIP: session initiation protocol. *RFC 3261, Internet Engineering Task Force.*

The Open Grid Forum (OGF). (2006). *The Open Grid Services Architecture, Version 1.5.* Retrieved from http://www.ogf.org/documents/GFD.80.pdf

Tierney, K. J. (2007). Testimony on Needed Emergency Management Reforms. *Journal of Homeland Security and Emergency Management, 4*(3). Retrieved from http://www.bepress.com/jhsem/vol4/iss3/15. doi:10.2202/1547-7355.1388.

Tosic, V., Kruti, P., & Pagurek, B. (2002). WSOL – Web Service Offerings Language. In *Proceedings of WES 2002* (LNCS 2512, pp. 57-67). Berlin: Springer Verlag.

Tosic, V., Lutfiyya, H., Yazhe, T., Sherdil, K., & Dimitrijevic, A. (2005, September 28-30). On requirements for management of mobile XML Web services and a corresponding management system. In *Proceedings of the 7th International Conference on Telecommunications in Modern Satellite, Cable and Broadcasting Services* (Vol. 1, pp. 57- 60).

UK Cabinet Office. (2004). *Civil Contingencies Act 2004: Emergency Preparedness.* Retrieved from http://www.ukresilience.info/preparedness/ccact/eppdfs.aspx

UK National Grid Service (NGS). (2010). *Homepage.* Retrieved from http://www.grid-support.ac.uk

University of Virginia. (2005). *WSRF.net.* Retrieved from http://www.cs.virginia.edu/~gsw2c/wsrf.net.html

Waldburger, M. (2007). Grids in a Mobile World: Akogrimo's Network and Business Views. *Praxis der Informationsverarbeitung und Kommunikation, 30*(1), 32–43. doi:.doi:10.1515/PIKO.2007.32

Waldburger, S., & Stiller, B. (2005). *Toward the Mobile Grid: Service Provisioning in a Mobile Dynamic Virtual Organization* (Tech. Rep. No. 2005.07). Zurich, Switzerland: University of Zurich

Weissman, J. B., Kim, S., & England, D. (2005). Supporting the Dynamic Grid Service Lifecycle. In *Proceedings of the IEEE/ACM CCGrid International Symposium on Cluster Computing and the Grid* (Vol. 2, pp. 808-815).

Welch, V., et al. (2003, June). Security for Grid Services. In *Proceedings of the Twelfth International Symposium on High Performance Distributed Computing (HPDC-12).* Washington, DC: IEEE Press.

Wesner, S., Dimitrakos, T., & Jeffery, K. (2004). Akogrimo - The Grid goes Mobile. *ERCIM news* (No. 59).

ENDNOTE

1. In the sequence diagrams that follow, the symbol "R" is used to indicate that the component featuring it is a WS-Resource.

This work was previously published in International Journal of Systems and Service-Oriented Engineering, Volume 1, Issue 3, edited by Dickson K.W. Chiu, pp. 39-59, copyright 2010 by IGI Publishing (an imprint of IGI Global).

Chapter 5
Developing Local Association Network Based IoT Solutions for Body Parts Tagging and Tracking

Zongwei Luo
The University of Hong Kong, China

Martin Lai
The University of Hong Kong, China

Mary Cheung
The University of Hong Kong, China

ShuiHua Han
Xiamen University, China

Tianle Zhang
BUPT, China

Zhongjun Luo
FDA, USA

James W. M. Ting
The University of Hong Kong, China

Patrick W. L. Wong
The University of Hong Kong, China

Sam K. Y. Chan
The University of Hong Kong, China

Kwok Fai So
The University of Hong Kong, China

George L. Tipoe
The University of Hong Kong, China

ABSTRACT

Traditional Internet is commonly wired with machine to machine persistent connections. Evolving towards mobile and wireless pervasive networks, Internet has to entertain dynamic, transient, and changing interconnections. The vision of the Internet of Things furthers technology development by creating an interactive environment where smart objects are connected and can sense and react to the environment. Adopting such an innovative technology often requires extensive intelligence research. A major value indicator is how the potentials of RFID can translate into actions to improve business operational efficiency (Luo et al., 2008). In this paper, the authors will introduce a local association network with a coordinated P2P message delivery mechanism to develop Internet of Things based solutions body parts tagging and tracking. On site testing and performance evaluation validate the proposed approach. User feedback strengthens the belief that the proposed approach would help facilitate the technology adoption in body parts tagging and tracking.

DOI: 10.4018/978-1-4666-1767-4.ch005

INTRODUCTION

Traditional Internet commonly is wired with machine to machine persistent connections. Evolving towards mobile and wireless pervasive networks, Internet has to entertain dynamic, transient, and changing interconnections. The vision of the Internet of Things furthers technology development by creating interactive environment where smart objects are connected and can sense and react to the environment. The resulted event flooding in such Internet of Things (IOT) environment has aroused interest in research in network architecture and topologies where the events can be filtered to meet event intensive application requirements. In this paper, we will introduce a local association network (LAN) with a coordinated P2P message delivery mechanism. This LAN is tested and validated as suitable building block for the Internet of Things.

Radio Frequency Identification(RFID) as a promising technology to revolutionize the things to be identified and managed, has shown great potentials and has demonstrated good competitiveness in improving visibility on things or items to be managed. The identification power of RFID has been observed from tagging the Internet of Things to human bodies. Possible use of RFID to tag body parts donated to science was already discussed in 2005 (Associated Press, 2005). To combat theft of donated body parts, RFID was considered to put in cadavers that can then be read by a handheld device. The U.S. Disaster Mortuary Operational Response Team (DMORT) and health officials in Mississippi's Harrison County were reported to implant human cadavers with RFID chips to help identify victims of Hurricane Katrina. In Louisiana, it is also expected to begin using the RFID systems to help officials cope with the estimated 500 unidentified bodies in that state (Kanellos, 2005).

In Hong Kong, there is an increasing interest in tagging human body to help manage the body or body parts either in a hospital or a mortuary. University medical school laboratories are also looking into this potential technology to help track and manage their laboratory body part assets in their everyday operation trainings. Is RFID a feasible candidate in such environment? How shall RFID technology be deployed in such environment? a key value indicators approach is proposed for performing value analysis in RFID adoption (Luo et al., 2008). The key value indicators serve as value metrics to evaluate perceived value from different adoption parties. The key value indicators for RFID technology adoption in body tagging would then relate to metrics indicating its value whether the body parts visibility can translate into actions to improve laboratory operation efficiency and to reduce cost. While the value promised by the RFID technology is eminent, the adoption level very much counts on the perceived value by the potential adoption parties – the labs, hospital, and many others.

There are a number of challenges in the value perceived by various stakeholders regarding operational efficiency in respect of body parts visibility, including:

- Functions of medical school training operations are highly fragmented, associating with a very large number of individual training laboratories. Very often they have their own training operation processes. The proliferation of diversified operation processes, if overlooked, makes it difficult for operational efficiency evaluation for RFID adoption in body part tagging.
- There is a lack of quality and consistent information on laboratory training operations (especially from RFID technology experts and laboratory management), leading barriers for setting focus on tackling body parts visibility. This makes it hard to identify how efficient of training operations. This makes it difficult to establish trends, make comparisons and manage cost with respect of RFID adoption in body part tagging.

- Limited information and mechanisms are available for reviewing training operational effectiveness. This would lead to limited adoption scrutiny and cross-checking for operational efficiency and costs savings with respect of RFID adoption in body parts tagging.
- There is a lack of procedures and methodologies in adopting RFID in body part tagging in medical school operation training laboratories, putting cost penalty for the lack of information visibility of body parts. This would result in great barriers in providing greater responsiveness in the provision of laboratory operation training services.

Thus, in this paper, we would tackle the challenges by forming a collaborative team with domain experts related to this RFID adoption in body part tagging, including RFID engineers, medical academics and professionals, and laboratory management teams. A local association network (LAN) approach is adopted to provide a cost effective means to deploy RFID in such a laboratory environment. This LAN approach can be generalized for other deployment environment including hospitals or mortuaries.

Organization of this paper is as follows. First we present an introduction to Internet of Things. Next is the LAN based approach for adopting RFID in body part tagging. Then, a prototype and solution deployment is described. We present the solution validation processes and tests conducted. Next, we present LAN integration and message delivery in LANs. Following this, a message delivery scenario is presented to illustrate LAN integration. Next, we present related two classical message delivery mechanisms. After the approach analysis, we present the performance evaluation, followed by an introduction of potential service offerings and applications. Finally, the evaluation of the deployed solution is conducted with users' feedback and then our conclusion.

INTERNET OF THINGS

Electronic tags, usually refereed as RFIDs, sensors, wireless VoIP terminals are likely to create a technological revolution similar to the one initiated by the Internet technology in the early nineties. These very cheap components are manufactured by billions, and are going to be inserted in quite all our everyday objects (Pujolie, 2006). The combination of the Internet and emerging technologies such as near-field communications, real-time localization, and embedded sensors let us transform everyday objects into smart objects that can understand and react to their environment (Kortuem et al., 2010). Such objects are building blocks for the Internet of Things and enable novel computing applications.

The vision of an Internet of Things (IOT) built from smart objects raises several important research questions in terms of system architecture, design and development. For example, what is the right balance for the distribution of functionality between smart objects and the supporting infrastructure? And how can people make sense of and interact with smart physical objects? In an IOT environment, as the number of nodes becomes large, routing information becomes a huge burden. The overhead involved in computing, storing, interacting and communicating the routing information becomes prohibitive.

In this paper we would develop a local association network (LAN) for Internet of Things. In this LAN, item tracking would be based on the association information identified and established. Further, an algorithm is introduced to guarantee a reasonable level of message delivery based on a delay tolerant mechanism in this LAN.

Architectures of IoT

Currently, there are several proposed architectures of Internet of Things (IoT), such as Web of Things (WoT) (Duquennoy et al., 2009). It introduces a kind of application-oriented IOT, which web service is embedded in, and can be used like a simple web service. Such architecture is user-oriented which aims to simplify information distribution and access of IOT using the web application architecture.

Autonomic architecture of IOT (Pujolie, 2006) is another kind of architecture designed for wireless communication environment, which uses automatic communication technology. Autonomic communication is the vision of next-generation networking, which will be a self-behaving system with properties such as self-healing, self-protection, self-configuration, and self- optimization (International Business Machines, 2003).

The autonomic architecture of IOT is showed as following, which is composed of 4 planes:

- The data plane to forward the packets.
- The control plane to send configuration messages to the data plane in order to optimize the throughput, and the reliability.
- The knowledge plane to provide a global view of all the information concerning the network.
- The management plane to administrate the three other planes (Figure 1).

Here automatic feature comes from the fact that TCP / IP protocol stack is replaced by Smart Transport Protocol and Smart Protocol (STP/SP) stack. However the protocol involved by automatic architecture is too complex, it only fits for IOT with a lot of computing resource

Figure 1. Autonomic networking architecture

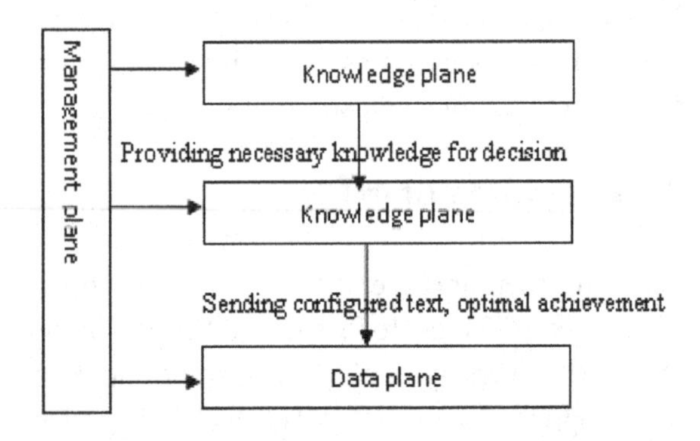

EPCglobal Network is another approach to realization of The Internet of Things. This EPCglobal Network is composed of five key elements:

- The electronic product code (EPC) is a unique number that identifies a specific item.
- The ID System consists of EPC tags and readers.
- EPC middleware manages real-time read events and information, provides alerts, and manages the basic read information for communication to EPC information services (EPC IS) and a company's other existing information systems.
- Discovery services are a suite of services that enable users to find data related to a specific EPC and to request access to that data. Object naming service (ONS) is one important component of discovery services. ONS provides a mapping between the EPC and the owner of the database that has the data.
- EPC information service (EPC IS) enables users to exchange EPC-related data with trading partners through the EPCglobal Network.

Figure 2 shows the EPCglobal Network architecture (Armenio et al., 2009). EPCglobal Network architecture is motivated mainly from the Internet concept dealing with certain applications such as supply chain management. The network is constructed with less dynamic connection support for meeting the dynamic, transient, and changing connections needs in many IOT applications.

Thus, EPC technology is often applied in a narrower capacity with efforts necessary to generalize leading to a general architecture for IoT.

Challenges

To propose a general architecture for IOT, there are a few challenges ahead. First, Internet of Things and its applications are likely to evolve

Figure 2. Pervasive RFID-based Infrastructure

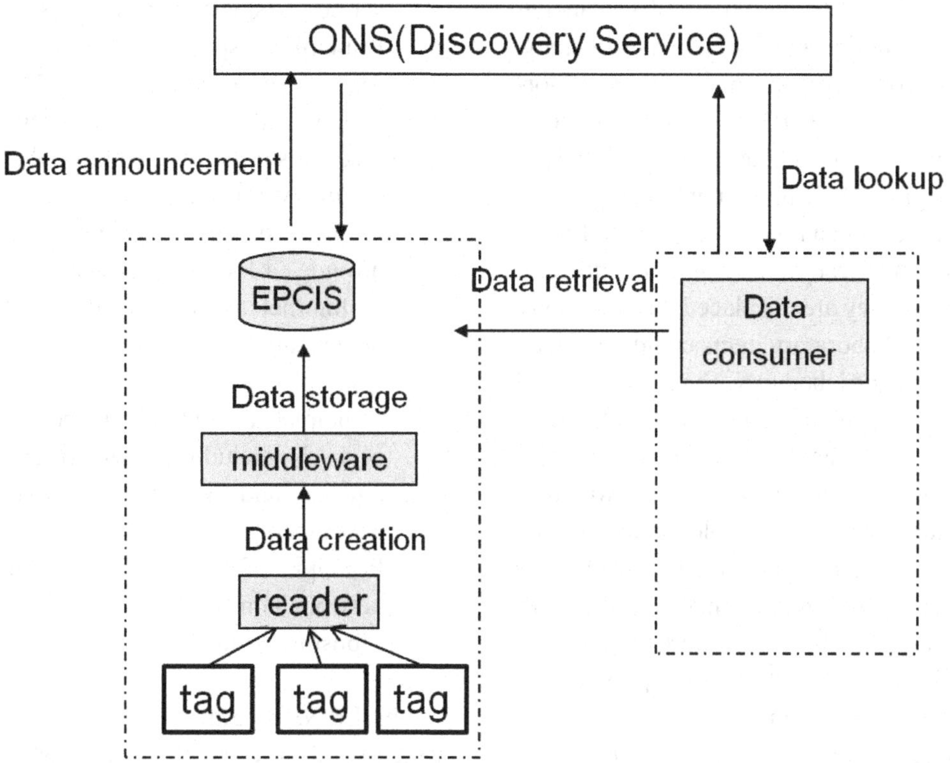

and grow over time. We already see RFID in elder care and object finding applications, each of which requires a flexible infrastructure that facilitates provisioning. Second, because a pervasive application will typically track people and belongings rather than items in inventory, privacy issues must be considered much more carefully. Finally, people are less predictable than goods moving through established distribution patterns in a supply-chain. As such, we must develop fundamentally new ways to deal with the variable-rate, partial, and noisy data likely to be generated by human activities.

Thus, we believe a good architecture of IOT should observe the following five principles:

- Principle of diversity (1), the architecture of IOT should be classified into various architectures according to different kind of nodes, or smart objects;
- Principle of space and time (2), the architecture of IOT should satisfy with requirement of time, space and energy efficiency,
- Principle of interconnection (3), the architecture of IOT should be able to connect with Internet smoothly,
- Principle of security (4), the architecture should defense a wide range of network attacks.
- Principle of robustness (5), the architecture of IOT should be with robustness and reliability.

LAN BASED APPROACH

In a laboratory environment, operation automation and efficiency improvement demands good position sensing technologies to track and manage body and parts. Like any technology or solutions, their adoption is very often under scrutinized with performance consideration for trade offs balancing out the cost reduction and operational efficiency. Performance consideration is one of the key performance indicators for RFID and other technology adoption exercises. Cost reduction often leads to a "line of sight" RFID implementation. That is, when you see, you can then approach and read. This would often lead to missing items difficult to locate when they are misplaced. Further, body parts and other laboratory items could be placed around any area in a laboratory. They can be moved and replaced as well. In order to track all the items where they could reach in such an environment, area of coverage becomes a concern, which is directly linked with network deployment cost. The needed area coverage and positioning accuracy for laboratory operation automation often comes with a corresponding level of expenditure for positioning network and RFID technology deployment to cover the laboratory premise.

A local association network (LAN) is proposed as a building block for IOT in the area of item tracking and situation monitoring. This local association network is developed in E-Business Technology Institute, the University of Hong Kong to develop cost effective technologies and solutions for item tracking to facilitate operation automation, improve operational efficiency and help cost reduction. In this LAN approach, the tracking accuracy delivery could be optimized by balancing the cost considerations and often expected operational efficiency. LAN is proposed and developed fitting with the five principles for IOT architecture development. In the following sections, we will mainly focus on first two principles although it preserves good properties for all the five.

- Principle of diversity (1), LAN forms a small area network in a P2P manner, in which a coordinator can be elected dynamically. Thus it offers diversified architectures and topologies accommodating various sensor nodes, smart objects, and RFID readers.
- Principle of space and time (2), the LAN, formed in a coordinated P2P network, adopts a delay tolerant message delivery mechanism, helping tackling the problems of energy consumption.
- Principle of interconnection (3), the LAN provides a local access networking means to Internet to allow sensor nodes, smart objects and RFID readers to be connected smoothly,
- Principle of security (4), security features of LAN including authentication, authorization, and privacy protection are being developed.
- Principle of robustness (5), the dynamic network construction of LAN offers both robustness and reliability.

In a LAN, there are typically multiple sensors which would present P2P relationships to each other. This setup is necessary as the tracking requirement vary from application to application. And different sensors will be needed as well from application to application. Such a P2P construction of a LAN would help ensure a flexible deployment method to meet different application requirements.

A LAN covers a small area within the line of sight. Multiple LANs can then be integrated to form a whole solution for item tracking in an IOT environment. The LAN would be the basic unit for providing positions of items. When RFID is used for the positioning purpose, a positioning Grid (Wang et al., 2007) would be set up for the LANs. The LAN integration will then link up all the LANs in a premise to form a complete item tracking and asset management IOT network.

A mobile asset tracking method (Zhang et al., 2008) can be adopted in the LAN integration based on delay tolerant routing principles. In an RFID only environment, a mobile handheld can also be utilized to track those items on the move. Interferences among different LANs are resolved via data processing and in a collision avoidance manner (Fei et al., 2009; Peng et al., 2008; Zhou et al., 2007).

In the LAN integration, a signing up and off protocol has to be developed for tracking those items being transferred from one LAN to another. Thus a message delivery mechanism has to in place for a LAN and LAN integration.

SOLUTION PROTOTYPING AND DEPLOYMENT

Operation counters or tables would act as placing tools where body parts are placed or by which body parts are carried. They will form a local association network. With these tools, association based tracking method will be able to locate a body part through the placing tools. That is, by tracking the placing tools, positions of the body parts placed on or carried by the tool will be known. The following is the association based tracking for such position sensing of body parts.

- The LAN would be able to identify the placing tool by the ID in the RFID tag.
- The LAN then associates the tag with the RFID reader in the LAN.
- The RFID reader will be able to read all the body parts with RFID tags placed or carried.
- The LAN then will associate the placing tool with those RFID tagged body parts placed or carried.
- The LAN then can calculate the placing tool's position leveraging available positioning infrastructure, e.g., provided by RFID.
- The position information will then be applied to all those items placed or carried.

When a body part is put into the placing tool, an association action will be detected. The position of the placing tool will be calculated to record the position where the body part is put into the placing tool. Then the position of the body part will now be associated with the placing tool. When a body part is removed, a disassociation action will be detected. The position of the body part now is taken out and disassociated from the placing tool. The position of the Placing tool will then be calculated to record the position where the body part is removed. This position will be used to update where the body part is being placed thus the whereabouts of the body part is known.

A prototype of the solution is developed and deployed. The following is the figure (Figure 3) which shows a typical operation training environment in a medical laboratory.

Figure 3 also shows a typical dissecting room layout (a partial drawing, 8m*35m) with LAN setup. The main checking zones are the dissecting tables, doors, sinks and benches. Each checking zone could form a LAN. Different LANs integrated would form another larger local laboratory LAN. The LAN is also able to set up as a mobile LAN. That is, a LAN can be moved to RF scan another checking zone.

The body parts, RFID tagged, will be read by the LAN, establishing the belonging association relationships with the LAN. The LAN will also form an association with the trainee according to the operation training schedule. Those associations are kept and managed by the LAN visibility module, integrated with the laboratory management systems.

SOLUTION VALIDATION

Our proposed solution is validated on site in the laboratory of the Department of Anatomy, LKS Faculty of Medicine, The University of Hong Kong. Areas of concerns for establishing the associations in a LAN are the following:

- Body part to tag. The part of the body to tag would affect the solution performance.

Figure 3. Dissecting room layout

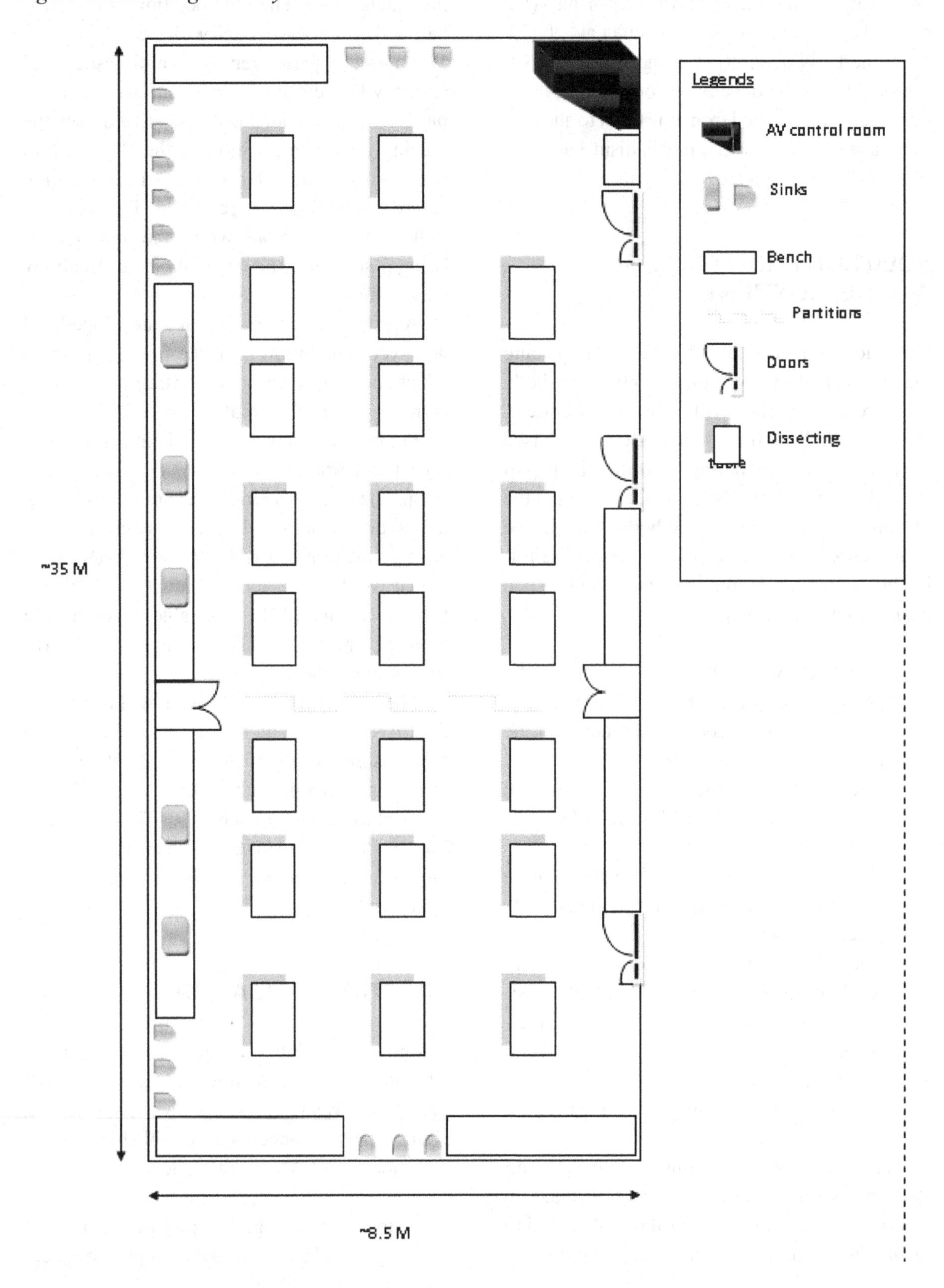

- Orientation. This orientation includes the body parts and the RFID reader.
- Distance. To construct a LAN, effective reading distance is a key performance indicator.
- Room environment. The room environment for operation and for storage will have impact upon the solution performance.

Available Validation Scenarios

Thus, the validation scenarios can be described as a combination of possible factors affecting the test result that is of interest to us. As the test objective is to see how RFID tag can help us to track bodies or body parts in the working environment, we mainly want to measure how well the RFID performs in reading RFID tags under different environmental conditions. As such, the factors, derived from the area of concerns, are:

- RFID reader and antenna
- RFID tag, model of chip used, packaging material
- Room temperature and humidity
- Position and orientation of RFID antenna
- Part of body tagged.
- Position and orientation of RFID tag relative to the dead body and the metal tray
- Room environment, including neighboring metal door, metal cabinet door, concrete wall, etc.
- Distance between RFID tag and antenna
- Storage condition of bodies

Validation Planning

The validation is planned with the following factors being considered.

1) RFID reader and antenna

We tested EPC Class 1 Gen 2 RFID tags using compatible readers and antennae. The following equipments are used.

- One model of handheld reader (Psion Teklogix WorkaboutPro)
- One model of fixed reader and antenna (CSL CS461)

2) RFID tag

Tags from one brand are tested (Avery Dennison AD-826). The tag has been packaged in a special format to allow it to attach to either the wrist or ankle of the body and stay in an almost upright position. This allowed us to easily test different orientation on dead bodies, and also allow us to somehow stay away from the metal tray below the body.

3) Room temperature and humidity

As room temperature and humidity of the validation lab is controlled. We only tested one set up based on actual lab environment. These are recorded as environmental factor.

4) Position and orientation of RFID antenna

The orientation is facing the metal tray, firing RF signal horizontally from around 1 m above ground.

5) Part of body tagged.

We plan to test body neck, wrist and ankle.

6) Position and orientation of RFID tag relative to the dead body and the metal tray

We plan to test the following:

- Tag above body (about 10cm separation from metal tray, edge of tag touching the body, tag parallel to antenna plane)

- Similar to a, with tag laying down, forming an angle greater than 70 degree (estimated) from the antenna plane.
- Tag lying on metal tray.
- Tag with one edge touching metal tray.
  7) Room environment, including neighboring metal door, metal cabinet door, concrete wall, etc.

We avoided facing squarely on a large metal surface, such as metal door of the cabinets and freezers. The path between antenna and the dead body were not obstructed. There is either a concrete wall or at least 2 meter of free space on the other side of the body. Due to difficulty in moving dead bodies around and setting up different room environment, only normal room environment would be tested.

8) Distance between RFID tag and antenna

The distance is adjusted on the fly to capture the maximum reading distance in each test case.

9) Storage condition of bodies

We tested bodies stored in 4 degree Celsius and -25 degree Celsius.

Validation to Carry Out

The following factors are decided to carry out the solution validation.

10) RFID reader and antenna

The RFID reader and tag chosen represent typical EPC RFID implementation.

11) RFID tag

The special tag packaging allows us to test the RFID tag in simulated working condition when attached to dead bodies. Metal tags are considered but was not made available during the validation period.

12) Room temperature and humidity

As we want to perform test in actual working environment, we are not allowed to adjust the temperature and humidity of the validation area. Although only one parameter is available, this represents common working condition of similar institutions.

13) Position and orientation of RFID antenna

The tested orientation is typical of real life installation. However, installation of reader under the ceiling, facing directly downward; and also at about 2 meters above ground level, facing downward at an angle; are also considered. These orientations were not tested due to difficulty in setting up the antenna in the lab of the department.

14) Part of body tagged.

The three body parts are the positions of our interest, where tag can be secured attached and easily reached by fixed and mobile reader.

15) Position and orientation of RFID tag relative to the dead body and the metal tray

The tag position was adjusted by hand to represent different ways the tag ends up after the dead body is moved in and out of the cabinet for some time.

16) Room environment, including neighboring metal door, metal cabinet door, concrete wall, etc.

We were limited to the set up of the room where validation was conducted.

17) Distance between RFID tag and antenna

The distance recorded represents the longest reading distance. It also represents how good the tag was detected by the fixed readers in real life operation and shows the feasibility of using the tag in each case.

18) Storage condition of bodies

The two tested conditions are typical of storage conditions of similar institutes, such as public mortuary.

Validation Results

The validation result is shown in Table 1. Number equations consecutively with equation numbers in parentheses flush with the right margin, as in (1).

LAN INTEGRATION AND MESSAGE DELIVERY

In a LAN and LANs, energy conservation to extend the network lifetime becomes a key issue in designing a message delivery protocol. Energy aware communication protocols for LAN help reduce the energy consumption by adopting sleep scheduling.

A LAN or LAN integration, integrated via mobile/wireless means, at a particular moment, could be not fully connected with each other. Traditional schemes have to ensure high duty cycle to maintain connectivity of network and derogate the energy efficiency. To help energy efficiency, a random sleep scheduling mechanism can be used to minimize the need to maintain the network connectivity (Pujolie, 2006). It employs routing for partially connected network to tackle the network connectivity deficiency. In comparison with fully connected routing, it achieves both higher energy efficiency and higher packet delivery ratio.

In a LAN, there are typically multiple sensors which would present P2P relationships to each other. To simplify the message delivery protocol design, we intend to select a coordinator in a LAN. This coordinator would be selected dynamically or through manual setup. This setup is necessary as the tracking requirement vary from application to application. And different sensors will be needed as well from application to application. In a LAN, a coordinator will make sure it can be identified and found by other peers in the LAN. Multiple coordinators are also possible if the partially connected LAN is too sparse and performance can be boosted.

A coordinator will have directory denoting connection status of its peers. A peer can scan/search the directory for other peers who can meet its particular criteria, and from the list they can determine the current connection and routing information. With that, the peer can then establish a direct peer-to-peer connection with a target peer. The target peer will have to register itself as active and known to be online, so it can respond right way. Otherwise, the coordinator has to wake up the target peer. Depending on the scope of the coordinator, the connection with the directory might remain, either to exchange supplementary information or to track current P2P transfer status.

In this peer-to-peer LAN network, whenever a new peer appears in the network, it first uses a broadcast protocol to send a query to join and wait for a response. Once the coordinator gets the message, it will give the response and sends its ID to the new peer. In this way the new peer has to register its information in the coordinator's directory.

At the same time, each peer who has already registered in the directory may send message to update its current information. Therefore, the coordinator keeps track of peers current information. Whenever peers switch to a new state, the peers send a new registration message, indicating the new status, the transmitted speed, buffer storage and so on.

When a local peer has to send message to a remote peer (a peer in another LAN), it will have to first contact its coordinator in the LAN. The

Table 1. Validation result

Body	Tag age (day) **	Body temp (°C)	Body part where tag attached	Tag position and orientation *	Reader	Distance (cm)
08-16	0	4	Neck	3	Fixed	180
08-16	0	4	Wrist	1	Fixed	190
08-16	0	4	Wrist	1	Handheld	70
08-16	0	4	Wrist	2	Fixed	180
08-16	0	4	Wrist	2	Handheld	12
08-16	0	4	Ankle	1	Fixed	116
08-16	0	4	Ankle	1	Handheld	50
08-16	0	4	Ankle	4	Fixed	60
X	0	20	Wrist	1	Fixed	250
X	0	20	Wrist	1	Handheld	100
X	0	20	Wrist	5	Fixed	80
X	0	20	Wrist	5	Handheld	40
09-37	0	-25	Neck	1	Fixed	240
09-37	0	-25	Neck	1	Handheld	50
09-37	0	-25	Neck	2	Fixed	200
09-37	0	-25	Neck	2	Handheld	20
09-37	0	-25	Wrist	1	Fixed	200
09-37	0	-25	Wrist	1	Handheld	80
09-37	0	-25	Wrist	2	Fixed	140
09-37	0	-25	Wrist	2	Handheld	20
09-37	0	-25	Ankle	1	Fixed	200
09-37	0	-25	Ankle	1	Handheld	10
09-37	0	-25	Ankle	2	Fixed	120
09-37	0	-25	Ankle	2	Handheld	20
08-16	11	4	Neck	3	Fixed	170
08-16	11	4	Neck	3	Handheld	30
08-16	11	4	Wrist	1	Fixed	170
08-16	11	4	Wrist	1	Handheld	40
08-16	11	4	Wrist	6	Fixed	No signal
08-16	11	4	Wrist	7	Fixed	No signal
08-16	11	4	Ankle	2	Fixed	170
08-16	11	4	Ankle	2	Fixed	40
X	11	-25	Wrist	5	Fixed	200
X	11	-25	Wrist	5	Handheld	40
09-37	11	-25	Neck	2	Fixed	200
09-37	11	-25	Neck	2	Handheld	30
09-37	11	-25	Wrist	2	Fixed	200
09-37	11	-25	Wrist	2	Handheld	30

continued on following page

Table 1. Continued

Body	Tag age (day) **	Body temp (°C)	Body part where tag attached	Tag position and orientation *	Reader	Distance (cm)
09-37	11	-25	Ankle	1	Fixed	200
09-37	11	-25	Ankle	1	Handheld	30

* Tag position on and orientation
1. upright, parallel to antenna
2. lay down, not parallel to antenna
3. lay down, with space between tag and chest
4. lay down, touching metal tray (which is covered by plastic body bag)
5. between wrist and trunk, not touching tray, almost parallel to antenna
6. reading tag on opposite limb (antenna on right hand side of body trying to read tag on left hand)
7. tag lying down on metal tray

** Tag age number of days after tagging of body

coordinator checks its directory to select three prior paths and sends the information of them to the local peer.

If the message is too large, then the peer can split the message into blocks according to the target peers' buffer storage information. If the peer decides to split the message into blocks it may set the sequent number for each block so the remote peer can re-sequence the sequences that it has yet to receive.

A MESSAGE DELIVERY SCENARIO

The following is a working scenario to show how the message is delivered in the peer to peer LAN networks. In Figure 4, there are peers and the coordinator in the LAN network. The message to send would include different multimedia including bits, text, audio, picture, and so on depending on the sensors deployed. In this scenario, we suppose that peer Andy wants to send video content to Ada in the LAN networks.

Figure 4. Message delivery in LAN

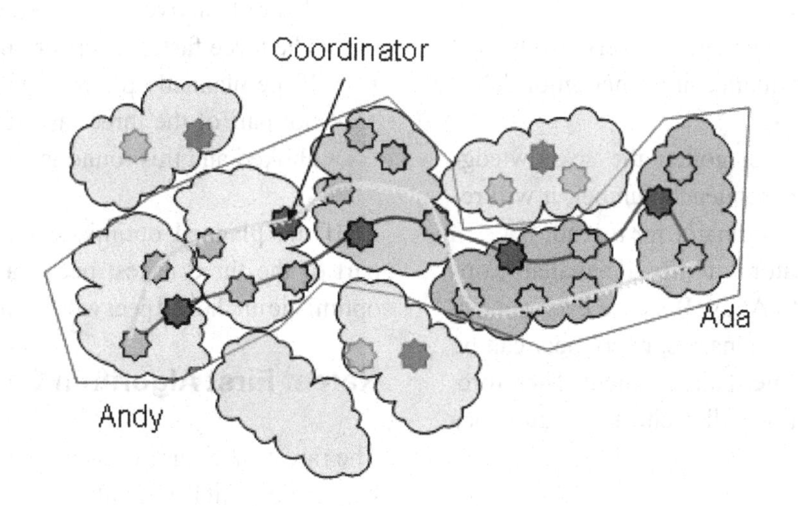

- Andy first uses a broadcast protocol to send a query to join the LAN network and await a response;
- When the coordinator gets this query, it will send the response and sends its registry information to this new peer, as well as inform the peer "I am the coordinator of the LAN network";
- Once Andy gets this response from the coordinator, he will register its information into the coordinator's directory;
- Coordinator will send back an acknowledgement to the new peer;

So far the registered process has completed.

- If Andy wants to send a message to Ada through the LAN network, Andy first connects the coordinator, and said "I want to send message to Ada, please help me";
- When coordinator receives the message, it may first check the peers' information between Ada and Andy;
- After Andy receives the message from coordinator, Andy finds that the buffer in the peers is too small for its message.
- So Andy splits the message into two blocks, then he sets the sequence number for each block;
- Andy picks the paths and sends the split messages out;
- In a reliably message delivery, Andy will receive confirmation information of delivered messages.
- If Andy has not gotten the acknowledgment of some sequence number, it will recontact the coordinator for resend;
- The coordinator will provide available optimal paths to Andy. Because each peer in the network is dynamic, every peer can be online or offline at any moment. Therefore, the coordinator will recheck the directory for Andy.

RELATED WORK IN MESSAGE DELIVERY

In this section, we will review two most relevant classical methods for message delivery in a P2P manner, i.e., choke algorithm and rarest first algorithm (Legout et al., 2006).

Choke Algorithm

The choke algorithm was introduced to guarantee a reasonable level of upload and download reciprocation (Legout et al., 2006). As a consequence, free peers that never contribute to uploading should be penalized. Each time the choke algorithm is called, we say that a new round starts, and the following steps are executed.

- At the beginning of every three rounds, i.e., every 30 seconds, the choke algorithm chooses one peer at random that is choked and interested. This peer is called the planned optimistic unchoked peer.
- The algorithm orders peers that are interested and have sent at least one block in the last 30 seconds according to their downloading rate (to the local peer).
- A peer that has not sent any block in the last 30 seconds is called snubbed. Snubbed peers are excluded in order to guarantee that only active peers are unchoked.
- The three fastest peers are unchoked.
- If the planned optimistic unchoked peer is not part of the three fastest peers, it is unchoked and this round is completed.

If the planned optimistic unchoked peer is part of the three fastest peers, another planned optimistic unchoked peer will be chosen randomly.

Rarest First Algorithm Description

The rarest first algorithm (Legout et al., 2006) is very simple BitTorrent algorithm. The local peer maintains the number of copies in its peer set of each content piece. It uses this information to define a rarest pieces set. Let m be the number of copies of the rarest piece, then the ID of all the pieces with m copies in the peer set is added to the rarest pieces set. The set of rarest is updated each time a copy of a piece is added to or removed from the peer set of the local peer. If the local peer has downloaded strictly less than 4 pieces, it chooses the next piece to request at random. This is called the random first policy. Once it has downloaded at least 4 pieces, it chooses the next piece to download at random in the rarest pieces set. BitTorrent also uses a strict priority policy, which is at the block level. When at least one block of a piece has been requested, the other blocks of the same piece are requested with the highest priority.

APPROACH ANALYSIS

Although the choke algorithm and rarest first algorithm are the key algorithms of BitTorrent, and it performs remarkably well, but that still suffers from several problems, such as traffic load problem. This could become more severe in a LAN as sensors would report events more frequently and due to the time, space and energy constraint, the routing would be quite different in a LAN as well.

Our coordinated message delivery mechanism not only avoids the problem, but also makes the retransmission more immediate by the coordinator's re-direction. By adding a role of coordinator, we can establish a more effective routing for message delivery by considering memory information in the peers.

In fact, the choke and other P2P message delivery protocols utilize a memory-to-memory topology established in the message delivery. That is, after a peer sends a message to a remote peer via the best paths suggested by the coordinator, the peer will put the path into its memory, and the peer can directly send the message next time if needed. But one important particular aspect is neglected, the peers in the network are dynamic, so the best path in the memory of a peer will be disabled, when some peers are offline.

The coordinator's centralized management is more effective in our LAN environment, a small scale network. This is one of the important reasons why we introduce a coordinator into LAN. The advantages of our approach are summarized as follows, compared with other P2P message delivery mechanisms:

- The peer finds the transmitted paths by coordinator, it saves much time, comparing with each peer routing;
- The peer can split messages into blocks for sending via the appropriate paths, it avoids some peers which has small storage buffering capacity;

There are potential pitfalls in a coordinator based message delivery:

- Single point of failure. If the coordinator crashes, then the entire P2P application crashes.
- Performance bottleneck. In a large P2P system, with hundreds of thousands of connected peers, a coordinator must maintain a huge database and must respond to thousands of queries per second.

But in our coordinated message delivery mechanism, the coordinator is selected dynamically. When a coordinator crashes, another peer would act as the new coordinator. In a LAN, there could be multiple coordinators as well.

The coordinator is relative to a LAN. That is, the coordinator is selected among peers in a LAN. The LAN is a small scale network, with little performance bottleneck introducing a coordinator. In a large scale network, we would have to introduce a scale distributed coordination mechanism.

PERFORMANCE EVALUATION

This LAN was tested and validated in a body parts tagging application in a local laboratory in Hong Kong. A laboratory visibility solution is studied and tested based on the LAN. In this application, the solution, prototyped through RFID only enabled system and tools, was validated via on site testing. Results from the validation show that the proposed LAN approach and solution are feasible, supported by a number of available RFID readers and tags in the market. The test shows, with a mobile handheld, the effective distance between the mobile reader and the tagged RFID on the body varies from 10cm to 80 cm for reliable reading. With a fixed reader, the effective distance varies from 40cm to 200cm for reliable reading.

Our LAN also presents similar features to wireless sensor network (WSN). However, in a WSN, energy conservation to extend the network lifetime is one of the primary issues. Energy aware communication protocols for WSN help reduce the energy consumption by adopting sleep scheduling. Traditional schemes have to ensure high duty cycle to maintain connectivity of network and derogate the energy efficiency. To tackle the problem of limited network connectivity in a pure P2P sensor network due to the random sleeping of nodes, message routing for partially connected network is employed. A Continuous Time Markov Chains model of the network is introduced to analyze the end to end responsiveness of multi-hop networks (Pujolie, 2006) (see Figure 5 and Figure 6).

Results show that the packet delivery ratio decreases rapidly with increasing E[TOFF] for fully connected routing protocol because the slightly low duty cycle (below 80%) will incur unacceptably frequent routing failures due to the impairment of the connection. While the packet delivery ratio is still above 55% (compared with 0% of counterpart) when E[TOFF]= 78s, for the routing for partially connected network, because the routing will resume after the wakeup of the downstream node thus reducing the routing failures ratio. By conducting series of simulation and comparing simulation results with the analytical results, we find the calculation error of our derived formula is no more than 5%.

Still many P2P sensor network deployments take the form of fully connected network topology. In these types of sensor networks, many of them adopt a sink node concept (e.g., ZigBee), similar to the coordinator in our LAN. Performance of these sensor networks is acceptable. The concern is with the energy consumption. A study of relationship between energy and package delivery performance is studied (Pujolie G, 2006) (see Figure 7). As we know, there is a conflict between the energy consumption and the responsiveness of network. According to the derived model and formulas, we can work out the optimal policy for deployment of WSN satisfying timeliness and duty cycle requirements. For example, for a WSN with dimension of 15 hops, state transition frequency of 1/120 Hz, we can find the reasonable resolution set to achieve reliable communication within a tolerant delay. According to computed result shown in the Fig. 5 and 6, to achieve packet delivery ratio above 80% fewer than 40% duty cycle, we have to afford the delay no less than 800s.

SERVICES OFFERINGS

To fully understand the capabilities of our approach, it is helpful to consider how the technology can be beneficial in real business situations. The following examples illustrate how the technology can impact throughout service sector (Lee et al., 2008).

1) Patient tracking

This LAN approach can be used for patient tracking in the check-in process. Patient tracking can result in a reduction in the number of staff required to manage the patient check-in process. For example, without a patient- tracking system, a

Figure 5. Packet delivery ratio vs. mean duration of time in ON phase

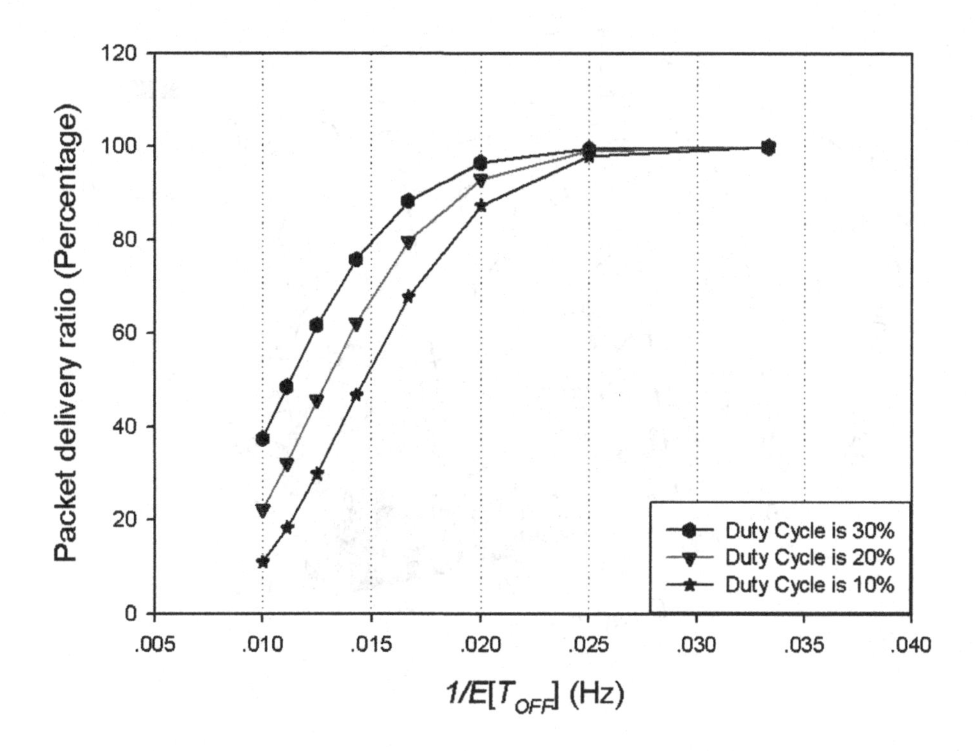

Figure 6. Packet delivery ratio vs. mean duration of time

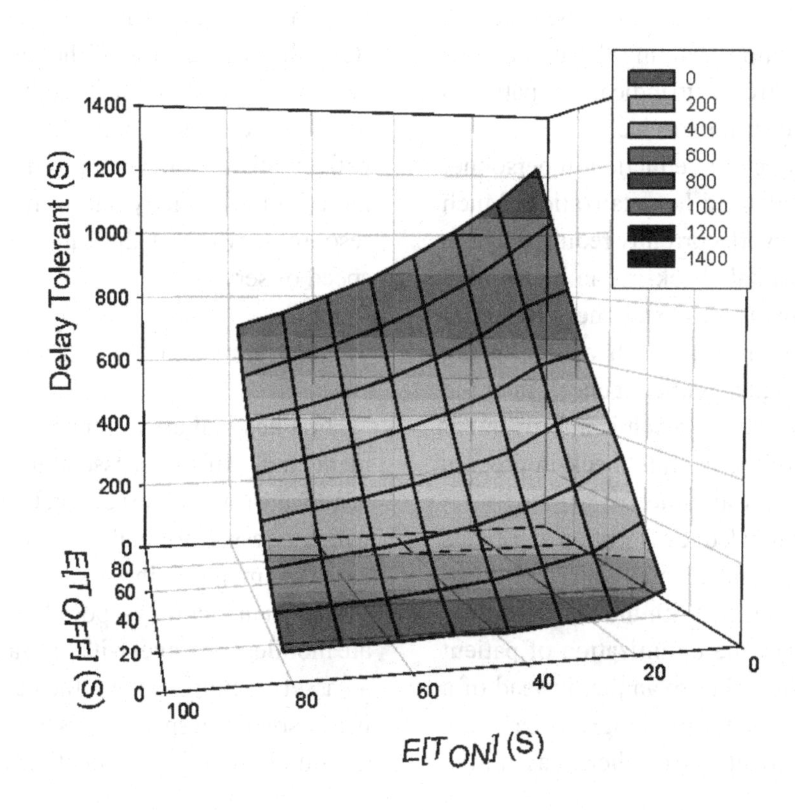

Figure 7. Trade off between energy and performance

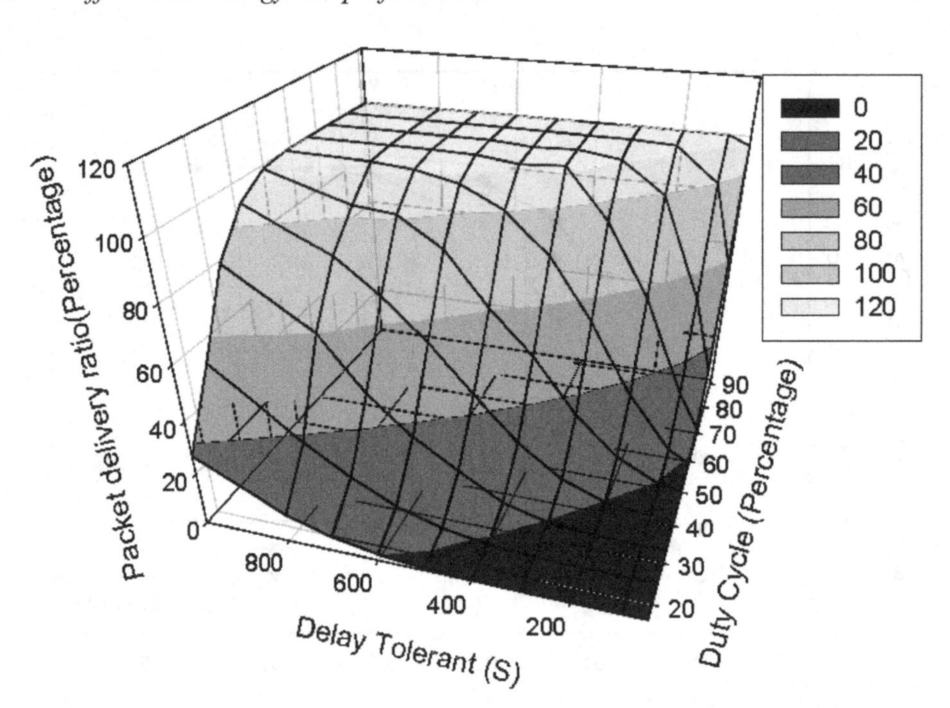

clerk in each department is responsible for managing the patient queue for service. With RFID, the role of the clerk in many cases can be reduced, with the patient simply walking through a door containing an RFID reader to indicate the patient's presence or readiness for service.

Another example of a reduction in personnel is in the case of planned hospitalization. Much of the initial paperwork currently required by the patient during hospital check-in can be handled before the patient arrives, and the patient ID bracelet can be embedded with an RFID tag distributed to the patient either at check-in or via mail before check-in. The result is improvement in the efficiency of the overall check-in process and reduction in the wait time.

The real-time knowledge of location of patients enables a hospital to fundamentally alter traditional ways of doing business. With the knowledge of location of patients, the optimization of patient flow is now possible. For example, instead of a patient waiting in line for radiology, a clerk now has access to the patient load in other areas like the laboratory and can redirect patients to other treatment areas if the process is not order dependent. Instead of a rigid, serial flow, RFID can enable flexibility in the flow of the patient through the various hospital services, resulting in shorter wait times, which has been noted to positively affect patient satisfaction. Such a method can potentially increase the capacity and utilization of hospital resources, while at the same time increase the speed of service.

2) Asset tracking

In a hospital environment, assets (like equipment and staff) are essential in providing the healthcare service to a patient. Unlike the library in which the assets normally have a fixed and expected location, many of the assets of a hospital (e.g., Wheel chairs, gurneys, portable X-ray machines) are mobile, and travel with or to a patient. Without an asset-tracking system, the central supply staff in a hospital can spend hours each day performing a "round-up" of equipment, which is a manual searching of departments for unused equipment or for equipment that requires servicing.

With LAN solution applied to the asset tracking of a hospital, the labor-intensive "roundup" of assets can be automated. Readers (sensors) can be placed throughout the hospital so as an asset moves, the system is tracking its location. The search time is minimized, since the RFID system can identify the location of all tagged equipment in the hospital. This results in fewer staff necessary to perform the round-ups, as well as minimizing the time nurses must spend on equipment searches. Additionally, it enables a faster response to patient needs as equipment location is known and can be accessed easier.

Another current use of asset tracking is to include maintenance information with the asset location information. Many hospital assets (like iv-pumps) go through a cleaning, maintenance, and inspection process. By integrating the service record of an asset with the asset's location information, hospitals can optimize getting the right (working) equipment to the right patient.

3) Road running race

LAN approach can used to provide timing services for running, cycling, and triathlon sporting events. In our framework, the runner represents the moving asset that is being tracked. The use of the RFID has enabled the race to automate the timing process to a greater extent than possible with barcodes. With RFIDs, a runner wears a passive RFID chip on his or her shoe. One RFID reader is at the starting line, tracking the start time, while another reader is at the finish line to track the finish time. Also, readers can be placed at different parts of the race to determine the split times, as well as to verify that the runner passed through certain checkpoints.

Such an approach has already transformed the racing sport by fundamentally redefining the capacity of the race (in terms of number of runners supported) and dramatically changing the way a task is carried out (timing of the runners). Additionally, it has enabled a new service—the timing of walkers.

Through increasing the overall capacity of the event, RFIDs have increased the race's revenue generating capability, since a primary source of funds is the registration fee charged per participant. Therefore, the Road Run represents an example where the technology has enabled the organization to increase revenue while keeping consistent with its goal of offering the best race experience for a venue of its size.

In conclusion, LAN approach can improve the utilization and control of key assets (staff and equipment), as well as provides knowledge of location of the people themselves. This convergence in knowledge of location of both assets and customer enables the deployment of the right equipment and the right staff to the right patient at the right time.

FEEDBACK AND SOLUTION EVALUATION

This LAN approach is to identify the concerns and value of adopting RFID in body tagging by medical professionals and laboratory management. The value proposition comes from the inquiry into understanding of such end user's expectation and concerns.

1) What is normal practice?

The normal practice in taking records of human cadavers and body parts is to physically use a patient tag assigned with the corresponding inventory number. However, these plastic tags have the following disadvantages namely: 1) limited amount of space to input adequate patient information; 2) tags are easily destroyed and subject to lost of information; 3) difficult to track down the location and movements of body parts in particular.

2) Are there any measures for good practices?

The measures put in place to ensure that accurate information of the human cadavers' body parts are kept properly are: 1) input of the inventory number with photographs of the human cadavers and body parts ; 2) input of additional information such as the causes of death, source, status of the endpoint whether the human cadavers need to be returned as a whole or be cremated for collection of the relatives or not and any special signs of the any parts of the body ; 3) Input to where the human cadavers or body parts are kept in the laboratory; and 4) regular check of the inventory at least once a year for accuracy and replacement of deteriorated tags,

3) Are there any pain points?

There are some serious problems with the existing system such as labor intensive in implementing the measures for good practices since the amount of human cadavers and body parts are increasing on the yearly basis. In spite of these investment, there are occasions where inaccuracy of the labeling the body parts and difficulty in tracking the movements of human body parts after being used. The other important issue is the security of the body parts after they are being used for teaching or display.

4) Are you aware of any help envisioned by adopting RFID?

The RFID will definitely solve a major part of the problems mentioned in no.3. RFID can also handle large volume of information with accuracy and NOT increase the cost of human resources. RFID will also increase the degree of the security in the mortuary and laboratory.

While the solution is validated by performing on site testing, user feedback on validation results of the proposed solution is also used for evaluating the solution. We have asked the following questions requiring users to respond.

5) Are we focusing on the right problems to solve?

We are focusing on the right problems. The main factor that contributed to the reading performance difference is the position and orientation of the tags. Tags generally perform poorly when putting near to or touching a metal surface, or when touching the dead body, especially on the chest. Test results show what is possible and what is not in actual usage of the technology in mortuary operation environment.

6) Are we able to plan the test smoothly?

As validation was done in an actual operation environment, the environmental parameters that we can adjust are very limited. It is also hard to control test parameters to repeat same test cases. For example, tag orientation and position cannot be prepared and recorded at a very high accuracy as if it was performed in a lab. On the other hand, the data that we collected represents what is happening in typical working environment where dead body is stored and processed.

7) Does the solution validation deliver good results?

The test result is very good. It shows where and when the technology works and also shows its limitation under certain conditions.

8) Are there any experiences to learn?

We understand more about the amount of detuning effect when the RFID tag is placed near the human body and the metal tray.

Things that we wished to perform in subsequent tests will be to test the tag and reader in real life situations where the dead body is moved in and out of the chamber multiple times, and measure the reading performance for a longer period and for more reading events. We would also like to test whether more expensive metal RFID tags can alleviate the difficulty of reading the attached tag in the awkward positions, such as when the tag is put between the body and the tray.

As automatic reading is not always possible, due to the high variation of reading performance of tag position and orientation, other measures, such as tag packaging design and operational procedure design may also help to adopt the technology to actual mortuary operation. Nevertheless, the automatic data capture capability of RFID is found to be a suitable technology to be deployed to help tracking dead bodies in laboratories, hospitals and mortuaries.

CONCLUSION

In this paper, we study the feasibility of body tagging in a local laboratory in Hong Kong. A solution is proposed based on a Local Association Network (LAN) to develop the laboratory visibility system and tools to facilitate automation and operation efficiency improvement by tagging the body parts. The solution, prototyped through RFID only enabled system and tools, is validated via on site testing. Results from the validation show that the proposed approach and solution are feasible, supported by a number of available RFID readers and tags in the market.

We also reviewed current available IOT architecture and proposed five principles for developing IOT. A local association network is introduced for item tracking and monitoring in a small scale network. A coordinated message delivery mechanism in a peer to peer LAN environment application is introduced. Comparison with other relevant algorithms is made. Advantages of our approach are cost effective and flexible.

Message delivery in peer-to-peer network is worthy further research. Although the approach proposed is more appropriate to apply in a small scale network such as a LAN, it can be applied in a large scale network as well, by improving the method in several fronts, such as allocating the distributed coordinators in a large scale network and dynamically elect any peer in the network as the coordinator when integrating several LANs. Then it will be an effective message delivery mechanism in a large scale IOT network.

Current ongoing work is focused on two fronts: 1) the change management in laboratory automation, process improvement by introducing the proposed solution for body tagging, and 2) testing of the data processing techniques for data filtering in a dense RFID or sensor deployment environment. Real time tracking under Wi-Fi and other wireless infrastructure (e.g. ZigBee, UWB) will also be tested and validated on a site implementation. Focus will then be on deployment efficiency/convenience and cost/performance considerations.

ACKNOWLEGMENT

This paper is an extended version and consolidation of previously papers (Luo & Lai et al., 2010; Luo et al., 2010).

REFERENCES

Armenio, F., Barthel, H., Dietrich, P., et al. (2009). *The EPCglobal Architecture Framework*. Retrieved from http://www.epcglobalinc.org/

Associated Press. (2005). *Body ID: Barcodes for Cadavers*. Retrieved from http://www.wired.com/medtech/health/news/2005/02/66519

Duquennoy, S., Grimaud, J. J., & Vandewalle, G. (2009). Smart and Mobile Embedded Web Server. In *Proceedings of the International Conference on Comlex, Intelligent and Software Intensive Systems* (pp. 571- 576).

Fei, Y., et al. (2009). RFID Middleware Event Processing Based on CEP. In *Proceedings of the 2009 IEEE International Conference on e-Business Engineering.*

International Business Machines. (2003). *An architectural blueprint for autonomic computing*. IBM.

Kanellos, M. (2005). *RFID chips used to track dead after Katrina.* Retrieved from http://news.cnet.com/RFID-chips-used-to-track-dead-after-Katrina /2100-11390_3-5869708.html

Kortuem, G., & Kawsar, F. (2010). Smart Objects as Building Blocks for the Internet of Things. *IEEE Internet Computing, 1/2*, 44–51. doi:10.1109/MIC.2009.143

Legout, A., Urvoy-Keller, G., & Michiardi, P. (2006). Rare first and choke algorithm are enough. In *Proceedings of the IMC'06*, Rio de Jameiro, Brazil.

Lorraine, S., Lee, K., Fiedler, D., & Smith, J. S. (2008). Radio frequency identification (RFID) implementation in the service sector: A customer - facing diffusion model. *International Journal of Production Economics, 112*(2), 587–600. doi:10.1016/j.ijpe.2007.05.008

Luo, Z. (2008). Value Analysis Framework for Technology Adoption with Case Study on China Retailers. *Communications of the Association for Information Systems, 23*, 295–318.

Luo, Z., et al. (2010). A Coordinated P2P Message Delivery Mechanism in Local Association Networks for the Internet of Things. In *Proceedings of the IEEE Intl Conference on e-Business Engineering*, Shanghai, China.

Luo, Z., Lai, M., et al. (2010). Adopting RFID in Body Tagging: a Local Association Network Approach. In *Proceedings of the IEEE Intl Conference on RFID TA*, Guangzhou, China.

Peng, P., et al. (2008). A P2P Based Collaborative RFID Data Cleaning Model. In *Proceedings of the 2008 International Symposium on Advances in Grid and Pervasive Systems (AGPS 2008).*

Pujolie, G. (2006). An autonomic-oriented architecture for the Internet of Things. In *Proceedings of the IEEE 2006 International Symposium on Modern Computing* (pp. 163-168).

Sheng, M. (2009). Ubiquitous RFID: Where are We? In *Information Systems Frontiers*. New York: Springer.

Wang, J., et al. (2007). RFID Assisted Object Tracking for Automating Manufacturing Assembly Lines. In *Proceedings of the 2009 IEEE International Conference on e-Business Engineering (ICEBE 2007).*

Zhang, T., et al. (2008). Developing a Trusted System for Tracking Asset on the Move. In *Proceedings of the 9th International Conference for Young Computer Scientists* (pp. 2008-2013).

Zhang, T., Luo, Z., Wong, E. C., Tan, C. J., & Zhou, F. (2008). Mobile Intelligence for Delay Tolerant Logistics and Supply Chain Management. In *Proceedings of the 2008 IEEE International Conference on Sensor Networks, Ubiquitous, and Trustworthy Computing* (pp. 280-284).

Zhou, Z., et al. (2007). Interconnected RFID Reader Collision Model and its Application in Reader Anti-collision. In *Proceedings of the IEEE RFID Conference*, TX.

This work was previously published in International Journal of Systems and Service-Oriented Engineering, Volume 1, Issue 4, edited by Dickson K.W. Chiu, pp. 42-64, copyright 2010 by IGI Publishing (an imprint of IGI Global).

Section 2
Intelligence Computing for Systems and Services

Chapter 6
Secure Key Generation for Static Visual Watermarking by Machine Learning in Intelligent Systems and Services

Kensuke Naoe
Keio University, Japan

Hideyasu Sasaki
Ritsumeikan University, Japan

Yoshiyasu Takefuji
Keio University, Japan

ABSTRACT

The Service-Oriented Architecture (SOA) demands supportive technologies and new requirements for mobile collaboration across multiple platforms. One of its representative solutions is intelligent information security of enterprise resources for collaboration systems and services. Digital watermarking became a key technology for protecting copyrights. In this article, the authors propose a method of key generation scheme for static visual digital watermarking by using machine learning technology, neural network as its exemplary approach for machine learning method. The proposed method is to provide intelligent mobile collaboration with secure data transactions using machine learning approaches, herein neural network approach as an exemplary technology. First, the proposed method of key generation is to extract certain type of bit patterns in the forms of visual features out of visual objects or data as training data set for machine learning of digital watermark. Second, the proposed method of watermark extraction is processed by presenting visual features of the target visual image into extraction key or herein is a classifier generated in advance by the training approach of machine learning technology. Third, the training approach is to generate the extraction key, which is conditioned to generate watermark signal patterns, only if proper visual features are presented to the classifier. In the proposed method, this

DOI: 10.4018/978-1-4666-1767-4.ch006

classifier which is generated by the machine learning process is used as watermark extraction key. The proposed method is to contribute to secure visual information hiding without losing any detailed data of visual objects or any additional resources of hiding visual objects as molds to embed hidden visual objects. In the experiments, they have shown that our proposed method is robust to high pass filtering and JPEG compression. The proposed method is limited in its applications on the positions of the feature sub-blocks, especially on geometric attacks like shrinking or rotation of the image.

INTRODUCTION

In this article, we propose a method of key generation scheme (Figure 1) for static visual digital watermarking (Figure 2) by using machine learning technology, neural network as its exemplary approach for machine learning method.

The proposed method is to provide intelligent mobile collaboration with secure data transactions using machine learning approaches, herein neural network approach as an exemplary technology. First, the proposed method of key generation is to extract certain type of bit patterns in the forms of visual features out of visual objects or data as training data set for machine learning of digital watermark. Second, the proposed method of watermark extraction is processed by presenting visual features of the target visual image into extraction key or herein is a classifier generated in advance by the training approach of machine learning technology. Third, the training approach is to generate the extraction key which is conditioned to generate watermark signal patterns only if proper visual features are presented to the classifier. In our proposed method, this classifier which is generated by the machine learning process is used as watermark extraction key.

The proposed method is to contribute to secure visual digital watermarking without losing any detailed data of visual objects or any additional resources of hiding visual objects as molds to embed hidden visual objects. The proposed method has used neural network for its training approach not limited but open in its applications to other machine learning approaches including fuzzy, Bayesian network and others. In this article, the target content is a static visual data which are constructed with discrete data set and we have demonstrated the feasibility of solving this problem by using neural network model. We would enhance our method by using those other approaches, such as fuzzy for dynamic visual data like video stream data and Bayesian network for continuous data structures. This article is different from the previous work by Ando et al. (Ando, R. & Takefuji, Y., 2003) in terms of embedding size where this article does not embed any information to the target content and also implies that the machine learning algorithm is not limited only to the neural network model as proposed in our previous work (Naoe, K. & Takefuji, Y., 2008).

ISSUES AND DEFINITIONS

The Service-Oriented Architecture (SOA) demands supportive technologies and new requirements for mobile collaboration across multiple platforms. One of its representative solutions is intelligent information security of enterprise resources for collaboration systems and services (Chiu, & Leung, 2005; Kafeza, Chiu, Cheung, & Kafeza, 2004).

Digital watermarking became a key technology for protecting copyrights. Digital watermarking protects unauthorized change of the contents and assures legal user for its copyright. Meanwhile, steganography conceals a hidden messages to a content but the existence of a message is kept secret (Artz, 2001). For the purpose of digital watermarking, content should not be encrypted or scrambled. Digital watermark is often embed-

Figure 1. Key generation scheme in embedding procedure

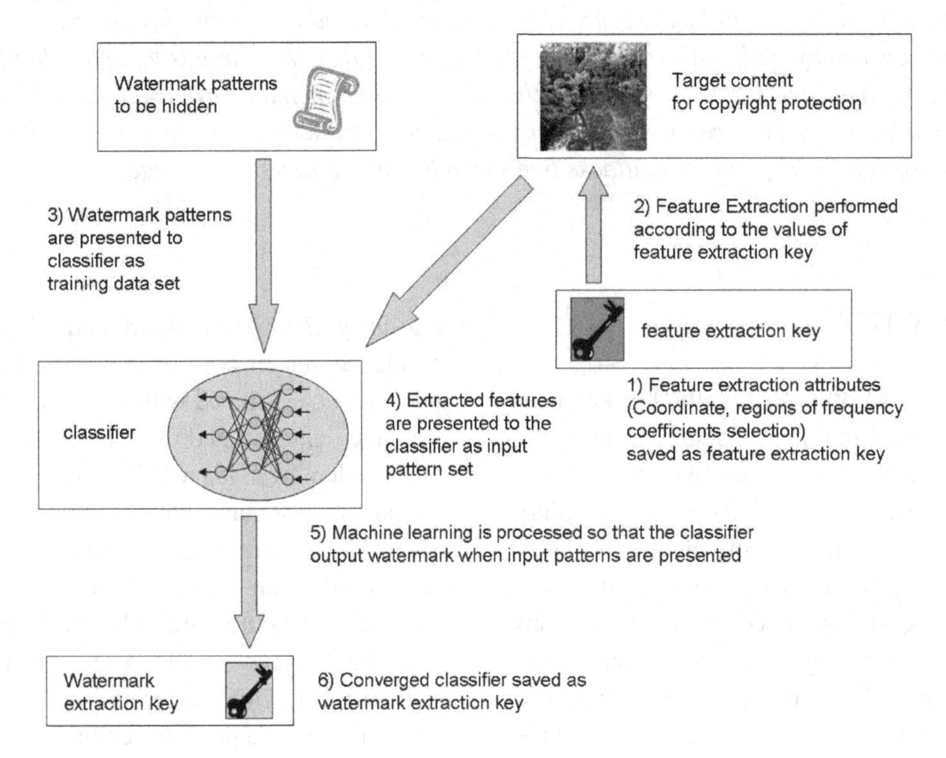

Figure 2. Watermark extraction scheme in extraction procedure

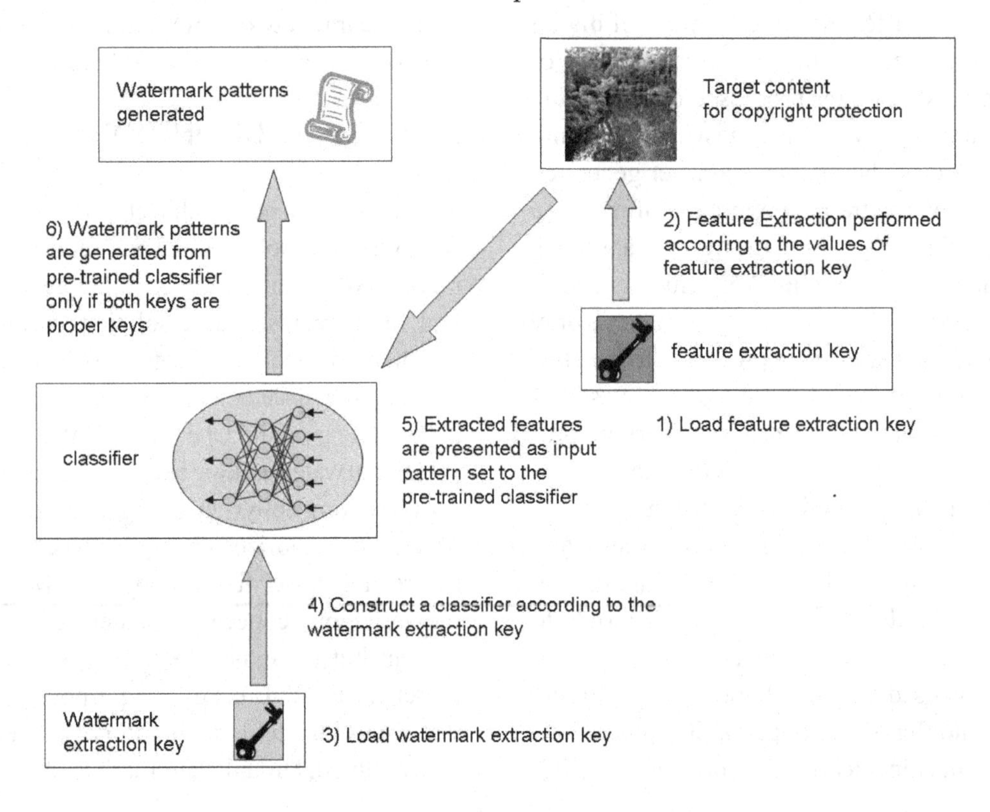

ded imperceptibly to human receptors to avoid contaminating the content and not to distract the content from the original expression. For imperceptible images, the human visual system (HVS) model is often used (Delaigle, Vleeschouwer, & Macq, 1998). Perceptible watermark are sometimes used, but it limits the use of the images. Therefore, main concern in this research area has focused on imperceptible watermark.

In general, the robustness and imperceptibility of digital watermarking take trade-off relationships. Embedding information must be placed in perceptually significant signal for it to be robust against removal attacks, but it is known that modifying these regions will lead to perceptual degradation of signal (Cox, Kilian, Leighton, & Shamoon, 1996). Therefore, an ideal digital watermarking algorithm should have minimal amount of embedding information. There is also a difficulty to fulfill the non-repudiation requirement by using a robust watermarking scheme alone. To address this problem, there must be a solid distribution protocol and secure infrastructure framework to put in practice. A contribution from previous work by Cheung, Chiu, & Ho, (2008) which proposed a distribution protocol and secure system framework addressed this problem and its watermarking algorithm can be replaced with another algorithms as long as the watermark can be inserted in the encrypted domain where digital contents are encrypted by public keys. Our proposed method generates pair of extraction keys where one of them can act as public key and also does not alter the target content. Therefore, our proposed method has an ability to collaborate with this previous work.

Backgrounds and Related Works

Here, we overview the backgrounds and related works in the research areas of digital watermarking and machine learning, neural network as an exemplary model. We discuss digital watermark, steganography, cryptography and machine learning.

The emergence of the Internet with rapid progress of information technologies, digital contents are commonly seen in our daily life distributed through the network. Due to the characteristics of digital contents are easy to make an exact copy and to alter the content itself, illegal distribution and copying of digital contents has become main concerns for authors, publishers and legitimate owners of the contents (Sasaki, 2007). There are several ways to protect digital content. One can protect the content by cryptographic-based schemes (Rothe, 2002), but this avoids free distribution and circulation of the content through the network because the decryption key must be shared, which most of the time not desirable for authors of the contents. To address this problem, the researches have developed "Public Key Infrastructure" (Adams, & Lloyd, 1999; Maurer, 1996) and "Secure Key Sharing" (Law, Menezes, Qu, Solinas, & Vanstone, 2003; Eschenauer, & Gligor, 2002). Moreover, digital watermarking technologies are noticed to substitute or to complement the conventional cryptographic schemes (Cox, Doerr, & Furon, 2006). Furthermore, in the era of mobile devices and technologies, conventional cryptographic schemes are highly power consuming and time consuming, which are not suitable for these types of devices (Prasithsangaree, & Krishnamurthy, 2003). Hence, necessary information must be obtainable with less calculation cost and power consumption.

Watermarking technique is one technique of an information hiding techniques. Information hiding provides a reliable communication by embedding secret code into a content for the various purposes: intellectual property protection, content authentication, fingerprinting, covert communications, content tracking, end-to-end privacy, secure distribution, etc. (Hung, Chiu, Fung, Cheung, Wong, Choi, Kafeza, Kwok, Pun, & Cheng, 2007; Wolf, Steinebach, & Diener, 2007; Kwok, Cheung, Wong, Tsang, Lui, & Tam, 2003). The researches in information hiding has a history (Kahn, 1996) and namely the researches in digital watermarking and steganography have been active (Katzenbeisser, & Fabien, 2000). Both are very similar but their applications are different (Cox, I., Miller, Bloom, Fridrich, & Kalker, 2007).

There are many digital watermarking and steganography algorithms, though it is difficult to use one algorithm together with another, because each other obstruct the embedded information and causing one to destroy another. Because our proposed method does not damage the target content, it has an ability to collaborate with another algorithm to strengthen the security of information hiding method. This characteristic is useful where one already manage digital rights using one watermarking algorithm or controls the file integrity using hash functions. If one wishes to strengthen the robustness of watermark using another algorithm, one must examine and assure that applying the algorithm will not affect embedded watermarking signals in advance. Furthermore, applying another watermarking algorithm will alter the fingerprint of the content managed by hash functions and forces administrator to recalculate a new hash values after applying new watermarking algorithm, which most of the time, result in higher calculation cost and time. Because our proposed method does not affect the target content at all, one can apply new watermark seamlessly without altering the fingerprint using the proposed method.

Machine learning and statistical analysis are very effective approach to approximate unknown class separating functions and to find potentially useful patterns in the data set (Bishop, 2006). Basically, machine learning algorithm uses training set to adjust the parameter of the model adaptively. When target vector or teacher patterns are presented for an training set, training or learning process is performed to condition the model to output a proper patterns which is same or close to as the target vector. Some of the noticeable models in machine learning are Bayesian network (Bernardo, & Smith, 2001; Jensen, 1996), fuzzy logic (Klir, & Yuan, 1995; Klir, & Yuan, 1996), support vector machine (Cortes, & Vapnik, 1995; Burges, 1998), and neural network (Kohonen, 1988; Bishop, 1995). The proposed method uses machine learning approach to generate watermark extraction keys and herein uses neural network model in this article.

The basic principle of neural network is that neuron, or most atomic unit of neural network, only has a simple function of input and output signal but is capable of complex function when these neurons are organically connected forming a network. Mathematical models of these networks are called neural networks, and processing these artificial neural networks on computers is called neural computing. Neural computing is a classic research topic and neural networks are known to be capable of solving various kinds of problems by changing the characteristics of neuron, synaptic linking and structure of the network.

The proposed method uses a multi-layered perceptron model for neural network model. Multi-layered perceptron basically has a synaptic link structure in neurons between the layers but no synaptic link between the neurons within the layer itself (Rosenblatt, 1958). Two layer perceptron can approximate data linearly but cannot classify data nonlinearly. Multilayered perceptron with more than three layers are known to have an approximation ability of a nonlinear function if properly trained, but then there was no ideal learning method for this kind of training. This model became popular when a training method called back propagation learning was introduced (Rumelhart, & McClelland, 1986). Other neural network models are RBF network model (Broomhead, & Lowe, 1988), pulsed neural network model (Johnson, 1994), Elman network model (Elman, 1990) and many others.

Structure

This article is organized as follows. Section 1 gives a background and some related works of digital watermarking as one application of information hiding technique, and general overview of machine learning technique especially focusing on neural network model. Section 2 explains our methodology for generating extraction key sets for extracting watermark patterns. Section 3 gives an experimental conditions and results of our experiments to examine the robustness of our watermarking method. Section 4 and 5 will give some discussions, conclusions and some future work to be addressed.

METHODOLOGY

In this section, we explain how our proposed method generates extraction keys from the target content and how to retrieve watermark information from the target content using the extraction keys generated in the watermark embedding procedure. With our method, the use of machine learning approach is the key methodology. Here, adjustment of neural network weights to output desired hidden watermark pattern by supervised learning of the neural network is performed. This conditioned neural network works as a classifier or watermark extraction key to recognize a hidden watermark pattern from the content. Therefore, extractor uses this neural network weights as extraction keys for extracting the hidden watermark patterns. Extractor must have proper visual features used for generation of extraction key and proper network weights of neural network which is the extraction key in our proposed method. Considering the difficulties for secret key transportation, this method should be applied in situations where the hider and the extractor is a same person or to use certification authorities to assure the integrity of the key as Cheung, Chiu, and Ho, (2008) proposed.

Our Proposed Method

Here, we explain the procedures for generation of extraction keys and extraction of watermark patterns. Generation of extraction key is included in the embedding process and extraction of watermark patterns is included in the extraction process. First, we simply demonstrate the procedures which must be taken for embedding process and extraction process.

Embedding process consists of following procedures:

1. Feature extraction by frequency transformation of target image
2. Selection of the feature regions according to the feature extraction attributes and saved as feature extraction key
3. Prepare watermark patterns to be embedded
4. Generation of extraction key to output watermark patterns by back propagation learning of neural network using feature extraction key
5. Save the generated neural network classifier as watermark extraction key

First, frequency transformation of the image is performed. There are several methods to transform an image to frequency domain, such as Discrete Fourier Transform (DFT), Discrete Cosine Transform (DCT) and Discrete Wavelet Transform (DWT). In our method, DCT is used to be robust against some image compression. Compression using DCT is known to be very effective, and is employed in MPEG format and JPEG format. DCT divides an image into N*N pixel small blocks, here we call sub-blocks. General DCT generates 8*8 pixel size sub-blocks. We must select certain amount of sub-blocks from frequency domain of target image and DCT coefficients are chosen diagonally from those sub-blocks. The same amount of unique sub-blocks must be chosen from the target image as number of classification patterns, which in this method is the watermark pattern. Sufficient number of neural networks must be prepared, which will be the number of binary digits to satisfy the watermark patterns. In case for choosing 32 watermark patterns, five networks are enough to represent 32 different classification values because five binary digits are

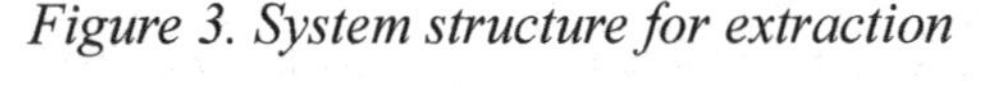

Figure 3. System structure for extraction

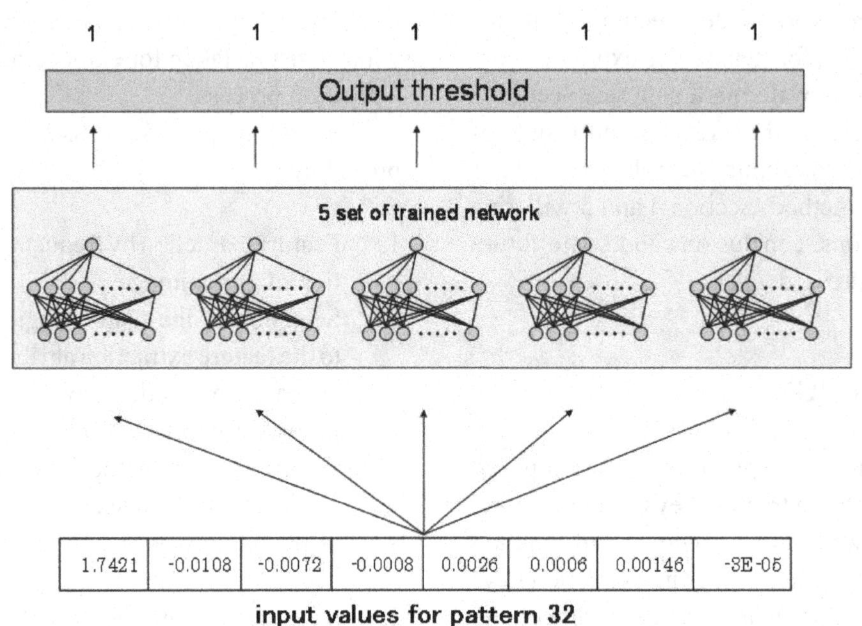

sufficient to distinguish for 32 patterns. Learning of all networks is repeated until the output value of neural network satisfies a certain learning threshold value. After all network weights are generated, the coordinates of sub-blocks and the values of network weights are saved. Extractor will use this information to extract hidden codes in the extraction process.

Extraction process consists of following procedures:

1. Obtain feature extraction key and watermark extraction key
2. Feature extraction of target image using the feature extraction key
3. Construct a neural network model using watermark extraction key
4. Observe the output patterns from neural network using both keys

Extractor must receive both proper feature extraction key and watermark extraction key to obtain proper watermark pattern. Only by having the proper feature extraction key will lead the user to the proper input patterns to the neural network. By knowing proper watermark extraction key, extractor can induce the structure of the neural network and only proper neural network is able to output the proper hidden watermark patterns. After constructing the neural network, extractor examines the output value from the network with the input values induced from the feature extraction key. This procedure is shown in Figure 3. Each network output either 1 or 0 with the aid of threshold for output unit.

Furthermore, here we explain each procedure more concretely. For key generation procedure, frequency transformation of the target content is processed for visual feature extraction of target content. Frequency transformation can be anything like DCT, DFT or DWT, here we use DCT because of its simplicity and its structure of feature vectors of a transformed image in frequency domain. This frequency transformation is done after converting the target content to YCbCr color domain. Basically, the transformation from RGB color signal to YCbCr signal is used to separate a luma signal Y and two chroma components Cb and Cr and mainly used for JPEG compression and color video signals. In our proposed method, instead of using RGB color space directly, YCbCr color space is used to make use of human visual system characteristic. The conversion from RGB to YCbCr is calculated using the following equation:

Y = 0.299R + 0.587G + 0.114B

Cb = -0.169R - 0.322G + 0.500B

Cr = 0.500 - 0.419G - 0.081B

Then, training of neural network is processed. For the training, one must decide the structure of neural network. The amount of units for input layer is decided by the number of pixels selected from target content data. In our proposed method, the feature values are diagonal coefficient values from frequency transformed selected feature sub-blocks. For better approximation, one bias neuron is added for input layer.

The neural network is trained to output a value of 1 or 0 as an output signal. In our proposed method, one network represents one binary digit for corresponding watermarking patterns. The adequate amount of neurons in the hidden layer, for back propagation learning in general, is not known. So the number of neurons in hidden layer will be taken at will. In our proposed method, ten hidden units are used. For better approximation, one bias neuron is introduced for hidden layer as well. Once network weights are generated to certain values, the proposed method use these values and the coordinates of selected feature sub-blocks as feature extraction key and watermark extraction key. These keys must be shared among the hider and the extractor in order to extract proper hidden watermark patterns from the contents.

Now, we demonstrate an overview of multi-layered perceptron model. In multilayered perceptron model, signals given to the input layer will propagate forwardly according to synaptic weight of neurons through the layers and finally reaches to the output layer as shown in Figure 4.

Figure 4. Multi-layered perceptron model

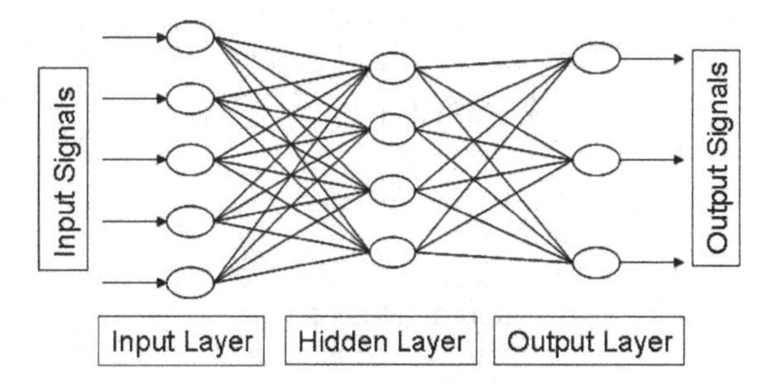

Signal that is being put to the neuron is converted to a certain value using a function and outputs this value as output signal. Normally sigmoid function is used for this model and this function is expressed as follows:

$$f(x) = \frac{1}{1 + e^{-x}}$$

Each synaptic link has a network weight. The network weight from unit i to unit j is expressed as W_{ij} and the output value for unit i is expressed as O_i. The output value for the unit is determined by the network weight and the input signal from the former layer. Consequently, to change the output value to a desired value for a certain input value patterns, the adjustment of these network weights must be conducted and this process is called learning. In our proposed method, we use back propagation learning as the learning method.

Back propagation learning is the process of adjusting the network weights to output a value close to the values of the teacher signal values which are presented to the neural network. Back propagation learning is a supervised learning (Rumelhart, & McClelland, 1986). This method tries to lower the difference between the presented teacher signal and the output signal dispatched for certain input value patterns by adjusting the network weight. The difference between the teacher signal values and the actual output signals are called as error

Figure 5. Embedding procedure

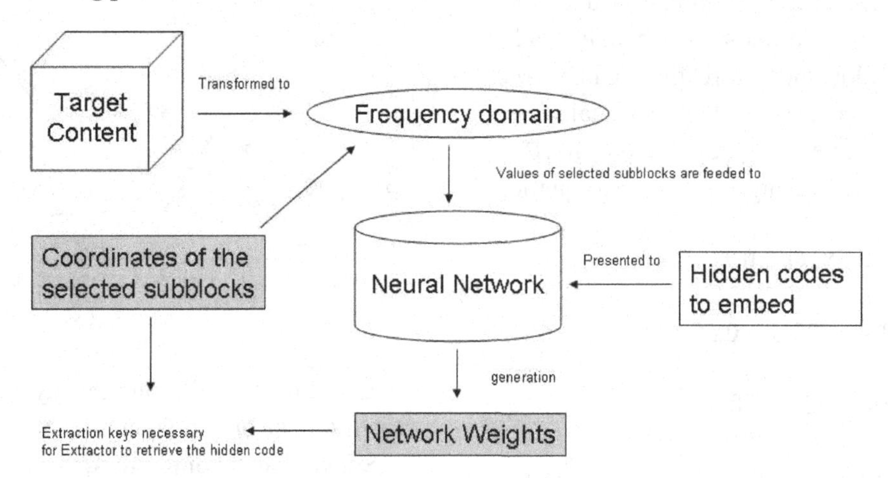

and often expressed as δ. The error will propagate backward to the lower layer and network weights are adjusted using these values. When teacher signal t_k is given to the unit k of output layer, the error δ_k will be calculated by following function:

$$\delta(x) = (t_k - O_k) \cdot f'(O_k)$$

To calculate the error value δ_j for hidden unit, error value δ_k of the output unit is used. The function to calculate the error value δ_j for hidden unit j is as follows:

$$\delta_j = (\sum_k \delta_k w_{jk}) \cdot f'(O_j)$$

After calculating the error values for all units in all layers, then network can change its network weight. The network weight is changed by using following function:

$$\Delta\delta_k w_{jk} = \eta\delta_j O_i$$

η in this function is called learning rate. Learning rate is a constant which normally has a value between 0 and 1 and generally represents the speed of learning process. The lower the learning rate is the more gradual the learning process will be, and the bigger the learning rate is the more acute the learning process will be. Sometimes, this parameter must be tuned for stable learning.

For extraction process, same neural network structure is constructed using the generated watermark extraction key. Only with the proper feature extraction key and watermark extraction key will be able to output the corresponding watermark patterns. These embedding and extraction procedure are shown in diagram in Figure 5 and 6.

We define necessary parameters for embedding and extracting. For embedding, there are two parameters to decide on. First is the number of watermarking patterns to embed. More the number of watermarking patterns, the more data can be embedded, but introducing large number of watermarking patterns will result in high calculation cost. Second parameter is the coordinate of the sub-blocks. Coordinates will determine the input patterns to the neural network for the embedding and extraction process. For extracting, there are two keys to be shared, that is the watermark extraction key and feature extraction key, between the embedding and extracting users. Former is the neural network weights created in the embedding process. Latter is the coordinates of feature sub-blocks. Only with the presence of the proper keys are able to generate proper watermark patterns as shown in Figure 5 and 6.

Figure 6. Extracting procedure

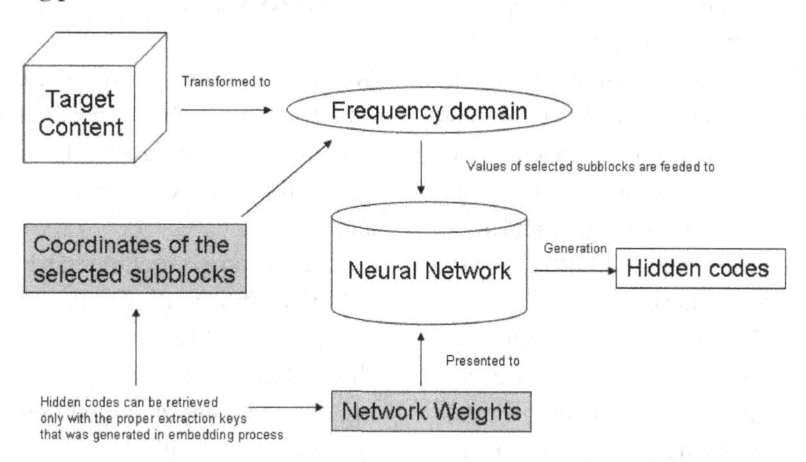

EXPERIMENTS

In this experiment, we will examine if we can retrieve a watermark patterns from both original image and graphically changed image. For the latter, we choose high pass filter and JPEG compressed image as the alteration method.

We used TIFF format Lenna image, which is 512*512 pixels in size, as target content data. Original and high pass filtered Lenna image is shown in Figure 7. We embedded 32 different watermark patterns as hidden signals. That is, hidden signals are [00000] for pattern 1, [00001] for pattern 2, ... [11111] for pattern 32. Five neural networks are used for classification of 32 patterns. Each network output value represents the binary digits of watermark patterns. In this experiment, network 1 represents the largest binary bit and network 5 represents the smallest binary bit. The number of hidden layer units is set to 11 including one bias unit. Learning process is repeated until the output values generate to a learning threshold of 0.1. Also, the output threshold in this experiment is set to 0.5. This means that if output value is larger than 0.5, output signal is set to 1, and if it is less than 0.5, output signal is set to 0. Again, this neural network structure is shown in Figure 3.

Figure 7. Original Lenna image and filtered image

With the threshold value of 0.5, the proposed method was able to extract proper signals for proper patterns. For example, output signals in binary bit pattern for pattern 1 is [00000], output signals for pattern 11 is [01010] and etc. The results of this experiment, output signals for original image, JPEG compressed image and high pass filtered image are shown in Figure 8, 9, 10, 11 and 12 for each neural network.

The output signals retrieved from high pass filtered image are shown to be slightly different compared to the output signals for the original image and also the output signals retrieved from JPEG compressed image is damaged more than of filtered image, but with the same output threshold of 0.5, we were able to retrieve same hidden watermark patterns for all 32 set from high pass filtered image and JPEG compressed image. These results showed the robustness to high pass filtering and JPEG compression alteration.

DISCUSSIONS

We would like to discuss the contributions and limitations on our propose method in its applications to intelligent information security of enterprise resources for collaboration systems and services.

Our proposed method has not embedded any additional information to the target images of the experiments. Also, in the experiment we have shown the possibility of a robust digital watermarking scheme without embedding any data into the target content. With the perspective of information security, this method has the possibility for the applications for both digital watermark and steganography.

Our proposed method is logically limited in its use for digital watermarking in the generation time of feature extraction key and watermark extraction key: the former feature extraction is a preprocess before the neural network learning;

Figure 8. Signal values for network 1

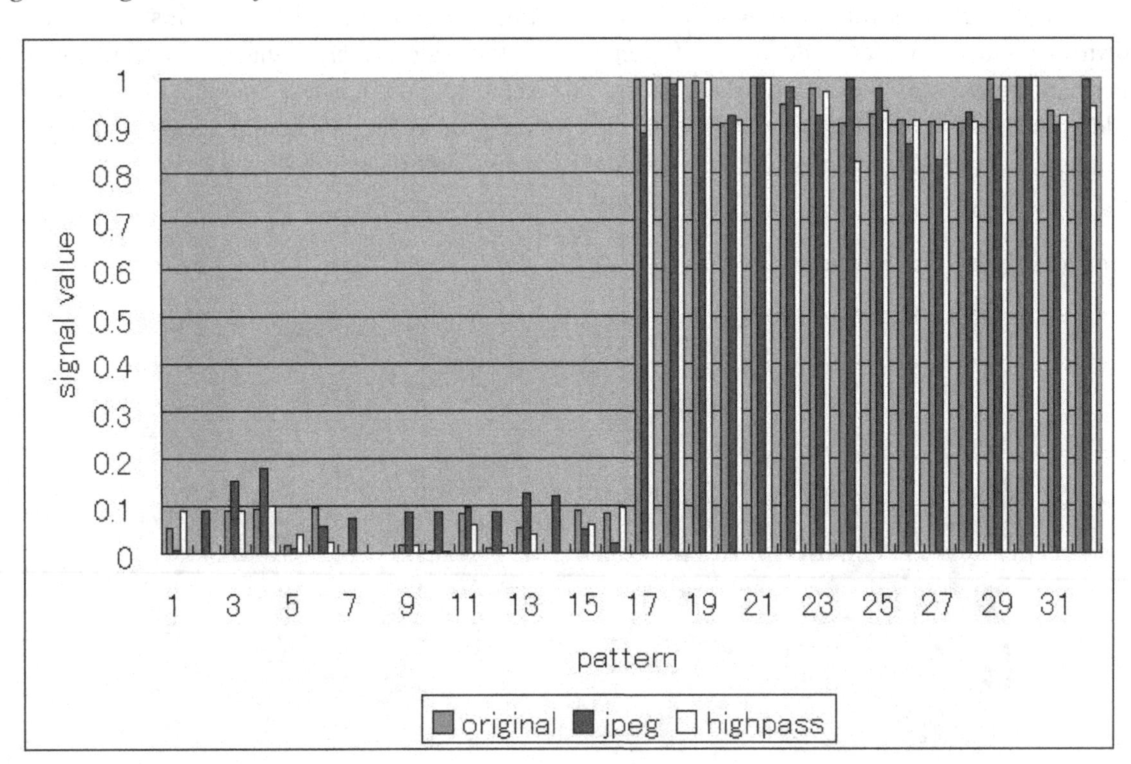

Figure 9. Signal values for network 2

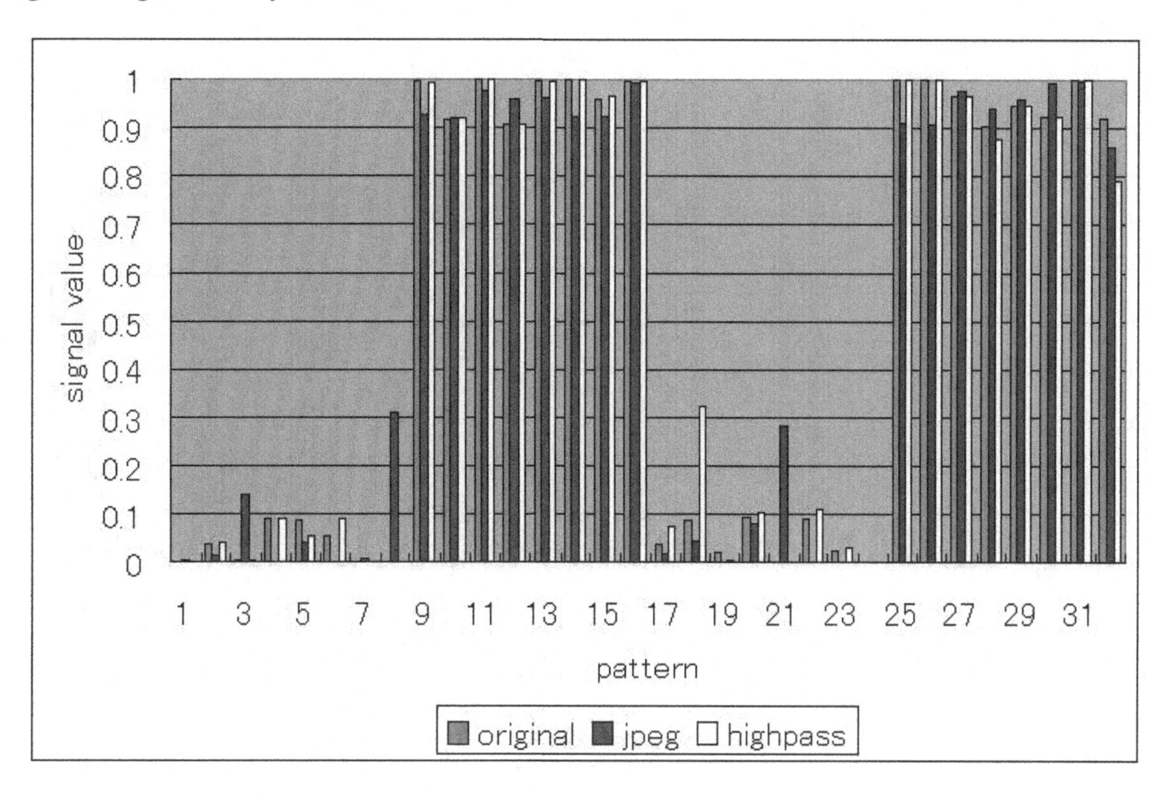

Figure 10. Signal values for network 3

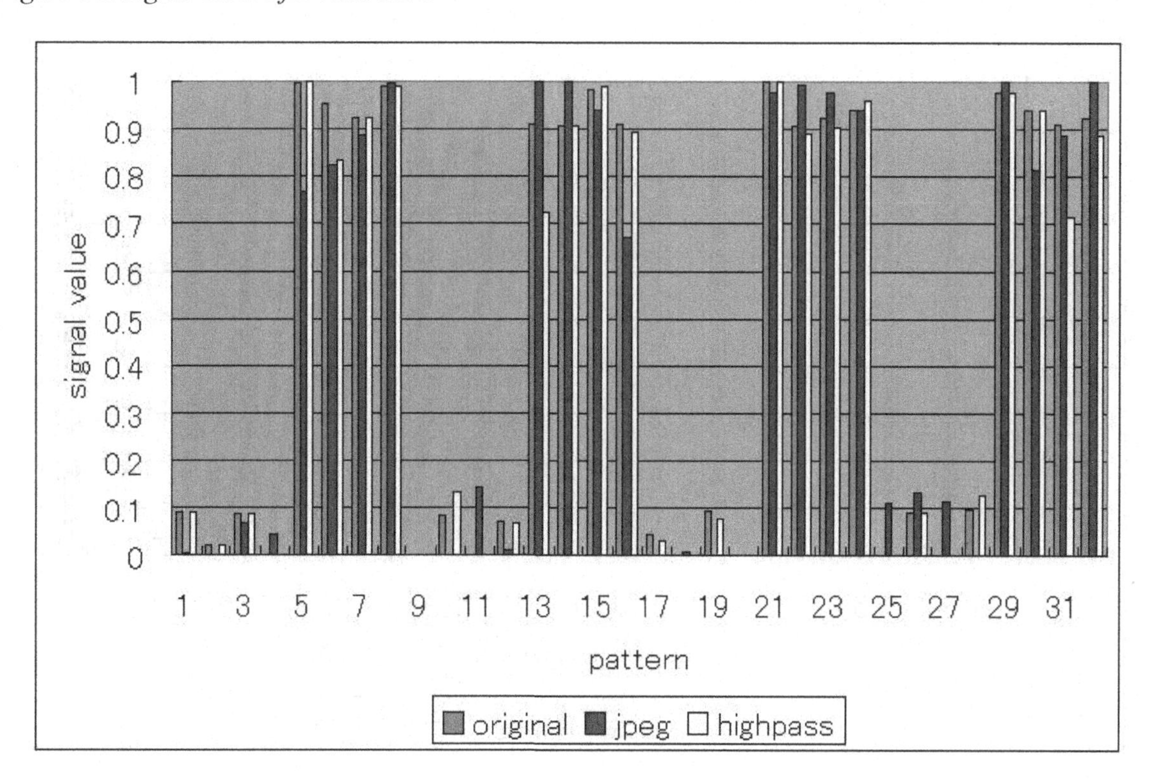

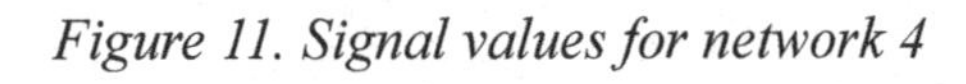

Figure 11. Signal values for network 4

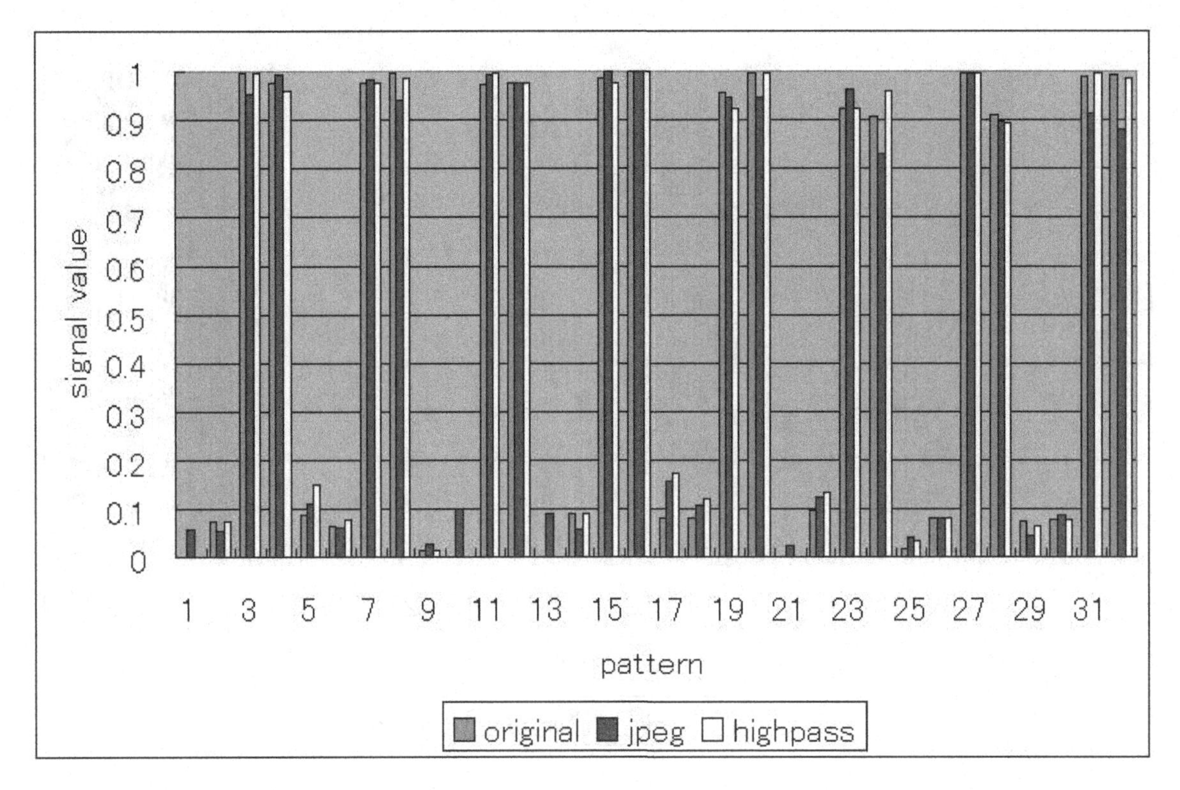

Figure 12. Signal values for network 5

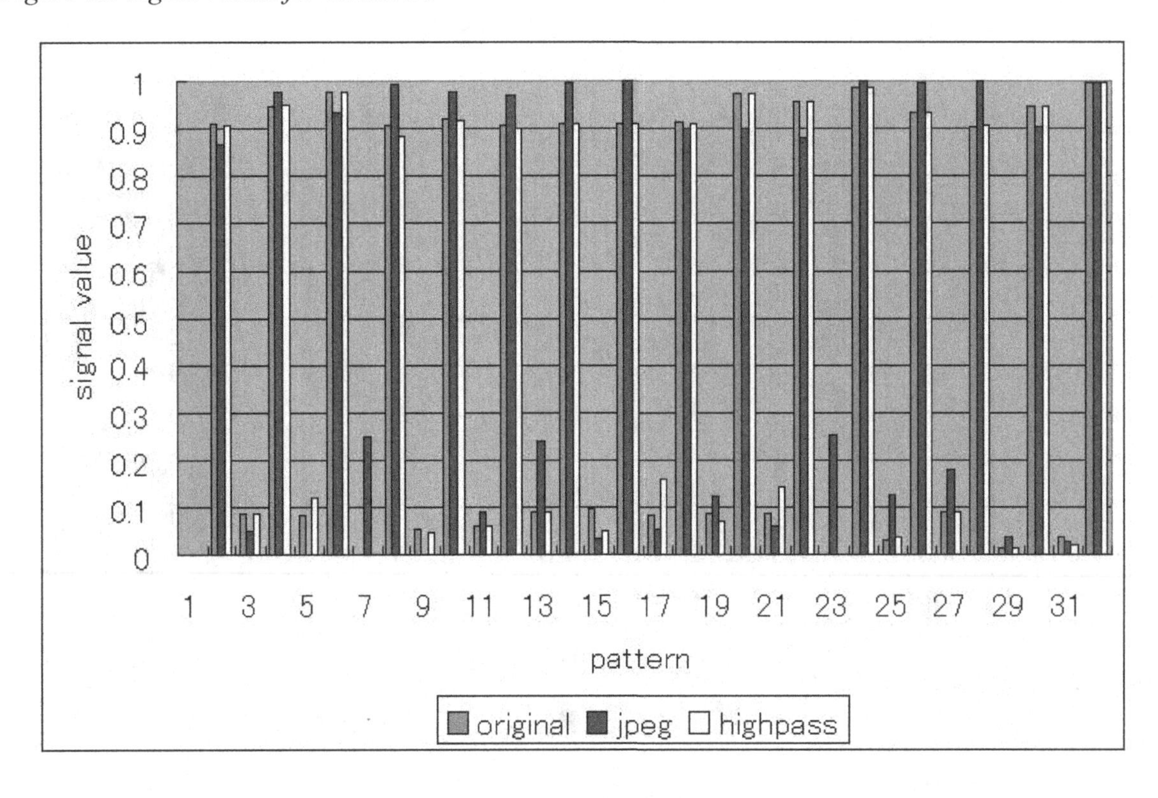

And, the latter processing time for watermark extraction is less than a second in practice of experiment. The generation of watermark extraction key is a learning process of neural network which is equivalent to a non-linear approximation of patterns. The elapsed time for the process of watermark extraction key is measured about a few minutes, although the generation of watermark patterns using watermark extraction key took only time of less than a second. The generation speed of key is not influential, and only the speed for generation of watermark pattern is important.

Meanwhile, our proposed watermark extraction method relies on the position of the feature sub-blocks. It is weak to geometric attacks like shrinking, expanding, and rotation of the image. This problem will be considered as future works of our proposed method.

CONCLUSION AND FUTURE WORK

The SOA demands supportive technologies and new requirements of which representative solution is intelligent information security of enterprise resources for collaboration systems and services. Digital watermarking became a key technology for protecting copyrights. In this article, we have proposed a key generation method for static visual watermarking by machine learning, neural network as an exemplary method. Key generation does not involve any embedding of data into target content which it means that it is a damage-less watermarking method. This characteristic is effective when user must not damage the content but must conceal a secret code into target content.

The proposed method uses multi-layered neural network model for classifying the input visual feature patterns to corresponding hidden watermark patterns. For input visual feature values, we used DCT coefficients on YCbCr domain. The proposed method does not limit to DCT as frequency transformation method only, one can use DFT and DWT otherwise. Also, machine learning method can be replaced with others such as Bayesian network and fuzzy.

In the experiment, we have shown that our proposed method is robust to high pass filtering and JPEG compression. Also, because our propose method does not alter the target content, it is applicable to steganography. Meanwhile, because the proposed method relies on the position of the feature sub-blocks, it is weak to geometric attacks like shrinking or rotation of the image. This must take into consideration for future work.

ACKNOWLEDGMENT

Kensuke Naoe is supported financially in part by the Mori Fund of Keio University, on this study. Hideyasu Sasaki is supported financially in part by the Grant-in-Aid for Scientific Research ("KAKENHI") of the Japanese Government: No. 21,700,281 (FY 2009-2011), on this study.

REFERENCES

Adams, C., & Lloyd, S. (1999). *Understanding public-key infrastructure: concepts, standards, and deployment considerations.* Macmillan Technical Publishing.

Ando, R., & Takefuji, Y. (2003, January). Location-driven watermark extraction using supervised learning on frequency domain. *WSEAS Transactions On Computers, 2*(1), 163–169.

Artz, D. (2001, May/June). Digital steganography: hiding data within data. *IEEE Internet Computing, 5*(3), 75–80. doi:10.1109/4236.935180

Bernardo, J., & Smith, A. (2001). Bayesian theory. *Measurement Science & Technology, 12*, 221–222.

Bishop, C. M. (1995). *Neural networks for pattern recognition.* Oxford University Press, USA.

Bishop, C. M. (2006). *Pattern recognition and machine learning (information science and statistics)* (New edition ed.). Springer-Verlag.

Broomhead, D. S., & Lowe, D. (1988). Multivariable functional interpolation and adaptive networks. *Complex Systems, 2*, 321–355.

Burges, C. (1998). A tutorial on support vector machines for pattern recognition. *Data Mining and Knowledge Discovery, 2*(2), 121–167. doi:10.1023/A:1009715923555

Cheung, S.-C., Chiu, D. K. W., & Ho, C. (2008). The use of digital watermarking for intelligence multimedia document distribution. *Journal of Theoretical and Applied Electronic Commerce Research, 3*(3), 103–118. doi:10.4067/S0718-18762008000200008

Chiu, D., & Leung, H. (2005). Towards ubiquitous tourist service coordination and integration: a multi-agent and semantic web approach. In *Proceedings of the 7th international conference on Electronic commerce* (pp. 574–581).

Cortes, C., & Vapnik, V. (1995). Support-vector networks. *Machine Learning, 20*(3), 273–297.

Cox, I., Doerr, G., & Furon, T. (2006). Watermarking is not cryptography. *Lecture Notes in Computer Science, 4283*, 1–15. doi:10.1007/11922841_1

Cox, I., Kilian, J., Leighton, T., & Shamoon, T. (1996). Secure spread spectrum watermarking for images, audio and video. *Image Processing, 1996. Proceedings., International Conference on, 3*.

Cox, I., Miller, M., Bloom, J., Fridrich, J., & Kalker, T. (2007). *Digital watermarking and steganography* (2nd ed.). Morgan Kaufmann.

Delaigle, J. F., Vleeschouwer, C. D., & Macq, B. (1998). Watermarking algorithm based on a human visual model. *Signal Processing, 66*(3), 319–335. doi:10.1016/S0165-1684(98)00013-9

Elman, J. (1990). Finding structure in time. *Cognitive Science, 14*(2), 179–211.

Eschenauer, L., & Gligor, V. (2002). A key-management scheme for distributed sensor networks. In *Proceedings of the 9th ACM conference on computer and communications security* (pp. 41–47).

Hung, P. C., Chiu, D. K., Fung, W. W., Cheung, W. K., Wong, R., & Choi, S. P. (2007). End-to-end privacy control in service outsourcing of human intensive processes: A multi-layered web service integration approach. *Information Systems Frontiers, 9*(1), 85–101. doi:10.1007/s10796-006-9019-y

Jensen, F. (1996). *Introduction to Bayesian networks*. Springer-Verlag New York, Inc. Secaucus, NJ, USA.

Johnson, J. (1994). Pulse-coupled neural nets: translation, rotation, scale, distortion, and intensity signal invariance for images. *Applied Optics, 33*, 6239–6253. doi:10.1364/AO.33.006239

Kafeza, E., Chiu, D., Cheung, S., & Kafeza, M. (2004). Alerts in mobile healthcare applications: requirements and pilot study. *IEEE Transactions on Information Technology in Biomedicine, 8*(2), 173–181. doi:10.1109/TITB.2004.828888

Kahn, D. (1996). The history of steganography. In *Proceedings of the First International Workshop on Information Hiding* (pp. 1–5). London, UK: Springer-Verlag.

Katzenbeisser, S., & Fabien, A. P. (Eds.). (2000). *Information hiding techniques for steganography and digital watermarking*. Artech House Publishers.

Klir, G., & Yuan, B. (1995). *Fuzzy sets and fuzzy logic: theory and applications*. Prentice Hall Upper Saddle River, NJ.

Klir, G. J., & Yuan, B. (Eds.). (1996). Fuzzy Sets, Fuzzy Logic, and Fuzzy Systems: Selected Papers by Lotfi A. Zadeh. World Scientific Publishing.

Kohonen, T. (1988). An introduction to neural computing. *Neural Networks*, *1*(1), 3–16. doi:10.1016/0893-6080(88)90020-2

Kwok, S., Cheung, S., Wong, K., Tsang, K., Lui, S., & Tam, K. (2003). Integration of digital rights management into the Internet Open Trading Protocol. *Decision Support Systems*, *34*(4), 413–425. doi:10.1016/S0167-9236(02)00067-2

Law, L., Menezes, A., Qu, M., Solinas, J., & Vanstone, S. (2003). An efficient protocol for authenticated key agreement. *Designs, Codes and Cryptography*, *28*(2), 119–134. doi:10.1023/A:1022595222606

Maurer, U. (1996). Modelling a public-key infrastructure. *Lecture Notes in Computer Science*, 325–350.

Naoe, K., & Takefuji, Y. (2008, September). Damageless Information Hiding using Neural Network on YCbCr Domain. *International Journal of Computer Sciences and Network Security*, *8*(9), 81–86.

Prasithsangaree, P., & Krishnamurthy, P. (2003). Analysis of energy consumption of RC4 and AES algorithms in wireless LANs. In *Proceeding of IEEE Global Telecommunications Conference* (pp. 1445-1449).

Rosenblatt, F. (1958). The perceptron: A probabilistic model for information storage and organization in the brain. *Psychological Review*, *65*(6), 386–408. doi:10.1037/h0042519

Rothe, J. (2002). Some facets of complexity theory and cryptography: A five-lecture tutorial. *ACM Computing Surveys*, *34*(4), 504–549. doi:10.1145/592642.592646

Rumelhart, D., & McClelland, J. (1986). *Parallel distributed processing: explorations in the microstructure of ognition, vol. 1: foundations*. MIT Press Cambridge, MA, USA.

Sasaki, H. (Ed.). (2007). *Intellectual property protection for multimedia information technology*. IGI Global.

Wolf, P., Steinebach, M., & Diener, K. (2007). Complementing DRM with digital watermarking: mark, search, retrieve. *Online Information Review*, *31*(1), 10–21. doi:10.1108/14684520710731001

This work was previously published in International Journal of Systems and Service-Oriented Engineering, Volume 1, Issue 1, edited by Dickson K.W. Chiu, pp. 46-61, copyright 2010 by IGI Publishing (an imprint of IGI Global).

Chapter 7
Modelling Self-Led Trust Value Management in Grid and Service Oriented Infrastructures:
A Graph Theoretic Social Network Mediated Approach

Antony Brown
University of Bedfordshire, UK

Nik Bessis
University of Bedfordshire, UK

Paul Sant
University of Bedfordshire, UK

Tim French
University of Reading and University of Bedfordshire, UK

Carsten Maple
University of Bedfordshire, UK

ABSTRACT

Current developments in grid and service oriented technologies involve fluid and dynamic, ad hoc based interactions between delegates, which in turn, serves to challenge conventional centralised structured trust and security assurance approaches. Delegates ranging from individuals to large-scale VO (Virtual Organisations) require the establishment of trust across all parties as a prerequisite for trusted and meaningful e-collaboration. In this paper, a notable obstacle, namely how such delegates (modelled as nodes) operating within complex collaborative environment spaces can best evaluate in context to optimally and dynamically select the most trustworthy ad hoc based resource/service for e-consumption. A number of aggregated service case scenarios are herein employed in order to consider the manner in which virtual consumers and provider ad hoc based communities converge. In this paper, the authors take the view that the use of graph-theoretic modelling naturally leads to a self-led trust management decision based approach in which delegates are continuously informed of relevant up-to-date trust levels. This will lead to an increased confidence level, which trustful service delegation can occur. The key notion is of a self-led trust model that is suited to an inherently low latency, decentralised trust security paradigm.

DOI: 10.4018/978-1-4666-1767-4.ch007

INTRODUCTION

The main motivation of this work is to provide a trust-based model that brings together ideas from previous research, and consolidates these in a graph-theoretic framework. The model proposed is simple and computationaly lightweight meaning that it has practical applications. The model describes a self-led trust approach, giving a level of autonomy to trust calculations, not simply being dependant on previous scores from a trust chain/neighbourhood, leading to more accurate trust levels. The trust scores should also be mediated, not just dependant on one trust score but allowing some level of control by enabling individuals to weight the available information in favour of more trusted sources and to better reflect the individuals own trust strategies. The extant trust related research literature is vast and highly diverse. Trust has been studied from each and every angle: in the philosophical, sociological, psychological, computer scientific, economic, and legal sense – just to name a few (Karvonen, 2007). One perennial barrier to synthesising a definitive e-service trust model and theory of trust from this work is the lack of agreement as to definitions of trust (Grabner-Krauter et al., 2005; Grabner-Krauter & Kaluschia, 2003). Thus, inadequate trust conceptualisations and a lack of a unifying viewpoint or paradigm have frequently lead to weak theoretical rationale for empirical studies and a consequent inability to develop coherent and efficient theories. Indeed, one of the central difficulties is that the notion of trust is closely related to other concepts such as *reliance, competence, trustworthiness* and *credibility*. Nevertheless, within the vast extant trust literature it is possible to identify a core body of classic work. Further, we have chosen to focus on the notion that trust as a definite measurable confidence level, mediated by inherent risk. Classic trust studies have firmly established the notion that trust and distrust are threshold points on a continuum of probability assessment (Gambetta, 1988). Thus, we trust an entity to perform a particular task if and only if the likelihood that the entity will fulfil its obligation lies above a particular threshold value. This notion is central to the consumer and provider e-service model that is presented later in this paper. A concise review of the relevant literature now follows so as to partially contextualise the model.

Classic Trust Models and Definitions

Deutsch (1958) was one of the first modern writers to seek to build a formal model of trust. Deutsch defined trust in terms of an individual confronted with an ambiguous path. Further, the path may lead to either an event leading to a beneficial outcome (Va+) to that individual or to an event perceived as being harmful (Va-). This individual perceives that the occurrence of Va+ or Va- is dependent on the behaviour of another human agent. Finally, the strength of Va- is greater than the strength of Va+. Essentially, his view of trust is of a trust relationship in which events are linked to other events, each of which has beneficial or non-beneficial paths. For a trust relationship to occur, the harmful path is more significant than the beneficial path. Deutsch goes on to explain that risk is an essential property of the environment within which a choice of paths occurs. The notion that trust building between individuals takes place within information spaces that are both potentially risky to the participants and where incomplete information is available to the human actors has been widely accepted and developed by many subsequent researchers (Corritore, Krasher, & Wiedenbeck, 2003; Mayer, Davis, & Schoorman, 1995). Within this information space the notion of expectation is central to many writers. For example, Gambetta (2000) provides us with a rich and potentially computationally useful definition that encapsulates the notion of trust as expectation:

Trust (or symmetrically, distrust) is a particular level of subjective probability with which an agent assesses that another agent or group of agents

Table 1. Trust studies: a summary of classic approaches

Typology of Trust Studies	Main focus	Selected Citation(s)
Prisoner's Dilemma (social) based games	Classic "prisoner's dilemma" trust game based studies, designed to probe trust building between individuals engaged in various "closed world" social situations. Pay-off and rewards are integral aspects.	Tucker (1950); Flood (1952).
Artefact field studies	Customer and user based studies within controlled settings. Artefacts may be real or artificial.	Kim and Moon (1998); Jarvenpaa et al. (2000); Egger (2003).
Social Capital studies	Studies of trust within large scale organisations, societies and social hierarchies.	Putnam (2000); Hardin (2002).
Trust propagation studies across social networks	Trust propagation across social networks of various kinds (virtual and real).	Fukuyama (1995); Golbek (2005); Ziegler (2005).
Typologies of trust	Literature survey and experiment.	McNnight et al. (2001, 2002).

will perform a particular action, both before he can monitor such action (or independently of his capacity ever to be able to monitor it) and in a context in which it affects his own action. (Gambetta, 2000, p. 217)

This idea of trust as an expectation (either *rational* or *affective* or a mixture of both) is a closely related concept to that of confidence levels. Confidence can be defined as a conscious or unconscious act or mental state involving placing confidence in something or someone. The idea that this confidence level can be formalised, hence measured is developed by Marsh (1994) in his much cited PhD thesis:

Trust is strongly linked to some thing, being it the person to be trusted, the environment, or whatever it is that the desirable outcome is contingent upon. We arrive at the concept of trust by choosing to put ourselves in another's hands, in that the behaviour of the other determines what we get out of the situation.(Marsh, 1994, p. 4.)

Many have stressed the importance of trust building over time (i.e. the temporal dimension of trust). Trust is not only clearly dependant on our past experiences but is also an expectation of reliability and confidence in future events too. These aspects have been incorporated into various formal and informal models of trust building (Dayal et al., 1999; Nikander & Karvonen, 2000). An important aspect of trust building is the degree to which *affective vs. rational* components are involved. It is clear from the literature that the affective component has been relatively under researched in comparison to the cognitive dimensions (Castelfranchi & Falcone, 2000) within the diverse collection of studies shown below in Table 1.

E-TRUST AS REPUTATION AND CONFIDENCE?

Reputation: Pioneering work has already been carried out by researchers in developing and applying various mathematical formalisms that can be used to design and implement trust models within autonomic systems. Many of these formalisms (Marsh, 1994) rely on the calculation of local trust thresholds of various kinds and their subsequent propagation across nodes via graph-theoretic models (Bistarelli & Santini, 2008). This approach has been extended so as to seek to enable MAS (Multi-agent Systems) with the power to investigate trust credentials, provenance and reputation.

A review of the relevant literature and MAS related theoretic reputation and several novel trust models, including agent based support can be found on the EU 6th Framework Project "*eRep*" site (http://megatron.iiia.csic.es/eRep/?q=node/37). The "*eRep*" trust model incorporates various dimensions including experience and hearsay evidence from trusted third parties (Sabater-Mir et al., 2007). This EU funded project has significantly highlighted the importance role of trust as reputation within the context of e-services of all kinds. However, the various models and technologies remain tentative and emergent.

Whereas there is much work on Reputation scores applied to "real" organisations, such as (Chiu et al., 2009) few have even tried to re-factor these for dynamic VO service contexts since, although reputation is seen as being highly relevant to the calculation of VO trust, the standard metrics are either too heavyweight and/or are not available dynamically and are in any case lagging not leading indicators. In addition, even if it could be done latency would be potentially high

Perhaps the best way forward is to seek to create a 'lightweight' trust index comprising a triangulation of a subset of corporate reputation metrics derived from well known and established sets of organisational performance metrics (Richard et al., 2009). This triangulation (correlation and weighting) remains speculative. However, if a small enough subset of metrics can be identified to be viable within high performance computing contexts, the resultant synthetic reputation index value could be used internationally. This would add value to the suggested use of Corporate Governance Scores.

One suggested mix (triangulation) derived from Richard's survey of some 722 academic management journals published in the years 2005-7 comprises: objective measures (accounting metrics); financial market measures (e.g., present value of future cash flows); mixed accounting/financial market measures (e.g., so called "Z" scores); subjective measures (e.g., Fortune 500 rankings etc.) and quasi objective measures (e.g., self reported sales growth figures). Richard reports that further basic research is needed to validate the proposed triangulation of these reputation measures.

Confidence: Confidence levels can be calculated in various ways. Within the specific context of the consumption and provision of Grid services from so-called Virtual Organisations (VO's) previously published work has shown the value of computationally heavyweight confidence engines as exemplified in the previous extensive work of the *TrustCom* community (Wilson, Arenas, & Schubert, 2007). Similarly, there have been recent calls for confidence levels to incorporate high-level proxy measures of VO reputation as well as using purely rational measures derived from previous performance history (Song, Hwang, & Kwok, 2003).

We seek to offer a distinctly computationally lightweight approach that is specifically designed to minimise run-time system latency, particularly within intensive high-performance Grid applications of the type exemplified by Monte-Carlo simulations used in the financial services industry "Super Quant Challenge, 2008" ("Super Quant Monte Carlo Challenge", 2008). Furthermore, we seek to incorporate and combine both *rational* confidence measures with *subjective* confidence measures so as to seek to more closely mimic human forms of trust formation. Indeed, the following working trust definition can be seen as the prime motivation for our confidence model: "*Trust of a party A to a party B for a service X is the measurable belief of A in that B behaves dependably for a specified period within a specified context in relation to service X*" (Olmedilla et al., 2006, p. 5). Here measurable "belief" means the derivation of a compound trust metric that combines and reflects both top-level VO trust as reputation with any known previous direct service history from an e-service provider node using a suitable set of weightings $\sum(x_{1*0.4}, x_{2*0.1} \ldots x_{n*y})$ so as to derive a synthetic trust metric scaled within the range 0 (distrust) to 1 (absolute trust). Clearly an agent based approach is implicit to our work. Specifically, the proposed MAS is tasked with seeking to check the trust credentials of a VO e-service provider just prior to invoking a request for a particular service. Within the agent community computational trust models are sufficiently advanced to give us confidence that an enabling technology is well within reach (Cahill et al., 2006).

In this paper, our approach seeks to support both the design of VO partnerships via graph theoretic models using a relatively simple and computationally lightweight service, designed to quantify "confidence level" using a set of simple algorithms and metrics. We achieve this by extending lightweight trust services work from Conrad et al. (2006); peer-to-peer identity and trust management work from Ion et al. (2007); and trust and contract management for dynamic VOs from Wilson et al. (2007). Specifically, our approach describes model's functionality using a number of aggregative service case scenarios. With this in mind, the contributions of this paper are to: i) familiarise and present readers with our methodological approach; and discuss consumer and service providers requirements via a number of aggregated service case scenarios; ii) present our model; and finally, iii) discuss the functionality of the produced model using the aggregated service case scenarios.

TRUST: A GRAPH THEORETIC SOCIAL NETWORK MEDIATED APPROACH

Our model development adopts a graph theoretic descriptive approach leading to an algorithm so that it can be used to help determine whether a particular consumer (of services) will use the services offered by a service provider. Graph theory offers a convenient, concise and expressive notation and is by now a well established formalism with respect to both trust propagation across social-networks (Golbek, 2005), and in tangible network security (Stang et al., 2003). With respect to Golbek's extensively cited work, it is clear that trust values can be systematically calculated (composed) across paths of nodes of individuals so as to derive personalised (local) trust values that are partially transitive. In our model trust values are composed in respect of particular e-service providers by e-service consumers in relation to a set of rational and subjective trust metrics that are personalised to each consumer. Thus, our model shares not only a common graph theoretic notation but also seeks to incorporate several principles derived from the social-network trust community. Let us now begin by providing the necessary background.

Self-led Trust Value Management

Our model is based upon the notion of self-led Trust Value Management (TVM). The main concept involved is that a community of users (encompassing service consumers and providers) can communicate within their own network, and manage their own perceptions of fellow users. In such a system, a user, for example, Alice, can decide (based upon limited local knowledge) how much trust to place in another user, Bob. The idea is that the system is dynamic, and based upon a series of interactions between users (represented by nodes in our graph theoretic model), the trust value between pairs of users can evolve over time. Such a notion provides a richness not seen in many other models.

The idea of self-led trusts mirrors the notion of trust relationships in the real world. If a person does not know someone, they may ask their friends about this person. Based upon the feedback they receive a judgement (personal, and self-led) is then made.

We purposely moved away from a centralised trust authority as this could lead to a single point of failure. To explain further, a centralised trust authority could be compromised by an external entity, and if all users are dependent upon this entity then the trustworthiness of the whole network would fail very quickly.

By using this idea of self-led trust values, we can ensure a degree of fault tolerance, and

Figure 1. A graph representation of consumers and service providers

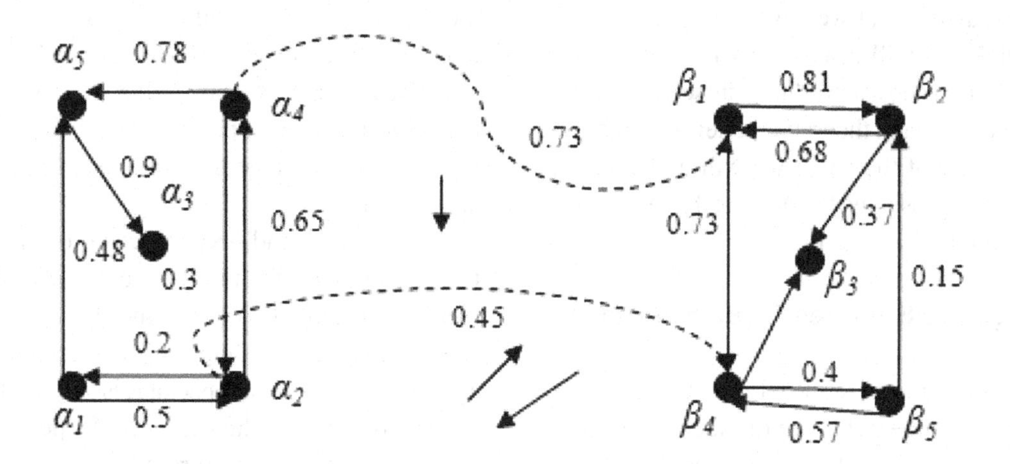

moreover, if a node within a network becomes compromised the potential impact is small (unless the network is particularly sparse). We are now in a position to introduce the necessary graph theoretic preliminaries used throughout the remainder of the paper.

Graph Theoretic Preliminaries

In describing our model we use standard graph theoretic notation (see, for example, Gibbons (1985)). A graph, $G = (V,E)$ consists of a set of *vertices* $V = \{v_1, v_2, \dots v_n\}$ and a set of edges, $E = \{(v_1,v_2), (v_i,v_j), \dots, (v_{n-1},v_n)\}$ representing connections between pairs of vertices v_i and v_j. In our case, we will be making reference to a *weighted* graph $G=(V,E,W)$ where V and E are as defined earlier, and W is a mapping of a *weight*, w_i to each edge, $e_i \in E$. This construct then allows us to define a value e_i in the range [0..1] that can be used to model the notion of *trust* within our graph theoretic framework.

Figure 1 provides an illustration of the graph G of the form that we will use in this paper. In this graph, the vertex set can be divided into three distinct subsets. The vertices α_1 to α_5 represent the consumers of services (such as, for example, Maria, insurance, banking etc.). Similarly, the set of vertices β_1 to β_5 represent service providers (such as banks and insurance companies).

At this point we would like to point out that we supplement our graph theoretic model with a number of additional (non-graph theoretic) concepts. The use of these additional concepts adds to the richness of model. We illustrate the usefulness of our model by referring to the scenarios described next.

As was illustrated by Golbeck and Hendler (2006), trust cannot simply be defined using numerical terms, and moreover, trust alone is not (at least in human terms) used by humans when deciding how trustworthy an individual is. The work of Goldbeck and Hendler (2006) introduced the notion of a confidence value which enhanced the original TidalTrust algorithm to produce a probabilistic algorithm called SUNNY (Kuter & Golbeck, 2007). With this in mind, we also incorporate a concept similar to confidence and this allows a consumer (or provider) to determine whether they would like to make use of the services provided (this can be from a consumer to service provider, consumer to consumer, or even a service provider to service provider perspective).

It is also important to note that trust relationships are (naturally) asymmetric. That is, two people, Alice and Bob, do not necessarily have the same level of trust in one another. This is an important aspect that we have incorporated into our model. We will discuss this further in later sections. Let us now introduce our scenarios and soon after our graph theoretic model. Throughout the remainder of the paper we will make use of the scenarios presented to illustrate the real-world application of our model.

Aggregated Services Case Scenarios

We are interested here in delegates who act as individual members (human or machine) of networked communities that are formed through ad hoc aggregation of services offered by various service providers. Five Aggregated Service (AS) case scenarios have been defined and analysed using graph theory.

AS1: The first AS case scenario involves Maria who has £15,000 in each of her three bank savings accounts. Maria wishes to invest her money in one or more companies or banks. She could go to her individual bank investors or to Marco who is an independent financial advisor. However, Maria realises that this is a rather limited service. She wishes to have access to a global system enabling her to valuate from a wider set of options. Specifically, Maria seeks to access past investment scenarios from others who have already invested their money in various companies or banks. She wants to see how others made such a decision and whether they were satisfied from their services. That is to say, Maria is looking for a dynamic service illustrating what benefits others had by joining with a particular company. In addition to this, Maria is also looking to find out what investment package is currently available and is relevant to her case. That is to say, Maria is looking for a dynamic service, which allows her to forecast what could be the benefits if she invests in their company now. This is an example of a consumer benefiting from the past experiences of other consumers they know to determine the trustworthiness of a variety of service providers, enabling them to choose between them. It benefits the user to be able to take the past experiences of other consumers they know and mediate them with their own past knowledge of how reliable those consumers have proved previously, meaning that they do not have to take others trust scores at 'face value'.

AS2: The second AS case scenario involves Y Ltd, a company like many others that wishes to identify and attract potential investors like Maria. Y Ltd realises that having a nice investment package is not enough. A nice package is only applicable as far it reaches all the right people, people who want to enter into this particular package. They also realise that making five investment packages available may be considered as compromised options in the eyes of the potential investors. Y would prefer if they were able to offer personalised investments packages as flexible as required by the investor. Their slogan is that every investor has different needs and therefore Y must offer them flexible and personalised options to choose from. They could go to Marco but also wish to have access to a global system enabling them to forecast who will be the best investors. This is due to the fact that they could spot Maria's willingness to invest and offer her a better deal/package than others. We need to understand that there are certain reasons that make someone like Maria join Y rather than Z, e.g., Y may offer a more stable return, part of the reputation. This scenario focuses on a service provider using pre-existing trust scores to identify potential consumers of their services and is an example of how providers also benefit from utilising the trust scores consumers generate about each other.

AS3: The third AS case scenario involves Y ltd from a different perspective. Y may not want Carlos investing his £10,000 because his record shows evidence that he enters the agreement and he leaves at a stage which may conflict with Y's interests. That is to say, Y wishes having access to past investment scenarios as a means to assess investors' profiles that are made available from companies who have already had Carlos as an investor in the past. Similarly, Y aims to use investors' funds (like Maria's) as to further invest them in other opportunities (other banks, trusted organizations, etc...) so they can generate a good enough profit for them and their investors (like Maria). Ideally, they need another (higher level) dynamic service, which will enable them to simulate where to invest. Y has a number of in-house financial advisors as analysts but they have also delegations with independent advisors like Marco. This scenario describes the situation where a service provider accesses the experiences of other services providers, to determine the trustworthiness of potential consumers. This example highlights the benefit for a provider to utilise the trust scores from other providers to identify untrustworthy consumers and so to help avoid risk.

AS4: The fourth AS case scenario involves Global Ltd, which is a Service Oriented Infrastructure (SOI) company enabling members of their multiple community networks (like Maria, Marco, Carlos and investment companies) to delegate (i.e., access and invest) via their available case scenarios and create their own personalised dynamic services on demand. They must also allow these companies making their investment offers available through their SOI. As these offers may change at any time, these must reside on each company's servers. Most importantly, Global also requires access to distributed computational power and data mining tools to model their complex calculations so they can render these ad hoc based dynamic services for the benefit of their clients. Thus, they need to join in partnerships with other companies offering them the computational power and the data mining tools required to model these ad hoc dynamic services. That is to say, Global requires access to services, which would allow them to evaluate and ultimately select and entrust automatically ad hoc based resources entering and leaving their domain of interest. An example of this type of scenario is grid computing and the need for automated entrustment within a constantly changing network. Distributed self-led trust values that can be computed quickly have an advantage in that they can reflect local changes in the network and reflect the differing risk strategies of individual nodes.

AS5: The fifth AS case scenario involves all delegates of this ad hoc based community network. For example, Maria needs to evaluate and trust Global (instead of other similar Global SOI companies) and its partner organisations prior to requesting a dynamic service (so what makes Y, Global and its partner organisations more trustful than other similar companies?). Y also needs to evaluate and trust Global (instead of other similar Global SOI companies) and its partner organisations prior to requesting a dynamic service or leverage of distributed computational power and mining tools (so what makes Global and its partner organisations more trustable than other similar companies?). Similarly, Anthony is one of the past investors who has shared his past investment scenarios but he needs to evaluate and trust Global for sharing his experiences prior to committing to sharing it (so what makes Y, Global and its partner organisations more trustable than other similar companies?). We must also appreciate that there may exist other companies like Global ltd offering similar services (which provides a greater horizon of what is available). Finally, we must take into account companies, which will need to evaluate and ultimately select and entrust automatically all companies joining the Global ltd partnerships (so what makes each individual partner organisation more trustable than other similar companies?).

In this paper, we are particularly interested with service agreements caused by ad-hoc based delegation services, which allow delegates rendering valuation and trust level service agreements in real time about the possible like-hood in joining an agreement (partnership or else) using distributed technologies including the Grid and Service Oriented Infrastructures.

SELF-LED TRUST VALUE MANAGEMENT GRAPH THEORETIC MODEL

Our graph theoretic model is based upon the concept of *Self-led Trust Value Management*. In our model both the consumers and the service providers are part of a social network. That is, the set of consumers form a social network that allows them to identify how trustworthy a particular (neighbouring) consumer is. This information is stored *locally* within a nodes *memory*. This can be thought of (*conceptually*) as a *virtualised* table of values that defines how trustworthy each neighbouring node is. Let us now describe our model, before going on to explain how our model works using the AS case scenarios presented earlier.

We represent the AS case scenarios presented earlier using graph theoretic notions. The problem can be represented by a graph $G = (V, E)$ where the vertex set V can be partitioned into two vertex-disjoint sets, V_1, and V_2 where $V_1 = \{ \alpha_1, \alpha_2, \ldots \alpha_n \}$ represents the set of *consumers* (of services) and $V_2 = \{\beta_1, \beta_2, \ldots, \beta_m\}$ represents the set of *service providers* (e.g., banks, insurance companies etc). The edge set, E is partitioned into three subsets E_1, E_2, and E_3. The set of edges E_1 represents relationships between consumers. Each edge $e \in E_1$ is labelled by a weight w_i representing the *trust* that two consumers share with one another. The component represented by this subgraph can be seen as a social network that evolves over time as more and more interaction takes place.

In a similar vain, the edge set E_2 represents relationships between service providers. These edges represent the level of trust between two service providers (based on previous encounters). In real-life terms this is akin to trading (of services) between different organisations (inter) or between departments within an organisation (intra). Again, this component can be seen as a social network that evolves as transactions take place. The third edge set, E_3 represents links between consumers and service providers. These links are generated when a consumer forms an opinion about a particular service providers service(s), through using those services. These links will also evolve over time as more and more interaction takes place.

We begin setting up our system by creating two distinct networks (which can be viewed as *social networks*). The first network represents the consumers, and contains links between pairs of vertices in V_1. This allows us to assign trust values between pairs of consumers. Each vertex maintains a finite set of information about each of its neighbours in this subgraph. This represents previous interactions and allows us to define an accurate trust value.

The second network represents the service providers, and in a similar manner to the consumer network, each vertex maintains a finite set of information about each of its neighbours. Figure 2 illustrates the components of our model.

We define the neighbourhood of a consumer node α_i, denoted $N(\alpha_i)$, as

$$N(\alpha_i) = \{\alpha_j : (\alpha_i, \alpha_j) \in E)\} \quad (1)$$

Similarly, we define the neighbourhood of a service provider node β_i, denoted N(βi) as

$$N(\beta_i) = \{\beta_j : (\beta_i, \beta_j) \in E \} \quad (2)$$

We define the customer relationships to service providers of a consumer as:

$$C(\alpha_i) = \{ \beta_j : (\alpha_i, \beta_j) \in E)\} \quad (3)$$

Similarly, the customers of a service provider node β_i, denoted C(βi) as

$$C(\beta_i) = \{ \alpha_j : (\beta_i, \alpha_j) \in E \} \quad (4)$$

$$N(\alpha_i, \alpha_k) = N(\alpha_i) \cap N(\alpha_k) \quad (5)$$

Figure 2. The graph theoretic model

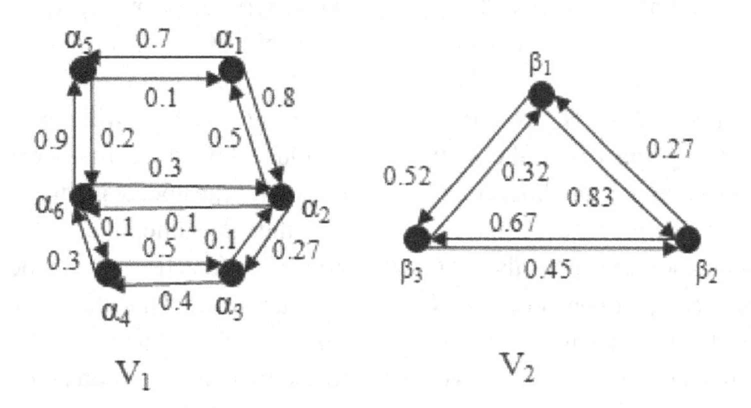

Therefore we can define the set of neighbours of consumer α_i who have knowledge of service provider β_i, denoted as $N(\alpha_i, \beta_i)$ as

$$N(\alpha_i, \beta_i) = N(\alpha_i) \cap C(\beta_i) \tag{6}$$

That is, all of the consumers liked to α_i who also have links to provider β_i

As our system evolves interaction between consumers and service providers (as, for example, in AS1 case scenario where Maria has £15,000 worth of savings and would like to find the best possible savings deal from a set of banks (service providers) that will give her the maximum Return On Investment). A natural starting point for Maria is to query her friends to see what their experience has been. Maria, node α_1, would obtain a list of all of the providers who provide the service she requires, X_n and the process of querying neighbours would then begin.

Each neighbour, α_i, of Maria (α_1) would be queried regarding a service provider, β_j's, Quality of Service when providing service X_n. Each neighbour, in turn, would then return a numerical rating of the ability of service provider β_j to provide service X_n. This allows Maria to gauge which β_j provides a better Quality of Service. Of course, when making a decision there are a number of additional factors to take into consideration.

DISCUSSION

Let us now explain this further using the aggregated service (AS) case scenarios presented earlier.

Consumer to Consumer Trust

This type of trust is relevant to our AS1 case scenario. As mentioned earlier, our graph, G is split into two distinct subsets, V_1 and V_2. These subsets represent our consumers and service providers, respectively. We start discussing our model by looking at the interactions between these subsets of users. More specifically, we begin by looking at calculating a level of trust between parties (consumers or service providers) within the same subset.

We will begin by looking at trust levels between consumers before moving on to look at trust levels between service providers. Once we have discussed interactions between pairs of consumers (alternatively, service providers) we will move on to interactions between consumers and service providers.

It is natural, when meeting someone that you have not had previous interaction with, to feel hesitant when seeking their opinion on a subject. The situation is slightly better when you have a network of friends with whom you interact – when asking someone's opinion you may factor in the judgement of your friends when deciding how much value to place in what has been said. In a similar manner, we incorporate such values into our graph theoretic model.

In order for our model to work, we require that each vertex in the graph can make some judgement about nodes in its neighbourhood. Such judgements are asymmetric (what one consumer thinks about another is not (always) a mirror image of what the other consumer thinks about them), and are likely to change over time as the number of interactions between the two increases. Each node will have an initial trust value, $TS_{default}(\alpha_i)$, which they will apply to any other node on their first encounter. This $TS_{default}(\alpha_i)$ value can vary from node to node, representing the optimism of each consumer. We can also gain indirect information by asking neighbours about another consumer. Obviously we need to balance out this information and make a final judgement. In our model, we define the Trust Score, $TS(\alpha_i,\alpha_j)$ as the weight of the edge between nodes α_i and α_j. When a consumer first forms a relationship with another consumer, the initial Trust Score is defined as follows:

$$TS_{initial}(\alpha_i,\alpha_j)= (TS_{default}(\alpha_i)+\Sigma_{\alpha k \in N(\alpha i, \alpha j) \setminus \alpha j}(TS(\alpha_i,\alpha_k)\, TS(\alpha_k,\alpha_j)))/(|N(\alpha_i, \alpha_j)|+1) \quad (7)$$

Here we ask each neighbour who knows them how much they trust α_j. Each of these trust scores must be mediated by how much we trust the node providing the trust score, so we take the product of our trust in their trust. In terms of the graph, we are multiplying the weights of the edges along the Path between α_i and α_j. We take the sum of these trust scores of all neighbours of a node α_i who currently have knowledge of α_j (including the default trust, $TS_{default}(\alpha_i)$), and then average these over the number of relevant neighbours (plus 1) to arrive at an overall (average) trust score. This seems natural and allows us to take account of all relevant judgements when arriving at a decision, mediating them into one usable trust score. This initial weight is then assigned to the new edge from $\alpha_i \rightarrow \alpha_j$. Similarly, α_j will determine its own trust of α_i in the same manner, setting the weight from $\alpha_j \rightarrow \alpha_i$. Note that in the event that α_i has no links to any consumers who know α_j then $TS_{initial}$ will equal $TS_{default}$.

Once the initial Trust score is established, the consumer can update it after each interaction with the other consumer, increasing it in the event of a positive interaction and reducing in the event of a negative one. This allows the *TS* to evolve over time, and these changes will propagate through the network as other consumers seek guidance on how much to trust.

Of course, consumers are not the only entities in our model – we also have service providers. In the next section we will discuss a similar mechanism for inferring trust between service providers.

Trust between Service Providers

This type of trust is relevant to our AS3 and AS4 case scenarios. Trust does not only occur between consumers, but also between service providers. A service provider β_i may provide services to other service providers. As an example, credit card companies may provide services for banks, but the credit card company can be seen as a separate service provider. As such, a bank may be able to choose between several credit card companies, and so there is a need for evaluation of trust between service providers as well as between consumers. In a process similar to that of providers, a service provider β_i will calculate a trust value, $TS(\beta_i,\beta_j)$ for each node β_j in its neighbourhood, by using the following equation:

$$TS_{initial}(\beta_i, \beta_j)= (TS_{default}(\beta_i)+\Sigma_{\beta k \in N(\beta i, \beta j) \setminus \beta j}(TS(\beta_i, \beta_k)TS(\beta_k, \beta_j)))/(|N(\beta_i, \beta_j)|+1)$$

Which means that when determining a trust value for a neighbour, β_j, β_i will be mediated by taking into account the trust values of all its neighbours (with respect to their trust of β_j) when calculating its own trust value for β_j.

Now that we have discussed calculating trust scores between nodes within the same subset, we are now ready to discuss the calculation of trust between nodes in different subsets, specifically, between consumers and service providers.

Consumer and Service Provider Interactions: Calculating Trust

This type of trust is relevant mainly to the interactions described in our AS2, AS4 and AS5 case scenarios but also applicable to AS1 and AS3 case scenarios. In our model, when a consumer node (in this case, Maria) is making a decision on which service provider to choose, it needs to combine knowledge about each of its neighbours in order to make a decision. Therefore, each node α_i, maintains (and dynamically updates) a so-called trust score for each of its neighbours, α_j. This represents, on the basis of previous interactions, whether α_i's evaluation of service provider β_j is trustworthy. Based upon this combination of trust scores, α_i can make a decision about which service provider to choose.

In terms of our model, we represent these trust scores using a series of equations. Firstly, each time a node, α_i, uses a service provider, β_j, it calculates a value *QoS* representing the Quality of Service provided (in other words, it is the trust in that provider to provide that service reliably). This information is then stored in a table so that when queries are received regarding β_j, α_i can provide the necessary information. Each provider can provide a variety of different services, which can be categorised into different types. Each of these can be labeled X_n and trust values can be stored and calculated for each service, $QoS_{xn}(a_i, \beta_j)$ by assigning values in the range [0..1], representing poor quality of service (0% satisfaction) to very high quality of service (100% satisfaction). Each time α_i interacts with β_j for service X_n it updates $QoS_{xn}(a_i, \beta_j)$ *to* reflect the most recent Quality of Service score.

In order for a node, α_i, to make a fully informed decision about a particular service provider β_j, we need to combine the QoS scores of each neighbour α_j, and to define a trust score, *TS*, for each of α_i's neighbours. As defined earlier, we represent the trust score, of a neighbour α_j (of another consumer, α_i) by $TS(\alpha_i,\alpha_j)$.

In order to obtain an overall QoS score for a provider of β_j we mediate the scores available by taking the average of the scores over all neighbours, α_j, of node α_i. Equation 8 defines the calculation of a QoS score for a service provider β_j:

$$QoS_{xn}(\alpha_i,\beta_j) = \left(\Sigma_{\alpha j \in N(\alpha i, Bi)} TS(\alpha_i,\alpha_j) QoS_{xn}(a_j, \beta_j)\right) / |N(\alpha_i,\beta_j)| \quad (8)$$

We calculate $QoS_{xn}(\alpha_i,\beta_j)$ for each node that is providing the service, X_n that we require, and store this at the node. Once all calculations are completed then we can create a table of scores for a service.

If we think of the equations above in real terms, then Maria is likely to ask her friends (Nik, Kostas and Pavlos) what they think of the banks ATE, National Bank of Greece and Citibank. Each of her friends may have had different experiences, so it is important to be able to combine the views of the friends together when making an overall decision about which bank to choose for her savings. Of course, it is important to be able to compare different banks, and it is likely that she will choose the bank that has, overall, provided the best QoS.

Returning to our model, once the calculations for each service provider β_j have been made, node α_i can then make an informed decision about which service provider to choose. Of course, once the service provider has been chosen then α_i will update its own perception of the service provider chosen, and in this way the system can continue evolving. This dynamic nature allows us to incorporate a temporal aspect into our model. After each interaction between a consumer, α_i, and a service provider, β_j, we recalculate the QoS values, and naturally, this requires us to update $QoS_{xn}(\alpha_i,\beta_j)$for the service provider concerned.

This dynamic component allows us to re-evaluate service providers. Such a process is natural in real-life. For example, initially a consumer could experience a poor level of service from a service provider at the beginning of their relationship, but over time the Quality of Service could increase. In the following section we will discuss the initial trust value calculations, and also an indicative model for dynamically recalculating trust as our system evolves.

Dynamic Calculation of Trust

One of the most important aspects of our model is how it calculates trust and more specifically, how it recalculates trust as the system evolves. This is important because it allows our model to evolve trust relationships in a manner similar to the real world.

Let us begin our discussion with an example. Imagine that we have two people, Maria and Kostas. In the beginning, Maria and Kostas do not know each other, but they meet at a mutual friend's party. Initially, Maria and Kostas may have no trust, or very little trust in one another (based, for example, upon the intersection of friends that they share). In our model, we would represent this as a small value (say, between 0 and 0.2). Of course, as their relationship develops, the level of trust that they have in one another will change. If, for example, Maria helps Kostas to sort out a problem with his bank (or vice versa) then the level of trust that Kostas has in Maria will increase. Conversely, if Maria lets Kostas down on a number of occasions, he may begin to distrust her, and the relationship between them may dissolve. In our model, we mimic this fluctuation of trust values by recalculating trust levels each time an interaction takes place. This ensures that we have the most up-to-date vision of the world, and allows accurate trust relationships (based on real information) to evolve and adapt.

Of course, deciding exactly how to alter trust values is an important question. In addition, we also need to think about the scenario when two (previously close) friends grow apart, and then reconvene some time later – at this point, what level of trust will they have in one another?

In our model we look at trust not as a static value at each point in time, but as a function. Trust naturally contains a temporal dimension, and also has a number of variables to take into consideration (such as a time lapse, malicious events occurring etc.). Figure 3 shows how trust can evolve over time (including the degradation of trust if there is a lack of interaction).

Figure 3 illustrates a number of important elements. Firstly, we begin with a level of trust that is low. In real terms, this represents the initial meeting of two people who may be meeting each other for the first time. In this case, the level of trust is low (this is commensurate with two people who may have a mutual friend in common) as they have little or no previous knowledge of each other. In some situations (where two strangers meet) the initial level of trust may be zero, but in general it will be 0+*e*, where *e* is a small (negligible) value representing some distant relationship. As time passes there are a series of positive interactions between

Figure 3. A view of trust over time

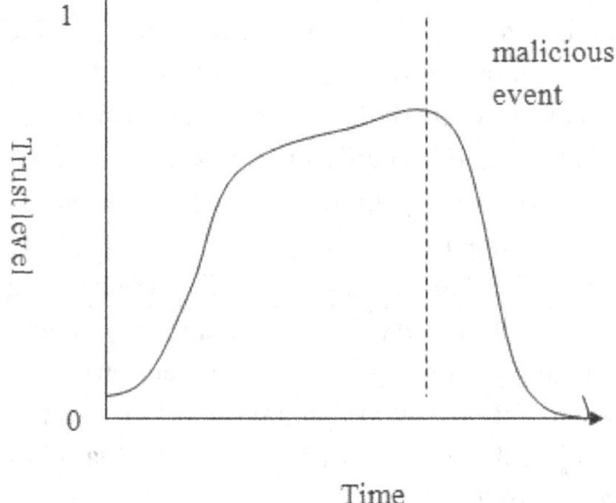

the parties, and as a result the trust value between the two (Figure 3 only represents one half of the relationship) will increase. As long as the interactions are positive the trust relationship will become stronger (indicated in Figure 3 by the upward slope of the curve) until a plateau is reached (either a maximum level of trust has been achieved, or there is a break in interaction between the two parties). Of course, at the point of reaching a maximum trust level there are two options – the plateau (if interactions continue after some small break), or a gentle downward slope (representing degradation of trust due to lack of interaction). These represent the normal case of trust relationships. There is one further factor to consider – malicious events. In this case, the level of trust decays rapidly (indicated by the sharp downward slope of the curve at the dotted line). This may result from a severe act being performed. It is important to note here that although the malicious act results in a sharp downward trend, it is not impossible (going beyond the rightmost boundary of Figure 3) to begin to climb again, although it is likely to be at a slower rate. By modelling our trust relationship as a function we allow our system to evolve over time, and this adds to its richness.

Timeliness of Trust

When asking a friend how much to trust a service provider, we would prefer that they had had recent experience of that provider. Knowing the level of service 6 months ago is obviously of less importance than knowing the level of service yesterday and so, trust gained recently should be given more weight.

We can make a modified version of our trust model that allows trust to decay over the time between interactions. We define $S(\alpha_i,\alpha_j)$ as the time since the last interaction between two consumers (similarly, $S(\alpha_i,\beta_j)$ is the time between interactions between a consumer and service provider). Each node can keep track of the last time it interacted with the other node. We can also define S_i as the half-life time of α_i, such that it is the time in which α_i's trust will halve, without any further interactions.

Thus we can define the Timely Trust Score, $TTS(\alpha_i, \alpha_j)$ as:

$$TTS(\alpha_i,\alpha_j) = \mathrm{TS}(\alpha_i,\alpha_j)/(2^{(S(\alpha_i,\alpha_j)/S_i)}) \quad (9)$$

Similar functions, $TTS(\beta_i,\beta_j)$ and $TQoS_{xn}(\alpha_i,\beta_j)$, can be defined for *TTS* between two providers and a consumer and provider, respectively.

$$TTS(\beta_i,\beta_j) = TS(\beta_i,\beta_j)/(2^{(S(\beta_i,\beta_j)/S_i)}) \quad (10)$$

$$TQoS_{xn}(\alpha_i,\beta_j) = QoS_{xn}(\alpha_i,\beta_j)/(2^{(S(\alpha_i,\beta_j)/S_i)}) \quad (11)$$

An example: Maria, α_1, that her trust will halve over a period of 3 months (S_1 = 90 days). She hasn't interacted with Marco, α_2, for 15 days and her trust score was 0.8, so her timely trust is $TT(\alpha_1,\alpha_2) = 0.8/2^{(15/90)} = 0.713$. If she does not interact with Marco for a further 75 days, her timely trust will drop to $0.8/2^{(90/90)} = 0.4$

However, each node will want to evaluate the trust of other nodes based upon their own threshold value, rather than that of the node supplying the *TS*. Therefore it is useful to define the Timely Trust Score, $TTS(\alpha_i, \alpha_j, \alpha_k)$ as:

$$TTS(\alpha_i, \alpha_j, \alpha_k) = TS(\alpha_j,\alpha_k)/(2^{(S(\alpha_j,\alpha_k)/S_i)}) \quad (12)$$

Here, validity of the timeliness of the trust between α_j and α_k is valuated based on the criteria of α_i

An example: Maria, α_1, trust Marco, α_2 with a value of 0.8 and last interacted with him 15 days ago. John, α_3, asks her how much he should trust Marco but John loses trust more easily than Maria and has a trust half-life of 30 days (S_3 = 30 days). Therefore, from Johns point of view $TTS(\alpha_3, \alpha_1, \alpha_2) = 0.8/(2^{(15/30)}) = 0.566$

Weighting by Interactions

Another factor to take into consideration is confidence of Trust. Trust based upon 2-3 interactions would often be judged to be more questionable to trust based on 200-300 interactions. When forming Trust for the first time, a node should give more weight to trust judgements that are based on more interactions. Each node, α_i, can track the total number of interactions there have been with another node, α_j, as $I(\alpha_i, \alpha_j)$ and we can define the total interactions for all nodes in their neighbourhood $NI(\alpha_i, \alpha_j)$ as:

$$NI(\alpha_i, \alpha_j) = (\Sigma_{\alpha k \in N(\alpha i, \alpha j)} (I(\alpha_k, \alpha_j)) \quad (13)$$

This allows the calculation of a weighted trust value, incorporating a weighted average of Trust scores:

$$TWTS_{initial}(\alpha_i, \alpha_j) = (TS_{default}(\alpha_i) + \Sigma_{\alpha k \in N(\alpha i, \alpha j) \backslash \alpha j} (TWTS(\alpha_i, \alpha_k)\, I(\alpha_k, \alpha_j)\, TTS(\alpha_i, \alpha_k, \alpha_j))) / (|NI(\alpha_i, \alpha_j)| + 1) \quad (14)$$

Here for each neighbour, α_k, that has experience with the node of interest, α_j, we get their timely trust score for the target node, $TTS(\alpha_i, \alpha_k, \alpha_j)$), then multiply by the number of interactions the trust is based on, $I(\alpha_k, \alpha_j)$, to weight the trust value. Then we multiply by our trust for the node supplying the information, $TWTS(\alpha_i, \alpha_k)$, to determine the final trust value from that node. This is repeated for all relevant nodes, the values are summed (including our initial trust value, α_i) and divided by the total number of interactions that all of the trust values are based on, $(|NI(\alpha_i, \alpha_j)|+1)$. This gives one final weighted average of all relevant time weighted trust score.

Similarly, for consumer to service provider relationships, we can define a time weighted quality of service value, *TWQoS*, as:

$$TWQoS_{xninitial}(\alpha_i, \beta_j) = (TS_{default}(\alpha_i) + \Sigma_{\alpha k \in N(\alpha i, \alpha j) \backslash \alpha j} (TWTS(\alpha_i, \alpha_k) I(\alpha_k, \beta_j) TQoS(\alpha_i, \alpha_k, \beta_j))) / (|NI(\alpha_i, \beta_j)| + 1) \quad (15)$$

Here we find the time weighted average of the Trust Values of our neighbours to use for the initial trust for service X_n. Such values can be calculated for all providers of service X_n and comparison can be made to determine which one α_i should choose as the most likely to provide good service.

Comparison with Centralised Model

The Self-led model presented allows each consumer or provider to make their own individual trust judgements, relying on both their own knowledge and the knowledge of others, rather than relying on a centralised trust authority. With a centralised trust model, the trust values represent an overall trust of all nodes in the network whereas the self-led model enables fine tuning of trust on a local level, as well as allowing individual nodes to determine the extent to which they trust other nodes to provide trust valuation of neighbouring nodes and services. This mediation of the information provided by other nodes ensures that individuals prioritise the views of their 'friends' (nodes they trust highly) over 'strangers' or nodes they know to be unreliable. Another feature of the self-led model is that the computation of trust scores is spread across the network, reducing the load to any individual node.

The self-led model also allows for the one or more centralised trust authorities to exist as service providers (providing the service of trust evaluations), which each node can choose to employ to help determine the trustworthiness of nodes in the network, mediated by their own trust of the reliability of the authorities. This allows for the strengths of the centralised model to be leveraged within the self-led trust framework presented.

CONCLUSION

In this paper, we extended lightweight trust services work from Conrad, et al. (2006); peer-to-peer identity and trust management work from Ion *et al.,* (2007); and trust and contract management for dynamic VOs from Wilson et al. (2007). Specifically, our approach utilises the notion of decentralised and dynamic *self-led* trust management. Central to this notion is that nodes continuously update their confidence level as regards other nodes using a measurable confidence level that is both lightweight, easily formalised using graph-theoretic notation and that is relevant to a number of aggregative service (AS) case scenarios, relevant to the grid and SOI.

We were particularly interested in developing a relatively simple and computationally lightweight dynamic trust model, which is capable of encompassing both consumer and provider trust inputs. Further, these inputs should reflect not only rational trust measures but also intangible trust measures so as to more fully reflect ways in which human collaborators build trust amongst partners. The ability to have individual self-led trust scores, as opposed to a centralised trust score, allows a level of autonomy for determination of trust and leads to more accurate trust levels (for the individual). The mediated aspect allows for an individual to avoid being dependant on scores given to it, but to prioritise information from more trusted sources over others and to combine all of the available trust information into a score that reflects that individuals overall strategies (in regards to reputation and risk). Trust measures whether rational or affective in origin have a lifecycle of their own. That is to say, from an initial value trust is built and then perhaps slowly decays in the absence of new information. Alternatively, initial trust confidence levels need to be updated in the light of service experience, service delegation and service reliability. Our set-theoretic notation allows for a fully transparent articulation of trust measures between nodes, including temporal aspects. The present model, though abstract and simplified can potentially aid in the development of richer, yet lightweight trust models that are both capable of being understood (hence checked) by human designers and also operationalised by MAS, in a manner that seeks to partially mimic human-to-human trust formation without inducing undue system latency. This is particularly suited to situations where automated trust decisions are required in networks comprised of nodes that constantly change and that encompass a range of organisations, individuals and VOs where trust in the provision of services is of key importance.

REFERENCES

Bistarelli, S., & Santini, F. (2008). Propagating Multitrust within Trust Networks. In *Proceedings of the Special Interest Group on Applied Computing 2008, Track on Trust, Recommendations, Evidence and other Collaboration Know-how*, Fortaleza, Ceara, Brazil. New York: ACM Publications.

Cahill, V., Gray, E., Seigneur, J. M., Jensen, C. D., Chen, Y., & Shand, B. (2007). A Survey of Trust and Reputation Systems for on-line Service Provision. *Decision Support Systems, 43*(2), 618–644. doi:10.1016/j.dss.2005.05.019

Cahill, V., Gray, E., Seigneur, J.M., Jensen, C.D., Chen, Y., Shand, B., Dimmock, N., Twigg, A., Bacon, J., English, C., Wagealla, W., Terzis, S., Nixon, P., Serugendo, G., Bryce, C., Carbone, M., Krukow, K., & Nielsen, M. (2006). *Pervasive Computing*, 52-61.

Castelfranchi, C., & Falcone, R. (2000). Trust is much more than subjective probability: Mental components and sources of trust. In *Proceedings of the 33rd Annual Hawaii International Conference on System Sciences* (Vol. 1, pp. 4-7).

Chiu, D. K. W., Leung, H., & Lam, K. (2009). On the making of service recommendations: An action theory based on utility, reputation, and risk attitude. *Expert Systems with Applications, 36*(2), 3293–3301. doi:10.1016/j.eswa.2008.01.055

Conrad, M., French, T., Huang, W., & Maple, C. (2006). A Lightweight Model of Trust Propagation in a Multi-Client Network Environment: To What Extent Does Experience Matter? In *Proceedings of the 1st International Conference on Availability, Reliability and Security* (pp. 482-487). Washington, DC: IEEE Publications.

Corritore, C., & Krasher, B., & Wiedenbeck. (2003). On-line trust: concepts, evolving themes, a model. *International Journal of Human-Computer Studies*, *58*, 737–758. doi:10.1016/S1071-5819(03)00041-7

Dayal, S., Landesberg, H., & Zeisser, M. (1999). How to build trust online. *Marketing Management*, *8*(3), 64–69.

Deutsch, M. (1958). Trust and suspicion. *The Journal of Conflict Resolution*, *2*(3), 265–279. doi:10.1177/002200275800200401

Egger, F. (2003). *From Interactions to Transactions: designing the Trust Experience for Business-to-Consumer Electronic Commerce*. Unpublished doctoral dissertation, Technical University Eindhoven, The Netherlands.

Flood, M. (1952). *Some Experimental Games. Research Memorandum RM0789*. Santa Monica, CA: RAND Corporation.

Fukuyama, F. (1995). *Trust: The Social Virtues and the Creation of Prosperity*. London: Hamish Hamilton.

Gambetta, D. (1988). Can we trust trust? In Gambetta, D. (Ed.), *Trust: Making and Breaking Cooperative Relations* (pp. 213–237). London: Basil Blackwell.

Gambetta, D. (2000). *Trust: Making and Breaking Cooperative Relations*. Retrieved from www.sociology.ox.ac.uk/papers/trustbook.html

Gibbons, A. (1985). *Algorithmic Graph Theory*. Cambridge, UK: Cambridge University Press.

Golbeck, J., & Hendler, J. (2006). FilmTrust: Movie Recommendations using Trust in Web-based Social Networks. In *Proceedings of the IEEE Consumer Communications and Networking Conference*.

Golbek, J. A. (2005). *Computing and applying Trust in Web-based Social Networks*. Unpublished doctoral dissertation, University of Maryland, College Park, MD.

Grabner-Krauter, S., & Kaluschia, E. A. (2003). Empirical research in on-line trust: a review and critical assessment. *International Journal of Human-Computer Studies*, *58*, 783–812. doi:10.1016/S1071-5819(03)00043-0

Grabner-Krauter, S., Kaluschia, E. A., & Fladnitzer, M. (2005). Perspectives of Online Trust and Similar Constructs – A Conceptual Clarification. In *Proceedings of the 8th International Conference on Electronic Commerce,* Fredericton, Canada (pp. 235-243).

Hardin, R. (Ed.). (2002). *Trust and Trustworthiness*. New York: Russell Sage Foundation.

Ion, M., Telesca, L., Botto, F., & Koshutanski, H. (2007). An Open Distributed Identity and Trust Management Approach for Digital Community Ecosystems. In *Proceedings of the International Workshop on ICT for Business Clusters in Emerging Markets.* Retrieved from http://www.create-net.org/osco/publications/ion-tele-bott-kosh-eBusiness-07.pdf

Jarvenpaa, S. L., Tractinsky, J., & Vitale, M. (2000). Consumer Trust in an Internet Store. *Journal of Information Technology Management*, *1*(1/2), 45–71. doi:10.1023/A:1019104520776

Karvonen, K. (2007). Enabling Trust between Humans and Machines. In *Proceedings of the European e-Identity Conference, Eema's 20th Annual Conference,* Paris, France. Retrieved from http://www.tml.tkk.fi/~kk/publications.html

Kim, J., & Moon, J. Y. (1998). Designing Towards Emotional Usability in Customer Interfaces: Trustworthiness of cyber-banking system interfaces. *Interacting with Computers*, *10*, 1-29. doi:10.1016/S0953-5438(97)00037-4

Krukow, K., & Nielsen, M. (2003). *Pervasive Computing*, 52-61.

Kuter, U., & Golbeck, J. (2007). SUNNY: A New Algorithm for Trust Inference in Social Networks, using Probabilistic Confidence Models. In *Proceedings of the Twenty-Second National Conference on Artificial Intelligence (AAAI-07)*, Vancouver, BC.

Marsh, S. P. (1994). *Formalising Trust as a Computational Concept*. Unpublished doctoral dissertation, Stirling University, UK.

Matthews, B., Bicagregui, C., & Dimitrakos, T. (2003). *Building Trust on the Grid: Trust Issues Underpinning Scalable Virtual Organizations*. Retrieved from http://epubs.cclrc.ac.uk/bitstream/643/trustedgridERCIM.pdf

Mayer, R. C., Davis, J. H., & Schoorman, F. D. (1995). An Integrative model of organisational trust. *Academy of Management Review*, *20*(3), 709–734. doi:10.2307/258792

McKnight, D. H., Choudhury, V., & Kacmar, C. (2002). Developing and validating trust measures for e-commerce: an integrative typology. *Information Systems Research*, *13*(3), 334–359. doi:10.1287/isre.13.3.334.81

Nikander, P., & Karvonen, K. (2000). Users and Trust in Cyberspace. In *Proceedings of the Cambridge Security Protocols Workshop 2000*.

Olmedilla, D., Rana, O., Matthews, B., & Nejdi, W. (2006). Security and Trust Issues in Semantic Grids. In *Proceedings Semantic Grid: The Convergence of Technologies*. Retrieved from http://drops.dagstuhl.de/popus/volltexte/2006/408

Putnam, R. D. (2000). *Bowling Alone: The Collapse and revival of American Community*. New York: Touchstone Press.

Richard, P., Devinney, T., Yip, G., & Johnson, G. (2009). Measuring Organisational Performance: Towards Methodological Best Practice. *Journal of Management*, *35*(3), 718–804. doi:10.1177/0149206308330560

Sabater-Mir, J., Pinyol, I., Villatoro, D., & Cuní, G. (2007). Towards Hybrid Experiments on reputation mechanisms: BDI Agents and Humans in Electronic Institutions. In *Proceedings of the 12th Conference of the Spanish Association for Artificial Intelligence (CAEPIA-07)* (Vol. 2, pp. 299-308).

SECURE. *Secure Environments for Collaboration among Ubiquitous Roaming Entities*. (2001). Retrieved from http://www.dsg.cs.tcd.ie/dynamic/?category_id=-30

Song, S., Hwang, K., & Kwok, Y. (2005). Trusted Grid Computing with Security Binding and Trust Integration. *Journal of Grid Computing*, *3*(1/2), 53–73. doi:10.1007/s10723-005-5465-x

Super Quant Monte-Carlo Challenge - V Grid Plugtests. (2008). Retrieved October 1, 2009, from http://www-sop.inria.fr/oasis/plugtests2008/ProActiveMonteCarloPricingContest.html

Tucker, A. (1950). A two-person dilemma. Lecture at Stanford University, Palo Alto, California. Stanford University Press. In Poundstone, W. (Ed.), *Prisoner's Dilemma* (2nd ed.). New York: Anchor Books.

Wilson, M., Arenas, A., & Schubert, L. (2007). *TrustCOM Framework Version 4*. Retrieved from http://www.eu-trustcom.com

Ziegler, C. N., & Lausen, G. (2005). Propagation models for trust and distrust in social networks. *Information Systems Frontiers*, *7*(4/5), 337–358. doi:10.1007/s10796-005-4807-3

This work was previously published in International Journal of Systems and Service-Oriented Engineering, Volume 1, Issue 4, edited by Dickson K.W. Chiu, pp. 1-18, copyright 2010 by IGI Publishing (an imprint of IGI Global).

Chapter 8
Autonomic Business-Driven Dynamic Adaptation of Service-Oriented Systems and the WS-Policy4MASC Support for Such Adaptation

Vladimir Tosic
NICTA, The University of New South Wales, Australia and The University of Western Ontario, Canada

ABSTRACT

When a need for dynamic adaptation of an information technology (IT) system arises, often several alternative approaches can be taken. Maximization of technical quality of service (QoS) metrics (e.g., throughput, availability) need not maximize business value metrics (e.g., profit, customer satisfaction). The goal of autonomic business-driven IT system management (BDIM) is to ensure that operation and adaptation of IT systems maximizes business value metrics, with minimal human intervention. The author presents how his WS-Policy4MASC language for specification of management policies for service-oriented systems supports autonomic BDIM. WS-Policy4MASC extends WS-Policy with new types of policy assertions: goal, action, probability, utility, and meta-policy assertions. Its main distinctive characteristics are description of diverse business value metrics and specification of policy conflict resolution strategies for business value maximization according to various business strategies. The author's decision making algorithms use this additional WS-Policy4MASC information to choose the adaptation approach best from the business viewpoint.

DOI: 10.4018/978-1-4666-1767-4.ch008

INTRODUCTION

In the modern world, technical and business changes are frequent and increasingly common. Additionally, the modern globalized economy increasingly supports and often requires various business relationships between diverse companies. These circumstances place important requirements on enterprise information technology (IT) systems: the ability to seamlessly interconnect with IT systems of diverse business partners irrespective of the implementation of these IT systems and the ability to handle various technical and business changes (e.g., temporary computer/network failures and establishment of new business alliances).

Service-oriented computing (SOC) was developed to answer these challenges. In service-oriented systems (Papazoglou & Georgakopulos, 2003), such as Web services and their compositions, parts of internal IT (e.g., software) systems are exposed as implementation-independent services, which are then composed in a loosely-coupled manner, possibly dynamically, i.e., during run-time (instead of during software/system design). However, just implementing and composing service-oriented systems is not enough to fully address diverse technical and business changes that can affect these systems. To discover and address changes, IT systems have to be managed (Sloman, 1995).

Management of IT systems, including service-oriented systems, is the process of their monitoring and control to ensure correct operation, enforce security, discover and fix problems (such as faults or performance degradations), maximize quality of service (QoS), accommodate changes, and achieve maximal business benefits. Monitoring determines the state of the managed system, e.g., by measuring or calculating QoS metrics, determining presence of problems, evaluating satisfaction of requirements and guarantees, accounting of consumed resources, and calculating prices/penalties to be paid. A QoS metric, such as response time or availability, is a measure of how well a system performs its operations. On the other hand, control puts the managed system into the desired state, by performing run-time adaptation of the system to ensure its correct operation, in spite of changes or run-time problems. For example, control of a service-oriented system includes its re-configuration, re-composition of services, and re-negotiation of contracts between the composed services and between the system and other parties. Control actions result in changes of monitored conditions (e.g., QoS metrics), which closes the management loop.

The majority of past IT system management products act as support tools for human system administrators – these products present summaries of monitored information and often automate some simpler control actions, but it is ultimately human responsibility to make more complex decisions about execution of control actions. Since the complexity of modern IT systems is rapidly increasing, human system administrators exhibit difficulties in making optimal decisions. Furthermore, human system administrators are expensive and might not be available at all times. Therefore, minimizing human involvement in IT system management has been a research goal for several decades and was made prominent by the vision of autonomic computing (Kephart & Chess, 2003). In autonomic computing, IT systems are self-managing, i.e., they manage (e.g., adapt) themselves using configurable policies, with minimal human intervention.

A prerequisite for performing IT system management activities is existence of a machine processable and precise format for specification of the monitored values/conditions and the control actions (Keller & Ludwig, 2003; Tosic, Pagurek, Patel, Esfandiari & Ma, 2005). Policies (Sloman, 1994) are a frequent approach to IT system management, not limited to autonomic computing. A policy formally specifies a collection of high-level, implementation-independent, operation and management goals and/or rules in a human-readable form. To improve flexibility, maintainability, reusability, and simplicity of specifications, policy description of monitoring and control aspects is separated from descriptions of the managed IT system. During run-time, a policy-driven management system refines these high-level goals and rules into many low-level, implementation-specific, actions controlling operation and management of the managed system and its components. Another format for specifying what should be achieved by IT system management is a service level agreement – SLA (Keller & Ludwig, 2003). An SLA is a special type of contract (a binding and enforceable formal agreement between two or more parties) that specifies QoS requirements and guarantees and, often, prices and monetary penalties to be paid. It can be used as an alternative or as a complement to policies. While information specified in policies and SLAs is similar in content, SLAs require two or more parties (while policies can be specified for one party only). Traditionally, SLAs and policies also differ in the formatting of represented information and in the architectures of management infrastructures (middleware).

The past IT system management (including autonomic computing) solutions were mostly focused on functional correctness, security, and optimization of technical QoS metrics (e.g., response time, availability). They provide only a simple treatment of financial business value metrics (e.g., prices and monetary penalties), without addressing non-financial business value metrics. A business value metric is any measure of business worth of an item or a situation to a particular business party. (Instead of the term "business value metric" some authors use the similar terms "key performance indicator – KPI", "business performance metric", "business QoS", and "quality of business – QoBiz".) Some business value metrics are financial, such as earned income, costs, profit, and return on investment (ROI). While financial business value metrics are important for all companies, they are not the only measures of business worth. Examples of non-financial business value metrics are the number of customers, market share, and customer satisfaction. Note that business value metrics are relative to business parties, e.g., concrete numerical quantities and possibly types of business value metrics for a consumer and a service provider can be different.

Business value metrics are more important to business users than technical QoS metrics. For example, a business user is usually not very interested in the fact that availability dropped from 99% to 95%, but wants to know how much this change costs her/his business. Unfortunately, the past practice has shown that mapping between technical and business models and metrics is difficult (Bartolini, Sahai & Sauve, 2006; Bartolini, Sahai & Sauve, 2007; Biffl, Aurum, Boehm, Erdogmus & Gruenbacher, 2006). For example, increasing availability need not lead to increases in business profits, because the costs to provide higher availability (e.g., through replication, partitioning, load balancing, or other means) could outweigh the increased income from customers.

The goal of business-driven IT management (BDIM) research community is to determine mappings between technical QoS and business value metrics and leverage them to make run-time IT system management decisions that maximize business value metrics (Bartolini, Sahai & Sauve, 2006; Bartolini, Sahai & Sauve, 2007). For example, it tries to quantify impact on business profits of increased/decreased availability. Integration of autonomic computing and BDIM was identified in (Bartolini, Sahai & Sauve, 2007, pp. 23-24) as an area of open research challenges. One of the limitations of the past BDIM works is that most of them focused on maximizing profit. However, human managers use many business strategies that differ (among many other aspects) in how they prioritize different business value metrics in different time frames (e.g., long-term vs. short-term). As will be illustrated later in the paper, maximization of short-term profit is not always appropriate. Business strategies are a major differentiator of companies in a market, so many diverse business strategies will exist in the market of (Web) services and should be supported by autonomic BDIM solutions.

We present our results on developing autonomic BDIM solutions, based on the WS-Policy4MASC language (Tosic, Erradi & Maheshwari, 2007; Tosic, 2008) for specification of management policies for service-oriented systems. WS-Policy4MASC can be used for monitoring and control activities focused on functional correctness and technical QoS metrics (Erradi, Tosic & Maheshwari, 2007), but its main original contributions are the description of diverse business value metrics (both financial and non-financial) and the specification of policy conflict resolution strategies for maximization of business value metrics according to various business strategies. We developed algorithms that use this additional WS-Policy4MASC information to provide unique support for autonomic business-driven dynamic adaptation of service-oriented systems. This paper complements our previous publications on the WS-Policy4MASC language, particularly (Tosic, Erradi & Maheshwari, 2007) and (Tosic, 2008), by providing additional explanations of some of the challenges for autonomic BDIM and how WS-Policy4MASC supports autonomic BDIM.

In the next section, we present motivating examples that illustrate both the need for autonomic BDIM software and some of the difficulties in developing such software. Then, we give a brief survey of the major related work. Next, we provide a high-level overview of the main features of WS-Policy4MASC. The main section of the paper contains detailed discussion of WS-Policy4MASC support for autonomic business-driven adaptation of service-oriented systems. In the final section, we summarize conclusions and items for future work.

EXAMPLES OF SOME CHALLENGES FOR AUTONOMIC BUSINESS-DRIVEN IT MANAGEMENT

Figure 1 shows an example situation of a distributed system that requires dynamic adaptation and which we will use to illustrate the need for autonomic business-driven adaptation of service-oriented systems. Party X implements some business process as a composition of 3 (Web) services: A, B, and C. (Note that A-B-C can be, in principle, any distributed computing system, but we will assume a service-oriented system.) Party B has a contractual guarantee (specified in an SLA or a policy agreed with the other parties) to provide 99% availability every day and charge AU$3 per invocation. If it does not meet the availability guarantee in a particular day, it pays the penalty of AU$10. Parties A and C have their own QoS guarantees and prices, which we will abstract. Party X gives an availability guarantee to its own consumers of 98% per day, with unavailability penalty of AU$50 and price of $20 per invocation. This service composition works without problems for some time, but then a problem happens with B that prevents it to keep it its guarantee of 99% availability. For example, B could become overloaded due to requests that it receives from some other consumers (not shown in Figure 1) – it is not uncommon that a service is concurrently part of several service compositions. We will assume that the problem is of such nature that B has to be replaced with an alternative service. After a search of all service directories, no exact replacement (in terms of functionality, QoS, and price) for B was found. Instead, 2 services, D and E, with the same functionality as B, but with different QoS and price were found. D guarantees 95% availability (lower than B) with AU$1 penalty but with the price of AU$1 per invocation (cheaper than B), while E guarantees 99.99% availability (higher than B) with $80 penalty with the price of AU$5 per invocation (more expensive than B).

The composing party X now has to make a decision which of these alternatives (D or E) to choose. We would like that software at X could make this decision without direct human involvement, i.e. autonomically. The traditional IT systems management (including Web service management) software would maximize technical metrics, so it would choose E. However, this need

Figure 1. A motivating scenario

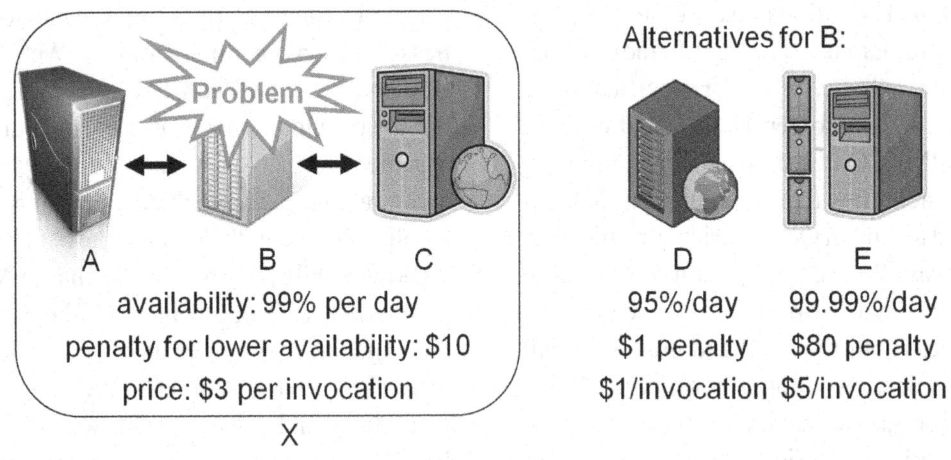

not be the best decision from the business viewpoint. The execution of the composition provides some business value to X – for simplicity, we will assume that this is profit. It is possible that the higher price of E would significantly reduce X's profit from the composition. On the other hand, it is likely that lower availability of D would cause X to pay availability penalty more often, which might lower X's profit in spite of D's lower price. The decision of whether to choose D or E also depends on a number of other factors (not shown in Figure 1), such as probabilities that particular parties will meet their availability guarantees.

Note that situations in real service-oriented systems are often much more complicated than the relatively simple example in Figure 1. For example, a party often offers multiple QoS guarantees (e.g., about response time or throughput, in addition to availability), guarantees/prices/penalties might differ between operations of the same service, there can be various payment models (e.g., per invocation, subscription, per information volume), guarantees/prices/penalties can differ based on context (e.g., X could charge consumers from developing countries less), etc. Additionally, service compositions are often more complicated than in Figure 1, both in terms of the number of composed services and in terms of the control structure (e.g., there could be parallel branches or conditional executions). Further, in some circumstances, many different control actions exist, with different consequences. For example, when a (Web) service does not respond on time, the composing party could decide to: wait for the reply a bit longer, resend the request to this service (if it is idempotent), cancel this request, simply skip this request (and ignore its results when/if they come later), replace the faulty service with a known alternative, search for an alternative in a service directory and then use this replacement, etc. Moreover, not only single control actions, but also groups (e.g., sequences) of actions could be appropriate. Consequently, even when only financial business values are considered and the business strategy is to maximize profit, there are many complexities that have to be taken into account.

The situation is further complicated by the fact that businesses often have non-financial business value metrics and business strategies that consider such metrics. Assume that in the example from Figure 1, choosing D gives X a bit higher profit, but significantly lowers satisfaction of X's customers. If X has the business strategy of maximizing short-term profits, it should choose D.

However, if X has the business strategy "exceptional customer satisfaction", then it has sense for X to choose E that brings lower profits (or even short-term loses), in order to maintain/enhance X's reputation and competitive advantage in the marketplace. In the long term, the reputation and the customer satisfaction will influence customer retention and recruitment, and, therefore, future profits. In the existing markets (e.g., the car market), there are many examples of companies with the "exceptional customer satisfaction" business strategy (e.g., high-end car manufacturers) and other business strategies that use non-financial business value metrics. We argue that solutions for autonomic BDIM (for IT systems in general, not only for service-oriented systems) should be able to use non-financial business value metrics and business strategies that consider such metrics.

RELATED WORK

Many languages for specification of policies, SLAs, and related IT system management constructs for service-oriented systems have been developed. The most important among these languages are accompanied by management-related tools, e.g., middleware. Some examples are the Dynamo middleware using the combination of the Web Service Constraint Language – WSCoL and the Web Service Recovery Language – WS-ReL (Baresi, Guinea & Plebani, 2007), the Web Service Level Agreement – WSLA language and middleware (Keller & Ludwig, 2003), the Web Service Offerings Infrastructure – WSOI using the Web Service Offerings Language – WSOL (Tosic, Pagurek, Patel, Esfandiari & Ma, 2005), and the Cremona (CReation and MONitoring of Agreements) toolkit using the Web Services Agreement Specification – WS-Agreement (Ludwig, Dan & Kearney, 2004). There are also additional research projects, such as JOpera (Pautasso, Heinis & Alonso, 2007), and industrial products that provide management infrastructures for service-oriented systems. While significant, these past results exhibit several limitations. They predominantly approached monitoring or QoS-based selection of Web services and did not sufficiently address the more challenging issues of control (adaptation), particularly for composite service-oriented systems. Only a few recent works (notably, Dynamo and JOpera) provide support for autonomic management of service-oriented systems.

Another important limitation of these past works is their focus on technical QoS metrics, with very limited support for simple financial business value metrics (e.g., pay-per-use and subscription prices, simple monetary penalties) in SLAs/policies. Only some of the past management infrastructures (usually, the commercial ones) contain accounting/billing subsystems that monitor/calculate monetary amounts to be paid. While business-driven management of service-oriented systems was mentioned in the past literature (Casati, Shan, Dayal & Shan, 2003; Tosic, Erradi & Maheshwari, 2007; Tosic, 2008), there are still many open research issues, particularly related to the use of diverse business strategies (beyond short-term profit maximization) and non-financial business value metrics.

Business literature provides evidence that modeling of non-financial business value metrics in addition to financial ones is beneficial for long-term strategic management of companies. A widely accepted approach that represents well-being of a company beyond financial metrics is the balanced scorecard – BSC (Kaplan & Norton, 1996). It organizes company-specific business value metrics along 4 standard perspectives: financial, customer, internal business processes, and learning & growth. Unfortunately, there are difficulties in direct reuse of balance scorecards for autonomic BDIM, because they are intended for human use. One of the issues is how to map from the 3 non-financial perspectives (particularly, learning & growth) into run-time BDIM activities at the IT level. Nevertheless, the balanced scorecard concepts inspired and significantly influenced several BDIM projects, in addition to our work. A notable example is Management by Business Objectives – MBO (Bartolini, Salle & Trastour, 2006). Its information model contains objectives, business value metrics (the authors use the term "KPI"), and perspectives (inspired by the balanced scorecard). Importance weights are assigned to these objectives and perspectives – this is a limitation, because in practice it can be difficult to determine precise weights. Aline is the corresponding engine that calculates alignment between different actions and objectives, defined as the likelihood of meeting the objectives. In spite of many contributions, this work did not address detailed modeling of business value metrics and business strategies.

(Tosic, 2008) examined some of the issues in modeling of financial and non-financial business value metrics: the need for explicit description of various characteristics of these metrics; advantages and disadvantages of monetization of non-financial business value metrics (e.g., replacing 80% customer satisfaction with the monetary amount AU$300), the need for modeling of possible (but not certain) business value metrics, and accounting for inflation and interest rates using the "time value of money" formulae from economics. The same paper also examined some of the issues in modeling of business strategies that maximize business value metrics: the need for explicit description of various characteristics of business strategies, options for limiting temporal scope of relevant business value metrics, calculation of overall (total, summary) business value metric associated with execution of a control action, and possible complications (e.g., cost limits, tiebreakers) in comparing these overall business value metrics of various control actions.

Many high-quality BDIM papers were published in proceedings of the BDIM workshops, e.g., (Bartolini, Sahai & Sauve, 2006; Bartolini, Sahai & Sauve, 2007), organized annually since 2006. However, important BDIM papers, such as (Salle & Bartolini, 2004), were published before the term "BDIM" was coined and BDIM-related papers are still published in a wide variety of venues (e.g., autonomic computing conferences), sometimes without using the "BDIM" term. Furthermore, BDIM is related to the other research areas that try to improve alignment between business and IT. (Salle, 2004) overviewed relationships between BDIM and IT governance, which is a set of human activities aiming to achieve that IT systems support business objectives and strategies. (Tosic, Suleiman & Lutfiyya, 2007) discussed the need for integrating BDIM with value-based software engineering – VBSE (Biffl, Aurum, Boehm, Erdogmus & Gruenbacher, 2006), which is a software engineering approach that explicitly considers value issues (e.g., value-based prioritization), in order to make the resulting software more useful. Integration of results from various communities will be needed to achieve autonomic BDIM.

OVERVIEW OF THE WS-POLICY4MASC LANGUAGE

WS-Policy4MASC extends the Web Services Policy Framework – WS-Policy (W3C Web Services Policy Working Group, 2007), an industrial standard by the World Wide Web Consortium (W3C). Analogously to the other Web service standards, the syntax of WS-Policy is specified in the Extensible Markup Language (XML) and its grammar is defined in XML Schema. In WS-Policy, a policy is defined as a collection of policy alternatives, each of which is a collection of policy assertions. WS-PolicyAttachment defines a generic mechanism that associates a policy with subjects to which it applies. Various policy subjects are possible, such as service-level constructs (e.g., operation, message) in the Web Services Description Language (WSDL) and process-level constructs (e.g., flow, link) in the Web Services Business Process Execution Language (WSBPEL). A policy scope is a set of policy subjects to which a policy may apply. WS-

Figure 2. The main concepts and relationships in WS-Policy4MASC

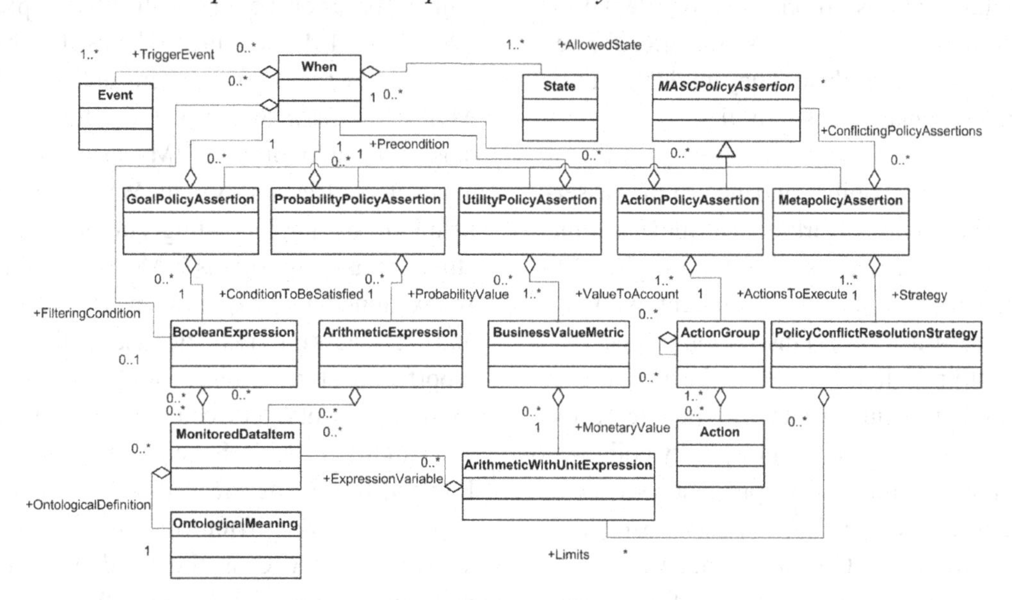

Policy has many good features, such as simplicity, extensibility, and flexibility. However, it is only a general framework, while details of specification of particular types of policy assertions are left for specialized languages. There are several recent academic WS-Policy extensions for specification of QoS, such as WS-QoSPolicy (Rosenberg, Enzi, Michlmayr, Platzer & Dustdar, 2007) and WS-CoL (Baresi, Guinea, Plebani, 2007), but not for business value metrics and many other issues relevant in autonomic BDIM.

WS-Policy4MASC defines 5 new types of WS-Policy policy assertions, as well as a number of auxiliary (supporting) constructs. The most important constructs and relationships in the current version 0.9 of WS-Policy4MASC are shown in Figure 2. The 5 new policy assertion types inherit from the abstract construct "MASCPolicyAssertion" and can be used (e.g., specified in WS-Policy files, attached using WS-PolicyAttachment) in the same way as other WS-Policy policy assertions. The new policy assertion types are:

1. Goal policy assertions specify requirements and guarantees (e.g., response time of an activity has to be less than 1 second) to be met in desired normal operation. They guide monitoring activities.
2. Action policy assertions specify actions (e.g., replacement/skipping/retrying of a sub-process, process termination) to be taken if certain conditions are met (e.g., some goal policy assertions were not satisfied). They guide control (adaptation) activities. As elaborated in (Tosic, Erradi & Maheshwari, 2007), WS-Policy4MASC has built-in support for specification of a diverse range of common service composition and business process adaptation actions.
3. Utility policy assertions specify diverse business value metrics assigned to particular run-time situations (e.g., non-satisfaction of some goal policy assertion, execution of some action, another event). They guide accounting/billing and provide information for business-driven management.
4. Probability policy assertions specify probabilities that particular situations will occur. They guide management of various forms of uncertainty (e.g., risks, trust in various parties).

5. Meta-policy assertions specify adaptation alternatives and business strategies for selection between them. They guide business-driven management activities.

The utility, probability, and meta-policy assertions will be explained further in the next section.

The above definitions of WS-Policy4MASC policy assertion types are consistent with the (Kephart & Walsh, 2004) discussion of goal, action, and utility policies, as well as the literature on policy conflict resolution with meta-policies (Lupu & Sloman, 1999). We added probability policy assertions after examination of application scenarios that contained uncertainties, risks, and trust issues.

WS-Policy4MASC enables specification of detailed information that is necessary for run-time policy-driven management and that overcomes some other limitations of WS-Policy (e.g., imprecise semantics of policy assertions' effects on policy subjects). Some of this information (e.g., which party performs evaluation/execution of a policy assertion, which party is responsible for meeting a goal) is specified in attributes of the above-mentioned 5 policy assertion types. Much more information is specified in additional auxiliary WS-Policy4MASC constructs, specifying ontological meaning, monitored QoS metrics, monitored context properties (e.g., geographic location), states, state transitions, events, schedules, applicability scopes, and various types of expression (Boolean, arithmetic, arithmetic-with-unit, string, time/date/duration). Not all auxiliary constructs are shown in Figure 2. "When" is the most important auxiliary construct. It specifies 1 or more states in which something (e.g., evaluation of a goal policy assertion) can happen, 1 or more events (e.g., Web service operation executed) that can (mutually independently) trigger this, and an optional filtering Boolean condition to be satisfied.

WS-Policy4MASC was used in the Manageable and Adaptable Service Compositions (MASC) middleware for management of service-oriented systems (Erradi, Tosic & Maheshwari, 2007), which is based on the Microsoft .NET 3.5 platform. (WS-Policy4MASC can also be used with different middleware and we are currently developing MiniMASC, a Java re-implementation and extension of the autonomic BDIM aspects of MASC.) We evaluated feasibility of the WS-Policy4MASC solutions by implementing corresponding data structures and algorithms in MASC. Additionally, we examined their expressiveness, effectiveness, and usefulness on 2 realistic case studies (weather report and stock trading) and several smaller examples. Syntax correctness (well-formedness and validity) of XML Schema definitions of WS-Policy4MASC was checked with XML tools.

One of the difficulties in working with WS-Policy4MASC is the need to write long and complicated XML files (WS-Policy files with WS-Policy4MASC policy assertions and WS-PolicyAttachment). Oftentimes, some of the information in policy files is known during design-time (software engineering). To facilitate writing of WS-Policy4MASC files and to help bridge design-time and run-time management issues, we developed the Unified Modeling Language (UML) profiles for WS-Policy4MASC, as well as the Extensible Stylesheet Language Transformations (XSLT) rules that generate WS-Policy4MASC and WS-PolicyAttachment files from information in these UML profiles (Tosic, Suleiman & Lutfiyya, 2007). Furthermore, we defined XSLT rules that annotate UML diagrams with run-time management information collected by the MASC middleware. The latter helps in design-time software adaptation (re-engineering, improvement) by software engineers and complements the autonomic run-time adaptation.

THE WS-POLICY4MASC SUPPORT FOR AUTONOMIC BUSINESS-DRIVEN ADAPTATION

WS-Policy4MASC enables specification of both financial and non-financial business value metrics in utility policy assertions. A utility policy assertion references a "When" construct describing situations in which business value metrics should be calculated, has attributes describing the management party (performing calculation and accounting) and the beneficiary party (e.g., the provider Web service), and lists 1 or more business value metrics. Each business value metric contains an arithmetic-with-unit expression and a set of attributes and sub-elements determining the optional paying party (e.g., a consumer), the optional payment recurrence (e.g., for subscription prices), and the business value metric category. For simplicity, all non-financial business value metrics are monetized (i.e., represented with an estimated monetary amount) and future payments are given in their net present value. The arithmetic-with-unit expression can contain constants, variables, standard mathematical operators, and function calls. Thus, business value metrics can be defined not only as absolute (e.g., "FutureSales = AU$700"), but also as relative (e.g., "FutureSalesToReturningCustomers = 0.7*FutureSales"). This expression is evaluated at run-time to produce a single number with a currency unit. Currency units can be associated with ontological definitions and linked to currency conversion services.

Note that in practice it can be difficult to monetize contributions of individual non-financial aspects. For example, if both customer satisfaction and market share (somewhat, but not completely, dependent on customer satisfaction) contributed to past profit, it can be difficult to estimate how customer satisfaction on its own impacts profit. Therefore, the WS-Policy4MASC approach is to define business value metric categories that are easier to estimate. There are 3 mutually orthogonal dimensions of properties that classify these business value metric categories: i) financial or non-financial, ii) agreed or non-agreed (i.e., only possible), iii) benefit or cost. A business value metric category is a combination of 1 characteristic from each of these 3 dimensions, so there are 8 business value metric categories. For example, "non-financial non-agreed benefit" characterizes customer satisfaction, while "financial non-agreed benefit" characterizes estimates of future sales. In different circumstances (e.g., business strategies), different business value metric categories are relevant for autonomic BDIM decision making (we will explain later how WS-Policy4MASC addresses this). A WS-Policy4MASC business value metric can have an optional array sub-element describing aspects (e.g., "customer satisfaction") that it represents, but this information is currently only for documentation purposes.

WS-Policy4MASC enables dealing with uncertainty through probability policy assertions. A probability policy assertion references a "When" construct, has attributes describing the management party as well as the trusting party (believing that the "When" condition will happen with the given probability) and the optional trusted party (promising that the "When" condition will happen), and includes an arithmetic expression. The expression is evaluated during run-time and results in a single number (probability) in the interval [0, 1]. For example, a situation that leads to the utility policy assertion with future sales of AU$700 could be assigned probability of 0.7, meaning that there is 70% chance that this situation will happen. It is also possible to assign probabilities that different estimates of the same business value metric will be correct (the sum of these probabilities has to be 1). An example is when satisfaction of some goal policy assertion GPA1 is associated with 2 utility policy assertions UPA1 and UPA2 and there are 2 probability policy assertions: PP1 specifies that there is 0.6 (60%) probability that UPA1 will be the correct estimate, while PP2 specifies that there is 0.4 (40%) probability that UPA2 will be the correct estimate.

One of the goals of BDIM is to model relationships between technical QoS metrics and business value metrics. In WS-Policy4MASC this is done through specification of utility (and probability) policy assertions for various conditions modeled in "When" constructs. For example, satisfaction/

non-satisfaction of a goal policy assertion or execution of an action policy assertion generates an event, which can be used in "When" constructs for utility policy assertions. Technical QoS metrics are specified in "MonitoredDataItem" constructs that can be used within any expression, such as Boolean expressions in goal policy assertions and filtering conditions of "When" constructs, or arithmetic-with-unit expressions in utility policy assertions (e.g., for modeling of pay-per-volume prices). In the motivating example from Figure 1, Party B has a contractual guarantee to provide 99% availability every day, with AU$10 penalty for not meeting this guarantee. This is modeled in WS-Policy4MASC by defining "AvailabilityPerDay" QoS metric as a "MonitoredDataItem", specifying in the Boolean expression of a goal policy assertion (GPA-B) that "AvailabilityPerDay" has to be >= 99%, specifying in the "When" construct of the same goal policy assertion (GPA-B) that evaluation is performed every day at a particular time (e.g., midnight), defining an event (E1) triggered when this goal policy assertion (GPA-B) is not satisfied, specifying in the arithmetic-with-unit expression of a utility policy assertion (UPA-B) the constant "AU$10", and specifying in the "When" construct of the same utility policy assertion (UPA-B) that accounting is performed whenever this event (E1) is triggered. Note that the "MonitoredDataItem" construct need not be used only for low-level technical QoS metrics – it can be also used for composite technical QoS metrics (e.g., average availability in a week), as well as for business process metrics that describe quality of business service, such as delivery time of a physical product (e.g., an ordered book) to a customer. Special action "MonitoredDataCollection" describes details of how, where, when, and by which party measurement or calculation of values for "MonitoredDataItem" instances is performed. All these features (plus several others) enable definition of complex, multi-level dependencies between technical QoS metrics and business value metrics. While this increases complexity of WS-Policy4MASC, it is appropriate because in practice it is difficult to determine direct mappings between technical QoS metrics and business value metrics and multiple intermediate layers and probability calculations are needed often.

A common issue in specification of business value metrics is how to determine particular numerical quantities, especially for estimates of non-agreed business value metrics (e.g., that future sales will be AU$600) and for monetization of non-financial business value metrics (e.g., that monetary value for customer satisfaction in some situation is AU$200). Unfortunately, there is no simple answer to this issue, which is even more difficult than determining numerical quantities for technical QoS metrics. The process of determining these numerical quantities should be based on different inputs: expert opinion, market research and analysis, economic models, simulations, impact analysis, historical information from the same or similar systems, context in various historical situations, anticipated developments, and others. Historical information can be very useful, but must be used carefully because changes in context can make it inappropriate and it can be difficult to predict future context. Accounting/billing subsystems in management infrastructures can provide historical financial information. Further, business intelligence (BI) and business activity monitoring (BAM) solutions can be useful for determining historical information about both financial and non-financial business value metrics, as well as context. We had to leave integration of our research with solutions for estimation of numerical values for technical QoS and business value metrics for future work. We simply assume that humans who write WS-Policy4MASC files somehow determined appropriate numerical quantities.

As discussed earlier in the paper, business strategy is important for autonomic BDIM solutions and modeling of influence of diverse business strategies on adaptation decisions is a still underexplored research area. WS-Policy4MASC enables specification of business strategies in meta-policy assertions, as a means of deciding which among alternative adaptation approaches to take. Each adaptation approach is modeled as an action policy assertion. When several policy assertions can be applied in the same situation, but cannot or should not be applied together (or at least not at the same time), there is a policy conflict. In the example from Figure 1, "replace B with D" and "replace B with E" are 2 conflicting adaptation options. Our work enables autonomic choice of 1 "best" conflicting action policy assertion.

Meta-policies (policies about policies) are often used in policy-driven management (Lupu & Sloman, 1999) to decide which among conflicting policies to apply. WS-Policy4MASC uses meta-policy assertions to describe business strategy criteria for: a) calculating conflicting option's overall business value metric that contains some (possibly all) of the above-mentioned 8 business value metric categories, and b) comparing these overall business value metrics to choose 1 "best" option that meets all specified constraints and results in the highest overall business value metric among the eligible options. WS-Policy4MASC also enables specification of priorities of policy assertions, but they are used only if overall business value metrics of several options are equal, because it can be difficult to a priory define appropriate priority levels. Compared with the approach that uses only priorities of the conflicting policies, as adopted in WS-ReL (Baresi, Guinea & Plebani, 2007) and other works, our approach to policy conflict resolution provides not only better alignment of service-oriented systems with business issues, but also more flexible handling of situations when priority of an option is not known beforehand (e.g., because it depends on some run-time information). It is central to the WS-Policy4MASC support for autonomic BDIM.

A WS-Policy4MASC meta-policy assertion references at least 2 conflicting action policy assertions and exactly 1 policy conflict resolution strategy, and has attributes describing the management party (executing our policy conflict resolution algorithm) and the party for which the overall business value metric is maximized (e.g., the provider Web service). The sub-elements and attributes of a policy conflict resolution strategy describe business strategy characteristics. Four among these attributes describe which properties (financial and/or non-financial, agreed and/or non-agreed) of business value metric categories should be used for calculation of the overall business value metric. For example, specifying "only non-financial" along with "both agreed and non-agreed" leads to the use of business value metrics in the categories "non-financial agreed benefit", "non-financial agreed cost", "non-financial non-agreed benefit", and "non-financial non-agreed cost" – these 4 categories are relevant for maximization of customer satisfaction. Further, there is an attribute that determines whether probabilities of occurrence should be taken into consideration. Then, there is a sub-element specifying common currency into which all currencies are converted and an optional currency conversion service. An optional sub-element specifies (as an arithmetic-with-unit expression evaluated at run-time) the length of time during which future recurring payments (e.g., subscriptions) should be considered. Another optional sub-element specifies (as an arithmetic-with-unit expression) cost limits that must not be exceeded. For example, if a policy conflict resolution strategy says that the cost limit is AU$1000 and an adaptation option requires investment with the total cost (for all 4 cost business value metric categories) of AU$1500, this option will be discarded.

Finally, 2 sub-elements and 1 attribute of a policy conflict resolution strategy are used to specify tiebreaking. Tiebreaking is used when overall business value metrics of 2 or more conflicting options are close enough so that the difference is considered negligible. For example, when one option provides profit of AU$500 and the other option provides profit of AU$499. In such cases, additional, previously excluded business value metric categories are used for calculation of new overall business value metrics. In the above example, if the original calculation of overall business value metrics did not take into consideration non-financial non-agreed benefits and costs, they can be added as tiebreakers. This can result in a situation when the new overall business value metric for the first option is AU$720, while for the second option is AU$750. Consequently, the second option will be chosen (in spite the fact that in the original calculations it was negligibly worse than the first option). WS-Policy4MASC supports up to 2 rounds of tiebreaking. An attribute of a policy conflict resolution strategy specifies whether the first tiebreaker will be additional business value metric categories along the financial/non-financial dimension (e.g., using "both financial and non-financial" instead of "only non-financial") or along the agreed/non-agreed dimension. Two sub-elements of a policy conflict resolution strategy specify (as arithmetic-with-unit expressions) 2 limits of what should be consider "negligible" difference between overall business value metrics. One of these limits is for the financial/non-financial dimension and the other is for the agreed/non-agreed dimension.

Our algorithm for selection of 1 "best" option among conflicting action policy assertions is outlined in Figure 3. Here, "best" means "highest overall business value metric, while meeting all constraints" and depends on which combinations of the 8 business value metric categories are used in the calculations. The algorithm is a refinement and improvement of the high-level algorithm presented in (Tosic, 2008). Versions of this algorithm were successfully prototyped, first in the original MASC middleware and recently in the new MiniMASC middleware we are developing. Further details, explanations, and illustrations of this algorithm will be given in our forthcoming publications.

CONCLUSION AND FUTURE WORK

Due to the rapidly increasing complexity of service-oriented (and other IT) systems, their management is becoming a serious problem. On one side, there is a need to perform management with minimal human intervention, i.e., autonomically. On the other, there is a need to align management decisions with company's business strategies and maximize relevant financial and non-financial business value metrics, by using BDIM approaches. Support for autonomic BDIM requires novel modeling of policies that guide management, additional monitoring solutions (e.g., if customer satisfaction is strategically important, it should be monitored), innovative decision making algorithms that maximize business value metrics important for the used business strategies, as well as management tools that provide the corresponding policy-driven monitoring and control. There are numerous challenges, particularly when non-financial business value metrics and corresponding business strategies are considered. In this paper, we mainly discussed the difficult challenges associated with: a) modeling of financial and non-financial business value metrics and their relationships with technical QoS metrics, and b) modeling of business value metric maximization strategies and their impact on adaptation decisions. We presented WS-Policy4MASC solutions to these challenges, as well as decision making algorithms that use this additional information for policy conflict resolution.

WS-Policy4MASC extends a widely-used industrial standard, WS-Policy, with information necessary for run-time management, including the unique support for autonomic BDIM. The specifications of diverse financial and non-financial business value metrics and business strategies that guide business-value driven selection among alternative control (adaptation) actions are the main distinctive characteristics and contributions of WS-Policy4MASC. WS-Policy4MASC also supports other management aspects, such as fault

Figure 3. Our algorithm for selection among conflicting action policy assertions

Inputs: A set of conflicting action policy assertions and the meta-policy assertion that describes their policy conflict resolution.

Step 1. *(Calculation of the overall business value metric for each conflicting action policy assertion)*

For each given conflicting action policy assertion:

1.1. Determine the set of all relevant utility policy assertions directly associated with execution of any of the actions contained in the current action policy assertion.

1.2. For each relevant utility policy assertion from Step 1.1:

1.2.1. If the policy conflict resolution strategy uses probabilities:

1.2.1.1. Find the probability policy assertion (if any) whose "When" condition is the same as for the currently processed utility policy assertion and calculate this probability (default: 1.0) – this is the probability that preconditions (e.g., correct execution of an action in the current action policy assertion) for the utility policy assertion will be satisfied.

1.2.1.2. Find the probability policy assertion (if any) whose "When" condition contains only the triggering event that the currently processed utility policy assertion is the correct estimate and calculate this probability (default: 1.0) – this is the probability that the estimate in the utility policy assertion will be correct.

1.2.2. For each business value metric in the current utility policy assertion:

1.2.2.1. Calculate the monetary amount and convert it to the common currency.

1.2.2.2. If the policy conflict resolution strategy uses probabilities, then multiply the converted monetary amount from Step 1.2.2.1 with the probabilities from Steps 1.2.1.1 and 1.2.1.2.

1.3. Sum all converted monetary amounts (for all business value metrics in all relevant utility policy assertions) determined in Step 1.2.2.2 (1.2.2.1, if probabilities not used), separately for each of the 8 business value metric categories – these are the 8 category-specific business value metric sums.

1.4. If the policy conflict resolution strategy specifies a cost limit constraint:

1.4.1. Sum all results from Step 1.3 for the 4 "cost" business value metric categories.

1.4.2. If the sum from Step 1.4.1 is higher than the given cost limit,

Then eliminate this action policy assertion,

Else sum all results from Step 1.3 for the business value metric categories relevant for the given policy conflict resolution strategy – this is the overall (summary) business value metric of the currently processed action policy assertion.

Step 2. *(Ordering of action policy assertions that satisfy all given constraints)*

If all conflicting action policy assertions were eliminated,

Then notify humans that no option satisfies all given constraints and exit this algorithm,

Else list the remaining action policy assertions in the descending order from the highest to the lowest result from Step 1.4.2.

Step 3. *(Tiebreaking)*

If the policy conflict resolution strategy specifies the first tiebreaker, then:

3.1. Determine the tiebreaking limit for the first tiebreaker.

3.2. Shorten the list from Step 2 to include only those conflicting action policy assertions for which the difference from the first element of the list (with the highest result from Step 1.4.2) is less or equal to the tiebreaking limit from Step 3.1.

3.3. For each short-listed conflicting action policy assertion, recalculate the overall business value metric (from Step .1.4.2) by summing all category-specific business value metric sums (from Step 1.3) that are relevant for the initial policy conflict resolution strategy (used in the "else" part of Step 1.4.2) or the first tiebreaker.

3.4. Re-order the short-listed action policy assertions in the descending order from the highest to the lowest result from Step 3.3.

3.5. If the policy conflict resolution strategy specifies the second tiebreaker, then repeat Steps 3.1-3.4 for the second tiebreaker, but use the results from Step 3.3 (instead of Step 1.4.2) and the list from Step 3.4 (instead of Step 2).

Step 4. *(Selection of the action policy assertion best from the business viewpoint)*

If 2 or more action policy assertions at the beginning of the list remaining from Step 3.5 have the same results for the overall (summary) business value metric,

Then use priorities of these action policy assertions to select the action policy assertion with the highest priority (if several action policy assertions have the same priority, use the one most recently added to the policy repository),

Else, select the first element of the list remaining from Step 3.5 – it will have the highest overall business value metric satisfying all given constraints and tiebreaking rules.

Outputs: The action policy assertion that should be executed.

management and maximization of technical QoS metrics. It has built-in constructs for specification of a wide range of adaptations and events common in management of service-oriented systems and business processes they implement. The WS-Policy4MASC language design and the algorithms using this policy information were successfully prototyped in the MASC middleware and the tools for annotation of UML models with WS-Policy4MASC policy assertions and run-time monitored data. In addition to demonstrating feasibility, we examined their correctness, expressiveness, effectiveness, and usefulness on realistic examples.

Some of the common issues for practical adoption of management solutions in general, and particularly autonomic BDIM solutions, are complexity and overhead of management activities. When WS-Policy4MASC policy assertions are written, in addition to writing policy assertions that exist in the traditional QoS-driven management, there is also a need to model business value metrics, determine numerical values for them, and map business strategies into policy conflict resolution strategies. At run-time, there can be an additional overhead of monitoring supplementary information (e.g., non-financial business value metrics) and executing the business-driven policy conflict resolution algorithm. In some situations, all this complexity and overhead will not provide sufficient benefits, because the traditional approaches maximizing technical QoS metrics will provide good-enough results. On the other hand, when impact of a wrong adaptation decision (e.g., in the example from Figure 1, choosing D when the business needs require choosing E) can have significant (or even catastrophic) negative business consequences, so the complexity and overhead of the presented solutions become worthwhile. Therefore, we treat the presented autonomic BDIM solutions as an optional layer beyond the traditional IT system management approaches. If they are unnecessary and too complicated for particular management situations, they should not be used.

While many challenges remain for research of autonomic BDIM and WS-Policy4MASC can be improved in different ways, several items have priority in our ongoing and future work. Most importantly, we continue with implementation of our solutions in the new MiniMASC middleware and plan additional evaluations, hopefully on real-life scenarios with industrial partners. We also work on facilitating practical adoption of WS-Policy4MASC, e.g., by improving the tools for translation of annotated UML models into WS-Policy4MASC files. For practical adoption of autonomic BDIM solutions, it would be also beneficial to relate the existing mechanisms for specification of business strategy with the mechanisms for specification of policies in autonomic BDIM solutions. Therefore, we have developed and currently evaluate integration of Business Motivation Model (BMM) hierarchies of ends (e.g., goals) and means (e.g., strategies) into WS-Policy4MASC policy assertions. Apart from making adoption by human managers easier, this extension broadens the range of strategies that can be modeled in WS-Policy4MASC. This work will be described in a forthcoming publication. We have also been working on addressing situations when multiple adaptation decisions can be chosen concurrently (e.g., different decisions for different classes of consumer), instead of choosing only 1 among several options. Integration with business intelligence and/or business activity monitoring solutions and design-time analysis of policies to preventively detect and resolve policy conflicts are some of the topics that we plan to study in the future.

ACKNOWLEDGMENT

NICTA is funded by the Australian Government as represented by the Department of Broadband, Communications and the Digital Economy and the Australian Research Council through the ICT Centre of Excellence program. Abdelkarim Erradi and Piyush Maheshwari contributed to the development of early versions of WS-Policy4MASC (up to the version 0.8), within the research project "Building Policy-Driven Middleware for QoS-Aware and Adaptive Web Services Composition" sponsored by the Australian Research Council (ARC) and Microsoft Australia. Rasangi Pumudu Karunaratne contributed to the development of WS-Policy4MASC version 0.9. Nahid Ebrahimi Nejad implemented and tested the described version of the algorithm for selection among conflicting action policy assertions.

REFERENCES

W3C Web Services Policy Working Group. (2007). *Web Services Policy 1.5 – Framework.* (W3C Recommendation 04 September 2007). Retrieved June 1, 2009, from http://www.w3.org/TR/ws-policy/

Baresi, L., Guinea, S., & Plebani, P. (2007). Policies and aspects for the supervision of BPEL processes. In J. Krogstie, A.L. Opdahl, & G. Sindre (Eds.), *Proceedings of CAiSE 2007, LNCS, Vol. 4495* (pp. 340-395). Berlin, Germany: Springer.

Bartolini, C., Sahai, A., & Sauve, J. P. (Eds.). (2006). *Proceedings of the First IEEE/IFIP Workshop on Business-Driven IT Management, BDIM'06*. Piscataway, USA: IEEE Press.

Bartolini, C., Sahai, A., & Sauve, J. P. (Eds.). (2007). *Proceedings of the Second IEEE/IFIP Workshop on Business-Driven IT Management, BDIM'07*. Piscataway, USA: IEEE Press.

Bartolini, C., Salle, M., & Trastour, D. (2006). IT service management driven by business objectives: An application to incident management. In J.L. Hellerstein & B. Stiller (Eds.), *Proceedings of NOMS 2006* (pp. 45-55). Piscataway, USA: IEEE Press.

Biffl, S., & Aurum, A. Boehm, B., Erdogmus, H., & Gruenbacher, P. (Eds.). (2006). *Value-based software engineering*. Berlin, Germany: Springer.

Casati, F., Shan, E., Dayal, U., & Shan, M.-C. (2003). Business-oriented management of Web services. *Communications of the ACM, 46*(10), 55–60. doi:10.1145/944217.944238

Erradi, A., Tosic, V., & Maheshwari, P. (2007). MASC -. NET-based middleware for adaptive composite Web services. In K. Birman, L.-J. Zhang, & J. Zhang (Eds.), *Proceedings of ICWS 2007* (pp. 727-734). Los Alamitos, USA: IEEE-CS Press.

Kaplan, R. S., & Norton, D. P. (1996). Using the balanced scorecard as a strategic management system. *Harvard Business Review, 74*(1), 75–85.

Keller, A., & Ludwig, H. (2003). The WSLA framework: Specifying and monitoring service level agreements for Web services. *Journal of Network and Systems Management, 11*(1), 57–81. doi:10.1023/A:1022445108617

Kephart, J. O., & Chess, D. M. (2003). The vision of autonomic computing. *Computer, 36*(1), 41–50. doi:10.1109/MC.2003.1160055

Kephart, J. O., & Walsh, W. E. (2004). An artificial intelligence perspective on autonomic computing policies. In M. Lupu & M. Kohli (Eds.), *Proceedings of Policy 2004* (pp. 3-12). Los Alamitos, USA: IEEE-CS Press.

Ludwig, H., Dan, A., & Kearney, R. (2004). Cremona: An architecture and library for creation and monitoring of WS-Agreements. In B.J. Kraemer, K.-J. Lin, & P. Narasimhan (Eds.), *Proceedings of ICSOC 2007* (pp. 65-74). New York, USA: ACM.

Lupu, E., & Sloman, M. (1999). Conflicts in policy-based distributed systems management. *IEEE Transactions on Software Engineering, 25*(6), 852–869. doi:10.1109/32.824414

Papazoglou, M. P., & Georgakopulos, D. (2003). Service-oriented computing. *Communications of the ACM, 46*(10), 25–28. doi:10.1145/944217.944233

Pautasso, C., Heinis, T., & Alonso, G. (2007). Autonomic execution of Web service compositions. In C.K. Chang & L.-J. Zhang (Eds.), *Proceedings of ICWS 2005* (pp. 435-442). Los Alamitos, USA: IEEE-CS Press.

Rosenberg, F., Enzi, C., Michlmayr, A., Platzer, C., & Dustdar, S. (2007) Integrating quality of service aspects in top-down business process development using WS-CDL and WS-BPEL. In M. Spies & M.B. Blake (Eds.), *Proceedings of EDOC 2007* (pp. 15-26). Los Alamitos, USA: IEEE-CS Press.

Salle, M. (2004). *IT service management and IT governance: Review, comparative analysis and their impact on utility computing*. (Technical Report HPL-2004-98). Palo Alto, USA: HP Laboratories. Retrieved June 1, 2009, from http://www.hpl.hp.com/techreports/2004/HPL-2004-98.pdf

Salle, M., & Bartolini, C. (2004). Management by contract. In R. Boutaba & S.-B. Kim (Eds.), *Proceedings of NOMS 2004* (pp. 787-800). Piscataway, USA: IEEE Press.

Sloman, M. (1994). Policy driven management for distributed systems. *Journal of Network and Systems Management, 2*(4), 333–360. doi:10.1007/BF02283186

Sloman, M. (1995). Management issues for distributed services. In N. Davies (Ed.), *Proceedings of SDNE'95* (pp. 52-59). Los Alamitos, USA: IEEE-CS Press.

Tosic, V. (2008). On modeling and maximizing business value for autonomic service-oriented systems. In D. Ardagna, M. Mecella, & J. Yang (Eds.), Proceedings of *BPM 2008 Workshops, LNBIP, Vol. 17* (pp. 407-418). Berlin, Germany: Springer.

Tosic, V., Erradi, A., & Maheshwari, P. (2007). WS-Policy4MASC - A WS-Policy extension used in the Manageable and Adaptable Service Compositions (MASC) middleware. In van der Aalst, W.M.P., L.-J. Zhang, & P.C.K. Hung (Eds.), *Proceedings of SCC 2007* (pp. 458-465). Los Alamitos, USA: IEEE-CS Press.

Tosic, V., Pagurek, B., Patel, K., Esfandiari, B., & Ma, W. (2005). Management applications of the Web Service Offerings Language (WSOL). *Information Systems, 30*(7), 564–586. doi:10.1016/j.is.2004.11.005

Tosic, V., Suleiman, B., & Lutfiyya, H. (2007). UML profiles for WS-Policy4MASC as support for business value driven engineering and management of Web services and their compositions. In M. Spies & M.B. Blake (Eds.), *Proceedings of EDOC 2007* (pp. 157-168). Los Alamitos, USA: IEEE-CS Press.

This work was previously published in International Journal of Systems and Service-Oriented Engineering, Volume 1, Issue 1, edited by Dickson K.W. Chiu, pp. 79-95, copyright 2010 by IGI Publishing (an imprint of IGI Global).

Chapter 9
ODACE SLA:
Ontology Driven Approach for Automatic Establishment of Service Level Agreements

Kaouthar Fakhfakh
LAAS-CNRS, Université de Toulouse, France and ReDCAD, University of Sfax, Tunisia

Said Tazi
LAAS-CNRS and Université de Toulouse, France

Tarak Chaari
ReDCAD and University of Sfax, Tunisia

Mohamed Jmaiel
ReDCAD and University of Sfax, Tunisia

Khalil Drira
LAAS-CNRS and Université de Toulouse, France

ABSTRACT

The establishment of Service Level Agreements between service providers and clients remains a complex task regarding the uninterrupted growth of the IT market. In fact, it is important to ensure a clear and fair establishment of these SLAs especially when providers and clients do not share the same technical knowledge. To address this problem, the authors started modeling client intentions and provider offers using ontologies. These models helped them in establishing and implementing a complete semantic matching approach containing four main steps. The first step consists of generating correspondences between the client and the provider terms by assigning certainties for their equivalence. The second step corrects and refines these certainties. In the third step, the authors evaluate the matching results using inference rules, and in the fourth step, a draft version of a Service Level Agreement is automatically generated in case of compatibility.

INTRODUCTION

According to the OASIS[1] definition "the Service Oriented Architecture (SOA) is a paradigm for organizing and utilizing distributed capabilities that may be under the control of different ownership domains". Thus, the ownership holds a key-role in realizing a SOA: for instance, who builds and makes a service available might be different from who consumes the service. As a consequence, service providers try to identify what are the potential requirements of their clients and develop the service accordingly. In the service oriented computing paradigm,

DOI: 10.4018/978-1-4666-1767-4.ch009

an SLA is a collection of service level requirements that have been negotiated and mutually agreed upon by the provider and the consumer. The Tele Management Forum worked out (in its SLA Management Handbook (TeleManagement Forum, 2004)) a split-up SLA lifecycle, consisting of six distinct phases: *Development, Negotiation, Implementation, Execution, Assessment, and Decommission (*Peer et al., 2006). Whilst the major part of SLA research concentrated on the development and the implementation issues of Service Level Agreements, the negotiation between customer and provider and therefore the creation of an agreement itself was insufficiently considered. Negotiation is one of the most important phases of an e-Business based collaboration, since it will define the conditions and terms that the service provider (and potentially customer) has to maintain during the lifetime of the collaboration. Obviously, one critical issue in the SLA lifecycle is to determine the Quality of Service (QoS) constraints in order to fulfill the client request. In fact, this request could be expressed using different words than the provider's technical language. For example, in the scope of the ITEA2's UseNet project[2], the public transportation network is equipped with high end communication facilities. A user of this network may want to download recent films on his netbook while waiting for the bus. Knowing that the waiting is about ten minutes, the download time of the film should not exceed this time slot. He also wants to pay less than 3 euros per film. On the other side, a provider has offers that can meet the client requirements but they are expressed in a different technical language. For example, the offers of the provider are based on the bandwidth that will be given for each user. The latter can be a non expert in the IT field and he may not understand the technical aspects of the bandwidth term and how it is computed. In this kind of situations, a fair and "intelligent" negotiation process is needed to match the client needs with the provider offers. The usual negotiation process consists in selecting a subset of clauses and values among predefined choices by the provider. However, the client may not understand these offers and, thus, has not the opportunity to express his needs with his own knowledge and language.

The challenge of our work is to propose an approach that helps the provider in analyzing a considerable number of client requirements (expressed in different words) and in identifying the suitable services or products that effectively meet their needs. In case of compatibility, we also aim to automatically generate an SLA between them. Consequently, our first objective in this work is to establish the necessary semantic enabled models that facilitate capturing the client's requirements and the provider's offers expressed by their own knowledge and their own languages. Our second objective is to analyze the correspondence between these models by defining an automatic semantic enabled matching process between them. In case of compatibility, our third objective consists in automatically generating a complete draft of an SLA between the client and the provider.

This paper reports innovative research on SLA with a focus on autonomous matching and establishment of QoS constraints. Our novel approach is composed of four steps. The first step consists in generating the correspondences between the client and the provider terms by assigning certainties to their similarity. The second step consists in refining and stabilizing these certainties in order to reduce the similarity measurement errors. The third step is a matching evaluation step in which we proceed to check the global similarity of the created correspondences. In this same step, we also use these correspondences to verify if the client constraints are satisfied according to the provider offers. Finally, in case of compatibility, our approach uses semantic inference techniques to automatically generate a draft version of an SLA in order to send it to the client after its validation by the commercial expert of the provider.

In Section "Modeling client intentions and provider offers" of this paper, we present the semantic enabled models that we have defined to capture the client's intentions and the provider's offers. In Section "A semantic matching approach between client intentions and provider offers", we detail our automatic matching approach between these intentions and offers: *Ontology Driven Approach for automatiC Establishment of Service Level Agreements* (*ODACE SLA*). In the next Section, we present the implementation and the testing results of our matching process. Before concluding, in Section "Related work", we present some existing negotiation solutions between clients and service providers and we compare them with our approach. We also cite some existing matching approaches and their inadequateness to resolve the issue of this paper.

MODELING CLIENT INTENTIONS AND PROVIDER OFFERS

In this Section, we present the high level semantic enabled models that we developed to facilitate the expression of client intentions and provider offers. These models are based on ontologies (Studer et al., 1998) offering reasoning and inference techniques that help in automating contract negotiation between service providers and clients. Ontologies provide a formal, syntactic, and semantic description model of concepts, properties and relationships between concepts. They also provide inference languages that can be used for reasoning about concepts and for data sources matching (Doan et al., 2004). We defined the *ClientOnto* ontology to describe the client intention and the *ProviderOnto* ontology to describe the provider offers. In our approach, we model an ontology O as: O = {C, R} where

- $C = \{C_i\}$: is the set of all the concepts C_i of the ontology O
    - $Ci = \{C_{i,k}\}$: is the set of all the individuals of C_i
- $R = \{R_n\}$: is the set of all the relationships $R_n = (C_i, C_j)$ between the concepts of the ontology
    - $R_{n,k} = (C_{i,k}, C_{j,m})$: is the set of the R_n relationship instances

ClientOnto: High Level Model of Client Intentions

The intention corresponds to a mental state of any actor who executes an action. Several works (Toma et al., 2006) attempted to account the relations between an action undertaken by a human being and the mental state which guides this action. To be able to use these relations in programs and information systems, it is necessary to formalize the notion of intention.

Many models tried to represent the intentions of the users and theirs preferences. We can cite (Garcia et al., 2008; Lamparter et al., 2007; Wang et al., 2006). The major drawback of such models is their complexity. In fact, they use the same models as the provider offers representations that contain complex technical terms that cannot be understood by the client. We adapted and enriched the intentional model of (Kanso et al., 2007) by adding service level agreements negotiation special elements. Figure 1 shows our client intentional model. This latter is characterized by a goal to achieve by performing an action. This goal can be for entertainment or educational purposes for example. The client wants to perform an action and to obtain an intended result. An action is defined by a *subject* to describe the searched products by the client according to constraints in order to express requirements and conditions. Each constraint is defined by a *property*, a comparison *threshold* and an *operator* as shown in the dashed rectangle in Figure 1. For example, the client can formulate a constraint on a service price. In this case, the *property* is the 'price',

Figure 1. The structure of the client model: ClientOnto

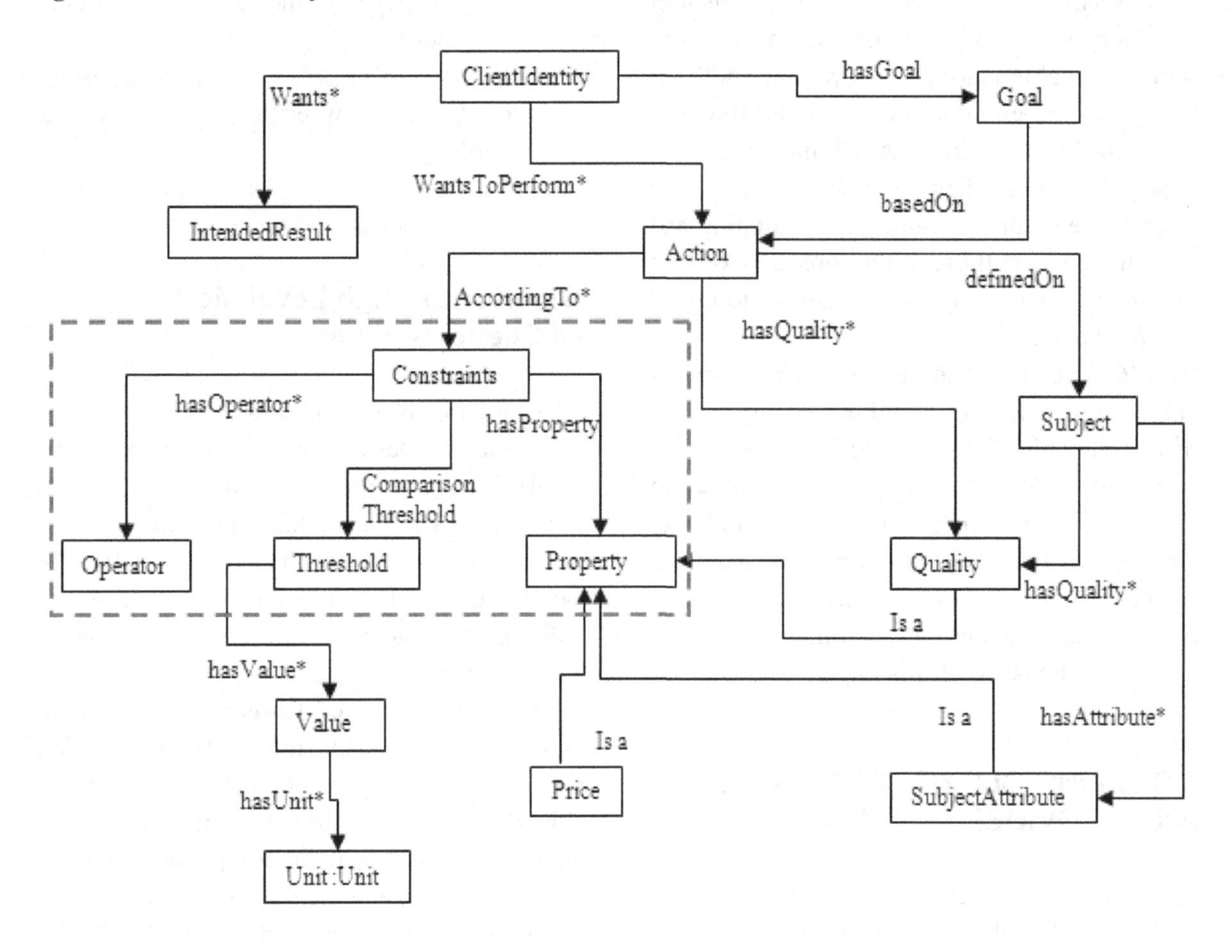

the *threshold* is the maximum 'value' fixed by the client and the *operator* is 'inferior'. We give a complete example of a *ClientOnto* instance in Section "Implementation and experimentation".

ProviderOnto: Business and Commercial Provider Offers Model

In the literature, we noticed that there are interesting description languages for provider offers. The most known and used language is WSOL (Web Services Offerings Language) (Tosic et al., 2004). It is an XML based language that allows specifying QoS levels attached to WSDL service descriptions. WSOL can define constraints on QoS dimensions. However, information extraction from this model is difficult to perform due its complexity.

For example, constraints are expressed as strings in this language. Consequently, WSOL does not help in matching the provider offers with the client requirements using semantic approaches. Figure 2 presents the simplified provider offer model[3] that we have defined to facilitate the matching process with the client intention. The root concept of this ontology is *ProviderIdentity* that identifies any provider. Each provider has offers that are defined by products, constraints and guarantees. A product can be described by a service or a service output. Service definitions are expressed by OWL-S (OWL-S Coalition, 2004) imports in the *ServiceDefinition* concept. A constraint is defined by a property (that can be a quality, a product feature or a price), a threshold and an operator. The property and the threshold

Figure 2. The structure of the provider model: ProviderOnto

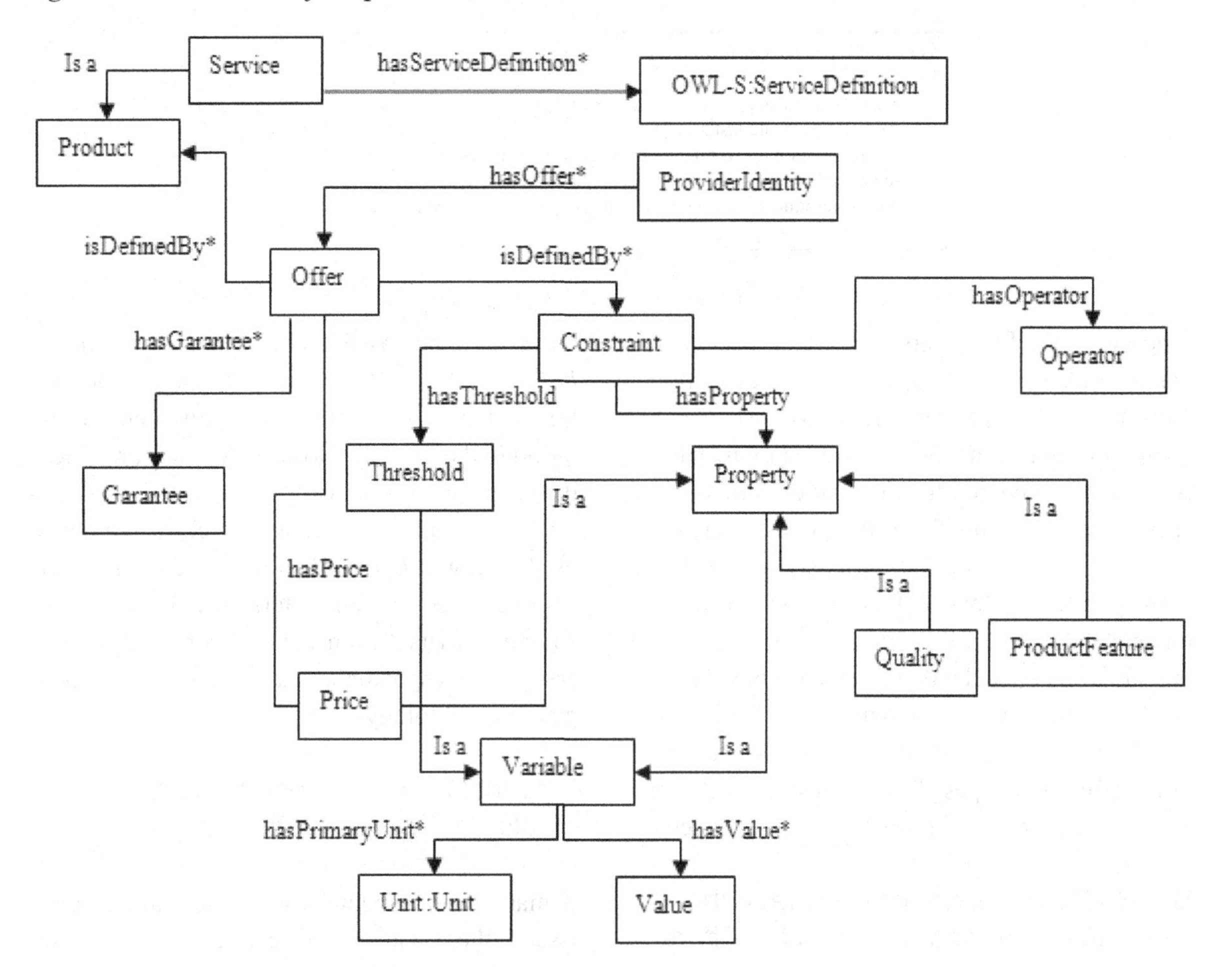

are defined as variables. Each variable has a measurement unit and a value. The constraints defined by the provider can be functional constraints, non functional constraints, price constraints or access right constraints. We give a complete *ProviderOnto* instance in Section "Implementation and experimentation".

A Semantic Matching Approach between Client Intentions and Provider Offers

In this Section, we present our semantic matching approach *"Ontology Driven Approach for automatiC Establishment of Service Level Agreements (ODACE SLA)"* between the client intentions and the provider offers according to the models that we presented in Section "Modeling client intentions and provider offers" of this paper.

Our approach is composed of four main steps: the correspondences generation step, the correspondences refinement step, the correspondences evaluation step and the SLA generation step in case of successful matching. In the remaining parts of this Section, we give more details about each step.

Correspondences Generation

The first step of our matching process consists in searching all the possible correspondences between the two ontologies (client intention and provider offers). We define a correspondence as

Listing 1 Correspondence algorithm between the client terms and the provider terms

```
generateCorrespondences(rootTerm1, rootTerm2)
{
1 For each individual1 instanceOf rootTerm1Do
2 For each individual2 instanceOf rootTerm2Do
3 semanticSimilarity:= wordnet::similarity::vector(individual1, individual2);
4 if semanticSimilarity > MIN_SIMILARITY
5 createCorrespondence(individual1, individual2, semanticSimilarity);
}
```

a possible similarity between two terms. Each correspondence has a source term, a destination term and a certainty represented by a float between 0 and 1. It represents the semantic similarity between the source and the destination terms. For each couple of terms from these ontologies, we evaluate their semantic similarity and we create a correspondence instance if the similarity value is higher than a selected threshold. In fact, below this threshold, we consider that the correspondence is weak and the terms are semantically disjoint. To link the client intention with the provider offers we have defined two types of correspondences: direct correspondences and indirect correspondences.

Direct Correspondences between the Client Intention and the Provider Offers

To compute the semantic similarities between the source terms and the destination terms of the correspondences, we used WordNet::Similarity (Pedersen et al., 2004) which is a Perl tool providing several similarity measures between words. This tool supports three main semantic similarity measures: hso (Hirst et al., 1998), lesk (Banerjee et al., 2003), and Gloss vector (Patwardhan, 2003). (Steffen et al., 2006) proved with their experiments that the Gloss Vector gives the most accurate measures using words meanings. Consequently, we used this measure to develop an algorithm that creates the direct correspondences between the client and the provider ontologies.

The first step of this algorithm (as presented in Listing 1) consists in browsing all the terms of the client intentions and the provider offers to retrieve all the possible terms pairs *(individual1, individual2)* from their corresponding ontologies. For each pair, we evaluate the equivalence of its individuals using the Gloss Vector measure. Then, if the resulting similarity is not too low (line 4 in Listing 1), we create a new correspondence by defining *individual1* as its source term, *individual2* as its destination term and the computed similarity as its certainty (line 5 in Listing 1). The created correspondences are then stored in a new matching ontology.

Indirect Correspondences Using a Quality of Service Ontology

In many cases, especially when we deal with QoS issues, there can be indirect relations between terms. For example, if the client has defined a constraint on the *download time* of a film, this can be mathematically computed using the *film size* and the *throughput* offered by the provider. In the literature, we find some existing QoS ontologies like QoSOnt (Dobson et al., 2005) and SL-ontology (Steffen et al., 2006). However, they don't express mathematical dependencies between QoS metrics when they are composite. Consequently, we can't compute the values of these composite metrics.

To automatically detect this kind of correspondences, we defined a QoS ontology named *QoSOnto* (see Figure 3) that can hold all the basic mathematical relations between QoS terms. In this ontology, we extended the OWL-QoS model by further elements to be able to compute composite metrics. In fact, we have defined an implement-

Figure 3. The QoS model:QoSOnto

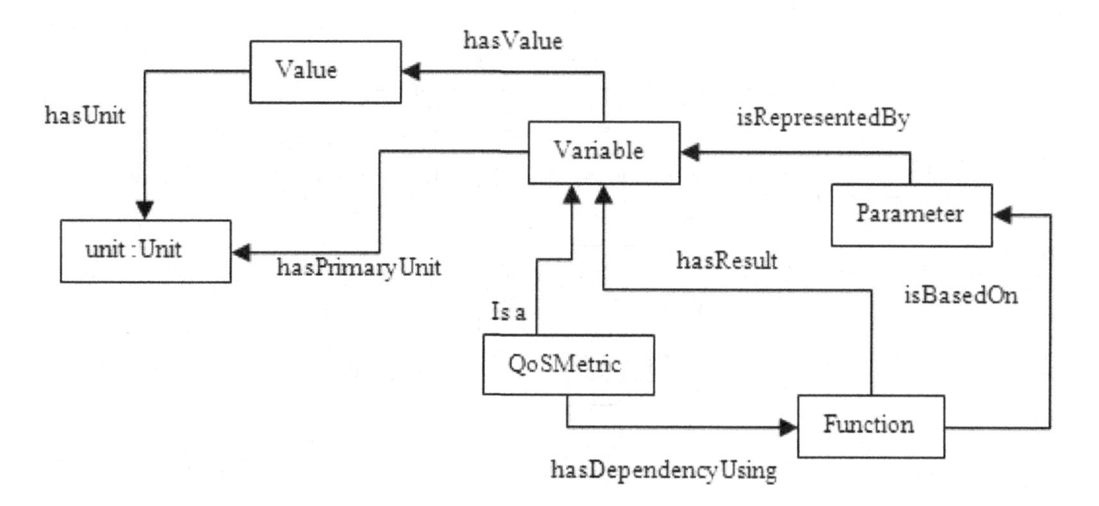

ing class for each dependency function between metrics. This class implements a super interface based on an abstract function which computes the value of the composite metric. We have also defined a range for the operands of each function to specify the order that has to be respected to compute the composite values correctly. Composite QoS Metrics can be specified using our QoSOnto model. In fact, a function output can constitute an input for another function. This composition is illustrated in Figure 3 by the "Is a" relation between the *QosMetric* and the *Variable* concepts.

To detect the indirect correspondences, we use the algorithm presented in the direct correspondences Section (Listing 1) two times. In the first time, we apply it on the *ClientOnto* terms and the *QoSOnto* terms. In the second time, we use it between the *QoSOnto* terms and the *ProviderOnto* terms. Then, we store all the created correspondences in the matching ontology. Figure 4 shows an example of an indirect correspondence using *QoSOnto*. The *Film download time* term has a correspondence with a *transfer time* term in the QoS ontology. The corresponding value of this term depends on the *Transfer size* and the *transfer bandwidth*. These two terms have correspondences with other terms in the provider ontology.

Correspondence Refinement

In this step, we proceed to the refinement of the generated correspondences in the previous step in order to obtain more correct semantic similarities between client and provider terms. Indeed, due to the existence of several meanings for a same term, some isolated errors can occur in the similarity measurements between the client and the provider terms. To correct these errors, we have defined a refinement algorithm in our global matching approach. In this algorithm, we start by generating adjacencies between the created correspondences according to their linguistic relatedness and we finish by applying a stabilization process to correct the wrong isolated values.

Adjacency Generation between Correspondences

We define an adjacency as two linguistically related correspondences according to their source terms or destination terms. An adjacency (see Figure 5) is characterized by a source correspondence, a destination correspondence and a certainty (float between 0 and 1) for its existence. This certainty expresses the linguistic relatedness measure between the source or destination terms

Figure 4. Indirect correspondence example

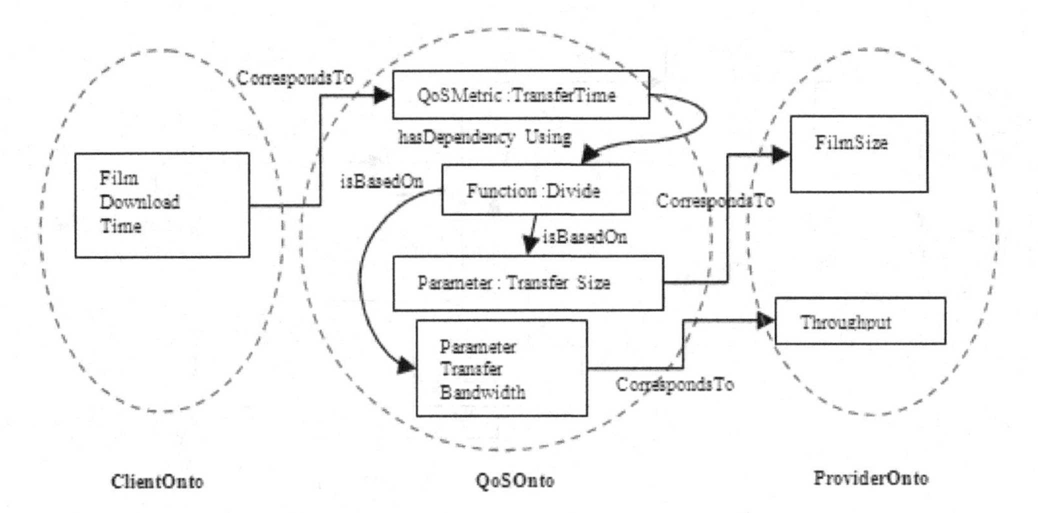

of the two considered correspondences. After comparing the accuracy of the existing linguistic relatedness measures in the Wordnet::Similarity Perl module Lch (Leacock et al., 1998), Wup (Wu et al., 1994), and Path (Ted et al., 2004), we choose the Path measure to assign weights to the adjacencies linking the created correspondences of the previous step in our global matching approach.

The algorithm of this step consists in browsing all the created correspondences in the previous steps (lines 1 to 6 in Listing 2) and in creating adjacencies (line 17 in Listing 2) between them when they have directly linked source or destination terms in the client, provider or QoS ontologies. For each created adjacency, we compute its average linguistic relatedness (line 16 in Listing 2) between the source terms and the destination terms of its source and destination correspondences.

Stabilization Process

To correct and refine the calculated semantic similarity measures between the correspondences, we defined a stabilization algorithm. In this algorithm (Listing 3), each correspondence will be affected by its adjacent correspondences to calculate its new

Figure 5. Correspondence stabilization formula

$$pi' = \frac{pi + \sum_{j=1}^{n}(\frac{Wij}{n} * pj)}{1 + \sum_{j=1}^{n}(\frac{Wij}{n})}$$

(1)

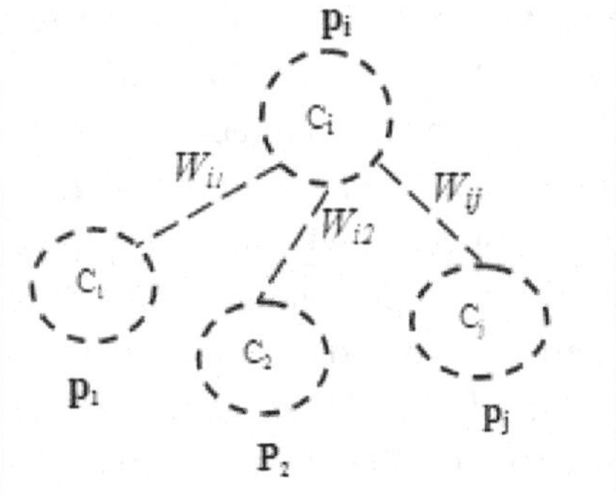

Pi C_i W_i1 W_i2 W_ij C_1 C_2 C_j P_1 P_2 Pj

Listing 2 Algorithm for creating correspondence adjacencies between client and provider terms

```
createCorrespondenceAdjacencies()
{
1 For each correspondence1 instanceOf Correspondence Do
2 sourceTerm1:= getSourceTerm(correspondence1);
3 destinationTerm1:= get DestinationTerm(correspondence1);
4 For each correspondence2≠correspondence1 instanceOf Correspondence Do
5 sourceTerm2:= getSourceTerm(correspondence2);
6 destinationTerm2:= getDestinationTerm(correspondence2);
7 linguisticRelatedness:=0;
8 adjacencySides:=0;
9 If ∃ R ∈ R_client OR R ∈ R_qos | R=(sourceTerm1, sourceTerm2)
10 linguisticRelatedness:= wordnet::relatedness(sourceTerm1, sourceTerm2);
11 adjacencySides++;
12 If ∃ R ∈ R_qos OR R ∈ R_provider | R=(destinationTerm1, destinationTerm2)
13 linguisticRelatedness+= wordnet::relatedness(destinationTerm1, destinationTerm2);
14 adjacencySides++;
15 If adjacencySides ≠ 0
16 linguisticRelatedness:= linguisticRelatedness / adjacencySides;
17 createAdjacency(correspondence1, correspondence2, linguisticRelatedness);
```

certainty (pi' in Figure 5) using a weighted average formula (1). This latter is compared repetitively with the old certainty until its stabilization.

The correspondence is considered as stabilized when two successive certainty values are too close, i.e., the difference between them is inferior to a predefined certainty precision (0.001 for example) like mentioned in line 11 of Listing 3. The "evidence" correspondences having a maximum certainty (equals to 1) are not modified and are considered as stabilized by default.

Our average formula is based on the similarity flooding approach of (Melnik et al., 2002). This approach states that if a node C_i has adjacent nodes C_1, C_2 … C_j with high certainties for their existence, then C_i should have a high certainty too. The principle of our average formula (1) is: for each correspondence C_i which has a certainty p_i (as shown on the right part of Figure 5), we get all the neighbor correspondences C_j and the weight W_{ij} of their corresponding adjacencies. W_{ij} is calculated using the Path measure of WordNet::Similarity applied on the source terms and the destination terms of the correspondences C_i and C_j. These weights are used to calculate the new certainty value p_i' of the correspondence C_i according to the formula of Figure 5. In this formula, n is the number of the adjacent correspondences of C_j.

Correspondence Evaluation

The third step of our global matching approach consists in evaluating the correspondences generated by the first two steps. It is composed of two types of evaluations: *structural evaluation* and *constraints evaluation*. The first one is considered to test if a sufficient number of the terms defined in the client intention have correspondences with the provider offers terms. It consists in calculating the global semantic correspondence degree with the provider ontology. This calculation is an optimistic average of the correspondence certainties attached to the client terms. We take the maximum of the correspondence similarity values for each client term. Then, we calculate the global average of these maximum values. If the result is less than a fixed threshold by the provider's commercial expert, we consider that there is no compatibility between the client intention and the provider offer. In this case, there is no need to drive further analysis to check the client constraints because his needs are totally different from the provider offers. The second

Listing 3 Correspondence stabilization algorithm

```
correspondenceStabilization()
{
1 For each correspondence1 instanceOf Correspondence Do
2 oldCertainty:= correspondence1.certainty;
3 ExtraCertainties:=0;
4 AdjacenciesRelatedness:=0;
5 For each adjacency instanceOf Adjacency | adjacency.hasExtremity(correspondence1) Do
6 correspondence2:= adjacency.hasExtremity | correspondence2≠correspondence1;
7 AdjacenciesRelatedness+= adjacency.relatedness / correspondence2.adjacenciesCount;
8 ExtraCertainties+= adjacency.relatedness*correspondence2.certainty / correspondence2.adjacenciesCount;
9 newCertainty:= (oldCertainty + ExtraCertainties)/(1 + AdjacenciesRelatedness);
10 correspondence1.certainty:= newCertainty;
11 If newCertainty-oldCertainty<CERTAINTY_PRECISION mark(correspondence1,"stabilized")
12 If ∃ correspondence instanceOf Correspondence | correspondence.isMarked("unstabilized")
13 correspondenceStabilization();
}
```

evaluation consists in verifying the client constraints with their corresponding available values at the provider side. For each client constraint, we consider its *Property* and we try to get its corresponding value (or interval value) from the provider ontology. For example, in Figure 4, the property *Film Download Time* is computed using the *Film Size* and the *Throughput* values at the provider side. These computed values are stored in the matching ontology after their direct extraction or indirect calculation (like the *Film Download Time* property) according to their corresponding values from the provider offers. We defined a SWRL (Semantic Web Rule Language) rule for each of these two cases.

Listing 4 presents an automatically generated SWRL rule that calculates the possible *download times* according to the provider's *throughputs* and *film sizes*. Unit conversion is not presented in this rule for clarity reasons.

After computing all the possible values for each client property, we check if there are combinations that can satisfy all the constraints. Listing 5 illustrates a simplified version of the generated SWRL rule that checks the compatibility of the provider offers with the client constraints.

If there are client constraint properties without direct or indirect relations with the provider ontology or if the client constraint is not verified by any provider value, we affirm that there is an

Listing 4 Calculate Download Time SWRL Rule

```
qos:hasDependencyUsing(qos:transferTime, ?function) ∧ qos:hasImplementingClass(?function, ?impClass) ∧ hasCorrespondingValue(TransferSizeQosValues, ?transferSizeQoSValue) ∧ hasCorrespondingValue(ThroughputQoSValues, ?throughputQoSValue) ∧
matchingTools:calculateFunction(?impClass, ?transferSizeQoSValue, 1, ?throughputQoSValue, 2, ?result)
→hasCorrespondingValue (downloadTimeProviderValues, ?result)
```

Listing 5 Check Constraints SWRL Rule

```
hasCorrespondingValue (downloadTimeProviderValues, ?downloadTimeValue) ∧ hasUnit(?downloadTimeValue,?providerUnit) ∧ hasUnit(client:downloadTime,?clientUnit) ∧hasConvertionFactor(?providerUnit,?clientUnit,?factor) ∧ swrlb:multiply(?providerValue,?downloadTimeValue,?factor)
∧ <other constraints can be added here> → matchingTools:checkConstraints(client:DownloadTime, ?downloadTimeOperator,
?downloadTimeValue <other properties can be added here...>)
```

Listing 6 Generate predicate SWRL rule

```
client:Constraints(?c)∧client:hasProperty(?c, ?property)
→ query:select(?c, ?property)∧matchingTools:createPredicate(?c, ?property)
```

incompatibility between the client intention and the provider offers. In case of compatibility, we automatically generate a complete SLA draft that should be validated by the commercial expert before its proposal to the client.

SLA Generation in Case of Successful Matching

If the previous steps give a successful compatibility between the client intention and the provider offers, we proceed to the generation of an SLA draft based on the *SLAOnt* (Service Level Agreements Ontology) structure (Fakhfakh et al., 2009). *SLAOnt* defines an ontology describing the concepts and the properties needed in a QoS contract. It's a semantically rich SLA model that we have developed in a previous work to automate SLA monitoring and to take concrete actions in case of violations.

The SLA generation principle is based on SWRL rules (Horrocks et al., 2004) to retrieve the necessary information from the client and the provider ontologies. Indeed, the measurement directive instances, their associated protocols and their operations in the provider ontology will constitute respectively the *measurement directive*, the *protocols* and the *operation* values in the *SLAOnt* ontology instance. In case of direct correspondence between the client and the provider ontologies, the *metrics* definition will be extracted from the *provider properties* which have a valid correspondence with the client property. The correspondence is considered as valid if its certainty is above than a threshold fixed by a commercial expert at the provider side (we took a 0.6 threshold for our experiments). In case of indirect correspondence, the metric definition will be extracted from the provider property which has a valid correspondence with a variable in the QoS ontology. Each variable must have a valid correspondence with a client property. Each property instance of the client ontology will constitute an SLA parameter in the generated *SLAOnt* instance. We also defined a SWRL rule (*generatePredicate rule*) as shown in Listing 6 to get the *property*, the *threshold* value and the *operator* of each client constraint. For each constraint, this rule will create a corresponding predicate (contract clause) that should be respected by the provider. When the predicate is not satisfied, the associated guarantee (in the provider ontology) is automatically triggered.

Listing 7 presents a concrete example of a generated SLA predicate in our approach. If the download time parameter is greater than or equal to 10 minutes, the action *disseminateViolation* is triggered.

The automatic generation of the SLA is based on the client terms (that can be directly or indirectly deducted from the provider offers) and not on the provider technical terms. Consequently, the SLA can be easily understood by the client. He is not obliged to sign a contract containing complex technical clauses predefined by the provider.

Listing 7 Download time predicate evaluation rule

```
slaont:hasEvaluation(downloadTimeParameter, ?x) ∧swrlb:greaterThanOrEqual(?x, 10)
→ slaActions:disseminateViolation(dowloadTimeConstraint, false, downloadTimeParameter, ?x)
```

Figure 6. The technical architecture of our matching approach

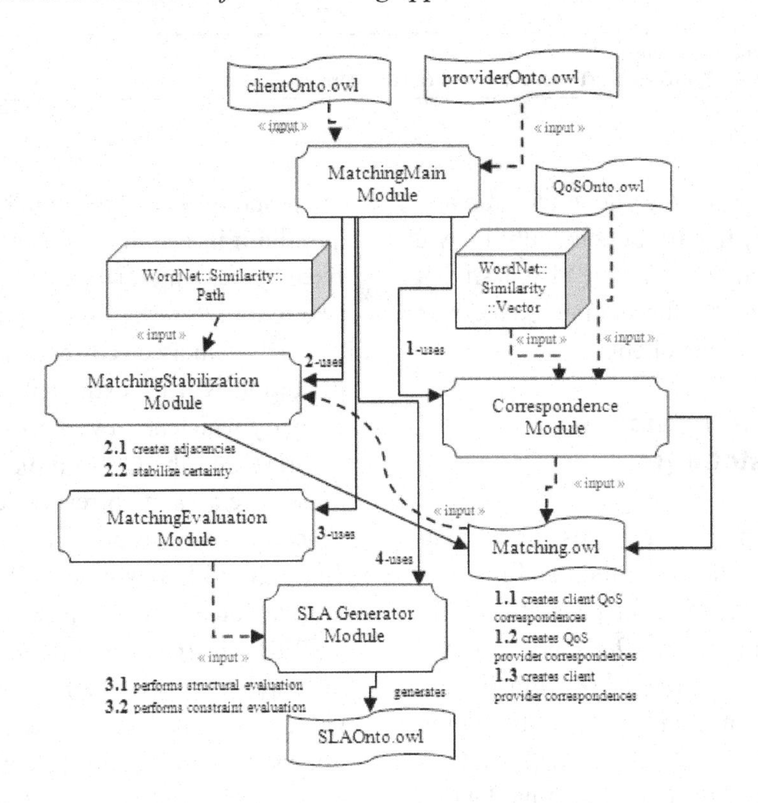

IMPLEMENTATION AND EXPERIMENTATION

In this Section, we present the Java tool that we have developed to implement all the steps of our matching approach. We start by describing the technical architecture of our tool and its implementation. Then, we detail the results of our approach in a business use case in the scope of European UseNet project.

Technical Architecture of Our Matching Approach

Our matching tool comprises four modules orchestrated by a fifth matching main module. Figure 6 shows the technical architecture that performs the matching process between the client intentions and the provider offers.

The matching main module of our prototype takes in input the client ontology instance and the provider ontology instance as owl (Mc Guiness et al., 2004) files. Then, in a first step, it uses the correspondence module to create the possible correspondences between the client, the provider and the QoS ontologies according to the WordNet::Similarity::Vector measure from Ted pederon's Perl module. In a second step, the matching main module uses the matching stabilization module (which is based on the WordNet::Similarity::Path measure) to create adjacencies between the identified correspondences in the previous step and to stabilize their certainties. In the third step, we use the matching evaluation module to perform the structural and the constraints evaluation in order to verify the compatibility of the client intention with the provider offers. Finally, in case of compatibility, the SLA generator module automatically generates a draft version of an SLA as an owl file.

Our matching tool is mainly based on (1) the OWL language to implement the client intentions and the provider offers, (2) SQWRL language to query the used ontologies and (3) SWRL rules (Horrocksb et al., 2004) to evaluate the compatibility of the client intention with the provider offers.

We used the protégé OWL API to execute these rules and queries. This API has several core software components, including (1) an editor that supports interactive creating, editing, reading, and writing SWRL rules; (2) a rule engine bridge that provides the necessary infrastructure to interoperate with third-party rule engines and OWL reasoners; (3) a bridge that provides a mechanism for defining built-in libraries; and (4) an extensive set of predefined built-in libraries. To perform all the steps of our matching approach, we extended the predefined libraries by personalized built-ins in our SWRL rules. We used the Jess rule engine[4] to evaluate these rules and to produce the inferred knowledge.

Figure 7 shows the technical layers of our matching tool modules. Each layer is used by its next upper one. In the lowest layer, the Xerces XML API[5] is used to read raw XML data from the ontologies (which are implemented using the owl language). Above this layer, the Protégé OWL API[6] is used to directly handle owl data in the ontologies. The SWRL Jess API[7] is used to make inferences and reasoning on the ontologies instances. In the upper layer, we have developed java classes that extract the SWRL rules defined in the OWL ontologies and use the SWRL engine Bridge to execute them. All these layers are implemented by each module of our approach.

Business Use Case

In this Section, we detail the results of our approach in the case study that we mentioned in the introduction of this paper in the scope of the ITEA2's UseNet project. A client (who is not an expert in IT field) wants to download comedy films with a download time less than ten minutes and a maximum price of three euros per film as described in the *clientOnto*[8] instance. On the other side, the provider proposes a comedy film gold offer among several other offers. This offer is defined in the *ProviderOnto*[9] instance by (1) a comedy film type constraint in which the property "*film type*" is equal to "*comedy*", (2) a comedy film gold price which is equal to two euros per film and (3) a gold throughput constraint (throughput equals to 8 Mbits per second).

We notice that this provider can satisfy the film type constraint and the client price constraint. Concerning the download time constraint, our approach uses the QoS ontology (*QosOnto*) to calculate the client download time which is equal to the division of the *transferred size* value by the *transfer bandwidth* value. This provider has films of 300 MB. Therefore, he can satisfy all the client constraints which are: *DownloadTimeConstraint*, *PriceConstraint*, and *FilmTypeConstraint (*in *clientOnto)*. Consequently, our tool generates a complete SLA automatically. The property of each client constraint is transformed to an *SLAParameter* in the generated SLA. Each parameter is attached to a set of metrics in two possible ways: directly from the provider (the case of *PriceParameter* for example) or indirectly using the QoS ontology (the case of *DownloadTimeParameter*). These parameters and predicates will be used in a future SLA monitoring stage (Fakhfakh et al., 2009).

Figure 7. Implementation layers of our matching tool modules

Java Program (matching module)	SLAMatchingMain SWRLBuildingLibraryIMPL
SWRL Jess API	SWRLEngineBridge SWRLFactory
Protégé OWL API	OWL model
Xerces XML API	XML Parser
Ontologies OWL (ClientOnto.owl, ProviderOnto.owl, QoSOnto.owl, SLAOnt.owl, SLAMatching.owl,)	

Figure 8. Samples of correspondences before and after stabilization

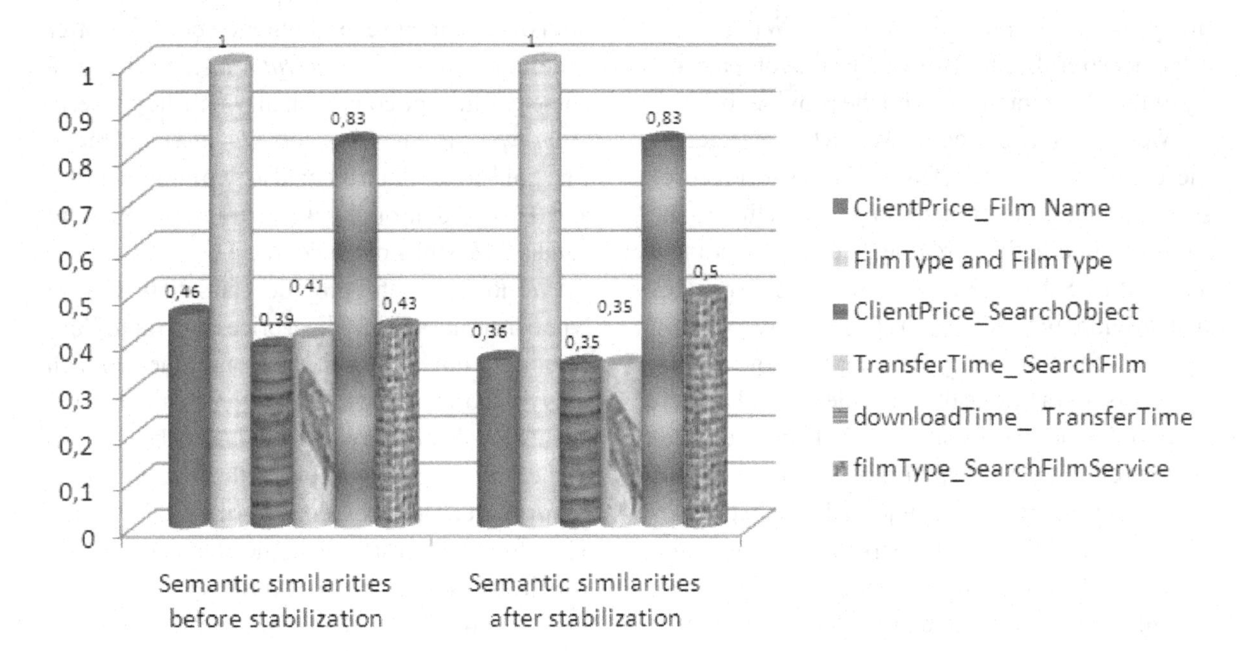

The generated OWL contract of this case study is available at (http://www.redcad.org/members/kaouthar.fakhfakh/negotiation/Contract.owl).

Results and Discussions

To give an idea about the generated results of our approach, we present in Figure 8 some samples of correspondences before and after the stabilization step. We notice that the stabilization process produces very good global results. However, we noticed that there are some values that should decrease after the stabilization but they increased (like the *[filmType, SearchFilmService]* correspondence). This is mainly due to the high certainties of the adjacent correspondences. It is important to mention that this increased difference is minimal and does not affect the constraint evaluation step which is the most important part of our matching process. In these values, the evidence correspondences having a "1" certainty are kept as they are. Consequently, they are not included in the stabilization process. They can influence their adjacent correspondences but they can't be influenced (like the evidence correspondence [*FilmType, FilmType*] in Figure 8).

Overall, the precision of the resulting correspondences of our approach is very promising (93% in our case study). This precision is calculated using the following formula (Do et al., 2003) (2):

$$Precision = \frac{|TP|}{|TP + FP|} \quad (2)$$

where TP and FP are the count of true positive and false positive correspondences compared with the human evaluations. However, our approach is time consuming. Its complexity is about O(c*p) where 'c' and 'p' are respectively the terms count of the client intention ontology instance and the provider offers ontology instance. In our case study, the number of client terms is 23 (c=23) and the number of provider terms is 70 (p=70). To reduce this complexity, we have defined some stop terms (like '*the*', '*of*' etc) and excluded classes (like the '*ClientIdentity*' and the '*ProviderIdentity*' classes) in the two models. These terms and classes do not affect the matching result and reduces the terms count of the client intention and the provider offers.

The matching process took 10 minutes and 13 seconds on a PC equipped with an I5 processor and 4GB of RAM. This is mainly due to the heavy communication with the Wordnet::Similarity Perl module that computes the semantic similarity between words. This module cannot handle parallel requests and sends a "busy" message if we don't inject some delays between the semantic similarity requests. After some changes in the communication with this module, we succeeded to reduce the execution time to 3 minutes and 18 seconds by running the first two matching steps two times: the first one to extract the pairs of terms that have to be semantically compared and the second time for the effective evaluation using the Wordnet::Similarity values. In fact, we noticed that Ted Pedersen's Perl module responds faster if all the terms pairs are given together in a file. We also stored the computed measures by this module in a hash-coded cache to accelerate their extraction by our semantic matching tool. If all the necessary values exist in the cache, our matching tool took 61 seconds on the same PC to run all the steps in our digital video library case study including the generation of a complete SLA. A video file (available at http://www.redcad.org/members/kaouthar.fakhfakh/negotiation/OdaceSLA.avi) gives an idea about the execution of our developed tool in this case study. We have made several tests in this application by changing the client intentions and the provider offers values. Our prototype can effectively detect partial compatibilities between these intentions and offers. It can also indicate the corresponding provider's values for the unsatisfied client constraints.

Actually, we are applying our approach on another use case for collaborative field activities where members (clients) want continuous video streams without lags from the providers.

Related Work

Automating the contract negotiation belongs to research issues pursued by multiple projects. They are mainly related to architectural solutions (Hasselmeyer et al., 2006; Keller et al., 2003) for negotiation or models for protocols. Our approach focuses mainly on the negotiation related protocols. The main protocols that we have identified in the literature are *WS-Agreement* protocol (Andrieux et al., 2007), *BEinGRID* protocol (Dimitrakos, 2010) and *NextGRID* protocol (Francis, 2008). *WS-Agreement* protocol is based on the principle of "take it or leave it", i.e., "approve or decline" completed by a signature phase if it is approved by the client. It is based on an XML structure to represent contracts. The problem of such solutions remains in the comprehension of the negotiation protocol which can be not understood by the client. The second protocol named *BEinGRID* protocol is based on templates defined by the provider. In *BEinGRID*, the expression of the client's needs is limited by the template structure. The third protocol named *NextGRID* protocol is specific to the grids. It is based on a predefined template for grids and its negotiation vision is very limited.

Overall, a negotiation protocol describes the actions, messages and rules to be applied by the participants. We noticed that there is a big interest to the actions (such as sending, receiving and accepting) but the exchanged messages semantics remain insufficiently explored. In fact, the majority of the existing QoS negotiation approaches use technical and complex terms in the proposed offers or contract clauses that cannot be understood by the client. The latter can have limited technical knowledge on the required services or products.

We have made a comparison of the existing negotiation protocols with our approach regarding four criteria: (1) the client request model complexity (2) the semantic richness of the negotiation language (3) the adequateness of the negotiation type and (4) the possibility of computing composite QoS metrics in the negotiation process. On the first criterion, the existing protocols use a same complex model for client requests and provider offers. On the second criterion, the used negotiation language in the existing solutions does not contain enough semantics to automate the client request correspondence with the provider offers. On the third criterion, we notice that the existing approaches are based on accept/reject exchanges between the providers and the clients; while we propose an automatic semantic matching process between their offers and requests. On the last criterion, the existing works do not allow to compute QoS aggregation functions to check the client request when it is composite. Overall, unlike the existing solutions, our matching approach (*ODACE SLA*) between client intentions and provider offers satisfies all these criteria which are fundamental to facilitate QoS negotiation and to fulfill the SLA establishment process.

We also compared our approach to other existing semantic matching techniques like S-Match (Giunchiglia et al., 2003) and Glue (Doan et al., 2004). These techniques are based on algorithms that generate a list of possible correspondences between the terms of two different data sources in order to merge them. However, this requires human actions to validate and to filter these correspondences. Moreover, the existing matching approaches only consider semantic correspondences between their several terms. In our case, this cannot be sufficient because we have to additionally evaluate QoS constraints expressed by the client and the provider to decide whether the matching between their requirements and offers is positive or negative. Our goal consists also in automating as much as possible the matching process between the client and the provider terms in order to automatically establish an SLA between them in case of compatibility.

CONCLUSION

In Service Oriented Computing paradigm, SLAs are commonly signed agreements forming an e-contract between a service provider and a service consumer defining the obligations of the two parties. The client has often predefined choices by the provider. In many cases, he may not understand these choices as he can be a non specialist. In this paper, we presented *ODACE SLA*, an automatic semantic matching approach between the client intentions with the provider offers to bridge the knowledge gap between them. In this approach, we started by modeling client intentions and provider offers using ontologies to facilitate their automatic matching. This matching is based on four steps. The first step consists in generating correspondences between the client and the provider terms by assigning certainties for their equivalence. The second step corrects and refines these certainties. In the third step, we evaluate the client constraints with the provider offers using the correspondences generated in the previous steps. In case of compatibility, a draft version of an agreement is automatically generated in a complete SLA. This draft should be verified by the commercial expert at the provider side before it proposal to the client. We have developed a Java tool that implements all the steps of *ODACE SLA* and we used it in a digital video library use case within the ITA2's UseNet European project. Our approach gave promising results and a high precision (93%) in this case study. We believe that our approach is beneficial for the providers by automating the analysis of the client requirements and by helping them finding the services and the products they effectively want. Our approach can be also beneficial for the clients to compare different offers from many providers for a same intention by defining semantic scoring functions for these offers. This can automatically lead to a complete generation of an SLA in order to send it to the selected provider. This issue is an ongoing work of a master thesis in our research team.

A PHD thesis is also proposed to automatically: (1) analyze text written in natural language by clients and (2) transform it to our *clientOnto* intentions structure. Actually, we continue testing our semantic-enabled matching approach on other business case studies.

ACKNOWLEDGMENT

This work has been partially funded by the ITEA2's UseNet (Ubiquitous M2M Service Networks) European project.

We would like to address our special thanks to Professor Ted Pedersen for his helpful advices on the implementation and the evaluation of our approach.

REFERENCES

Andrieux, A., Czajkowski, K., Dan, A., Keahey, K., Ludwig, H., Nakata, T., Pruyne, J., Rofrano, J., Tuecke, S., & Xu, M. (2007). *Web Services Agreement Specification (WS-Agreement)* (No. GFD-R-P. 107). Global Grid Forum.

Banerjee, S., & Pedersen, T. (2003). Lesk: Extended gloss overlaps as a measure of semantic relatedness. In *Proceedings of the 18th International Joint Conference on Artificial Intelligence*, Acapulco, Mexico (pp. 805-810). Menlo Park, CA: AAAI Press.

Coalition, O. W. L.-S. (2004). OWL-S: Semantic markup for web services. *W3C Member Submission*. Retrieved April 05, 2010, from http://www.w3.org/Submission/OWL-S/

Dimitrakos, T., Martrat, J., & Wesner, S. (Eds.). (2010). *A selection of common capabilities validated in real-life business trials by the BEinGRID consortium XV* (p. 210). New York: Springer.

Do, H. H., Melnik, S., & Rahm, E. (2003). Comparison of Schema Matching Evaluations. In A. B. Chaudhri, M. Jeckle, E. Rahm, & R. Unland, Eds. *Revised Papers from the Node 2002 Web and Database-Related Workshops on Web, Web-Services, and Database Systems* (LNCS 2593, pp. 221-237). London: Springer Verlag.

Doan, A., Madhavan, J., Domingos, P., & Halevy, A. (2004). Ontology matching: a machine learning approach. In Staaband, S., & Studer, R. (Eds.), *Handbook of ontologies: International handbookon informations systems* (pp. 385–404). Berlin: Springer Verlag.

Dobson, G., Lock, R., & Sommerville, I. (2005). QoSOnt: a QoS Ontology for Service-Centric Systems. In *Proceedings of the 31st EUROMICRO Conference on Software Engineering and Advanced Applications*. Washington, DC: IEEE Computer Society

Fakhfakh, K., Chaari, T., Tazi, S., Drira, K., & Jmaiel, M. (2009). Semantic Enabled Framework for SLA Monitoring. *International Journal on Advances in Software*, *2*(1), 36–46.

Francis, W. (2008). *The NextGRID Final Report (Project Final Report)*. Edinburgh, UK: The University of Edinburgh.

Garcia, J. M., Toma, I., Ruiz, D., & Ruiz-Cortes, A. (2008, November). A service ranker based on logic rules evaluation and constraint programming. In *Proceedings of the 2nd ECOWS Non-Functional Properties and Service Level Agreements in Service Oriented Computing Workshop*, Dublin, Ireland.

Giunchiglia, F., & Shvaiko, P. (2003). Semantic matching. *The Knowledge Engineering Review*, *18*(3), 305–332. doi:10.1017/S0269888904000074

Hasselmeyer, P., Qu, C., Schubert, L., Koller, B., & Wieder, P. (2006). Towards Autonomous Brokered SLA Negotiation. In Cunningham, P. M. (Ed.), *Exploiting the Knowledge Economy - Issues, Applications, Case Studies* (*Vol. 3*). Amsterdam: IOS Press.

Hirst, G., & St-Onge, D. (1998). Hso: Lexical chains as representations of context for the detection and correction of malapropisms. In Fellbaum, C. (Ed.), *WordNet: An Electronic Lexical Database* (pp. 305–332). Cambridge, MA: MIT Press.

Horrocks, I., Patel-Schneider, P. F., Boley, H., Tabet, S., Grosof, B., & Dean, M. (2004). *SWRL: A semantic web rule language combining OWL and RuleML*. W3C Member Submission.

Kanso, H., Soulé-Dupuy, C., & Tazi, S. (2007, June). Representing Author's Intentions of Scientific Documents. In *Proceedings of the International Conference on Enterprise Information Systems, 3*, 489–492.

Keller, A., & Ludwig, H. (2003). The WSLA Framework: Specifying and Monitoring Service Level Agreements for Web Services. *Journal of Network and Systems Management, 11*(1), 57–81. doi:10.1023/A:1022445108617

Lamparter, S., Ankolekar, A., Studer, R., & Grimm, S. (2007). Preference-based selection of highly configurable web services. In *Proceedings of the 16th international Conference on World Wide Web (WWW '07)*, Banff, Canada (pp. 1013-1022). New York: ACM.

Leacock, C., & Chodorow, M. (1998). Lch: Combining local context and WordNet similarity for word sense identification. In Fellbaum, C. (Ed.), *WordNet: An Electronic Lexical Database* (pp. 265–283). Cambridge, MA: MIT Press.

Mc Guiness, D., & van Harmelen, F. (2004). *OWL Web Ontology Language Overview, W3C Recommendation*. Retrieved April 05, 2010, from http://www.w3.org/TR/owl-features/

Melnik, S., Garcia-Molina, H., & Rahm, E. (2002). Similarity Flooding: A Versatile Graph Matching Algorithm and Its Application to Schema Matching. In *Proceedings of the 18th international Conference on Data Engineering (ICDE)* (p. 117). Washington, DC: IEEE Computer Society.

Patwardhan, S. (2003). *Vector: Incorporating dictionary and corpus information into a context vector measure of semantic relatedness*. Unpublished master's thesis, University of Minnesota, Duluth, MN.

Pedersen, T., Patwardhan, S., & Michelizzi, J. (2004) WordNet:Similarity - Measuring the Relatedness of Concepts. In *Proceedings of Fifth Annual Meeting of the North American Chapter of the Association for Computational Linguistics* (pp. 38-41). Boston: Association for Computational Linguistics.

Peer, H., Changtao, Q., Bastian, K., Lutz, S., & Philipp, W. (2006, October). *Towards Autonomous Brokered SLA Negotiation Exploiting the Knowledge Economy: Issues, Applications, Case Studies (eChallenges 2006)* (pp. 44-51). Barcelona, Spain: IOS Press.

Steffen, B., Thomas, W., & Kurt, G. (2006). An Ontology for Quality-Aware Service Discovery. *International Journal of Computer Systems Science & Engineering, 21*(5).

Studer, R., Benjamins, V. R., & Fensel, D. (1998). Knowledge engineering: principles and Methods. *IEEE Transactions on Data and Knowledge Engineering, 25*(1/2), 161–197.

TeleManagement Forum. (2004). *SLA Management Handbook, TeleManagement* (Tech. Rep. No. GB917, Version 2, Vol. 4). Berkshire, UK: The Open Group.

Toma, I., Foxvog, D., & Jaeger, M. C. (2006). Modeling QoS characteristics in WSMO. In *Proceedings of the 1st Workshop on Middleware for Service Oriented Computing MW4SOC 2006*, Melbourne, Australia (Vol. 184, pp. 42-47). New York: ACM.

Tosic, V., Ma, W., Pagurek, B., & Esfandiari, B. (2004). Web Service Offerings Infrastructure (WSOI) - A Management Infrastructure for XML Web Services. In *Proceedomgs of NOMS 2004, Seoul, Korea* (pp. 817–830). Washington, DC: IEEE.

Wang, X., Vitvar, T., Kerrigan, M., & Toma, I. (2006). A QoS-aware selection model for semantic web services. In *Proceedings of Service Oriented Computing (ICSOC 2006)* (LNCS 4294, pp. 390-401). New York: Springer.

Wu, Z., & Palmer, M. (1994). Verbs semantics and lexical selection. In *Proceedings of the 32nd Annual Meeting on Association For Computational Linguistics* (pp 133-138). Las Cruces, NM: Association for Computational Linguistics

ENDNOTES

1. The OASIS Consortium. http://www.oasis-open.org
2. The ITEA2 Europeen project. https://usenet.erve.vtt.fi
3. The provider offers ontology structure http://www.redcad.org/members/kaouthar.fakhfakh/negotiation/ProviderOntoClasses.gif
4. Jess, the Rule Engine for the JavaTM Platform. http://www.jessrules.com/
5. The Xerces API. http://xerces.apache.org/xerces-j/apiDocs/index.html
6. The protégé-owl API. http://protege.stanford.edu/plugins/owl/api/
7. The SWRL Jess API. http://protege.cim3.net/cgi-bin/wiki.pl?SWRLJessTab
8. A client intention example. www.redcad.org/members/kaouthar.fakhfakh/negotiation/clientOntoInstances.gif
9. A provider offer example. http://www.redcad.org/members/kaouthar.fakhfakh/negotiation/providerOntoInstances.gif

This work was previously published in International Journal of Systems and Service-Oriented Engineering, Volume 1, Issue 3, edited by Dickson K.W. Chiu, pp. 1-20, copyright 2010 by IGI Publishing (an imprint of IGI Global).

Chapter 10
Using Description Logics for the Provision of Context-Driven Content Adaptation Services

Stephen J.H. Yang
National Central University, Taiwan

Jeff J.S. Huang
National Central University, Taiwan

Jia Zhang
Northern Illinois University, USA

Jeffrey J.P. Tsai
The University of Illinois at Chicago, USA

ABSTRACT

This article presents our design and development of a description logics-based planner for providing context-driven content adaptation services. This approach dynamically transforms requested Web content into a proper format conforming to receiving contexts (e.g., access condition, network connection, and receiving device). Aiming to establish a semantic foundation for content adaptation, we apply description logics to formally define context profiles and requirements. We also propose a formal Object Structure Model as the basis of content adaptation management for higher reusability and adaptability. To automate content adaptation decision, our content adaptation planner is driven by a stepwise procedure equipped with algorithms and techniques to enable rule-based context-driven content adaptation over the mobile Internet. Experimental results prove the effectiveness and efficiency of our content adaptation planner on saving transmission bandwidth, when users are using handheld devices. By reducing the size of adapted content, we moderately decrease the computational overhead caused by content adaptation.

INTRODUCTION

Mobile computing poses big challenges to Web content delivery services in several significant ways. First, increasing volumes of handheld devices (e.g., Personal Digital Assistants (PDAs) and mobile phones) have been used to access Web content nowadays; however, most of the existing Web content is originally designed for desktop devices instead of handheld devices. Second, mobile users usually move constantly; their residing environments thus may change accordingly and Web content delivery should also subject to the changes for a better performance. For example, if a user moves into a blurred environment (e.g., due to a sunny or gloomy weather), the content should

DOI: 10.4018/978-1-4666-1767-4.ch010

be consequently enlarged or the background color should be turned brighter. Third, people's status may change dynamically, which may consequently request adjusted content delivery. For example, if a user on a multimedia phone conversation walks into a room for another physical meeting running in parallel, the audio should be turned off. The corresponding audio transmission thus becomes unnecessary.

Therefore, tools and mechanisms are in need to provide mobile users with transparent and seamless content delivery services. To achieve this ultimate goal, it is essential to deliver personalized and adaptive content according to users' situated environments. In this article, the two terms "*situated environment*" and "*context*" are used interchangeably, both referring to content receivers' surrounding information that has an impact on content delivery and presentation such as receivers' personal profiles, receiving devices, communication network, location, activity, and time (Schilit, Adams et al. 1994; Dey & Abowd 1999; Satyanarayanan 2004; Mukherjee, Delfosse et al. 2005; Julien & Roman 2006).

The conventional approach to provide Web content supporting various types of computational devices is to prepare and maintain different versions (formats) of the same Web content for different devices. For example, a Web page typically holds one HTML version supporting desktop devices and one Wireless Markup Language (WML) version supporting wireless devices. This approach is straightforward but labor-intensive yet error-prone. Content providers have to prepare different layouts and formats for the same Web content, which results in tremendous overhead. To support a new device, all previous Web pages have to support a new format. Even worse, any change in the Web content may result in consequent changes in every related version, which is highly inflexible. Obviously, this approach is neither practical nor feasible for providers of a large volume of Web content.

To bridge the gap between content providers and mobile consumers, content adaptation refers to a technique that provides the most suitable content presentation by means of transformation. While some researchers focus on content adaptation techniques on multimedia types, such as image and video adaptation (Mohan, Smith et al. 1999; Mukherjee, Delfosse et al. 2005; Nam, Ro et al. 2005; Vetro & Timmerer 2005; Xie, Liu et al. 2006; Wang, Kim et al. 2007), some other researchers focus on exploring how to conduct proper content adaptation based on receiving contexts (Lum & Lau 2002; Hua, Xie et al. 2006; He, Gao et al. 2007). Although the literature has witnessed these effective content adaptation efforts and techniques, they typically do not support automatic content adaptation decision; nor do they support configurable and extensible contextual environment specifications.

In contrast with the previous works lacking a clear semantic basis, this research intends to study a semantic foundation for content adaptation. Specifically, this research aims to investigate solutions to four research challenges:

C1: How to formally detect and represent mobile user contexts?

C2: How to design configurable and reconfigurable adaptation rules?

C3: How to manage and automate content adaptation management?

C4: How to automatically generate adaptation output format?

To the best of our knowledge, our research is the first effort to apply description logics (DLs) to formally define context profiles and requirements, and to automate content adaptation decision. We also introduce a Meta Medium Object concept and an Object Structure Model (OSM) to formally model an adaptable medium object. Then we construct a context-driven content adaptation planner that is designed and developed to automatically transform Web content to an appropriate format based on users' surrounding contexts, especially when users are using handheld devices over the mobile Internet. Another benefit of our method it that it may reduce content access time. For example, if a user is accessing a film while driving, then our context-driven content adaptation planner automatically turns off the video for safety. This strategy could potentially save a significant amount of bandwidth by not transferring video clips (or other unnecessary data) in the already crowded mobile Internet.

The remainder of this article is organized as follows. Related research regarding context and content adaptation planning is presented in Section 2. We formalize context profile and requirement definition using description logics in Section 3. We introduce a Meta Medium Object concept and an Object Structure Model (OSM) in Section 4. We present the design of or context-driven content adaptation planner in Section 5. Implementation considerations and solutions will be presented in Section 6. Experiments and evaluations are discussed in Section 7. We conclude the article in Section 8.

RELATED WORK

He et al. (He, Gao et al. 2007) identify three types of objects in an HTML page, namely, structure, content, and pointer objects. Mukherjee et al. (Mukherjee, Delfosse et al. 2005) propose to associate content with metadata defining adaptation choices and their resulting media characteristics. In contrast with their work, we propose a concept of Meta Medium Object (MMO) as a self-containing entity encapsulating presentation objects and their metadata. Structural information of a Web page and the inter-relationships between presentation objects are managed by interactions and relationships between identified MMOs. Instead of having to manage two individual objects for each content entity, our approach only has to manage one unified type of object. The maintainability and integrity become higher.

Hua et al. (Hua, Xie et al. 2006) propose a technique of supporting dynamic Web content adaptation based on semantic block identification, adaptation algorithms, as well as a caching strategy. While both partition a Web page based on the W3C's Document Object Model (DOM) (W3C) technique, our work differs from theirs after a DOM tree is constructed. Their strategy then tries to detect semantic blocks for page recreation based on predefined receiving devices. In contrast, our approach then generates a lazy-initiated and dynamically managed Object Structure Model (OSM). In addition to supporting semantic block scaling (Hua, Xie et al. 2006), our research introduces the OSM as a flexible and adaptable foundational model to support context-driven content adaptation.

Adaptation rules, or policies, are typically used to guide content adaptations. Kinno et al. (Kinno, Yukitomo et al. 2004) adopt policy descriptions to designate how an adaptation engine should behave according to changing environments. Lemlouma and Layaida (Lemlouma & Layaida 2004) and Phan et al. (Phan, Zorpas et al. 2002), on the other hand, both design application-specific adaptation engines without employing adaptation policy descriptions. In contrast with their works based on predefined content adaptation rules, our research applies DLs to formalize context and requirement specifications and automate content adaptation decision making.

Mohan et al. (Mohan, Smith et al. 1999) introduce a rate-distortion framework equipped with a subjective fidelity measure for the analysis of adaptation policies. A representation scheme named InfoPyramid provides a multi-modal and multi-resolution representation hierarchy for multimedia. An intelligent engine is introduced to help select the best content presentation style according to residing contexts, by modeling the selection process as a resource allocation problem in a generalized rate-distortion framework. Lum and Lau (Lum & Lau 2002) propose an adaptation system based on predefined decision policies.

Upon receiving a request, the system first assigns a score to each possible content version based on various factors such as processing overhead and the volume of the content. Afterwards, a decision engine selects the optimal version by searching the score tree. In contrast with their work with high-level divisions of presentation qualities and immutable adaptation policies, our research proposes a fine-grained presentation Object Structure Model (OSM) supported by adaptable, configurable, and re-configurable adaptation policies.

Wang et al. (Wang, Kim et al. 2007) propose a utility function that considers video entity, adaptation, resource, utility, and their inter-relationships. Utility values are dynamically calculated to predict and prepare real-time video transcoding. In contrast with their work, we propose a rule-based decision engine equipped with configurable and re-configurable adaptation rules and lookup tables.

Many content adaptation prototypes have been built in recent years. Among them, Phan et al. (Phan, Zorpas et al. 2002) propose a middleware, called Content Adaptation Pipeline (CAP), to perform content adaptation on any complex data type, not only text and graphic image. XML is used to describe all the elements in a content hierarchy. Berhe et al. (Berhe, Brunie et al. 2004) present a service-based content adaptation framework. An adaptation operator is introduced as an abstraction of various transformation operations such as compression, decompression, scaling, and conversion. A logic adaptation path is determined by associating adaptation constraints to proper combinations of adaptation operators. To determine the optimal service, a path selection algorithm is proposed based on cost and time considerations. Their work shows a proof-of-concept of Web-based content adaptation; however, their implementations stay in a preliminary phase. How to map from constraints to adaptation operators is unsolved.

Lee et al. (Lee, Chandranmenon et al. 2003) develop a middleware-based content adaptation server providing transcoding utilities named GAMMAR. A table-driven architecture is adopted to manage transcoding services located across a cluster of network computers. Their approach allows incorporation of new third-party transcoding utilities. Lemlouma and Layaida (Lemlouma & Layaida 2004) propose an adaptation framework, which defines an adaptation strategy as a set of description models, communication protocols, and negotiation and adaptation methods. In contrast with their work, our content adaptation planner is founded on a theoretical model of OSM with an adaptation rule engine, as well as formal design patterns (Gamma, Helm et al. 1995).

Some researchers focus on content decomposition methods. Chen et al. (Chen, Xie et al. 2002a) propose a block-based content decomposition method, DRESS, for quantifying the content representation. An HTML page is factorized into blocks, each being assigned a score denoting its significance. DRESS selects the block with the highest score to represent the content. This method prevents from missing significant information. It also enables content layout to become adjustable according to the region of interest, attention value, and minimum perceptible size (Chen, Xie et al. 2002b). Ramaswamy et al. (Ramaswamy, Iyengar et al. 2005) propose an efficient fragment generation and caching method based on detection of three features: shared behavior, lifetime, and personalization characteristic. The smallest adjustable element in these two approaches is a composite of objects (i.e., text, image, audio, and video). This granularity of decomposition is too large for mobile device screens; therefore, they are not suitable for mobile content adaptation.

Xie et al. (Xie, Liu et al. 2006) propose a decomposition and re-composition technique for facilitating browsing large pictures on small-screen devices. In contrast with their approach focusing on building a visual model of attention objects in an image, our research intends to establish a visual model of presentation objects in a Web page (i.e., HTML page).

MPEG-21 Digital Item Adaptation (DIA) adopts metadata to guide transformation of digital items for supporting various levels of Quality of Services (QoS) in accordance with a set of constraints such as storage, transmission and consumption (Vetro & Timmerer 2005). In contrast with their work focusing on generic content adaptation for providing different QoS features, our research focuses on Web page transformation oriented to mobile computing scenarios.

A number of context-based adaptation methods (Lei & Georganas 2001; Lum & Lau 2002; Pashtan, Kollipara et al. 2003; Toivonen, Kolari et al. 2003; Kurz, Popescu et al. 2004; Lemlouma & Layaida 2004; Krause, Smailagic et al. 2006) are proposed to customize Web content according to client contextual environments, including personal preferences, device capabilities, and access environments. Julien and Roman (Julien & Roman 2006) introduce a *view* concept to represent application-specific contextual information. Bellavista et al. (Bellavista, Corradi et al. 2003) adopt metadata for representing context characteristics at a high level of abstraction. Cabri et al. (Cabri, Leonardi et al. 2003) propose a two-dimensional model to describe the location information in a mobile context: a physical location in space and a logical location within a distributed group or application. Nam et al. (Nam, Ro et al. 2005) propose to consider both user preferences and perceptual characteristics as a basis for effective visual content adaptation. Mukherjee et al. (Mukherjee, Delfosse et al. 2005) take into consideration run-time conditions about terminal, network, user preference, and rights. Krause et al. (Krause, Smailagic et al. 2006) believe that contextual information includes users states and surroundings. A user's state can be extracted from the user's activity, location, schedule, and physiological information. In contrast with their context models, in this research we propose a multi-dimensional context model. Along each dimension, users can configure attributes, each being associated with a list of configurable adaptation rules.

Some researchers focus on dynamic content adaptation based on users' changing contextual environments. CB-SeC framework (Mostefaoui, Tafat-Bouzid et al. 2003) and Gaia middleware (Roman, Hess et al. 2002) provide functionalities to help users find content based on their contexts or enable content adaptations according to user's contextual information. Mostefaoui and Hirsbrunner propose a formal definition of context to model content's contextual information (Mostefaoui & Hirsbrunner 2004) Lemlouma and Layaida propose a framework to assert metadata information of Web content (Lemlouma & Layaida 2001). All of these works use Composite Capabilities/Preferences Profiles (CC/PP) as interoperable context representation carriers to enable communication and negotiation of device capabilities and user preferences in terms of a service invocation. Zhang et al. further propose extensions to CC/PP to enable transformation descriptions between various receiving devices (Zhang, Zhang et al. 2005). Besides, several Web Ontology Language (OWL)-based context models are presented (Khedr & Karmouch 2004; Yang, Zhang, Chen 2008; Yang, Zhang et al 2009) to provide high-quality results of Web content presentation beyond the expressive limitations of CC/PP. These researchers all utilize ontology to describe contextual information including location, time, device, preference, and network. In addition, some researchers pursue quantitative approaches to decide adaptation at run time. Among them, Wang et al. (Wang, Kim et al. 2007) formalize the decision process as a classification and regress problem. One major difficulty of these adaptation approaches is that content providers have to predefine exhaustive adaptation rules for adapting the contents for every contextual situation.

Another limitation of aforementioned content adaptation methods is that, they do not take users' changing contextual information into full account, especially when users are using handheld devices in a mobile environment. Users may receive some invalid content, and they have to manually verify received content. In contrast, our context-driven content adaptation rules can provide automatic and better contextually matched Web content to meet user' contextual changes in the mobile Internet.

As a fundamental technique in the field of semantic Web, Description logics (DLs) support formal knowledge representation in a structured manner (Baader, Calvanese et al. 2003). W3C-endorsed OWL (W3C 2004) is built on top of DLs. However, OWL is established to serve for generic Web services delivery. Compared to OWL, we apply DLs to study context-aware content adaptation. OWL can be used as one tool to specify contextual environments and content delivery requirements.

FORMALIZATION OF CONTEXT SPECIFICATIONS AND DECISIONS

To automate content adaptation decision, we apply description logics (DLs) to formally define context profiles and requirements.

Description Logics

Description logics (DLs) refer to a family of knowledge representation languages that can be used to express knowledge of an application domain in a structured manner (Baader, Calvanese et al. 2003). Equipped with formal logic-based semantics, DLs are capable of describing hierarchical notions of concepts (classes) and roles (relations) and formal reasoning about concepts and roles. It is a fundamental technique in the field of semantic Web; several widely used semantic Web languages are based on DLs such as W3C-endorsed OWL (W3C 2004).

Syntax of DLs contains three major components: predicate, relation, and constructor. A unary predicate symbol denotes a concept name; a binary relation denotes a role name; and a constructor is a recursive definition that defines comprehensive concepts and roles from atomic ones. In DLs, TBox (terminological box) is used to represent sentences describing concept hierarchies (e.g., roles between concepts); ABox (assertional box) is used to contain "ground" sentences stating to which in the hierarchy individuals belong (e.g., roles between individuals and concepts).

We adopt DLs to establish a hierarchical ontology to enable and facilitate context-aware dynamic content adaptation. Three layers are identified: a context requestor layer for describing receiving contextual environments, a context provider layer for defining contextual requirements and constraints, and a content planner layer for managing content adaptation definitions and matchmaking.

Content Planner Layer

In the content planner layer, we define ContentPlanner as a superclass for content adaptation measurements and matchmaking. It has five subclasses formulated as follows: ContextProfile, ContextRequirement, ContextInquiry, AdaptationTemplate, and Metric.

ContentPlanner ⊆ T
ContextProfile ⊆ ContentPlanner
Context Requirement ⊆ ContentPlanner
AdaptationTemplate ⊆ ContentPlanner
Metric ⊆ ContentPlanner

All content planner layer ontologies are described in DLs' TBox to formalize the descriptions and facilitate semantic matchmaking. ContextProfile defines the receiving contextual environments of the service requestor; ContextRequirement defines the required contexts of a content delivery service; ContextInquiry defines a service requestor's inquiry Ontology; and AdaptationTemplate defines some predefined content adaptation templates (e.g., in a format of tables or rules).

Metric defines an abstract template for the context requirement layer to properly define context attributes and their semantic meanings.

For each attribute, an instance of the Metric class defines a 3-tuple (metricName, unit, value). The item metricName is a user-defined identifier for a metric. The item unit defines how to measure an attribute and implies its semantic meaning. For simplicity reason, we allow two types of unit: &xsd;#nonNegativeInteger and &xsd;#string (&xsd; is the entity macro delimiting an XML Schema namespace). The former declares a numeric measurement as a cardinality constraint for a context attribute; the latter allows users to define application-specific measurements.

Metric is further divided into AtomicMetric and ComplexMetric: the former defines a metric over a single context attribute; the latter defines a metric over multiple context attributes through operators. The operands of an operator in a ComplexMetric definition can be either a single attribute or a composite attribute. An operator defines a function of how to process operand metrics. As a proof of concept, we define three operators: BooleanFunction, ArithmeticFunction, and AggregateFunction. A BooleanFunction allows three operations ($\vee, \wedge, \neg$). An ArithmeticFunction allows two operations (+,-). An AggregateFunction allows accumulating multiple metrics.

AtomicMetric $\subseteq$ Metric
ComplexMetric $\subseteq$ Metric

Each context metric is a subclass of either AtomicMetric or ComplexMetric. The taxonomy of metrics is designed by content service provider in the format of TBox; assertions on metrics are defined in the format of ABox. Proper definition of metrics is a key to appropriate content adaptation and delivery.

Context Requestor Layer

Without losing generality, in this research, we consider four aspects of a receiving context environment: receiver identity, receiving condition, network bandwidth, and receiving device.

Definition 1. The context profile of a receiver (ContxtProfile) is denoted as a 4-tuple:

ContextProfile = <I, C, N, D>, where:

I denotes the receiver's identity; C denotes the receiver's access condition; N denotes the receiver's communication network; and D denotes the receiver's receiving device. In the context requestor layer, we define ContextProfile as a superclass defining receiving contextual environments. ContextProfile has four subclasses: Identity, Condition, Network, and Device.

ContextProfile $\subseteq$ T
Identify $\subseteq$ ContextProfile
Condition $\subseteq$ ContextProfile
Network $\subseteq$ ContextProfile
Device $\subseteq$ ContextProfile
What $\subseteq$ Condition; Where $\subseteq$ Condition; What $\subseteq$ Condition; Location $\subseteq$ Condition; CommunicationProtocol $\subseteq$ Network;

Each dimension of ContextProfile is further refined into contextual attributes at a finer granularity. For example, as shown below, condition information can be further described by when, where, and what activities a person is involved, as well as the location identified by Global Positioning System (GPS), sensor networks, Radio Frequency Identification (RFID), and so on. Network information can be further described by communication protocols such as General Packet Radio Service (GPRS), Third-Generation Technology (3G), Voice over Internet Protocol (VoIP), Wireless Fidelity (WiFi), and so on.

Furthermore, device information can be further described using Composite Capability/Preference Profiles (CC/PP), User Agent Profile (UAProf), and so on. Please note that the assumed information discussed in this article is selected for ease of illustration; the actual contextual information in the real world could be much more complicated.

Context Requirement Layer

This layer defines suitable contexts for a content delivery service by the corresponding service provider. Existing DL reasoners typically provide better support for subsumption reasoning in TBox than datatype reasoning in ABox (Zhou, Chia et al. 2005); therefore, we adopt attribute-oriented cardinality to define contextual requirements and constraints. Note that contextual attributes are domain specific; it is up to content delivery providers to establish the attributes and cardinality specifications.

From content adaptation perspective, context requirements can be viewed as context input (ContextInput), and receiving context definitions can be viewed as required context output (ContextOutput). In addition, some other constrains may be defined. Some content might require external conditions to be satisfied to ensure that it can be properly displayed. For example, a video clip may require that a specific browser or software is installed for proper display. Such a specification is defined in ContextPrecondition. Furthermore, some side effects might be stated in (ContextEffect). For example, the display of some video clip might have the effect of lower throughout. Moreover, we use ContextDefault to represent no contextual requirements defined.

Thus, we define five ContextProfile classes for context requirements: ContextDefault, ContextInput, ContextOutput, ContextPrecondition, and ContextEffect. They are expressed as follows:

ContextDeFault $\subseteq$ ContextProfile
ContextInput $\subseteq$ ContextProfile
ContextOutput $\subseteq$ ContextProfile
ContextPrecondition $\subseteq$ ContextProfile
ContextEff ect $\subseteq$ ContextProfile

In this layer, content service providers define the metrics for required context attributes. Supported by the definitions of the context attributes, cardinality constraints are ready to be added on the defined attributes in context planner layer for matchmaking purpose.

Our context ontology allows extensibility. Context attributes can be customized and specified. New attributes can be defined; existing attributes can be updated; and constraints on attributes can be specified and updated in context metrics hierarchy taxonomy. The metrics definitions and verifications in ABox are left to the corresponding domain and organization. Proper definition of this context requirement layer is the key to context matchmaking and content adaptation.

An Example of Using Dls To Define Context Requirements And Environments

In this section, let us take access condition as an example to show how we apply DLs to define context requirements and environments.

As shown in Figure 1, a metric is defined to describe an attribute "condition" in ContextProfile. According to the abstract template described in section 3.2, condition is defined as a ComplexMetric:

ComplexMetric(condition):= ("condition", &xsd;#string, value)

Condition is defined to describe people's access condition including when, where, and what activities they are involved. In other words, condition is a composite attribute that comprises three atomic attributes, each being defined using a string ("when", "where", and "what"). The three atomic attributes are aggregated into a composite attribute. For example, if a person intends to access films when she is involved in an activity "*meeting*" (what) on "03/09/2008" (when) at a place "*meeting room A306*" (where), the receiving condition is defined as ("Condition", (when, where, what), (meeting, 03/09/2008, meeting room A306)). Using DL syntax, the profile can be represented as follows:

Figure 1. Part of context ontology

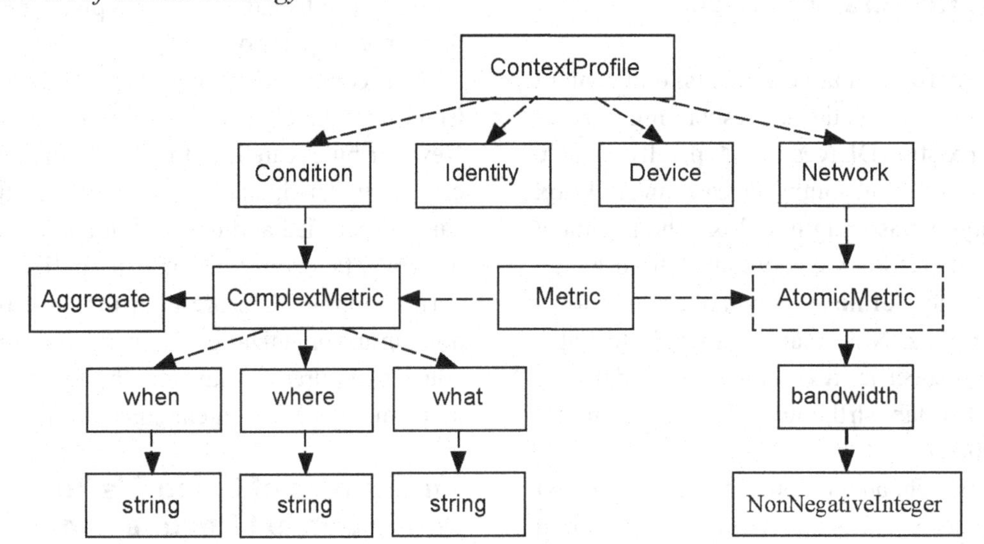

aProfile = ContextProfile
I(*"meeting" what.WhatStringMetric*)
I(*"03/09/2008" when.WhenStringMetric*)
I(*"meetingroomA306" where.WhereStringMetric*)
As shown in Figure 1, an AtomicMetric is defined for the context attribute *network* as follows:
AtomicMetric(network):= ("bandwidth", &xsd;#NonNegativeInteger, value)

A network connection can be measured by its bandwidth. For example, using DL syntax, a content service provider may define a suitable network connection as a bandwidth 400KB and 2.5MB in the following TBox definition:

aService = ContextRequirement
I($\geq$ 400*bandwidthBandwidthKBMetric*)
I($\leq$ 2500*bandwidthBandwidthKBMetric*)

The following ABox definition declares an assertion of such a context requirement published on 03/08/3008. The first line is a concept assertion; the rest lines are role assertions.

aService(aServiceInstance)
hasServiceProfile(aServiceInstance, "&provider_service;#")
startTime(aServiceInstance, 03/09/2008)
endTime(aServiceInstance, 03/09/2008)

Content Adaptation Decision

By applying DL-based ontology, the published requirements of a content delivery service define a set of constraint-specified context metrics in the format of ABox; the profile of a receiving party defines a set of constraint-specified context metrics in the format of TBox. Whether content adaptation is needed can be decided by comparing the two sets of specifications, which can be formalized into a problem of deciding ontology subsumption relationship. To automate the decision process, we define the following operator.

Definition 2. An isStronger operator ($\succ$) is denoted as an order operation between two DL specifications. For two constraints, x and y, on the same context attribute *i*, $x(i) \succ y(i)$ indicates that x(i) is a stronger specification than y(i).

Recall that we define two types of units for metrics: &xsd;#nonNegativeInteger and &xsd;#string. For the first type of unit, the applied domain is nonnegative integer that is an ordered set, meaning that any two values using this type of measurement unit are comparable. A general definition of a constraint (x) on a context attribute (i) specifies a range of variable values: $a \leq x(i) \leq b$, where $a \leq b$, and a and b are both non negative integers.

Axiom 1. For two constraints, x and y, on the same context attribute *i*,

$x \leq x(i) \leq b,\ c \leq y(i) \leq d,\ a \leq b,\ c \leq d$

$x(i)$ f $y(i)$, iff $c \leq a$ and $b \leq d$

For simplicity, we consider that one constraint on a context attribute only specifies one range. For the second type of unit (&xsd;#string), the applied domain is string. A general definition of a constraint (x) on a context attribute specifies an enumeration of string values: $x(i) \in A$, where $A = \{a_1, a_2, ...a_m\}$, and $a_l (1 \leq l \leq m)$ is a string. We define an isStronger operator ($\succ$) between them as follows:

Axiom 2. For two constraints, x and y, on the same context attribute *i*, $x(i) \in A, y(i) \in B, A = \{a_1, a_2, ...a_m\}, B = \{b_1, b_2, ...b_n\}$, $x(i) \succ y(i)$, iff $\forall a_k \in A, \exists p \Rightarrow a_k = b_p$.

Based on our definition of $\succ$ operator between two constraints on the same context attribute, we can define a compatible relationship ($\lhd$) between two context ontology descriptions, R and P (R is a content delivery service requirement, P is a receiving profile) as follows:

Definition 3. For $P = x(C) \cap x(D) \cap x(N)$, $R = y(C) \cap y(D) \cap y(N)$, $P \lhd R$, iff $(x(C) \prec y(C)) \wedge (x(D) \prec y(D)) \wedge (x(N) \prec y(N))$

This means that unless a receiving profile is compatible with the required contexts, content adaptation is needed. Thus, we transform the problem of context constraint comparison into the problem of judging ontology subsumption relationship. The two operators both exhibit transitive properties, which are useful in content adaptation decision making.

Proposition 1. The isStrong operator $\succ$ has transitive property:

$(x(i) \succ y(i)) \wedge (y(i) \succ z(i)) \Rightarrow (x(i) \succ z(i))$

Proposition 2. The compatible operator $\lhd$ has transitive property:

$(P \lhd R) \wedge (R \lhd Q) \Rightarrow (P \lhd Q)$

Proof: These propositions are straightforward derived from their definitions.

Content Adaptation Rules

If a context comparison concludes a required content adaptation, an adaptation strategy is needed comprising a set of adaptation rules that guide proper adaptation of contents to fit users' contexts. We formally define an adaptation strategy as follows:

Definition 4. $AS_{personID} = \prod_{i=1}^{n} AR_i$, where:

AR_i is an adaptation rule, $i \in [1,n]$, n is the number of dimensions defined in the user's situated environment profiles.

For example, if people access films when they are involved in an activity "*meeting*" at a place "*meetingroomA306*," then the content adaptation should automatically turn off the sound to avoid making noise. If people access films when they are involved in an activity "*driving*" at a place "*car*," then the content adaptation should automatically turn off the video for their safety. Surely, people could disable content adaptation if they decide to do so. For example, they may insist on listening to content during meetings (e.g., via ear phones),

Table 1. Rules

Rule#	original object	Network	adapted object
N1	Object(video,original)	Network(2M+)	Object(video,original)
N2	Object(video,original)	Network(2M-) and ObjectSize(2M+)	Object(video,low_resolution)

or reading content while they are driving (e.g., by stopping the car and read).

We formalize an adaptation rule as an event controller that automatically responds to changes of users' situated environments.

Definition 5. An adaptation rule can be represented by a 6-tuple as follows:

r = <id, o, r, p, g, tS>, where:

id denotes the identifier of the rule; *o* denotes the original content; *r* denotes the resulting content; *p* denotes the prerequisite of the rule; *g* denotes the objectives of the rule; *tS* denotes the transformation strategy of the rule.

We use the following two rules in Table 1 as examples to explain the major components of a content adaptation rule.

The two rules can be represented as:

r1 = <"N1", Object(video.original), Object(video.original), "ContextRequirement $\sqcap$ (≥ 2*bandwidth. BandwidthMBMetric*), null, null>

r2 = <"N2", Object(video.original), Object(video.low_resolution), "ContextRequirement $\sqcap$ (≥ 2*bandwidth. BandwidthMBMetric*) $\sqcap$ (≥ 2*objectSize. ObjectSizeMBMetric*)", "Lower video resolution", "videoTransMethod">

The first rule (N1) indicates that if the network condition is fast enough (>2M), video content does not have to be transformed. The second rule (N2) indicates that if the network condition is slow (a TBox definition <2M) and a video object is large (a TBox definition >2M), then the video object should be transformed into a lower resolution using the specified method (videoTransMethod).

All rules can be stored in a rule base incrementally: new facts can be asserted when new contextual information is acquired; new adaptation rules can be inserted when new context types are considered. Meanwhile, facts and rules can be removed when they are no longer suitable in a user's situated environment. Any insertion and deletion of facts and rules to an existing rule base may result in inconsistency or conflicts. Therefore, truth maintenance is required to ensure the consistence of the rule base. We have developed a truth maintenance mechanism for verifying rule base concerning redundancy, inconsistency, incompleteness, and circularity to ensure the incrementally constructed rule base is sound and complete (Yang, Tsai et al. 2003).

META MEDIUM OBJECT

Presentation content typically comprises different types of presentation objects, in the format of text, video, audio, image, and so on. In this section, we will focus on transforming presentation objects in different medium types; therefore, throughout the rest of the article we will use the terms "presentation objects" and "medium objects" interchangeably. Our previous work yields algorithms to automatically identify semantically coherent presentation units (Unit of Information or UOI) in a Web page that have to be displayed on the same screen (Yang, Zhang et al. 2007). Each UOI may also comprise different medium objects. Since different transformation strategies and procedures are needed to serve different medium types, our adaptation process is focused on various medium objects.

MMO Concept

To formalize the content adaptation process, we introduce a concept of *MetaMediumObject* (MMO) that represents a self-contained meta concept of a presentation object encapsulating various presentation formats controlled by adaptation rules. As a proof of concept, we consider two major aspects of presentation formats: modality and fidelity. Modality indicates the type of a medium object such as text, video, audio, and image. Each modality is associated with a certain fidelity indicating the quality of the presentation of the objects, such as image resolution, color depth, and video bit-rate. Thus, a content adaptation rule can be viewed as a pattern for transforming a medium object's modality and fidelity. We also simplify the description of modality and fidelity in this article. In the real world, modality information can be further classified into flash or streaming media, for example; fidelity information can be further described by image types (e.g., BMP or JPEG), image resolution (e.g., 24 bit), color depth (e.g., 32bit), video bit-rate, and so on.

Figure 2 illustrates our idea of MMO by using a UML Class Diagram. An MMO represents an individual presentation object. Instead of being a presentation object with actual code, it is a meta-object containing the high-level information of the described presentation object.

Figure 2. Concept of MetaMediumObject (MMO)

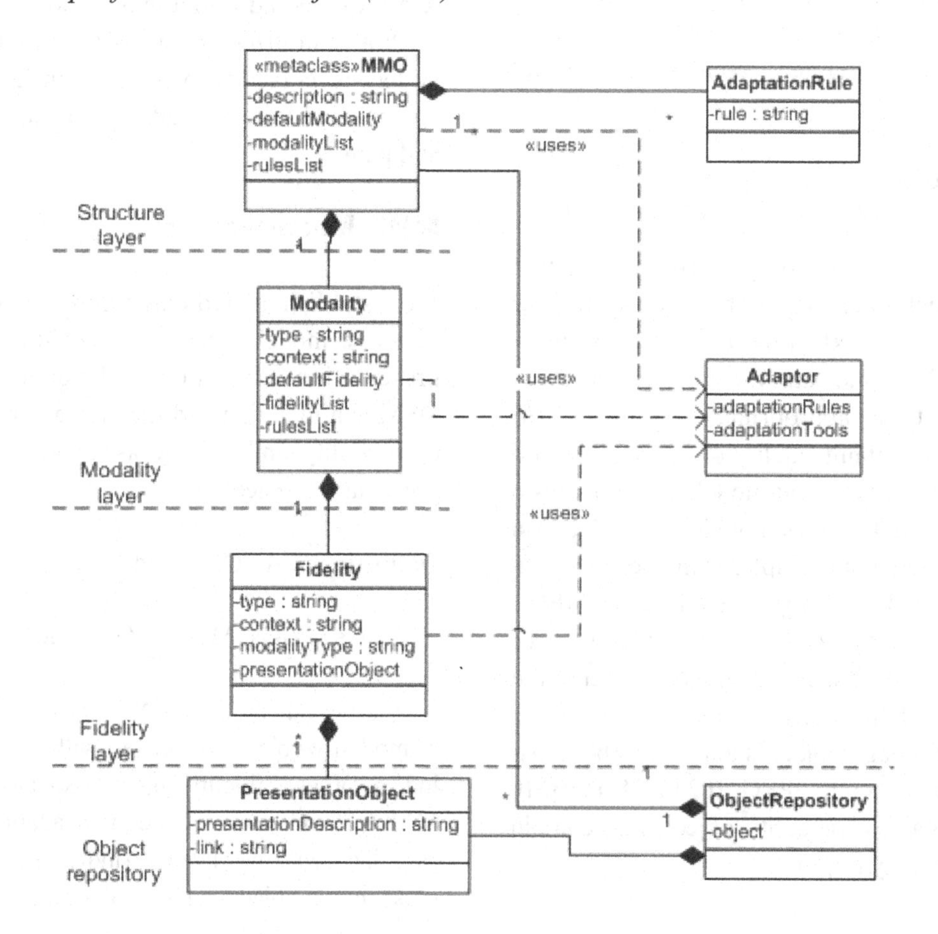

Definition 6. An MMO is denoted as a 4-tuple:

aMMO = <d, dM, M, R>, where:

d contains the basic descriptive information of the object such as its goal and short description; *dM* specifies the default modality of the presentation object. Without losing generality, the value of *dM* is one of the four items from the enumeration (text, image, audio, video): $dM \in \{t, i, a, v\}$. *L* contains a list of possible modalities in which the object can be represented. For example, a text object can also be represented as a piece of audio: $L = \{t, a\}$. As shown in Figure 2, an MMO may contain multiple modality objects. *R* contains a list of content adaptation rules associated with the presentation object for modality transformation, for example, whether the presentation object is allowed to be transformed from text to audio.

As shown in Figure 2, each modality object is a 5-tuple:

m = <t, c, dF, F, R>, $\forall m \in M$, where:

t indicates the presentation type of the presentation object, which is one of the four items from the enumeration (text, image, audio, video): $t \in \{t, i, a, v\}$. *c* indicates the contexts for which the modality is appropriate: $c \in TBox$. *dF* specifies the default fidelity of the presentation object for the associated modality. *F* contains a list of possible fidelities in which the object can be represented. For example, an image object can be represented in two resolutions, either JPEG 1024x728 or BMP 16-bits. *R* contains a list of adaptation rules associated with the presentation object for fidelity transformation, for example, whether an image object is allowed to be transformed from format JPEG 1024x728 to BMP 16-bits. Again, a modality object may contain multiple fidelity objects.

As shown in Figure 2, each fidelity object is a 4-tuple:

f = <t, mT, c, o>, $\forall f \in F$, where:

t indicates the user-specified fidelity type. *mT* indicates the category of presentation modality, which is one of the four items from the enumeration (text, image, audio, video): $mT \in \{t, i, a, v\}$. *c* indicates the contexts for which the fidelity is appropriate. *o* points to the actual presentation object.

As shown in Figure 2, an independent component *adaptor* is used by MMOs, modality objects, and fidelity objects to transform an existing presentation format into another. An *adaptor* typically is a tuple: (R, T). *R* define configurable and re-configurable adaptation rules for the adaptor; *T* links to a set of available content adaptation tools (e.g., text-to-audio tool and audio-to-text tool). Such a centralized adaptor allows us to build a system-level facility to manage configurable and reconfigurable content adaptation based on rules and patterns.

MMO Management

We establish a dedicated three-layer object structure model (OSM) to manage MMOs for presentation adaptation and version control. OSM can maintain and manage all possibilities of modality and fidelity associated with each presentation object.

Definition 7. An OSM is denoted as a 5-tuple:

OSM = <MMOs, M, F, R, O>, where:

MMOs indicate a set of MMOs; *M* indicate a set of modality objects associated with the MMOs; *F* indicate a set of fidelity objects associated with the MMOs; *R* indicate a set of content adaptation rules associated with the MMOs; *O* indicate a set of actual presentation objects associated with the MMOs.

Figure 3. Object structure model

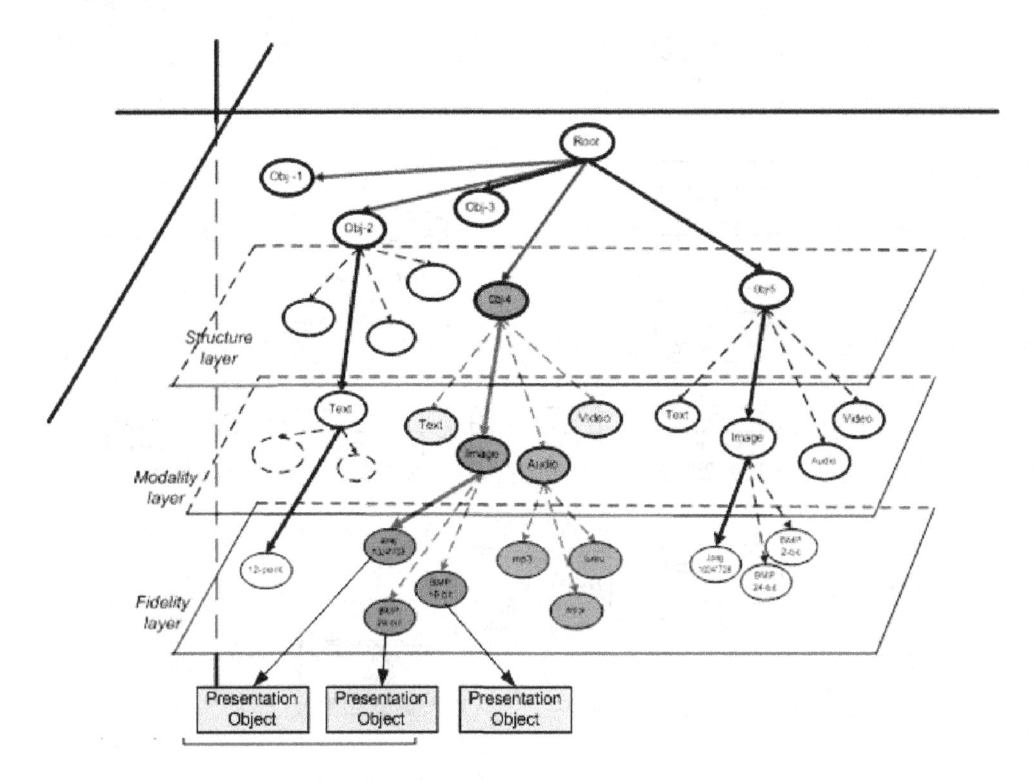

As shown in Figure 3, OSM implies a three-layer framework: structure layer, modality layer, and fidelity layer. The structure layer holds a tree-like structure of MMOs decomposed from a Web page. For each MMO (leaf node) in the structure layer, each of its modality attributes is mapped into a child node in the modality layer. For each modality node in the modality layer, each of its fidelity attributes is mapped into a child node in the fidelity layer. Each fidelity node maps to an actual presentation object (i.e., adapted version according to specified attributes).

Each Web page can have an instance of OSM associated with it, representing in a tree-like structure. Figure 3 shows such an illustrative OSM tree. Taking *Obj4* as an example, it can be represented in one of the four types of modality (text, image, video, and audio); its modality of image supports three types of fidelity (JPEG 1024x728, BMP 24-bits, and BMP 16-bits). Therefore, *Obj4* in the structure layer is associated with four child nodes in the modality layer, in which *Image* node is further associated with three child nodes in the fidelity layer. Each fidelity node eventually points to a presentation object. *Obj4* and all of its descendent nodes form an object cluster that refers to a sub-tree of the OSM, indicating all possible combinations of modality and fidelity associated with the object (*Obj4*).

CONTEXT-DRIVEN CONTENT ADAPTATION PLANNER

Based on our DL-based context formalization and MMO presentation, we have designed a rule-based planner for supporting and automating context-driven content adaptation, as presented in Figure 4. The design of the planner comprises five major components: (1) Web page decomposition, (2) object management, (3) dynamic content adaptation including dynamic transcoding and cache

Figure 4. Design of content adaptation planner

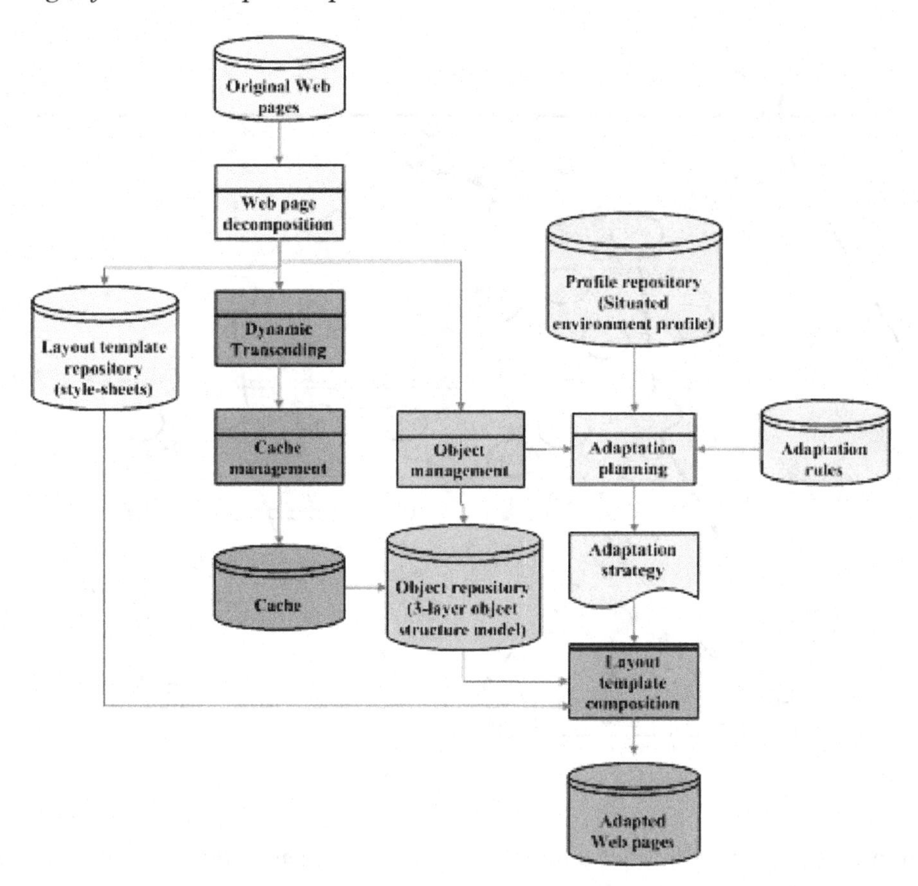

management, (4) adaptation planning, and (5) layout template composition. Each component also represents a step (phase) in the corresponding content adaptation process. As shown in Figure 4, seven database repositories are included in the planner: original Web pages, layout template repository, profile repository, adaptation rules, object repository, cache, and adapted pages. We will discuss the five components together with their associated repositories (highlighted in different colors) in detail in the following sections.

Web Page Decomposition

A Web page typically comprises a set of interrelated presentation objects, or simply objects. Objects on a Web page are characterized by their modality indicating their types such as text, video, audio, and image. Each modality is associated with fidelity indicating the object's presentation quality such as image resolution, color depth, and video bit-rate. In order to render the same object on various devices, content adaptation may have to perform transcoding and change object's modality and fidelity accordingly. For example, if a mobile phone can only play images with a low resolution, the fidelity of an image with a high resolution has to be adapted to a lower level. In this article, we focus on the design of content adaptation by providing rules for transforming objects' modality and fidelity to fit in with users' situated environment.

The input of the page decomposition step is a Web page in HTML format from the *original Web pages* repository. The goal of this step is to decompose the HTML page into encompassed

Figure 5. Example of a decomposed Web page and its corresponding object tree

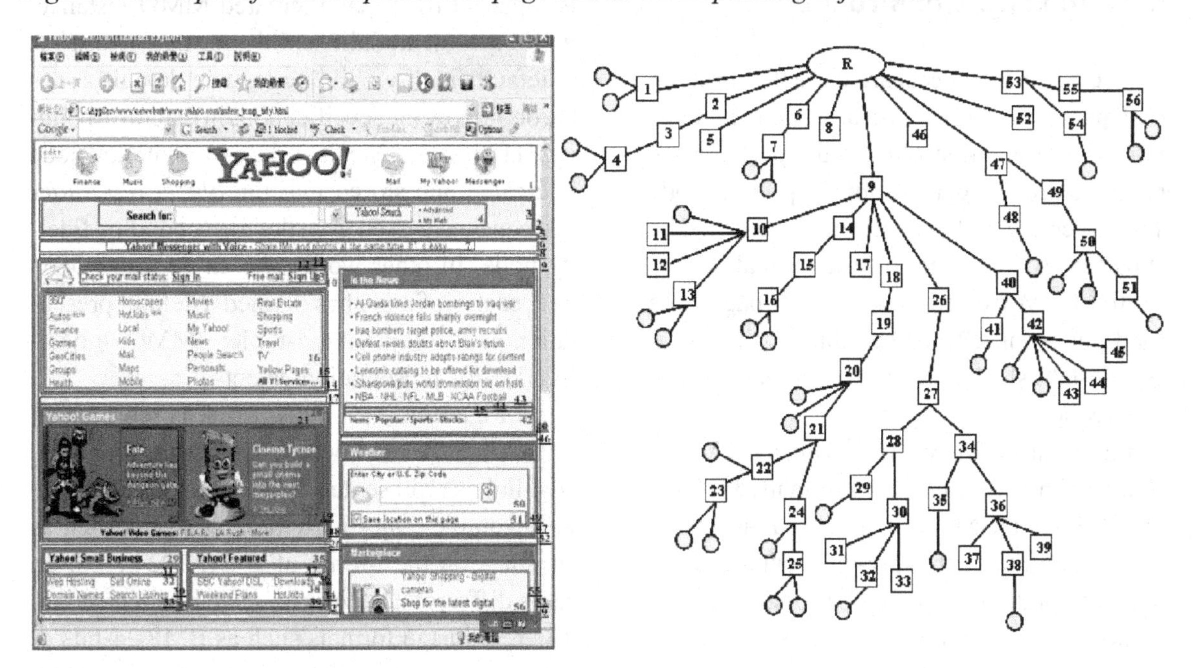

presentation objects. To enable automatic page decomposition, a non-well-formed Web page has to be first transformed into a well-formed format. We adopt an open-source software tool Tidy (Bose & Frew 2005)to perform the task due to its simplicity and our familiarity with it.

Then we decompose the formatted Web page by identifying presentation objects based on W3C's Document Object Model (DOM) (W3C). As an *ad hoc* standard for Web components, DOM proposes a platform- and language-independent model representing the content, structure, and style of documents. The resulting objects are stored in the *object repository* through the *object management* phase.

Besides identifying individual objects as instructed by the DOM framework, we also detect and identify the inter-object relationships during the Web page decomposition phase. Relationships, such as spatial and temporal relationships, can be used to describe or decide the layout and presentation sequence of objects during Web page rendering and re-rendering. For example, for two objects that are rendered side-by-side spatially in a desktop's screen, they could be adapted to be displayed sequentially from top to bottom on a mobile phone's screen.

In order to maintain inter-object relationships, we construct an object tree for each decomposed Web page and record encompassed objects' spatial and temporal relationships in the tree. Figure 5 illustrates how a commonly accessed Web page (http://www.yahoo.com) is transformed into an object tree. As shown in Figure 5, objects can be displayed individually or aggregated into segments. The root of an object tree represents the Web page; the intermediate nodes represent segments; and the leaf nodes represent individual presentation objects. The details of how to generate such an object tree will be discussed in later sections. An object tree can be traversed in any order, either pre-order or post-order, where a node is accessed before any of its child nodes or after any of its child nodes. The resulted linear sequence will be used to restore inter-object relationships when we perform page layout template composition at a later stage. It can also be used to serialize the structure into the *object repository* to save memory space.

OSM and MMO Construction

After presentation objects are extracted from an HTML page, we can establish an OSM to manage identified presentation objects for presentation version control. Figure 6 shows pseudo code of how to create and manage an OSM tree. The first part is an OSM tree constructor that creates an initial OSM tree from an HTML page. The second part is an OSM tree adjustor that manages an established OSM tree upon receiving an access request.

To create an OSM tree, first of all, a DOM tree is generated by decomposing a formatted HTML page. Then the DOM tree is traversed, using any traversal algorithm such as pre-order or post-order, where each node is visited before or after any of its child nodes. Each leaf node in the DOM tree is replaced by a newly created MMO instance, with a default modality child node that in turn has a default fidelity child node. The default nodes are defined by the specifications in the corresponding HTML page. Meanwhile, a presentation node is created and stored into the *object repository* that can be accessed from the corresponding fidelity node. In other words, for an initiated OSM tree, each leaf node has defined a default presentation path comprising 4-tuple: <MMO, modality, fidelity, presentationObject>.

In Figure 4, each default link is represented by a solid directed line; each dynamically generated link is represented by a dotted directed line. For example, *Obj4* is by default represented as an image in the format of JPEG 1024x728. To serve different devices, *Obj4* can be adapted into different image formats such as BMP 24-bits or

Figure 6. Pseudo code of OSM tree creation and management

```
//1. Constructor that creates an initiate OSM tree
void OSMTree(HTML page) {
   createDOMTree(HTML page);
   traverse(DOMTree);
   for (each leafnode) {
      createMMO();
      createDefaultModalityNode();
      createDefaultFidelityNode();
      createPresentationObject();
   }
}

//2. Upon receiving a presentation request
void adjustOSMTree (request) {
   gatherContext(request);
   traverse(OSMTree);
   for (each MMO) {
      if (!satisfy(defaultModality, context) && !satisfy(anyExistingModality, context)) {
            createModalityNode(context);
            createFidelityNode(context);
            createPresentationObject(context);
      }
      if (!satisfy(defaultFidelity, context) && !satisfy(anyExistingFidelity, context)) {
         createFidelityNode(context);
         createPresentationObject(context);
      }
   }
}
```

BMP 16-bits. To adapt to different user environments, *Obj4* can be adapted into other presentation types, such as text, audio, or video. Note that for simplicity reason, Figure 3 skips internal nodes in the structure layer representing HTML tags.

Upon receiving an access request, the receiving contextual information has to be first gathered. Then the OSM tree is traversed in any order. For each MMO object, we check whether the default modality or existing modalities can satisfy the receiving contexts. If the answer is a "no," a new modality node will be created; then in turn a new fidelity node will be created. A new presentation object will also be created using adaptation tools, which will be discussed in detail in later sections. If the answer is a "yes," however, we need to first check whether its default fidelity or existing fidelities can satisfy the receiving contexts. If not, a new fidelity node will be created, together with a new presentation object. The dynamic content adaptation process will be discussed in the next section.

Dynamic Content Adaptation with Transcoding and Cache Management

If existing modality or fidelity formats cannot be found in the *object repository* for a presentation object, a *dynamic transcoding* process is invoked. As shown in Figure 4, the transcoded objects are cached and stored into *object repository* for future reuse. The process may employ a set of transcoding apparatus dedicated for transforming different types of presentation objects into various modalities and the fidelities. We have experimented a variety of third-party transcoding tools. Among them, we found that the following four tools show satisfactory results for proprietary presentation transformation: VCDGear as a toolbox for video transformation, PictView as a command-line-driven image converter, and JafSoft (Chebotko, Lu et al. 2007) and Microsoft Reader (Microsoft) as text-to-audio (and vice versa) converters.

The design of our framework allows any third-part transcoding service to be easily plugged into the system, by adopting the strategy design pattern (Gamma, Helm et al. 1995). Any transcoding tools and services can be registered at the transcoder with specific functionalities. Upon receiving a transcoding request, the transcoder dynamically decides the most appropriate service and redirects the request to the selected service.

We enable two types of content adaptation, either static or dynamic. For static content adaptation, contents are transcoded and prepared before users' requests. Thus, content adaptation can only occur when applicable objects can be found in the *object repository*. No computational overhead is imposed in this manner because the actual transcoding process is performed offline. The problem of the static approach, however, is that the system has to predict all potential Web page users beforehand. For dynamic content adaptation, the system performs dynamic transcoding and caching if applicable objects cannot be found in the *object repository*. The advantage of this dynamic approach is the flexibility it offers: users can access any Web page and obtain adapted content upon requests. Its problem is that the performance of Web access may degrade due to additional transcoding overhead at run time. Users probably have to wait a significant amount of time when they access adapted Web content at the first time. Nevertheless, the performance will be enhanced the next time when the same user or other users access the same adapted Web content. The reason is that we have applied a caching mechanism presented by Kinno et al. (Kinno, Yukitomo et al. 2004), so that the transcoding overhead can be reduced. Note that instead of caching the entire adapted Web page, only the dynamically transcoded objects will be cached by our *cache management*. When dynamic content adaptation begins, the dynamically generated layout template and the associated objects will also be cached.

Because of the cache management, when users re-visit the same Web page, the access performance with dynamic content adaptation is almost the same as that without content adaptation. For PDA and mobile phone users, the performance of dynamic content adaptation is even better. This is because the size of the transcoded objects will be reduced, which results in tremendous saving of transmission bandwidth.

Page Layout Template Composition

The *layout template composition* phase is used to decide presentation styles (i.e., page layout templates) at run time. To enable fully personalized content adaptation, it is critical to allow end users to customize their own style sheets, to a certain level of granularity. To achieve this goal, we design an Extensible Stylesheet Language (XSL) style sheets-based strategy.

As shown in Figure 7, we have designed different page layout templates, for NBs, for PDAs, and for wireless phones. For each presentation type, each receiving device is associated with a proprietary style sheet. For example, since all three devices can present images and texts, their corresponding style sheets for the two presentation types are the same. However, wireless phones may not be able to play video clips; PDAs may not intend to show all video clips unless users explicitly decide to do so. Therefore, as shown in Figure 7, the style sheet for NBs allows the source of a video clip to be embedded into the delivered content; while the style sheets for PDAs and phones only embed a link to the source of the video clip.

As shown in Figure 7, each receiving device is also associated with some other utility style sheets. For example, in order to present a table, NBs are associated with a "table_nb.xsl"; PDAs are associated with a "table_pda.xsl"; phones are associated with a "table_phone.xsl." We design a set of utility style sheet templates for all receiving device types. Some other examples are illustrated in Figure 7, such as "tr" (row), "td" (column), "image_play," "audio_link," and "video_link."

These individual style sheets are configurable and re-configurable by different end users for different receiving devices. At the phase of *layout template composition*, as shown in Figure 6, an aggregated XSL style sheet file is created, by importing all related style sheets serving various purposes (for both presentation types and supporting utilities).

Our strategy of separation of concerns allows configuration and re-configuration of personalizable/customizable presentation styles at a fine-grained granularity. Our dynamic aggregation strategy ensures a format management solution, with high flexibility and extensibility.

As mentioned in *Web page decomposition* phase in section 5.1, the inter-object relationships (pre-order or post-order sequences) are used to restore an object's presentation sequence and generate an adapted Web page. We adopt post-order sequence to describe parallel order of objects. For PC and NB, one object can be displayed beside or next to another object in terms of their spatial relationships, or one object can be rendered concurrently with another object in terms of their temporal relationships. In contrast, pre-order sequence is used to describe sequential order of objects. For smaller screen size of PDAs and mobile phones, one object can be displayed above or below another object in terms of their spatial relationships, or one object can be rendered before or after another object in terms of their temporal relationships.

IMPLEMENTATION DETAILS

In this section, we discuss some of our implantation details, focusing on the aspects of context definition and design pattern usages. In our research, we adopt OWL to build our ontology, as OWL is based on DLs and is natural for us to define receiving contexts and requirements.

Figure 7. An example of XSL temples for NBs, PDAs, and phones

Template	XSL for NB	XSL for PDA	XSL for phone
image	... <xsl:template match="**image**"> <img src="**{@src}**" /> </xsl:template> ...	... <xsl:template match="**image**"> <img src="**{@src}**" /> </xsl:template> ...	... <xsl:template match="**image**"> <img src="**{@src}**" /> </xsl:template> ...
Video	... <xsl:template match="**video**"> <embed src="**{@src}**" autostart="**false**" /> </xsl:template> ...	... <xsl:template match="**video**"> <a href="**{@src}**"> **Video_Link**</a > </xsl:template> ...	... <xsl:template match="**video**"> <a href="**{@src}**">**Video_Link**< /a> </xsl:template> ...
Text	... <xsl:template match="**text**"> <font style="**{@size}**"> <xsl:value-of select="**.**" /> </font> </xsl:template> </xsl:stylesheet> ...	... <xsl:template match="**text**"> <font style="**{@size}**"> <xsl:value-of select="**.**" /> </font> </xsl:template> ...	... <xsl:template match="**text**"> <font style="**{@size}**"> <xsl:value-of select="**.**" /> </font> </xsl:template> ...
Aggregated XSL file	<?xml version="1.0" encoding="UTF-8" ?> - <xsl:stylesheet version="**1.0**" xmlns:xsl=http://www.w3.org/1999/XSL/Transform/> <xsl:import href="**body.xsl**" /> <xsl:import href="**p.xsl**" /> <xsl:import href="**table_nb.xsl**" /> <xsl:import href="**tr_nb.xsl**" /> <xsl:import href="**td_nb.xsl**" /> <xsl:import href="**text.xsl**" /> <xsl:import href="**image_play.xsl**" /> <xsl:import href="**audio_play.xsl**" /> <xsl:import href="**video_play.xsl**" /> - <xsl:template match="**html**"/> - <html> <xsl:apply-templates. /> </html> <xsl:template> </xsl:stylesheet>	<?xml version="1.0" encoding="UTF-8" ?> - <xsl:stylesheet version="1.0" xmlns:xsl=http://www.w3.org/1999/XSL/Transform/> <xsl:import href="**body.xsl**" /> <xsl:import href="**p.xsl**" /> <xsl:import href="**table_pda.xsl**" /> <xsl:import href="**tr_pda.xsl**" /> <xsl:import href="**td_pda.xsl**" /> <xsl:import href="**text.xsl**" /> <xsl:import href="**image_play.xsl**" /> <xsl:import href="**audio_link.xsl**" /> <xsl:import href="**video_link.xsl**" /> - <xsl:template match="**html**"/> - <html> <xsl:apply-templates. /> </html> <xsl:template> </xsl:stylesheet>	<?xml version="1.0" encoding="UTF-8" ?> - <xsl:stylesheet version="**1.0**" xmlns:xsl=http://www.w3.org/1999/XSL/Transform/> <xsl:import href="**body.xsl**" /> <xsl:import href="**p.xsl**" /> <xsl:import href="**table_phone.xsl**" > <xsl:import href="**tr_phone.xsl**" /> <xsl:import href="**td_phone.xsl**" /> <xsl:import href="**text.xsl**" /> <xsl:import href="**image_play.xsl**" /> <xsl:import href="**audio_link.xsl**" /> <xsl:import href="**video_link.xsl**" /> - <xsl:template match="**html**"/> - <html> <xsl:apply-templates. /> </html> <xsl:template> </xsl:stylesheet>

Rule-Based Adaptation Management Using Jess

We have adopted JESS (JESS) (Jess Expert System Shell), a rule engine for the Java platform, to design and implement content adaptation rules. Being a superset of CLIPS programming language, JESS provides a rule-based paradigm suitable for automating an expert system. In contrast to a procedural programming language where a single program is activated at one time, the JESS' declarative paradigm allows us to apply a collection of rules to a collection of facts based on pattern matching. In addition, JESS is Java-based and conforms to our implementation environment.

Examples of rules over conditions are listed in Figure 8. The JESS rules define the template of an object as a 3-tuple (object-ID, modality, fidelity), and then define the template of a situated environment as a 4-tuple (pserson-ID, condition, network, device). Figure 8 defines 5 rules over different conditions. Take rule 1 as an example. The receiving environment is normal; thus, rule 1 requires no special action except printing out a message at DOS prompt. Take rule 5 as another example. The receiving environment is noisy and it allows video and audio objects. Thus, rule 5 requires checking object identifier, verifying the modality of the presentation objects in the format of video or audio; and turning the fidelity of the presentation objects to be loud.

Design Pattern Usages

In the implementation of our content adaptation planner, we adopted three design patterns (Gamma, Helm et al. 1995) to enhance design and code reusability, flexibility, and extensibility. The four patterns are: lazy initiation, Model-Viewer-Controller (MVC), mediator, and strategy.

In the previous section, we discussed our strategy of adopting the lazy initiation design pattern in constructing an OSM tree. In the initialization phase, only the default presentation path is detected and created. All other possible presentation paths (including modality, fidelity, and presentation object) are created at an on-demand basis.

In order to offer dynamic content adaptation to provide various presentation formats, we adopted the Model-View-Controller (MVC) design pattern. The MVC pattern has been known to facilitate maintenance by separating code for modeling the application domain from its GUI code. Figure 9 illustrates how we use the MVC pattern. As discussed earlier, we use our OSM tree to enable and manage multiple presentation formats of an HTML page. Inside of an OSM tree, each MMO represents a model for a specific presentation object; its associated adaptation rules represent a controller; any of its presentation paths (either the default path or dynamically generated paths) represents a view (presentation) of the modeled presentation object. Note that the multiplicity relationship between MMO and adaptation rule is a one-to-many relationship; the multiplicity relationship between MMO and presentation path is also a one-to-many relationship. This means that under different adaptation rules, a presentation object may show different presentation formats.

When we implemented the dynamic content adaptation engine, we adopted the mediator pattern and the strategy pattern, as shown in Figure 10. Both of them are in the category of behavioral patterns. The mediator pattern intends to offer a unified interface to a set of interfaces in a subsystem; the strategy pattern intends to provide an agile technique so that algorithms can be selected on-the-fly at runtime.

As mentioned earlier, without losing generality, in this research we consider four modalities: text, image, audio, and video. Based on various receiving contexts, a presentation object in one modality may need to be transformed into another modality. For example, a text message can be transformed into an audio clip for a user to listen instead of reading, assuming the user is driving. The maximum number of categories on content modality adaptation can be calculated as: $P_4^2 = 12$. In addition,

Figure 8. Examples of content adaptation rules written in JESS

```
;; Facts for describing object
 (deftemplate Object "define object modality and fidelity"
                    (slot object-ID)
                    (slot Modality)
                     (slot Fidelity))
;; Facts for describing users situated environment
(deftemplate Situated-Env " define users situated environment "
                    (slot Person-ID)
                    (slot Condition)
                    (slot Network)
                     (slot Device))
;;   Condition Rules
(defrule Condition-r1   (Situated-Env (Condition normal))
                        =>
                        (printout t "no adaptation while at normal condition" crlf))
(defrule Condition-r2   ((Object ((ObjID ?objID) (Modality ?mod& image | video)))
                       (Situated-Env (Condition?cond & dark | driving))
                       =>
                       (assert (Object (ObjID ?objID) (Modality ?mod) (Fidelity blank))))
(defrule Condition-r3   ((Object ((ObjID ?objID) (Modality ?mod& image | video | text))
                       (Situated-Env (Condition blurred)))
                       =>
                       (assert (Object (ObjID ?objID) (Modality ?mod) (Fidelity bright))))
(defrule Condition-r4   ((Object ((ObjID ?objID) (Modality ?mod& video | audio)))
                       (Situated-Env (Condition?cond & quiet | meeting)))
                       =>
                       (assert (Object (ObjID ?objID) (Modality ?mod) (Fidelity mute))))
(defrule Condition-r5   ((Object ((ObjID ?objID) (Modality ?mod& video | audio)))
                       (Situated-Env (Condition noisy)))
                       =>
                       (assert (Object (ObjID ?objID) (Modality ?mod) (Fidelity loud))))
```

even staying with the same modality, a presentation object can change its fidelity. For example, an image object can be adapted from a high resolution into a lower resolution.

In order to enable one interface to serve all possible content adaptation requests, we adopted the mediator design pattern. As shown in Figure 10, an adaptation controller provides a unified interface for any content adaptation requests. Based on its carried algorithms and parsers, the adaptation controller examines an incoming request and analyzes its details such as its goal, input parameters, output parameters, context, and so on. Then the adaptation controller directs the request to proper executor.

To design the executor, we adopted the strategy design pattern. As discussed in Section 4.3, any transcoding tools and services can be registered at the executor with specific functionalities. Upon receiving an adaptation request, the executor dynamically decides the most appropriate service and redirects the request to the selected service.

Figure 9. Model-view-controller design pattern adopted

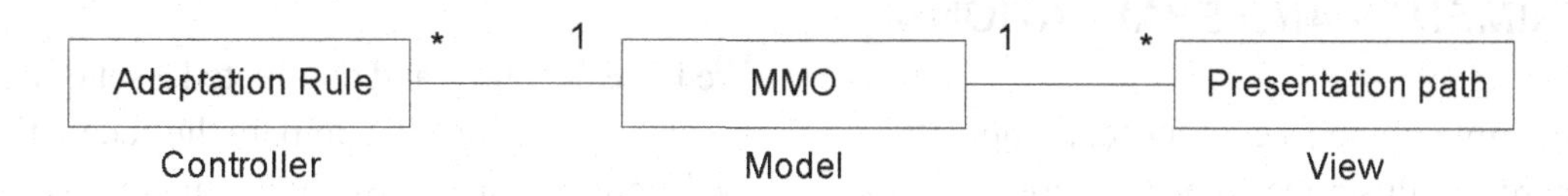

Figure 10. Mediator and strategy patterns adopted

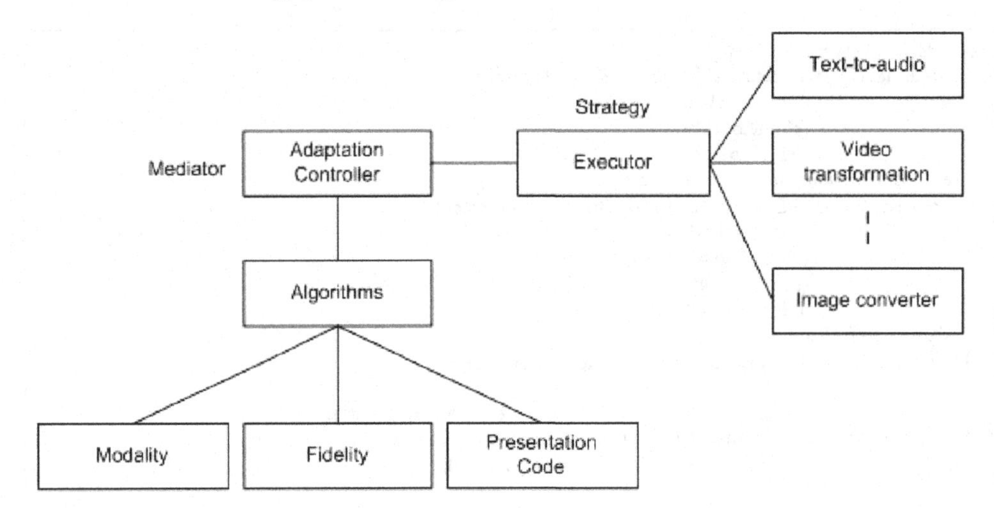

Lazy Initialization of OSM Trees

An OSM tree is dynamically created and maintained. A *lazy instantiation* design pattern (Gamma, Helm et al. 1995) is used here. At the initial construction, only the default presentation path is created. Other presentation paths are created based on an on-demand basis. All dynamically created MMO objects, modality objects, fidelity objects, and presentation objects are stored in the *object repository* for later reuse. The OSM tree is also stored in the *object repository*. Upon the first access, an OSM tree is created for a specific Web page. We realize their linkage through a hyperlink. The hyperlink concept was originally created to represent a link embedded in an HTML page to redirect access to another URL. We extend the hyperlink concept to realize association of an OSM tree to its HTML page. In other words, a hyperlink pointing to an OSM tree is embedded into each Web page to manage its dynamic content adaptation and presentation in its entire life cycle.

EXPERIMENTS AND EVALUATIONS

We have implemented a content adaptation planner. Figure 11 illustrates some execution results of our content adaptation planner through a comparison of Web browsing between conventional and handheld devices. The middle of Figure 11 shows the Yahoo home page (http://www.yahoo.com/) on a desktop browser. The left-hand side of Figure 8 shows how this page is shown on a PDA screen without content adaptation (upper-side) and with content adaptation (lower-side). The right-hand side of Figure 11 shows how the same page is shown on a wireless phone screen without content adaptation (upper-side) and with content adaptation (lower-side). As shown in Figure 11, without content adaptation, users of PDAs and wireless phones have to move scroll bars left and right, up and down in order to view the whole Web page. In contrast, our content adaptation obviously provides better content presentation to the users by transforming the Web page into a column-wise presentation, so that users only need to move scroll bars in one direction (up and down) instead of in two directions.

Measurement and Comparison of Access Time

We have designed and conducted a set of experiments to measure and compare the access time of Web content with and without our context-driven

Figure 11. Snapshots of adapted Web page for NB, PDA, and phone

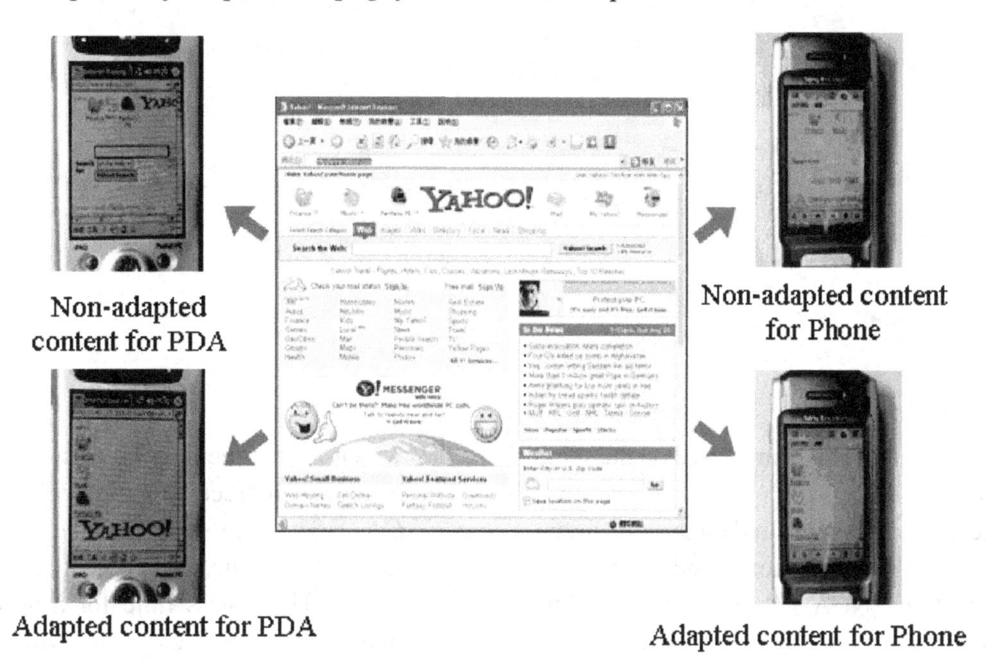

content adaptation planner. The goal is to evaluate whether adopting our content adaptation planner could enhance mobile Web accessibility for users with handheld devices. The detailed experimental set up is shown in Figure 12. Yahoo (http://www.yahoo.com) was chosen as the content provider and test bed. The reasons why we chose this commonly accessed Web site are as follows. First, Yahoo's site is well known and its overall structure (i.e., its division of sections) is stable. Therefore, our tests can be easily repeated by third parties, and the test results are more trustworthy. Second, Yahoo's site comprises a comprehensive set of sections, each containing significantly different content types with proprietary static attributes and dynamic features. Therefore, our experiments and results will be more useful and convincing.

This experiment of content adaptation (CA in short) was designed to measure the content access time for a user to browse different types (i.e., sections or categories) of Yahoo Web pages. Without losing generality, we selected five sections of Yahoo pages, namely *Home*, *News*, *Astrology*, *Auto*, and *Movie*. Each section exhibits some specific features. For example, the *Home* section maintains a relatively stable structure; the *News* section is under frequent updates so that its structure is volatile; the *Astrology*, *Auto*, and *Movie* sections contain a lot of images, photos, audio, and video clips. In short, we used Yahoo's diversified sections as the test bed to evaluate the effectiveness and efficiency of our content adaptation planner.

As shown in Figure 12, our content adaptation planner is implemented on an IBM server machine equipped with an Intel 1.0GHz CPU and a 1GB memory. The receiving devices we tested include a Compaq 3870 PDA with Microsoft Windows C.E. 3.0 PPC, a Sony Eriksson P900 smart phone with Symbian, and a notebook (NB) with Microsoft Windows XP installed. The connections between the receiving devices and the content adaptation planner are via a wireless network; the connection between the content adaptation planner and the Yahoo Web site is through the wired Internet. The contrast tests without using

Figure 12. Experimental set up

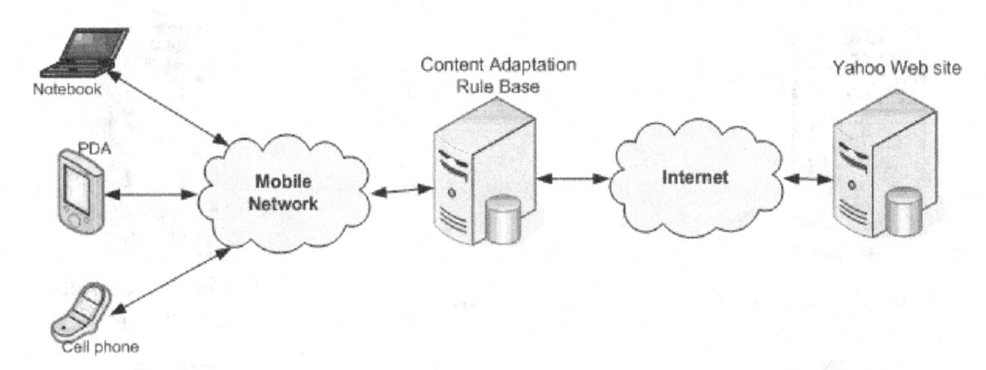

content adaptation planner are based on the same environmental settings.

The access time is measured for three different devices (NB, PDA, and phone) under three different conditions (NCA, CA_1, and CA_2). The experiment results are summarized in Table 2, 3, and 4. Various kinds of access time measured and denoted in this section are defined as follows:

- NCA: The access time for a user's visiting a Web page without content adaptation, which is the time interval between sending a request and receiving a reply.
- CA_1: The access time for a user's first visiting a Web page with content adaptation.
- CA_2: The access time for a user's second visiting a Web page with content adaptation.
- pre-CA process: The process before content adaptation, including Web page decomposition, trans-coding, object repository construction, and context acquisition

Table 2. Second-time visiting the same Web page with content adaptation

Yahoo page	device	adaptation planner (sec.)	adaptation engine (sec.)	total access time (sec.)	adaptation engine%
home	NB	0.2723	0.7928	1.0651	74.43%
news	NB	0.3727	1.0151	1.3878	73.14%
astrology	NB	0.2138	0.5842	0.7980	73.21%
auto	NB	0.1587	0.4749	0.6336	74.95%
movie	NB	0.3739	0.7615	1.1354	67.07%
home	PDA	0.7561	7.0414	7.7975	90.30%
news	PDA	0.8510	10.3604	11.2114	92.41%
astrology	PDA	0.2195	7.8263	8.0458	97.27%
auto	PDA	0.1617	6.7837	6.9454	97.67%
movie	PDA	0.8858	10.3503	11.2361	92.12%
home	phone	1.7266	21.8300	23.5566	92.67%
news	phone	0.8473	26.7388	27.5861	96.93%
astrology	phone	0.2185	15.7217	15.9402	98.63%
auto	phone	0.6791	10.4697	11.1488	93.91%
movie	phone	1.7736	22.0621	23.8357	92.56%

Table 3. Second-time visiting the same Web page with content adaptation

Yahoo page	device	adaptation planner (sec.)	adaptation engine (sec.)	total access time (sec.)	adaptation engine%
home	NB	0.2723	0.7928	1.0651	74.43%
news	NB	0.3727	1.0151	1.3878	73.14%
astrology	NB	0.2138	0.5842	0.7980	73.21%
auto	NB	0.1587	0.4749	0.6336	74.95%
movie	NB	0.3739	0.7615	1.1354	67.07%
home	PDA	0.7561	7.0414	7.7975	90.30%
news	PDA	0.8510	10.3604	11.2114	92.41%
astrology	PDA	0.2195	7.8263	8.0458	97.27%
auto	PDA	0.1617	6.7837	6.9454	97.67%
movie	PDA	0.8858	10.3503	11.2361	92.12%
home	phone	1.7266	21.8300	23.5566	92.67%
news	phone	0.8473	26.7388	27.5861	96.93%
astrology	phone	0.2185	15.7217	15.9402	98.63%
auto	phone	0.6791	10.4697	11.1488	93.91%
movie	phone	1.7736	22.0621	23.8357	92.56%

Table 4. Overhead of the first-time visit and second-time visit on the same Web page

Yahoo page	device	access time for NCA (sec.)	access time for CA_1 (sec.)	access time for CA_2 (sec.)	$CA\_1_{overhead}$	$CA\_2_{overhead}$
home	NB	1.6470	12.0281	1.0651	630.30%	-35.33%
news	NB	3.6064	15.8398	1.3878	339.21%	-61.52%
astrology	NB	2.2178	7.3857	0.7980	233.02%	-64.02%
auto	NB	2.1850	7.4502	0.6336	240.97%	-71.00%
movie	NB	2.5448	13.2938	1.1354	422.39%	-55.38%
home	PDA	10.4060	18.7605	7.7975	80.29%	-25.07%
news	PDA	28.8180	25.6634	11.2114	-10.95%	-61.10%
astrology	PDA	12.4760	14.6335	8.0458	17.29%	-35.51%
auto	PDA	16.4620	13.7620	6.9454	-16.40%	-57.81%
movie	PDA	21.1840	23.3945	11.2361	10.43%	-46.96%
home	phone	43.8100	34.5196	23.5566	-21.21%	-46.23%
news	phone	119.7560	42.0381	27.5861	-64.90%	-76.96%
astrology	phone	68.5200	22.5279	15.9402	-67.12%	-76.74%
auto	phone	83.9000	17.9654	11.1488	-78.59%	-86.71%
movie	phone	120.3380	35.9941	23.8357	-70.09%	-80.19%

(please refer to Figure 1 for content adaptation process).

- pre_CA%: The percentage of pre-CA processing time over the total access time.

When a user first visits a Web page, the content adaptation process has to go through three steps: first is the pre-CA process; then the adaptation planner decides an adaptation strategy; afterwards the adaptation engine executes the adaptation accordingly.

Table 2 shows the performance measurement of a user's first-time visit of a Web page with content adaptation. The total access time is the sum of the execution time from three parts: pre-CA process, adaptation planner, and adaptation engine. For each device's accessing one of the five Yahoo sections, we recorded the processing time (in seconds) of its three individual steps and the total access time, as well as the percentage of the pre-CA processing time over the total access time. As shown in Table 2, the pre_CA% for NB is significantly higher than that for PDA and for phone. This observation reveals that the bottleneck of the content adaptation process for NB is at the pre-CA process (around 91%).

Table 3 shows the performance measurement of the second-time visiting of the same Web page with content adaptation. When a user visits the same Web page for the second time, the content adaptation process can skip the pre-CA process so the total access time will be reduced. For each device's accessing one of the five Yahoo sections, we recorded the processing time (in seconds) of its two individual steps and the total access time, as well as the percentage of the processing time of the adaptation engine over the total access time. As shown in Table 3, the processing time of the adaptation engine is the bottleneck of the entire content adaptation for all three kinds of devices, especially for devices with less computational power (PDA and cell phone).

Definition 8. $CA\_1_{overhead}$ denotes the overhead due to content adaptation at user's first-time access, $CA\_1_{overhead}$ is the overhead due to content adaptation at user's second-time access:

$$CA\_1_{overhead} = \frac{(access\_time_{CA\_1} - access\_time_{NCA})}{access\_time_{NCA}}$$

$$CA\_2_{overhead} = \frac{(access\_time_{CA\_2} - access\_time_{NCA})}{access\_time_{NCA}}$$

The higher the index, the more overhead it represents. If the value of an index is negative, it means that the content adaptation results in time saving instead of overhead. To better illustrate the overhead comparison, we further use a line chart in Figure 13 to represent the data shown in Table 4.

As shown in Figure 13 and Table 4, content adaptation does bring overhead to NB at the first time of content access, as indicated by the column $CA\_1_{overhead}$. However, when accessing the same content at the second time, the values of the corresponding $CA\_2_{overhead}$ column become negative, meaning that it takes less time to access the same content. This is because the adapted content has been processed and cached by our adaptation planner. The time for accessing adapted content is significantly less than that for accessing the original content due to the significant reduction in content size.

It should be noted that such phenomenon of saving content access time due to content adaptation becomes more significant when users are using portable devices. As shown in Table 4, a phone's content access time is reduced dramatically with content adaptation, no matter whether it is the first-time or the second-time visit. Similar results are observed from PDAs. This is mainly because the content adaptation process has utilized our adaptation planners to change the content's modality and fidelity based on the users' situated environment, which results in a tremendous saving of bandwidth. In other words, our experimental

Figure 13. CA_1 overhead vs CA_2 overhead

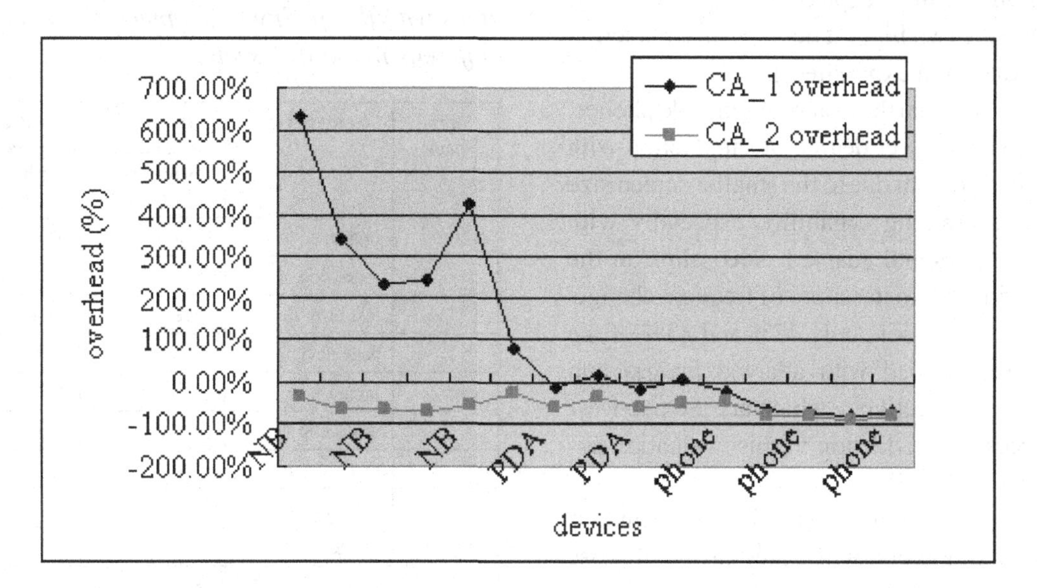

results show that the computational overhead due to the context-driven content adaptation rules can be moderately decreased through the reduction of the size of adapted content.

Qualitative Evaluation of Users' Satisfactions

Under the same experimental set up as described in Figure 11, we further designed a qualitative evaluation process to test users' satisfaction levels on the quality of adapted contents using our context-driven content adaptation planner. The motivation is that we realize content adaptation might result in degradation of content quality, which may affect people's comprehension of the adapted content. The reasons of quality degrade might include: reading difficulties due to either a device's screen size or improper content resizing or content reorganization, degrade in fidelity due to either devices' computing and bandwidth limitation or improper trans-coding, and so on. Thus, we designed four evaluation questions to evaluate users' satisfactions with the quality of adapted images, video clips, texts, and the experience of page navigation.

1. Are testers satisfied with the quality of adapted image?
2. Are testers satisfied with the quality of adapted video clip?
3. Are testers satisfied with the quality of text browsing on the adapted page?
4. Are testers satisfied with the quality of page navigation on the adapted page?

43 randomly selected student testers participate in this evaluation. Each of them was asked to use PDAs and mobile phones to access the aforementioned five sections of Yahoo pages (*home*, *news*, *astrology*, *auto*, and *movie*) after content adaptation, and then record their answers to the four questions above. Based on their responses, we summarize our findings in Table 5. When testers use PDAs, the *home* section provides the best results of content adaptation because it has the most stable Web page structure. 83% of the testers are satisfied with the adapted image on the *home* section. Regarding the experience of text browsing and page navigation on the *home* section, we found that most of the testers are satisfied with the adaptation. (96% and 94% of the testers have good experience of text browsing and page

navigation on the adapted page, respectively, which are the two highest degrees of satisfaction in this evaluation experiment).

However, when the testers use mobile phones, they showed a lower degree of satisfaction with the adapted content due to the smaller screen size and less computing capability, especially with adapted images and adapted video clips on the *news* section due to its nature of frequent change. (On the *news* section, only 47% and 43% of the testers are satisfied with adapted *images* and *video clips*, respectively, which are the two lowest degrees of satisfaction in this evaluation experiment).

The testers also expressed their strong demands of content adaptation when they browsed those image-oriented Web pages such as the sections of *astrology*, *auto*, and *movie*. They indicated that a flexible and readable content presentation will be one of their major concerns when they use handheld devices for mobile Internet surfing.

Table 5. Evaluation results of testers' satisfactions with the quality of adapted content on five different Yahoo Web sections

Yahoo page	Content types	Degree of satisfaction with PDA (%)	Degree of satisfaction with phone (%)
home	image	83%	67%
	video clip	80%	56%
	text browsing	96%	87%
	page navigation	94%	85%
news	image	68%	47%
	video clip	63%	43%
	text browsing	75%	78%
	page navigation	71%	72%
astrology	image	73%	56%
	video clip	68%	53%
	text browsing	90%	80%
	page navigation	92%	73%
auto	image	79%	57%
	video clip	75%	51%
	text browsing	87%	82%
	page navigation	89%	74%
movie	image	80%	60%
	video clip	81%	55%
	text browsing	86%	81%
	page navigation	85%	77%

Further Discussions

Although content adaptation can provide better content presentation based on people's contextual environment, it may also cause significant computational overhead. Therefore, one might argue that there is no need to perform content adaptation such as turning off the video while a person is driving. However, one major reason why such adaptation is valuable is that it could largely improve transmission performance, especially when transmitting rich media over wireless networks. For example, removing unnecessary video may largely save transmission time.

Although our experiments test access time for one user and prove that content adaptation overhead can be reduced after the user's first access, the observation can be extended to multi-user situation. The reason is that our content adaptation is conducted at the content adaptation planner acting as a proxy serving multiple users. After any user accesses a page, its adapted content will be cached, and all later users will benefit as long as they have similar receiving conditions. Thus, the more popular a Web page is, the more users will benefit from our technique.

In addition, not every handheld device can play all types of media. For example, a non-multimedia cell phone cannot play video clips. Hence, by removing such a multimedia component if it is un-playable on receiving devices, content adaptation could save unnecessary network bandwidth. Another alternative is to perform transcoding to downgrade such a multimedia component into a playable format, e.g., downgrade a video clip to static images. Nevertheless, there is always a tradeoff between quality and performance, and this is one of the motivations why we design a dynamic content adaptation planner.

Furthermore, in a mobile computing environment, context is typically a generalized concept and requires specialization based on specific application logic (Julien & Roman 2006). Our study here provides a guidance and prototype for people to build domain-specific content adaptation and delivery services.

CONCLUSION

In this article, we present a context-driven content adaptation planner concerning users' situated contextual environments. Our experimental results show that content adaptation can dramatically reduce the access time to Web content, especially when users are using handheld devices over the mobile Internet.

Our content adaptation planner may also benefit people who are with disability such as deaf and blind. For people who are deaf, they are in the same condition as people who are sitting in a meeting because both of them will not hear the audio. For people who are blind, they are in the same condition as people who are driving because both of them will not read the content. For those who suffer from weak vision, they are in the same condition as people who are in a blurred environment (e.g., due to a sunny or gloomy weather); the content should be consequently enlarged or the background color should be turned brighter. For people who suffer from weak hearing, they are in the same condition as people who are in a noisy environment (such as in a marketplace); then the audio volume should be turned louder.

The main contribution of this article is the development of a context-driven content adaptation planner to enhance the Web page content accessibility using handheld devices over the mobile Internet. We apply description logics (DLs) to formally define context profiles and requirements and automate content adaptation decision. Experimental results show that the computation overhead caused by the context-driven content adaptation planner can be moderately decreased through multiple accesses and through the reduction of the size of adapted content. The saving of transmission bandwidth becomes more significant when users are using handheld devices over the mobile Internet.

Our presented content adaptation planner provides some solutions to the four research challenges we identified at the beginning of this research project. For the challenge of *detection and representation of mobile user contexts*, we present a multi-dimensional formal model (including device, environment, status, and end user profile) that is then used to guide content adaptation. For the challenge of *adaptation rule design*, we present a JESS-enabled rule engine with an incrementally constructed rule base. For the challenge of *content adaptation management*, we present the design and implementation of content adaptation planner that balances the tradeoffs of various concerns. A formal OSM model is proposed as the fundamental basis for content adaptation management with high reusability and adaptability. Algorithms and detailed designs are presented for guiding automatic and dynamic content adaptation. For the challenge of *format (e.g., style sheet) generation*, we present our fine-grained style sheet management solution to enable configurable and re-configurable presentation style control.

Nevertheless, we also found several barriers existing in our current adaptation techniques, such as lacking the capability of processing script languages (e.g., JavaScript and VBScript), lacking the mechanism of processing image without alternative descriptions and voice recognition.

In addition, we found that many Web pages have similar layout structures, even though their contents are significantly different. Some similar and rarely changed portions (e.g., header fragment, navigation fragment, and advertisement fragment) occupy significant amount of storage space and consume computing resources to process. To further speed up the decomposition process and reduce required storage space, we are currently exploring a dedicated fragment detection method for automatically detecting similar fragments by identifying the coding patterns (Gamma, Helm et al. 1995) among Web pages.

Furthermore, currently we allow individual end users to configure and re-configure their dedicated style sheets. However, this extremely fine-grained solution is impractical since the cost of style sheet management is propositional to the number of end users. It may become too costly to maintain a large amount of style sheets. A more feasible solution is to attach a spread sheet with a group of users with common features. However, how to control the granularity of user groups remains challenging and will be investigated in our future research.

ACKNOWLEDGMENT

This work is supported by National Science Council, Taiwan under grants NSC 95-2520-S-008-006-MY3 and NSC96-2628-S-008-008-MY3.

REFERENCES

W3C. (2004). "*OWL-S: Semantic Markup for Web Services*." from http://www.w3.org/Submission/OWL-S/.

W3C. "*Document Object Model (DOM)*." from http://www.w3.org/DOM/.

Baader, F., Calvanese, D., McGuinness, D. L., Nardi, D., & Patel-Schneider, P. F. (2003). *The Description Logic Handbook: Theory, Implementation, Applications*. Cambridge, UK, Cambridge University Press.

Bellavista, P., Corradi, A., Montanari, R., & Stefanelli, C. (2003). Context-Aware Middleware for Resource Management in the Wireless Internet. *IEEE Transactions on Software Engineering*, *29*(12), 1086–1099. doi:10.1109/TSE.2003.1265523

Berhe, G., Brunie, L., & Pierson, J. M. (2004). Modeling Service-Based Multimedia Content Adaptation in Pervasive Computing. *the First Conference on Computing Frontiers on Computing Frontiers*, ACM Press, 60-69.

Bose, R., & Frew, J. (2005). Lineage retrieval for scientific data processing: a survey. *ACM Computing Surveys*, *37*(1), 1–28. doi:10.1145/1057977.1057978

Cabri, G., Leonardi, L., Mamei, M., & Zambonelli, F. (2003). Location-Dependent Services for Mobile Users. *IEEE Transactions on Systems*, Man and Cybernetics. *Part A.*, *33*(6), 667–681.

Chebotko, A. (Manuscript submitted for publication). S, Lu., & Fotouhi, F. (2007). Semantics preserving SPARQL-to-SQL query translation. *Data & Knowledge Engineering*.

Chen, L. Q., Xie, X., Fan, X., Ma, W. Y., Zhang, H. J., & Zhou, H. Q. (2002b). *A Visual Attention Model for Adapting Images on Small Displays*. Technical report MSR-TR-2002-125, Microsoft Research.

Chen, L. Q., Xie, X., Ma, W. Y., Zhang, H. J., Zhou, H. Q., & Feng, H. Q. (2002a). *DRESS: A Slicing Tree Based Web Representation for Various Display Sizes*. Technical report MSR-TR-2002-126, Microsoft Research.

Dey, A. K., & Abowd, G. D. (1999). *Toward A Better Understanding of Context and Context-Awareness*. GVU Technical Report GIT-GVU-99-22, College of Computing, Georgia Institute of Technology.

Gamma, E., Helm, R., Johnson, R., & Vlissides, J. (1995). *Design Patterns: Elements of Reusable Object-Oriented Software*. Addison Wesley, Boston, MA, USA.

He, J., Gao, T., Hao, W., Yen, I.-L., & Bastani, F. (2007). A Flexible Content Adaptation System Using a Rule-Based Approach. *IEEE Transactions on Knowledge and Data Engineering*, *19*(1), 127–140. doi:10.1109/TKDE.2007.250590

Hua, Z., Xie, X., Liu, H., Lu, H., & Ma, W.-Y. (2006). Design and Performance Studies of an Adaptive Scheme for Serving Dynamic Web Content in a Mobile Computing Environment. *IEEE Transactions on Mobile Computing, 5*(12), 1650–1662. doi:10.1109/TMC.2006.182

JESS. "Jess Expert System Shell." from http://www.jessrules.com/.

Julien, C., & Roman, G.-C. (2006). EgoSpaces: Facilitating Rapid Development of Context-Aware Mobile Applications. *IEEE Transactions on Software Engineering, 32*(5), 281–298. doi:10.1109/TSE.2006.47

Khedr, M., & Karmouch, A. (2004). Negotiating Context Information in Context-Aware Systems. *IEEE Intelligent Systems, 19*(6), 21–29. doi:10.1109/MIS.2004.70

Kinno, A., Yukitomo, H., & Nakayama, T. (2004). An Efficient Caching Mechanism for XML Content Adaptation. *the 10th International Multimedia Modeling Conference* (MMM'04), Jan., IEEE Press, 308-315.

Krause, A., Smailagic, A., & Siewiorek, D. P. (2006). Context-Aware Mobile Computing: Learning Context- Dependent Personal Preferences from A Wearable Sensor Array. *IEEE Transactions on Mobile Computing, 5*(2), 113–127. doi:10.1109/TMC.2006.18

Kurz, B., Popescu, I., & Gallacher, S. (2004). FACADE - A Framework for Context-Aware Content Adaptation and Delivery. *Second Annual Conference on Communication Networks and Services Research*, IEEE CS Press, 46-55.

Lee, Y. W., Chandranmenon, G., & Miller, S. C. (2003). *GAMMA: A Content Adaptation Server for Wireless Multimedia Applications*. Bell-Labs, Technical Report.

Lei, Z., & Georganas, N. D. (2001). Context-based Media Adaptation in Pervasive Computing. *Canadian Conference on Electrical and Computer Engineering*, May

Lemlouma, T., & Layaida, N. (2001). The Negotiation of Multimedia Content Services in Heterogeneous Environments. *The 8th International Conference on Multimedia Modeling (MMM)*, Amsterdam, The Netherlands, Nov. 5-7, 187-206.

Lemlouma, T., & Layaida, N. (2004). Context-Aware Adaptation for Mobile Devices. *2004 IEEE International Conference on Mobile Data Management*, IEEE CS Press, 106-111.

Lum, W. Y., & Lau, F. C. M. (2002). A Context-Aware Decision Engine for Content Adaptation. *IEEE Pervasive Computing / IEEE Computer Society and IEEE Communications Society, 1*(3), 41–49. doi:10.1109/MPRV.2002.1037721

Microsoft. "*Microsoft Reader*." from http://www.microsoft.com.

Mohan, R., Smith, J. R., & Li, C. S. (1999). Adapting Multimedia Internet Content for Universal Access. *IEEE Transactions on Multimedia, 1*(1), 104–114. doi:10.1109/6046.748175

Mostefaoui, S. K., & Hirsbrunner, B. (2004). Context Aware Service Provisioning. *the IEEE International Conference on Pervasive Services (ICPS)*, Beirut, Lebanon, Jul. 19-23, 71-80.

Mostefaoui, S. K., Tafat-Bouzid, A., & Hirsbrunner, B. (2003). Using Context Information for Service Discovery and Composition. *5th International Conference on Information Integration and Web-based Applications and Services (iiWAS)*, Jakarta. *Indonesia*, (Sep): 15–17, 129–138.

Mukherjee, D., Delfosse, E., Kim, J.-G., & Wang, Y. (2005). Optimal Adaptation Decision-Taking for Terminal and Network Quality-of-Service. *IEEE Transactions on Multimedia, 7*(3), 454–462. doi:10.1109/TMM.2005.846798

Nam, J., Ro, Y. M., Huh, Y., & Kim, M. (2005). Visual Content Adaptation According to User Perception Characteristics. *IEEE Transactions on Multimedia, 7*(3), 435–445. doi:10.1109/TMM.2005.846801

Pashtan, A., Kollipara, S., & Pearce, M. (2003). Adapting Content for Wireless Web Service. *IEEE Internet Computing*, *7*(5), 79–85. doi:10.1109/MIC.2003.1232522

Phan, T., Zorpas, G., & Bagrodia, R. (2002). An Extensible and Scalable Content Adaptation Pipeline Architecture to Support Heterogeneous Clients. *the 22nd International Conference on Distributed Computing Systems*, IEEE CS Press, 507-516.

"PictView." from http://www.pictview.com/pvw.htm.

Ramaswamy, L., Iyengar, A., Liu, L., & Douglis, F. (2005). Automatic Fragment Detection in Dynamic Web Pages and Its Impact on Caching. *IEEE Transactions on Knowledge and Data Engineering*, *17*(6), 859–874. doi:10.1109/TKDE.2005.89

Roman, M., Hess, C. K., Cerqueira, R., Ranganathan, A., Campbell, R. H., & Nahrstedt, K. (2002). Gaia: A Middleware Infrastructure to Enable Active Spaces. *IEEE Pervasive Computing / IEEE Computer Society and IEEE Communications Society*, *1*(4), 74–83. doi:10.1109/MPRV.2002.1158281

Satyanarayanan, M. (2004). The Many Faces of Adaptation. *IEEE Pervasive Computing / IEEE Computer Society and IEEE Communications Society*, *3*(3), 4–5. doi:10.1109/MPRV.2004.1321017

Schilit, B. N., Adams, N. I., & Want, R. (1994). Context-Aware Computing Applications. *IEEE Workshop on Mobile Computing Systems and Applications*, Santa Cruz, CA, USA, 85-90.

Toivonen, S., Kolari, J., & Laakko, T., USA. (2003). Facilitating Mobile Users with Contextualized Content. *Artificial Intelligence in Mobile System Workshop*, Seattle, WA, USA, October,

"VCDGear." from http://www.vcdgear.com.

Vetro, A., & Timmerer, C. (2005). Digital Item Adaptation: Overview of Standardization and Research Activities. *IEEE Transactions on Multimedia*, *7*(3), 418–426. doi:10.1109/TMM.2005.846795

Wang, Y., Kim, J.-G., Chang, S.-F., & Kim, H.-M. (2007). Utility-Based Video Adaptation for Universal Multimedia Access (UMA) & Content-Based Utility Function Prediction for Real-Time Video Transcoding. *IEEE Transactions on Multimedia*, *9*(2), 213–220. doi:10.1109/TMM.2006.886253

Xie, X., Liu, H., Ma, W.-Y., & Zhang, H.-J. (2006). Browsing Large Pictures Under Limited Display Sizes. *IEEE Transactions on Multimedia*, *8*(4), 707–715. doi:10.1109/TMM.2006.876294

Yang, S. J. H., Tsai, J. J. P., & Chen, C. C. (2003). Fuzzy Rule Base Systems Verification Using High Level Petri Nets. *IEEE Transactions on Knowledge and Data Engineering*, *15*(2), 457–473. doi:10.1109/TKDE.2003.1185845

Yang, S. J. H., Zhang, J., & Chen, I. Y. L. (2008). A JESS enabled context elicitation system for providing context-aware Web services. *Expert Systems with Applications*, *34*(4), 2254–2266. doi:10.1016/j.eswa.2007.03.008

Yang, S. J. H., Zhang, J., Chen, R. C. S., & Shao, N. W. Y. (2007). A Unit of Information-Based Content Adaptation Method for Improving Web Content Accessibility in the Mobile Internet. *Electronics and Telecommunications Research Institute (ETRI). Journal*, *29*(6), 794–807.

Yang, S. J. H., Zhang, J., Tsai, J. J. P., & Huang, A. F. M. (2009). SOA-based Content Delivery Model for Mobile Internet Navigation. *International Journal of Artificial Intelligence Tools*, *18*(1), 141–161. doi:10.1142/S0218213009000081

Zhang, J., Zhang, L.-J., Quek, F., & Chung, J.-Y. (2005). A Service-Oriented Multimedia Componentization Model. *International Journal of Web Services Research*, *2*(1), 54–76.

Zhou, C., Chia, L.-T., & Lee, B.-S. (2005). Web Services Discovery on DAML-QoS Ontology. *International Journal of Web Services Research*, *2*(2), 47–70.

This work was previously published in International Journal of Systems and Service-Oriented Engineering, Volume 1, Issue 1, edited by Dickson K.W. Chiu, pp. 96-129, copyright 2010 by IGI Publishing (an imprint of IGI Global).

Section 3
Knowledge Engineering for Systems and Services

Chapter 11
Two Heads Are Better Than One:
Leveraging Web 2.0 for Business Intelligence

Ravi S. Sharma
Nanyang Technological University, Singapore

Dwight Tan
Nanyang Technological University, Singapore

Winston Cheng
Nanyang Technological University, Singapore

ABSTRACT

This paper examines how Web 2.0 may be used in organizations to support business intelligence activities. Five leading professional services firms in the Energy, IT, software and health industries were used as the field research sites and action research performed on their Web 2.0 tools and environment. Business intelligence was the most significant driver of service value to their clients. From the data, five key findings were observed on the strategic use of Web 2.0 in the leading services firms. Firstly, the firm is aware that social networking tools can improve employees' performance. Secondly, there are more tools for tacit-to-tacit and tacit-to-explicit knowledge transfer than explicit-to-explicit and explicit-to-tacit. Thirdly, the firm has a higher number of tools where knowledge flows within itself and almost none for external knowledge flows. Fourthly, social network is part of normal work responsibilities. Finally, among KM tools that were most recognized as assisting social network use were of the Web 2.0 genre such as wikis, RSS feeds and instant messaging and blogging. The authors show that using Web 2.0 improves social networking and may be linked to a service professional's individual performance.

DOI: 10.4018/978-1-4666-1767-4.ch011

INTRODUCTION

Web 2.0 technologies have emerged as the next level of innovation in web-based applications that facilitate collaboration and sharing between users and the use of these technologies have been growing rapidly in recent years (O'Reilly, 2005). Typically, Web 2.0 platforms comprise technologies such as messaging, blogs, wikis, Really Simple Syndication (RSS) feeds, and social networking tools. Whereas the first generation of web tools were asymmetric in terms of users receiving information and knowledge from various websites, Web 2.0 is a euphemism for a more bi-directional, and one might say democratic, flow which often leads to peer-to-peer knowledge sharing and dissemination. This has enabled knowledge workers to interact and socialize more intuitively through online forums, directly exchanging know-how and experience with one another and creating content online. Web 2.0 technologies have been effectively used in the business environment where keeping increasingly widely distributed employees in contact with one another is vital. Their value lies in creating communities, collaboration, co-creation and connections, that when applied to the workplace can impact the way employees work, innovate, engage and deliver to their clients, partners and other colleagues (Xarchos & Charland, 2008). Hence, many leading edge businesses are using Web 2.0 as part of their knowledge management strategies to improve performance and to find new opportunities for knowledge creation and diffusion.

Arguably, the most intense of such knowledge management initiatives is business intelligence which is often misunderstood as data mining but also about keeping abreast of the customer, competitor and the market through intelligence (cf. Foo, 2008; Tiwana, 2002). More specifically, whereas knowledge management is about exploiting intellectual capital, business intelligence is more focused on knowing about the business conditions that an organization operates in. The sources of intelligence is not confined within research analyst reports, databases, statistics, annual reports, government reports, catalogues, directories and publications. It can also be in the form of observations such as price lists, advertisements, financial data, patent applications and commercials. Finally, intelligence can be gathered by attending seminars, information trade events, social contacts, sales force meetings, meeting and discussions with suppliers, distributors and partners. Through a case study, this paper attempts to examine how Web 2.0 social networking tools can be strategically used to improve business intelligence which ultimately leads to better company's performance and results.

Sveiby (1997) created the first intellectual capital framework and defined it in terms of three elements, namely, employee competence, internal structure and external structure. These three elements comprise the hardware and software of an organization from human and organization capital, processes, data, to leadership, vision, policies and relationship capital. Therefore, a primary objective of KM and BI must be to discover and analyze an organization's strengths and weaknesses along the 3 elements and benchmark them with competencies of competitors (Pan & Scarbrough, 1999).

Richard & Nory (2005) have further described how BI may help firms analyze transactions within each element. Using the SECI framework of Nonaka & Takeuchi (1995) to described the knowledge creation spiral, they also highlighted that BI may directly affect *combination* more than the other three conversion processes – *socialization, externalization and internalisation*. However this provides an incomplete view as it focuses on internal processes to the exclusion of external interactions with the market.

This paper describes an investigation of how a knowledge-intensive professional services firm leverages social interactions among employees, business partners, customers and its environment in order to support business intelligence activities that contribute to knowledge management effectiveness and corporate performance. We believe that business intelligence initiatives are particularly vital for services as both quality and productivity are directly aided by what is known about the operating environment. In this study, the scope of business intelligence transcends data mining and analytics and includes other sources of social intelligence about the customer, market and the competitor. The source of intelligence hence shifts from the explicit and into the realm of the tacit – mainly social interactions that are difficult to capture and mobilize. As opposed to the view of Richard & Nory (2005), we look at how organisations utilize "social capital" through employee interactions to capture knowledge from the marketplace and stay competitive. More specifically, we ask whether and how Web 2.0 tools help to support the four knowledge conversion processes in order to derive business intelligence. Note that the focus is on the more strategic business intelligence rather than a more operationally driven knowledge management perspective.

The remainder of this paper is organized as follows. In the next section we describe how Web 2.0 integrates with BI in general and present a framework based on Boisot's (1998) I-Space model which helps examine the impact of social capital on KM strategy. Section 3 outlines the field research methodology which includes the case background of five professional services firms. In Section 4, we analyze and discuss findings from our field investigation, again using the adopted framework from Boisot. We conclude with some implications for BI research and practice.

THEORETICAL BACKGROUND

We conjecture that Web 2.0 supports business intelligence by enabling the synergistic relationship and interplay between tacit and explicit knowledge specifically, through the four-step process of socialization, articulation, integration, and internalization (the SECI model of Nonaka & Takeuchi, 1995). In this section, we elaborate how this is accomplished.

Socialization requires individuals to share their tacit knowledge (Nonaka & Konno, 1998). It is during such interactions that tacit knowledge are transferred. It is well accepted that knowledge workers learn from their mentors by observation and practice under guidance rather than through codification alone (Davenport et al., 1998). With Web 2.0 tools, business colleagues routinely share their information and insights through blogs and wikis. As experts reply with their knowledge and experiences, the novices invariably learn from them. These tools hence create a common space or *ba* (Nonaka & Konno 1998) for individuals to share and diffuse knowledge.

Articulation is the process of converting tacit knowledge to explicit knowledge. A common example of this is thinking aloud and seeking consensus in the decision-making process. Web 2.0 tools, such as blogs, wikis and discussion forums provide the platform suitable for in-depth discussions and enable users to articulate their thoughts into words that allow others to understand.

Integration according to Nonaka & Konno (1998), involves three steps: 1) to combine and collect data from both internal and external sources; 2) diffusion of the valuable knowledge; and 3) processing of the knowledge into patterns and relations that make it more usable. In BI, integration of explicit knowledge can be in the form of "analysis of multiple, related what-if cases that show how these key factors interact to influence the decision" (Richard & Nory, 2005). Web 2.0 tools which support repositories as well as comments and referrals are an effective means to share explicit knowledge such as videos and presentation slides while social bookmarking tools creates cluster of tags that may be shared among the subscribers who are regularly informed of new tags.

Internalization is the process of using new patterns and relations together with the arguments of why they fit the purpose. This is followed by

Figure 1. Knowledge strategy formulation (adapted from Zack, 1999)

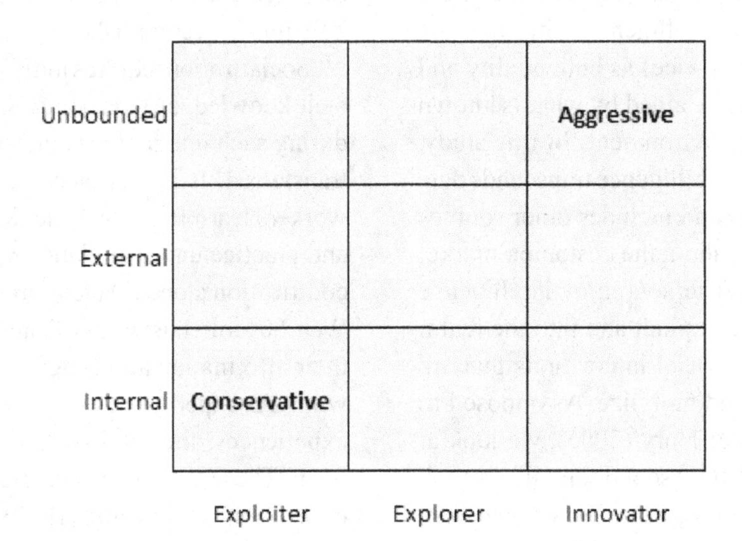

an update that extends the decision maker's own tacit knowledge base, thus creating a spiral of learning and knowledge that begins and ends with the knowledge worker. Although Web 2.0 is not able to directly support internalization, tools such as blogs and e-diaries provide an environment for users to internalize their knowledge when writing about their experiences. During the process of writing a blog, authors go through a thinking process of clarifying, refining and then sharing insights, analyses and story-telling which of course are meant to be challenged, validated or augmented with the accompanying social networks.

Juxtaposing the SECI model, Cross et al. (2001) have also identified four dimensions that are key characteristics of social relationships that were effective for acquiring and sharing knowledge, solving problems and learning. These are: (1) knowing what another person knows and thus when to turn to them, (2) being able to gain timely access to that person, (3) willingness of the person sought out to engage in problem solving rather than a knowledge dump, (4) a degree of safety in the relationship that promoted learning and creativity. Web 2.0 and social networks provide the platform for a knowledge worker to know where to go through wikis, forums and blogs contributed by experts. Although Web 2.0 is not able to entirely replace face-to-face meetings for such social discourse, they nevertheless support knowledge flows especially in cases of geographic dispersion. Although the use of Web 2.0 tools may not create an environment of trust and sharing that promotes learning and creativity, it enables the collective knowledge that generates new business insights when sharing experiences in blogs, wikis and other forums.

To sustain competitive edge in the marketplace, it is vital to know how Web 2.0 tools that support BI may be linked to knowledge strategies through the charting of what is known as a knowledge map knowledge map is performed. Figure 1 shows a framework adapted from Zack (1999) that allows some thought on what type of knowledge assets an organization should capture as a KM strategy and hence in the course of its BI pursuits.

The vertical axis reveals the primary sources of knowledge in an organization - internal or external or both (ie unbounded). On the horizontal axis, knowledge is categorized into those that are exploited, explored or innovated. Where and how knowledge is created and diffused determines whether an organization is taking an aggressive or conservative approach. An aggressive approach is needed if a firm's knowledge is lagging far behind its competitors or is defending its knowledge position (Kim, 1998) and BI should calibrate this need. A conservative approach is needed if a firm is in a position of market leadership and self-reliance in terms of knowledge assets. The framework in Figure 1 may be used to map out the Web 2.0 tools that support the internal versus external and exploiter versus explorer. This will make visible the areas of a firm's weaknesses and strengths with respect to the industry. Zack (1999) hence concluded that "knowledge is the fundamental basis of competition" and that organizations must know their capabilities, resources so as to strategically address knowledge gap to gain competitive advantage. In short, BI must unearth such knowledge. Building on what the firm already knows avoids re-inventing the wheel and not repeating mistakes makes it effective. The impact of BI is that it provides a firm an edge over the competitors.

Besides the abilities to transfer and create knowledge, an organization needs to know what to create and when to diffuse the knowledge that is created. This strategy enables the firm to build up its own knowledge-base and enter into a realm of increasing returns (Romer & Teece, 1998). These will require input in the areas of market, competitors and customers intelligence. By understanding what is happening in the market and changing customer needs allows the firm to decide when to invest in the right technology. As risk is imminent when investing in new technologies, companies competing in knowledge intensive industries inherently face higher risk (Boisot, 1998). Hence it is important for the company to execute its business strategy at a time such that it can maximize the profit by leveraging the knowledge when its value is at the maximum.

We develop a theoretical lens to investigate how Web 2.0 can affect the network structures and knowledge flows within a firm by mapping Web 2.0 tools to Boisot's (1998) I-Space model. Tacit knowledge is traditionally acquired through face-to-face interactions or could be transformed into explicit knowledge by capturing them in a directory of expertise. The introduction of Web 2.0 tools have large corporations thinking about decentralizing their knowledge and business processes which is bureaucratic and hierarchical in nature to informal networks where trust and reputation could be cultivated. However, as Boisot pointed out, the presence of a formal structure is still essential to engineer knowledge flows through social and other interactions.

The I-Space model suggests that diffusion can take in environments of abstract or concrete problems with explicit or tacit knowledge as the basis. Whereas Boisot's argued that maximum value is derived in a region of abstract, codified and undiffused information while minimum value is derived from concrete, un-codified and diffused information, we take a different view. As suggested in Figure 2, we posit that Web 1.0 tools serve a context of concrete, diffused and codified knowledge resulting in less value than Web 2.0 tools which support in the mobilization of abstract, un-codified and undiffused knowledge. Knowledge flows best between sources and targets that are at the same point in the scale in terms of codification, abstraction and diffusion. Table 1 below captures the essential characterizations of low, medium and high diffusion.

We may also examine how Web 2.0 supports the firm's overall business strategy through the use of Drew's (1999) "Boston Box". This identifies four different types of business knowledge: 1) what we know we know, 2) what we know we don't know, 3) what we don't know we know, and 4) What we don't know we don't know. Each of these occupy a quadrant of the Boston Box along the dimensions of procession and awareness of various knowledge assets. In "what we know we know", Web 2.0 can support the process of sharing and distribution of existing knowledge. This could include sharing best practices across internal and external boundaries. In "what we know we don't know", Web 2.0 tools combined with data mining

Figure 2. Adaptation of Boisot's I-space model

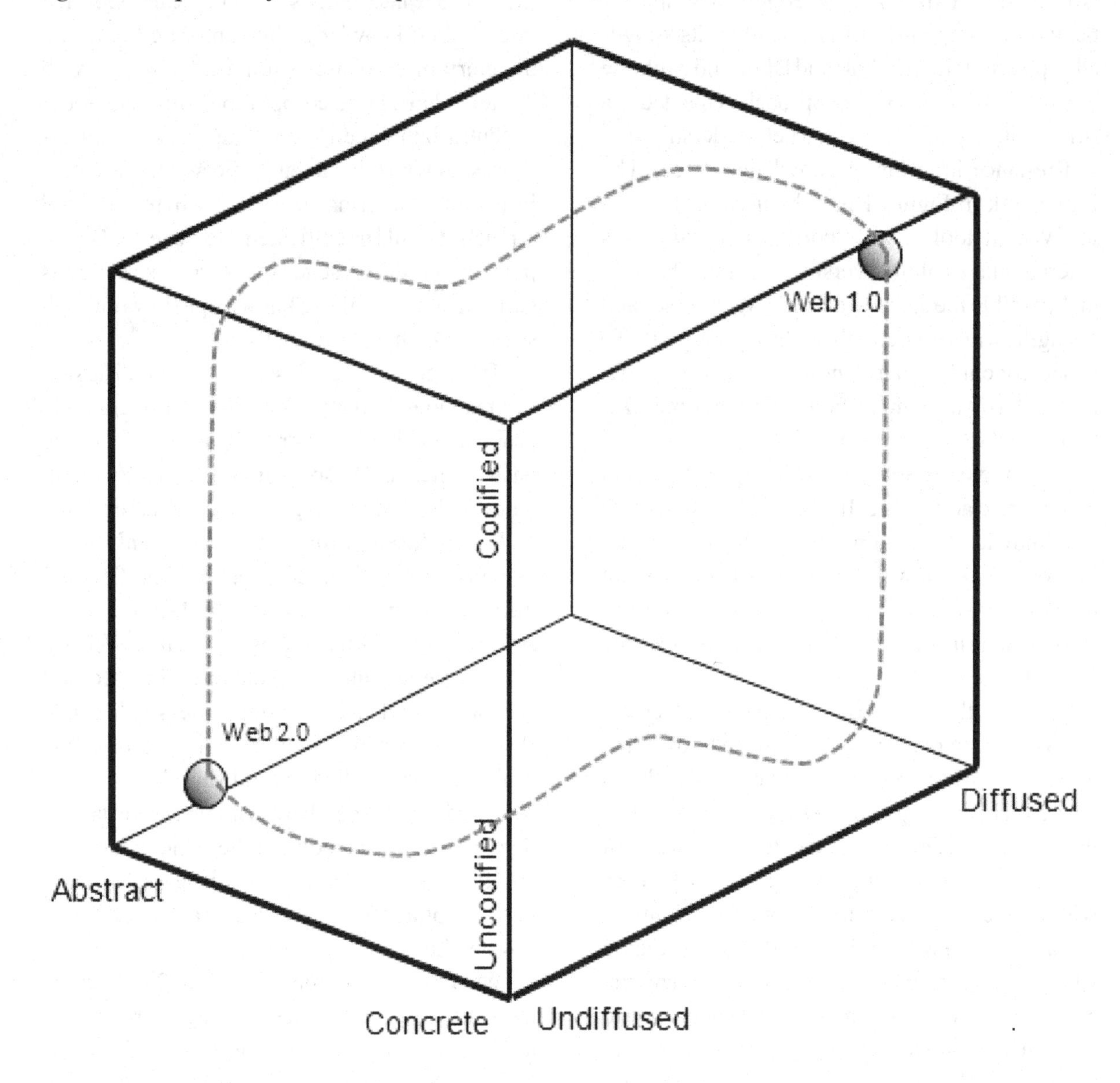

Table 1. Mapping Web 2.0 tools to I-space model

Position on Scale	Codification	Abstraction (within the firm)	Diffusion
High	Need to be captured in figures and formulae? Does it lend itself to standardization and automation?	Generally applicable to all agents whatever departments they are in?	Readily available to employees as well as external partners?
Medium	Can it be described in words and diagrams? Can it be readily understood by others from documents and written instructions alone?	Does it need to be adapted to the context in which it is applied?	Readily available to employees?
Low	Hard to articulate? Is it easier to show someone than to tell them about it?	Does it need extensive adaptation to the context in which it is applied?	Restricted to targeted or invited personnel?

tools supports intelligence-gathering and market research activities. It could be an EKP platform that contains blogs, forums, wikis, dash-boards with drill-down capabilities. In "what we don't know we know" past knowledge may be rediscovered and used through wikis and forums which store past contributions from historical users. From a strategy perspective perhaps the type of knowledge that poses the greatest threats and opportunities is "what we don't know we don't know". RSS Feeds may be suitable to 'push' critical information to the users quickly in order to keep them aware of new risks and possibly new threats and opportunities that the firm may be encountering.

As the Boston Box helps firms to identify which types of knowledge are keys to sustain their existence, it also shows which tools are useful to capture the necessary business intelligence. A knowledge gap framework such as the template shown in Figure 3, allows firms to align its knowledge strategy to business goals. It can also be used to show whether the Web 2.0 tools available in the firms are able to support their business intelligence needs for the present and future. The ability to know what the company must know prepares the organization for the future in terms of market's movement, competitors and customers' reactions (cf. Alavi & Leidner, 2001; Alvesson & Kaarreman, 2001; Anantatmula & Kanungo, 2005).

To conclude this background section, we also need to mention Gupta & Govindarajan's (2000) theoretical framework of intra-corporate knowledge transfers within multinational corporations using five factors that affect knowledge flow into and from a subsidiary within an organization. Their findings shall be applied in our study to understand how knowledge flows into an organization and how Web 2.0 can support this.

FIELD RESEARCH APPROACH

In order to investigate the research question of how Web 2.0 and other social networking tools may be used to support the business intelligence activities of firms, field research was performed on five large, multi-national, professional services firms with a strong global footprint and significant presence in Singapore. One was in the Energy Sector, two in IT Services, one in Software as a

Figure 3. Knowledge strategy gap analysis (adapted from: Tiwana, 2001)

Timeline	Types of Intelligences		
Future: What your company must know			
Now: What your company knows			
	Market	Competitor	Customer

Service (SaaS) and the remaining one in Health Services. More specifically, an immersive ethnographic study was conducted over a 12 month period using surveys, interviews, observation and record reviews. The Energy services firm was most cooperative and was used as the anchor site, the IT and software services firms were used for corroboration and consistency, whereas the Health Services firm (having given us much less time for our study) was used to replicate a quick series of interviews. All five sites demanded strict confidentiality and possessed the following characteristics that were particularly appropriate for our field study:

- Operated in knowledge intensive domains with critical BI activities in the delivery of professional services.
- Strong KM practices and a presence of knowledge sharing culture.
- Made use of Web 2.0 tools in their business functions.
- Well placed with regards to peer implementation of KM.
- Industry leaders in terms of KM and BI tools and processes.

An internal KM survey conducted in 2007 at the anchor site was used in understanding the Web 2.0 culture in the delivery of professional services. The survey of over 30000 service professionals was designed to assess the importance of social networking for this population on two axes - grade and performance. It explored how and why each population made use of its social network to achieve work objectives. The result of this enquiry was designed to understand the conditions that favored successful use of social networking to achieve exceptional career performance. The survey also probed the population on its understanding social networking tools.

Web 2.0 and social networking tools used in the professional services firm were identified and mapped into knowledge and business strategy models to determine how these tools help the organization to gain business intelligence. Following this, interviews with the IT teams were conducted to understand the extent to which social networking initiatives are aligned to the company's organizational goals. In a manner outlined in Sharma & Chowdhury (2007) we determined how well Web 2.0 tools supported service professionals (particularly the high performers) in terms of BI activities. Finally an action plan was proposed for each of the organizations to consider when planning for their future knowledge management initiatives. A subsequent half-yearly review and re-assessment of the initiatives to be conducted as part of the iterative process of action research was mapped out.

Case Background

The anchor subject of our case was founded 80 years ago and is today, one of the leading providers of engineering in the energy industry. The company is a knowledge intensive firm which supplies technology, information solutions and integrated project management that increase customers' performance.

Divided into two different business divisions, it has employees from more than 100 nationalities located around the world. The business is managed by different geographical areas. That the organization values people is reflected on its employment and development policies that constantly train and develop employees to excel in their skill sets. One aspect of training and development is the constant knowledge transfer by relocation personnel. Relocating employees facilitates tacit knowledge transfer by placing employees of diverse experiences and knowledge together.

The Energy Services firm is a believer of technological innovations as it strives to continue to lead in the industry. It currently has a network of Research and Development Centers around the world to drive innovation and state of the art technology and solutions into the marketplace.

While new knowledge was being created, the company recognized the importance of exploiting them to monetise their inventions into profits. Without downplaying the contribution of the knowledge and experience gained throughout the 80 years since it was founded, the tacit knowledge in the employees has a vital role in servicing their customers with their high value expertise. To enable the transfer of knowledge, the firm has 200 million dollars over the past 5 years in software and technology.

The IT services firms are two of the biggest from India with an excess of 100 000 employees in over 100 locations across the globe. Exploiting knowledge while being innovative is a core competency at both firms and this accounts for their SEI / CMM Level 5 status. The Software services firm was a leader in the enterprise space until it was recently acquired by a much larger rival (… hence the field data collection was left incomplete). Finally, the Health services firm was an independent business unit extracted from a fairly large health-care network in order that best practices and benchmarks may be identified and diffused throughout its regional outreach. Whilst the anchor case provided the most valuable field test for our research questions, our intention was to generalize the findings with quicker replicated studies in the other 4 cases. Hence, using the opportunity to work with each of these organizations for 6 months to a year, we extrapolated our findings on the effective use of Web 2.0 in BI.

KM Strategy and Approach

The ability to harness the existing knowledge and technical innovation has been the key reason why professional services firms have been able to supply technology and information solutions that increase customers' performances successfully. The successful implementation of KM tools, practices and processes have since evolved to include knowledge transfer within the company across time (history such as corporate history, lessons learnt etc), space (geographical locations) and cultural backgrounds effectively and efficiently.

One example of the Energy firm's successful KM initiatives was an engineering query support system which reduced the average time to resolve engineering modifications and technical queries by 75% and 97% respectively.

The timeline shown in Figure 4 of the KM tools and processes implemented throughout the years is indicative of the maturity of KM within the organization culture. In 2001, the Energy Services firm was nominated as one of the Most Admired Knowledge Enterprises (MAKE). Since then, it has always been nominated every year. One of the two IT services firms has been nominated for the Asian MAKE award since 2006.

The Web 2.0 aware organization is constantly exploiting and exploring knowledge to support its business vision and strategic objectives. In order to support these knowledge processes, it is imperative to understand how Web 2.0 tools could be leveraged strategically to better manage their knowledge.

Use of Social Networking Tools

In this section, we will cover the data collection processes and their results. The data collection in our study comprises two parts:

1. Collecting information about social networking tools by browsing EKP resources.
2. Collecting information about how social networking initiatives are aligned to organizational goals by interviewing the IT and KM teams in the organization.

A compilation study on most prevalent Web 2.0 tools was carried out on all five case sites and Table 2 is a generalization and summary of the most impactful of Web 2.0 and social networking tools that were used in the five field sites. This is followed by an itemized description. A more extensive list may be found in Appendix I.

Figure 4. Timeline of KM tools implemented (corporate internal source)

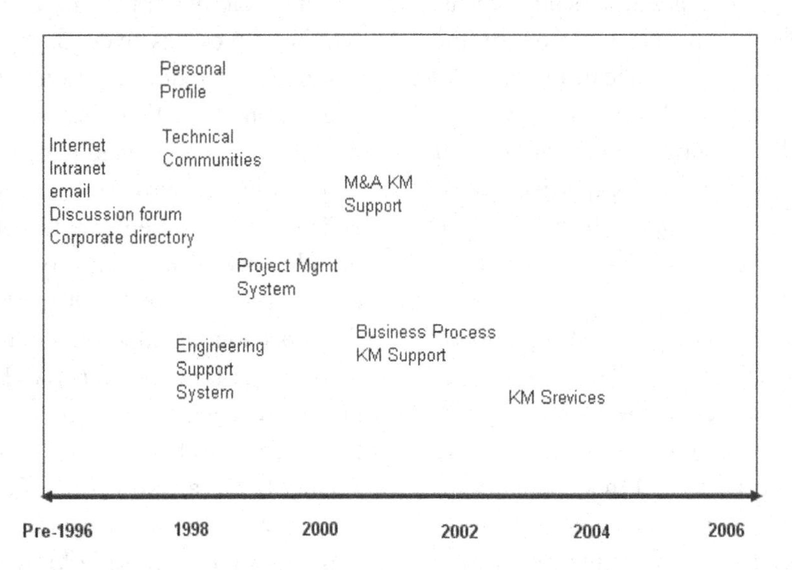

Bulletin Boards, Video and Photo Sharing

The typical services organization has hundreds of bulletin boards, for topics as varied as tips when moving into a new city, to domain of science discussions, passing through management topics. The technical community uses this technology as its principal means of maintaining contacts inside the Community of Practices and Special Interest Groups. Video and photo sharing is a site where people can post videos and images, which can later be viewed by all subscribers, searched for and referred to via addresses.

Blogs

A blog or weblog is basically a diary recorded on a web site, with simple editing functions. It is said that more than anything else, blogs are changing the way media and marketing operate today, with

Table 2. Web 2.0 tools used in professional services

Active Sharing	Collaboration Management	Computerized Communication	Contact Management	Conventional Telecommunication	Knowledge Bases	Audiovisual	Web Publishing
Shared Whiteboard	Workflow System	E-mail	Profile	Phone	Bulletin Board	Video Sharing	Blog
Application Sharing	Document Management	Chat	Relationship Handler	SMS	Unstructured Repositories	Photo Sharing	Wiki
Game Rooms	Shared Calendar	Internet Telephony		Teleconference	Structured Repositories		Podcst
				Video Conferencing			RSS
				Fax			Social Bookmarking

more news found and exposed today by bloggers, and brands being made or broken by good or bad exposition on those blogs.

Wikis

In a Wiki everybody can edit the original post or add other ones. This is considered as one of the best way to emulate the wisdom of crowds, although in reality there is a large risk of groupthink. It is likely that the concept might extend to be a repository of histories and snippets from organization history.

Podcasts, RSS and Social Bookmarking

Podcast is associated with sound files (not necessarily for the iPod) that can be played online or downloaded for future listening. This is often used for informal e-learning (the podcast can also have video) or quick news and live reports. RSS feeds provide a way for users to be notified when a page is modified and to get the update. Social bookmarking enables sharing user's bookmarks, sometimes adding comments, ratings, taxonomy etc.

Profile and Relationship Handlers

People share their profile in a common database and can find each other, sometimes with restrictions depending on their subscription. This is basically a web version of the yellow and white pages – public or private directories where people can look for other persons or companies. Relationship handler is used for many reasons: dating, recruiting, school alumni, etc. Unlike a directory where anybody can meet everybody, in this case it is all about who knows who, and about maintaining the links between them.

Shared Whiteboards and Application Sharing

Shared whiteboard emulates the classroom whiteboard, members can write at the same time on the board on their computer and view "live" the result of their writing and their co-writers. Application sharing enables more than one user to work at the same time on the same application, where usually one person controls while other view and the hand can be passed around, for example, NetMeeting.

Alignment with Organizational Goals

Follow-up interviews with relevant members of the IT and KM teams in each organization were conducted over a 6 month period. The measurement of success in social networking initiatives were not intended to be linked to innovation or financial performance. This was out of scope. What was investigated was the professional service employee's work performance that was impacted by the extent of social networking culture in the firm. The performance factors for BI were grouped into: Personal, Managing the Assigned Tasks, Working with Others and Environment, Health and Safety. Precise definitions of these BI performance indicators may be found in Appendix II. These were useful to target the usage of Web 2.0 tools by high performers undertaking BI activities.

Personal

- Knowledge of Work – familiarity with individual task or tasks performed by business unit.
- Decision-making – Both willingness to make decisions and the quality of decisions (judgement).
- Creativity – Seeks innovative ways to solve problems.
- Initiative –Ability to think and act without being directed.

Managing the Assigned Task

- Understanding of Product or Service Technology – Knowledge of engineering principles or field technology.

- Short-range planning – Markets, products, facilities, people.
- Long-range planning – Emphasis on the new markets, products, facilities.
- Implementation of plans – Puts short and long range plans to practical use.

Working with Others

- Peers – Respected by and can influence.
- Teamwork – Works effectively on teams by actively contributing to the accomplishment of goals.
- Clients/Others – Works effectively and successfully with clients and others.
- Knowledge Sharing – Shares own knowledge, learns from others and applies knowledge in daily work. Open to New ideas and continuous learning.

Environment, Health and Safety

- Effectiveness of Prevention–Achieves continuous improvement in accidental risk reduction.

FINDINGS AND DISCUSSION

The fifteen (15) Web 2.0 tools that were identified as common and impactful within the five organizations are tabulated below in Table 3. Of the 15 tools, Intern Alumni was the only one that was not universally present in all five sites. For the purpose of discussion, each tool is given a unique number which serves as its identifier.

Mapping Web 2.0 Tools into Knowledge Creation

To aid our conceptual understanding, the BI-specific KM tools are mapped into Nonaka and Takeuchi's knowledge creation model to understand how the application of Web 2.0 supports the knowledge creation process.

Figure 5 was derived by consensus in our field interviews and observations. As evident from Figure 5, most Web 2.0 tools supporting BI are for linking of explicit to explicit knowledge although some of them also support the other three processes. However, the SECI model has its limitation and is unable to show clearly the continuity between the transformations from socialization, externalization, combination to internalization. Hence, we analyzed gaps in knowledge capture and reuse.

Mapping Web 2.0 Tools into Knowledge Gaps

From observation and confirmation, it was found that the social networking tools may be mapped into the knowledge strategy gap framework in order to understand how these tools can help the firm exploit or explore the knowledge derived from BI.

The figure above was derived by consensus in our field interviews and observations. Figure 6

Table 3. List of business intelligence tools investigated

No.	Tool	No.	Tool	No.	Tool
1	Application Sharing	6	Technical Community	11	Intern Alumni
2	Blog	7	Enterprise Knowledge Portal	12	Internal Wikibook
3	Internal Youtube	8	RSS	13	Internal Wikipedia
4	Expert Finder	9	Sales & Marketing Sharing Tool	14	Webcast
5	CRM	10	Employee Alumni	15	Internal Wikis

Figure 5. Web 2.0 tools in the SECI framework

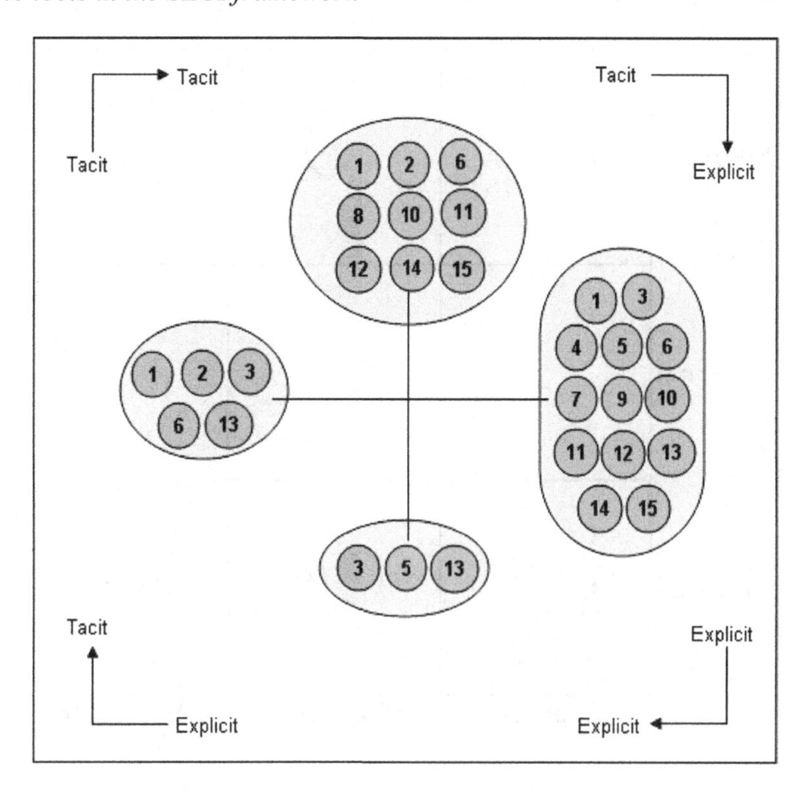

suggests that the tools that support the exploration of knowledge are those that are used to create new knowledge from BI gathered to keep pace with competition while tools that allow the organization to exceed its requirement are considered exploiters. When a tool supports exploitation and exploration, it is considered to be an innovator.

On the vertical axis, internal refers to tools that have knowledge residing in systems, repositories and processes within the organization, while external refers to external partners such as customers, vendors, and consultants. Their positions within each grid quadrant are not relative to one another as there is insufficient quantitative data from our case studies to indicate their exact positions. However, the Knowledge Gap model shows that the firm has various Web 2.0 tools to explore and exploit the firm's own knowledge.

Figure 6 also shows that the majority of the tools in the organization could be used for exploiting knowledge in the firm. Although there are no specific Web 2.0 tools for gathering BI from external sources, the unbounded row indicates that there are tools for both internal and external sources. Webcast is used as an innovator tool to aggressively create new knowledge from both internal and external resources.

Again, both figures above were derived by consensus in our field interviews and observations. Figure 7a shows that most of the tools used require a certain level of codification and abstraction prior to transfer except for Application Sharing, Internal YouTube and Expert Finder tools. Although Application Sharing is for invited personnel only, they are open for employees to participate when required. Most of the tools aid the transfer of knowledge to the employees from the undiffused state to diffused state. This is consistent with Boisot's argument that electronic revolution has pushed the social learning cycle towards the right hand side of the I-Space thus more people can be reached within

Figure 6. Web 2.0 tools for exploitation vs exploration

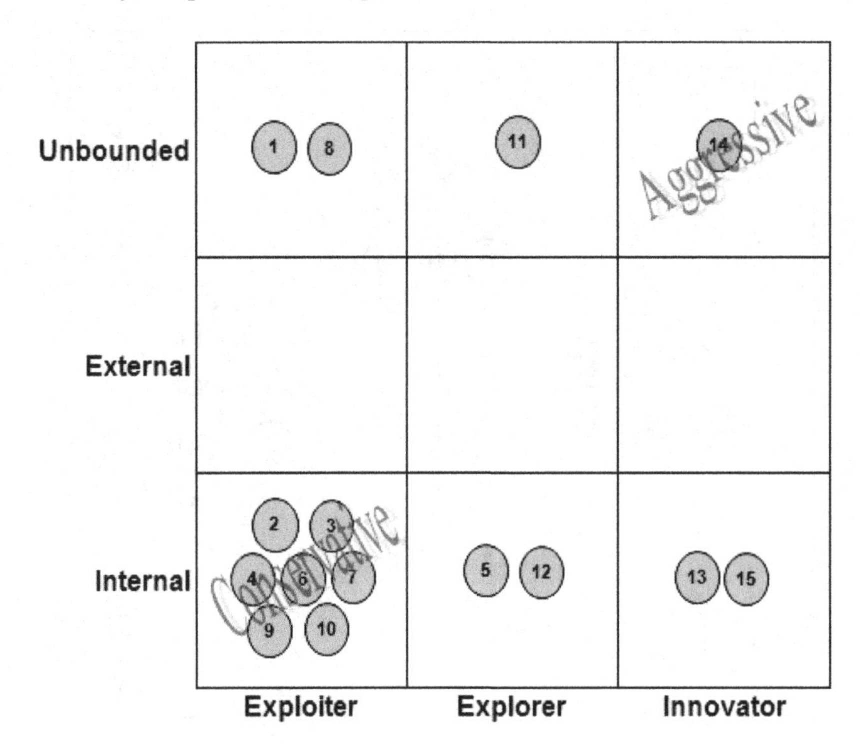

the same time. This can result in difficulties to retain knowledge in the V_{max} region as knowledge diffuses even further.

Figure 7b shows that the knowledge shared via the Application Sharing tools (cf balloon 1) need not be well codified and abstract as tools like video conferencing can help transfer tacit knowledge to employees and external personnel without face-to-face meetings. For the Technical Community (cf balloon 6), the social learning cycle indicates that as the communities evolve, they begin to function like clans where transfer of knowledge can be highly tacit and concrete. As core members and experts continue to stay committed yet others do not see added values being part of the communities. Hence, new knowledge created will only flow to a small group of employees. One option is to end the communities that are no longer bringing value while new ones are being created. The Employee Alumni and Intern Alumni tools in red show similar trends as the Technical Community except that the knowledge transferred does not need to be highly codified prior to transfer as the number of tools available is vast. However, they also face the risk of losing "audience" should there be no blockage put in the path and the participants choose to remain comfortable and cliques are formed.

Figure 8 summarizes our findings from the field along the aspect of types of BI. The matrix shows that the five firms have various social networking tools that are used to collect different types of BI related knowledge - Market, Competitor and Customer. Web 2.0 tools are mapped into the matrix based on their functions. The matrix also shows how the tools are deployed to help the company to collect BI in the two areas: knowledge they already have and the knowledge they must have.

Mapping Web 2.0 Tools into the Boston Box

By considering how Web 2.0 tools support knowledge awareness and knowledge content in the area

Figure 7. Web 2.0 tools in I-space (a) Figure Web2.0 tools' social learning cycle (b)

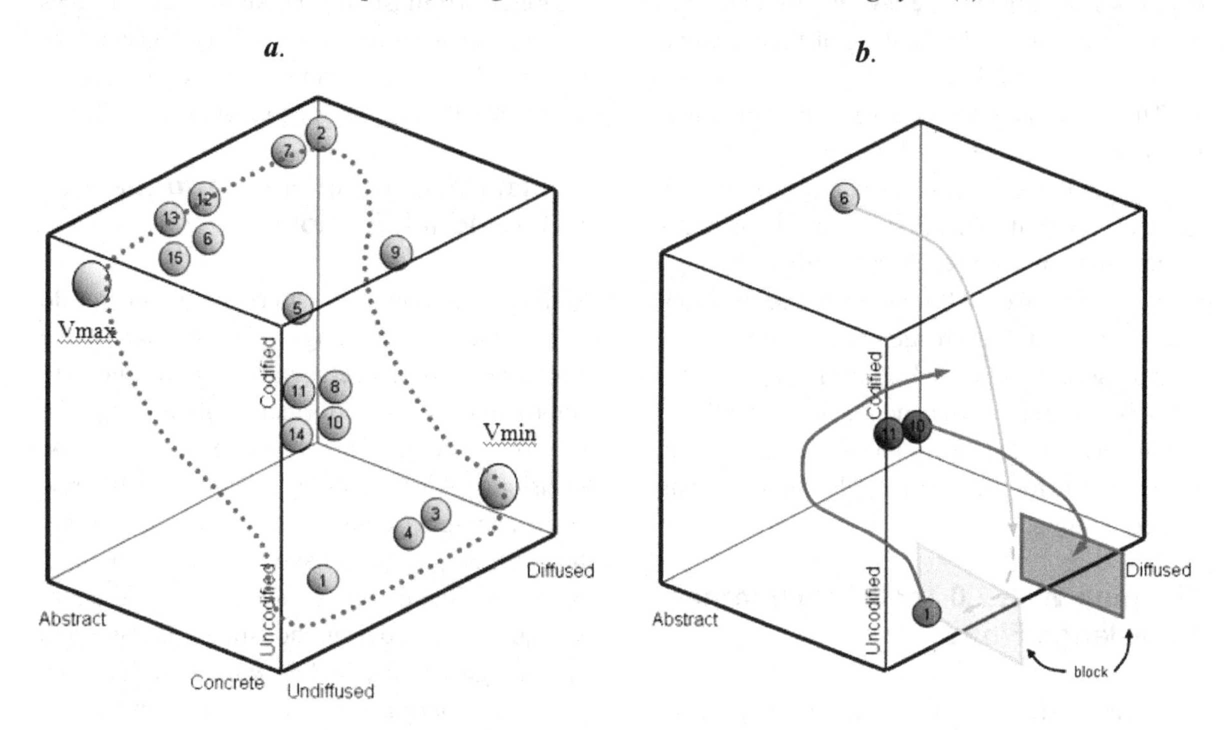

Figure 8. Web 2.0 tools in BI type vs timeline

Timeline	Types of Intelligences		
Future: What your company must know	8 11 13	2 8 11	2 5 8 11 13
Now: What your company knows	1 4 5 6 9 10 12 14 15	1 2 4 6 9 10 12 14 15	1 2 4 5 9 10 12 14 15
	Market	Competitor	Customer

of business intelligence, we have extrapolated from the five sites an understanding of their context within the Boston Box.

The figure above was derived by consensus in our field interviews and observations. Figure 9 suggests that the competitive services firm is more established in the first and third quadrants where the knowledge exists internally. Although there are few Web 2.0 tools in the second and fourth quadrants, the creation and capture of new knowledge cannot be solely judged by the number of tools but their impact and reach. Web 2.0 and social networks are an obvious case for reaching to the outside for the purpose of learning, sharing and combining.

Mapping Web 2.0 Tools into Internal Knowledge Flows

It was a consistent finding that a leading firm's knowledge sharing culture and ability to adopt new technology such as Web 2.0 tools is important for knowledge to flow to throughout. In terms of value of knowledge stock, the project deliverables and needs specified in the employee's job expectations require employees to pursue greater success in their own functions. Motivations to share and acquire knowledge are driven by distinct characteristics of the organization. The diverse cultural background of its employees increases the motivation to acquire knowledge for breakthroughs for innovation ideas. Employees are encouraged to share knowledge as part of their performance evaluation. Our results revealed that the social network plays a critical role in career progress and the employees are willing to use social network tools if they are made more available and easy to use. The presence of a great variety of social networking tools also improved the existence and richness of transmission. Finally, new subsidiaries as a result of mergers and acquisitions (M & A) are better equipped through Web 2.0 tools to acquire and impart the necessary knowledge between them and HQ.

Figure 10 illustrates the knowledge outflows and inflows between the subsidiary, HQ and new entrants that are consistent with the research on knowledge flows by Gupta & Govinderajan (2000).

Social Networking and Performance of Individual Employee

Although leading professional services firms do not see social networking as directly linked to the financial performance, social networking helps to improve the work performance of an individual employee in terms of his or her knowledge of work, decision-making, creativity and initiative. In terms of managing assigned tasks, social networking helps an employee to have a better understanding of product or service technology and better business insights during short-range planning and long-range planning. When working with others, social networking helps to facilitate interaction and knowledge sharing between peers, improve teamwork, and more effectively with clients. All these factors are taken into consideration when assessing the performance of the employees as it helps to increase the business intelligence of the organization.

CONCLUSION, IMPLICATIONS AND FURTHER WORK

Based on the extrapolations derived from our immersive ethnographic studies, we unearthed the following implications for organizations planning for seeking strategic impact from Web 2.0 tools.

- Current social networking tools lack user-friendliness as reported in the internal survey. A technology or product that provides a suite of social networking tools and a portal that enables personalization would encourage more users to be effective.
- The knowledge of ex-employees and now alumni could be further tapped on by en-

Figure 9. Web 2.0 tools mapped onto the Boston box

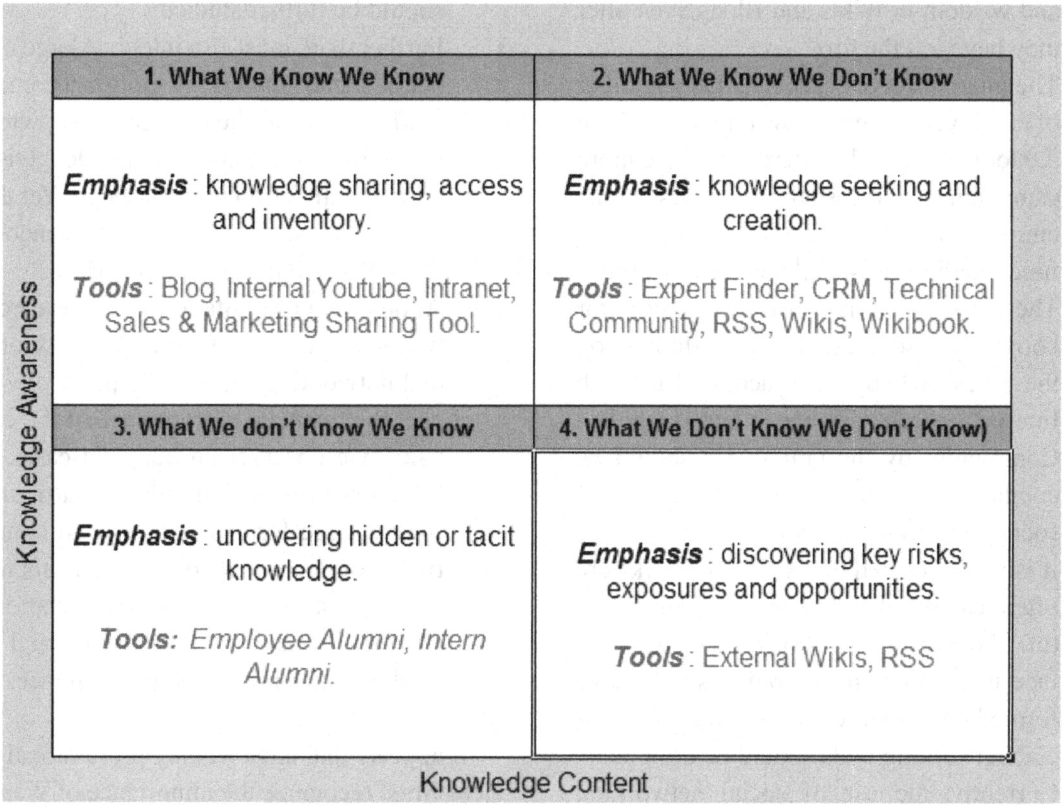

1. What We Know We Know	2. What We Know We Don't Know
Emphasis: knowledge sharing, access and inventory. ***Tools***: Blog, Internal Youtube, Intranet, Sales & Marketing Sharing Tool.	***Emphasis***: knowledge seeking and creation. ***Tools***: Expert Finder, CRM, Technical Community, RSS, Wikis, Wikibook.
3. What We don't Know We Know	**4. What We Don't Know We Don't Know)**
Emphasis: uncovering hidden or tacit knowledge. ***Tools:*** *Employee Alumni, Intern Alumni.*	***Emphasis***: discovering key risks, exposures and opportunities. ***Tools***: External Wikis, RSS

Knowledge Awareness (vertical axis); Knowledge Content (horizontal axis)

Figure 10. Web 2.0 tools in corporate knowledge flows

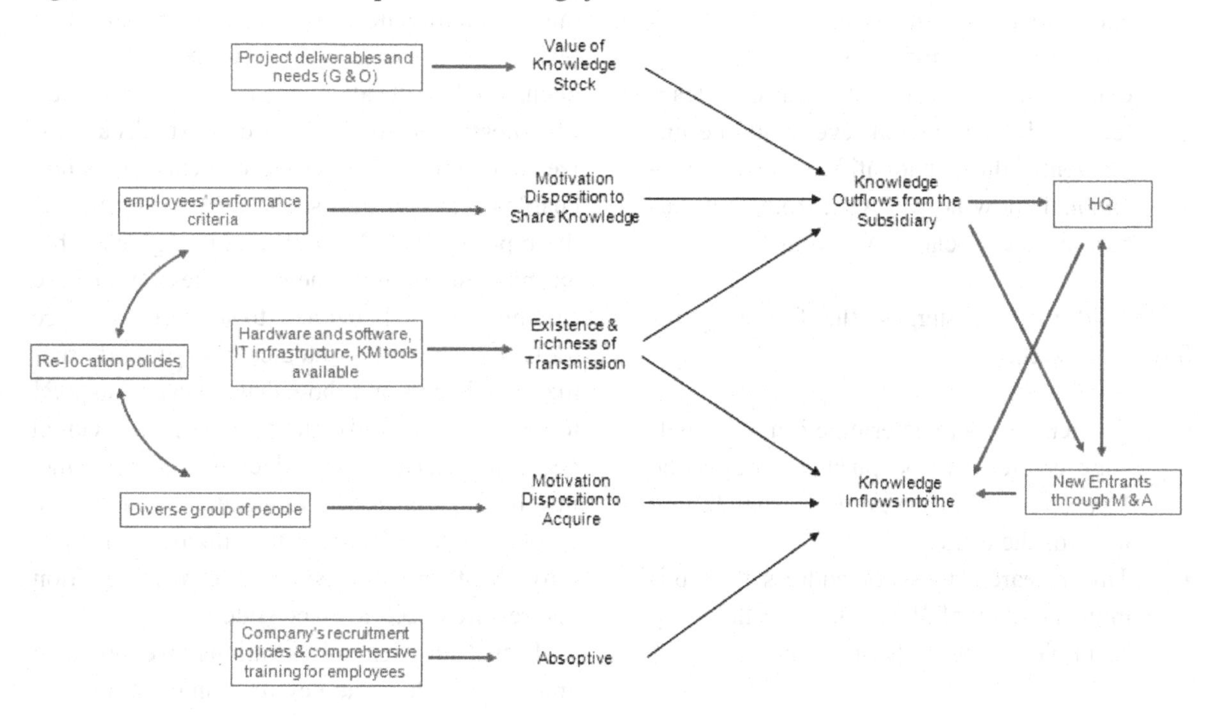

couraging them to share their experience and wisdom in Wikis and Blogs even after they have left the firm.

- The internal survey revealed that a number of employees do not know how to use Web 2.0 tools effectively. There should be more campaigns to increase the awareness of the employees in terms of the presence and functionalities of social networking tools.
- There is also a lack of recognition for knowledge sources. Further studies on the types of knowledge acquired through Internet search, databases and Technical Community by the employees should be conducted to improve the functionalities of social networking tools.
- It is well accepted that social networks are often catalyzed by personal interactions (cf. Wenger et al., 2002). More face-to-face interactions for employees who have formed communities of practice using social networking tools should be designed.
- Currently, the use of social networking tools is only seen as positively impacting the performance of individual employees and not sufficient to bring about a radical change in the company's social networking culture. Recognition and rewards at team level and departmental level could be implemented through identifying new organizational know-how flowing from activities as a result of social networking.

In addition, we suggest the following for further research.

- Further studies to determine if new knowledge created can be pushed towards the Vmax region and prevent it from falling towards the diffused state.
- This research does not address the optimum number of Web 2.0 tools that may be implemented without a firm's IT infrastructure being overloaded. This aspect should be further studied.
- Further work must also integrate large number of social networking tools in a common platform such as the Enterprise Knowledge Portal that allows flow of knowledge is vital to the diffusion of knowledge. Yet there does not exist an architecture that incorporates this social aspect into KM.
- Another major limitation in our research is the lack of data to prove that the use of social networking tools could positively impact the financial performance of the company. We observed in our field cases that high performing individuals particularly sought large forums and external connectivity for social networking. On this note, we suggest an empirical determination of whether the size and reach of a social network impacts individual performance.

It appears that most leading professional services firms recognize the importance of Web 2.0 and specifically, social networking to support the company's business strategy in terms acquring market intelligence, especially in the areas of technological expertise, long-term planning and client trends. After all, a unique selling proposition of professional services is the contextual, customized deliverables that accrue to clients. Our study enables us to assess the social networking culture, the types of Web 2.0 tools commonly available and more importantly, their usefulness. With these findings, we are better able to suggest how these tools could be used strategically to support organizational goals and show that the use of such BI tools could be linked to the employee's individual performance. However, inducing a radical change in a firm's social networking culture to encourage employees to participate in utilizing such tools through other means such as team recognition and rewards need to be considered.

From our field observations, we are also able to provide some key recommendations for service firms to improve their social networking culture through better technological platform and awareness campaigns. Business managers and KM champions of social networking initiatives would also be able to have a better understanding of how Web 2.0 social networking tools fits into the overall picture of a company's business strategy. Recent anecdotal evidence seems to suggest that social networks are productivity killers with employees wasting up to 45 minutes each day (cf. http://getahead.rediff.com/report/2009/dec/22/are-social-networking-sites-killing-productivity.htm). Our in-depth field studies have enabled us to refute this and suggest that the most valuable intelligence of the market may be gleaned by employees active in social networks with prospective customers, regulators, innovation centres and business partners. Such BI support better allows a professional services firm to position itself in the market and make the tradeoffs on explored vs. exploited (or alternately innovative vs. core) knowledge strategies. We conclude with some conviction that Web 2.0 is integral to business intelligence, an essential component of knowledge management strategy. Insights and experience are intuitively tacit and hence better suited to leverage Web 2.0 than codified knowledge repositories for sharing and knowledge transfer.

ACKNOWLEDGMENT

The authors are part of an informal, irreverent knowledge research factory (styled on the Bourbaki group) at the Nanyang Technological University, Singapore. The findings reported in this article are part of the on-going efforts to develop a formal understanding of knowledge management methods and policies. The authors are grateful to the participating organizations for allowing the immersive ethnographic field studies. Many thanks are also due to the guest editors and the anonymous reviewers for their thoughtful review of the article which has led to a much improved paper.

REFERENCES

Alavi, M., & Leidner, D. E. (2001). Knowledge management and knowledge management systems: conceptual foundations and research issues. *Management Information Systems Quarterly*, *25*(1), 107–136. doi:10.2307/3250961

Alvesson, M., & Kaarreman, D. (2001). Odd couple: Making sense of the curious concept of knowledge management. *Journal of Management Studies*, *38*(7), 995–1018. doi:10.1111/1467-6486.00269

Anantatmula, V., & Kanungo, S. (2005). Establishing and structuring criteria for measuring knowledge management efforts. In *Proceedings of the 38th Hawaii International Conference on System Sciences (HICSS-38)*. Washington, DC: IEEE Press.

Bartlett, C. (1998). *McKinsey & Company: Managing knowledge and learning* (Case 9-396-357). Cambridge, MA: Harvard Business School Press.

Bierly, P., & Chakrabarti, A. (1996). Generic knowledge strategies in the U.S. pharmaceutical industry. *Strategic Management Journal*, *7*(4), 123–135.

Bohn, R. (1994). Measuring and managing technological knowledge. *Sloan Management Review*, *36*(1), 61–73.

Boisot, M. (1999). *Knowledge Assets - The Paradox of Value*. Oxford, UK: Oxford University Press.

Cross, R., Parker, A., Prusak, L., & Borgatti, S. (2001). Knowing What We Know: Supporting Knowledge Creation and Sharing in Social Networks. *Organizational Dynamics*, *30*(2), 100–120. doi:10.1016/S0090-2616(01)00046-8

Davenport, T. H., Long, D., & Beers, M. C. (1998). Successful knowledge management projects. *Sloan Management Review*, *30*(2), 43–57.

Drew, S. (1999). Building Knowledge Management into Strategy: Making Sense of a New Perspective. *Long Range Planning*, *32*(1), 130–136. doi:10.1016/S0024-6301(98)00142-3

Foo, S., Sharma, R., & Chua, A. (2007). *Knowledge Management Tools and Techniques* (2nd ed.). Upper Saddle River, NJ: Prentice Hall.

Gupta, K., & Govindarajan, V. (2000). Knowledge Flows Within Multinational Corporations. *Strategic Management Journal*, *21*(4), 473–496. doi:10.1002/(SICI)1097-0266(200004)21:4<473::AID-SMJ84>3.0.CO;2-I

Kim, L. (1995). Crisis Construction and Organisational Learning: Capability Building in Catching-up at Hyundai Motor. *Organization Science*, *9*(4), 506–521. doi:10.1287/orsc.9.4.506

Nahapiet, J., & Ghoshal, S. (1998). Social capital, intellectual capital and organizational advantage. *Academy of Management Review*, *23*(2), 242–266. doi:10.2307/259373

Nonaka, I., & Konno, N. (1998). The Concept of "Ba": Building a Foundation for Knowledge Creation. *California Management Review*, *40*(3), 40–54.

Nonaka, I., & Takeuchi, H. (1995). *The Knowledge-Creating Company*. New York: Oxford University Press.

Nonaka, I., Toyama, R., & Konno, N. (2000). SECI, Ba and Leadership: A unified model of dynamic knowledge creation. *Long Range Planning*, *33*(1), 5–34. doi:10.1016/S0024-6301(99)00115-6

O'Reilly, T. (2005). What is Web 2.0? Design Patterns and Business Models for the Next Generation of Software. *O'Reilly Media, Inc*. Retrieved June 15, 2008, from http://oreilly.com/pub/a/oreilly/tim/news/2005/09/30/what-is-web-20.html

Pan, L., & Scarbrough, H. (1999). Knowledge Management in Practice: an Exploratory case study. *Technology Analysis and Strategic Management*, *11*(3), 370–387.

Richard, T., & Nory, E. (2005). Knowledge Management and Business Intelligence: The importance of integration. *Journal of Knowledge Management*, *9*(4), 45–55. doi:10.1108/13673270510610323

Romer, T., & Teece, D. (1998). Capturing Value from Knowledge Assets: The New Economy, Markets for Know-How, and Intangible Assets. *California Management Review*, *40*(3), 55–79.

Sharma, R., & Chowdhury, N. (2007). On the Use of a Diagnostic Tool for Knowledge Audit. *Journal of Knowledge Management Practice*, *8*(4).

Skyrme, D. (2000). *Knowledge Management Assessment: A Practical Tool*. London: David Skyrme Associates Limited.

Sveiby, K.-E. (1997). *The New Organizational Wealth*. San Francisco, CA: Berrett-Koehler.

Tiwana, A. (2001). *The Essential Guide to Knowledge Management: E-Business and CRM Applications*. New York, NY: Prentice Hall.

Wenger, E., McDermott, R., & Snyder, W. (2002). *Cultivating communities of practice: a guide to managing knowledge*. Boston, MA: Harvard Business School Press.

Xarchos, C., & Charland, B. (2008). Innovapost uses Web 2.0 tools to engage its employees. *Strategic HR Review*, *7*(3), 13–18. doi:10.1108/14754390810865766

Zack, M. (1999). Developing a Knowledge Strategy. *California Management Review*, *41*(3), 125–145.

APPENDIX I – BI SUPPORT TOOLS FOR PROFESSIONAL SERVICES

1. Application Sharing

Application sharing tools such as NetMeeting and WebEx are used to conduct discussions, meetings and collaborate. These applications enable decision making possible without users being in the same physical location. Although geographical location does not affect the possibility of decision making, users are to be in the same meeting at the same time. To overcome this problem, some meetings are recorded and disseminate to the directed employees for update. This is especially useful when no decision making is required from the users who are viewing through the recordings.

2. Blog

The organization's blogs feature news and updates of the events from within the firm and market. The organization allows individual employees to create blogs of their own profiles as well as for projects. The blogs for projects are used to share information without flooding colleagues' inbox. History, progress and update of projects are captured in the project blogs. There are also blogs created for sharing of competitor, market movement information, and internal processes.

3. Internal Youtube

Internal Youtube is equvilant to the Youtube in the public web. It is used for the organization's internal video broadcasting forum. Internal Youtube allows employees to capture their thoughts on videos and broadcast them. This tool was launched for four weeks in the month of September 2008 and has since been discontinued.

4. Expert Finder

Expert Finder is a career networking tool for employees to input their career profiles such as bio-data, experiences, past and current assignments, expertise and subject interests. This information is also used by personnel managers internally as a CV to identify candidates for vacancies within the organization. Expert Finder is linked to the Technical Community where details of expertise can be found. It provides a place to have an overview of the personnel in the company, be it new project members or someone they are collaborating with but not within physical proximity.

5. Customer Relationship Management (CRM)

CRM is a web-based tool that manages customer accounts and contract information, sales plans, leads and opportunities, market share, and competitor data. It also generates reports and contains training files on CRM. Although it allows contribution and extraction of data for business analysis studies, it typically restricted to select employees within the firm.

Figure 11.

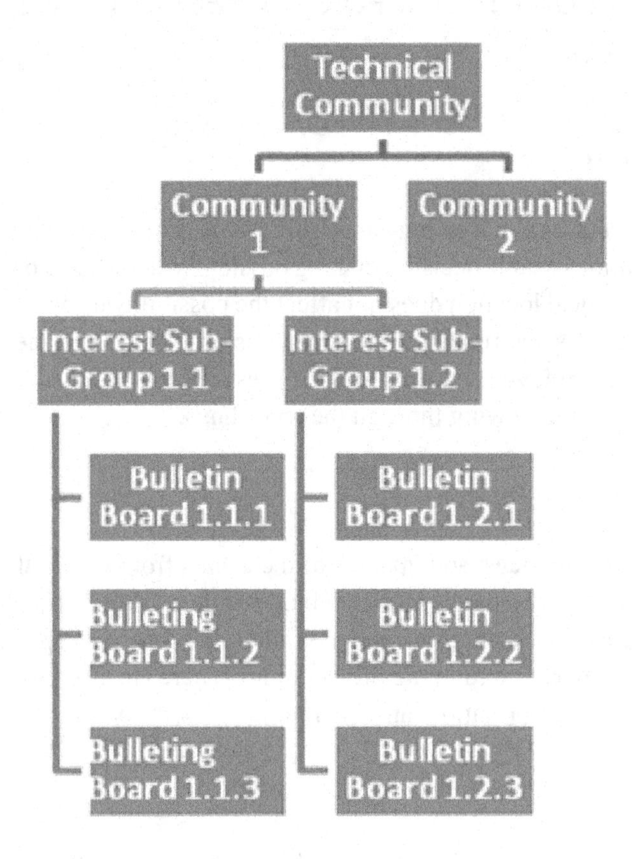

6. Technical Community (Forums, CoPs)

Technical Community is communities of practice used within the organization and is one of its knowledge management tools that focus on people. It is a self-directed tool where members form their own communities, nominate leaders and pursue their own common interests voluntarily. These communities will then form their own different sub-groups based on their interests and most of the communication in the sub-group takes place in the bulletin boards. To date, there are more than 20 and 100 communities and sub-groups respectively. The relationships between CoPs, Sub-groups and Bulletin Boards (BB) can be mapped out in Figure 11.

In Technical Community, the virtual infrastructure connects people-to-people and people-to-information. Interest Sub-Group members can find the experts in the particular areas and contact them directly or look at their past and present assignments for more information. Employees can also find opportunities to develop and share knowledge; apply it for customer and R&D needs within the organization. From business' point of view, this is an ideal tool when technical knowledge, input and experts in specific fields are required by management to make strategic decisions. The experts locating tool allows search by name, community types, technical expertise, qualifications and societies. As a Technical Community member, employees are mapped into the organization's technical career advancement roadmap which recognizes and rewards technical leadership for their contribution through peer and management review.

7. Enterprise Knowledge Portal (EKP- for log, stories posting etc)

EKP is used as "a central feature of the communications and knowledge management strategy" said the CEO during an internal interview. EKP links to all the other tools in the organization. It also shows the top 10 Best Practices periodically, ranking the most searched words in weekly basis, latest industry news and updates. EKP also provides language translation tool to assist personnel in a multi-cultural company. To facilitate ease of information and knowledge retrieval, different business entity can create their portal within the EKP specifically for their business functions. Another type of portal created within the EKP is a Product Portal that links all the product offerings from various business entities together. These portals are built based on Wikis, see item 15, Internal Wikis, below.

8. Really Simple Syndication (RSS)

RSS news feed is available for use in the organization. It allows users to subscribe to websites of interest so as to get updated of any changes to the web pages.

9. Sales & Marketing Sharing Tool

Sales & Marketing Sharing Tool is a web-based tool where marketing materials are managed in a powerful database which allows employees to search from various catalogues of files from advertising to case studies, brouchres, product and presentations.

10. Employee Alumni

This social network tool is for employees with three years seniority and ex-employees only. Employee Alumni serves as an intermediate point between EKP and public website where non- members can access to news and information not readily available elsewhere. This online community tool sees an estimate of 80,000 memberships from to-be-retirees, clients, consultants, competitors and distributors and employees. By keeping contact with these groups of people, employees can tap into the knowledge and experiences to deal with challenges at work. This tool can be used as an ambassador for the organization's products and services.

11. Intern Alumni

Intern Alumni is a web-based social network tool for those who are going to join, have joined and had joined the organization as interns only. As the Employee Alumni is used to create networks with those who have corporate experiences, Intern Alumni aims to target the future of the organization.

12. Internal Wikibook

Using the concept of the public Wikibook, this tool is used to generate larger and more structured bodies of knowledge on a single subject from the community. Often a corporate taxonomy helps organize bodies of knowledge into hierarchies and clusters that facilitate storage, searching and browsing.

13. Internal Wikipedia

Internal Wikipedia, modeled after the Wikipedia project, provides definitions of terms and acronyms, their meanings and uses within the organization. It is also an entry point where references, official sites, extra information such as competitors information and publish industrial papers etc can be found. This tool is opened to all employees to contribute and edit the articles. In the anchor case, since the tool went into production on 1 October 2007, 11,823 articles were submitted when research for this report was done.

14. Webcast

Live and recorded webcast used for seminar, conferences and broadcasting of news. This is a commonly used tool when decision making between parties from geographically different locations. Sometime also known as webinars, a range of audio visual devices, compression algorithms and application sharing tools are configured for customized usage in the field.

15. Internal Wikis

Professional services firms adopt Wikis as one of the tools which different business entities can create their own to taps into the wisdom of crowds throughout the organization. These portals reside within the EKP of the organization. An example is the Singapore Engineering Portal (SEP) created by its engineering office located in Singapore. SEP was created to share information, knowledge and experience for that specified product only. Personnel within the office can contribute and search for product related issues. The engineers are using blogs in this portal to update their weekly reports and track their planned and unplanned Process Authorisation issued.

APPENDIX II - KEY PERFORMANCE INDICATORS FOR BI

Definition of Performance Factors used in the Study.

1. Individual
    a. Knowledge of work – Familiarity with individual task or tasks performed by business unit.
    b. Creativity – Seeks innovative ways to solve problems.
    c. Communication – Makes accurate and thorough reports. Persuasive in speaking and on paper.
2. Task Management
    a. Understanding of Product or Service Technology – Knowledge of engineering principles or field technology.
    b. Shor-Range Planning – Markets, products, facilities, people.
    c. Implementation of Plans – Puts short and long range plans to practical use.
3. Managing Subordinates
    a. Selecting – Matches job requirements with skills and abilities.
    b. Training & Development – Works to upgrade skills and abilities.
4. Working with Others
    a. Peers – Respected by and can influence.
    b. Teamwork – Works effectively on teams by actively contributing to the accomplishment of goals.
    c. Customers – Works effectively and successfully with customers.
    d. Knowledge Sharing – Shares own knowledge, learns from others and applies knowledge in daily work. Open to new ideas and continuous learning.
5. Development Plan
    a. Competency & Area of Need – Development of action plans for identified areas to improve.

This work was previously published in International Journal of Systems and Service-Oriented Engineering, Volume 1, Issue 2, edited by Dickson K.W. Chiu, pp. 1-24, copyright 2010 by IGI Publishing (an imprint of IGI Global).

Chapter 12
Knowledge Capture in E-Services Development: A Prosperous Marriage?

Eva Söderström
University of Skövde, Sweden

Lena Aggestam
University of Skövde, Sweden

Jesper Holgersson
University of Skövde, Sweden

ABSTRACT

In this paper, the authors examine whether the union of Knowledge Management with e-services development would be successful in performing as a collaborative functioning unit. The focus of this research is examining the potential for using Knowledge Management as a means for improving research and practice in e-services development. The authors analyze a real-life case against the Knowledge Capture model and its associated knowledge loss. The results show that KM theory has definite potential to elevate e-services research and practice, for example, by adding analysis and decision points concerning what knowledge to use and how to collect it. This is particularly relevant when collecting requirements, information, and desires from potential users of an e-service at the start of a development project.

INTRODUCTION

We have heard it before. Technology *is* developing at rapid speed, and organizations *are* facing great challenges given the increasingly competitive and global world. Although these may be "buzz-sentences", they are nevertheless true. One of the recent developments is the emergence of electronic services, e-services, and one of its technical implementations in the form of web services and service-oriented architectures. Web Services is a technology for publishing, identifying and calling services in a network of interacting computer nodes (Barry, 2003), while SOA can be defined as "*a framework for integrating business processes and supporting IT infrastructure as*

DOI: 10.4018/978-1-4666-1767-4.ch012

secure, standardized components (services) that can be reused and combined to address changing business priorities" (Bieberstein et al., 2005, p. 4). While these two focus on the technical level, this paper will take a higher level perspective and focus on e-services that enable service improvement to customers, citizens, and other organizations. Knowledge is a critical asset in modern organizations, not the least when developing e-services. For example, knowledge about who the users of the e-service are is critical for successfully developing the e-service. Hence, in order to gain and sustain competitive advantage, organizations must manage their knowledge resources, i.e., they need to perform Knowledge Management (KM) work. KM includes both knowledge reuse and knowledge creation (Davenport et al., 1996), and the organization must support and stimulate the knowledge-creating activities of individuals (Nonaka & Takeuchi, 1995). Therefore, the potential marriage of e-services research with Knowledge Management (KM) is highly relevant to investigate.

The aim of this paper is to demonstrate how KM theory and research can be applied in e-services development as a means of improving e-services usability and applicability, with particular emphasis on how and what knowledge is captured from and about the users. We do so by using a model of the knowledge capture process with its identified knowledge loss types, and use a real-life case to illustrate findings on how the e-services development process can be improved. The results show that KM theory has definite potential to elevate e-services research and practice, for example by adding analysis and decision points concerning what knowledge to use and how to collect it. This is particularly relevant when collecting requirements, information, and desires from potential users of an e-service at the start of a development project. The paper is structured as follows: Knowledge Management is introduced and described in Chapter 2, before e-services are explained in Chapter 3. The real-life case (Chapter 4) precedes the analysis of the case against the Knowledge Capture process (Chapter 5), and the paper ends with a discussion in Chapter 6.

KNOWLEDGE MANAGEMENT

Successful Knowledge Management (KM) that contributes to improved organizational effectiveness requires that the appropriate knowledge is provided to those that need it when it is needed (Jennex et al., 2007). One way to do this is to implement Electronic Knowledge Repositories (EKR). EKR is a key form of KM (Kankanhalli et al., 2005), and EKR is also the focus in this paper. EKR make it possible to store and provide the right knowledge to those when they need it, but EKR also prevent knowledge from being lost when a specific employee leaves the organization. However, knowledge sharing through the use of EKRs must be regarded as a means, not an end, to the purposes for sharing knowledge (Carlsson & Kalling, 2006).

KM Introduction

There are different types of KM with regard to how organizations accumulate knowledge, insights, and valuable expertise over time (Wiig, 1994). One type accumulates knowledge outside people in order to disseminate knowledge to support learning (Wiig, 1994); this is the type to which EKR refers. EKR enable both individual and organizational learning, and hence support the other two types of KM identified by Wiig (1994): to accumulate knowledge inside people and to embed knowledge in processes, routines etc. With respect to Binney's (2001) six elements, developing EKR includes both a product and a process perspective. There must be processes associated with the management of the knowledge repository and improvements of work processes in order to support different types of knowledge conversions as described by Nonaka and Takeu-

Figure 1. Different parts of KM and their relations (developed from Aggestam & Backlund, 2007)

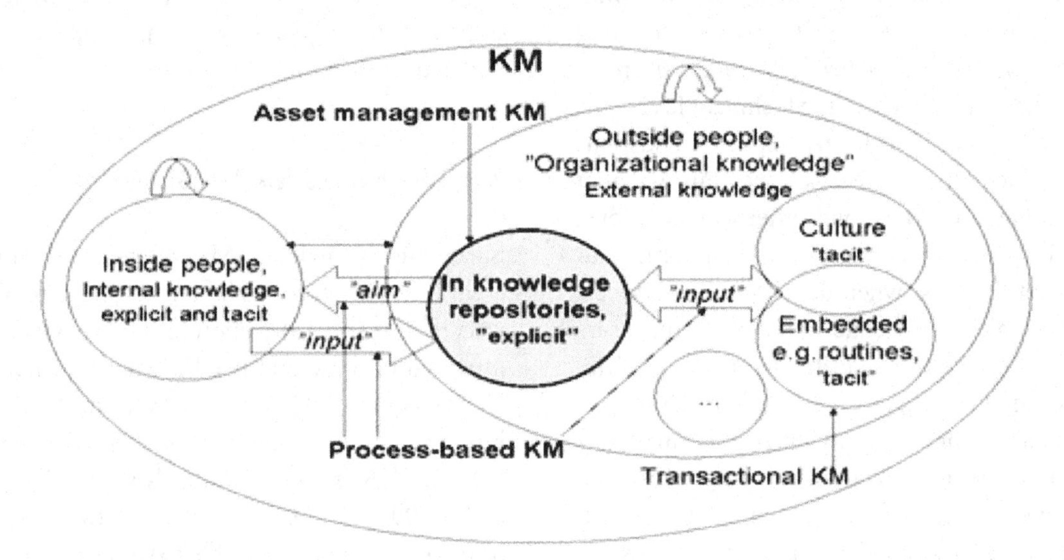

chi (1995). The application of technology when building the repository embeds knowledge in the application and the use of it. Binney (2001) terms this transactional KM, which is a side-effect of building knowledge repositories.

There are different types of knowledge that have to be managed. Wiig (1993) terms knowledge that people hold in their minds internal knowledge. Knowledge in e.g., books and IT systems is referred to as external knowledge. From the perspective of an employee, external knowledge is organizational knowledge, i.e. knowledge that remains in the organization even if employees quit. An EKR is a part of the organizational knowledge. Another common distinction in the literature is between tacit and explicit knowledge (e.g., Gore & Gore 1999; Loermans, 1993; Nonaka & Takeuchi, 1995; Wiig, 1993). Tacit knowledge is difficult to identify and to express since it is highly personal and concerns insights and intuition (Nonaka & Takeuchi, 1995; Blodgood & Salisbury, 2001). Explicit knowledge is easier to express and can, in contrast to tacit knowledge, also be processed by a computer (Blodgood & Salisbury, 2001; Nonaka & Takeuchi, 1995). From an organization's perspective organizational knowledge stored in a repository can be regarded as explicit and organizational knowledge stored in the culture and embedded in work routines as tacit. Figure 1 summarizes our conceptualization of KM.

Learning and accumulation of (new) knowledge always start from the perspective of an individual (Jensen 2005). Furthermore, knowledge derives from information (Davenport & Prusak, 1998; Wiig, 1993), and produces new information (Schreiber et al., 1999). Knowledge-creating activities for creating knowledge take place within or between people (Davenport & Prusak, 1998), but the real transformation process, when information changes to knowledge, is individual. Thus, it is impossible to store "knowledge"; it is information that supports knowledge transformation that is stored. However, we have experienced that people regard stored information as knowledge, and thus we can also refer to such stored information as external knowledge.

An EKR requires capturing, packaging and storing of relevant knowledge. These processes take place when a knowledge repository is created for the first time in a KM project, as well as every time new knowledge is generated that has potential relevance for incorporation in an existing EKR.

The latter is critical for ensuring that the EKR is updated and for maintaining usefulness and trust in the repository over time. EKR success depends on whether or not the repository is actually used in which case the users must perceive that its usage will greatly enhance their work performance (Sharma & Bock, 2005). Hence, *what* is stored in the repository is critical for success. Before storing knowledge in the repository, it must be captured. Managing knowledge thus requires the ability to capture it (Matsumoto et al., 2005).

The Knowledge Capture Process and Knowledge Loss Types

The process of capturing new knowledge starts when knowledge with the potential to be incorporated in the repository is identified and closes when the identified knowledge is passed onto the process of packaging and storing knowledge, or when a decision is made that the knowledge should not be stored. It is crucial to understand that new knowledge is not regularly generated, and, accordingly, knowledge to be *continuously* captured. This ability to continuously capture new, relevant and correct knowledge challenges the long-term survival of a repository, since failure will eventually result in a repository that is out-dated and irrelevant. Thus, users will most likely abandon the use of the repository. Therefore, developing and implementing an IT-supported knowledge repository does not only involve building the repository itself, but also involves implementing processes which ensure that the repository will be maintained.

Capturing the *right* knowledge is necessary for KM success (Jennex et al., 2007). Sharma and Bock (2005) manifest that quality, for example reliability and relevance, in the knowledge repository has to be high for knowledge re-use to take place. Capturing the *right* knowledge means that "all" knowledge should be identified and evaluated for possible storage. However, in the capture process there are different types of knowledge loss, which means that no new knowledge is stored in the repository. The effective management of knowledge loss concerns decreasing the amount of knowledge that is unidentified i.e., decrease unwanted knowledge loss, and increasing the amount of knowledge that not should be stored, i.e., increase wanted knowledge loss. Hence, building awareness of knowledge loss is important (Hari et al., 2005), and KM work related to e-services is no exception.

Seven different types of knowledge loss (Figure 2) can be identified in the capture process (Aggestam & Söderström, 2010). The following description is a summary from Aggestam and Söderström (2010). Five of these can be regarded as *unwanted*, because no employee has consciously chosen not to capture the actual piece of knowledge. Two types of knowledge loss can be regarded as *wanted*, because an employee has, for some reason/s, consciously chosen not to pass this knowledge on to the package and store process.

Figure 2 shows the seven types of knowledge loss, depicted by the numbers. Numbers 1-5 concern unwanted knowledge loss, because no employee consciously decided not to capture the knowledge. Numbers 1-3 refer to loss meaning that knowledge did not even reach the capture process, and the need of a structured approach to identify when new knowledge has been created is obvious. Numbers 4-5 refer to loss meaning that the actual piece of knowledge was identified, but for different reasons (e.g., no employees were responsible for relevant KM tasks), it was not passed onto the package and store process. Numbers 6-7 concern wanted knowledge loss, that an employee chose not to pass the knowledge onto the package and store process, for example for irrelevance or legal reasons.

Knowledge loss 1: The first knowledge loss concerns knowledge of which corresponding information is already stored somewhere, e.g. in documents, books and/or protocols. Even if it is already stored, it will probably

Figure 2. Different types of knowledge loss in the capture process (Aggestam & Söderström, 2010)

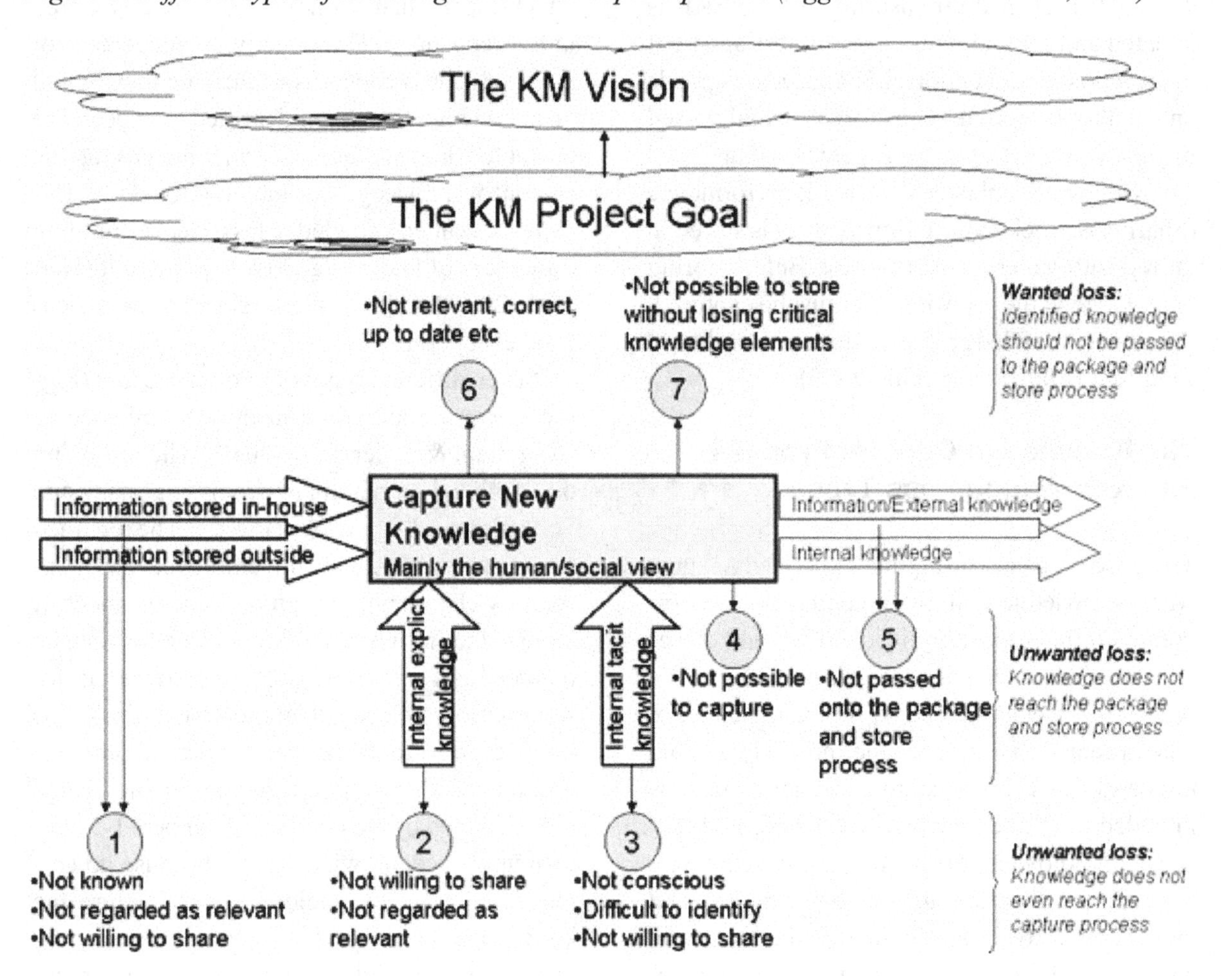

be hard to find and reuse unless it is integrated and related to other organizational knowledge. This stored knowledge can be found inside the organization, in the actual department or in another one, or outside the organization, and it is probably structured and stored for other purposes compared to the repository. If no one knows about these different sources of knowledge, or do not think about them from the perspective of the actual KM work, they will not be identified, and accordingly, never reach the capture process. Furthermore, there is also stored knowledge which is known, but the person who knows about it does not want to disseminate it.

Knowledge loss 2: The second knowledge loss concerns internal explicit knowledge. The person who knows is the knowledge owner, and with regard to the fact that actual knowledge is explicit, it must be known. Thus, the main challenge concerning explicit knowledge is to increase employees' willingness to contribute it to the repository. Reasons for not wanting to contribute are, for example, that employees want to stay where they are and do not experience that they receive any benefits from sharing the knowledge. One further problem that may result in explicit knowledge not being identified is that the knowledge owner does not think about it as being relevant for the repository.

Knowledge loss 3: Knowledge loss 3 concerns internal tacit knowledge. Because the knowledge is internal, there are similarities to knowledge loss 2. The person who knows is always the knowledge owner, and if this person is conscious of the knowledge, reasons for this type of knowledge loss concern willingness or ignorance of its relevance for the repository. However, with regard to the fact that the knowledge is tacit may mean that it varies if the knowledge owner is conscious of the knowledge or not. Furthermore, tacit knowledge is more difficult to identify compared to explicit knowledge, and the need for suitable approaches to identify this type of knowledge as well as work procedures to try to convert tacit knowledge to explicit is clear. These conditions are also relevant for loss 4.

Knowledge loss 4: Knowledge loss 4 concerns internal tacit knowledge that is identified, but which could not be captured. Compared to number 3, the person who knows is conscious about it and willing to share the knowledge, but the organization lacks suitable tools, competences etc. to capture it. An alternative is to store information about the owner. If this type of knowledge loss is significant, there is also a risk that the organization lacks methods, work procedures, and so on, that enhance converting tacit knowledge to explicit knowledge.

Knowledge loss 5: This knowledge loss concerns all knowledge that has been identified and valued as important to disseminate through the repository. A critical issue is to pass the knowledge on to the package and store process. Employees must have time for performing these tasks, as well as IT systems' support. Furthermore, if this is not integrated in daily work processes, the risk that the knowledge will never be stored increases.

Knowledge loss 6: The sixth knowledge loss concerns knowledge that is identified, but should not, for different reasons, be passed onto to the package and store process. Reasons for not wanting to store identified knowledge are, for example, that it does not contribute to the knowledge goal, that it already is stored, or that storing is prohibited. Furthermore, this type of knowledge loss also includes deleting already stored information, which can be the result of revising the content. Documented evaluation criteria adapted to the organization's needs support this decision process. This is in accordance with the Ev-CSF described in (Aggestam et al., 2009).

Knowledge loss 7: There is tacit knowledge that cannot be stored without losing critical knowledge, which is what the seventh knowledge loss is about. If there is a risk of losing knowledge during storage, it is important to judge if it is worth it or not. An alternative is to store information about who has this particular knowledge. Some of these aspects have been discussed in knowledge loss number 4.

E-SERVICES

E-services are used in many different settings and there is no commonly agreed definition (Rowley, 2006). There are, however, some general characteristics of e-services. For example, they are based on electronic interactions (Javalgi, 2004; Liao et al., 2007; Rowley, 2006). Furthermore, e-services have a lot in common with traditional services and share, to some extent, the same properties, such as intangibility, inseparability and heterogeneity (Javalgi, 2004; Järvinen & Lehtinen, 2004; Edvardsson & Larsson, 2004; Johannesson et al., 2008). Some authors also stress the importance of the creation of value, for the provider, the consumer, or for both (Priest, 2004; Hultgren, 2007; Edvardsson & Larsson, 2004). In this paper we view e-services as artefacts for the electronic delivery of services, i.e. e-services are applications making it possible to offer and use services via electronic communication channels, such as the Internet.

E-Services Development

The development of e-services (ESD) poses somewhat different challenges and perspectives compared to traditional information systems development. This section introduces some of these issues: public vs. private e-services (section 3.1.1), ESD user focus (section 3.1.2), and the uniqueness of ESD compared to traditional development (section 3.1.3).

Public vs. Private E-Services

E-services can be viewed from two perspectives: commercial and public (e-government services). The most important difference between these two is that e-government services are and should not be based on an ability or desire to pay for using the service, since public administrations do not aim to make a profit (Henriksen, 2004). The driving force for public administrations is instead to save money by increased internal efficiency combined with a more efficient communication with citizens (Axelsson & Melin, 2007; Charabalidis et al., 2006). Commercial organizations use e-services for the same tasks as public administrations, but the target user groups differ. While users of commercial e-services may be targeted in specific directions, e-government services must be universal to all citizens and consider a wide range of aspects such as different disabilities for users, geographical limitations, language limitations and such (van Velsen et al., 2009; Henriksen, 2004).

Additionally, e-government services must consider, or are affected by, laws and regulations to a much greater extent than private ones (Hung et al., 2006; Teicher et al., 2002). E-government services are unique, and are in many cases infrequently used by citizens, for example, an e-service for applying for a drivers license (van Velsen, 2009). This means that the citizen must adapt to new e-services, and hence be supported in how to use them. This is different for private e-services, which often are similar and more frequently used. The private e-service user therefore generally finds it significantly easier to handle these types of e-services (van Velsen, 2009).

E-Services Development User Focus

Until recently, ESD has focused on automating manual processes within organizations (Asgarkhani, 2005) with little or no consideration to the users (Andersen & Medaglia, 2008; Anthopoulos et al., 2007; Melin et al., 2008). At best, user needs are guessed and not thoroughly analyzed by the developing party (Jupp, 2003). However, the focus is now shifting to the users, since it has been found that increased focus on the users will boost the probability for success once an e-service has been deployed (Axelsson & Melin, 2008; Lindblad-Gidlund, 2008; Melin et al., 2008). The main arguments is that user needs are more likely to be met, since both the providing government and the e-service user experience benefits from said e-service (van Velsen et al., 2008; Carroll & Rosson, 2007; Andersen & Medaglia, 2008). It should not be a surprise that the incorporation of the user in the development shows good results. In traditional information systems development (ISD) it has been known for years that user participation has some general positive effects, such as better information quality resulting in more consistent and accurate requirements, more realistic expectations on the upcoming system from the users point of view etc.

Recently the use of e-services has shifted focus from company-internal to company external users. This means that ESD is different from traditional ISD. By using for example the Internet for communicating, organizations today are trying to shift focus from internal user groups with distinct needs and easy to target, to a situation with external users with dispersed needs that are hard to target (Albinsson et al., 2007). Furthermore, it is challenging to target users for internal ISD projects, but even more challenging to target external users of e-services, i.e., customers, citizens, other businesses etc. (Melin et al., 2008; Axelsson & Melin, 2008).

The Uniqueness of E-Services Compared to Traditional IT Systems

E-service applications are no different from other IT systems. However, traditional IT systems have an internal focus whereas e-services also have an external focus. A majority of users then reside outside the e-service providers organizational boundaries. By using the Internet, e-services organizations are trying to shift focus to a situation with external users with dispersed needs that are hard to target (Albinsson et al., 2007). As with the previous issue, it is more challenging to target external users than internal ones (Melin et al., 2008; Axelsson & Melin, 2008).

Added Complexity of User Participation in E-Services

Getting users to participate in the ESD process differs from that of traditional ISD for internal use. User participation of any kind is not easy to organize, for example due to time and resource constraints (Melin et al., 2008). However, getting users to participate can provide beneficial results (Lindblad-Gidlund, 2008; Melin et al., 2008; Axelsson & Melin, 2008). There is some complexity to overcome, and this section will describe three such issues: knowing the e-service market segment (section 3.2.1), identifying users for development efforts (section 3.2.2), and getting users to participate in e-service development (section 3.2.3).

Knowing the E-Services Market Segment

An e-service should address a clear "target audience", i.e., it should be clear for what purpose and for what user it is developed. The *target service market segment* for the e-service should therefore be identified and analyzed. Depending on the degree of participation, or the type of users developers have access to, different approaches should be selected. For example: is an organization primarily seeking to attract new customers or do they wish to provide better e-services for existing customers? Is the organization commercial and aiming for higher revenues or is it a government striving for better services to citizens? In this sense, e-government services complicate things since they have to be offered universally to *all* citizens (Henriksen, 2004), and therefore have to be usable for many different types of users. Identifying new e-services for commercial reasons may call for involvement of innovation-driven users, whereas advisory users may be more suitable for developing basic e-government services.

Identifying Users for Development Efforts

The decision of design approach is steered by the amount of available resources to allocate to a development project. Users of both commercial and e-government e-services are often external (Albinsson et al., 2007). This will make it more complicated to find the *appropriate users* to participate with compared to traditional ISD within organizational boundaries. Finding lead users may be even more complicated since not any representative user will do. Incorporating innovative users often seem like an appealing design approach, but the problem of finding the desired users may call for other design approaches to be considered.

Getting Users to Participate in E-Services Development

It may be straight forward to get in-house users residing in-house to *participate* in the design process, since they can be obligated to participate and may see the benefits more clearly (Albinsson et al., 2007). E-services for commercial as well as governmental usage poses greater challenges since participation must be based on free will, i.e., the users must be persuaded to participate which most likely will reduce the number of suitable participants. Restricted user access may influence the type of interaction techniques used, that is type of seminars for eliciting requirements etc. This problem concerns degree of participation, and is a crucial question to ask at this stage. If active user participation is sought, part of team or advisory participation could be suitable approaches.

REAL-LIFE CASE

As a means to demonstrate the applicability and usefulness of KM in ESD, a real-life case will be used as a basis of analysis. The case is based on a real company, and all the background information is genuine. Since the KM usefulness is the primary focus, the identity of the company is irrelevant and all actors will therefore be protected by anonymity. The company will from here on be referred to as "Company X". It is a large, international company in the travel industry with branches in several countries, and which offers and provides different types of services all over the world. In order to protect its anonymity, no further details will be provided.

Renewal via E-Services

One main concern in Company X is to continuously develop and enhance their customer support. They therefore launched a membership program some years ago, to keep and increase customer loyalty. One consequence, however, was that the manual handling of member bookings increased, which became very expensive. There was also a need to increase the number of ways in which to make offers, and to remove some limitations in the service structure. Online functionality in terms of an e-service was considered a highly suitable approach to address these problems, and it was decided that a new e-service should be developed. This e-service would reduce the number of contacts made to the customer support center, and thereby relieve the strain on administrative functions.

The desired solution should be generic enough to suit different member needs concerning variety of services provided by Company X. Connections to surrounding internal and external systems had to be established, and appropriate follow-up procedures concretized. Company X determined that unless the e-service was developed, customer support demands would be continuous and Company X would fail to compete with other similar membership programs provided by competitors. Profitability and opportunities could thereby be lost.

E-Services Development Process and Methodology

The basic development methodology used for developing e-services is Rational Unified Process (RUP), with some modifications and adjustments to Company X's needs. In summary, the development processes can be divided into four steps or phases, which are iterated until a control gate has been approved:

1. Inception: primarily focused on clarifying and analysing current and desired future states. There are three mandatory control gates, all of which are focused on specifying what the project will include and result in, and whether or not it will be performed at all or stopped for further processing. The main activities include business modelling and requirements specification.
2. Elaboration: concerns processing and refining the requirements developed during the Inception phase, with an emphasis on how to transfer the requirements to a practical level. Before the phase is completed, a decision is made whether or not the project will be

implemented. The main activities include analysis and design, and implementation.

3. Construction: focuses on technical realisation combined with testing. The main activities are construction, testing, implementation and configuration.
4. Transition: focuses on implementing the project in the organisation. The phase ends with a decision on project closure or a need for further iterations to achieve a more satisfying result. The main activities include implementation and configuration.

The main phase in use is the first, *Inception*, since external suppliers and developers are mainly used for the next-coming phases. During development projects, a number of control gates have been pre-determined where human actors approve project continuation.

Architectural Context

The e-service will store information about all members in the membership program, including track records of what they do and how they travel. When customers are to book for a specific service and pay using the points gathered, a booking is made in Amadeus, which in turn communicates with the e-service. With the new e-service environment, the user (Company X's customer) is supposed to work via a booking dialogue connected to the Amadeus e-Retail system. There, they can pay with their customer loyalty points or with money. Amadeus, which is not a property system of Company X but a standalone product, has thus far sent information to some of the old systems in Company X's architecture, and these systems will now be replaced. One problem is that the old systems are so tightly coupled that if changes are made to one, all need to be tested for functionality. Many old systems are also written in the old programming language FORTRAN, which few of today's programmers have knowledge of. With the new e-service, a hub will be created to which Amadeus will send all agreed upon information.

ANALYSIS OF THE CASE AGAINST THE KNOWLEDGE CAPTURE PROCESS

In the line of the paper aim, to demonstrate how KM theory and research can be applied in e-services development, we will now use the knowledge gained from Company X and make a theoretical analysis of the fit of the Knowledge Capture model into the situation at hand, and in what way this model may have been beneficial for Company X. The basis of the analysis is a real problem situation. It should be noted, however, that concrete details may be lacking due to the theoretical nature of the analysis. The analysis is focused on how and what knowledge is captured from and about the users. The results are presented in Table 1, where the flows of the Knowledge Capture model are assigned one row each. For each flow, a match is described for Company X. For example, information stored in-house corresponds to Company X's customer database. This description is thereafter matched to an associated knowledge loss, which is then described as well. The remainder of this chapter is devoted to describe the analysis results.

The need for the e-service was discovered by the increasing strain on the customer service department in Company X. Availability of new technology also affected the choice of an online solution. In order to show how KM research fits into and can affect service-orientation, we will use the basic information about Company X and theorize how the existing Knowledge Capture model could have been used during this development.

When beginning to capture knowledge required to develop a usable and useful e-service, Company X first needs to assess which information they have access to. *In-house information* consists of the customer database, to which the e-service will be connected. It contains all customer information,

Table 1. Relations between Company X and the Knowledge Capture model

Knowledge Capture flows	Description	Associated knowledge loss	Description
Information stored in-house	Customer database	1	Which data should be stored? Required time = knowledge selection (desired loss)
Information stored outside	Data stored by partners, online social communities	1	Which data do partners store?
Internal explicit knowledge	Documentation from past projects, systems architecture descriptions	2	Important experiences have not been documented
Internal tacit knowledge	Experienced developers, commercial department, customer support	3	Important information is withheld or regarded as irrelevant
Capture new knowledge	Knowledge associated to the entire Knowledge Capture process	4	Lack of tools for expressing traceability and dependencies between systems and e-services
		6	Customer knowledge that is irrelevant for e-services Service information outside of the e-services scope
		7	Lack of evaluation procedures
Internal knowledge	Project documentation on customer and company requirements, development team experience	5	Control gates are in place, methodology components may complement
Information/ External knowledge	Competitor information, RUP methodology documentation	5	Control gates are in place, methodology components may complement

and stored actions. Knowledge about customer preferences, behavior, demography etc are sample input to the process. *Information stored outside* Company X may refer to data stored in partner organizations. This information may contribute to requirements for what the users want the e-service to do and how it should work with the partner systems. Another external information source is the recent boom in online social communities, including discussion forums concerning Company X and its membership program. It is here necessary for Company X to pay attention to potential knowledge loss, such as lack of knowledge of what information partners store, and keeping track of which customer information that is stored in other internal systems. Company X has an extensive IT architecture, and there is always a risk of missing something. There may be a conscious choice not to take in all available information, since that requires time, while time-to-market is essential in the travel industry. This makes up desired knowledge loss.

Company X's internal knowledge is either *explicit or tacit.* Employees possess a lot of knowledge about the customers (users of the e-service) and the systems, and one problem (but still an essential task) is therefore to identify who to talk to and where to search for relevant information and knowledge. Experienced developers, the commercial department and customer support are important sources. The latter two have great knowledge about user preferences and requirements because of their direct user connection. Systems experts and architects also possess valuable knowledge in terms of what the technology can do. The knowledge loss risk is that important information is withheld or not regarded as relevant. Capturing information that its possessors are unaware of is the most challenging task, however, and careful preparations and analyses are necessary. For example, since the developers are customers themselves, they can observe one another when acting as customers. Thereby, tacit knowledge can be detected. Alternatively, real customers can be observed when booking and acting towards the system. This is difficult face-to-face due to customer lack time, and geographical distances. However, the tracking Company X does of their customers is a good substitute that enables analysis of knowledge gaps and system flaws. The extensive customer behavior knowledge will enable Company X to simulate future scenarios to predict future behavior. This and the mere size of their collected customer knowledge enable pattern analysis for improving, renewing and innovating the e-services. One or more employees in Company X may have important knowledge that *cannot be captured* due to a lack of tools. For example, e-services bring a dramatic increase in the number of systems and services, making it more difficult to keep track of traceability and dependencies between systems. Awareness of this problem will put the issue on the agenda.

Company X has captured a lot of knowledge and now needs to evaluate it to ensure that all relevant information is passed on, but also that *unwanted or irrelevant* knowledge is not. In our case, irrelevant knowledge may concern the frequency with which a user travels, which is not directly affecting the e-service features. If limitations have been made concerning what travel services the e-service should comprise, knowledge falling outside these limitations should be omitted as well. Or if Company X was to cut down on the number of travel service it provides, this information should be deleted from the e-service. Company X should also evaluate if there is a risk of losing knowledge during storage, and take appropriate action accordingly. The final step in the Knowledge Capture process is to *pass the knowledge on* to the package and store process. Company X has a set development process and control gates along the way to ensure viability, relevance, and quality. In a sense, they already evaluate the knowledge and requirements gathered as part of this process. However, explicit attention to the knowledge loss risks embedded into RUP as an additional methodology component may increase the likelihood of employees being given sufficient time to address these issues.

DISCUSSION

The aim of this paper has been to demonstrate how knowledge management theory and research can be applied in ESD, with particular emphasis on how and what knowledge is captured from and about the users. Our results show that the marriage between these areas has great potential for improving ESD processes and thus e-services themselves. For example, we can see that e-services and the complexity added by user availability and development participation emphasizes the need for evaluating external knowledge sources. The sheer number of potential knowledge sources makes the task increasingly complex. The travel industry is very large with numerous organisations of varying sizes. All cannot be scanned, but a selection is needed. Conscious ways of searching and selecting relevant sources are needed, and could be beneficial for Company X and its peers. Identification of knowledge sources is also complex given that social online networks can be created by anyone and placed anywhere. Future research is thus needed concerning to identify the sources that may have a bearing on the e-services relevant to a particular company, not to mention how to get access to these sources. There is a much larger base to sort through and analyze. Company X has identified one particularly relevant online community. Excluding some may mean that relevant user/customer issues, problems and opinions are lost. It is true that traditional ISD had these problems too, but not to the same extent. Additional mechanisms are needed for ascertaining that only relevant knowledge has been captured.

In addition to external knowledge, internal tacit knowledge is at least as difficult to identify.

Capturing information that its possessors are unaware of is a challenging task requiring careful preparations and analyses. Company X has the advantage that many of its employees use the services and is part of the membership program themselves, and therefore also have ideas, requirements and desires for how to improve and innovate. Still, there may be ideas that are not captured and which would be of great gain if identified. Company X should therefore strive even harder to develop ways of working that enable employee ideas to become explicit and communicated. It is a large company, and may as such have personnel resources to assign one person the role of "idea collector" or facilitator. This person's job would include communicating with employees and to develop ways and templates for how to document expressed ideas. Many similar companies have implemented "walls" on the Internet pages allowing customers to express opinions and ideas. However, these ideas are not by definition sufficiently detailed, and a "template" ("fill out this form") approach could therefore be used to enable users to be more detailed. Noteworthy is that such "walls" or template implementations always need to be someone's responsibility, or they will be left unhandled.

In this paper, we have demonstrated that conscious awareness of the Knowledge Capture process and its knowledge losses can help Company X and its peers to enhance their work and their e-services development methodology. Examples include adding extra control gates that focus on knowledge capture and knowledge loss, and addressing the need for additional methodological components towards the maintenance phases where system dependencies and traceability becomes crucial. The likelihood of a successful marriage between e-services research and practice with Knowledge Management research and practice is therefore great, and both areas should continue to strive for a long-lasting and fruitful union.

REFERENCES

Aggestam, L. (2006, May 21-24). Wanted: A Framework for IT-supported KM. In *Proceedings of the 17th Information Resources Management Association (IRMA'06)*, Washington (pp. 46-49).

Aggestam, L., & Backlund, P. (2007). Strategic knowledge management issues when designing knowledge repositories. In H. Österle, J. Schelp, & R. Winter (Eds.), *Proceedings of the Fifteenth European Conference on Information Systems*, University of St. Gallen, St. Gallen, Switzerland (pp.528-539).

Aggestam, L., Backlund, P., & Persson, A. (in press). Supporting Knowledge Evaluation to Increase Quality in Electronic Knowledge Repositories. *International Journal of Knowledge Management*.

Aggestam, L., & Söderström, E. (2010). Seven types of knowledge loss in the knowledge capture process. In *Proceedings of the 18th European Conference on Information Systems (ECIS 2010)*.

Albinsson, L., Forsgren, K., & Lind, M. (2007). *Towards a Co-Design Approach for Open Innovation. Designed for Co-designers workshop.* Paper presented at the meeting of the Participatory Design Conference (PDC 2008), School of Informatics, Indiana University, Bloomington, IN.

Andersen, V. K., & Medaglia, R. (2008) eGovernment Front-End Services: Administrative and Citizen Cost Benefits. In M. A. Wimmer & E. Ferro (Eds.), *Proceedings of EGOV 2008*. Berlin: Springer.

Anthopoulos, L. G., Siozos, P. S., & Tsoukalas, I. A. (2007). Applying participatory design and collaboration in digital public services for discovering and re-designing e-Government services. *Government Information Quarterly*, *2*(24), 353–376. doi:10.1016/j.giq.2006.07.018

Asgarkhani, M. (2005). The effectiveness of e-Service in Local Government: A Case Study. *Electric journal of e-government, 3*(4), 157-166.

Axelsson, K., & Melin, U. (2007). Talking to, not about, citizens – Experiences of focus groups in public e-service development. In M. A. Wimmer, H. J. Scholl, & A. Grönlund (EdS.) *Proceedings of EGOV, 2007*, 179–190.

Axelsson, K., & Melin, U. (2008). Citizen Participation and Involvement in eGovernment Projects – An emergent framework. In M. A. Wimmer & E. Ferro (Eds.), *Proceedings of EGOV 2008* (pp. 207-218). Berlin: Springer.

Barry, D. K. (2003). *Web Services and Service-Oriented Architectures: The savvy manager's guide*. San Francisco, CA: Morgan Kaufmann Publishers.

Bieberstein, N., Bose, S., Fiammante, M., Jones, K., & Shah, R. (2005). *Service-Oriented Architecture (SOA) Compass: Business Value, Planning, and Enterprise Roadmap (Developerworks)*. Indianapolis, IN: IBM Press.

Binney, D. (2001). The knowledge management spectrum – understanding the KM landscape. *Journal of Knowledge Management, 1*(5), 33–42. doi:10.1108/13673270110384383

Blodgood, J. M., & Salisbury, W. D. (2001). Understanding the influence of organizational change strategies on information technology and knowledge management strategies. *Decision Support Systems, 1*(31), 55–69. doi:10.1016/S0167-9236(00)00119-6

Carlsson, S. A., & Kalling, T. (2006). Why is it that a knowledge management initiative works or fails. In *Proceedings of the Fourteenth European Conference on Information Systems*, Gothenburg, Sweden.

Carroll, J. M., & Rosson, M. B. (2007). Participatory design in community informatics. *Design Studies, 3*(28), 243–261. doi:10.1016/j.destud.2007.02.007

Charalabidis, Y., Askounis, D., Gionis, G., Lampathaki, F., & Metaxiotis, K. (2006). Organising Municipal e-Government Systems: A Multi-facet Taxonomy of e-Services for Citizens and Businesses. In W. E. Al (Ed.), *Proceedings of EGOV 2006* (pp. 195-206). Berlin: Springer.

Davenport, T. H., Jarvenpaa, S. L., & Beers, M. C. (1996). Improving knowledge work processes. *Sloan Management Review, 4*(37), 53–65.

Davenport, T. H., & Prusak, L. (1998). *Working Knowledge*. Boston: Harvard Business School Press.

Edvardsson, B., & Larsson, P. (2004). *Service guarantees*. Lund: Studentlitteratur.

Gore, C., & Gore, E. (1999). Knowledge management: The way forward. *Total Quality Management, 4/5*(10), 554–560.

Hari, S., Egbu, C., & Kumar, B. (2005). A knowledge capture awareness tool: An empirical study on small and medium enterprises in the construction industry. *Engineering, Construction, and Architectural Management, 6*(12), 533–567. doi:10.1108/09699980510634128

Henriksen, Z. H. (2004). The diffusion of e-services in Danish municipalities. In Traunmüller (Ed.), *Proceedings of EGOV 2004* (pp.164-171). Berlin: Springer.

Hultgren, G. (2007). *E-services as social interaction via the use of IT systems: a practical theory*. Unpublished doctoral dissertation, University of Linköping, Sweden.

Hung, S., Chang, Y., & Yu, T. J. (2006). A classification of business-to-business buying decisions: Risk importance and probability as a framework for e-business benefits. *Government Information Quarterly, 1*(23), 97–122. doi:10.1016/j.giq.2005.11.005

Järvinen, R., & Lehtinen, U. (2004). Services, e-Services and e-Service innovations – combination of theoretical and practical knowledge. In Hannula, M., Järvelin, A.-M., & Seppä, M. (Eds.), *Frontiers of e-Business research* (pp. 78–89).

Javalgi, R., Martin, C., & Todd, P. (2004). The export of e-services in the age of technology transformation: challenges and implications for international service providers. *Journal of Services Marketing*, *7*(18), 560–573. doi:10.1108/08876040410561884

Jennex, M. E., Smolnik, S., & Croasdell, D. (2007). Towards Defining Knowledge Management Success. In *Proceedings of the 40th Hawaii International Conference on the System Sciences*, Waikoloa, HI (p. 193).

Jensen, P. E. (2005). A Contextual Theory of Learning and the Learning Organization. *Knowledge and Process Management*, *1*(12), 53–64. doi:10.1002/kpm.217

Johannesson, P., Andersson, B., Bergholtz, M., & Weigand, H. (2008). Enterprise Modeling for Value Based Service Analysis. In Stirna, J., & Persson, A. (Eds.), *Proceedings of PoEM 2008*. Berlin: Springer.

Jupp, V. (2003). Realizing the vision of e-government. In Curtin, G., Sommer, M., & Sommer-Vis, V. (Eds.), *The world of E-government*. New York: Haworth Press.

Kankanhalli, A., Tan, B. C. Y., & Wei, K.-K. (2005). Contributing knowledge to electronic knowledge repositories: an empirical investigation. *Management Information Systems Quarterly*, *1*(29), 113–143.

Liao, C., Chen, J.-L., & Yen, D. (2007). Theory of planning behavior (TPB) and customer satisfaction in the continued use of e-service: An integrated model. *Computers in Human Behavior*, *6*(23), 2804–2822. doi:10.1016/j.chb.2006.05.006

Lindblad-Gidlund, K. (2008). Driver or Passenger? An analysis of Citizen-Driven eGovernment. In M.A. Wimmer & E. Fererro (Eds.), *Proceedings of EGOV 2008* (pp.267-278). Berlin: Springer.

Loermans, J. (2002). Synergizing the learning organization. *Journal of Knowledge Management*, *3*(6), 285–294. doi:10.1108/13673270210434386

Matsumoto, I. T., Stapleton, J., Glass, J., & Thorpe, T. (2005). A knowledge capture report for multidisciplinary design environments. *Journal of Knowledge Management*, *3*(9), 83–92. doi:10.1108/13673270510602782

Melin, U., Axelsson, K., & Lundsten, M. (2008). Talking to, not about, Entrepreneurs – Experiences of Public e-service Development in a Business Start Up Case. In Cunningham, P., & Cunningham, M. (Eds.), *eChallenges*. Amsterdam: IOS Press.

Nonaka, I., & Takeuchi, H. (1995). *The Knowledge-creating Company*. Oxford, UK: Oxford University Press.

Preist, C. (2004). A Conceptual Architecture for Semantic Web Services. In F. Van Harmelen, S. A. McIlraith, & D. Plexousakis (Eds.), *Proceedings of ISWC 2004*. Berlin: Springer Verlag.

Rowley, J. (2006). An analysis of the e-service literature: towards a research agenda. *Internet Research: Electronic Networking Applications and Policy*, *6*(16), 879–897.

Schreiber, G., Akkermans, H., Anjewierden, A., de Hoog, R., Shadbolt, N., Van de Velde, W., & Wielinga, B. (2000). *Knowledge Engineering and Management: The CommonKADS Methodology*. Cambridge, MA: MIT Press.

Sharma, S., & Bock, G.-W. (2005). Factor's Influencing Individual's Knowledge Seeking Behaviour in Electronic Knowledge Repository. In D. Bartmann, F. Rajola, J. Kallinikos, D. Avison, R. Winter, P. Ein-Dor, Jr. Becker, F. Bodendorf, & C. Weinhardt (Eds.), *Proceedings of the Thirteenth European Conference on Information System*, Regensburg, Germany (pp.390-403). ISBN 3-937195-09-2

Teicher, J., Hughes, O., & Dow, N. (2002). E-government: a new route to public sector quality. *Managing Service Quality*, *6*(12), 384–393. doi:10.1108/09604520210451867

Van Velsen, L., Van der Geest, T., Ter Hedde, M., & Derks, W. (2008). Engineering User Requirements for e-Government Services: A Dutch Case Study. In M. A. Wimmer, H. J. Scholl, & E. Fererro (Eds.), *Proceedings of EGOV 2008* (pp.243-254). Berlin: Springer Verlag.

Van Velsen, L., Van der Geest, T., Ter Hedde, M., & Derks, W. (2009). Requirements engineering for e-Government services: A citizen-centric approach and case study. *Government Information Quarterly*, *3*(26), 477–486. doi:10.1016/j.giq.2009.02.007

Wiig, K. M. (1993). *Knowledge Management Foundations – Thinking About Thinking – How People and Organizations Create, Represent, and use Knowledge*. Arlington, TX: Schema Press LTD.

Wiig, K. M. (1994). *Knowledge Management The Central Management Focus for Intelligent-Acting Organizations*. Arlington, TX: Schema Press LTD.

This work was previously published in International Journal of Systems and Service-Oriented Engineering, Volume 1, Issue 2, edited by Dickson K.W. Chiu, pp. 25-39, copyright 2010 by IGI Publishing (an imprint of IGI Global).

Chapter 13
Knowledge Extraction from a Computational Consumer Model Based on Questionnaire Data Observed in Retail Service

Tsukasa Ishigaki
National Institute of Advanced Industrial Science and Technology, Japan

Yoichi Motomura
National Institute of Advanced Industrial Science and Technology, Japan

Masako Dohi
Otsuma Women's University, Japan

Makiko Kouchi
National Institute of Advanced Industrial Science and Technology, Japan

Masaaki Mochimaru
National Institute of Advanced Industrial Science and Technology, Japan

ABSTRACT

In service industries, matching the level of demand of the consumer and the level of service of the provider is important because it requires the service provider to have knowledge of consumer-related factors. Therefore, an intelligent model of the consumer is needed to estimate such factors because they cannot be observed directly by the service provider. This paper describes a method for computational modeling of the consumer by understanding his or her behavior based on datasets observed in real services. The proposed method constructs a probabilistic structure model by integrating questionnaire data and a Bayesian network, which incorporates nonlinear and non-Gaussian variables as conditional probabilities. The proposed method is applied to an analysis of the requested function from customers regarding the continued use of an item of interest. The authors obtained useful knowledge for function design and marketing from the constructed model by a simulation and sensitivity analysis.

DOI: 10.4018/978-1-4666-1767-4.ch013

INTRODUCTION

The concepts of values and lifestyles of consumers have become increasingly diverse. The demand of a consumer can vary over time depending on the situation or the state of mind of the consumer. A good situation for the consumer and the service provider is a matching of the level of demand of the consumer and the level of service of the provider. This matching requires the service provider to have knowledge of consumer-related factors, such as the satisfaction level or the concept of value of the consumer. In order for the service provider to understand consumer-related factors, a consumer model that sufficiently explains the behavior of the consumer is necessary because these factors cannot be observed directly by the service provider. If we can clarify such factors, then the behavior of a consumer can be simulated using a consumer model in order to predict the behaviors of consumers. As such, the service provider is able to use the simulation results to match the demand and level of service at the contact point of the consumer and the service.

In marketing research, mass-marketing or mass-production methods are not useful in such cases. Marketing to such consumers requires focusing on small communities or individuals, for example, by segment marketing or one-to-one marketing. The realization of such marketing first requires an understanding of the consumer. An understanding of the behavior, satisfaction level, or values of the customer can lead to an improvement in customer satisfaction and productivity in service industries (Japanese Ministry of Economy, Trade and Industry, 2007).

Moreover, in service industries, customer relationship management (CRM), the goal of which is continued use of a product, brand, or service, is a significant concern. In order to achieve efficient CRM, it is necessary to understand consumer related factors. Such considerations have been examined, in a general manner, through questionnaire data or interviews, and massive amounts of such data has been collected and stored for the purposes of assessing customer satisfaction or understanding customer behavior. The useful and practical estimation of such considerations requires effective modeling of the customer based on such data. However, in a practical marketing situation, it is difficult to extract useful knowledge using conventional consumer behavior models because most conventional models are qualitative models, as described in detail later herein. Since the structure or property of a qualitative model often depends on the designer. On the other hand, quantitative models of customer behavior, as typified by statistical modeling, have also been investigated using a probabilistic choice model or a logistic model (Luce, 1959; Domencich & McFadden, 1975). However, addressing the consumer behavior and/or decision making, including nonlinear or non-Gaussian variables, as well as the interaction between variables, is difficult by conventional statistical modeling, which assumes linearity, a Gaussian distribution, and independence among variables. The structural equation model (Fornell & Larcker, 1981) has been actively applied to research on consumer behavior. However, this model is not able to handle the nonlinearity and non-Gaussian property of the variables.

The present paper describes a method for clarifying consumer-related factors by computational modeling of the consumer based on questionnaire datasets for real services. The proposed method constructs probabilistic consumer models by combining a questionnaire data and a Bayesian network (Jordan, 1998; Pearl, 2000; Motomura & Kanade, 2005). The Bayesian network can deal with nonlinear and non-Gaussian variables as conditional probabilities. The model structure can be constructed automatically based on information criteria and can embed some of the experiences of the model designer and/or physical or social rules in advance. The proposed method is applied to an analysis of the requested function from customers regarding a product (customizable shoe insert) with a focus on the continued use of the product.

We attempt to obtain useful knowledge for function design and marketing from the constructed model through a simulation and sensitivity analysis.

Figure 1 shows a schematic diagram of the proposed method for matching optimization at the contact point between the consumer and the provider. Service providers can use the consumer model to clarify consumer-related factors if the model sufficiently explains the behavior of the consumer.

The remainder of the present paper is organized as follows. In the following section, we review research on consumer behavior modeling. In the following section, we explain the computational consumer behavior model. The statistical modeling and a brief overview of the Bayesian network are presented in the Bayesian Network section. The construction of the probabilistic cognitive structure model using the Bayesian network and questionnaire data are described in the Model Construction section. In the Evaluation of the Model section, we describe the evaluation of the constructed model. In the Knowledge Extraction section, knowledge extraction from the constructed model is described. Finally, a discussion of the experiments and the results of the present study are presented in the Discussion section, and conclusions are presented in the Conclusion section.

CONSUMER BEHAVIOR THEORY

In this section, a small number of consumer behavior models in consumer behavior theory are described. Figure 2 shows a schematic diagram of trends in consumer behavior modeling. Consumer behavior models have been investigated extensively. The goal of such models is to explain common phenomena associated with consumption-related human behaviors. Most of the models that are considered in the study of consumer behavior theory tend to be qualitative.

Initially, consumer behavior was investigated primarily with respect to economics or psychology. In economics consumption theory, the concept of utility has been developed and has become influential in marketing research. On the other hand, the psychological state of the consumer with respect to consumption behavior was investigated from a cognition or attitude viewpoint. From these beginnings, the study of consumer behavior theory began to flourish in the 1950s. In the following, we introduce four conventional consumer behavior models that have a significant influence on consumer behavior theory, namely, the process model, the probabilistic choice model, the S-R model, and the information processing model.

The process model represents the psychological process of the consumer from the point at which the consumer first notices the product to the point at which the consumer purchases the product. Process models, as typified by the Attention, Interest, Desire, and Action (AIDA) model and the Attention, Interest, Desire, Memory and Action (AIDMA) model, were proposed for efficient advertisement. Most process models are expressed as a simple unidirectional process of the psychological state upon the decision to purchase a product. As such, the behavior of the consumer is not sufficiently explained.

The probabilistic choice model developed in the field of mathematical psychology was introduced to consumer behavior theory in the 1960s. Probabilistic choice models attempt to provide an understanding of the consumer or to predict the selection behavior of the consumer using mathematical consumer models. Learning theory (Massy et al., 1970), the constant ratio model (Luce, 1959), and the probabilistic utility model (Domencich & McFadden, 1975) are well-known probabilistic choice models. By the 1990s, as a result of poor computational power, models such as logistic models or Markov models, in which linearity and independence are assumed, were generally used. However, increased computational

Figure 1. Schematic diagram of the concept of efficient use of the computational consumer model

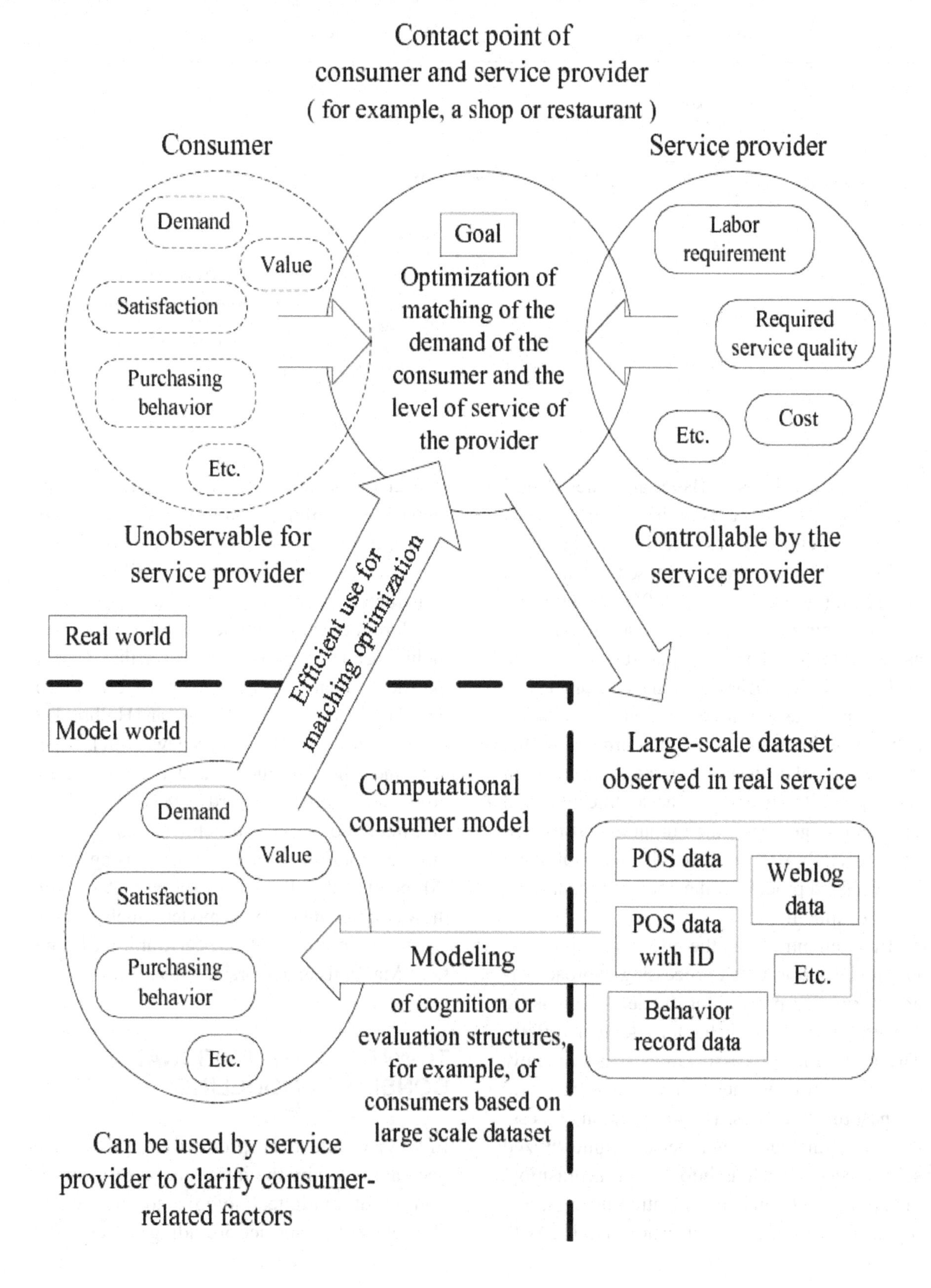

Figure 2. Trend in consumer behavior modeling

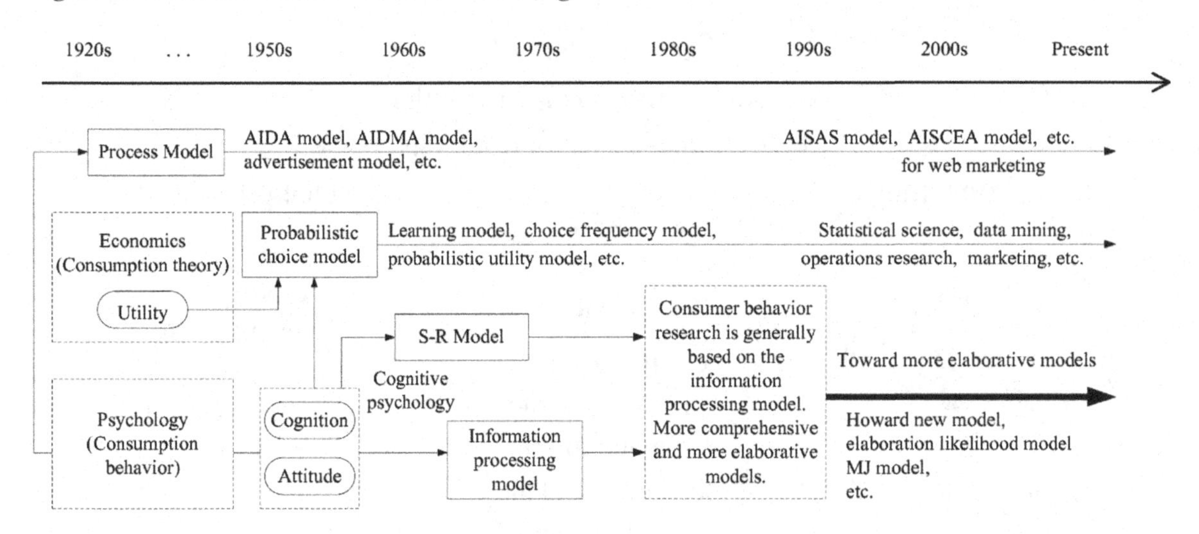

power has led to the use of Bayesian models (Rossi et al., 2005), such as the empirical Bayes model or the hierarchical Bayes model, in marketing.

The S-R model was proposed by Howard and Sheth (Howard & Sheth, 1969) to represent a relationship between an input (stimulus), such as advertisement, and an output (response), such as the purchasing behavior of the consumer. The S-R model focuses on the correspondence relation between the input and the output in order to predict the response of the behavior of the consumer with respect to advertisement. The architecture and process that generates the output is constructed as a black box. Therefore, the psychological situation and process of the consumer cannot be treated explicitly.

Bettman introduced the information processing model (information processing approach) as an information process for the decision making process of consumer behavior (Bettman, 1970). The information processing model has frequently been used in consumer behavior research over the past three decades. The information processing model includes a number of segmentalized sub-processes, such as information acquisition, memory search, and an evaluation process, and represents a process for decision making with respect to the purchase of a product from the viewpoint of information processing. However, most information processing models tend to be excessively complex. Therefore, verification of a constructed model based on real data is difficult.

Moreover, various models have been proposed, including an experimental approach that presents questions regarding the concept of utility proposed by Holbrook and Hirschman (Holbrook & Hirschman, 1982), Howard's new model, which combines the S-R model and the information processing model (Howard, 1989), the elaboration likelihood model, which was designed as a more elaborative information processing model (Mitchell, 1981; Petty & Cacioppo, 1986), and the MacInnis and Jaworski model, which provides greater elaboration with respect to attitude formation (MacInnis & Jaworski, 1989)

TOWARD COMPUTATIONAL CONSUMER MODELING

In service industries, productivity growth has been achieved by the introduction of information or automatic technologies to improve the efficiency of the service providing process, such as a central kitchen, an operation process, or a distribution center. Such a method can reduce labor and financial costs, but cannot deal directly with the improvement of customer satisfaction or value-added service quality. Matching consumer demand and the level of service of the provider is important for improving service productivity.

In order to improve service quality and provide higher customer satisfaction, clarification of consumer-related factors and the use of a consumer model are helpful. If a useful model of the customer can be constructed, then knowledge about the customer can be extracted in order to better understand customer behavior. However, in a practical situation in a real service, the extraction of useful knowledge using a consumer model is difficult because most conventional models, such as process models, S-R models, and information processing models, are qualitative models. Only skilled and experienced designers can construct useful and practical models, because the structure or property of a qualitative model often depends on the designer's sense (rather than on actual data). Moreover, such models are deterministic, which means that treating the uncertainty of consumer behavior is difficult. On the other hand, probabilistic choice models or structural equation models are quantitative and probabilistic models. However, addressing consumer behavior, including nonlinear or non-Gaussian variables, is difficult by conventional statistical modeling, which assumes linearity, a Gaussian distribution, and independence among variables. A number of studies on consumer behavior, including studies on prospect theory (Kahneman & Tversky, 1979), indicate that consumer cognition and decision making are influenced by the nonlinearity and/or interaction of variables.

In order to realize consumer modeling that is useful for a real service, the model must have the following properties:

A) The model must be quantitative.
B) The model must be constructed based on actual data observed in a real service.
C) The model must be able to deal with uncertainness of consumer behavior, including nonlinear or non-Gaussian variables and interactions between variables.

BAYESIAN NETWORK

A nonlinear and non-Gaussian probabilistic model satisfies the above three conditions. Recently, nonlinear/non-Gaussian problems have been investigated in the fields of statistical science, machine learning, and data mining (Bishop, 2006). The kernel method (Muller et al., 2001), as typified by the Support Vector Machine (SVM), can realize nonlinear calculation with relatively low computational cost and has been applied successfully in various fields. In the present study, we consider large-scale data in service industries and expect that such data will have strong nonlinearity, because this data incorporates the results of decisions made by customers. Therefore, in the present study, a Bayesian network is used to model customer behavior.

Figure 3 shows a schematic diagram of a Bayesian network model. The Bayesian network has good visual comprehensibility. The variables treated in the present paper are limited to discrete variables. The Bayesian network constructs a probability network model using variable nodes, arrows, and a conditional probability table (CPT). The nodes and arrows define the relationships between variables, and the relationship between two linked variables is defined by the CPT. The Bayesian network can represent the nonlinear and non-Gaussian distribution given in the CPT and interactions between variables as a model structure. The model structure can be constructed automatically based on information criteria. In addition, a number of the experiences of the model designer and/or physical or social rules

Figure 3. Example of a Bayesian Network

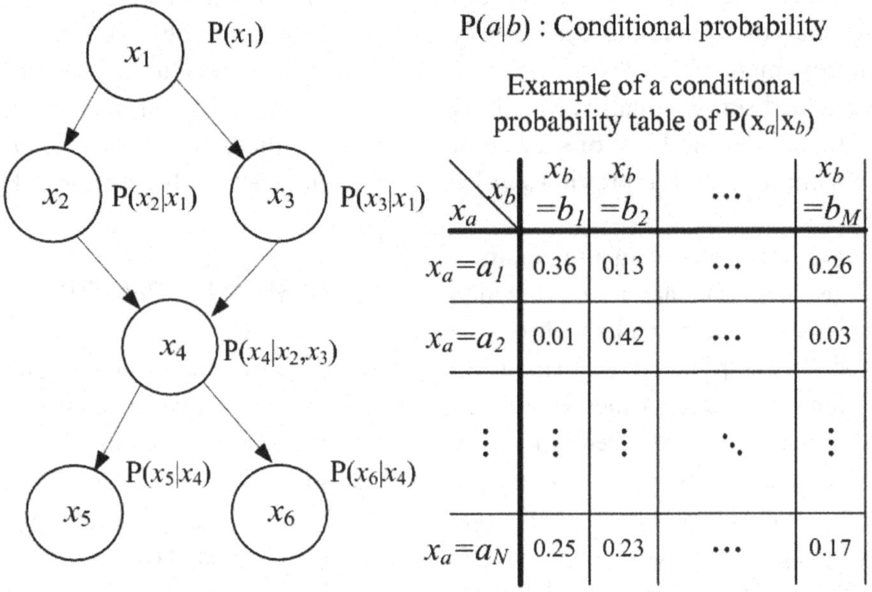

x_1 $P(x_1)$

x_2 $P(x_2|x_1)$ x_3 $P(x_3|x_1)$

x_4 $P(x_4|x_2,x_3)$

$P(x_5|x_4)$ $P(x_6|x_4)$

x_5 x_6

P(a|b) : Conditional probability

Example of a conditional probability table of P(x_a|x_b)

x_a \ x_b	$x_b=b_1$	$x_b=b_2$	…	$x_b=b_M$
$x_a=a_1$	0.36	0.13	…	0.26
$x_a=a_2$	0.01	0.42	…	0.03
⋮	⋮	⋮	⋱	⋮
$x_a=a_N$	0.25	0.23	…	0.17

Joint distribution of the model

$$P(x_1,x_2,x_3,x_4,x_5,x_6)=P(x_1)P(x_2|x_1)P(x_3|x_1)P(x_4|x_2,x_3)P(x_5|x_4)P(x_6|x_4)$$

can be embedded in the model in advance. The probabilities of each variable are calculable using efficient probabilistic computational algorithms, such as belief propagation or loopy BP (Murphy et al., 1999). These properties are convenient for expressing complex phenomena, such as human behavior and decision making. Therefore, Bayesian networks have recently been applied to recommendation systems (Blodgett & Anderson, 2000; Sebastiani et al., 2000; Rosis et al., 2006; Ono et al., 2007). In the following experiments, sensitivity analysis, and simulations, we use belief propagation for probability inference.

MODEL CONSTRUCTION

The questionnaire data used to construct the model were obtained in 2008 from 311 customers who had purchased customizable shoe inserts in 2007. The questions were designed to elicit demographic information (five variables: "have/do not have children", "marital status", "career", "age", and "area of residence"), reasons for purchase (12 variables: "price", "size", "function", "comfort", "fatigue free", "relief of pain", "customizable functionality", "regulation", "new item", "appearance", "new item", "advertisement", and "color"), information related to evaluation after use (three variables: "reasonability of price", "comfort", and "overall satisfaction"), expectations before use (one variable: "expectation"), and intent to continue using the product (one variable: "continue use").

In the present study, we use the Akaike Information Criterion (AIC) (Akaike, 1974) for probability structure search. Optimization structure search based on the AIC can avoid over-fitting the constructed model to the given data. The joint distribution of a Bayesian network given a model structure S is

Figure 4. Notation used in the present study

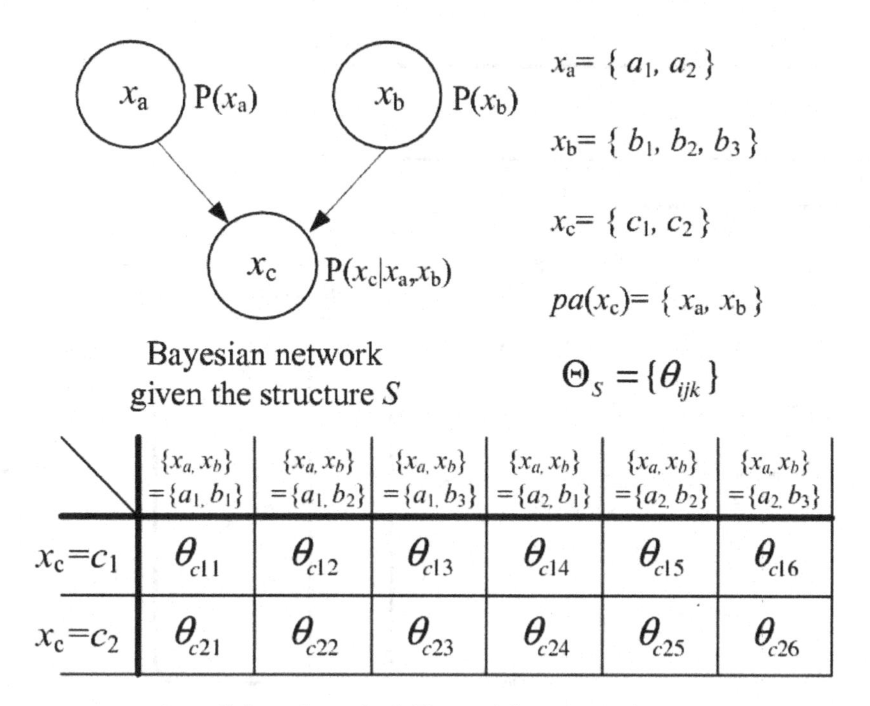

	$\{x_a, x_b\} = \{a_1, b_1\}$	$\{x_a, x_b\} = \{a_1, b_2\}$	$\{x_a, x_b\} = \{a_1, b_3\}$	$\{x_a, x_b\} = \{a_2, b_1\}$	$\{x_a, x_b\} = \{a_2, b_2\}$	$\{x_a, x_b\} = \{a_2, b_3\}$
$x_c = c_1$	θ_{c11}	θ_{c12}	θ_{c13}	θ_{c14}	θ_{c15}	θ_{c16}
$x_c = c_2$	θ_{c21}	θ_{c22}	θ_{c23}	θ_{c24}	θ_{c25}	θ_{c26}

Conditional probability table of P($x_c|x_a, x_b$)

$N_v = 3$

$N_{pa(c)} = 2 \times 3 = 6$

$N_{K(c)} = 2$

$$P(\mathbf{x} \mid S) = \prod_{i=1}^{Nv} P\left(x_i \mid pa\left(x_i\right), S\right),$$

where N_v is the dimension of the explaining variables, $\boldsymbol{x}$ is a variable set having elements x_i ($\boldsymbol{x} = \{x_1, x_2, \ldots, x_{Nv}\}$), $P(\boldsymbol{x})$ is a joint distribution of a probabilistic model, and $pa(x_i)$ is a set of parent nodes of x_i. In addition, $N_{pa(i)}$ is the number of patterns of $pa(x_i)$, $N_{K(i)}$ is the number of patterns of x_i given $pa(x_i)$, $\Theta_S = \{\theta_{ijk}\}$ is a parameter set given S with respect to $x_i = k$ when $pa(x_i)$ indicates a pattern j, and N_{ijk} is the number of data that include a pattern of $x_i = k$ when $pa(x_i)$ indicates a pattern j on dataset D (see Figure 4.) The log-likelihood of the probabilistic model can then be approximated as follows:

$$l\left(\Theta_S \mid D\right) = \sum_{i=1}^{N_v} \sum_{j=1}^{N_{pa(i)}} \sum_{k=1}^{N_{K(i)}} \left(N_{ijk}\right) \log \theta_{ijk},$$

and the AIC score of the Bayesian network can be calculated as follows:

$$AIC = -2l\left(\Theta_S \mid D\right) + 2\sum_{i=1}^{N_v} N_{pa(i)}.$$

A model structure search of the Bayesian network is carried out in order to minimize this score.

Here, we set a constraint whereby not all of the directed arrows of the variables point toward the demographic variables, because it is not natural

Figure 5. Bayesian network model constructed using questionnaire data

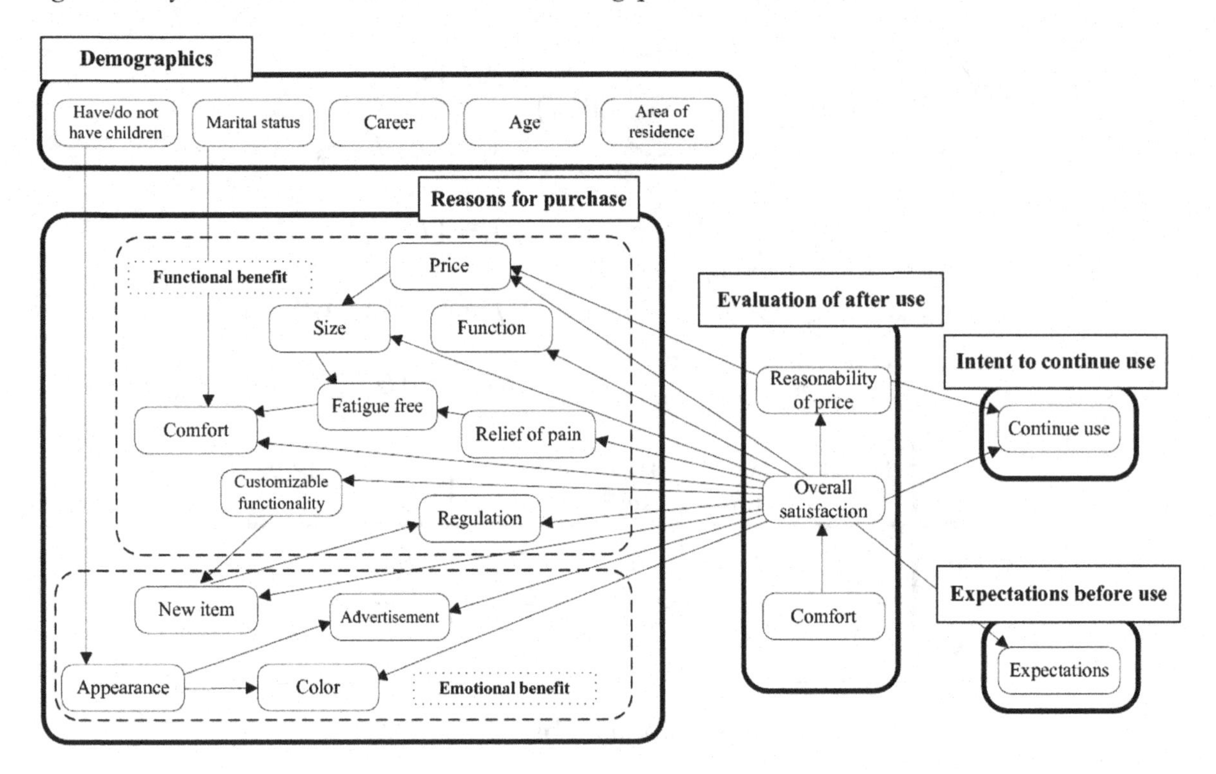

for the demographic variables to be influenced by other variables. Under this condition, an optimized Bayesian network structure in terms of the AIC is constructed by a greedy search (Motomura, 2001), as shown Figure 5.

Links between variables indicate a strong probabilistic relation according to the AIC. For example, the links illustrate that the intent to "continue use" is influenced by "reasonability of price" and "overall satisfaction". This structure is intuitive.

EVALUATION OF THE MODEL

In this section, we describe the validity of the constructed model. Here, the correct answer rate is used as a classifier for quantitative evaluation. The performance of the constructed model as a prediction model can be evaluated using the correct answer rate.

The Bayesian network model can be regarded as a classifier (Friedman et al., 1997). Consider a training set consisting of N pairs of inputs and outputs ($\boldsymbol{x}^{(i)}$, c_i) in $X \times C$, $C = \{ c_1, c_2, \ldots, c_{Nc} \}$ ($I = 1, \ldots, N$), where C is a classification class and $\boldsymbol{x}^{(i)}$ is a feature vector of sample i. The posterior distribution of C given $\boldsymbol{x}^{(i)}$ is expressed as

$$P\left(C \mid \mathbf{x}^{(i)}\right) = \alpha P\left(\mathbf{x}^{(i)} \mid C\right) P\left(C\right)$$

$$\alpha = \left\{ \sum_{i=1}^{Nc} P\left(\mathbf{x}^{(i)} \mid c_j\right) \right\}^{-1}$$

where $P(\boldsymbol{x}^{(i)}|C)$ is a vector of likelihood, $P(C)$ is a prior distribution with respect to the classification class, and α is a normalization term. The posterior probabilities of each class $\{ c_1, c_2, \ldots, c_{Nc} \}$ given $\boldsymbol{x}^{(i)}$ are calculated using the above equation, and the class that has the highest posterior probability is regarded as an output class with respect to an input $\boldsymbol{x}^{(i)}$.

We compare the prediction performance of the constructed model with the naive Bayes classifier, quantification theory (Imai et al., 2004), and the SVM. The naive Bayes classifier assumes local independence between the input variables and has one parent node (output variable). Other variables (input variables) are assumed to be child nodes of the parent node. The following equation is a posterior distribution of the naive Bayes classifier:

$$P\left(C \mid \mathbf{x}^{(i)}\right) = \alpha P\left(C\right) \prod_{k=1}^{Nv} P\left(\mathbf{x}^{(i)} \mid C\right).$$

Quantification theory is used to perform linear discriminant analysis of categorical data using dummy variables. The SVM is a popular kernel machine that can treat the nonlinearity of data. Over the past decade, the combination of the SVM and the kernel method, which realizes nonlinear prediction and classification, has been successfully implemented in various fields as a result of excellent prediction performance. In the present study, we use the Gaussian kernel with the SVM, and the parameters of the SVM are decided by a cross-validation test.

We performed classification experiments to evaluate the prediction performance of the constructed model. The sample number is N = 311, and the output class is C = {c_1 = {"continue use": Yes}, c_2 = {"continue use": No answer} }. We used two-thirds of the samples of each class c_1 and c_2 as the training data, and other samples were used as the test dataset. In other words, we evaluated the prediction performance of the model by the average correct answer rate of a three-fold cross-validation. Here, the 12 variables describing the reasons for purchase are used as the input variables, and the posterior probabilities of the "continue use" variable (c_1 and c_2) are estimated in the model in order to determine which classes have a high probability. We then examined the correct answer rates. The test results are shown in Table 1 (NBC: naive Bayes classifier, QT: quantification theory, BN: Bayesian network).

Table 1. Results of the cross-validation test

	NBC	QT	SVM	BN
Accuracy rate	66.8%	71.1%	74.1%	74.9%

The Bayesian network demonstrated a higher performance than the other methods. The results of the evaluation indicate that the Bayesian network model constructed from the questionnaire data has significant prediction power as a cognitive structure model. Since the performance of the naive Bayes classifier was low, the independence of the relationships between input variables is a poor assumption in this case. The computation times of all methods used herein were short (less than one second), because the scale of the models and data was small.

KNOWLEDGE EXTRACTION

This section describes the extraction of knowledge by sensitivity analysis (Saltelli et al., 2008) using the constructed model. The sensitivity analysis investigates the effect on outputs with respect to the variation of input variables. We focus on the "continue use" variable in order to find variables that affect the intent to continue using the customizable shoe insert. Using this model and the sensitivity analysis, we obtained the following results:

1) Demographic factors do not affect other factors.
2) The intent to "continue use" is greatly influenced by the "reasonability of price" and the "overall satisfaction" (see Figure 6).
3) The functional benefit factors and emotional benefit factors among the reasons for purchase variables have different influences on the intent to "continue use" (see Figure 7).

Figure 6. Posterior probability of the "continue use" variable with respect to the satisfaction variables

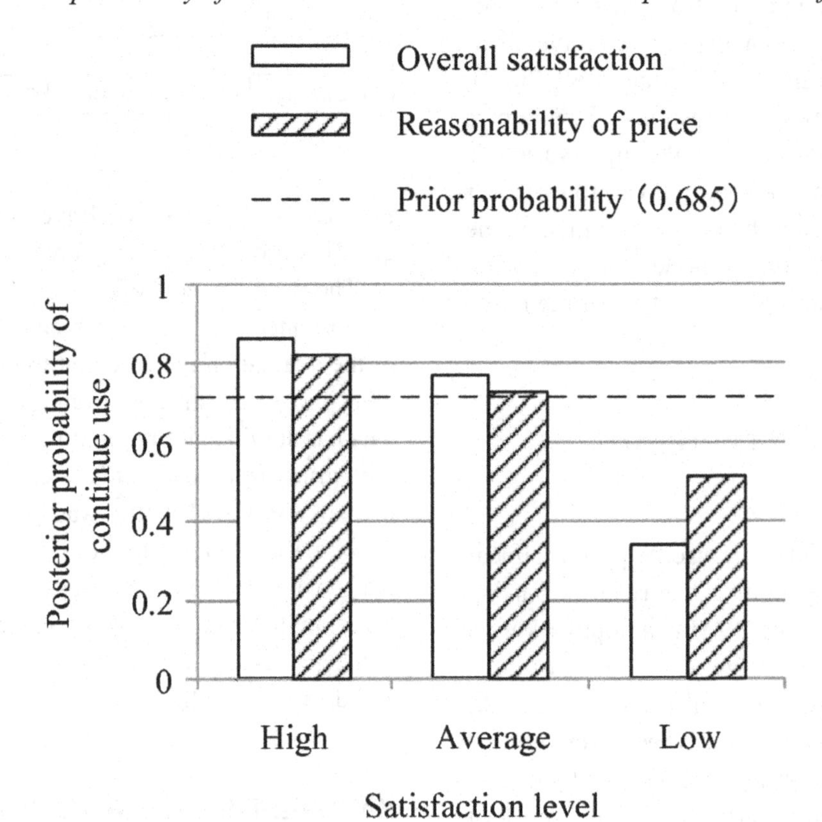

In Figure 6, an intuitive trend occurs whereby the intent of the customer to "continue use" is high when the satisfaction level after use is high. The validity of the constructed cognitive structure model is also confirmed by the results shown in Figure 6.

Figure 8 shows the prior probabilities of the "reasons for purchase" variables. In Figure 8, the probabilities of the "function", "fatigue free", "relief of pain", and "comfort" variables of customers who specified these variables as reasons for purchase were over 90%. However, these variables have little influence on the probability of "continue use", as shown in Figure 7. Moreover, the posterior probability of "continue use" by customers who did not specify "fatigue free" or "relief of pain" as reasons for purchasing the product decreases by more than 10%, as compared with the prior probability of "continue use". Therefore, customers who purchased the customizable insert considered the functional benefits of the product, such as customizable functionality, fatigue mitigation, comfort, or relief from pain, when deciding whether to purchase the product. In addition, the emotional benefit variables shown in Figure 7 have little influence on the probability of "continue use" with respect to the reasons for purchase variables. These results indicate that planners or designers of the product should incorporate the emotional functions of customers in advertisements.

Figure 9 shows the posterior probabilities of the intent to continue use given the expectation variable. The posterior probability of "continue use" of customers who had a "high" expectation before use was approximately 70%. The posterior probability of "continue use" of customers who had a "reasonable" expectation before use was approximately 25%, and the posterior probability of "continue use" of customers who had a

Figure 7. Posterior probability of the "continue use" variable with respect to the reasons for purchase variables

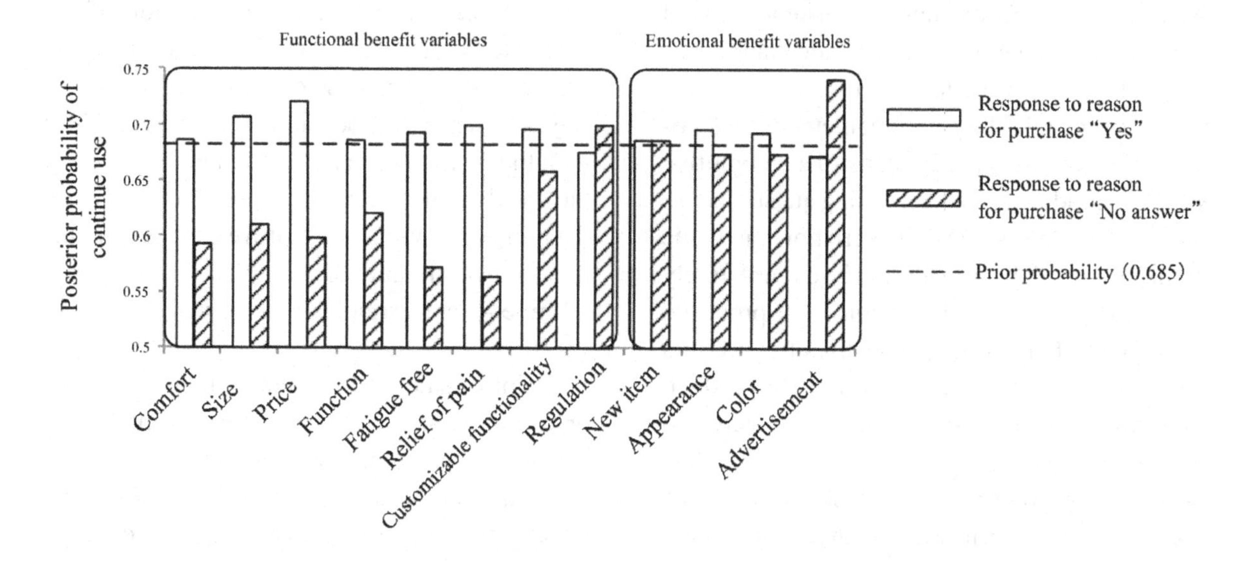

"low" expectation before use was approximately 5%. However, in Figure 9, customers who had a "reasonable" expectation before use had a higher intent to continue use than customers who had "high" or "low" expectations before use. In addition, the results of the sensitivity analysis of the "advertisement" variable in Figure 7 indicate that the posterior probability of "continue use" by customers who did not indicate certain variables as reasons for purchase were 8% higher than customers who indicated that certain variables were reasons for purchase. This indicates that advertisement likely provides insufficient or useless information to the customers.

Figure 8. Prior probability of reasons for purchase variables

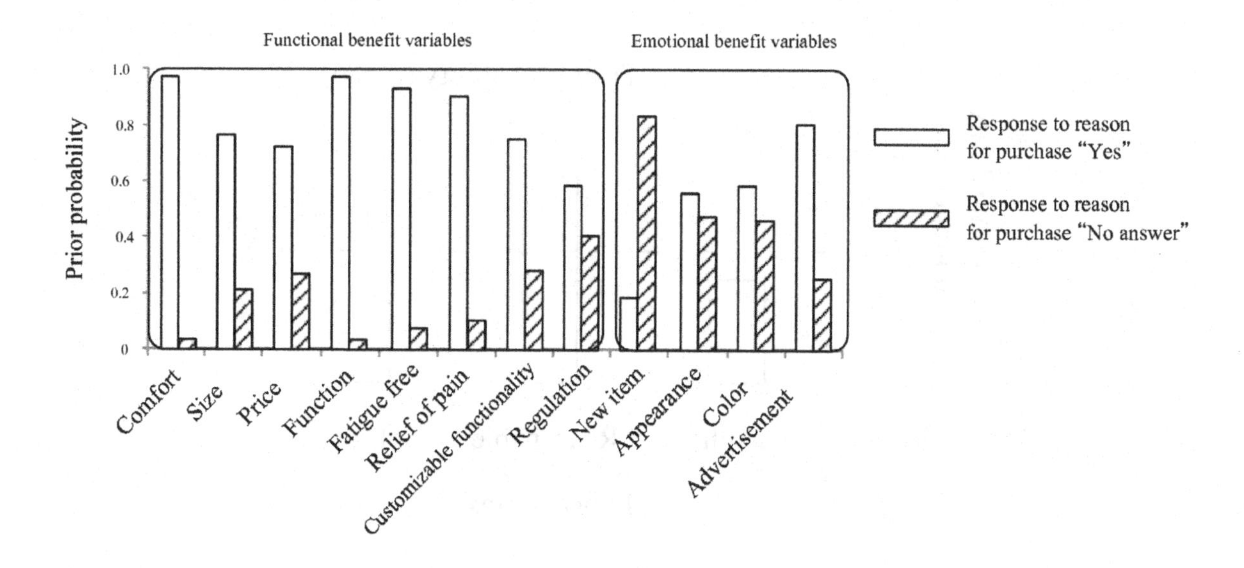

DISCUSSION

An adequate understanding of consumer related factors requires a more accurate understanding of the customer in order to match the dynamically varying demand. The proposed method can automatically and dynamically construct a cognitive structure model of customers using questionnaire data and a computer, and the variation and value of the output of all variables can be quantitatively evaluated through a simulation. The proposed method can help service industries to ascertain the attitude of the customer toward a product. Moreover, the proposed method is widely applicable to other items if the questionnaire data includes objective variables such as "continue use". Therefore, we must design the questionnaire so that objective variables are included as question items in the questionnaire in order to apply the proposed method.

The Bayesian network constructed in the present paper is constructed using discrete variable data obtained from questionnaire answers. If the data includes continuous variables, then we must discretize the variables. The structure and/or performance of constructed models depend on the discretization method. Therefore, we must introduce the multi-resolution method (Margaritis & Thrum, 2001) or some prior knowledge to the variables discretization. In addition, if we calculate the AICs of all structures strictly, then the computational cost of construction of the Bayesian network model increases exponentially with the number of variables. Therefore, we use the greedy structure search algorithm. However, the number of variables that we can treat is limited to several hundred variables. The handling of larger-scale variables will be investigated in the future.

Point of sale (POS) data, weblog data, and/or questionnaire data have been compiled in various service industries using sensor networks or ubiquitous computing. Various large-scale datasets have been compiled from data collected as consumers shop or browse web pages using POS systems, RFID systems, or Internet technology. Such datasets include records of consumer behavior and may potentially be capable of clarifying consumer demand, satisfaction level, and the concept of value with respect to purchased products or services. The efficient use of such datasets for service design by clarifying customer-related factors is of growing interest to service providers.

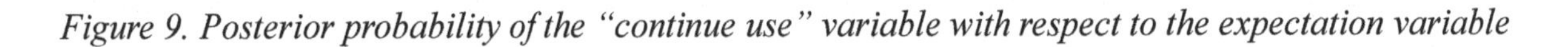

Figure 9. Posterior probability of the "continue use" variable with respect to the expectation variable

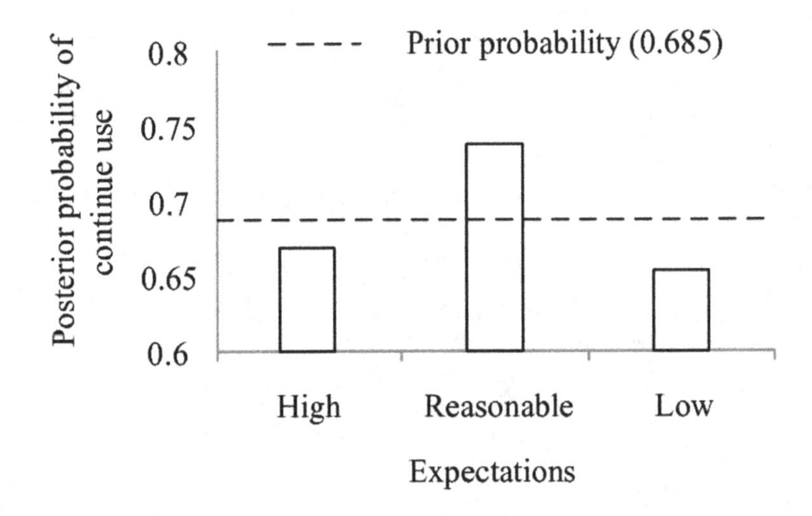

CONCLUSION

We described a method for understanding consumer-related factors by computational modeling of consumer behavior based on questionnaire data observed in real services. The contributions of the present paper are as follows: (i) the proposal of modeling of consumer behavior for the prediction of the consumer behavior through simulation, and (ii) the introduction of automatic computational modeling of the consumer based on datasets in order to clarify consumer behaviors. The proposed method can contribute to matching between the demand of the consumer and the level of service of the provider at the contact point. Factors such as the concept of value, satisfaction level, and the lifestyle of the customer change over time. An adequate understanding of such factors requires a more accurate understanding of the customer in order to match the dynamically varying demand. The proposed method can automatically and dynamically construct a model of customers using a large-scale dataset and a computer, and the variation and values of the output of all variables can be quantitatively evaluated through simulation. The proposed method is efficient for several other services that use a variety of data. However, the models described herein are not control models, but rather prediction models. In order to better understand the consumer, the realization of a control model is necessary. In the future, we intend to construct a control model.

ACKNOWLEDGMENT

The authors would like to thank Sensyukai Co. Ltd. for providing the questionnaire data. This study is supported in part by the Project for the Promotion of Service Research Center of the Japanese Ministry of Economy, Trade and Industry (METI) in 2008.

REFERENCES

Akaike, H. (1974). A new look at the statistical model identification. *IEEE Transactions on Automatic Control*, *19*(6), 716–723. doi:10.1109/TAC.1974.1100705

Bettman, J. R. (1970). Information processing models of consumer behavior. *JMR, Journal of Marketing Research*, *7*(3), 370–376. doi:10.2307/3150297

Bishop, C. M. (2006). *Pattern recognition and machine learning*. London: Springer.

Blodgett, J. G., & Anderson, R. D. (2000). A Bayesian network model of the consumer complaint process. *Journal of Service Research*, *2*(4), 321–338. doi:10.1177/109467050024002

Domencich, T., & McFadden, D. (1975). *Urban Travel Demand: A Behavioural Analysis*. Amsterdam: North-Holland.

Fornell, C., & Larcker, D. F. (1981). Evaluating structural equation models with unobserved variables and measurement error. *JMR, Journal of Marketing Research*, *18*(1), 39–50. doi:10.2307/3151312

Friedman, N., Geiger, D., & Goldszmid, M. (1997). Bayesian Network Classifiers. *Machine Learning*, *29*(2-3), 131–163. doi:10.1023/A:1007465528199

Holbrook, M. B., & Hirschman, E. C. (1982). The Experiential Aspects of Consumption: Consumer Fantasies. *Feelings and Fun*, *9*(2), 132–140.

Howard, J. A. (1989). *Consumer behavior in marketing strategy*. Upper Saddle River, NJ: Prentice Hall.

Howard, J. A., & Sheth, J. N. (1969). *The theory of buyer behavior*. New York: John Wiley & Sons.

Imai, H., Izawa, D., Yoshida, K., & Sato, Y. (2004). On Detecting Interactions in Hayashi's Second Method of Quantification. *Modeling Decisions for Artificial Intelligence*, *3131*, 1–22.

Japanese Ministry of Economy. Trade and Industry. (2007). *Towards Innovation and Productivity, Improvement in Service Industries*. Retrieved March 27, 2009, from http://www.meti.go.jp/english/report/downloadfiles/0707ServiceIndustries.pdf

Jordan, M. I. (1998). *Learning in graphical models*. Cambridge, MA: MIT Press.

Kahneman, D., & Tversky, A. (1979). Prospect theory: An analysis of decision under risk. *Journal of the Econometric Society*, *47*(2), 263–291. doi:10.2307/1914185

Luce, R. D. (1959). *Individual choice behaviour: a theoretical analysis*. New Yokr: Wiley.

MacInnis, D. J., & Jaworski, B. J. (1989). Information processing from advertisements: Toward and integrative framework. *Journal of Marketing*, *53*, 1–23. doi:10.2307/1251376

Margaritis, D., & Thrun, S. (2001). A Bayesian Multiresolution independence test for continuous variables. In *Proceedings Seventeenth Conference on Uncertainty in Artificial Intelligence*.

Massy, W. F., Montgomery, D., & Morrison, D. G. (1970). *Stochastic Models of Buyer Behavior*. Cambridge, MA: MIT Press.

Mitchell, A. A. (1981). The dimensions of advertising involvement. *Advances in Consumer Research. Association for Consumer Research (U. S.)*, *8*, 25–30.

Motomura, Y. (2001). BAYONET: Bayesian Network on Neural Network. *Foundations of Real-World Intelligence*, 28-37.

Motomura, Y., & Kanade, T. (2005). Probabilistic human modeling based on personal construct theory. *Journal of Robot and Mechatronics*, *17*(6), 689–696.

Muller, K. R., Kika, S., Ratsch, G., Tsuda, K., & Scholkopf, B. (2001). An introduction to kernel-based learning algorithms. *IEEE Transactions on Neural Networks*, *12*(2), 181–202. doi:10.1109/72.914517

Murphy, K., Weiss, Y., & Jordan, M. I. (1999, July 30-August 1). *Loopy belief propagation for approximate inference: An empirical study.* Paper presented at Uncertainty in Artificial Intelligence, London.

Ono, C., Kurokawa, M., Motomura, Y., & Aso, H. (2007). A context-aware movie preference model using a bayesian network for recommendation and promotion. In *Proceedings of User modeling 2007, 4511*, 257-266.

Pearl, J. (2000). *Causality: Models, Reasoning, and Inference*. Cambridge, UK: Cambridge University Press.

Petty, R. E., & Cacioppo, J. T. (1986). *Communication and persuasion: central and peripheral routes to attitude change*. Berlin: Springer Verlag.

Rosis, F., Novielli, N., Carofiglio, V., & Cavalluzzi, A. (2006). User modeling and adaptation in health promotion dialogs with an animated character. *Journal of Biomedical Informatics*, *39*(5), 514–531. doi:10.1016/j.jbi.2006.01.001

Rossi, P. E., Allenby, G., & McCulloch, R. (2005). *Bayesian statistics and marketing*. New York: John Wiley and Sons.

Saltelli, A., Chan, K., & Scotte, E. M. (2008). *Sensitivity Analysis*. New York: Wiley.

Sebastiani, P., Ramoni, N., & Crea, A. (2000). Profiling your customers using Bayesian networks. *ACM SIGKDD Explorations Newsletter*, *1*(2), 91–96. doi:10.1145/846183.846205

This work was previously published in International Journal of Systems and Service-Oriented Engineering, Volume 1, Issue 2, edited by Dickson K.W. Chiu, pp. 40-54, copyright 2010 by IGI Publishing (an imprint of IGI Global).

Chapter 14
Non-Monotonic Modeling for Personalized Services Retrieval and Selection

Raymond Y. K. Lau
City University of Hong Kong, China

Wenping Zhang
City University of Hong Kong, China

ABSTRACT

With growing interest in Semantic Web services and emerging standards, such as OWL, WSMO, and SWSL in particular, the importance of applying logic-based models to develop core elements of the intelligent Semantic Web has been more closely examined. However, little research has been conducted in Semantic Web services on issues of non-mono-tonicity and uncertainty of Web services retrieval and selection. In this paper, the authors propose a non-monotonic modeling and uncertainty reasoning framework to address problems related to adaptive and personalized services retrieval and selection in the context of micro-payment processing of electronic commerce. As intelligent payment service agents are faced with uncertain and incomplete service information available on the Internet, non-monotonic modeling and reasoning provides a robust and powerful framework to enable agents to make service-related decisions quickly and effectively with reference to an electronic payment processing cycle.

INTRODUCTION

The growing interests in Semantic Web services has led to the development of emerging standards such as OWL, WSMO, and SWSL in recent years. In particular, the importance applying logic-based models to develop the core elements of Semantic Web services has been realized by many researchers (Diaz et al., 2006; Guo, 2008; Liu et al., 2008; Roman & Kifer, 2007; Steller & Krishnaswamy, 2009). In the context of Semantic Web services, the classical logic-based models have been applied to services discovery (Steller & Krishnaswamy, 2009), services choreography

DOI: 10.4018/978-1-4666-1767-4.ch014

(Roman & Kifer, 2007), services enactment (Guo, 2008), and services contracting (Liu et al., 2008). However, classical logic such as description logic which is the basis of the OWL reasoning framework of the Semantic Web is monotonic (e.g., old knowledge can never be retracted from a knowledge base). Such an assumption does not really modeling the reality well because when one finds a piece of new information which contradicts the previous information archived on the Web, s/he will probably discard the old information. Moreover, classical logic is also weak in dealing with the uncertainty present in many real-world applications. Unfortunately, little research work has examined the issue of non-mono-tonicity and uncertainty for Web services retrieval and selection. Because of the large number of web services that exist nowadays, discovering and invoking one or several web services to fulfill the functional requirements of a user, becomes a very complex (e.g., non-mono-tonicity and uncertainty) and time consuming activity for application developers. One possible solution to alleviate such a problem is to develop a sound inference mechanism to autonomously deduce the most suitable services to fulfill the partially defined functional requirements of the user.

The exponential growth of the Internet has rapidly changed the way businesses are performed, and the way consumers do their shopping (He et al., 2003). Recently, a number of electronic payment services have emerged on the World Wide Web (Web) (Song et al., 2006). No doubt, electronic payment services will grow very rapidly and several large online organizations such as Yahoo, Amazon, eBay, etc. have already joined in the race. There are several advantages for the consumers to make use of electronic payment services on the Web. Firstly, these payment services autonomously and accurately process the incoming bills on behalf of the consumers. Therefore, consumers do not need to worry about the tedious job of keeping track of their bills, and perhaps paying penalties for late payments. Secondly, there could be financial advantage of using electronic payment services. For example, the total transaction cost may be reduced by having a single bulk payment from the payment service rather than having several individual micro-payments settled between a consumer and their biller. Security is also an important issue for electronic commerce. Payment service providers take care of all the security issues with other payment service providers, or the ultimate financial institutes where bill settlements take place. Therefore, the consumer's risk of conducting on-line shopping and payment is reduced to a minimum. In fact, similar advantages are also brought to the virtual stores or utility companies (e.g., electricity, gas, etc.). For instance, they do not need to spend a lot of money to set up their own payment services on the Web. Moreover, they are guaranteed by the payment service providers that outstanding bills will be settled once the services or goods are delivered to their clients.

The market of Internet-based electronic payment services is highly competitive, and there are new payment services brought to the Web every day. In order to make profit and survive under such a keen competition, payment service providers must be able to select the best settlement options (e.g., cheap, secured, and reliable settlement services) according to the latest market information such as the transaction costs and the service qualities of the external payment or settlement service providers. An agent-based electronic payment service is appealing because payment agents, a kind of service agents, can *proactively* monitor the latest market information on the Internet (Lau, 2007; Sim & Wang, 2004). Moreover, they can *autonomously* keep track of bill payments on behalf of each registered client and make sensible decisions regarding optimal settlement options available in a payment cycle. Some business agents can even learn the consumers' shopping requirements and recommend appropriate products for purchasing (Lau et al., 2000; Lau et al., 2008; Maes et al., 1999). Nevertheless, one difficulty that payment agents need to deal with is that market information (e.g., prices and service qualities of settlement services) available on the Internet is highly volatile. For instance, even though a settlement Web service was up and running few hours' ago, it might be out of service at the current payment processing cycle. Transaction cost pertaining to a settlement Web service may also vary quite frequently. This trend has already been revealed in nowadays telecommunication market. When a new payment or settlement Web service is first introduced to the Internet, it may even be difficult to have full knowledge about its service characteristic (e.g., reliability).

Therefore, payment agents are faced with the challenge of making good settlement decisions under the constraint of uncertain and incomplete information. Though sophisticated quantitative methods have been develop to model the decision making processes (Edwards, 1992; Keeney & Raiffa, 1993), there are weaknesses of these methods. For example, those axioms underlying a particular decision theory may not be valid under all decision making situations (Edwards, 1992). Bayesian network has widely been used for uncertainty analysis. However, it is difficult to obtain all the conditional probabilities required in a large network model. Though assuming independencies among events may simplify the calculations, the resulting model may not reflect the realities of the underlying problem domain. In the context of agent-mediated electronic commerce, it is desirable for business agents to explain and justify their decisions so that consumer trust and satisfaction can be obtained (Guttman & Maes, 1998; He et al., 2003; Hoffman et al., 1999). It is not easy to explain an agent's decisions purely based on a quantitative model where the relationships between various decision factors are buried by numerous conditional probabilities or weight vectors.

The main contribution of this paper is the development of a non-monotonic modeling and reasoning framework for Semantic Web service agents in general and payment agents in particular so that uncertain and incomplete information of a service market can explicitly be represented and reasoned about during agents' decision making processes. Moreover, under such an integrated framework of quantitative and symbolic approaches, the explanatory power of these agents can also be enhanced. The reasons are that the relationship between an agent's decision and its justifications can be represented by human comprehensible causal rules, and the decision making process can be explained based on formal deduction. Enhanced explanatory capabilities of payment agents will subsequently lead to improved consumer trust, and the deployment of these agents to the intelligent Semantic Web.

The rest of the paper is structured as follows. The next section gives an overview of our proposed agent-based electronic payment service. The preliminaries of non-monotonic reasoning are then provided in the following section. The formulation of a payment agent's decision rules and the representation of uncertain market information of settlement Web services will then be discussed. Afterward, how non-monotonic reasoning can be applied to the payment service agent so that optimal settlement decisions are made will be illustrated. Finally, future work is proposed, and a conclusion summarizing our findings will be given.

AN OVERVIEW OF INTELLIGENT PAYMENT SERVICES

An overview of the proposed Intelligent Agents Based Web Payment Service (IAWPS) is depicted in Figure 1. The Web site housing the proposed Semantic Web service is certified digitally so that all the related parties can verify its identification. Moreover, encrypted transmissions are used to exchange information with external agents (e.g. customers, billers, banks, external payment services, etc.). A user must first register with IAWPS by providing their personal information such as

Figure 1. An overview of Intelligent Agent Based Web Payment Service (IAWPS)

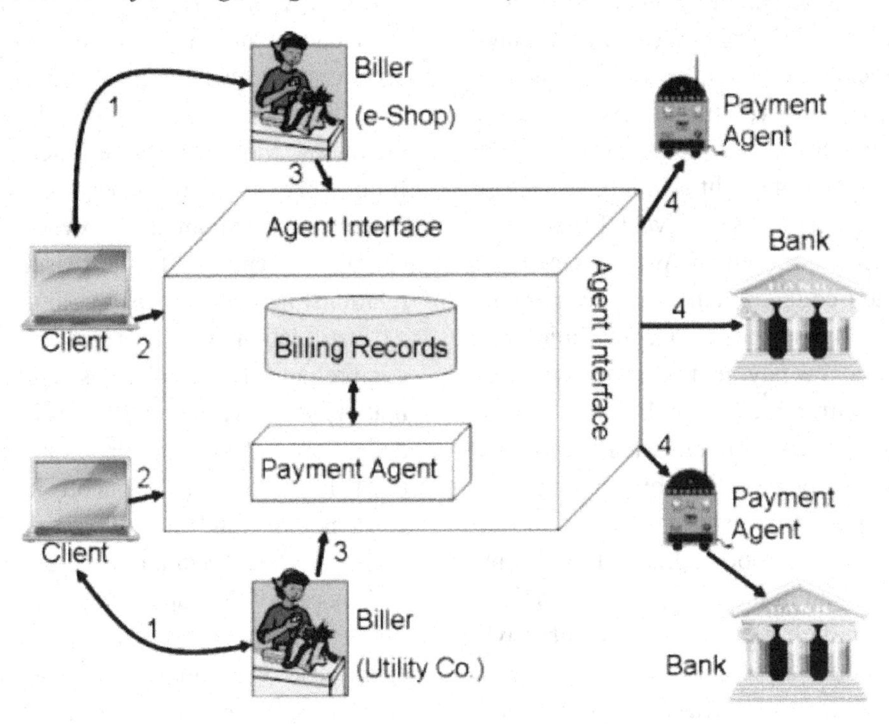

their digital certification and bank details. After acquiring utility services (e.g., electricity) or conducting on-line shopping (i.e., step 1 in Figure 1), registered users or their agents can authorize IAWPS to pay the related parties (i.e., step 2). The payment agent of IAWPS can then handshake with the retailers or their agents to obtain bills or settlement instructions (i.e., step 3) for personalized payment settlement. It is assumed that either the SOAP or REST based communication protocol (Shi, 2006) is used to facilitate the exchange of such information. Finally, the payment agent settles the bills via the Web services provided by financial institutes (e.g., commercial banks) or other payment services providers (i.e., step 4). Communications with external agents or Web services can be conducted via dedicated lines, Internet connection with Secure Sockets Layer (SSL) or Secure Electronic Transaction (SET). These external financial institutes or their agents may pass the payments through the settlement chain until the payments can finally reach the bankers of the retailers or the utility companies. For the discussion in this paper, these external financial institutes or payment services are collectedly referred to as *settlement services* or *settlement agents*. The focus of this paper is on step 4 from which the payment agent selects the optimal settlement options according to the service characteristics of pertaining settlement services during a payment processing cycle. Because of the dynamic nature of the Internet, the characteristics of these settlement services are uncertain or incomplete. Therefore, a formalism that allows explicit representation of uncertainties and sound reasoning about these uncertainties is required for the development of the effective and efficient payment services.

Since all the billing and payment records are stored in IAWPS, a registered user can enjoy personalized payment service and review their billing and payment histories from time to time. Moreover, they can specify regular payment requirements (e.g., direct debits) and set maximum payment limit. In each payment processing cycle (e.g., every 2 hours or half day), the payment agent will scan through the billing database and extract billing items which are going to be due. Billing items are then grouped into bulk payments according to the settlement instructions pertaining to these bills. Then, the payment agent needs to decide which settlement services are more suitable (e.g., being able to meet the payment deadline, cheap and secured) to settle each bulk payment. It is assumed that there could be more than one settlement options for each bulk payment processed in a payment processing cycle.

PRELIMINARIES OF NON-MONOTONIC REASONING

Non-monotonic reasoning enables an intelligent agent to make plausible decisions and retract fact and knowledge stored in a knowledge base (Dubois & Prade, 1994). More specifically, this paper refers to necessity-valued possibilistic logic (Dubois et al., 1991; Dubois et al., 1993), which is one class of non-monotonic logic (Dubois & Prade, 1994). Let L be a classical first-order language and Ω be the set of classical interpretations for L. A possibilistic formula is of the form (α, m), where $\alpha \in L$ is a closed formula and $m \in (0,1]$ is a positive number. The intuitive meaning of the formula (α, m) is that the truth of α is certain at least to the degree m (i.e., $N(\alpha) \geq m$, where N is the necessity measure of α). The semantics of a set of classical formulae Δ is defined by a subset $M(\Delta) \subset \Omega$ that satisfies all formulae in Δ. Each interpretation $\omega \in M(\Delta)$ is called a model. For a set of possibilistic formulae $F = \{(\alpha_i, m_i), i = 1, ..., n\}$, a *possibility distribution* π over Ω that characterises the *fuzzy set of models* $M(F)$ of F is introduced. In other words, F induces a preference ordering over Ω via π_F. The necessity measure N induced by π is then defined as (Dubois et al., 1993):

$$\forall \alpha \in L, N(\alpha) = \inf\{1 - \pi(\omega), \omega \models \neg\alpha\}$$

In other words, the necessity measure (i.e., the lower bound certainty) of a formula α equals to the complement of the highest *possibility* attached to an interpretation ω where $\neg\alpha$ is classically satisfied. A possibility distribution π on Ω is said to satisfy the necessity-valued formula (α, m) (i.e., $\pi \models (\alpha, m)$), iff $N(\alpha) \geq m$. $\pi \models F$, iff $\forall i = 1, ..., n, \pi \models \alpha_i, m_i$. If $\forall \pi, \pi \models (\alpha, m)$, (α, m) is said to be valid and is denoted $\models (\alpha, m)$. A possibilistic formula Φ is a logical consequence of a set of possibilistic formulae F (i.e. $F \models \Phi$), iff $\forall \pi, (\pi \models F) \rightarrow (\pi \models \Phi)$. The consistency of F (i.e., $Cons(F)$) is a measure of to what degree that there is at least one completely possible interpretation for F and is defined by $Cons(F) = sup_{\omega \in \Omega} \pi_F(\omega)$. Then, the inconsistency degree of F is defined by: $Incons(F) = 1 - Cons(F) = 1 - sup_{\omega \in \Omega} \pi_F(\omega)$.

The deduction problem in possibilistic logic is taken as finding the best valuation m (i.e. $Val(\beta, F)$) such that F entails (β, m). Based on the classical resolution, a sound and complete proof method has been developed (Dubois et al., 1993; Dubois et al., 1994) as follows:

1. A possibilistic knowledge base $F = \{(\alpha_i, m_i), i = 1, ..., n\}$ is converted to possibilistic clausal form (i.e. conjunctive normal form) $C = \wedge_{i,j} \{(c_{i,j}, m_i)\}$ where $c_{i,j}$ is a universally quantified classical first-order clause.
2. The negation of a first-order formula β, which is to be deduced from F, is also converted to clausal form $c_1, \&, c_n$.
3. Let $C' = C \cup \{(c_1, 1), ..., (c_n, 1)\}$.
4. Search for the empty clause $(\perp, \bar{m})$ by applying the *possibilistic resolution rule* (R):

$$(c_1, m_1), (c_2, m_2) \vdash (R(c_1, c_2), \min(m_1, m_2))$$

Table 1. Characteristics of settlement services

Service	Service time	Operability	Risk	Service Cost
A	2 hours	Operating	Low	60 dollars/transaction
B	2 hours	Operating	Medium	60 dollars/transaction
C	1 hour	Operating	Low	90 dollars/transaction
D	1 hour	Likely Operating	High	30 dollars/transaction

from C' repeatedly until the maximal $\overline{m}$ is found. $R(c_1, c_2)$ is the classical resolvent of c_1 and c_2.

5. Let $Val(\beta, F) = \overline{m}$.

In addition, it has been shown that the inconsistency degree $Incons(F)$ of any possibilistic knowledge base F is the least certain formula $\min\{m_1, ..., m_n\}$ involved in the strongest contradiction (i.e., $(\perp, \overline{m})$). Based on these propositions, *nontrivial possibilistic deduction* ($|\sim$) is defined as [4]:

$$F \mid\sim (\beta, m) \text{ iff } F \vDash (\beta, m) \text{ and } m > Incons(F)$$

The intuition behind nontrivial possibilistic deduction is that the proof of a formula β is restricted to the subset F' of a partially inconsistent possibilistic knowledge base F where the information is more or less certain. For example, $\forall(\alpha_i, m_i) \in F' : m_i > Incons(F)$. In other words, the certainties attached to the formulae in F' must be strictly higher than the certainty of the formula in F that causes the inconsistency. Therefore, β is proved with the most reliable information (i.e. the most certain piece of knowledge) from the knowledge base F (Dubois et al., 1994).

KNOWLEDGE REPRESENTATION

It is believed that the first step towards decision making is to produce a quantitative description of the problem domain (Holtzman, 1989) (e.g., identifying decision variables and estimating uncertainties). This is also applied to a possibilistic-based framework of knowledge representation and reasoning. For easy of illustration, only the following decision variables are discussed in this paper: *service time*, *availability*, *risk*, and *transaction cost*. Service time and availability are considered as *hard constraints*, whereas transaction cost and risk are treated as *soft constraints* (Holtzman, 1989). For example, even though a particular settlement service is the most expensive one, the payment agent may still consider it because of its ability to meet the payment deadline and relatively low risk. Table 1 is a simplified example for four settlement services with corresponding service characteristics.

To transform the service characteristics depicted in Table 1 to possibilistic formulae, the certainty value associated with each characteristic (i.e., service value) pertaining to a settlement service is computed. For example, even though a settlement service announces that its service time for a transaction is 2 hours on average, there is uncertainty related to this quoted figure because of the dynamics of the Internet (e.g., certain sites along the settlement chain are down). Similarly, even though the payment agent can be "handshake" with a settlement service (e.g., by using the "ping" command), the service may go down when a transaction is finally routed there.

Table 2 depicts the possibilistic formulation of the service characteristics for these settlement services. Under each service attribute (e.g., service time), the left hand column is the predicate formulae of corresponding service values, and the right hand column is the associated certainty values. It is assumed that a history file is kept for each settlement service. If a settlement service is brand new, default values and associated certainties could be applied. Based on the history file, if a settlement agent A announces an average service time of 2 hours, and nine out of ten times that a payment can be settled within 2 hours, the certainty attached to the predicate $Servtime(A, 2)$ will be 0.9. The predicate $Servtime(x, y)$ means that the average service time of the settlement agent x is y hour. Availability of a settlement agent is modeled by the predicate $Down(x)$, which indicates if a service is down or not. If a settlement agent does not respond to the initial "handshake" command, the certainty of $Down(x)$ is 1.0. However, even though the remote agent responds, it is still possible that it will be out of service when a transaction arrives later on. The certainty m of $Down(x)$ is computed based on the history records (e.g., frequency of out of service given that it responds at the "handshake" stage). It is assumed that routing payments to external settlement agents is risky in nature. The certainty of the predicate $Risk(x)$ is derived subjectively according to certain basic features of a settlement service x. For example, it can be assessed based on whether it is a certified site, transaction history, kinds of secured transmission protocols supported (SSL/SET), etc. If a settlement service A is housed at a certified site and it supports highly secured transmission protocol, it is not as risky as another service D which is not certified at all. The cost factor is modeled as a *fuzzy predicate* $Expensive(x)$ of a settlement service x. Expensive is a relative term for the evaluation of settlement services. There is not a definite value which will be considered as expensive. Among a group of settlement services, the one with the highest transaction cost will be considered as expensive with certainty 1.0, whereas the cheapest service will be assigned a certainty of 0.0. The rest of the settlement services can be evaluated according to a *fuzzy membership function* μ (Zadeh 1978). For the discussion in this paper, it is assumed that $\mu_{Expensive} = \frac{x - lowest}{highest - lowest}$, where $highest$ and $lowest$ represent the highest and the lowest transaction costs as quoted by the settlement agents, and x is the transaction cost of a particular settlement service.

The payment agent of IAWPS will only select a settlement service x as a candidate $Cand(x)$ for payment processing if it is likely to satisfy the hard constraint such as service time. The payment agent prefers settlement services which are cheap and less risky. The following formulae represent the payment agent's personalized evaluation criteria of a good candidate service $Cand(x)$ and a bad candidate service $\neg Cand(x)$:

$$Servtime(x, y) \wedge Reqtime(z) \wedge lq(y, z) \rightarrow Cand(x), 1.0$$
$$Down(x) \rightarrow \neg Cand(x), 1.0$$
$$Risky(x) \rightarrow \neg Cand(x), 0.6$$
$$Expensive(x) \rightarrow \neg Cand(x), 0.8$$
$$Expensive(x) \rightarrow \neg Risky(x), 0.4$$
$$Reqtime(2), 1.0$$

Table 2. Possibilistic formulation of service characteristics

Services	Service Time		Operability		Risk		Cost	
A	*Servtime*(*A*, 2)	0.9	*Down*(*A*)	0.2	*Risky*(*A*)	0.3	*Expensive*(*A*)	0.5
B	*Servtime*(*B*, 2)	0.8	*Down*(*B*)	0.3	*Risky*(*B*)	0.4	*Expensive*(*B*)	0.5
C	*Servtime*(*C*, 1)	0.8	*Down*(*C*)	0.3	*Risky*(*C*)	0.3	*Expensive*(*C*)	1.0
D	*Servtime*(*D*, 1)	0.5	*Down*(*D*)	0.5	*Risky*(*D*)	0.7	*Expensive*(*D*)	0.0

The first formula means that if a settlement service x 's service time is y hour and it is less than or equal to (i.e., predicate $lq(y,z)$) the required service time z of the current processing cycle, it is a candidate service (i.e., $Cand(x)$) with certainty 1.0; If a service x is unavailable (i.e., $down(x)$), it is not a candidate service (i.e., $\neg Cand(x)$) with certainty 1.0; A risky service is not a candidate service with certainty 0.6; An expensive settlement service is not a candidate service with certainty 0.8; An expensive service may imply that it is not risky with certainty 0.4, a low confidence level. It is assumed that the deadline of processing the bulk payments in the current processing cycle is within 2 hours, and so $Reqtime(2), 1.0$ is added to the payment agent's knowledge base K.

NON-MONOTONIC REASONING FOR INTELLIGENT PAYMENT SERVICES

The payment agent in IAWPS employs possibilistic deduction to infer if a settlement service is a feasible solution (i.e., candidate) according to its preference (e.g., cheap, less risky, meeting deadline, etc.). When a settlement service is evaluated, its characteristics are encoded as possibilistic formulae and added to the payment agent's personalized knowledge base as *assumptions*. If the payment agent's knowledge base K entails a settlement service x as a candidate service (i.e., $Cand(x)$), x will be added to the set S of feasible solutions. This scenario is similar to the symbolic approach of finding a feasible plan or schedule through *assumption based reasoning* (Kraetzschmar, 1997). In MX-FRMS (Kraetzschmar, 1997), each possible plan is added to the agent's knowledge base as assumption. If the plan is consistent with the knowledge base where the constraint about the planning process is stored, the plan is considered feasible. However, for the payment agent, some of the characteristics of a service x may lead to its consideration as a candidate service, whereas other service qualities may remind the agent to exclude it from consideration. This is a general problem in multiple criteria decision making (Zeleny, 1982). In logical term, inconsistency ($\perp$) may exist in the resulting possibilistic knowledge base K. Therefore, the nontrivial possibilistic deduction $|\sim$ is used so that the payment agent can draw sensible conclusions based on the most certain part of K even though some of its elements are contradictory to each other. To choose the optimal settlement option(s) from S, the valuation $Val(Cand(x), K)$ will be compared for each $x \in S$. A candidate settlement service that receives the highest valuation will be chosen as the settlement service since the payment agent is most certain that it should be a candidate service.

With reference to our example, the formulae representing the characteristics of each settlement service will be added to the payment agent's knowledge base K as assumptions. This process creates K_A, K_B, K_C, and K_D respectively. By possibilistic resolution, the inconsistency degree of each revised knowledge base can be computed (e.g., $Incons(K_A)$). The third step is to compute the valuation $Val(Cand(x), K_x)$. According to the definition of nontrivial possibilistic deduction $|\sim$ described in Section 3, if $Val(Cand(x), K_x)$ $>$ $Incons(K_x)$ is true, K_x $|\sim$ $Cand(x)$ can be established. To conduct possibilistic resolution, all the formulae must be converted to their clausal form first. The following is a snapshot of K_A after the characteristics of settlement service A is added to K:

$$\neg Servtime(x,y) \vee \neg Reqtime(z) \vee \neg lq(y,z) \vee Cand(x), 1.0$$
$$\neg Down(x) \vee \neg Cand(x), 1.0$$
$$\neg Risky(x) \vee \neg Cand(x), 0.6$$
$$\neg Expensive(x) \vee \neg Cand(x), 0.8$$
$$\neg Expensive(x) \vee \neg Risky(x), 0.4$$
$$Reqtime(2), 1.0$$
$$Servtime(A,2), 0.9$$
$$Down(A), 0.2$$
$$Risky(A), 0.3$$
$$Expensive(A), 0.5$$

It can easily be observed that K_A is partially inconsistent (e.g., $Cand(A) \in K_A$, $\neg Cand(A) \in K_A$). Therefore, $Incons(K_A) > 0$ is derived. In addition, there are several valuations of $Val(\bot, K_A)$. The maximal valuation $\bar{m} = 0.5$ is derived from the resolution path depicted in Figure 2. It is assumed that standard unification procedure in Prolog and the possibilistic resolution procedure described in Section 3 are used to derive the refutation. As can be seen, $Incons(K_A) = 0.5$ is derived. To find an optimal refutation without traversing a large number of nodes along the resolution paths, efficient resolution strategies have been proposed (Dubois et al., 1987). To see how strong the payment agent's knowledge base K_A entails $Cand(A)$ (i.e., how strong the payment agent believes that service A is a candidate for processing payments), the valuation $Val(Cand(A), K_A)$ is computed. According to the resolution procedure described in Section 3, $(\neg Cand(A), 1)$ is added to K_A, several refutations can be found again. Figure 3 depicts the resolution path with the strongest refutation. Therefore, $Val(Cand(A), K_A) = 0.9$ and $K_A \models (Cand(A), 0.9)$ are established. According to the definition of nontrivial possibilistic deduction $|\sim$ described in Section 3, $K_A \mid\sim Cand(A)$ since $Val(Cand(A), K_A) = 0.9 > Incons(K_A) = 0.5$. In other words, settlement service A is a feasible solution.

Similarly, $(Servtime(B,2), 0.8)$, $(Down(B), 0.3)$, $(Risky(B), 0.4)$, and $(Expensive(B), 0.5)$ are added to the possibilistic knowledge base K as assumptions. By means of possibilistic resolution, $Incons(K_B) = 0.5$ and $Val(Cand(B), K_B) = 0.8$ are derived. There-

Figure 2. Resolution path for the maximal Val(⊥, K_A)

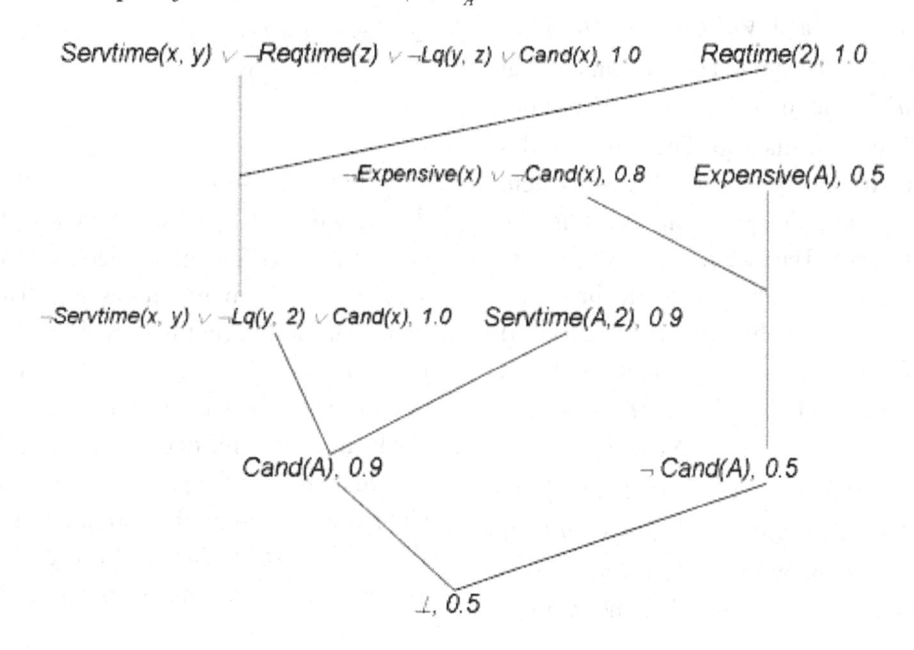

Figure 3. Resolution path for the maximal Val(Cand(A), K_A)

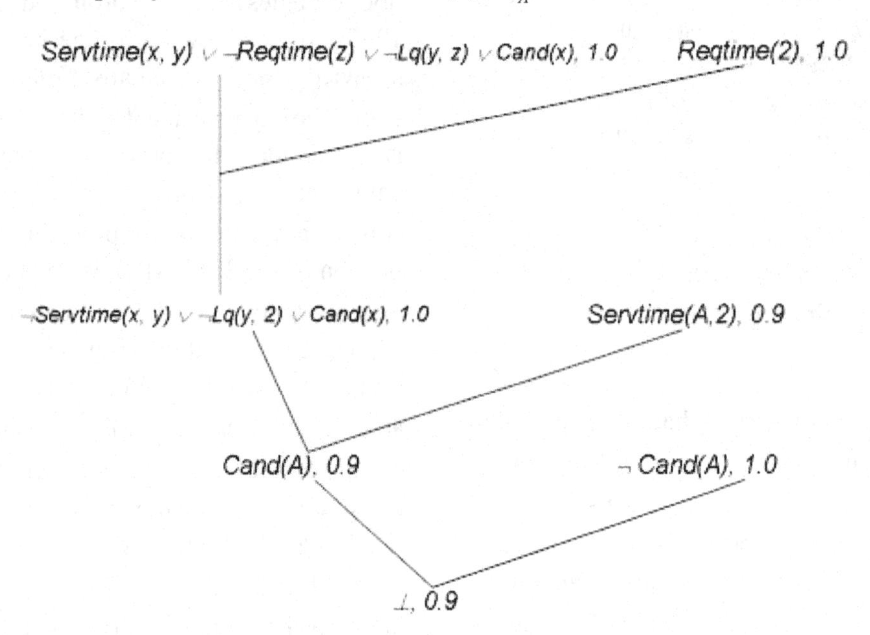

fore, the payment agent can also deduce that $K_B \mid\sim Cand(B)$. Consequently, settlement service B is also a potential solution. According to the characteristics of the settlement services described in Table 2, $(Servtime(C,1),0.8)$, $(Down(C),0.3)$, $(Risky(C),0.3)$, and $(Expensive(C),1.0)$ are also added to K as assumptions. By means of possibilistic resolution, $Incons(K_C) = 0.8$ and $Val(Cand(C), K_C) = 0.8$ are derived. The payment agent deduces that $K_C \not\approx Cand(C)$ based on the definition of nontrivial possibilistic deduction. Therefore, settlement agent C will not be considered for payment processing because the payment agent is not certain that C should be a candidate service (e.g., it is an expensive service). To evaluate the settlement service D, the assumptions of $(Servtime(D,1),0.5)$, $(Down(D),0.5)$, $(Risky(D),0.7)$, and $(Expensive(D),0.0)$ are used to derive K_D. Accordingly, $Incons(K_D) = 0.5$ and $Val(Cand(D), K_D) = 0.5$ are computed. Therefore, $K_D \not\approx Cand(D)$ is also established. In other words, it is uncertain that settlement service D can fulfill the requirements of being a candidate service because of its frequent down time and relatively high risk. In summary, the payment agent draws the following conclusions based on nontrivial possibilistic deduction:

$$K_A \mid\sim Cand(A)$$
$$K_B \mid\sim Cand(B)$$
$$K_C \not\approx Cand(C)$$
$$K_D \not\approx Cand(D)$$

Since $Val(Cand(A), K_A)$ is higher than $Val(Cand(B), K_B)$, the payment agent is more certain that settlement service A is a candidate service for payment processing. The intrinsic nature of the Internet is very dynamic, and so is the Semantic Web. A full set of service attributes pertaining to a particular settlement service may not be available for decision making. Therefore, it is desirable for the payment agent to draw sensible conclusions with incomplete information. Possibilistic reasoning is also useful under such a circumstance. For instance, if the only information available for service A is service time and availability, $K_A \mid\sim Cand(A)$ is still maintained according to the given evidence. So, A is still considered as a feasible solution because of its likelihood to meet the payment deadline. However, further investigation is required to study the expected rational behavior of payment agents when they are faced with uncertain or missing market information. According to the above example, it also sheds light on the explanation ability of payment agents. For example, the reason why settlement service A is considered as a candidate $Cand(A)$ is that there is high certainty (0.9) of it being able to meet the service deadline. Comparatively speaking, it is unlikely (0.5) that A should not be a candidate service $\neg Cand(A)$. The reason are that there are low certainties for out of service (i.e., $Down(A), 0.2$) and risk (i.e., $Risky(A), 0.3$), and medium certainty for high cost (i.e., $Expensive(A), 0.5$). Possibilistic logic provides an explicit representation of these causal relationships and a gradated assessment of the likelihood of these relationships given the available evidence.

CONCLUSION AND FUTURE WORK

The proposed agent-based payment service sheds light on a formal approach towards Web services selection in general and electronic payment on the Internet in particular. Because of the intrinsic dynamics of the Internet-based settlement services, payment agents are faced with the challenge of making good decisions under the constraint of uncertain and incomplete service market information. Non-monotonic logic provides an expressive language to formalize non-monotonicity and uncertainty that are often associated with Web services retrieval and selection. The explicit modeling of the causal relationships between a decision and its justifications also facilitates the explanation of the agent's decision making process. Although we illustrate our non-monotonic reasoning framework using the Web payment service as an example, the proposed framework can be applied to Web services retrieval and selection in general. Our research work opens the door to the development and deployment of intelligent Web services to support electronic commerce and the Semantic Web.

The effectiveness and computational efficiency of the proposed services selection agent, which is underpinned by non-monotonic logic, will be further examined in the future. One limitation of the possibilistic resolution method illustrated in this paper is that it is applied to non-fuzzy predicates only. Since payment agents also deal with fuzzy predicates (e.g., $Expensive(x)$), further work includes applying the fuzzy possibilistic resolution method (Dubois & Prade, 1990) to the Web service agents. Apart from intelligently selecting the optimum settlement service, a step further for the development of the agent-based payment service is to utilize automated negotiation (Lau et al., 2006) between the payment agents and external settlement services for making better deals (e.g., cheaper cost per transaction for larger amounts of payments) in our future work.

ACKNOWLEDGMENT

The work reported in this paper has been funded in part by the Strategic Research Grant of City University of Hong Kong (Project No.: 7008018).

REFERENCES

Diaz, O. G. F., Salgado, R. S., & Moreno, I. S. (2006). Using case-based reasoning for improving precision and recall in web services selection. *International Journal of Web and Grid Services*, *2*, 306–330. doi:10.1504/IJWGS.2006.011358

Dubois, D., Lang, J., & Prade, H. (1987, August). Theorem proving under uncertainty - a possibility theory based approach. In J. McDermott (Ed.), *Proceedings of the Tenth International Joint Conference on Artificial Intelligence* (pp. 984-986). San Francisco, CA: Morgan Kaufmann Publishers Inc.

Dubois, D., Lang, J., & Prade, H. (1991, October). A brief overview of possibilistic logic. In R. Krause & P. Siegel (Eds.), *Proceedings of Symbolic and Quantitative Approaches to Uncertainty (ECSQAU '91)* (LNC 548, pp. 53-57). New York: Springer.

Dubois, D., Lang, J., & Prade, H. (1993). Possibilistic Logic. In Gabbay, D. M., Hogger, C. J., Robinson, J. A., & Nute, D. (Eds.), *Handbook of Logic in Artificial Intelligence and Logic Programming* (*Vol. 3*, pp. 439–513). Oxford, UK: Oxford University Press.

Dubois, D., Lang, J., & Prade, H. (1994). Automated Reasoning using Possibilistic Logic: Semantics, Belief Revision, and Variable Certainty Weights. *IEEE Transactions on Knowledge and Data Engineering, 6*(1), 64–71. doi:10.1109/69.273026

Dubois, D., & Prade, H. (1990). Resolution principles in possibilistic logic. *International Journal of Approximate Reasoning, 4*(1), 1–21. doi:10.1016/0888-613X(90)90006-N

Dubois, D., & Prade, H. (1994, August). Possibility theory, belief revision and nonmonotonic logic. In A. L. Ralescu (Ed.), *Proceedings on the IJCAI-93 Workshop on Fuzzy Logic in Artificial Intelligence* (LNCS 847, pp. 51-61). New York: Springer.

Edwards, W. (1992). *Utility Theories: Measurements and Applications*. Norwell, MA: Kluwer Academic Publishers.

Guo, W.-Y. (2008). Reasoning with semantic web technologies in ubiquitous computing environment. *Journal of Software, 3*(8), 27–33. doi:10.4304/jsw.3.8.27-33

Guttman, R., & Maes, P. (1998). Cooperative vs. Competitive Multi-Agent Negotiations in Retail Electronic Commerce. In *Proceedings of the Second International Workshop on Cooperative Information Agents (CIA'98)*, Paris (LNAI 1435, pp. 135-147). New York: Springer Verlag.

He, M., Jennings, N. R., & Leung, H. (2003). On agent-mediated electronic commerce. *IEEE Transactions on Knowledge and Data Engineering, 15*(4), 985–1003. doi:10.1109/TKDE.2003.1209014

Hoffman, D., Novak, T., & Peralta, M. (1999). Building consumer trust online. *Communications of the ACM, 42*(4), 80–85. doi:10.1145/299157.299175

Holtzman, S. (1989). *Intelligent Decision Systems*. Reading, MA: Addison-Wesley Publishing Co.

Keeney, R., & Raiffa, H. (1993). *Decisions with Multiple Objectives: Preferences and Value Tradeoffs*. Cambridge, UK: Cambridge University Press.

Kraetzschmar, G. (1997). *Distributed Reason Maintenance for Multiagent Systems*. Unpublished doctoral dissertation, School of Engineering, University of Erlangen-Nürnberg, Erlangen, Germany.

Lau, R. Y. K. (2007). Towards A Web Services and Intelligent Agents Based Negotiation System for B2B eCommerce. *Electronic Commerce Research and Applications, 6*(3), 260–273. doi:10.1016/j.elerap.2006.06.007

Lau, R. Y. K., Tang, M., Wong, H., Milliner, S., & Chen, Y. (2006). An Evolutionary Learning Approach for Adaptive Negotiation Agents. *International Journal of Intelligent Systems, 21*(1), 41–72. doi:10.1002/int.20120

Lau, R. Y. K., ter Hofstede, A. H. M., & Bruza, P. D. (2000, April). Adaptive Profiling Agents for Electronic Commerce. In J. Cooper (Ed.), *Proceedings of the 4th CollECTeR Conference on Electronic Commerce*, Breckenridge, CO.

Lau, R. Y. K., Wong, O., Li, Y., & Ma, L. C. K. (2008). Mining Trading Partners' Preferences for Efficient Multi-Issue Bargaining in e-Business. *Journal of Management Information Systems*, *25*(1), 81–106. doi:10.2753/MIS0742-1222250104

Liu, H., Li, Q., Gu, N., & Liu, A. (2008, April 21-25). A logical framework for modeling and reasoning about semantic web services contract. In J. Huai, R. Chen, H.-W. Hon, Y. Liu, W.-Y. Ma, A. Tomkins, & X. Zhang (Eds.), *Proceedings of the 17th International Conference on World Wide Web (WWW 2008)*, Beijing, China (pp. 1057-1058). New York: ACM.

Maes, P., Guttman, R., & Moukas, A. (1999). Agents that buy and sell. *Communications of the ACM*, *42*(3), 81–91. doi:10.1145/295685.295716

Roman, D., & Kifer, M. (2007, September 23-27). Reasoning about the behavior of semantic web services with concurrent transaction logic. In C. Koch, J. Gehrke, M. N. Garofalakis, D. Srivastava, K. Aberer, A. Deshpande, D. Florescu, C. Y. Chan, V. Ganti, C.-C. Kanne, W. Klas, & E. J. Neuhold (Eds.), *Proceedings of the 33rd International Conference on Very Large Data Bases*, University of Vienna, Austria (pp. 627-638). New York: ACM.

Shi, X. (2006). Sharing service semantics using SOAP-based and REST Web services. *IT Professional*, *8*(2), 18–24. doi:10.1109/MITP.2006.48

Sim, K. M., & Wang, S. Y. (2004). Flexible negotiation agent with relaxed decision rules. *IEEE Transactions on Systems, Man, and Cybernetics. Part B, Cybernetics*, *34*(3), 1602–1608. doi:10.1109/TSMCB.2004.825935

Song, W., Chen, D., & Chung, J.-Y. (2006). Web services: an approach to business integration models for micro-payment. *International Journal of Electronic Business*, *4*, 265–280. doi:10.1504/IJEB.2006.010866

Steller, L., & Krishnaswamy, S. (2009). Efficient mobile reasoning for pervasive discovery. In S. Y. Shin & S. Ossowski (Eds.), *Proceedings of the 2009 ACM Symposium on Applied Computing (SAC)*, Honolulu, HI (pp. 1247-1251). New York: ACM.

Zadeh, L. A. (1978). Fuzzy sets as a basis for a theory of possibility. *Fuzzy Sets and Systems*, *1*, 3–28. doi:10.1016/0165-0114(78)90029-5

Zeleny, M. (1982). *Multiple Criteria Decision Making*. New York: McGraw-Hill.

This work was previously published in International Journal of Systems and Service-Oriented Engineering, Volume 1, Issue 2, edited by Dickson K.W. Chiu, pp. 55-68, copyright 2010 by IGI Publishing (an imprint of IGI Global).

Chapter 15
Adaptable Services for Novelty Mining

Flora S. Tsai
Nanyang Technological University, Singapore

Wenyin Tang
Nanyang Technological University, Singapore

Agus T. Kwee
Nanyang Technological University, Singapore

Kap Luk Chan
Nanyang Technological University, Singapore

ABSTRACT

Novelty mining is the process of mining relevant information on a given topic. However, designing adaptable services for real-world novelty mining faces several challenges like real-time processing of incoming documents, computational efficiency, multi-user working environment, diverse system requirements, and integration of domain knowledge from different users. In this paper, the authors bridge the gap between generic data mining methodologies and domain-specific constraints by providing adaptable services for intelligent novelty mining that model user preferences by synthesizing the parameters of novelty scoring, threshold setting, performance monitoring, and contextual information access. The resulting novelty mining system has been tested in a variety of performance situations and user settings. By considering the special issues based on domain knowledge, the authors' adaptable novelty mining services can be used to support a real-life enterprise.

INTRODUCTION

With the fast growth of technology, the Web is changing from a data-centric Web into Web of semantic data and Web of services (Yee, Tiong, Tsai, & Kanagasabai, 2009). Moreover, the demand for Web services that enable users to run offline standalone applications over the Internet has increased rapidly. The World Wide Web Consortium (W3C) defines a Web service as "a software system designed to support interoperable machine-to-machine interaction over a network" (Hugo & Allan, 2004). As a result, recently more and more software applications are available online and can be accessed on the remote system at the client side. To easily identify those online applications, each of them is assigned with the unique URI (Uniform Resource Indicator), which serves as the address of the individual application or service. Individual services can be built and combined with each other to create other services with more comprehensive functionality at little

DOI: 10.4018/978-1-4666-1767-4.ch015

additional cost (Zheng & Bouguettaya, 2009). The use of these Web services has significance in the business domain, where they are used as means of communication or exchanging data between businesses and clients (Kwee & Tsai, 2009).

Another consequence of the rapid growth of technology is the information overload from news articles, scientific papers, blogs (Chen, Tsai, & Chan, 2007), social networks (Tsai, Han, Xu, & Chua, 2009), and mobile content (Tsai et al., 2010). Because of the vast amount of information available, people tend to suffer from information overload because of irrelevant and redundant information in these documents. Thus, novelty mining (NM), or novelty detection, is a solution to this phenomenon. A novelty mining process is a process of retrieving novel yet relevant information, based on a topic given by the user (Ng, Tsai, & Goh, 2007; Ong, Kwee, & Tsai, 2009). Novelty mining can be used to solve many solve many business problems, such as in corporate intelligence (Tsai, Chen, & Chan, 2007) and cyber security (Tsai, 2009; Tsai & Chan, 2007). The pioneer work of novelty mining was proposed by Zhang et al. at the document level (Zhang, Callan, & Minka, 2002). They defined the definition of "novelty" which was the opposite of "redundancy". Given any set of documents, a document which was less similar to its history documents was regarded as a "novel" document. Although users can retrieve all the novel documents, each document still needs to be read to find the novel sentences within these documents (Tsai & Chan, 2010). Therefore, to serve users better, later studies of novelty mining were performed at the sentence level (Allan, Wade, & Bolivar, 2003; Kwee, Tsai, & Tang, 2009; Tang & Tsai, 2010; Zhang, Xu, Bai, Wang, & Cheng, 2004; Zhang & Tsai, 2009b). To the best of our knowledge, no previous work has been reported in designing and developing adaptable services for novelty mining system in the business enterprise. For enterprise users, this service-oriented novelty mining system conveniently helps them to retrieve new information about certain events of interest. They do not need to read through all documents or passages in order to find the novel information. Creating Web services for the novelty mining components allows for the rapid deployment and availability of these services for these diverse set of users, which can balance technical significance and business concerns in business processes and enterprise systems.

This paper addressed the problem of *intelligent* novelty mining, which is a big challenge in the data mining and business community. Intelligent novelty mining solves the domain-specific problem of mining novel information from text data with specific regard to the user context. Traditional novelty mining focuses on optimizing the retrieval of novel information with a predefined evaluation measure. This tends not be useful in realistic situations where a fixed unit of measure is unable to adapt to the performance requirements of different users. Intelligent novelty mining, on the other hand, aims to balance the technical significance and business concerns to create techniques that are useful in real-world scenarios. We describe the adaptable services used in an intelligent novelty mining system that is used to detect and retrieve any new information of an incoming stream of text. More specifically, the contributions are to propose services for each aspect of novelty mining, which includes novelty scoring service for mixed metrics, adaptable threshold setting service, novelty feedback service for contextual information, and document-to-sentence service for improved document-level detection.

LITERATURE REVIEW

Document-Level Novelty Mining

Zhang et al. (2002) performed document-level novelty mining based on the distance between the new document and the previously delivered documents (in history), i.e., document-document distance. This method simulates the way users view novelty because it is easier for an assessor to identify a new document as a redundant one with a specific document, and harder to identify it as redundant with a set of previously seen documents. Five novelty measures were used to evaluate the similarity, including set difference, geometric distance (cosine similarity), distributional similarity with Dirichlet smoothing, distributional similarity with shrinkage smoothing, and a mixture model. In their experimental study, all the relevant documents have been assumed to be known. The experimental results on TIPSTER data showed that the cosine similarity metric was very effective and robust.

Sentence-Level Novelty Mining

Allan et al. (2003) performed sentence-level novelty mining using several commonly used novelty measures to show that the relevant analysis appeared to be the more difficult part of this task. The performance of novelty measures could be very sensitive to the presence of non-relevant sentences. In other words, if non-relevant sentences have not been filtered out at first, they could be regarded as novel sentences in novelty mining. Seven novelty measures have been compared with each other, including the vector space model measures: simple new word count, set difference, cosine similarity and the language model measures: interpolated aggregate smoothing, Dirichlet smoothing, "shrinkage" smoothing and sentence core mixture model. Their paper reviewed both situations: (i) all the relevant sentences were known, and (ii) only the relevant sentences were unknown but were given by the top 10% relevant sentences retrieved. In the first situation, all the novelty measures performed very well. The cosine similarity and language model measures worked better than the simple new word count and set difference. However, in the second situation, the performance degraded a lot. The performance of simple new word count and set difference would work better than others.

Kwee et al. (Kwee, Tsai, & Tang, 2009) evaluated the performance on sentence-level novelty mining in both English and Malay languages, and found that the same novelty mining algorithm can be used for both languages if the preprocessing was sufficient. The results show that by replacing English stop words with their proposed Malay stop words and English stemming algorithm with their adapted Malay stemming algorithm, existing algorithms that were originally developed for English can also be used for Malay sentences, without much loss in precision (<2%) and recall (<1%), and can possibly be extended to other alphabet-based languages.

Tang et al. (2010) first analyzed the potential pitfalls for two typical novelty metrics, cosine and new word ratio. Then, by testing these novelty metrics on topics with different ratios of novel sentences, they found that each metric had its own weak zone, while the weak zones of different metrics did not overlap. The non-overlapping weak zones provided a basis for creating a blended metric. The experimental results showed that, by blending both cosine similarity and new word ratio metrics, a generally better performance than single component metrics can be obtained on both TREC 2003 and 2004 Novelty Track data (Soboroff, 2004).

Zhang and Tsai (2009b) applied named entity recognition (NER) and part-of-speech (POS) tagging for novel sentence detection. They discussed three methods to combine the two techniques, and proposed a new method that treats the sentence novelty score as two parts: the novelty score between entities of sentences and the novelty score between other significant words. Experimental results showed that their combined method can improve the novel sentence mining performance.

Although previous studies have analyzed text documents and sentences for novelty based on optimizing fixed measures such as precision or F Score, none have yet to provide a unified approach to intelligent novelty mining based on the adapting to different performance requirements. Our proposed services aim to leverage the adaptability of the algorithms to accommodate the changing

Figure 1. Our novelty mining framework

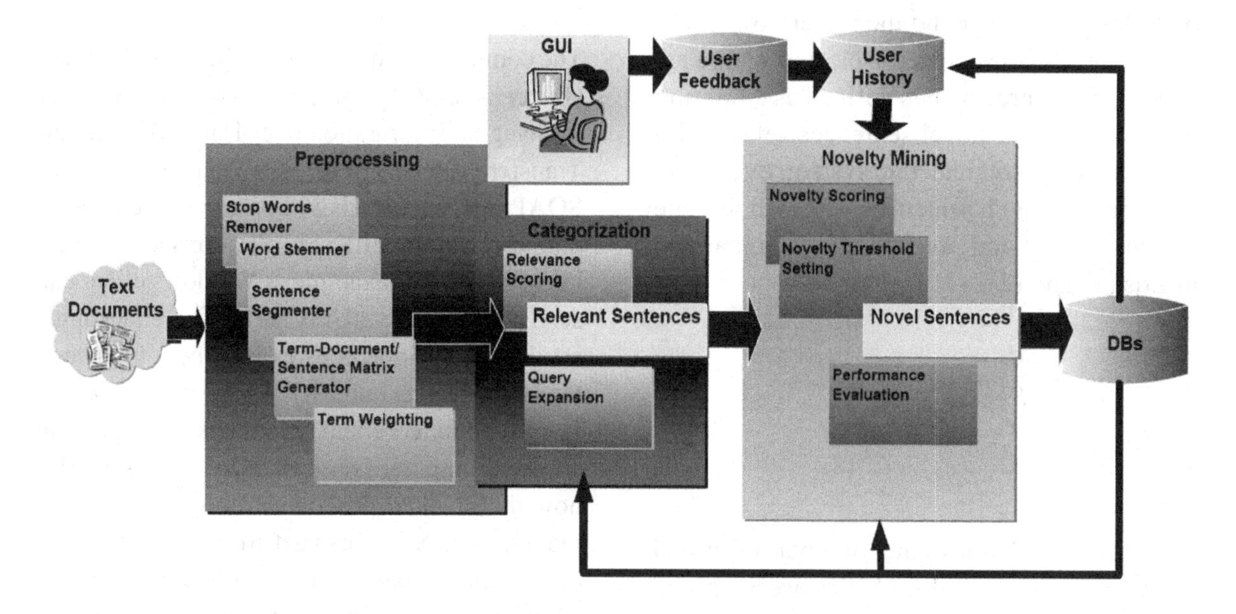

conditions for real-world situations, which has rarely been explored for novelty mining.

NOVELTY MINING FRAMEWORK

Our novelty mining framework contains three main steps: (i) preprocessing, (ii) categorization, and (iii) novelty mining (Liang, Tsai, & Kwee, 2009). In the first step, text documents are preprocessed by removing stop words, performing word stemming, etc. In the second step, each incoming sentence or document (later, we only refer to sentences without loss of generalization) is categorized into the relevant topic bin (Zhang, Kwee, & Tsai, 2010). Finally, within each topic bin, novelty mining searches the time sequence of relevant sentences and retrieves only those with enough novel information (Zhang & Tsai, 2009a). An overview of the novelty mining framework is shown in Figure 1.

Novelty Mining Algorithm

Our novelty mining algorithm for the sentence level can be described as follows. Given a specific topic, all the relevant sentences are arranged in a chronological order, i.e., *s*1, *s*2,... *sn*. For each sentence *st* (*t* = 1...*n*), the degree of novelty of *st* is quantitatively scored by a novelty metric, based on its history sentences, i.e., *s*1 to *st*¡1. The final decision on whether a sentence is novel or not depends on whether the novelty score falls above or below a *novelty threshold* (Tang & Tsai, 2009). Finally, the current sentence is pushed into the history sentence list. This algorithm is also the same for document-level novelty mining. The two major components in our algorithm are (i) novelty scoring and (ii) novelty threshold setting. Before we describe the suitable methods for these two components, we will first describe the evaluation measures in the context of novelty mining. This has implications for different types of users with various performance requirements.

Evaluation Measures

Two types of measurements are used to evaluate the system performance. One is recall and the other is precision. Recall is defined as a fraction of the number of novel sentences selected by both the assessor and the system over total number of novel sentences selected by the assessor, where precision is defined as a fraction of the number of novel sentences selected by both the assessor and the system over total number of novel sentences selected by the system. Recall and precision are summarized in Eqs (1) and (2).

$$\Pr ecision = \frac{M}{S} \text{..................}(1)$$

$$\operatorname{Re} call = \frac{M}{A} \text{.......................}(2)$$

Where M denotes the number of novel sentences selected by both the assessor and the system; S denotes as the number of novel sentences selected by the system; and A denotes as the number of novel sentences selected by the assessor. It has been acknowledged that there is a trade-off between precision and recall, which can be adjusted by the novelty threshold. A novelty threshold is a value that is used to compare with the incoming sentence. If the novelty score of an incoming sentence is below this threshold value; then, this sentence is predicted as non-novel and vice versa. If the user does not know the performance requirements, our system defaults to optimizing for high F score. F score is the primary requirement of previous studies (Soboroff, 2004), which is assumed that the user wants to keep balance between recall and precision. The definition of F score is shown in Eq. (3).

$$Fscore = \frac{2 \times \Pr ecision \times \operatorname{Re} call}{\Pr ecision + \operatorname{Re} call} \text{..............................}(3)$$

In a real-life novelty mining system, enterprise users may not be able to judge all the incoming sentences. In this case, the precision, recall, and F score results will be calculated based on a subset of sentences.

SERVICES FOR NOVELTY MINING

The communication uses the XML based Simple Object Access Protocol (SOAP) protocol for exchanging information over HTTP (HyperText Transfer Protocol) to access a Web service. Since SOAP is a message-based protocol, it can communicate among different platforms and operating systems. For example, a Windows client can send a message to a Unix server for requesting a service. SOAP specifies exactly how to encode the HTTP header and XML file so that a program in one computer (client) can invoke a program in another computer (server). SOAP also specifies how the invoked program in the server returns the response. SOAP can get around firewall servers, so that it can communicate with programs anywhere in the server (Duraisamy, 2008). The overall process involves two steps. The first step establishes the protocol between Web service requester and service provider. It searches the requested Web service in the Universal Description, Discovery, and Integration (UDDI) database, which is a directory service to store information about Web services. UDDI allows users or business to identify themselves by name, product, location, or the Web services they offer (WhatIs.com, 2003). Web services are not tied to any specific operating system or programming language. As a result, different operating systems or programming languages can be used to develop services which later can communicate with one another. UDDI is used to find the location of requested Web Service Description Language (WSDL) document. WSDL is used to describe and locate a Web service. WSDL contains information about the location of the service, the function calls and how they can be accessed. The Web service requesters use this information to form the SOAP request to the server (James, 2001). Then, it uses the location of WSDL to get the service information, such as location of requested Web service, function calls, and how to access them. Once it obtains all this information, the requester makes

Table 1. Symmetric vs. Asymmetric Metrics

	Symmetric metric	Asymmetric metric
Definitions	A metric M yields the same result regardless the ordering of two sentences, $M(s_i, s_j)=M(s_j, s_i)$	A metric M yields different results based on the ordering of two sentences, $M(s_i, s_j)\neq M(s_j, s_i)$
Typical metrics in novelty mining	Cosine similarity, Jaccard similarity	New word count, Overlap

a SOAP request to invoke the Web service. The second step exchanges information between the requester and provider. After a request to a Web service has already been performed, the server responds by requesting all input documents or sentences that need to be predicted and all the settings necessary to perform the process. This information that is used in the SOAP protocol is sent over the network to the server. Upon receiving this information, all the necessary processes will be performed. The results are sent back via the Internet to the user and displayed on the client screen. In our system, we choose the SOA (Service-Oriented Model) of Web services, as it allows for looser coupling. The basic unit of communication of this style is the "message", and is sometimes referred to as "message oriented" services (Wikipedia, 2009). Other types of Web services, such as RPC (Remote Procedure Calls) and REST (Representational State Transfer) are either not as loosely coupled or not well-supported in standard software development kits, which often offer support only for WSDL 1.1 (Wikipedia, 2009). The service-oriented model solves the drawbacks in the traditional model, and developers can fully control the system since there is only one copy which resides on the server.

Novelty Scoring Service

For novelty scoring, many efforts have been placed on creating some new metrics or comparing different metrics across different corpora, in order to find the empirically suitable metrics in a very general situation (Zhang, Callan, & Minka, 2002; Allan, Wade, & Bolivar, 2003; Tsai, Tang, & Chan, 2010). Based on whether the ordering of sentences is taken into account, novelty metrics can be divided into two types, i.e. symmetric and asymmetric (Zhao, Zheng, & Ma, 2005), as summarized in Table 1.

Symmetric metric is a metric M yielding the same result regardless the ordering of two sentences, i.e., $M(s_i, s_j)=M(s_j, s_i)$. The cosine similarity metric is a typical example here. In novelty mining, the cosine similarity novelty metric first calculates the similarities between the current sentence s_t and each of its history sentences s_i ($1\leq i \leq t-1$), as defined in Table 2. Then, the novelty score is simply one minus the maximum of these cosine similarities. Another popular symmetric metric is Jaccard similarity. Jaccard similarity metric is used to measure the similarity between two sets, which is defined as the ratio of the intersection size and the union size of two sample sets. Similar to cosine similarity, the Jaccard similarity is converted to the novelty score by 1 minus the maximum Jaccard similarity between s_t and s_i $1\leq i \leq t-1$)), as defined in Table 2.

Since symmetric metrics are not able to take the ordering of sentences into account, they may cause some problems in the novelty scoring. For example, suppose there are two sentences s_1 and s_2 as follows.

s_1 = {A,B,C,D,E,F,G}

s_2 = {A,B,C}

where A, B, C, D, E, F, and G are seven different terms, and s_2 appears immediately after s_1.

Table 2. Definitions of novelty metrics

Novelty Metrics	Type	Definitions
Cosine Similarity (N_{cos})	Symmetric	$N_{\cos}(s_t) = \min_{1 \le i \le t-1} [1 - \cos(s_t, s_i)]$ $\cos(s_t, s_i) = \frac{\sum_{k=1}^{n} w_k(s_t) \cdot w_k(s_i)}{\lVert s_t \rVert \cdot \lVert s_t \rVert}$ Where $w_k(s_i)$ is the term weight of word *k* in sentence *i*. We use TF.ISF term weighting function as follow. $w_k(s_i) = tf_{w_k, d_i} \log\left(\frac{n+1}{sf_{w_k} + 0.5}\right)$
Jaccard Similarity ($N_{jaccard}$)	Symmetric	$N_{Jaccard}(s_t) = \min_{1 \le i \le t-1} [1 - Jaccard(s_t, s_i)]$ $Jaccard(s_t, s_i) = \frac{\lvert s_t \cap s_i \rvert}{\lvert s_t \cup s_i \rvert} = \frac{\sum_{k \in s_t \cap s_i} \min(w_k(s_t), w_k(s_i))}{\sum_{k \in s_t \cup s_i} \max(w_k(s_t), w_k(s_i))}$ For example, if we assume there are two sentences defined as $s_t = [1,0,1,0,0,2]^T, s_i = [0,1,2,0,1,1]^T$, $\lvert s_t \cap s_i \rvert$ that means the count of the overlapping words of s_t and s_i can be estimated by the sum of the overlapping words, i.e. $\sum_{k \in s_t \cap s_i} \min(w_k(s_t), w_k(s_i))$. In this case, it equals to: $\min(1,0) + \min(0,1) + \min(1,2) + \min(0,0) + \min(0,1) + \min(2,1) = 2$. The determinator $\lvert s_t \cup s_i \rvert$ is used to normalize the overlap measure by the combined length of the sentence s_t and s_i. In this case, it equals to: $\max(1,0) + \max(0,1) + \max(1,2) + \max(0,0) + \max(0,1) + \max(2,1) = 7$

continued on following page

Table 2. Continued

Overlap Metrics ($N_{overlap}$)	Asymmetric	$N_{overlap}(s_t) = \min_{1 \le i \le t-1} [1 - overlap(s_t \mid s_i)]$ $overlap(s_t \mid s_i) = \frac{\lvert s_t \cap s_i \rvert}{\lvert s_t \rvert} = \frac{\sum_{k \in s_t \cap s_i} \min(w_k(s_t), w_k(s_i))}{\sum_{k \in s_t} w_k(s_t)}$ For example, if we again assume there are two sentences defined as $s_t = [1,0,1,0,0,2]^T, s_i = [0,1,2,0,1,1]^T$, $\lvert s_t \cap s_i \rvert$ that means the count of the overlapping words of s_t and s_i can be estimated by the sum of the overlapping words, i.e. $\sum_{k \in s_t \cap s_i} \min(w_k(s_t), w_k(s_i))$. In this case, it equals to: $\min(1,0) + \min(0,1) + \min(1,2) + \min(0,0) + \min(0,1) + \min(2,1) = 2$. The determinator $\lvert s_t \rvert$ is used to normalize the overlap measure by the length of the sentence s_t, i.e., 4 in this case.
New Word Ratio (N_{nwr})	Asymmetric	$N_{newWord}(s_t) = \frac{newWord(s_t \mid \{s_i\}_{i<t})}{\lvert W(s_t) \rvert}$ $newWord(s_t \mid \{s_i\}_{i<t}) = \lvert W(s_t) \rvert - \lvert W(s_t) \cap [\cup_{i=1}^{t-1} W(s_i)] \rvert$ Where $W(s)$ is the set of word in the sentence s; $\lvert W(s) \rvert$ is number of words in the sentence s. For example, if we again assume there are two history sentences defined as $s_1 = [1,0,1,0,0,2]^T, s_2 = [0,1,0,0,1,1]^T$, the newWord count for a new incoming sentence $s_t = [0,1,0,\underline{2},0,1]^T$ is 2. The newWord Ratio is the normalized version of newWord count by dividing the length of the sentence s_t.

Measured by symmetric metrics, s_2 is likely to be predicted as "novel" due to the low similarity between s_1 and s_2. But in fact, s_2 should actually be predicted as "non-novel", based on the history information.

To alleviate the problems caused by symmetric metrics, some asymmetric metrics have been proposed to measure the "novelty" of sentences by considering the ordering between sentences, i.e. $M(s_i, s_j) \neq M(s_j, s_i)$. A typical asymmetric metric used in novelty mining is the overlap metric (Zhang, Callan, & Minka, 2002), as defined in Table 2. Another popular asymmetric novelty metric is new word count which counts the number of new words in the incoming sentence given its history sentences (Allan, Wade, & Bolivar, 2003), also defined in Table 2.

Because the strengths of symmetric metrics and asymmetric metrics complement each other, we propose a service for measuring the novelty by a mixture of both types of novelty metrics. The goal of the mixed metric is to integrate their merits and hence generalize better over different topics.

Two major issues for constructing a mixed metric are (i) solving the scaling problem to ensure different component metrics comparable and consistent and (ii) combining the strategy that defines the way of fusing the outputs from different component metrics. Having resolved the scaling problem, novelty scores from all novelty metrics range from 0 (i.e. redundant) to 1 (i.e. totally novel). Therefore, they are both comparable for having the same range of values and consistent. For the combining strategy, we combine two types of metrics linearly, one from each type, as shown in Equation 1.

Equation 1: $M(s_t) = \alpha N_{sym}(s_t) + (1\text{-}\alpha)N_{asym}(s_t)$

where α is the combining parameter ranging from 0 to 1. The bigger the α, the heavier the weight for the symmetric metrics. Without losing any generalization, we first choose equal weights of 50% (α =0.5) for both symmetric and asymmetric metrics and compare the performance of the mixed metric with those of individual metrics.

Four different mixed metrics can be generated from Equation 1, i.e., M_1 to M_4, as follows.

$$M_1 (s_t) = \alpha N_{cos}(s_t) + (1\text{-}\alpha)N_{newWord}(s_t)$$

$$M_2 (s_t) = \alpha N_{cos}(s_t) + (1\text{-}\alpha)N_{overlap}(s_t)$$

$$M_3 (s_t) = \alpha N_{Jaccard}(s_t) + (1\text{-}\alpha)N_{newWord}(s_t)$$

$$M_4 (s_t) = \alpha N_{Jaccard}(s_t) + (1\text{-}\alpha)N_{overlap}(s_t)$$

The series of mixed metrics from M_1 to M_4 have been tested on all 100 topics from TREC 2003 and 2004 Novelty Track data and compared with 4 individual metrics, cosine, Jaccard, overlap and new word count metrics, with the experimental results shown in Figure 2. The F scores of various metrics are shown in both the high-precision region (Figure 2 (a)) and the high-recall region (Figure 2 (b)). Since the highest F scores appear in the high-recall region, the global view is omitted. From Figure 2, within the individual metrics, asymmetric and symmetric metrics perform well in high-recall and high-precision cases respectively, but not both. On the other hand, the mixed metrics can perform well in both cases.

Threshold Setting Service

As an adaptive filtering algorithm, one primary challenge in novelty mining is to set the threshold of novelty scores adaptively. There are several motivations for designing an adaptive threshold setting algorithm. First of all, since novelty mining is a real-time system, there is little or no training information in the initial stages. Therefore, the threshold cannot quite be predefined with confidence. Secondly, the adaptive threshold setting is necessary to satisfy the diverse user requirements varying from high-precision to high-recall, and the diverse user definitions of "novelty", as discussed in user studies. As novelty mining is an accumulating process, more training information will be available for threshold setting via the user's feedback given over time. The adaptive threshold setting algorithm is also necessary to make good use of this information from the user and to satisfy the user's needs better.

To the best of our knowledge, few studies have focused on adaptive threshold setting in terms of novelty mining. (Zhang & Callan, 2001) proposed a simple threshold setting algorithm, which slightly increased the threshold if a novel document is retrieved as a non-novel one based on the user's feedback. Clearly, it is a weak algorithm because it can only increase the threshold. This section addresses the problem of setting an adaptive threshold to meet different system requirements.

Figure 2. Mixed metrics vs individual metrics for all 100 topics in high-precision (a) and high-recall (b) system in TREC 2003 and 2004 Novelty Track data

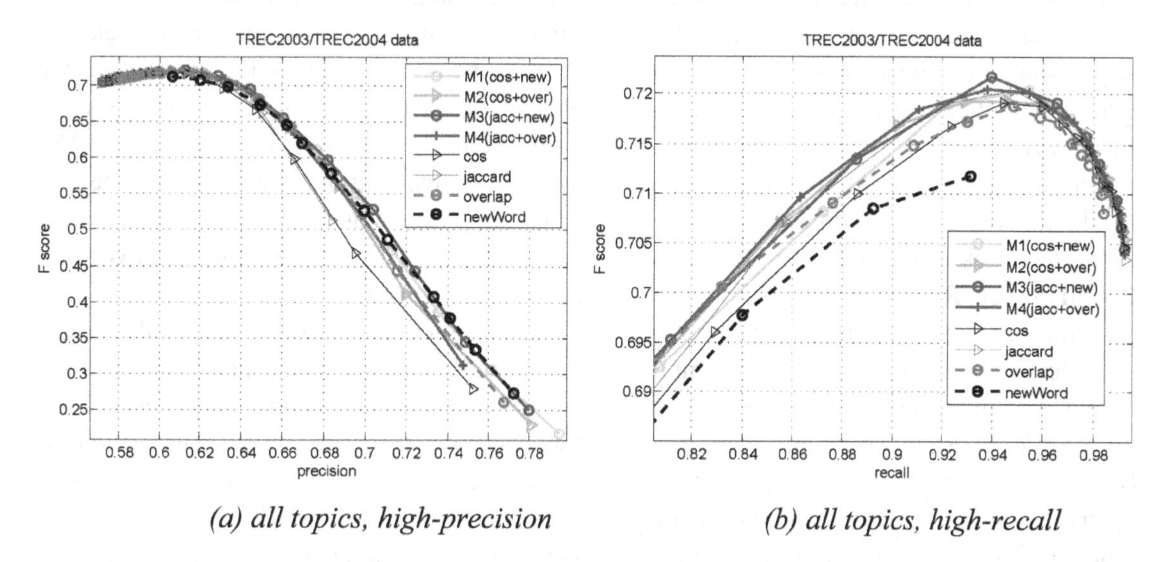

(a) all topics, high-precision *(b) all topics, high-recall*

Our work is related to the score distribution-based threshold setting algorithms that have been proposed for relevant sentence retrieval (Zhai et al., 1999; Zhang & Callan, 2001). Some ideas of score distribution modeling can be borrowed from the relevance retrieval area; however, the novelty score in novelty mining has its distinctive characteristics. First, we find that scores from the novel and non-novel classes heavily overlap. This is because novel and non-novel information are always interlaced in one sentence, while in the relevance retrieval problem; most of the non-relevant sentences show little similarity with the relevant ones. Secondly, we find that the score distributions for both novel and non-novel classes can be approximated by Gaussian distributions, while in the relevance retrieval problems, the scores of non-relevant sentences follow the exponential distribution (Zhang, Callan, & Minka, 2002),which actually implies that most of the non-relevant sentences are not similar to the relevant ones. Similar results can be observed on the document level.

The score distributions of classes provide the global information necessary for constructing an optimization criterion for threshold setting. The threshold optimizing this criterion is the best we can obtain so far until new user feedback is provided. The proposed service is based on the Gaussian-based adaptive threshold setting (GATS) algorithm, and can be tuned according to different performance requirements on documents, by employing different optimization criteria. For example, the F_β score, which is the weighted harmonic average of precision and recall, where β controls the tradeoff between them.

Novelty Feedback Service

Traditional novelty mining techniques cannot detect novelty in a sentence that looks similar to its history sentences, but contains information that is useful in the user's current context, such as an analyst interested in price fluctuations of oil. A user's context in novelty mining is defined as his/her specific requirements, which can be one or more of the following:

1) **Level of Novelty**: Different users have different definitions of "novel" information. For example, a user would regard a sentence with 60% novel information as a novel sen-

tence while another user would only regard a sentence with 90% novel information as a novel sentence. The threshold of novelty scores should be higher for the user with a stricter definition for the "novel" sentence. As novelty mining is an accumulating system, more training information will be available for threshold setting, based on the user's feedback given over time.

2) **Performance Requirement**: The user's performance requirements can tell us the specific concerns about the novelty mining output. For example, when the user does not want to miss any novel information, a high-recall system which only filters out the most redundant sentences is desired. On the other hand, when the user wants to read the sentences with the most significant novelty in a short time, a high-precision system which only retrieves highly novel sentences is preferred.

3) **Specific Domain Interest**: Last but not least, understanding the user's context is very helpful for the system to set the mining target. For example, a user who is interested in the financial news may be sensitive to different numbers, currency symbols, etc. To make use of the user's contextual information, we have to answer several questions, i.e., what contextual information can be acquired in novelty mining, how to make automatic annotation about the context, and how to design the user interface for the user's context.

In the previous studies, the contextual information accessed through relevance feedback has been successfully utilized for query expansion in relevant information retrieval (Robertson & Soboroff, 2002). In this case, the user may either provide the relevance feedback for documents/passages where the useful terms are extracted, or select the key terms directly.

In novelty mining, however, accessing the context information through user feedback has rarely been explored. One possible reason is that it is much more difficult to describe the novel information using key words. In the relevant document/passage retrieval, if "finance" is input as a key word by the user, documents/passages with "finance" are more likely to be retrieved by a search engine. But in novelty mining, the user cannot always use "finance" as novel term because the word "finance" may become non-novel after its first appearance.

In this section, we attempt to bridge a gap between the explicit user's preference and the advance of contextual information, by answering three questions: (i) what aspects of user's context can be acquired and how to guide user to interact with system efficiently? (ii) how could we utilize the acquired contextual information in the novelty mining system? and (iii) how could we evaluate the effectiveness of the resulting system?

Instead of terms feedback, which is not quite applicable in novelty mining systems, the high-level structures of terms, i.e., named entities (number, time, location, person, etc) are suggested in this paper. These high-level structures show very natural aspects of information that can be easily understood by the user to facilitate the novelty feedback. For example, readers of financial news may be more sensitive to different numbers. Given a pair of sentences s_1 and s_2 where s_1 is the nearest history sentence of s_2, s_2 contains novel information because they include different number information, i.e., US$75 and US$70.

Example 1:

s_1: Now the oil price is <u>US$**75**/barrel</u>.

s_2: Now the oil price is <u>US$**70**/barrel</u>.

However, s_2 may be predicted as non-novel by the conventional novelty mining system because it is very similar to s_1, according to the similarity score calculated by cosine similarity (Allan, Wade, & Bolivar, 2003). This example shows that the linguistic-level information, such us number, time, location and person, reflect the user's context and can potentially interest the user. The major issue here is designing a method for user feedback.

In the GUI design, the user is assumed to be quite clear about what his/her target of novel information retrieval. Since a specific topic usually contains many aspects of information, the system allows the user to feedback his/her preferred aspects of novel information simultaneously and hence to describe his/her context completely. Accordingly, in the GUI design, we use check-boxes with different aspects of information, as shown in Figure 3.

The user is able to specify his/her context for each topic separately via GUI.

Figure 4 shows the framework of a heuristic annotation-based novelty mining (HAN) system (Tang, Kwee, & Tsai, 2009). The basic idea of HAN is to retrieve the novelty information missed by the original novelty mining based on the annotation of named entities from novelty feedback. Therefore, a cascade framework is designed for HAN, where the original novelty mining algorithm classifies novel sentences at the first layer. The HAN algorithm retrieves the novel information missed by the original novelty mining system at the second layer, based on the user's context learnt from novelty feedback. In this system, any new incoming sentence s_t will be predicted in 4 steps, i.e.

- step 1: Record the corresponding contextual information specified by the user, if any.
- step 2: Run the original novelty mining on s_t. Go to step 3 if the novelty mining prediction is "non-novel", step 4 otherwise.
- step 3: Run HAN on s_t. Go to step 4 if the HAN prediction is "novel".
- step 4: Predict s_t as novel.

One of the most important issues in HAN is the annotation of named entities. In this paper, the named entities in sentences are identified by a commercial named entity extractor. We extract four types of entities including "number", "location", "person" and "time".

Figure 3. Snapshot of GUI for additional contextual information

Unlike other three types of named entities, numbers themselves are not that meaningful. Therefore, the first word after the number is also accessed by the HAN algorithm using part-of-speech tagging (POS tagging or POST). Only the number that is in conjunction with a noun (e.g., player, mile, etc) or cardinal number (e.g., million, billion, etc) is recorded as *number pattern*. For example, "20 people", "3 million", "45 people" will be recorded and compared in HAN, instead of numbers themselves like "20", "3" and "45". Other named entities such as time, location, and person show enough information by themselves and no additional information will be appended. For example, "Singapore", "London" and "Liverpool" clearly show three different locations and some sentences are prone to be novel due to some new location that has never appeared before.

After contextual information extraction, HAN also compares the incoming sentence with its history sentences, but makes the final prediction based on a set of predefined rules related to each kind of named entity. There are two strategies in sentence comparison, i.e. one-to-one and one-to-all comparisons. The rules using one-to-one sentence comparison are defined as below.

Number-rule: the nearest sentence pair s_1 and s_2 have <u>different NUMBER + same NOUN</u>

Figure 4. A heuristic annotation-based novelty mining system with the user feedback of "number" and "location"

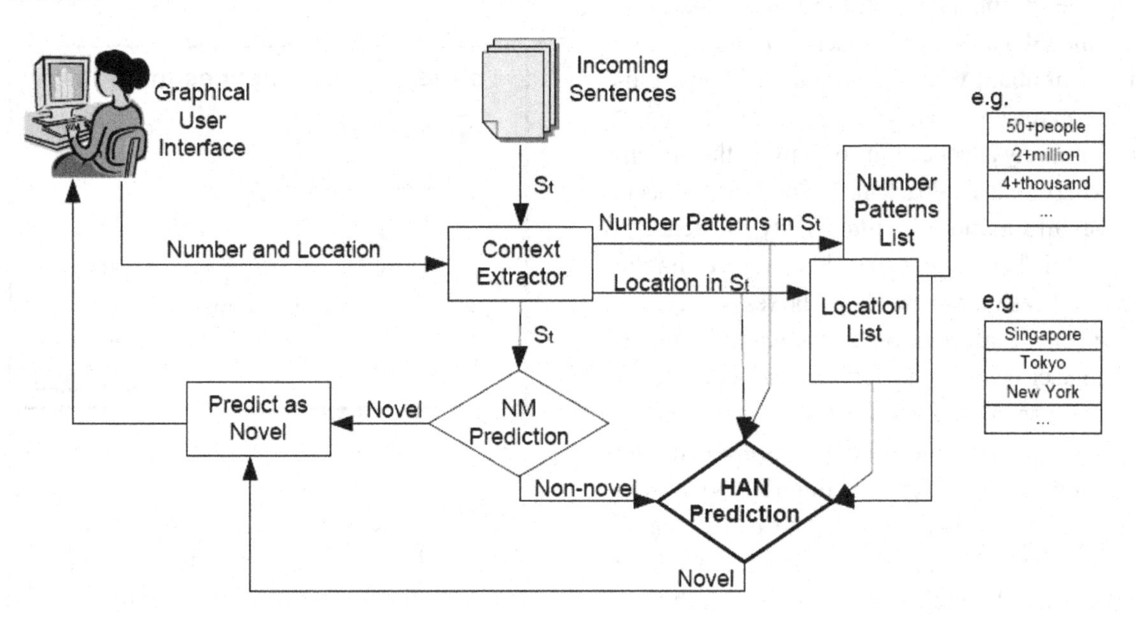

=> s_2 is novel

Location/Time/Person-rule: the nearest sentence pair s_1 and s_2 have different

LOCATION/TIME/PERSON => s_2 is novel

Using the above number-rule, the sentence s_2 in Example 1 can be predicted as "novel" as it is retrieved by the HAN. In the experimental study, however, we found that one-to-one comparison may not be sufficient for the prediction of novelty because the novel number pattern in terms of s_1 may have already occurred in other history sentences, i.e., the context of s_2. To consider the context of the incoming sentence, we accumulate all number patterns appeared in the history sentences (as shown in Figure 4, for example), and propose another set of rules based on one-to-all sentence comparison, denoted as ruleA, defined as below.

Number-ruleA: satisfy number-rule AND s_2 has different number pattern compared with the number pattern list in history => s_2 is novel

Location/Time/Person-ruleA: satisfy Location/Time/Person-rule AND s_2 has different LOCATION/TIME/PERSON compared with the LOCATION/TIME/PERSON list in history => s_2 is novel

The experimental result on TREC 2004 Novelty Track data shows that *-ruleA always outperforms *-rule in various situations in the HAN algorithm. We also compare these two strategies in the original novelty mining on the same data. Some interesting summaries are listed as follows.

- The one-to-one sentence comparison outperformed one-to-all sentence comparison in the original novelty mining. This indicates that one-to-one sentence comparison may be more effective for the novelty mining based on the bag-of-words model.
- One-to-all sentence comparison is more effective in HAN. This indicates that one-to-all sentence comparison may be more effective for detecting novel information based on named entities.

Figure 5. Framework of the HAN algorithm using one-to-all sentence comparison

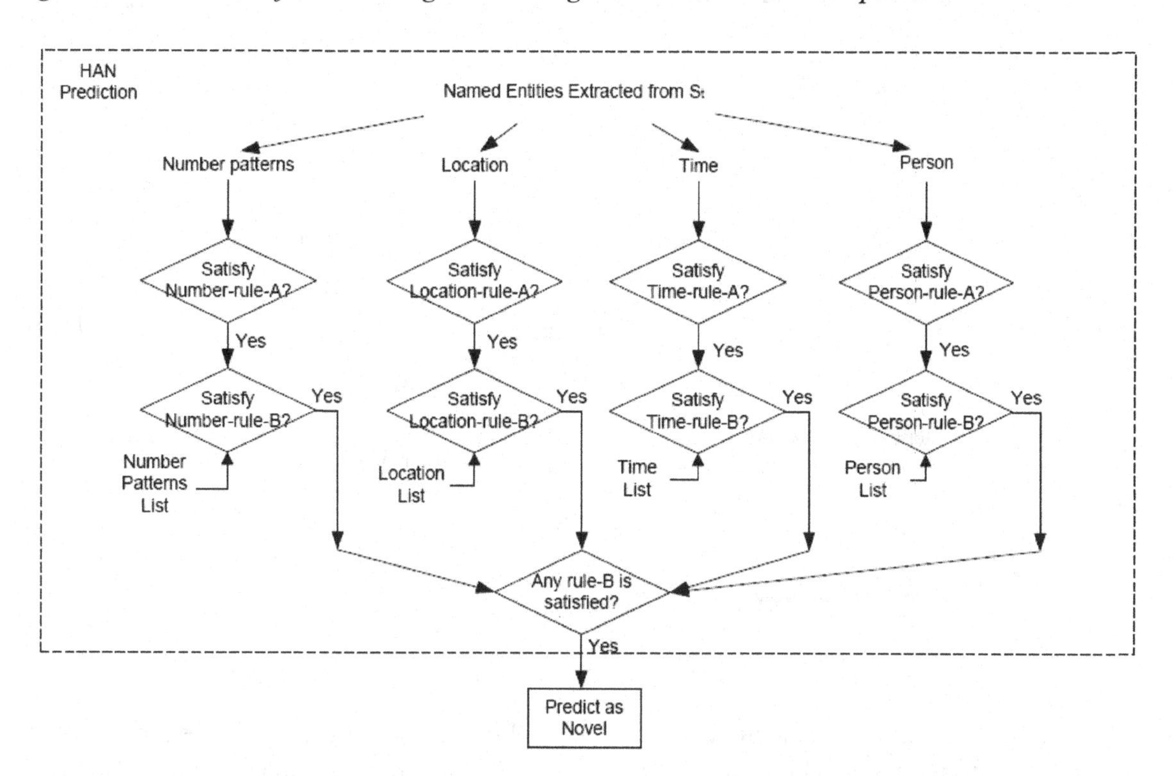

Obviously, one-to-all comparison strategy conducts a tighter criterion for novel sentence retrieval. This result also shows, for highly similar sentences that are predicted as "non-novel" by the original novelty mining, the tighter criterion is necessary because only the complete novel named entities may make a highly similar sentence novel. Figure 5 shows how the HAN algorithm works by using one-to-all sentence comparison strategy.

HAN attempts to retrieve the novel information hidden in the sentences missed by the original novelty mining algorithm, as shown in Figure 4. Another issue here is how to construct a searching pool of sentences for the HAN algorithm. One simplest way is to let HAN search all the sentences predicted as non-novel by the original novelty mining. However, this searching pool is fixed and unable to adjust the tradeoff between precision and recall for different users' needs.

Instead of a fixed searching pool for HAN, we defined a soft searching pool, by introducing the width of the searching pool, i.e., δ. Given a width parameter δ, the searching pool of the sentences with a novelty score from (θ- δ) to θ (θ is the novelty threshold in the original novelty mining) can be constructed accordingly. With the increasing value of the width of the searching pool δ, a larger pool of sentences are formed and then processed by the HAN algorithm.

From the physical point of view, the width of the searching pool reflects how badly the user does not want to miss any information. When δ is set as δ=θ, i.e., the maximum searching pool of sentences for the HAN algorithm, the precision will be sacrificed in some degree for a higher recall.

Document-to-Sentence Service

The idea of our document-to-sentence (D2S) service is to segment the document into sentences and perform novelty mining at sentence level (Zhang & Tsai, 2010). It is assumed that

Figure 6. Detection process when applying sentence novelty mining on document novelty mining

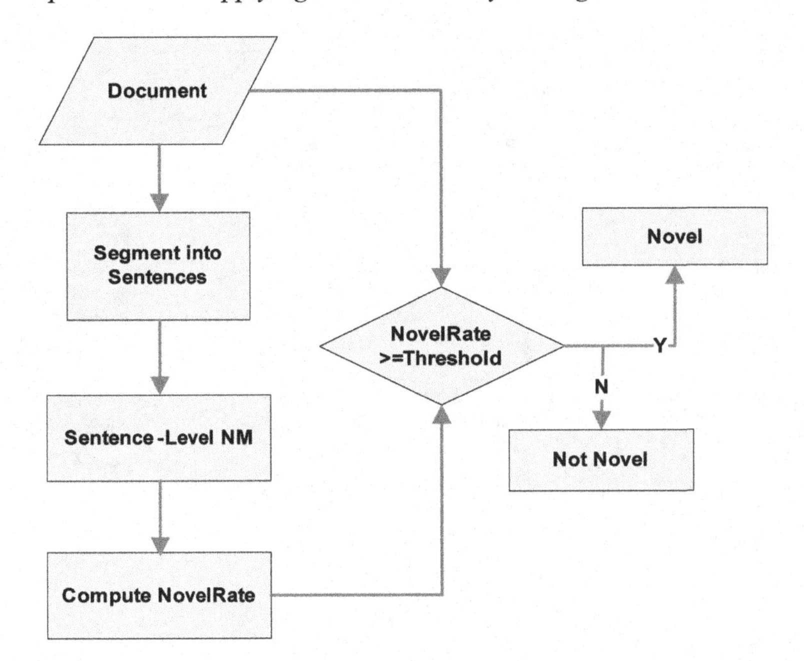

all sentences coming from one document are content-relevant to the topic. This assumption is the prerequisite for sentence-level novelty mining in D2S.

The novelty of the sentences in one document can reflect the novelty of the document. It simulates the way users view novelty. It is easier for an assessor to identify a new document as a redundant one with several specific previously seen sentences. Then, the novelty of a new document was set to the proportion of its novel sentences which are judged through comparing the similarity with all history sentences.

Therefore, we define NovelRate to measure the novelty of a document. If the NovelRate of one document is larger than a predefined threshold, we judge it to be a novel document. The definition of NovelRate is:

Equation 2

$$NovelRate = \frac{count(novelsentence)}{count(allsentence)}$$

Where *count(allsentence)* is the number of sentences in one document and *count(novelsentence)* is the number of novel sentences which are predicted by sentence-level novelty mining for the document.

The novelty mining process of D2S can be seen in Figure 6. First, we segment the documents into sentences and apply sentence-level novelty mining on these sentences. Then based on the prediction of each sentence, we can calculate NovelRate for each document. If the novel rate of one document is larger than the threshold, it will be predicted as a novel document.

CONCLUSION AND FUTURE WORK

This paper described the design of adaptable services for intelligent novelty mining which, to the best of our knowledge, has not been reported in previous studies. We described services for novelty mining, such as novelty scoring, novelty threshold, novelty feedback, and document-to-sentence technique that can meet the actionable demands of a real-life enterprise. By making this service-oriented application online, it enables users to access it from anywhere using the Internet protocol. Features of this service-oriented application were provided so that users can use this application effectively. The services can benefit the users who want to keep track of novel information of a given topic, but do not want to waste time going through all possible documents. To integrate the technical and business concerns between the classic novelty mining methodologies and business expectations, we performed human-centered user studies. The proposed services are thus adaptable to the diverse needs of a realistic enterprise application. By considering the domain-specific knowledge and real-world user requirements, our intelligent novelty mining services can help to discover novelty that can support enterprise users who are taking decision-making actions. Future work will focus on implementing and deploying the services in a real-world environment which can benefit users in a business context. Further studies can be conducted on the limitations of the proposed services in resource-constrained systems such as mobile devices, which may involve offloading the heavy computation onto the Web server instead of the user's system.

REFERENCES

Allan, J., Wade, C., & Bolivar, A. (2003). Retrieval and Novelty Detection at the Sentence Level. In. *Proceedings of SIGIR, 2003*, 314–321.

Chen, Y., Tsai, F. S., & Chan, K. L. (2007). Blog search and mining in the business domain. In *Proceedings of the 2007 International Workshop on Domain Driven Data Mining (DDDM '07)* (pp. 55-60).

Duraisamy, S. (2008). *SOAP*. Retrieved from http://searchsoa.techtarget.com/sDefinition/0,sid26gci214295,00.html

Hugo, H., & Allan, B. (2004). *Web Services Glossary*. Retrieved from http://www.w3.org/TR/wsgloss/

James, K. (2001). *Overview of WSDL*. Retrieved from http://developers.sun.com/appserver/reference/techart/overviewwsdl.html

Kwee, A. T., & Tsai, F. S. (2009). Mobile Novelty Mining. *International Journal of Advanced Pervasive and Ubiquitous Computing, 1*(4), 43–68.

Kwee, A. T., Tsai, F. S., & Tang, W. (2009). *Sentence-level Novelty Detection in English and Malay* (LNCS 5476, pp. 40-51).

Liang, H., Tsai, F. S., & Kwee, A. T. (2009). Detecting Novel Business Blogs. In *Proceedings of the Seventh International Conference on Information, Communications, and Signal Processing (ICICS)*.

Ng, K. W., Tsai, F. S., & Goh, K. C. (2007). Novelty Detection for Text Documents Using Named Entity Recognition. In *Proceedings of the 2007 Sixth International Conference on Information, Communications and Signal Processing* (pp. 1-5).

Ong, C. L., Kwee, A. T., & Tsai, F. S. (2009). Database Optimization for Novelty Detection. In *Proceedings of the Seventh International Conference on Information, Communications, and Signal Processing (ICICS)*.

Robertson, S., & Soboroff, I. (2002). *The TREC 2002 Filtering Track Report*. Paper presented at TREC 2002 - the 11th Text Retrieval Conference.

Soboroff, I. (2004). *Overview of the TREC 2004 Novelty Track*. Paper presented at TREC 2004 - the 13th Text Retrieval Conference.

Tang, W., Kwee, A. T., & Tsai, F. S. (2009). Accessing Contextual Information for Interactive Novelty Detection. In *Proceedings of the European Conference on Information Retrieval (ECIR) Workshop on Contextual Information Access, Seeking, and Retrieval Evaluation*.

Tang, W., & Tsai, F. S. (2009). Threshold Setting and Performance Monitoring for Novel Text Mining. In *Proceedings in Applied Mathematics 3 Society for Industrial and Applied Mathematics - 9th SIAM International Conference on Data Mining 2009* (pp. 1310-1319).

Tang, W., Tsai, F. S., & Chen, L. (2010). *Blended Metrics for Novel Sentence Mining*. Expert Syst. Appl.

Tsai, F. S. (2009). Network intrusion detection using association rules. *International Journal of Recent Trends in Engineering, 2*(1), 202–204.

Tsai, F. S., & Chan, K. L. (2007). Detecting cyber security threats in weblogs using probabilistic models. *Intelligence and Security Informatics, 4430*, 46–57. doi:10.1007/978-3-540-71549-8_4

Tsai, F. S., & Chan, K. L. (2010). *Redundancy and novelty mining in the business blogosphere*. The Learning Organization.

Tsai, F. S., Chen, Y., & Chan, K. L. (2007). *Probabilistic techniques for corporate blog mining* (. *LNCS, 4819*, 35–44.

Tsai, F. S., Etoh, M., Xie, X., Lee, W.-C., & Yang, Q. (2010). Introduction to Mobile Information Retrieval. *IEEE Intelligent Systems, 25*(1), 11–15. doi:10.1109/MIS.2010.22

Tsai, F. S., Han, W., Xu, J., & Chua, H. C. (2009). Design and Development of a Mobile Peer-to-peer Social Networking Application. *Expert Systems with Applications, 36*(8), 11077–11087. doi:10.1016/j.eswa.2009.02.093

Tsai, F. S., Tang, W., & Chan, K. L. (2010). *Evaluation of Metrics for Sentence-level Novelty Mining*. Information Sciences.

WhatIs.com. (2003). *UDDI*. Retrieved from http://searchsoa.techtarget.com/sDefinition/0,sid26gci508228,00.html

Wikipedia. (2009). *Web Service*. Retrieved from http://en.wikipedia.org/wiki/Webservice#Stylesofuse

Yee, K. Y., Tiong, A. W., Tsai, F. S., & Kanagasabai, R. (2009). OntoMobiLe: A Generic Ontology-centric Service-Oriented Architecture for Mobile Learning. In *Proceedings of the Tenth International Conference on Mobile Data Management (MDM) Workshop on Mobile Media Retrieval (MMR)* (pp. 631-636).

Zhai, C., Jansen, P., Bai, S., Stoica, E., Grot, N., & Evans, D. A. (1999). Threshold calibration in CLARIT adaptive filtering. In *Proceedings of Seventh Text Retrieval Conference TREC-7* (pp. 149-156).

Zhang, H.-P., Xu, H.-B., Bai, S., Wang, B., & Cheng, X.-Q. (2004). Experiments in TREC 2004 Novelty Track at CAS-ICT. In *Proceedings of TREC 2004 - the 13th Text Retrieval Conference.*

Zhang, Y., & Callan, J. (2001). Maximum Likelihood Estimation for Filtering Thresholds. In. *Proceedings of SIGIR, 2001*, 294–302.

Zhang, Y., Callan, J., & Minka, T. (2002). Novelty and Redundancy Detection in Adaptive Filtering. In. *Proceedings of SIGIR, 2003*, 81–88.

Zhang, Y., Kwee, A. T., & Tsai, F. S. (2010). Multilingual Sentence Categorization and Novelty Mining. *Information Processing and Management: an International Journal* .

Zhang, Y., & Tsai, F. S. (2009a). Chinese Novelty Mining. In *Proceedings of the Conference on Empirical Methods in Natural Language Processing (EMNLP '09)* (pp. 1561-1570).

Zhang, Y., & Tsai, F. S. (2009b). Combining Named Entities and Tags for Novel Sentence Detection. In *Proceeding of the WSDM '09 Workshop on Exploiting Semantic Annotations in Information Retrieval (ESAIR '09)* (pp. 30-34).

Zhang, Y., & Tsai, F. S. (2010). *D2S: Document-to-Sentence Framework for Novelty Detection*. Tech. Rep.

Zhao, L., Zheng, M., & Ma, S. (2006). The Nature of Novelty Detection. *Information Retrieval*, *9*, 537–541. doi:10.1007/s10791-006-9000-x

Zheng, G., & Bouguettaya, A. (2009). Service Mining on the Web. *IEEE Transactions on Service Computing*, *2*(1), 65–78. doi:10.1109/TSC.2009.2

This work was previously published in International Journal of Systems and Service-Oriented Engineering, Volume 1, Issue 2, edited by Dickson K.W. Chiu, pp. 79-95, copyright 2010 by IGI Publishing (an imprint of IGI Global).

Chapter 16
Engineering Financial Enterprise Content Management Services:
Integration and Control

Dickson K. W. Chiu
Dickson Computer Systems, China

Patrick C. K. Hung
University of Ontario Institute of Technology, Canada

Kevin H. S. Kwok
The Chinese University of Hong Kong, China

ABSTRACT

The demand is increasing to replace the current cost ineffective and bad time-to-market hardcopy publishing and delivery of content in the financial world. Financial Enterprise Content Management Services (FECMS) have been deployed in intra-enterprises and over the Internet to network with customers. This paper presents Web service technologies that enable a unified scalable FECMS framework for intra-enterprise content flow and inter-enterprise interactions, combining existing sub-systems and disparate business functions. Additionally, the authors demonstrate the key privacy and access control policies for internal content flow management (such as content editing, approval, and usage) as well as external access control for the Web portal and institutional programmatic users. Through the modular design of an integrated FECMS, this research illustrates how to systematically specify privacy and access control policies in each part of the system with Enterprise Privacy Authorization Language (EPAL). Finally, a case study in an international banking enterprise demonstrates how both integration and control can be achieved.

DOI: 10.4018/978-1-4666-1767-4.ch016

INTRODUCTION

Enterprise Content Management (ECM) refers to the management of textual and multimedia content across and between enterprises (Tyrväinen et al., 2003). In the context of the Financial Enterprise Content Management Services (FECMS), content refers to the pieces of information in the enterprise, including financial research, market commentary, calendar events, trading ideas, bond offerings, and so on. Recently, internal FECMS, as well as external content portals for customer access, have been deployed to replace the current cost ineffective and bad time-to-market hardcopy release of content delivery in the financial world. Published content contributes highly to customer relationship management (CRM) (Tiwana, 2001), as this is an important value-added service to clients in the financial industry, such as brokerage firms (Chiu et al., 2003). Content produced by an analyst of a financial enterprise often provides valuable advice for the decision making of client investors, and therefore has a high impact on the image, reputation, and professionalism of the enterprise. In addition, content received or composed is also used throughout the enterprise for internal decision making. Knowledge is power. As knowledge and organizational memory can be captured in enterprise content, access to content is an effective source of knowledge (Küng et al., 2001). A good ECM system can produce high return on investment, which is a valuable asset to the enterprise (McNay, 2002). Thus, this is especially important for financial enterprises.

Integration, instead of building from scratch, is the preferred strategy in building large enterprise information systems as demonstrated in our case study in a large international banking enterprise (Kitayama et al., 1999; Edwards et al., 2000). However, the management of such a large volume of content and such a complex system is non-trivial. For a global system with multiple sites, it is a big challenge to provide a mechanism for content analysts all over the world to contribute commentary that they will publish on the Web in a timely way. The maximum time to market a commentary should be within minutes, because its intrinsic value depreciates exponentially. Nevertheless, an important contradicting requirement is that editors and auditors must check content publication against any possibility of violating laws and regulations, which vary across countries, and even States. In this paper, we demonstrate how contemporary Web service technologies can facilitate such conflicting objectives of integration and control.

With an integrated FECMS deployed for both internal and external users, risks appear if there is inadequate control. In this context, privacy and access control is the focus of concern. For example, malicious or even un-intentional alternations to financial content may not only cause disasters to internal management decisions, but also affect valuable external client investors. The latter case might lead to severe damage of enterprise reputation or even legal responsibilities as FECMS contains a large amount of sensitive and confidential information. Access control technologies can also reinforce management control as demonstrated later in this paper, while privacy issues often go hand-in-hand with access control (Powers et al., 2002). In particular, there are usually additional legal and trade requirements for financial institutions, such as the U.S. Privacy Act of 1974 (Davis, 2002) as a result of the sensitivity and value of the customers' information.

To the best of our knowledge, no previous, comprehensive studies regarding FECMS reporting exist on how the conflicting requirements of integration and control can be facilitated with technologies. We present a holistic approach to the problem in this paper, based on the previous studies of Kwok and Chiu (2004) and Chiu and Hung (2005). The coverage of this paper is the description and analysis of the following: (i) requirements and technical problems of ECM in the financial industry, (ii) a methodology to elicit such requirements, (iii) an enhanced FECMS architecture for such an environment, (iv) the design of FECMS components for secured internal content flow management and external access, and (v) a comprehensive case study with detailed illustration of how various Web service technologies can streamline the main objectives of integration and control.

To reach these objectives, we organize our paper as follows. Section 2 introduces an overview of the FECMS background. Section 3 surveys related work. Section 4 presents the overall system architecture for integration, and Section 5 presents our approach to address the privacy requirements. Section 6 details the design and implementation of the FECMS components. Section 7 discusses how our approach facilitates the management's goals. We then conclude our paper in Section 8 with further research issues.

FECMS BACKGROUND AND OVERVIEW

First, we introduce some common terms used in an FECMS before discussing the main requirements for the stakeholders.

Tagging refers to the labeling of content for easy classification, search, and retrieval. Tags can be thought of as index entries (meta-data) with specified values linked to a piece of content. All content are tagged when it is created. Some tags can be defined automatically by inference (for example, Country=China implies Region=Asia) or by *templating*, while others may need to be selected from a list of valid tags or specified by the author or editor. Templating refers to functionality for an individual to be able to save any particular piece of content information template for future use by the individual or the group.

Taxonomy refers to the overall structure and organization of tags across the enterprise. It is the basic mechanism for tiering, entitlement, and filtering of content. The taxonomy should reflect the creators' view on what is important about any piece of content, as well as the users' view. In addition, it enables all content to be organized in a way that facilitates CRM activities, such as cross-selling, up-selling, and increase in customer orientation (Tiwana, 2001). While the enterprise should maintain a consistent global repository of taxonomy, different business units may also have their own local taxonomies (e.g., language, terminologies, and regulatory difference). Some sort of mapping is required before delivery to different business units or external parties. For example, in a securities' world, product is regional/exchange base, such as Japan/Nikkei, US/NASDAQ/NYSE, Hong Kong/HKSE, and so on. But in other business units, products normally mean the financial institution provided instruments, such as Foreign Exchange Swap and Corporate Bonds. So, we have to re-map these tags to maintain the taxonomy ontology.

Entitlement is the ability to ensure that different types of customers and customers of different values are offered appropriate levels of service. Tiering is the ability to offer different levels of service (by providing access to different sets of content) to customers of different values.

Based on a study of an international banking enterprise, Figure 1 depicts an overview of an FECMS, highlighting the main system components and stakeholders. The design of an FECMS must specifically match the need and interest of each stakeholder within and related to the enterprise (Chiu & Kwok, 2004). Besides the management, there are four main types of stakeholders involved, namely, *Content Creators*, *Content Providers*, *Content Distributors*, and *Content Users*.

Content Creators collectively refer to internal users who are involved in the content creation processes of the enterprise. The FECMS should be able to accommodate the different operational and administrative requirements of these different roles of internal users, and maintain appropriate security control. They interact mainly with Content Editorial Engines of the FECMS. Content Creators include the following roles:

Figure 1. Overview of an FECMS

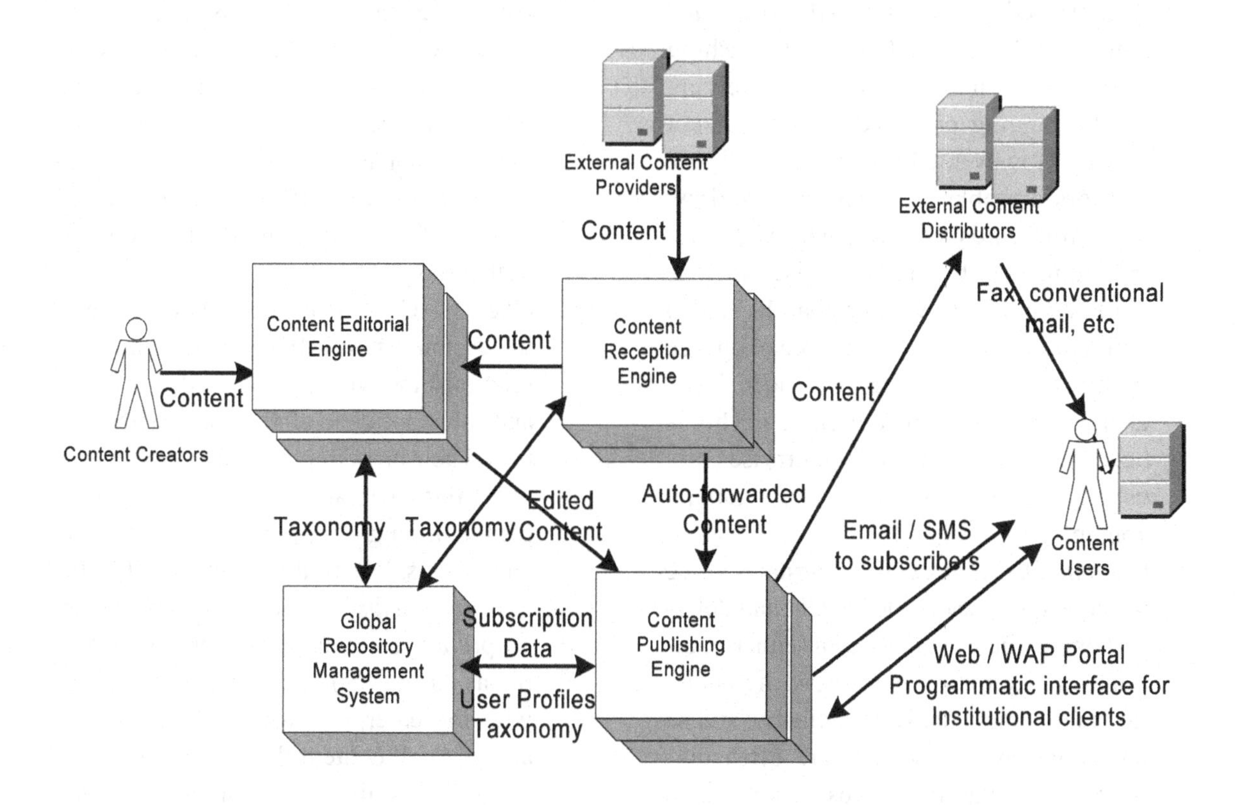

- *Authors* compose content or publish content for analysts, in addition to providing initial tiering and tagging of the content. Content creation privilege is limited according to different roles, and different users can create different sets of content as classified by tags. Also, content flow is based on the user privilege and the type of content. Some users (such as unit heads) may bypass the editorial or even the approval process but others cannot. Some content types allow straight-through processing but others may need multi-level approval. The system must be flexible enough to handle these variations in the content flow.
- *Editors* are power users who review content and tagging from authors or external sources. They also rectify this if necessary.
- *Approvers* review others' content. All approvers are categorized by a business unit, that is, content created by a certain business unit requires approval from a particular group of approvers.
- *Auditors* review the content for the company's interest, along with compliance to laws and regulations. This is different from approvers who can only stop pending content, auditors can pull any piece of content back even if it has already been published.
- *Administrators* are super users to manage the overall operation of content creation. Administrators also maintain local or global taxonomy.
- *Content Providers* are external sources (such as Reuters and Bloomberg) providing content (such as news, stock quotes, indices, and interest rates) to the enterprise through a Content Reception

Engine. To ensure timeliness, content from trusted sources is usually forwarded automatically to the Content Publishing Engine for immediate delivery, relying on the tagging provided by the content source. However, editors and compliance auditors are able to review or withdraw them afterwards. On the other hand, content composed by the enterprise (such as market commentary and research) is also delivered to these providers free of charge (public research), on a per piece basis charge, or as a lump sum charge. This is because major financial enterprises are usually an important source of financial content.

- *Content Distributors* are external service providers that render the content and deliver them to clients via different (either traditional or electronic) channels, such as mass fax, mail, email, hardcopy delivery, and so on. Nowadays, these jobs are often outsourced. Though this is costly, traditional services need to be maintained because of some clients' needs and their extra service payment.
- *Content Users* can be internal or external to the enterprise, and are classified into five tiers in our case. In particular, content services to these external users are very important CRM activities. Content Users obtain their access through a Content Publishing Engine. They are maintained by an enterprise-wide Global Repository Management System. Based on their subscription data, the Content Publishing Engines also actively send appropriate content to the subscribed users. The five tiers are:
- *Public Visitors* – Anonymous users are often allowed to access some limited amount of public content through a portal. This helps attract them to visit the enterprise's Web site.
- *Registered Visitors* – Potential customers who have not yet been using the enterprise's services are attracted to register by the usefulness of the content. After registration, the enterprise knows more of the details of potential customers and therefore can perform more effective service recommendations and other marketing activities to them.
- *Clients* – Customers (such as retail banking customers or SME) with basic business relationships who are allowed full access and subscription to all the unrestricted content. Their browsing and subscription provides further input to an analytical engine for the mining of opportunities for up-sale and cross-sale activities (Tiwana, 2001).
- *Priority Clients* – Premier customers (such as private banking customers or institutional customers) with deep relationships with the enterprise who are allowed full access to all content that are not classified as "internal only". Programmatic access of contents for institutional customers should be supported.
- *Internal Users* – Internal staff can access "internal only" content related to them, as well as all the content for external users. They are also automatically subscribed to relevant content, according to their job functions, market sector, geographical location, seniority, and so on. Based on similar criteria, further access control may be imposed.

LITERATURE REVIEW

Enterprise Content Management (ECM) is an emerging research area. Tyrväinen et al. (2003) give an excellent concise introduction to the research issues in this area, which mainly include technical, user, process, and content perspectives. McNay (2002) presents an overview of ECM and stresses the need of an ECM system with consistent tagging to ensure a timely-updated, well-organized Web site. However, the paper does not cover any design of such an ECM system.

Croll et al. (1997) point out that the trading of content between broadcasters requires descriptive data and some versions or illustrations of the content to be quickly assessed. The commitment should be confirmed and honored with minimal delay and administration, despite the complex content ownership and legal issues. Their Atman project attempts to model content trading using both archived programs and live events coverage as examples. Some of their requirements are similar to our FECMS but in a different application domain. However, available technologies nowadays can provide a much more sophisticated framework for similar applications.

Fensel (2001) and Omelayenko (2001) relate the challenges in inter-enterprise content management to business-to-business (B2B) electronic commerce in the context of product information integration and ontology in electronic marketplaces. Küng et al. (2001) relates knowledge management to enterprise Web content management with focus on superimposed information and domain ontology. They employ a *Topic Maps* approach in their system architecture because the underlying abstract model provides a high degree of power and flexibility to combine these approaches by supporting evolutionary construction of computer-based organizational memories. There is a large amount of research on the topic of ontology in the context of Semantic Web (Berners-Lee et al., 2001), and therefore, taxonomy ontology is not the focus of this paper.

Surjanto et al. (2000) introduce XCoP (XML Content Repository) as a repository based on an object-relational database management system to improve content management of eXtended Markup Language (XML) documents, thereby exploiting their structural information. Arnold-Moore et al. (2000) describe the data model for implementing an XML-native content management server and the requirements for supporting text-intensive applications. However, these works present mainly technical details of a content repository. Weitzman et al. (2002) present the Franklin Content Management System, developed by IBM's Internet Technology Group with XML technologies. Their goals are content reusability, simplified management of content and design that enforces integrity and consistency, the customization of content to individual users, and the delivery of content to a variety of display devices. However, multi-engine and heterogeneous engine integration issues essential for scalability and interoperability are not covered.

Chiu et al. (2003) discuss the requirements of customer relationship management for SME stock brokerage in Hong Kong, and propose an event-driven approach to ensure efficiency and timeliness in converting knowledge into business actions effectively. One such action is to relay received stock price and market news content to relevant customers. This means ECM helps CRM. This motivates a more in-depth research on a large-scale ECM context, as presented by Kwok and Chiu (2004) and Chiu and Hung (2005), as well as in this paper.

Only until recently have studies in RBAC for documents been started. Tiitinen (2003) proposes a methodology based on roles to analyze the requirements of individual and organizational users of documents as well as those of organizational needs related to security and access control. Bertino et al. (2002) describe Author-X, a Java-based system for discretionary access control to XML documents. Author-X supports a set-oriented and a document-oriented, credential-based document protection, a differentiated protection of document/document type contents through multi-granularity object protection, and positive/negative authorizations together with different access control strategies.

In the past few years, there are increasing demands and discussions about privacy access control technologies for supporting different business applications. For example, the Platform for Privacy Preferences Project (P3P) working group at the World-Wide-Web Consortium developed the P3P specification for enabling Web sites to express their privacy practices (Stufflebeam et al., 2004). On the other hand, a P3P user agent allows users to automatically be informed of site practices and to automate decision-making based on the Web site's privacy practices. Thus, P3P also provides a language called P3P Preference Exchange Language 1.0 (APPEL1.0), to be used to express user's preferences for making automated or semi-automated decisions regarding the acceptability of machine-readable privacy policies from P3P enabled Web sites. On the other hand, IBM proposes the Enterprise Privacy Authorization Language (EPAL) technical specification to formalize privacy authorization for actual enforcement within an intra- or inter- enterprise for business-to-business privacy control. EPAL services exchange privacy policies and make privacy authorization decisions. In particular, EPAL concentrates on the privacy authorization by abstracting data models and user-authentication from all deployment details. Similarly, eXtensible rights Markup Language (XrML) is used to describe the rights and conditions for owning or distributing digital resources. XrML concepts include license, grant, principal, right, resource, and condition (Wang et al., 2002). Based on the specification of licenses, the XrML agent can determine whether to grant a certain right on a certain resource to a certain principal or not.

Currently in the financial industry's widespread but scattered efforts of in-house development of ECM systems are emerging, because the value of FECMS has recently appreciated. Such valuable solutions, tailor-made for individual enterprises, have been treated as trade secrets, and therefore are rarely published. In summary, none of the existing research papers have discussed a detailed enterprise-wide architecture that can adequately support a complete content flow. Inter- and intra- enterprise ECM system integrations issues are almost unexploited. To the best of our knowledge, the study of privacy and access control in these systems has not been published.

SYSTEM ARCHITECTURE AND INTEGRATION

Based on the requirements discussed in Section 2, Figure 2 summarizes our Enterprise Content Model in a Unified Model Language (UML) (Carlson, 2001) class diagram, highlighting the main entities and their relationships in an FECMS. Figure 3 depicts an overview of an FECMS, highlighting the main privacy and access control over stakeholders, namely, *Content Creators*, *Content Providers*, *Content Distributors*, and *Content Users*. The main components of this architecture are three major types of engines: Content Editorial Engines, Content Reception Engines, and Content Publishing Engines. A Content Editorial Engine provides content and taxonomy maintenance and approval functions for different levels of administrators in the financial institute. A Content Reception Engine collects content from external sources, and then delivers it to different parts of the system for approval and publication. A Content Publishing Engine stores approved content, which it then sends to different parties via different channels (such as email, fax, and conventional mail). It also serves as the Web storefront of the enterprise for user enquiries. In addition, the Global Repository Management System provides backing support for user information and taxonomy.

The FECMS supports a highly heterogeneous environment with multiple data sources, external systems of business partners, and enterprise information systems (EISs). In order to allow e-commerce activities to be carried out based on a set of XML standards such as Simple Object Access Protocol (SOAP), Universal Description, Discovery and Integration (UDDI), and Web Services Description Language (WSDL) (Linthicum, 2003) we have employed Web services in information system integration (Aversano et al., 2002). The benefits of adopting Web services include faster time to production, convergence of disparate business functionalities, a significant reduction in the total cost of development, and

Figure 2. An enterprise content model in UML class diagram

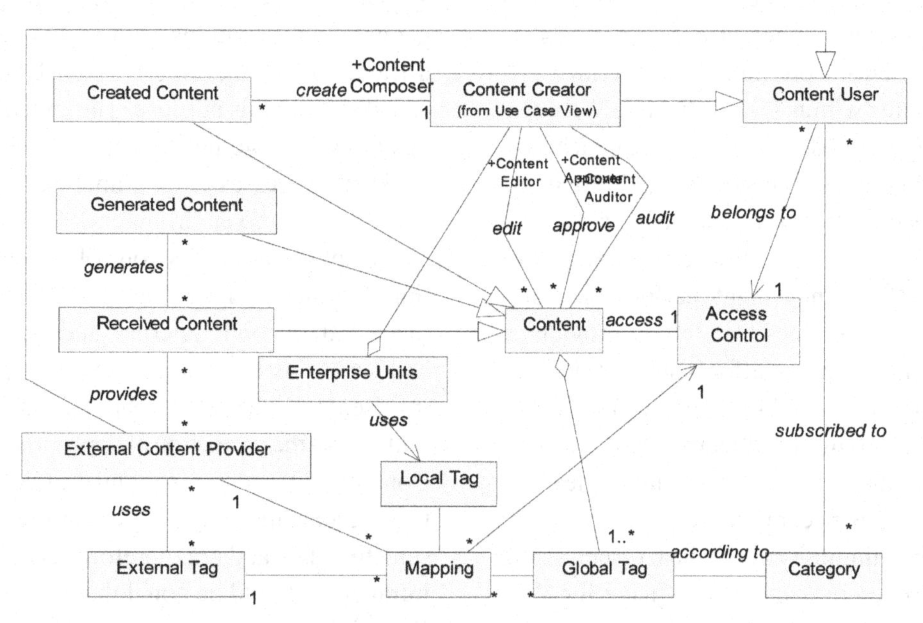

Figure 3. An overview of an FECMS architecture

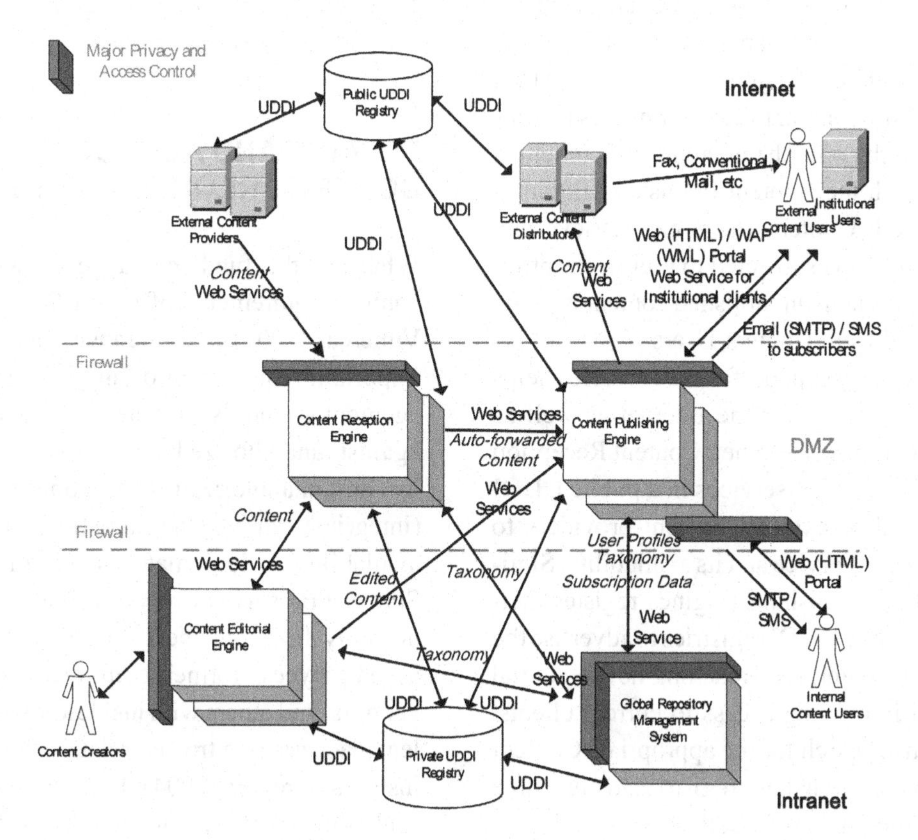

easy to deploy business applications for trading partners (Ratnasingam, 2002). In addition, Web services provide a convenient architecture to support both human and programmatic interfaces. At the same time, within the enterprise, Web services can provide loosely coupled communications among autonomous sub-systems. Multiple instances of each sub-system can be hosted at different sites (possibly in different geographical areas) for better management, performance, and scalability. Web services technology provides a firewall friendly, open platform supporting both synchronous (such as WS-Transaction) and asynchronous messaging (Linthicum, 2003). Further, the combination of workflow technologies and Web services has become more popular in both the research community and industry, such as the Business Process Execution Language for Web Services (Erl, 2006).

In contrast to traditional databases, enterprise content is much larger in volume but not all content is useful to all users. In particular, working, incomplete, unedited, unapproved content is often useless or could be dangerous if released. Thus, isolation of the Content Editorial Engines is well-justified. On the other hand, approved content is published with the Content Publishing Engines, through which Content Users can retrieve and share effectively is subjected to security control. With this approach, unpublished content is much more isolated from Content Users.

UDDI registry support for Web services helps in service location over the Internet as well as within the enterprise intranet. Content Reception Engines register their services in a public UDDI registry to allow external content providers to query its type of service and its availability. Similarly, Content Publishing Engines register their services in public UDDI registries to advertise the service of the enterprise, reaching new potential clients and improving access to current clients. They can also search for the appropriate Content Distributors through these registries to outsource any conventional content delivery.

In a global enterprise, a large number of internal systems, services from various units possibly at geographical locations, and their interfaces are difficult to keep track of. A private UDDI registry can also serve this purpose. Thus, UDDI technology helps manage and describe services as well as business processes programmatically in a single, open, and secure environment.

For physical access control, Content Reception Engines and Content Publishing Engines (accessed by both internal and external users and served as information gateways) are set up in a Demoralized Zone (DMZ) (Shimonski et al., 2003) with the protection of appropriate firewalls. Content Editorial Engines and the Global Repository Management System contain valuable and sensitive data and are therefore set up within the intranet. It should be noted that there is only one Global Repository Management System in place to maintain users' access to various global and regional Web sites as a single entity (compare airline companies' portals), together with globally consistent content taxonomy.

PRIVACY AND ACCESS CONTROL REQUIREMENTS

What are the fundamental privacy and access control requirements of FECMS? Samarati and Vimercati (2001) stated the answer as follows: "An important requirement of any information management system is to protect data and resources against unauthorized disclosure (confidentiality) and unauthorized or improper modification (integrity), while at the same time ensuring their availability to legitimate users." In the United States, FBI surveys reported that threats come not only from outsiders, but also insiders such as employees, former employees, contractors, vendors, and others with inside knowledge, privileged access, or a trusted relationship with other insiders (Gregory, 2004). In order to circumvent such threats, three fundamental requirements, confidentiality, integrity, and availability, have to be addressed:

- Confidentiality: Confidentiality is the assurance that sensitive information is not disclosed. The confidentiality of FECMS is violated when unauthorized parties obtain protected content over the Internet.
- Integrity: Integrity prevents the unauthorized modification of information. The integrity of FECMS is violated when the correctness and appropriateness of the content is modified, destroyed, deleted, or disclosed.
- Availability: Availability refers to the notion that information and services are not available for use when needed. The availability of FECMS is violated when the system is brought down or malfunctioned by attackers or intruders.

Privacy is a state or condition of limited access to a person (Schoeman, 1984). In particular, information privacy relates to an individual's right to determine how, when, and to what extent information about the self will be released to another person or to an organization (Leino-Kilpi et al., 2001). In general, privacy policies describe the following data practices of an organization: what information they collect from individuals (subjects), for what purpose the information (objects) will be used, whether the organization provides access to the information, who are the recipients of any result generated from the information, how long the information will be retained, and who will be informed in the circumstances of dispute. Therefore, information privacy is usually concerned with the confidentiality of sensitive information such as health and financial data, as well as personal identifiable information (PII). However, privacy control is more concerned with the policy enforcement than just the individual subjects because a subject releases his/her data to the custody of an enterprise while consenting to the set of purposes for which the data may be used (Fischer-Hübner, 2001).

Access control is the process of limiting access to the resources of a system to only those authorized users, programs, processes, or other systems on a need-to basis (Ferraiolo, 2001). In general, access control is defined as the mechanism by which users are permitted access to resources according to the authentication of their identities and the associated privileges authorization. Though access control technology can be directly applied in protecting PII data, privacy concepts, such as purpose and obligation, also have to be incorporated. The traditional view of an access control model should be extended with an enterprise wide privacy policy for managing and enforcing individual privacy preferences (Powers et al., 2002). In the U.S., the Privacy Act of 1974 (Davis, 2000) requires that federal agencies grant individuals access to their identifiable records that are maintained by the agency, ensure that existing information is accurate and updated in a timely manner, and limit the collection of unnecessary information and the disclosure of identifiable information to third parties. From a recent survey (Hinde, 2002), bank officers said that they had ongoing concerns, mostly procedural, about how to handle the anticipated privacy regulations of the U.S. Gramm-Leach-Bliley Act (GLB), which requires financial institutions to regularly communicate privacy policies to customers and provide adequate opportunities for "opting-out" of personal information disclosure to non-affiliated third parties.

The current FECMS's approach to access control with a simple entitlement and tiering is inadequate for the ever growing complexities especially after the system integration. Therefore, we proceed to an in-depth re-investigation of the privacy and access control issues in this paper. The steps taken are as follows:

1. Identify the new paradigm and technology required;

2. Identify the information entities to be protected;
3. Identify the entitlement and protection that should be imposed on the stakeholders;
4. By tracing the flow of the information entities to be protected, identify the processes during which such protection should be enforced and hence the detailed protection policies as well as the required enhancement to existing system components; and
5. Identify any modification of the existing content flow or content management process required.

Firstly, to overcome the limitations of simple entitlement and tiering, we employ a Role-based Access Control (RBAC) paradigm (Sandhu et al., 1999). The RBAC presents a conceptual model to describe different approaches such as base model, role hierarchies, constraint model, and consolidated model. The RBAC works by assigning content users to roles where roles are granted privileges for accessing different categories of content. The major advantages of the RBAC are its accuracy in specifying access control policies, and ease of management. To achieve this goal, the National Institute of Standards and Technology (NIST) conducted market analysis for identifying RBAC features into two layouts: the RBAC Reference Model and the RBAC Functional Specification (Ferraiolo, 2001). The RBAC Reference Model describes a common vocabulary of RBAC element sets and relations for specifying requirements together with the scope of the RBAC features included in the standard. The RBAC Functional Specification describes the requirements of administrative operations for creating and managing RBAC element sets and relations, as well as the system functions for creating and managing RBAC attributes on user sessions and making access control decisions. In particular, the proposed RBAC model with privacy-based extension in the next section is based on the core RBAC model discussed (Ferraiolo, 2003). To demonstrate this, we first present the core RBAC, which mainly includes the following entities:

- USERS, ROLES, OPS, and OBS represent users, roles, operations, and objects, respectively.
- UA $\subseteq$ USERS $\times$ ROLES is a many-to-many mapping between users and roles (that is, a user-to-role assignment relation).
- assigned_users: (r:ROLES) $\rightarrow 2^{USERS}$ is the mapping of role *r* onto a set of users. Formally, assigned_users(r) = $\{u \in USERS \mid (u,r) \in UA\}$
- PRMS = $2^{\{OPS \times OBS\}}$ is the set of permissions.
- PA $\subseteq$ PRMS $\times$ ROLES is a many-to-many mapping between permissions and roles (role-permission assignment relation).
- assigned_permissions(r:ROLES) $\rightarrow 2^{PRMS}$ is the mapping of role *r* onto a set of permissions. Formally, assigned_permissions(r) = $\{p \in PRMS \mid (p,r) \in PA\}$.
- SUBJECTS is the set of subjects.
- subject_user(s:SUBJECT) $\rightarrow$ USERS is the mapping of subject *s* onto the subject's associated user.
- subject_roles(s:SUBJECT) $\rightarrow 2^{ROLES}$ is the mapping of subject *s* onto a set of roles. Formally, subject_role(s_i) $\subseteq \{r \in ROLES \mid (subject\_user(s_i), r) \in UA\}$

Figure 4 presents an access control framework of the core RBAC with privacy-based extension such as purpose, recipient, obligation, and retention. When a request arrives at the access control system, in addition to a grant permission or a denial to the subject, the system returns a set of obligations and a retention policy according to privacy requirements as shown in Figure 4. The following sets of privacy-based entities are proposed to extend the core RBAC:

Figure 4. An access control framework of core RBAC with privacy-based extension

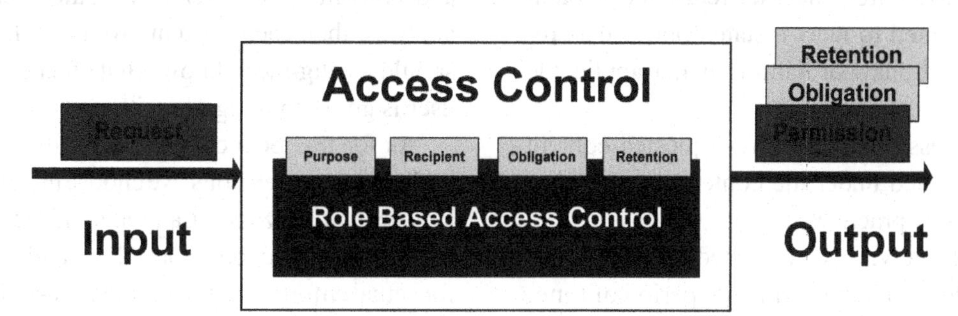

- PURPOSES = $\{pp_1, pp_2, \ldots, pp_n\}$ is the set of n purposes to describe a user's purpose(s) of content access to the FECMS. As most of the content contains sensitive information, the FECMS has to know the user's purposes of requesting datasets in order to make a decision of permission such as "Pseudo Financial Analysis," "Pseudo Financial Decision," and "Financial Research and Development."
- RECIPENTS = $\{rp_1, rp_2, \ldots, rp_n\}$ is the set of n recipients of the result generated by the set of collected object(s) such as the content. This is usually solicited by the content providers because each content provider has its own rules to guide the usage of datasets. Referring to the Platform for Privacy Preferences (P3P) specification (W3C, 2002), the recipient can belong to one of the following parties:
- Ours: the organization/user and/or entities acting as the content user's agents. An agent is defined as a third party that processes datasets only on behalf of the organization/user for the completion of the stated purposes.
- Other-recipient: legal entities constrained by, and accountable to the content user.
- Unrelated-third-parties: legal entities whose data usage practices are not known by the content user.
- Public-fora: any entity belonging to the public and able to access the result via different medium such as public directories and commercial directories.
- OBLIGATIONS = $\{obl_1, obl_2, \ldots, obl_n\}$ is the set of n obligations to be taken after the decision of permission is made. In general, an obligation is opaque and is returned after the permission is granted. The obligations describe the promises that a subject must make after gaining the permission. The contents are particularly related to the agreement or contract between the content user and provider. In an FECMS, an example obligation for a user that is granted with permission can be "the content must not be released to any unauthorized content user."
- RETENTIONS = $\{rt_1, rt_2, \ldots, rt_n\}$ is the set of r retention policies that are to be enforced in the object(s) in effect. Each content provider may have its own retention policy to enforce the usage of content. According to the P3P specification, the retention policy can be one of the following:
- No-retention: The requested content is not retained for more than a brief period of time necessary to make use of it during the session. The content must be destroyed following the session and must not be logged, archived, or otherwise stored.
- Stated-purpose: The requested content is retained to meet the stated purpose. This requires the content to be discarded at the earliest time possible.

- Legal-requirement: The requested content is retained to meet a stated purpose as required by law or liability under applicable law.
- Business-practices: The requested content is retained under the content user's stated business practices.
- Indefinitely: The requested content is retained for an indeterminate period of time.

In RBAC, a subject can never have an active role that is unauthorized for its user (Ferraiolo, 2001). With all the privacy-based extension (purposes, recipients, obligations, and retentions) discussed above, Ferraiolo (2003) revised the role authorization in the core RBAC model as follows:

- $\forall$ s:SUBJECTS, u:USERS, r:ROLES, op:OPS, $\{o_1, o_2, \ldots, o_i\} \subseteq$ OBS, $\{pp_1, pp_2, \ldots, pp_j\} \subseteq$ PURPOSES, $\{rp_1, rp_2, \ldots, rp_k\} \subseteq$ RECIPIENTS, $\{obl_1, obl_2, \ldots, obl_l\} \subseteq$ OBLIGATIONS, rt:RETENTIONS
- $r \in subject\_roles(s) \wedge u \in subject\_user(s) \Rightarrow u \in assigned\_users(r)$
- access: SUBJECTS $\times$ OPS $\times$ OBS $\times$ PURPOSES $\times$ RECEIPENTS $\rightarrow$ BOOLEAN $\times$ OBLIGATIONS $\times$ RETENTIONS
- access(s, op, $\{o_1, o_2, \ldots, o_i\}$, $\{pp_1, pp_2, \ldots, pp_j\}$, $\{rp_1, rp_2, \ldots, rp_k\}$) = (1, $\{obl_1, obl_2, \ldots, obl_l\}$, rt) if subject *s* can access any object in $\{o_1, o_2, \ldots, o_i\}$ using operation *op* for any purpose in $\{pp_1, pp_2, \ldots, pp_j\}$ with any recipient in $\{rp_1, rp_2, \ldots, rp_k\}$, (0, $\varnothing$, $\varnothing$) otherwise. If the access is granted, a set of obligations $\{obl_1, obl_2, \ldots, obl_l\}$ and also a retention policy *rt* are returned to the subject.

The main reason to extend the core access rule with a set of objects ($\{o_1, o_2, \ldots, o_i\} \subseteq$ OBS) is to deal with the specific operation "integrate" or "link" in FECMS. For example, if a content user requests to integrate content *A*, content *B*, and content *C* in order to execute a particular analysis, then each content provider (*A*, *B*, or *C*) should be eligible to know what other content the user is going to integrate with.

As for the technology for specifying RBAC with privacy extensions, we choose the Enterprise Privacy Authorization Language (EPAL). This is because EPAL not only formalizes authorizations for actual enforcement within an intra- or inter-enterprise for business-to-business privacy control, but also support privacy authorization. In addition, EPAL is an interoperability language for defining enterprise privacy policies on data handling practices in the context of fine-grained positive and negative authorization rights. Further, this choice is in line with the IBM dominated architecture of the financial enterprise that we studied.

In EPAL, there are two major components: vocabulary and policy. The EPAL vocabulary includes lists of hierarchies in data-categories, user-categories, and purposes, as well as sets of actions, obligations, and conditions. Data-categories are used to define different categories of collected data handled differently from a privacy perspective such as financial information. User-categories are used to describe the users or groups assessed collected data such as investors. Purposes are used to model the intended service for which data is used such as an investment. Actions are used to model how the data is used, such as buy and sell. Obligations are used to define actions that must be taken by the environment of EPAL such as "No personal data will be released to any unauthorized party." In particular, conditions are Boolean expressions, such as "all sellers must have signed the confidential agreement form." A vocabulary may be shared by more than one enterprise. Alternatively, the EPAL policy defines the privacy authorization rules that allow or deny the actions on data-categories, by user-categories for certain purposes under certain conditions while mandating certain obligations.

We propose to adopt EPAL as the language to specify the privacy access control policy in this application. Not only is EPAL one of the most promising privacy access control languages, but EPAL also satisfies the following properties of a privacy authorization model (Jones, 1995):

- Interoperable: A privacy authorization model should be able to interpret and use credentials issued by any other issuing authorities. In EPAL, the concepts of vocabulary and privacy policy are used to support the interoperability between a set of sector-specific enterprises. For example, enterprise A's privacy policy can refer to a vocabulary defined by enterprise B. Furthermore, enterprise A's privacy policy can always refer to more than one vocabularies from different enterprises.
- Expressive: Credentials should contain not only an individual identity but also other useful information. In EPAL, a vocabulary is used not only to describe the user's identity, but also to describe other privacy information in an XML format.
- Extensible: The credential system should be flexible enough to register new individuals and organizations with any new types of information. In EPAL, one can easily add new users or even define some new types of information in the vocabulary for implementing different requirements.
- Anonymous: An individual identifier should be kept secret in any circumstance. In EPAL, no user's personal identifier can be revealed in the vocabulary or the privacy policy.
- Scalable: Credential systems should be robust enough to handle the increasing number of users, service providers, and issuing authorities. In EPAL, administrators are always able to support a many-to-many relationship between vocabularies and privacy policies in a sector-specific environment.

Next, to streamline steps 2 and 3 and to later facilitate system maintenance, an effective approach is to elicit a comprehensive ontology for the information entities and stakeholders. Only after this can access control rules and privacy policies be specified systematically. In particular, the information entities to be protected are as follows:

- The major concern of an FECMS is naturally the vast amount of content;
- Almost equally important are the PII and profiles of content users (in particular customers);
- Users' activity records should also be protected because of privacy requirements. This is often inadequately handled in existing systems; and
- Content and user taxonomies though mostly visible to the content management software systems should be maintained only by specialists.

Referring to steps 4 and 5, different aspects of the information and content flows can be modeled in a single framework. As an indispensable part of any information system (Hung, 2001), information flow is the bridge between the information system and the users' activity model. Threats to privacy and access control are identified as events; an undesirable event that takes advantage of vulnerability. These threats can come from both insiders and outsiders within each organization (Fischer-Hübner, 2001). Based on the studies on workflow management of Hung (2001), possible threats to a content flow may include:

- Unauthorized disclosure, modification, and destruction of information;
- Unauthorized utilization and misuse of resources;
- Interruption, unknown status, and repudiation in content access;

- Denial of service from stakeholders or resources; and
- Corruption of stakeholders.

In the context of this paper, we concentrate on content access control as well as privacy protection regarding users' personal information and users' content access activity records. We identify the following strategies in our study regarding privacy and access control:

- Reception of contents into an FECMS should be adequately monitored and controlled;
- To maintain the privacy and integrity of content items, sophisticated content access control should be exercised over content creators and supervisors (such as editors, approvers, auditors, and administrators), according to the content flow and process requirements, based on a 'need-to-know' principle;
- Users' access to content should be managed with the role-based access control technology by matching users' roles and authorization with the classification of content items. Inference of tags should be supported in matching for ease of flexibility specification (e.g., subscription to Asia => China and HK, Stock => warrants);
- Users' personal information and profiles as well as their activity records should be protected for their privacy. Access control should be strictly restricted to the user himself and to user managers; and
- As for taxonomies' protection, the current approach of tight control for only specialists' access is adequate and therefore the issue is left for future research.

SUBSYSTEM DESIGN AND IMPLEMENTATION

In this section, we outline our design and implementation framework of the main system components of the FECMS. This includes the Content Reception Engines, Content Editorial Engines, Content Publishing Engines, and Global Repository. We focus on how to achieve integration while exercising privacy and access control in the content creation and flow.

Content Reception Engines

One main function of a Content Reception Engine is to receive content from various sources (with Java Messaging Services and Web services) and pre-processing (such as classification and machine analysis) as depicted in Figure 5. Receiving a content item (such as news or stock price update) is a typical event. In order to process the received content and generated information effectively, we use an event-driven approach in the design, centered on an Active Rule Module. The *event* is a significant occurrence that affects either the system or a user application. The approach we have chosen is motivated by the active database paradigm (Dayal, 1989; Chiu & Li, 1997).

The Environment Listener receives input from Content Providers. In addition to Web services, we have to support a common message system, the Java Message Service (JMS), for purposes such as connection to Reuter's services, until all information sources have been migrated to Web services. Traditionally, information of the Content Providers is only restricted to the Content Reception Administrators. Some content sources require customized programming on proprietary protocols. Worse still, some even restrict content reception with special clients only, and therefore, tricky wrapper programs are required for feeding them into the Content Reception Engine. With the gradual migration to Web services, a unified platform for reception can eventually be achieved. In addition, service from new Content Providers can be explored through public UDDI registries.

The Environment Listener mainly uses a publish-and-subscribe mechanism. Content Providers publish a summary of subscribe-able

Figure 5. Content reception engine and access control

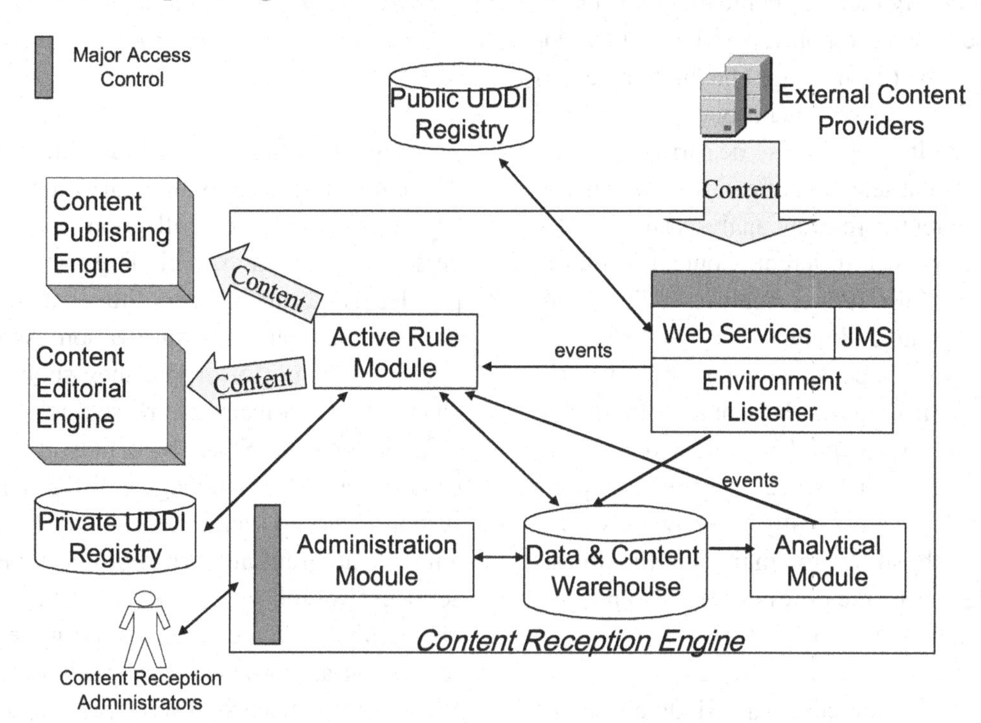

content at a public UDDI registry. This allows users to subscribe to content that is relevant or interesting to them, through the subscription Web service of the Content Provider. The Environment Listener's corresponding reception Web service is provided in order to receive the subscribed content. Thus, polling Content Providers is not necessary. Instead, when a new piece of content is ready, the Content Provider can actively send the content to the Environment Listener, generate an event to the Active Rule Module for processing, and store the received content into the Data and Content Warehouse.

The Data and Content Warehouse stores all the relevant information (including received content) and provides backing storage to the engine. The Analytic Module then analyzes the information in order to discover knowledge from the received content, such as summary reports calculations, market signal analysis, and so on. Since the knowledge discovered is very useful content, it is also stored in the Data and Content Warehouse and a corresponding event is also generated to the Active Rule Module for processing.

The Active Rule Module processes rules in the *event-condition-action* (ECA) format (Dayal, 1989), which specifies the constraints and actions to be taken upon reception and generation of new contents. When an event occurs, it triggers some rules with matching event specifications so that the model evaluates the *condition* parts of these rules. Conditions are logical expressions defined upon content status and information, such as tags and their values. Only if the *condition* is evaluated to be true, will the *action* part be executed, which could lead to other events. The semantics of ECA rules can be summarized by the following: *On* event *if* condition *then* action. As such, rules can be executed in a timely manner, avoiding the need of inefficient polling or ineffective batch processing. This also provides more flexible filtering and mapping capabilities than the traditional table-driven approach. The ECA rules can serve the following main purposes:

- Re-classify received content into the enterprise's own taxonomy, based on content information (such as tags and their values) as provided by the content source. Rule-based active filtering can also be carried out.
- Forward a selection of received or generated content to relevant analysts and Content Creators via different Content Editorial Engines for further analysis, editing, approval, and auditing.
- Forward selected content for immediate publishing via the Content Publishing Engine. Content from trusted Content Providers and some generated reports or signals are usually directly forwarded for publishing to maintain timeliness. Relevant Content Editors are also notified to continue the content flow.

We separate the Active Rule Module from the Analytical Module because the analytic engine should mainly deal with knowledge discovery, which tends to be resource intensive and computationally expensive. This cannot meet the timeliness requirements of urgent events for content processing. We also separate the Active Rule Module from other modules in order to manage the rules in a repository (the Content and Data Warehouse can also serve as the backing storage for the rules). This facilitates control and management of content sources for a particular Content Reception Engine.

The main concern in access control for a Content Reception Engine is that the reception of contents into an FECMS should be adequately monitored and controlled. We have to deal with protection against Content Providers. Strict verification and authorization before accepting new Content Providers is of vital importance. Here is the simplified EPAL code to illustrate:

```
<ALLOW
user-category = "Content_Creator"
data-category = "Any_Content"
purpose= "Distribution"
operation = "publish"
condition = "Authorization_Clearance
= TRUE">
```

Authorized Content Providers must connect through Web services with security tokens. Security tokens represent a collection of claims (i.e., personal information) including name, identity, privilege, and capability for the security services of authentication and authorization. As security tokens contain a lot of PII, they should be exchanged in a privacy-aware setting with other information in the SOAP body messages. OASIS proposes an XML language called Security Assertions Markup Language (SAML) for making authentication and authorization decisions at Web services (Rosenberg, 2004). Web services providers submit SAML messages to security servers for requesting authorization decisions. In addition, Web Services Security (WS-Security) describes enhancements to SOAP messaging to provide quality of protection through message integrity, message confidentiality, and single message authentication (Rosenberg, 2004).

In addition, Content Providers are authorized to provide only certain types of content based on tags. Content of unauthorized types are rejected. Sources maliciously flooding the system may even be totally rejected.

All accepted contents should normally be stored and logged. However, administrators can identify and mark problematic content providers. For example, when content approvers complain that they deliberately and repeatedly tag items wrong or send invalid content items, their subsequent content will be quarantined automatically and only accessible by designated specialists. This is because tags help classify content and therefore facilitate access control. For example, content items with specific tags that are related to sensitive topics, for example, politics and major market changes, are forwarded to, and only accessible by designated specialists for approval. Here is the simplified EPAL code for illustration:

Figure 6. Content editorial engine and access control

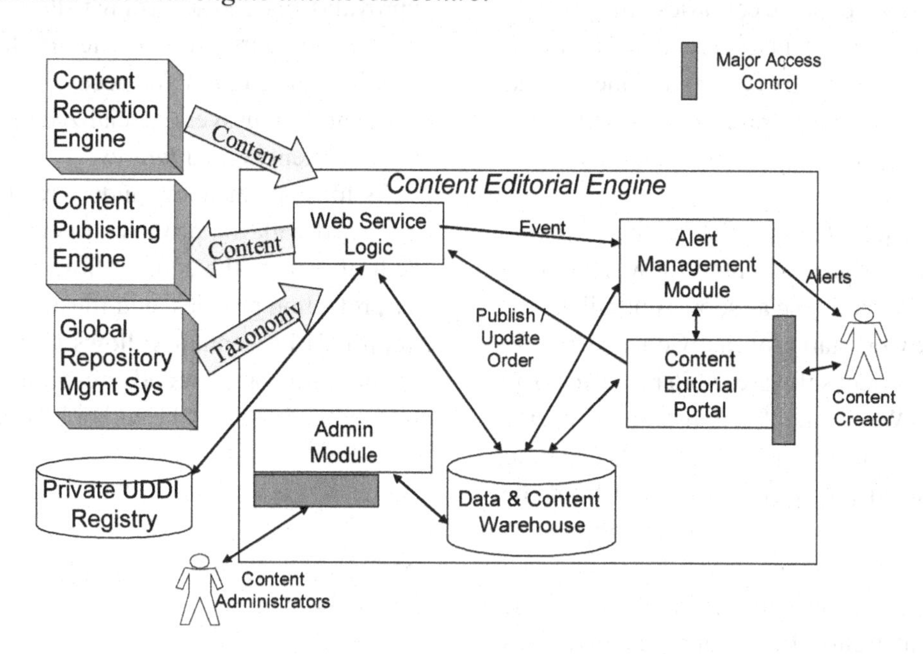


```
<DENY
user-category = "Content_Users"
data-category = "Politics_Content"
purpose= "Any"
operation = "access"
condition = "Designated_Specialists =
FALSE">
```

In addition, the Data and Content Warehouse stores all information, including received content, in addition to providing backing storage to the engine, such as active rules for filtering or forwarding content. Thus it must be protected. Only designated administrators are allowed to access such information and only through the administrative module.

Content Editorial Engines

The main function of the Content Editorial Engine is to support Content Creators in order to create new content and work with the received content, as depicted in Figure 6. The operations of the engine are mainly *alert* driven, as the timeliness requirement of financial content is crucial. *Alerts* are notification messages triggered by events, and managed by the Alert Management Module. The main function of the module is for alert routing and monitoring. Alerts further request the assigned user to carry out a job with time (urgency) constraints (Chiu et al., 2008). If the default Content Editor responsible for the next step of a job in the content flow is not available or too busy, it will attempt to route the job and alert a replacement Content Editor. It also monitors whether or not the assigned Content Editor begins working on, and finishing the job within a given timeframe. Otherwise, reminder messages will be sent to the assigned Content Editor. Further internal mechanisms of an alert management system can be found in the work of Chiu et al. (2008). A typical creation content flow is as follows.

1. A Content Author creates a piece of content, determines its tier and tags, and then sends it to a Content Editor for revision.
2. Normally after editing, the content is approved by a Content Approver.
3. However, if the Content Editor suspects that the content might violate the laws or

regulations of some countries, this piece of content will be sent to a Content Auditor of the affected country for a compliance check. Before the Content Auditor's approval, customers from those countries cannot receive or read it.

4. Once a job step is finished, the next one responsible in the content flow should be alerted to continue as soon as possible.
5. On the workstation of each Content Creator, special client software similar to *I Seek You* (ICQ) (Weverka, 2000) is installed to receive the alerts. Further details of alerts and their associated jobs can be retrieved from the Content Editorial Portal in the form of a job-list.
6. By opening an entry in the job-list, the Content Editor, by default, acknowledges that work has been started.
7. If the above does not occur, he/she may decline the job (say, because of conflict of interest or specialty mismatch) by pressing the cancel button. If so, the Alert Management Module will try to find another suitable Content Creator for the job.

The main concern in privacy and access control for a Content Editorial Engine is to maintain the integrity and privacy of content items. Sophisticated content access control should be exercised over content creators and supervisors (such as editors, approvers, auditors, and administrators), according to content flow and process requirements, based on the 'need-to-know' principle. Furthermore, all alerts, changes, removals, and approvals are logged.

Content received from Content Reception Engines are forwarded to appropriate content approvers by matching content tags with the capability of the content approvers. The advantage of using a capability matching approach (Chiu et al., 2001) is that classified materials can be handled with a tag (potentially called "classified"). To reinforce this policy, access control of content waiting for approval is guarded according to the capability of content approvers. At least one match in tag and capability is required for access. For example, only content approvers specialized in futures can access content tagged *futures*.

As for content items under creation, we are concerned with the privacy of content creation. We further trace the content flow, identify some key protection policies, and illustrate with some simplified EPAL code as follows. Content in progress may be incomplete and error prone. Thus, they are only accessible to the author before approval.

```
<DENY
user-category = "Content_Users"
data-category = "Any_Content"
purpose= "Any"
operation = "access"
condition = "Approval = FALSE">
```

Further, Content Creators cannot update content items submitted for editing, unless editors request their amendments, in the event that the content editor is updating it. Edited content can only be published after approval.

```
<DENY
user-category = "Content_Creator"
data-category = "Any_Content"
purpose= "Any"
operation = "update"
condition = "Approval_Status = SUB-
MITTED">
```

Content auditors can change or remove all content items classified under their capabilities plus regional restrictions.

```
<ALLOW
user-category = "Content_Auditors"
data-category = "Futures"
purpose= "Any"
operation = "update"
condition = "Restrictions = NONE">
```

Figure 7. Content publishing engine and access control

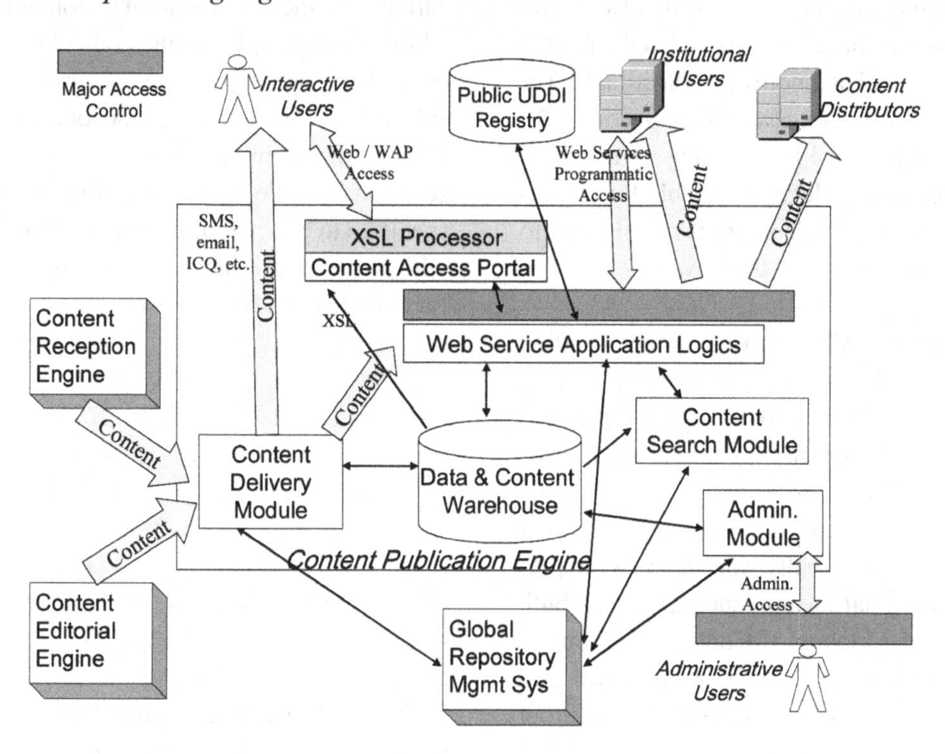

Supervisor override is necessary for work flexibility. Supervisors can be granted read access to all content items under their subordinates' work unless otherwise classified. However, updated access should require managerial approval because of accountability and possibility of confusion. The manager of a department can access all content items under work for that department.

```
<ALLOW
user-category = "Manager"
data-category = "Any_Content"
purpose= "Any"
operation = "access"
condition = "Departmental = TRUE">
```

The alert management module supports rerouting of work. For example, if the default Content Editor responsible for the next step of a job in the content flow is not available or too busy, it will attempt to route the job and alert to a replacement Content Editor. Supervisors can also manually reassign work. It should be noted that access rights of a rerouted content item must be updated accordingly.

Content Publishing Engines

The main function of a Content Publishing Engine is to send the new content to subscribed users and store them for later queries as depicted in Figure 7. In order to receive new content automatically, Content Users must subscribe to the relevant content categories beforehand. Interactive users may do this from the Content Access Portal, while programmatic subscription can be done through Web services interfaces. A Global Repository Management System maintains the user registrations and subscriptions so that users can interact with the enterprise as a single entity.

Upon receiving a new piece of content, the Content Delivery Module queries the User Registration and Subscription System for the list of relevant Content Users with Web services. The category of the content is determined by its tier and tags, which have been translated into the enterprise's global taxonomy (see Section 6.1) or defined by the Content Creators (see Section 6.2). Only users subscribed to the category and with adequate access privilege are legible for the delivery. In summary, the content is mainly sent to the user via the following three ways:

- Via email, SMS, and/or ICQ as specified by interactive users at subscription time.
- Via Web services to the access point as specified by programmatic (usually institutional) users.
- Indirectly through external Content Distributors, usually for those who prefer traditional means (such as fax or bulk mail). The lookup of Content Distributors is possible via public UDDI registries, and is communicated automatically without any human intervention.

Another main function of the engine is to support the Content Users' queries and browsing, which is mainly responsible by the Content Search Module. To maintain maximum code reusability and modularity, all external access to the engine is performed through the Web Service Application Logic. Even the Content Access Portal has to invoke functions through these service points. This approach is further justified because XML messages returned by Web services can immediately be rendered with XML Stylesheet Language (XSL) technologies for users on different platforms. For example, different Hypertext Markup Language (HTML) outputs are generated for Web browsers on desktop computers and Personal Digital Assistants (PDAs) respectively, while WAP Markup Language (WML) outputs are generated for mobile phones (Lin & Chlamtac, 2000).

The main concern of the Content Publishing Engine is control users' access to content. This could be effectively managed with the RBAC technology by matching users' roles and authorization with the classification of content items. In additional to simple tiering, content users' access control requires further considerations of their subscription payment, regional locale (because of legal requirements), and a more refined customer segmentation. For example, a content item can be specified to be accessible by a user in Hong Kong with *platinum* privilege and subscription fee paid for *premium research report*:

```
<ALLOW
user-category = "Content_Users"
data-category = "Any_Content"
purpose= "Any"
operation = "access"
condition = "Location = Hong Kong"
and "Privilege Class = Platinum" and
"Subscription = Premium Research" >
```

It should be noted that different parts of a content item often require different access control. For example, summaries usually have a "lower" access level than full content.

In addition, users should only allow subscribing for categories that they are authorized to, according to the specification above. After subscription, all the new content items of the subscribed categories will be delivered automatically to the user. Therefore, if a user's classification changes, conflicting subscription categories should be removed. To further enforce this, access control should be checked before the distribution of every content item.

Global Repository and Overall Integration

The Global Repository Management System provides backing support for user profiles and taxonomy. This includes all internal and external Content User registration, personal profiles, and their subscription data, in order to maintain users' access to various global and regional Web sites as a single entity. In addition, the global taxonomy is

Figure 8. Design summary of web services interfaces

Content Reception Engine	**Content Publishing Engine**	**Global Repository Management System**
External Service: receiveContent // for global taxonomy update Internal Service: receiveGlobalTag Internal Service: updateGlobalTag **Content Editorial Engine** Internal Service: receiveContent // for global taxonomy update Internal Service: addTag Internal Service: updateTag	Internal Service: publishContent Internal Service: checkDeliveryStatus Internal Service: updateContentStatus External Service: getSubcriptionCategories External Service: getSubcribedCategories External Service: updateSubscription External Service: searchContent **External Content Distributor** External Service: deliverContent External Service: checkDeliveryStatus	Internal Service: createUser Internal Service: updateUserInformation Internal Service: getSubcriptionCategories Internal Service: getSubcribedCategories Internal Service: updateSubscription Internal Service: searchTags Internal Service: addTag Internal Service: updateTag Internal Service: addCategory Internal Service: updateCategory Internal Service: getSubcribedUsers

maintained in this system for global consistency. The strategy is to keep minimal vital information in this repository to maintain its efficiency. Therefore, massive enterprise content is not stored here. In order to improve performance and reliability, replication techniques from contemporary relation databases such as Oracle (Garmany, 2003) may be used.

The overall system integration of the FECMS is based on a Web services interface to maintain autonomous sub-systems in various units of the enterprise. Figure 8 summarized the main Web services offered by Content Reception Engines, Content Editorial Engines, Content Publishing Engines, the Global Repository Management System, and external Content Providers. Both inter- and intra- enterprise interactions are implemented with Web services (labeled with *external* and *internal*, respectively). Let us examine the following typical use case of content publish-and-subscribe through Web services.

1. An institutional user may request Web services based content delivery. This is accomplished by submitting a request to the *updateSubscription* Web service of a Content Publishing Engine of the enterprise with the appropriate parameters, which include the categories of required content, along with the address of its own reception Web services access point.
2. The institution user must implement a Web service conforming to the specification of the *receiveContent* service of the Content Reception Engine.
3. The Content Publishing Engine verifies the request and relays successful request to the Global Repository Management System.
4. When new content arrives at the Content Publication Engine, the engine queries the Global Repository Management System through its *getSubscribedUsers* Web service, with the tier and tags of the new content as parameters.
5. If the institutional user is included in the list, the Content Delivery Module of the Content Publication Engine will invoke the user-specified Web service accordingly to deliver the piece of content.

Because of the importance of the information here, access controls restrict authorization through software systems only. In addition, we have to protect users' privacy in the FECMS. For example, users are allowed to view and update their profiles after authentication.

```
<ALLOW
user-category = "Content_Users"
data-category = "User_Profiles"
purpose= "Any"
```

```
operation = "update"
condition = "Authenticated = True" >
```

Only the broker or financial advisor, and the advisor's supervisors, of a user can read access a user's profile and update it only upon authorization.

```
<ALLOW
user-category = "Brokers" and "Finan-
cial_Advisors"
data-category = "User_Profiles"
purpose= "Any"
operation = "access" or "update"
condition = "Authorized = True" >
```

Thus, the system has to update access rules when supervisors assign temporary or alternate brokers or financial advisors. On the other hand, privacy also concerns the protection of identity of the content users and what content they have accessed. This should also be protected, similar to the policies listed above. However, supervisors of the employees should be able to know what their subordinates have accessed upon managerial authorization.

DISCUSSION

As pointed out by Kwok and Chiu (2004), the main goals of an FECMS are management, cost, value, and legal issues. In this section, we discuss how our approach helps archive these goals.

This case study demonstrates how a complex ECM system can be composed with a set of highly coherent but loosely coupled sub-systems that might be physically distributed within an enterprise. They are orchestrated by Web services technology to work together seamlessly for the enterprise. Thus, our approach enables seamless access to knowledge while integrating disparate business functions. Because of the power of knowledge, this not only enhances the management of the FECMS as a whole, but also assists the management and operations of the organization. As knowledge and organizational memory can be captured in enterprise content, access to content is an effective source of knowledge (Küng et al., 2001).

Existing content management systems in many such enterprises are semi-manual and not integrated, mainly due to the diversity of external information sources, heterogeneous platforms, and legacy systems. Some of the sub-systems are redundant and not unified; different sub-systems may exist in different units or in different geographical locations. In this case, chaos ensues, resulting in a high cost of system operations and maintenance. Thus, an integrated implementation framework for a new FECMS with standardized business processes and contemporary technologies help to reduce the maintenance costs by the re-use and integration of the existing sub-systems. Furthermore, switching as far as possible to channels that are free at the point of delivery, such as electronic, reduces the costs and unnecessary time delay.

As Web services are designed to support interoperable application-to-application interaction over the Internet, its goal is to overcome some of the main drawbacks of traditional business-to-business (B2B) applications that often result in complex, custom, one-off solutions that are not scalable, and are costly and time consuming. Some benefits of adopting Web services are that they are platform and vendor independent. Since they are based on a set of standards, they provide a means for the convergence of disparate business functionalities (Ratnasingam, 2002), and they make it easier to deploy business applications for trading partners. This results in a significant reduction in total cost of development.

Our proposed architecture based on Web services is also highly scalable and interoperable. There are no practical limitations in the implementation platform for each of these sub-systems as long as they support Web services and are programmed to be compliant with the enterprise's call conventions. For example, existing Java-2 Enterprise Edition (J2EE) enterprise applications can employ Sun Microsystems' Web service solutions, while current Content Editorial Engines using macros of Microsoft Office for its front-ends may be extended with the .NET framework (Bustos & Watson, 2003). In the case of legacy systems, wrappers may be built around them to enable compatibility with Web services. As such, upgraded sub-systems are able to join the FECMS gradually for adequate testing and streamlining the switch-over, which might otherwise cause a great impact for a major enterprise-wide system.

In addition, Web services serve as the middleware for interactions among business partners for sharing contents in both directions. Similar gradual migration strategies are also possible. In order to further streamline interactions among enterprises, application layer semantics (such as content taxonomy and category definitions based on Semantic Web research), protocols for interaction, and service-level standards are called for. Trade unions and regulatory bodies may help in such standardizations. If so, content *service grids* (Gentzsch, 2002) can be formed for seamless and effective inter-enterprise content sharing and management.

Besides timeliness, our architecture enables content delivery across a wide range of clients, both internal and external, and across multiple channels to increase its values in CRM. If necessary, the editors can tailor the content for a particular market sector. Support for different formats of information such as eXtended Markup Language (XML), Hypertext Markup Language (HTML), Portable Document Format (PDF), graphics, image, and multimedia is archived for the current Internet environment.

Standardized enterprise-wide policies and business processes provide a mechanism for various content creation and management functions, such as content flow, document lifecycle, and collaboration. In particular, supporting different access control levels as described in the previous sub-section is important to the integrity and control of all the enterprise content. On the other hand, similar to a library system, means for creating and maintaining the meta-data (taxonomy) around the content is vital for the correct distribution of content and facilitates the opportunities for further analysis. Furthermore, there is a long-term need for integration with third-party FECMS or information sources, as well as the definition and maintenance of third-party content references.

In addition, content distribution to clients is an important CRM activity, especially for financial enterprises. Thus, an FECMS provides a high value for both internal management and external marketing objectives. In addition to the usual aim of an information system to reduce costs and improve management, an FECMS also helps to ensure compliance with legal issues and maximize value extracted from the content.

Legal issues are vital because of the large amounts of money involved, and possible risks of damage to reputations. The FECMS assists management and the Content Creators to ensure compliance with relevant laws and regulations. For example, the enterprise should adopt a single set of approval policies and procedures covering all forms of distribution in order to ensure that content are only published through official distribution channels (and not through personal mail distribution lists). The FECMS also ensures the identity of the Content Users, as well as what information they receive. Access control can ensure that only legally audited content is published, and only through official distribution channels (not through personal mail distribution lists) to the correct set of content users. In addition, effective privacy protection is also mandatory because of privacy acts all over the world. Moreover, because different countries have different requirements, our proposed mechanism is essential for the customization of such policies.

Our methodology for tracing content flow in all system components ensures the coverage of privacy and access control policies for the entire enterprise content. Furthermore, a systematic review of the content management processes identifies not only possible loopholes, but also other flaws in these processes. In particular, the content creation and editorial process is complicated, sensitive, and difficult to manage. The policies suggested in this case study provide an effective solution, which also enforce the right personnel processing relevant content.

Our composition approach further facilitates a systematic specification of the privacy and access control policies. Based on central policies, subsidiaries and regional offices can customize such policies based on their individual requirements. EPAL is in a textual form, and easy to understand. As we can formulate EPAL based on the global enterprise ontology, exchange and customization of such policies is further streamlined. Because of the ease of development, customization, and maintenance, costs involved in privacy and access control can be reduced. To further this, as EPAL is a standard technology, investment can also be protected as opposed to proprietary ones. Please note that applying EPAL is only part of the solution to enforce privacy access control in the whole picture. Another important issue is to have legally binding stakeholders.

A well-controlled and secure system generally provides a higher value because of improved information integrity. As content service plays an important role in CRM, an image of a secure portal with strong privacy protection further enhance customers' confidence. Therefore, this helps improve customer relationships and possibly attracts paid content subscriptions. This is particularly important for financial enterprises because a significant amount of sensitive personal information is involved.

CONCLUSIONS AND FUTURE WORK

Throughout this paper, we have presented a flexible, scalable, and sophisticated architecture for a Financial Enterprise Content Management System with contemporary Web services technologies to support both internal content flow and external interactions. Based on a study of the requirements of a large international bank, we have demonstrated a pragmatic approach to address the conflicting objectives of an FECMS, in particular, integration and control. We have detailed the design of different components of an FECMS by systematically tracing content flow. Four major goals, management, cost, legal issues, and value of an FECMS have been considered during requirement elicitation. Further, we perceive that this pragmatic framework and methodology is applicable to a wide range of enterprises. Though the scenarios discussed above are not intended to be a completely accurate representation of all the complex requirements in an FECMS, nor does this paper necessarily provide a detailed plot of the solution, we believe that the presentation and discussions are sufficiently representative to introduce the relevant research issues and to be the starting point for future development in this area.

We have also considered the applicability of our ECM implementation framework to other industries highly positive and optimistic. This is due to the trend in which organizations are rapidly moving towards service-oriented operations. For large enterprises, our operation model for ECM based on the anticipated content flow is highly generic. Though some enterprises currently do not consider importing contents from other sources, they eventually need to do so in order to increase their competitiveness. In addition, they will eventually realize the value of providing their clients with content for CRM through a secured portal. As our case study may be an extreme one in the complexity, it may well be a reference model for smaller organizations and those with less stringent requirements.

For SME, though the complete architecture may be overkill and quite expensive, it is feasible to use some of these design concepts in their own architectures, especially because software houses may develop packages with our approach. Moreover, the external Web service interfaces presented in this paper are not complicated and can be easily programmed for content reception and delivery. This is as result of the main complexity (such as taxonomy and internal workflows) being encapsulated within the complete system, but not directly involved in the inter-enterprise interface.

In conclusion, to be able to trace content distributions, we would like to integrate the FECMS framework with a watermarking infrastructure to reinforce document management policies by supporting non-repudiation in the document distribution protocol (Cheung et al., 2008). We are also investigating the application of an advanced workflow management system, such as ADOME-WFMS (Chiu et al., 2001, 2004) in the FECMS for both inter- and intra- enterprise content flow management. In addition, we are interested in the application of Semantic Web technologies for the management of content taxonomy.

REFERENCES

Arnold-Moore, A., Fuller, M., Kent, A., Sacks-Davis, R., & Sharman, N. (2000). Architecture of a content management server for XML document applications. In *Proceedings of 1st International Conference on Web Information Systems Engineering (WISE 2002)*, Hong Kong, China (Vol. 1, pp. 97-108).

Aversano, L., de Canfora, G., Lucia, A., & Gallucci, P. (2002). Integrating document and workflow management tools using XML and web technologies: a case study. In *Proceedings of Sixth European Conference on Software Maintenance and Reengineering* (pp. 24-33).

Berners-Lee, T., Hendler, J., & Lassila, O. (2001). The Semantic Web. *Scientific American, 284*(5), 34–43. doi:10.1038/scientificamerican0501-34

Bertino, E., Castano, S., Ferrari, E., & Mesiti, M. (2002). Protection and administration of XML data sources. *Data & Knowledge Engineering, 43*(3), 237–260. doi:10.1016/S0169-023X(02)00127-1

Bustos, J., & Watson, K. (2002). *Beginning. NET Web Services using C*. New York: Wrox Press Ltd.

Carlson, D. (2001). *Modeling XML Applications with UML*. Reading, MA: Addison-Wesley.

Cheung, S. C., Chiu, D. K. W., & Ho, C. (2008). The Use of Digital Watermarking for Intelligence Multimedia Document Distribution. *Journal of Theoretical and Applied Electronic Commerce Research, 3*(3), 103–118. doi:10.4067/S0718-18762008000200008

Chiu, D. K. W., Chan, W. C. W., Lam, G. K. W., Cheung, S. C., & Luk, F. T. (2003). An Event Driven Approach to Customer Relationship Management in an e-Brokerage Environment. In *Proceedings of the 36th Hawaii International Conference on System Sciences*, Big Island, HI. Washington, DC: IEEE Computer Society Press.

Chiu, D. K. W., Cheung, S. C., Karlapalem, K., Li, Q., Till, S., & Kafeza, E. (2004). Workflow View Driven Cross-Organizational Interoperability in a Web Service Environment. *Information Technology and Management, 5*(3-4), 221–250. doi:10.1023/B:ITEM.0000031580.57966.d4

Chiu, D. K. W., & Hung, P. C. K. (2005). Privacy and Access Control in Financial Enterprise Content Management. In *Proceedings of the 38th Hawaii International Conference on System Sciences*, Big Island, HI. Washington, DC: IEEE Computer Society Press.

Chiu, D. K. W., Kafeza, M., Cheung, S. C., Kafeza, E., & Hung, P. C. K. (2009). Alerts in Healthcare Applications: Process and Data Integration. *International Journal of Healthcare Information Systems and Informatics, 4*(2), 36–56.

Chiu, D. K. W., & Li, Q. (1997). A Three-Dimensional Perspective on Integrated Management of Rules and Objects. *International Journal of Information Technology, 3*(2), 98–118.

Chiu, D. K. W., Li, Q., & Karlapalem, K. (2001). Web Interface-Driven Cooperative Exception Handling in ADOME Workflow Management System. *Information Systems, 26*(2), 93–120. doi:10.1016/S0306-4379(01)00012-6

Croll, M., Lee, A., & Parnall, S. (1997). Content Management - the Users Requirements. In *Proceedings of the International Broadcasting Convention,* Amsterdam, The Netherlands (pp. 12-16).

Davis, J. C. (2000). Protecting privacy in the cyber era. *IEEE Technology and Society Magazine, 19*(2), 10–22. doi:10.1109/44.846270

Dayal, U. (1989). Active Database Management Systems. In *Proceedings of the 3rd International Conference on Data and Knowledge Bases* (pp. 150-169).

Edwards, J., Coutts, I., & McLeod, S. (2000). Support for system evolution through separating business and technology issues in a banking system. In *Proceedings of the International Conference on Software Maintenance* (pp. 271-276).

Erl, T. (2006). *Service-Oriented Architecture: Concepts, Technology, and Design.* Upper Saddle River, NJ: Prentice-Hall.

Fensel, D. (2001). Challenges in Content Management for B2B Electronic Commerce. In *Proceedings of the 2nd International Workshop on User Interfaces to Data Intensive Systems (UIDIS 2001)*, Zurich, Switzerland (pp. 2-4).

Ferraiolo, D. F., Kuhn, D. R., & Chandramouli, R. (2003). *Role-based access control.* Norwood, MA: Artech House Publishers.

Ferraiolo, D. F., Sandhu, R., Gavrila, S., Kuhn, D. R., & Chandramouli, R. (2001). Proposed NIST Standard for Role-Based Access Control. *ACM Transactions on Information and System Security, 4*(3), 224–274. doi:10.1145/501978.501980

Fischer-Hübner, S. (2001). *IT-Security and Privacy – Design and Use of Privacy-Enhancing Security Mechanism* (LNCS 1958). New York: Springer.

Garmany, J. (2003). *Oracle Replication: Snapshot, Multi-master & Materialized Views Scripts.* Kittrell, NC: Rampant TechPress.

Gentzsch, W. (2002). Grid computing: a new technology for the advanced web. In *Proceedings of the NATO Advanced Research Workshop on Advanced Environments, Tools, and Applications for Cluster Computing* (LNCS 2326, pp. 1-15). New York: Springer.

Hinde, S. (2002). The perils of privacy. *Computers & Security, 21*(3), 424–432. doi:10.1016/S0167-4048(02)00508-4

Hung, P. C. K. (2001). *Secure Workflow Model. Unpublished doctoral disseration.* Hong Kong, China: Department of Computer Science, The Hong Kong University of Science and Technology.

Jones, V. E., Ching, N., & Winslett, M. (1995). Credentials for privacy and interoperation. In *Proceedings of the New Security Paradigms Workshop* (pp. 92-100).

Kitayama, F., Hitose, S., Kondoh, G., & Kuse, K. (1999). Design of a framework for dynamic content adaptation to Web-enabled terminals and enterprise applications. In *Proceedings of the Sixth Asia Pacific Software Engineering Conference* (pp. 72-79).

Küng, J., Luckeneder, T., Steiner, K., Wagner, R. R., & Woss, W. (2001). Persistent topic maps for knowledge and Web content management. In *Proceedings of the 2nd International Conference on Web Information Systems Engineering*, Kyoto, Japan (Vol. 2, pp. 151-158).

Kwok, K. H. S., & Chiu, D. K. W. (2004). An Integrated Web Services Architecture for Financial Content Management. In *Proceedings of the 37th Hawaii International Conference on System Sciences*, Big Island, HI. Washington, DC: IEEE Computer Society Press.

Leino-Kilpi, H., Valimaki, M., Dassen, T., Gasull, M., Lemonidou, C., Scott, A., & Arndt, M. (2001). Privacy: A review of the literature. *International Journal of Nursing Studies, 38*(6), 663–671. doi:10.1016/S0020-7489(00)00111-5

Lin, Y.-B., & Chlamtac, I. (2000). *Wireless and Mobile Network Architectures*. New York: John Wiley & Sons.

Linthicum, D. S. (2003). *Next Generation Application Integration: From Simple Information to Web Services*. Reading, MA: Addison-Wesley.

McNay, H. E. (2002). Enterprise Content Management: an Overview. In *Proceedings of the IEEE International Professional Communication Conference* (pp. 396-402).

Omelayenko, B. (2001) Preliminary Ontology Modeling for B2B Content Integration. In *Proceedings of the 12th International Workshop on Database and Expert Systems Applications*, Munich, Germany (pp. 7-13).

Powers, C. S., Ashley, P., & Schunter, M. (2002) Privacy promises, access control, and privacy management - Enforcing privacy throughout an enterprise by extending access control. In *Proceedings of the 3rd International Symposium on Electronic Commerce* (pp. 13- 21).

Ratnasingam, P. (2002). The Importance of Technology Trust in Web Services Security. *Information Management & Computer Security, 10*(5), 255–260. doi:10.1108/09685220210447514

Rosenberg, J., & Remy, D. (2004). *Securing Web Services with WS-Security: Demystifying WS-Security, WS-Policy, SAML, XML Signature, and XML Encryption*. Sams.

Samarati, P., & Vimercati, S. (2001) *Access Control: Policies, models, mechanisms* (LNCS 2171). New York: Springer.

Sandhu, R. S., Bhamidipati, V., & Munawer, Q. (1999). The ARBAC97 model for role-based administration of roles. *ACM Transactions on Information and System Security, 1*(2), 105–135. doi:10.1145/300830.300839

Schoeman, E. D. (1984). *Philosophical dimensions of privacy: an anthology*. New York: Cambridge University Press. doi:10.1017/CBO9780511625138

Shimonski, R. J., Schmied, W., Chang, V., & Shinder, T. W. (2002). *Building DMZs for Enterprise Networks*. Syngress.

Stufflebeam, W. H., Antón, A. I., He, Q., & Jain, N. (2004, October 28). Specifying privacy policies with P3P and EPAL: lessons learned. In *Proceedings of the 2004 ACM Workshop on Privacy in the Electronic Society*, Washinton, DC (pp. 35-35).

Surjanto, B., Ritter, N., & Loeser, H. (2000). XML Content Management Based on Object-Relational Database Technology. In *Proceedings of the 1st International Conference on Web Information Systems Engineering*, Hong Kong, China (Vol. 1, pp. 70-79).

Tiitinen, P. (2003). User Roles in Document Analysis. In *Proceedings of CAISE'03 Forum Short Paper Proceedings* (pp. 205-208). University of Maribor Press.

Tiwana, A. (2001). *The Essential Guide to Knowledge Management – E-Business and CRM Applications*. Upper Saddle River, NJ: Prentice Hall.

Tyrväinen, P., Salminen, A., & Päivärinta, T. (2003) Introduction to the Enterprise Content Management Minitrack. In *Proceedings of the 36th Hawaii International Conference on System Sciences*, Big Island, HI. Washington, DC: IEEE Computer Society Press.

Wang, X., Lao, G., DeMartini, T., Reddy, H., Nguyen, M., & Valenzuela, E. (2002, November 22). XrML - eXtensible rights Markup Language. In *Proceedings of the 2002 ACM Workshop on XML Security*, Fairfax, VA (pp. 71-79).

Weitzman, L., Dean, S. E., Meliksetian, D., Gupta, K., Zhou, N., & Wu, J. (2002). Transforming the content management process at IBM.com. In *Proceedings of the Conference on Human Factors and Computing Systems*, Minneapolis, MN (pp. 1-15).

Weverka, P. (2000). *Mastering ICQ: The Official Guide*. ICQ Press.

World Wide Web Consortium. (W3C). (2002, April 16). *The Platform for Privacy Preferences 1.0 (P3P1.0) Specification*. Retrieved February 21, 2010, from http://www.w3.org/TR/P3P/

This work was previously published in International Journal of Systems and Service-Oriented Engineering, Volume 1, Issue 2, edited by Dickson K.W. Chiu, pp. 86-113, copyright 2010 by IGI Publishing (an imprint of IGI Global).

Section 4
Systems Engineering and Service-Oriented Engineering

Chapter 17
RT-Llama:
Providing Middleware Support for Real-Time SOA

Mark Panahi
University of California, USA

Weiran Nie
University of California, USA

Kwei-Jay Lin
University of California, USA

ABSTRACTS

Service-oriented architectures (SOA) are being adopted in a variety of industries. Some of them must support real-time activities. In this paper, the authors present RT-Llama, a novel architecture for real-time SOA to support predictability in business processes. Upon receiving a user-requested process and deadline, our proposed architecture can reserve resources in advance for each service in the process to ensure it meets its end-to-end deadline. The architecture contains global resource management and business process composition components. They also create a real-time enterprise middleware that manages utilization of local resources by using efficient data structures and handles service requests via reserved CPU bandwidth. They demonstrate that RT-Llama's reservation components are both effective and adaptable to dynamic real-time environments.

INTRODUCTION

Service-oriented architecture (SOA) is the prevailing software paradigm for dynamically integrating loosely-coupled services into one cohesive business process (BP) using a standard-based software component framework (Bichler & Lin, 2006; Huhns & Singh, 2005). SOA-based systems may integrate both legacy and new services, created by either enterprises internally or external service providers.

However, current SOA solutions have not addressed the strict predictability demands that many enterprise applications require, from banking and

DOI: 10.4018/978-1-4666-1767-4.ch017

finance to industrial automation and manufacturing. Such enterprises, many of whom already embrace SOA for a large part of their systems, would greatly benefit from a comprehensive SOA solution that can also encompass their real-time applications. In other words, as SOA gains prominence in many domains, the confluence of real-time and SOA systems is inevitable. We must prepare SOA for meeting the predictability requirements of real-time enterprise systems.

For example, the auto industry is looking at applying SOA to their development of the AUTOSAR (AUTomotive Open System ARchitecture) standardization project (AUTOSAR, 2008) to address the needs of future automotive systems, called "service integrated systems". Such systems can facilitate: 1) the integration of various on-board services (e.g., cruise control, brakes, etc.), 2) both vehicle-to-vehicle and vehicle-to-roadside for traffic information, and 3) communication with the dealer about vehicle maintenance. This must be accomplished while also maintaining the real-time requirements of such systems, particularly with on-board service integration.

SOA brings both advantages and challenges to real-time integration. By its very nature, SOA was designed to be flexible to deal with dynamic environments. As such, different service component candidates may have very different timing behaviors. Moreover, the distributed nature of execution environments makes it almost impossible to make any kind of "guarantees." However, such flexibility and dynamism may be leveraged to allow 1) different choices of candidate services to better meet deadlines during BP composition and 2) dynamic run-time adjustments to reconfigure a BP to make up for lost time when some services have taken much longer than expected. We therefore want to study the real-time support in SOA both to realize its benefits and to meet its challenges.

In this paper, we present the RT-Llama project (as an extension of Llama (Lin, Panahi, Zhang, Zhang, & Chang, 2009; Zhang, Lin, & Hsu, 2007)) which meets the real-time enterprise challenge by enabling SOA users to schedule an entire BP, thus eliminating the risk of missed deadlines due to the over-utilization of resources. RT-Llama differs from previous service-oriented architectures (Panahi et al., 2008) in that it allows end-to-end BP deadline guarantees through advance reservations of local resources.

In order to make this work, we 1) design global resource management and composition components that reserve resources in advance for each service in a BP to help guarantee its end-to-end deadline; 2) leverage an efficient data structure for managing reservation utilization data, a dynamic adaptation of the TBTree (Moses, Gruenwald, & Dadachanji, 2008) called the dTBTree; 3) implement a CPU bandwidth management system for each host, dividing a CPU into multiple temporally-isolated virtual CPUs, allowing different classes of service with various levels of predictability; and 4) develop pre-screening mechanisms to decrease the likelihood of unsuccessful distributed service reservations.

The rest of the paper is organized as follows. We first review the challenges of bringing real-time to SOA and describe the RT-Llama RT-SOA architecture. We then present the performance study of the RT-Llama implementation. Finally, we compare our work on RT-Llama to related efforts.

BACKGROUND

Motivation

RT-SOA is a relatively new and challenging field of study. While some aspects of SOA make its transition to real-time simpler, still other aspects pose serious challenges. One real-world problem that can use an RT-SOA solution is algorithmic trading. Algorithmic trading is defined in Wikipedia as "a sequence of steps by which patterns in real-time market data can be recognized and responded to." The performance requirements of algorithmic trading demand not only fast transactions but also predictable ones (Martin, 2007). Therefore, it is an application which should be implemented as RT-SOA. Examples of algorithmic trading strategies include: 1) trading several positions in coordination and 2) breaking larger trades into smaller sequential trades to minimize market impact.

As an example scenario, we assume that a customer has a choice for each trade among several brokerage firms. Each brokerage firm offers three levels of service:

1. **institutional:** for high volume traders where small delays can result in large losses,
2. **premium:** for moderate volume investors, and
3. **individual:** for low volume casual investors.

Moreover, two operations are permitted. The getQuote operation requires a ticker symbol on an equity and returns its price per share. This operation is usually immediate and not reserved. The trade operation takes the quantity of shares and whether it is a buy or sell order. This type of operation has strict execution requirements and will usually be reserved in advance.

Ensuring sufficient availability of resources on a host to execute tasks is critical for any real-time application. As an example, Figure 1 shows the available utilization of a given host in 10 minute increments (note that embedded systems may record much smaller increments). The 5pm to 5:10 increments indicates 25% utilization availability whereas the subsequent increment shows 75% availability. Figure 1 then shows the result of composing a business process (BP) that requires one service from each of the three classes and has a total execution time of 35 minutes. The BP has an expected start time of 5pm and deadline of 5:50. Without proper consideration of execution times of the service candidates under the availability of resources on each host, we see that the composed process can complete no earlier than 5:57 and will miss the specified deadline of 5:50. Clearly, many other service combinations could have been selected to compose a feasible real-time BP.

To address the problems exposed in this example, we propose the RT-Llama framework to help SOA take into consideration the execution times of service candidates and availability of host resources to compose predictable BPs.

Scope of Real Time Applications

We make the following assumptions to guide us on our initial work. In the future, we plan to explore the effect of relaxing these assumptions.

Figure 1. Utilization of a host before and after a new reservation attempt

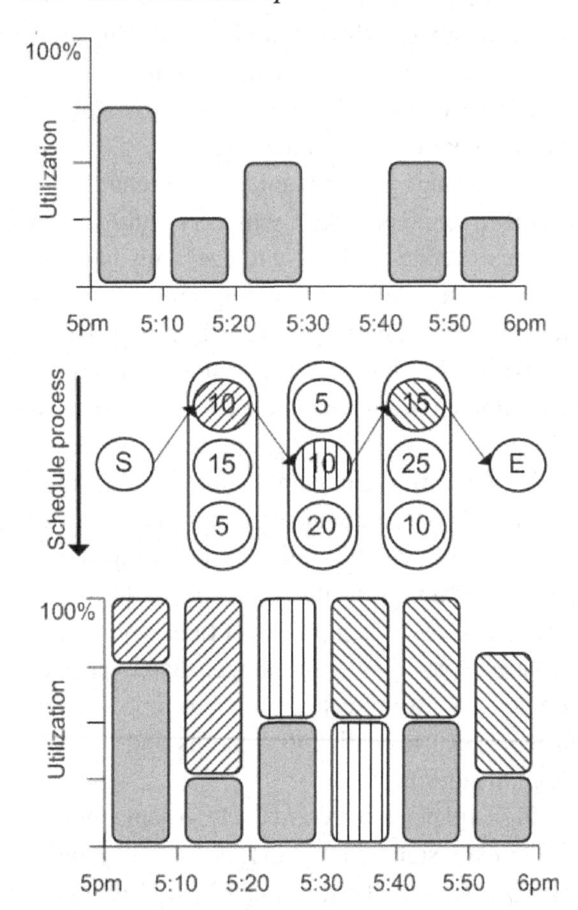

One-Shot Execution

Due to the variable availability of hosts for executing services, we currently assume a "one-shot" model for BP composition and execution. In some systems, BPs may be composed and then reused multiple times. However, since RT-Llama's composition process takes into account service availability, the composed BP may be guaranteed to run successfully only once. If a user has the same requirements, the BP must be regenerated, possibly selecting different services for each BP.

Advance Reservations

The RT-Llama architecture is based on the advance reservations of service executions to guarantee timeliness. Currently, we assume that reservation requests are made well enough in advance of the actual service execution such that there is plenty of time to send the request after the reservation. Therefore, reservation requests themselves are not subject to real-time requirements. In later work, we plan to explore "just-in-time" reservations.

Process Information

Some real-time systems determine admissibility of a job only when it arrives at a server (i.e., admission control). For RT-SOA applications, this may not be acceptable. It would take the rejection of only one component service or one service overrun to severely disrupt the entire BP's execution. However, the advantage of SOA is that a BP is usually encoded in a higher level language, such as BPEL. The execution path is assumed to be known before hand and can be reserved in advance. Therefore, we leverage this information in our RT-Llama framework and build advance reservation mechanisms for the whole process so that the risk of service rejections and cost overruns is eliminated.

Real-Time SOA Support

In order to promise end-to-end predictability for any real-time distributed system, subcomponent or dependency must also provide predictability. Furthermore, an RT-SOA framework poses even more challenges than any stand-alone or distributed real-time system. We have identified the following required support for an RT-SOA and discuss the subset of these issues we wish to address in this work.

Operating System

Any real-time middleware framework must be built atop an operating system that provides real-time scheduling and a fully preemptible kernel.

Communications Infrastructure

The communications infrastructure must provide predictability – a requirement that the existing Internet infrastructure currently does not provide. However, approaches like Differentiated Services (Blake et al., 1998) and Integrated Services (Braden, Clark, & Shenker, 1998), despite their lack of wide acceptance, provide a starting point for QoS-based communications.

BP Composition Infrastructure

The SOA composition infrastructure must be able to generate a BP that satisfies both a user's functional and timeliness requirements. It must be able to negotiate such timeliness requirements with distribution middleware to ensure predictable BP execution.

Distribution Middleware

The main purpose of a real-time middleware platform (like a real-time enterprise service bus (ESB)) is to ensure the predictable execution of

Figure 2. RT-SOA system using RT-Llama

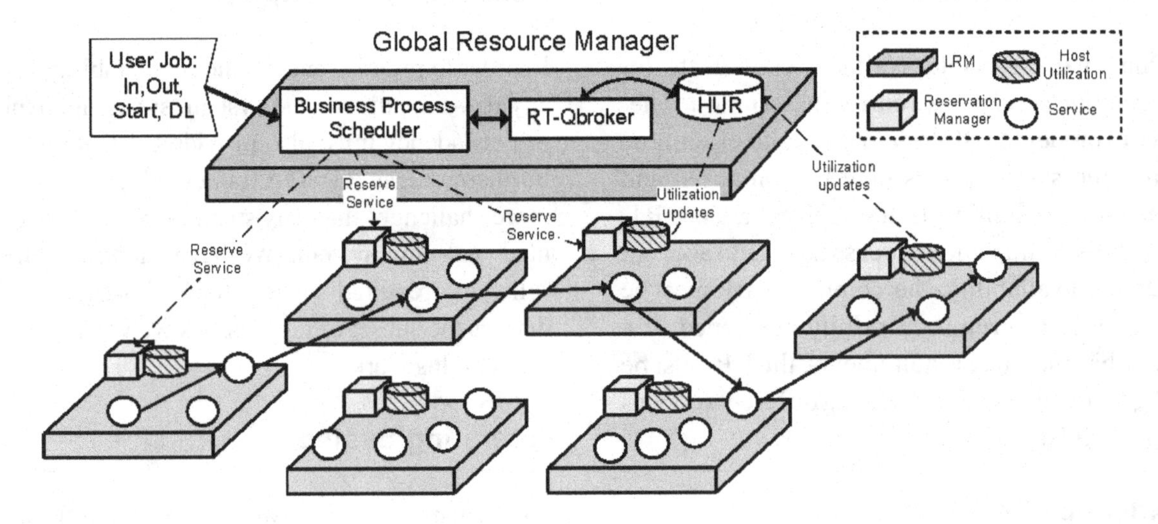

individual service requests. Therefore an RT-SOA middleware platform must provide support for advance reservations and avoid overloading or overbooking its host's resources.

Client Infrastructure

Many existing SOA deployments use a business process execution language (BPEL) engine, which is a centralized mechanism to coordinate all remote service interactions within the process. An RT-SOA solution must address any unpredictability that a BPEL engine may produce. Alternatively, there are also distributed coordination mechanisms that route from service to service without the intervention of a centralized apparatus, which may be more amenable to achieving RT-SOA.

In our current work, we focus on introducing predictability into two of the main areas mentioned above: BP composition infrastructure and distribution middleware. We leverage Real-time Java (Bollella et al., 2000) and the Solaris 10 OS to provide real-time scheduling capabilities to our middleware. We leave the issues of integrating real-time networking and predictable SOA client infrastructures to future work.

ARCHITECTURE

System Model

In contrast to some existing distribution frameworks, SOA can support multiple classes of service that can be offered by providers at different QoS levels. RT-Llama is comprised of several components that aid providers in offering predictable service execution and users in ensuring that their BPs meet their specified deadlines. A typical operating scenario is shown in Figure 2. From a user's perspective, all that is required of them is to specify input values, desired output, and timeliness parameters, including start time and deadline. Based on this information, RT-Llama selects and reserves a feasible BP that matches the user's requirements.

In RT-Llama, service providers can deploy services under two categories:

- **Unreserved:** Services deployed under this category accept requests without any prior workload reservation. There are two main subclasses:
  1. **Immediate** requests are serviced in first-in first-out order according to the system's underlying real-time

scheduler. There may be at most one immediate class. However, it may support several priorities.

2. **Background** requests are serviced in a best effort fashion according to the operating system's underlying non-real-time scheduler.

- **Reserved:** Services deployed under this category may only receive requests that have been reserved in advance. There may be any number of reserved classes. Reserved classes are intended to map to different levels of service that providers would like to offer. We identify three policies that may govern multiple classes:
  1. **Resource favorability.** Providers can give higher classes more bandwidth (a larger share of system resources) than lower ones.
  2. **Run-time spillover.** Providers can set policies such that higher class requests can steal bandwidth from lower classes in the case of cost overruns or tardy requests, in order to ensure that higher class requests meet their deadlines.
  3. **Reservation spillover.** Higher class users may be able to have a service reservation span into lower classes allowing it to finish sooner and thus increasing the likelihood of a successful reservation of the overall BP.

Real Time Model

Similar to the standard real-time task model, we define the real-time task model for SOA as follows: A business process BP_i is a workflow composed of sequential service invocations $S_{i,j}$. Each BP_i begins execution at time r_i, finishes at time f_i, with an execution time of c_i, and has a deadline of d_i, that is respected if $f_i <= d_i$. Similarly, each service invocation arrives at time $r_{i,j}$, finishes at time $f_{i,j}$ after executing for $c_{i,j}$, with a deadline of $d_{i,j}$.

Since the RT-Llama framework supports advance reservations, we have a_i as the scheduled start time for BP_i and $a_{i,j}$ as the scheduled start time for $S_{i,j}$. We thus have the relation $w_i = [a_i, d_i]$ as the time window for BP_i and $w_{i,j} = [a_{i,j}, d_{i,j}]$ as the time window for $S_{i,j}$. Moreover, each $S_{i,j}$ has a known worst case execution time $wcet_{i,j}$.

This system model must be supported by the underlying infrastructure. Therefore, we currently make use of hosts that have at least 2 CPUs or dual core, with one (or more) CPU/core completely devoted to servicing real-time tasks (RT-CPU) and at least one available for background tasks and other operating system tasks (non-RT-CPU) as shown in Figure 3. The RT-CPU will execute both the unreserved immediate class, as well as all reserved classes. These classes, although sharing the same CPU, must be temporally isolated from each other, meaning that cost overruns that may occur in one class cannot interfere with other classes. This is accomplished by creating multiple virtual servers out of the RT-CPU using the constant bandwidth server (CBS) framework (Abeni & Buttazzo, 2004). Therefore, each class is assigned to a virtual server, which in turn has a system defined utilization percentage. For example, if we have one immediate class, and three reserved classes (H, M, and L), we may assign utilization allocations of 20%, 40%, 30%, and 10%, respectively.

RT-Llama Architecture

The RT-Llama architecture is shown in Figure 3. Specifically, the components are as follows.

The **Global Resource Manager (GRM)** components are responsible for scheduling a user's BP based on their requirements. **QBroker/RT-QBroker** is a QoS broker designed to select a BP based on user specified constraints (previously studied in (Yu & Lin, 2005; Yu, Zhang, & Lin, 2007)). In addition to its original role for best-effort BPs, RT-QBroker performs feasibility checks on individual services during service

Figure 3. RT-Llama architecture

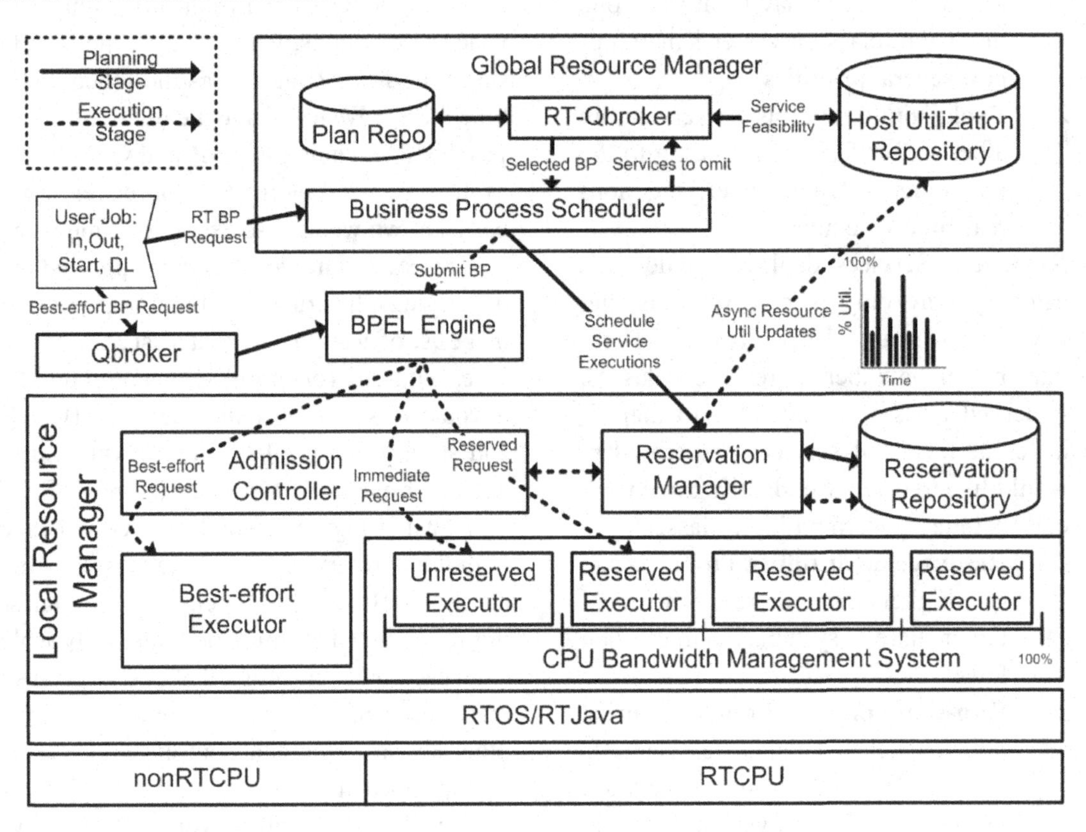

selection, by consulting the Host Utilization Repository, to determine if the host is likely to be available during the general span of time of the BP. The **Business Process Scheduler (BPS)** is responsible for reserving a BP selected by RT-QBroker according to either a concurrent or sequential mechanism (discussed later) by contacting the host of each service. In the event of a reservation failure (i.e., the reservation is not feasible or the request timeout is exceeded) the BPS requests a new BP from RT-QBroker, absent any unfeasible services. If RT-QBroker cannot find a feasible BP, the user's request is rejected. **The Host Utilization Repository (HUR)** stores cached future utilization information regarding each host using dTBTrees (discussed shortly). As a cache, it is updated at predefined intervals, and thus not always completely up-to-date. It may also have coarser grain information than that stored on individual hosts to save on space. However, it provides an efficient way of determining if a service is likely to be available at some specific future time.

The **Local Resource Manager (LRM)** is responsible for hosting services and ensuring that requests on such services are executed predictably by working with the GRM components. The **Reservation Manager (RM)** manages advance reservations for the host. It uses a dTBTree for each of the reserved classes to manage their overall utilization, as well as a simple hashmap to manage specific reservation information. Both data structures are represented by the **Reservation Repository (RR)**. The RM supports search, insert, and delete operations for reservations. Additionally, it sends asynchronous updates on recent utilization changes to the HUR. The **Admission Controller (AC)** accepts incoming service requests and routes them to an **Executor** responsible for servicing requests for the various classes of service. Each executor is mapped to one or more threads on the underlying operating system. The real-time executors include the **Unreserved Executor** and the **Reserved Executors** and are each given a guaranteed amount of bandwidth on the RT-CPU by promoting the FIFO priority of the real-time thread backing the Executor for an amount of time out of a period according to its bandwidth percentage. For reserved requests, it looks up the $a_{i,j}$, $wcet_{i,j}$, and $d_{i,j}$ according to the request's reservation ID in the RR. This information is necessary for the Reserved Executors to ensure that requests are serviced in an earliest deadline first (EDF) fashion and to properly accommodate tardy requests and deadline overruns. The **Best-effort Executor**, by contrast, executes on the non-RT-CPU along with other system threads and provides no guarantees for predictability.

Reservation Data Management

One of the unique features of the RT-Llama framework in contrast to other SOA systems is its ability to reserve service requests in advance. This avoids the pessimistic nature of on-demand admission control strategies and leads to potentially higher utilization rates as users are able to plan their BP executions ahead-of-time.

In the RT-Llama framework, we adopt the temporal bin tree (TBTree) data structure discussed in (Moses et al., 2008) due to its flexibility, efficiency in both time and space, and suitability for real-time planning applications. The TBTree uses a binary tree structure to store the total amount of time available within a time interval at different levels of granularity. Each node in the tree contains the sum of the available time in the left and right children, each of which covering half the time interval of the parent. A search for available time begins at the root and descends into the tree and identifies the leaf node that can completely accommodate the task, specified as either an earliest start time or latest finish time, and the expected execution time. For improved efficiency, the TBTree can implement forward pointers for each node, which indicate the next part of the tree to search in the event that the given node cannot accommodate the task, thereby avoiding costly backtracking operations.

This data structure was originally designed (Moses et al., 2008) to store advanced reservations for fixed but long time horizons in manufacturing systems where tasks may be planned out days or weeks in advance and take minutes or hours. By contrast, we adapt this data structure to the RT-SOA environment since we expect service executions to be planned out minutes (occasionally, seconds) in advance. Moreover, we expect individual service execution times to be also on the order of seconds.

These new requirements impel us to make the following two adaptations to the TBTree:

1. small task granularities and
2. subtree recycling. We call this new data structure the dynamic TBTree, or *dTBTree*.

dTBTree Leaf-node Granularity

In the design of the original TBTree, the capacity of each leaf node is greater or equal to the largest possible task size. Moreover, any task must be completely accommodated within a leaf node. In other words, tasks cannot span across multiple leaf nodes. However, the problem with this approach in its application to RT-SOA is that it leads to underutilization of resources. Instead, we make the capacity of each leaf node as the smallest possible task size and allow tasks to span across multiple leaf nodes. During the search operation, the dTBTree's forward pointers are utilized to laterally traverse the tree and available time slots are collected in a list. Since the HUR is used only to guide the reservation pre-screening process, the cached versions of the local dTBTrees here may have coarser granularities to manage space/accuracy tradeoffs.

dTBTree Subtree Recycling

Having a long time horizon with small task granularities would require a prohibitively large TBTree. Therefore, our second change is to enhance the original TBTree with dynamic capabilities to recycle subtrees whose right-most leaf node's time index is less than the current system time, called *subtree expiration*.

Figure 4 depicts a dTBTree and how it operates. The left and right subtrees are swapped since the left subtree has expired. Each node has two fields: (*subtree id, remaining capacity*). The dTBTree offers two recycling policies:

1. **eager reinitialization:** where the capacities of each of the nodes of the expired subtree are reinitialized during the recycling operation itself; and
2. **lazy reinitialization:** where the node capacities are only reinitialized during search operations.

Figure 4. dTBTree example with lazy reinitialization. Phase 2 shows the recycling operation commencing at time 4. A subsequent search operation is depicted in phase 3. Changes to the tree are highlighted in bold.

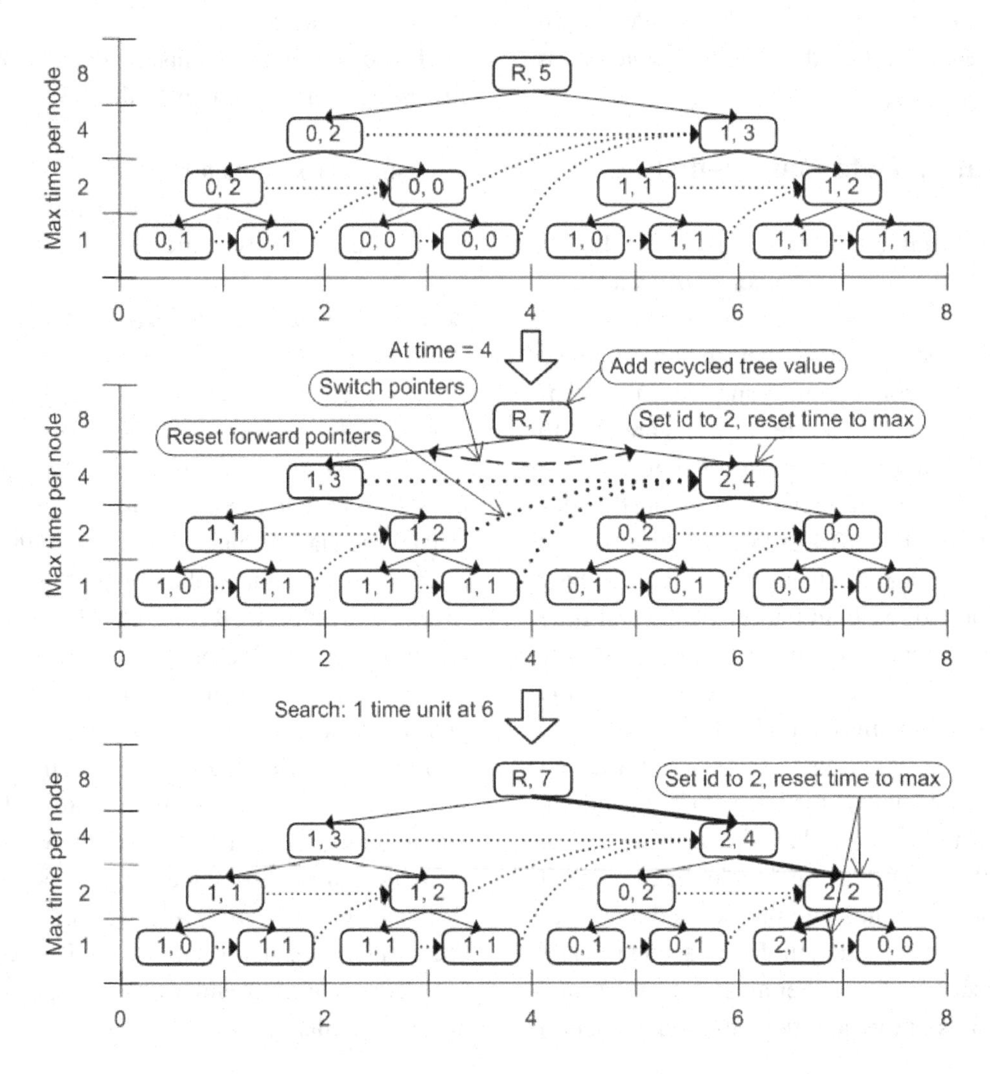

For the lazy reinitialization approach, while descending into the tree during a search operation, the subtree id is compared to that of its parent. If they are different, the subtree id is set to the parent's subtree id, and the time capacity is reset to the maximum (calculated based on the node's depth).

During a subtree recycling operation, several things need to happen; 1) the forward pointers along the right side of the non-expired subtree need to be set to point to the root of the recycled subtree, 2) the used capacity of the recycled subtree needs to be added back to the root, 3) the root's pointers to its left and right children must be swapped, 4) for the root node of the recycled subtree the subtree id must be given a new unique value and its capacity must be replenished. These steps are shown in phase 2 of Figure 4 where the recycling operation commences at time 4. The search operation is depicted in phase 3.

If we compare the running time of the lazy versus the eager approach, we see that with the lazy approach, the only non-constant-time operation is step 1, setting the forward pointers of the non-expired subtree, which will run in logarithmic time to the size of a subtree. The eager approach, on the other hand, will require linear time to reinitialize the subtree. The lazy approach will become especially important for future work when we explore adding real-time constraints on the reservation system itself (i.e., "just-in-time reservations").

dTBTree Usage in RT-Llama

The dTBTree is used in the LRM to store utilization information for each reserved executor. Moreover, there is a cached version of each local dTBTree in the HUR of the GRM. This cached dTBTree is an approximation of the original in both time and space in that it receives change updates only after specified intervals, and can have a height that is shorter than that of the original, if space is a constraint and can be traded for accuracy. Later we will study the impact of varying these two parameters on the effectiveness of the pre-screening process.

In terms of using a dTBTree for real systems, a typical scenario would involve creating the tree to accommodate tasks as far ahead as 1 hour, with 100 ms leaf node granularities. Using even powers of 2, we could create a tree of height 16 with 65,536 leafs nodes and 131,071 total nodes in the tree, yielding 1.82 hours of reservation time. These figures would be doubled in order to accommodate subtree recycling. In order to optimize on space, the subtree recycling procedure could be designed to recycle smaller parts of the tree at a time (e.g., a quarter or eighth, instead of half); this is to be explored in the future.

Reservation Mechanism

As shown in Figure 2, the reservation mechanism begins when a user submits the input, desired output, QoS requirements, start time and deadline to the GRM. The GRM in turn composes a suitable BP, schedules it for execution, and submits it to the BPEL engine at the specified start time. The main challenges during this process include 1) deciding the intermediate start times and deadlines for each service within the BP, and 2) appropriately reserving all the hosts involved in the BP using the two-phase commit transaction protocol. The exact difficulty with each challenge depends on the strategy being used.

When a BPS receives the initial input from a user, it then contacts RT-QBroker to construct the BP based on the functional requirements. During RT-QBroker's selection phase, it can pre-screen each service to be selected based on the service's utilization information in the HUR. The utilization of S_i's host over time t is given by $u_{i,j}(t)$, and let k be some utilization threshold. The service may be considered if its host's utilization over the BP's time frame is below the threshold, i.e., $u_{i,j}(w_i) <= k$.

Once service selection is complete and a BP is composed, there are two main strategies that can be used to perform service reservations:

- **Concurrent:** This strategy is based on first selecting the BP and then determining the start time and deadline for each service, i.e., $a_{i,j}$ and $d_{i,j}$, after which the reservation process may begin. Moreover, both RT-QBroker and the BPS can screen service availability against the HUR. This way, if a service is unlikely to be available, RT-QBroker may produce another BP without the overhead of unsuccessful distributed reservation procedure.

Before the reservation, the BPS must determine parameters $a_{i,j}$ and $d_{i,j}$ for each service. Under the concurrent strategy, we adopt two intermediate deadline assignment methods, proportional deadline (PD) assignment and normalized proportional deadline (NPD) assignment. For PD, the values of $a_{i,j}$, $d_{i,j}$, and $w_{i,j}$ are based on $wcet_{i,j}$ as a proportion of the total time available for the BP (i.e., w_i). Thus, we have $w_{i,j} = (wcet_{i,j} / \sum wcet_{i,j}) w_i$, $a_{i,j} = d_{i,j-1}$, and $d_{i,j} = a_{i,j} + w_{i,j}$. NPD is a more reasonable intermediate deadline assignment method in that it takes into account host utilizations. Under this scheme, we have $w_{i,j} = (wcet_{i,j} * u_{i,j}(w_i)) / (\sum_j wcet_{i,j} * u_{i,j}(w_i))$. Once the intermediate start times and deadlines are computed, the BPS may apply a more precise screening process according to the relation $u_{i,j}(w_{i,j}) <= k$, essentially checking the service's host's utilization over the service's own time window.

Now that the BP is ready to be reserved, asynchronous reservation requests are sent to the RM on each services' LRM, with a timeout attached to each request, using a transaction-based two-phase commit algorithm. If a timeout expires, or a reservation request is rejected, all pending service reservations must be undone, and the procedure repeats again with a new BP from RT-QBroker.

- **Sequential:** According to this strategy, a BP is reserved service-by-service from start to end using a greedy strategy in order to maximize the remaining time available for remaining services in the BP. RT-QBroker, and its screening option, is used in much the same way as in the concurrent strategy. However, the BPS does not find the intermediate start times and deadlines for each service. Timeout values are attached to each reservation request, and once a timeout expiration or reservation failure occurs, the BP reservation must be rolled back.

There are tradeoffs to both approaches. With the sequential approach, there will likely be a higher level of success with each attempt. Each attempt, however, will most likely take longer than the concurrent approach. But each attempt of the concurrent approach would be less likely to be successful as it relies at best on the cached information in the HUR.

Simulation Results

The main goal of our simulation study is to observe 1) the performance and tradeoffs of different reservation methods and 2) the effectiveness of the RT-QBroker pre-screening function since one unique feature of our RT-SOA architecture design is to use the cached host utilization in HUR as a hint in order to increase the reservation efficiency. To focus on the real-time requirements of BPs rather than the complexity of BP structures, all BPs in our simulation have only sequential structures.

We have defined the following system properties in our study. 1) The *workload factor* is the amount of workload generated for the simulation. A workload factor of 1.0 represents a workload exactly equal to the system capacity, which is the maximum amount of work that can be handled by the system. A workload factor of 0.8 represents a workload 80 percent of the system capacity. 2) The *success ratio* is the number of successful reservation requests divided by the total number of reservation requests. 3) In our design, the GRM has the option of rejecting a request after the first unsuccessful reservation attempt or trying to come up with another BP and repeat the reservation. In other words, there could be multiple reservation attempts by the GRM per user request. We define the *average attempts* to be the total number of attempts divided by total number of requests, which approximates the average response time to a user request. 4) The *efficiency* is the number of successful reservation requests divided by the number of reservation requests that passes pre-screening. Efficiency, in contrast with the success ratio, measures the success of reservations that only pass the pre-screening process, and thus captures the overhead spared by avoiding remote reservation operations.

In the following experiments, we simulate 5 hosts. Each host has a single reserved executor for simplicity. However, our simulation applies to multiple reserved executors, as they are temporally isolated from each other. A dTBTree with 1024 time units is associated with each host to record the utilization information. Services with the same function are replicated so that we can always switch to a host with a low utilization if there is one. In the simulation, we simulate a process containing 3 services with each service having an execution time of 1 time unit. Process instances differ in their start times and have a relative end-to-end deadline varying between 20 to 32 time units. To simulate a more realistic environment, we pre-loaded the system with 30% of system capacity before collecting data.

Evaluation of Reservation Methods

As discussed earlier, advance reservations can be made using: 1) concurrent with proportional deadline (PD) assignment, 2) concurrent with normalized proportional deadline (NPD) assignment, or 3) sequential (SEQ) schemes. We have designed experiments to see 1) the effect of the reservation methods and the *workload factor* on the success ratio, and 2) how many attempts on average are necessary to either accept or reject a reservation if we allow multiple attempts.

Figure 5 shows the success ratio of the three reservation methods for different workload factors under a monotonic-random pattern. In a monotonic-random pattern, reservation requests form clusters that increase monotonically along the time horizon. But within each cluster, the reservation requests are random in their start times. We believe this work pattern simulates realistic workloads.

In Figure 5, we can see that when the workload factor increases, the success ratio decreases because more requests are generated but the system capacity remains the same. For example, the maximum success ratio that can be achieved for a workload factor of 2.0 is 50%. Another observation is that when the workload factor increases, the success ratios achieved by all schemes approach the upper bound value, i.e., 50% for a workload factor of 2.0.

Figure 6 shows the effect on success ratio and avg. attempts if multiple reservation attempts can be made for a single business process request. As expected, avg. attempts (lines) increase for all three schemes when the max attempt value increases and the result for SEQ remains the lowest. For the success ratio (bars), concurrent schemes (PD and NPD) enjoy a relatively greater benefit from increasing the max attempts value, while SEQ remains high. Although SEQ outperforms both PD and NPD in terms of success ratio and average attempts, there is a tradeoff not shown in the figure:

Figure 5. Success ratio for different reservation schemes and workloads (MaxAttempts=1, monotonic-random pattern).

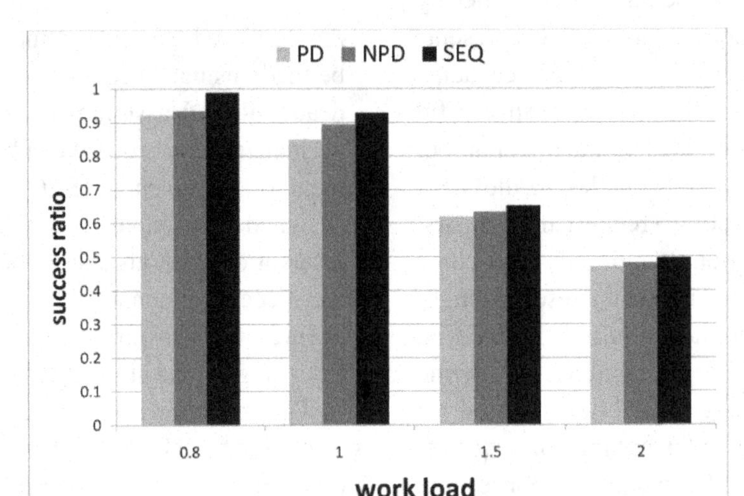

The response time for a sequential reservation is the sum of response times of individual reservations, whereas for the concurrent schemes, the response time is the maximum of response times of all individual service reservations. Therefore, the concurrent schemes may perform more efficiently where long communication latencies (between GRM and LRM) are expected.

Evaluation of Pre-Screening Functionality

As mentioned before, the GRM avoids selecting services with very high host utilization because that increases the likelihood of missing deadlines. Specifically, we set up a utilization threshold k. Whenever a service is to be selected to compose

Figure 6. Effects on success ratio and avg. attempts using multiple reservation attempts (workload factor=1.0, monotonic-random pattern).

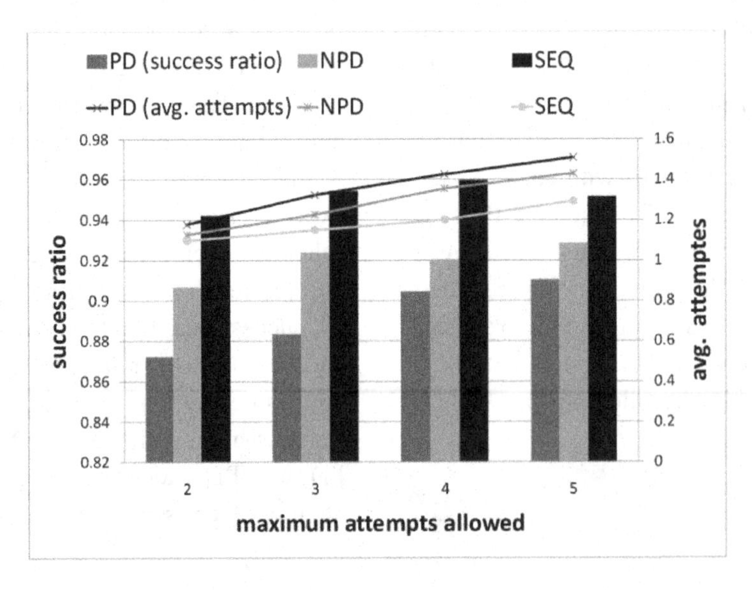

Figure 7. QBroker result (workload = 1.0 and HUR interval = 2)

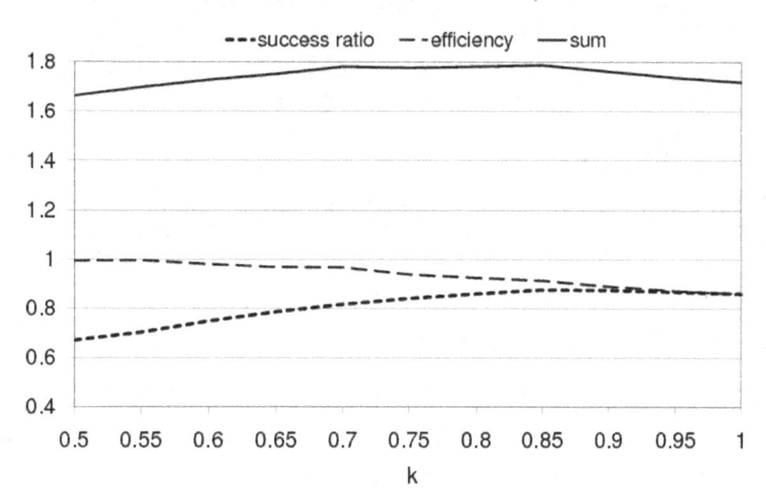

a process, we require that its host utilization be under a threshold k. If all candidate services cannot satisfy the condition, the GRM will notify the user about the reservation failure without actually performing the reservation on LRMs.

We have designed experiments to see the effect of host threshold k, workload factor, and HUR update interval on the RT-QBroker pre-screening success ratio and efficiency. Since our goal is to jointly optimize both the success ratio and efficiency, we also plot the sum of these two values as a general indicator of the joint optimum. As a point of reference, a k value of 1 represents no pre-screening. The results of our experiments are shown in Figures 7 and 8. Choosing the optimization criteria to be maximizing the sum of these two values, we see that when the workload factor is 1, the optimal k is around 0.85. However, when the workload factor increases to 1.5, the optimal k decreases to about 0.6. The implications for real systems is that the k value should be varied according to historical or expected workload to achieve the optimal level of reservation success ratio and efficiency.

Figure 8. QBroker result (workload = 1.5 and HUR interval = 2)

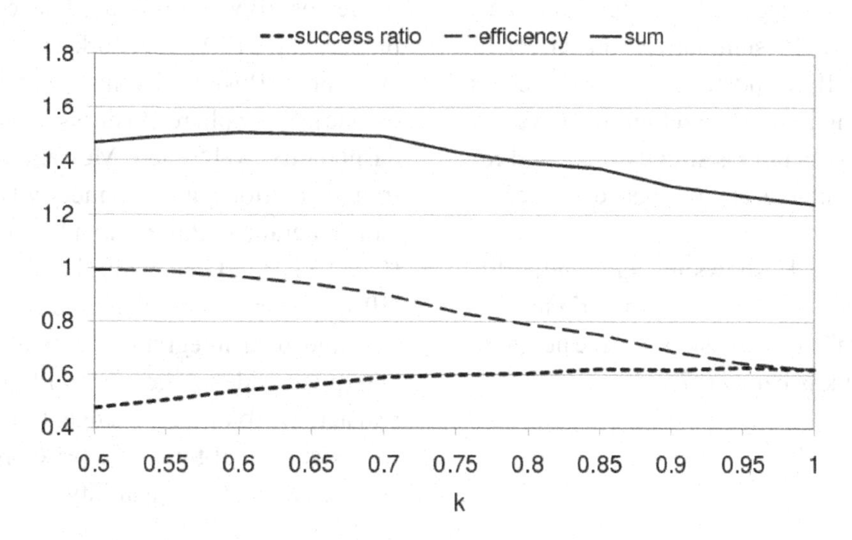

Figure 9. Effect of (workload, HUR interval) on efficiency (tree height = 10)

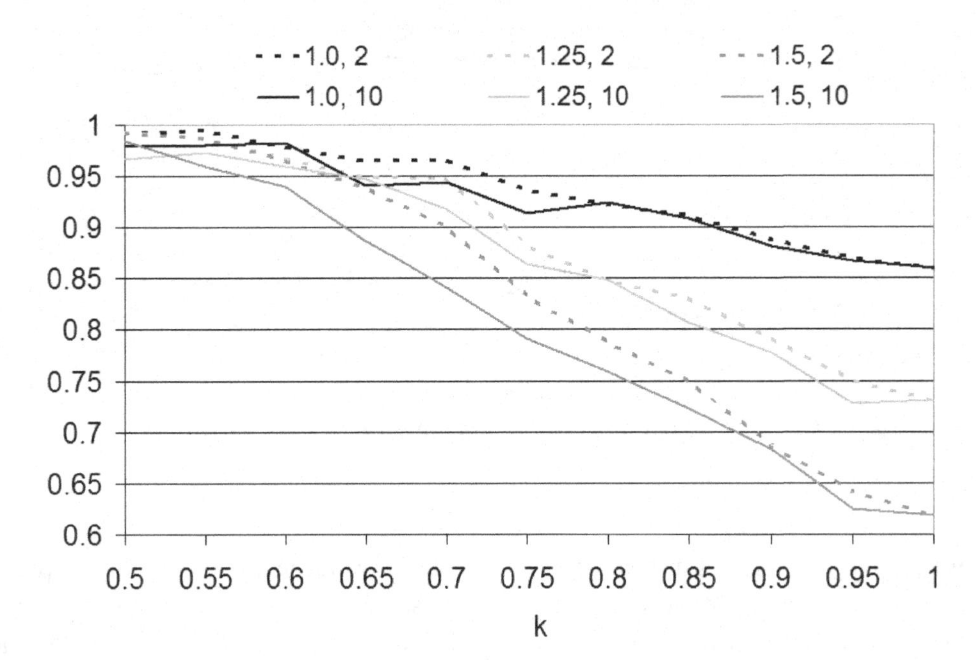

Figure 9 summarizes the effect that workload and HUR update interval has on efficiency and to see why it is important to have a lower value of k when the workload is high. In terms of the effect of the HUR update interval, we see that a larger interval does not have a noticeable impact on success and efficiency. Therefore, further network overhead can be saved by updating the HUR less frequently. Figure 9 also shows that a clear but small difference in the effect of the update interval only appears for higher workloads. Similarly, Figure 10 summarizes the effect of workload and HUR update interval on success ratio. Comparing Figure 9 and Figure 10, we can see that the success ratio is slightly increased for a higher update interval at the expense of reservation efficiency.

Finally, Figure 11 shows the sum under different HUR tree heights. There is a definite tradeoff that arises from decreasing the tree height in the HUR when k is below 0.7.

Related Work

Real-time enterprise is an attractive idea that has captured interest from many IT companies. Companies such as IBM, Microsoft, Sun Microsystems, and HP have all invested heavily on developing the technology. Microsoft has developed Microsoft Dynamics as a line of integrated business management solutions that automate and streamline financial, customer relationship, and supply chain processes. IBM has proposed the complex event processing (CEP) framework which is made up of WebSphere Business Events (the event processing engine), WebSphere Business Monitor (the rich dashboard), WebSphere Message Broker (event transformations and connectivity functions), and Generalized Publish and Subscribe Services (GPASS) (Bou-Ghannam & Faulkner, 2008). The HP ZLE framework claims to provide application and data integration to create a seamless, enterprise-wide solution for real-time information and action. To our knowledge, however, none of the existing real-time enterprise products offers service reservation capability.

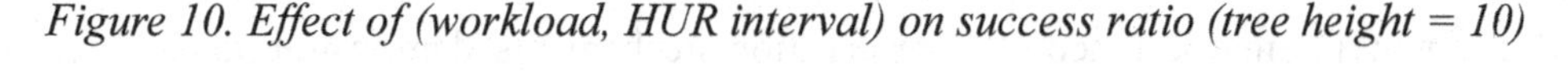

Figure 10. Effect of (workload, HUR interval) on success ratio (tree height = 10)

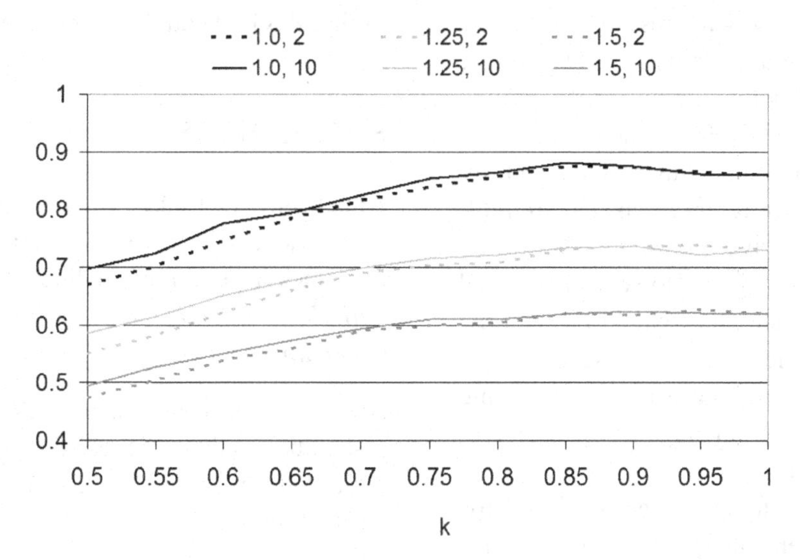

The common object request broker architecture (CORBA) (Object Management Group, 2004) is an object-oriented distributed architecture designed to provide location, platform, and programming language transparency. RT-CORBA (Object Management Group, 2003) brings real-time features to CORBA by specifying a priority-based scheme for handling object requests. RT-Llama architecture differs from RT-CORBA in that 1) RT-Llama is deadline-based rather than priority-based and 2) RT-Llama can reason about the predictability of an entire process rather than just one service at a time. In other words, based on the deadline of the entire process, RT-Llama manages to determine the intermediate deadline for and schedules each individual service that comprises the process.

Figure 11. Tree height result on sum factor (workload = 1.5 and HUR interval = 2)

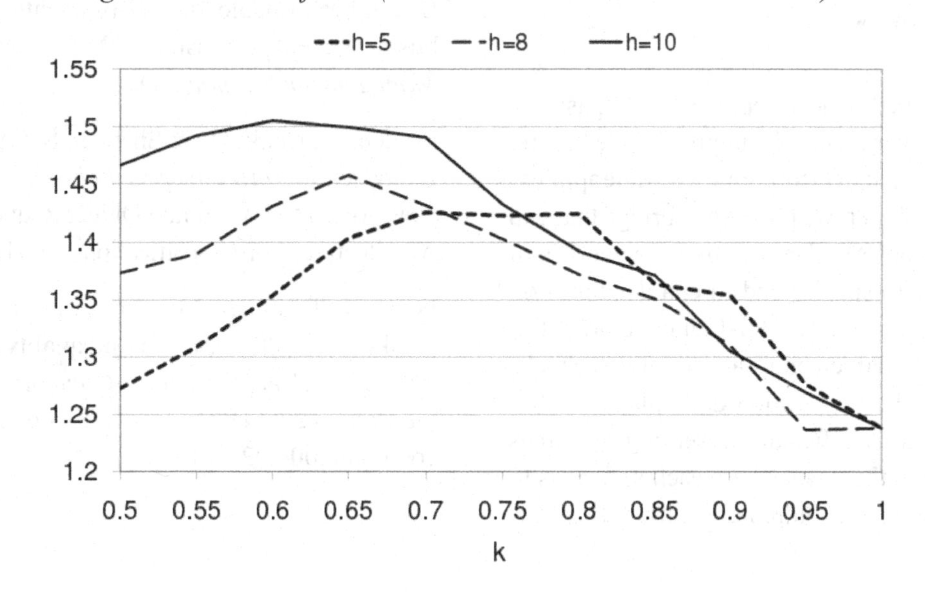

Advance reservation systems like that designed in RT-Llama is a growing research area. For example, Mamat et al. (Mamat, Lu, Deogun, & Goddard, 2008) discuss an advance reservation system for clusters, particularly to help manage I/O conflict among nodes in a cluster. They identify an advance factor for reservations that are multiples of task interarrival times, anywhere from immediate to a factor of ten. However, no special discussion of data structure would be required for such a short time range. Our research, however, requires efficient data structures for storing managing reservations on the order of seconds to minutes in advance.

The GARA system (Foster, Fidler, Roy, Sander, & Winkler, 2004) is another advance reservation system that is especially useful for reserving bandwidth for streaming applications. However, streaming media applications have coarser requirements than the individual service execution reservations required for RT-SOA. Many projects have studied distributed real-time resource scheduling (Abdelzaher, Thaker, & Lardieri, 2004; Zhang, Gill, & Lu, 2008), but they do not address service process composition for real-time applications and are not designed to use the flexibility of SOA.

CONCLUSION

SOA has gained wide acceptance in the past few years. However, due to its unpredictable nature, current SOA cannot incorporate real-time applications. In this paper, we propose to bridge that gap by creating an RT-Llama SOA framework that allows users to specify end-to-end deadlines on their business processes. RT-Llama in turn, after performing initial screening and feasibility checks based on cached utilization data, plans the full BP execution by efficiently reserving resources in advance on the hosts where each service is to execute. We have explored the merits of various reservation mechanisms, demonstrating the flexibility of RT-Llama for different environments.

REFERENCES

Abdelzaher, T., Thaker, G., & Lardieri, P. (2004). A feasible region for meeting aperiodic end-to-end deadlines in resource pipelines. In *Proceedings 24th International Conference on Distributed Computing Systems* (pp. 436-445).

Abeni, L., & Buttazzo, G. (2004). Resource reservation in dynamic real-time systems. *Real-Time Systems*, *27*(2), 123–167. doi:10.1023/B:TIME.0000027934.77900.22

AUTOSAR (2008). Specification of the virtual functional bus, v1.0.2, r3.1, rev 0001. *Automotive Open System Architecture*.

Bichler, M., & Lin, K.-J. (2006, March). Service-oriented computing. *IEEE Computer*, *39*(3), 99–101.

Blake, S., Black, D., Carlson, M., Davies, M., Wang, Z., & Weiss, W. (1998). *An architecture for differentiated services, RFC 2475.*

Bollella, G., Brosgol, D., Furr, H., et al. (2000). *The real-time specification for java*. Addison-Wesley. Bou-Ghannam, A., & Faulkner, P. (2008, December). Enable the real-time enterprise with business event processing. *IBM Business Process Management Journal*(1.1).

Braden, R., Clark, D., & Shenker, S. (1998). *Integrated services in the internet architecture: an overview, RFC 1633.* Common Object Request Broker Architecture: Core Specification. (2004, March).

Foster, I. T., Fidler, M., Roy, A., Sander, V., & Winkler, L. (2004). End-to-end quality of service for high-end applications. *Computer Communications*, *27*(14), 1375–1388. doi:10.1016/j.comcom.2004.02.014

Huhns, M. N., & Singh, M. P. (2005, January-February). *Service-oriented computing: Key concepts and principles.* IEEE Internet Computing.

Lin, K.-J., Panahi, M., Zhang, Y., Zhang, J., & Chang, S.-H. (2009). Building accountability middleware to support dependable SOA. *IEEE Internet Computing, 13*(2), 16–25. doi:10.1109/MIC.2009.28

Mamat, A., Lu, Y., Deogun, J., & Goddard, S. (2008, July). Real-time divisible load scheduling with advance reservation. In *Euromicro conference on real-time systems, ecrts '08* (pp. 37-46).

Martin, R. (2007, May). Data latency playing an ever increasing role in effective trading. *InformationWeek.*

Moses, S. A., Gruenwald, L., & Dadachanji, K. (2008). A scalable data structure for real-time estimation of resource availability in build-to-order environments. *Journal of Intelligent Manufacturing, 19*(5), 611-622. Object Management Group. (2003, November). *Realtime-corba specification, v 2.0.*

Panahi, M., Lin, K.-J., Zhang, Y., Chang, S.-H., Zhang, J., & Varela, L. (2008). The llama middleware support for accountable service-oriented architecture. *In Icsoc '08: Proceedings of the 6th international conference on service-oriented computing* (pp. 180–194). Berlin, Heidelberg: Springer-Verlag.

Yu, T., & Lin, K.-J. (2005, December). Service selection algorithms for composing complex services with end-to-end QoS constraints. In *Proc. 3rd Int. Conference on Service Oriented Computing* (ICSOC2005), Amsterdam, The Netherlands.

Yu, T., Zhang, Y., & Lin, K.-J. (2007, May). Efficient algorithms for web services selection with end-to-end QoS constraints. *ACM Transactions on the Web.*

Zhang, Y., Gill, C., & Lu, C. (2008). Reconfigurable real-time middleware for distributed cyber-physical systems with aperiodic events. *International Conference on Distributed Computing Systems* (pp.581-588).

Zhang, Y., Lin, K.-J., & Hsu, J. Y. (2007). Accountability monitoring and reasoning in service-oriented architectures. *Journal of Service-Oriented Computing and Applications, 1*(1).

This work was previously published in International Journal of Systems and Service-Oriented Engineering, Volume 1, Issue 1, edited by Dickson K.W. Chiu, pp. 62-78, copyright 2010 by IGI Publishing (an imprint of IGI Global).

Chapter 18
SLIM:
Service Location and Invocation Middleware for Mobile Wireless Sensor and Actuator Networks

Gianpaolo Cugola
Politecnico di Milano, Italy

Alessandro Margara
Politecnico di Milano, Italy

ABSTRACT

One of the main obstacles to the adoption of Wireless Sensor Networks (WSNs) outside the research community is the lack of high level mechanisms to easily program them. This problem affects distributed applications in general, and it has been replied by the Software Engineering community, which recently embraced Service Oriented Programming (SOP) as a powerful abstraction to ease development of distributed applications in traditional networking scenarios. In this paper, the authors move from these two observations to propose SLIM: a middleware to support service oriented programming in mobile Wireless Sensor and Actuator Networks (WSANs). The presence of actuators into the network, and the capability of SLIM to support efficient multicast invocation within an advanced protocol explicitly tailored to mobile scenarios, make it a good candidate to ease development of complex monitoring and controlling applications. In the paper the authors describe SLIM in detail and show how its performance easily exceeds traditional approaches to service invocation in mobile ad-hoc networks.

INTRODUCTION

After an initial period of research and experimentation, *Wireless Sensor Networks* (WSNs) (Sohraby, Minoli, & Znati, 2007) and their siblings: *Wireless Sensor and Actuator Networks* (WSANs) (Akyildiz & Kasimoglu, 2004) are entering a more mature phase, with several commercial companies offering products to support a wide range of application domains concerned with monitoring and control. On the other hand, to hold the promise of bridging the gap between the physical and the virtual world, WSANs have still to overcome

DOI: 10.4018/978-1-4666-1767-4.ch018

a major limitation: the difficulty of developing applications on top of a given WSAN platform.

Indeed, usually WSAN programming has to be carried out in a very low-level, system-dependent way. This requires programmers to have a strong technical background in several domains, from system to network, while also resulting in very long development and testing phases: something that is currently limiting the adoption of WSAN technology.

To overcome this limitation, the research community proposed various approaches to raise the level of abstraction for WSAN programming (Picco & Mottola, in press) but none of them has been widely adopted. Meanwhile, the Software Engineering community is proposing Service Oriented Programming (SOP) as a powerful approach to ease development of complex distributed applications. While others have already proposed using SOP in WSANs (see our section on related works), to the best of our knowledge none has offered a complete and integrated solution to bring the SOP world into the tiny scale of WSANs. In this paper we fill this gap by presenting SLIM: a middleware infrastructure to support service-oriented programming in *mobile WSANs*.

Issues in Mobile WSANs. Several applications of WSANs involve monitoring and controlling mobile entities like people, animals, or goods, through access and control points that can be also mobile, like PDAs in the hands of operators. This brings the need of considering mobility as a key aspect of a sensor network: an aspect such important because it has a great impact on the communication and coordination layers. Indeed, as the research on Mobile Ad-Hoc Networks (MANETs) shows, the communication layer of a mobile application has to include effective mechanisms to cope with unreliable links and routes. Similarly, ad-hoc coordination mechanisms are required to cope with nodes that may quickly become unreachable due to network partitions, while other nodes may appear as they approach the coordination area. Even the application layer is impacted by mobility, as the location, and more in general the *context* of nodes, becomes an important issue to consider in choosing which nodes to contact.

SLIM addresses these issues by adopting an advanced routing protocol explicitly designed to manage unreliable links and dynamic group membership, while it decouples the service matching policy from the communication layer, allowing applications to choose and install into the middleware the matching component that better fits their needs (e.g., one including contextual information as part of service descriptions).

SLIM also provides an efficient, peer-to-peer service discovery protocol fitting the needs of a mobile network in which nodes (including a registry) may easily become unreachable. It also covers the typical needs of a network designed for sensing, by providing two forms of invocation: unicast and multicast. The former allows service consumers to get data from a single sensor or to send a command to an actuator, while the latter allows to invoke all the services that satisfy a given query at once, e.g., to gather the data produced by a set of sensors in a single step.

Organization of the presentation. In the remainder of the paper we describe SLIM in details. In particular, the SLIM architecture and API is the topic of the next section, then we describe the routing protocol behind SLIM, while in section "Evaluation" we discuss SLIM performances and compare them with those attainable with more classical solutions for SOP. Finally, the "Related Work" section surveys other research results related with SLIM, while in the last section we draw some conclusions and describe our future research plans.

Figure 1. Reference scenario

THE SLIM ARCHITECTURE

The reference scenario for SLIM (see Figure 1) is that of a *Wireless Sensor and Actuator Network* (WSAN), composed of a set of nodes, including fixed or mobile sensor motes, fixed or mobile actuators, PDAs in the hand of operators, and gateways toward a traditional network (a LAN or the Internet). In a SOP style, every node may act as a *service provider*, a *service consumer*, or both. Service providers offer one or more *services* to consumers, each described through a *service descriptor*. Service consumers (or simply *clients*) locate the services they need by passing a *query* to a *lookup service*, and send *messages* to invoke them, receiving other messages back as replies.

As mentioned, SLIM clients may also send messages in multicast, to reach all those services that match a given query. This form of communication, which is not common in the panorama of SOP middleware, not only has the potential of reducing the cost of invocation, through an intelligent use of multicast routing, but it also allows avoiding the service location phase for all those cases, which are common in WSANs, in which the service consumer is interested in invoking all the service providers that offer the same service, e.g., that of temperature and humidity reading in a vineyard.

The kind of applications that may benefit of SLIM include environmental monitoring and actuating, e.g., for precision agriculture (Sikka et al., 2006); monitoring people at home or in hospitals, e.g., for elderly care (*Proceedings of the 3rd Int. Conf. on Pervasive Computing Technologies for Healthcare*, 2009); monitoring animals in farms (Andonovic et al., 2009); or monitoring goods moving around, e.g., for supply chain management (Evers & Havinga, 2007).

SLIM supports these scenarios through a middleware explicitly designed to be embedded in small scale nodes like sensor motes, which supports both service location (i.e., discovery) and invocation. Figure 2 shows the internal architecture of the SLIM middleware implemented into each node. On top of the MAC protocol (e.g., IEEE 802.15.4), SLIM includes an advanced routing protocol explicitly conceived to support the SLIM functionalities in a mobile, multi-hop WSAN. Next section describes this protocol in detail. Here we focus on the other layers that compose the SLIM middleware starting from the *Service Matcher*,

Figure 2. The SLIM internal architecture

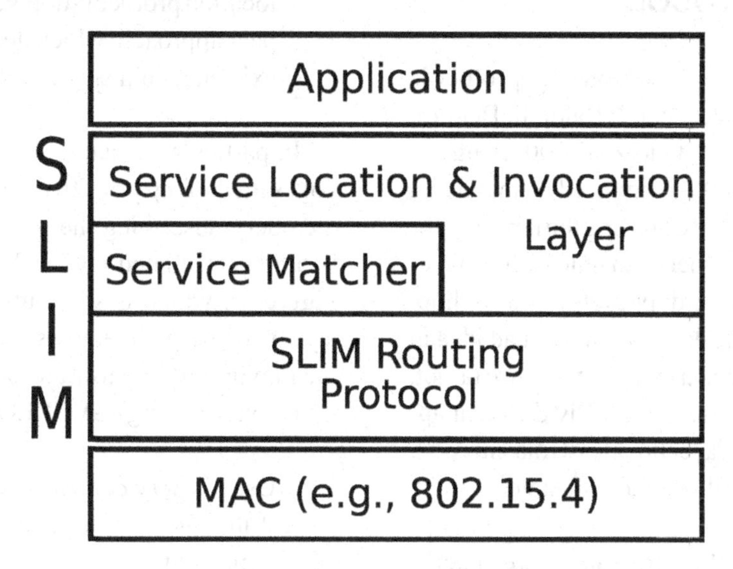

which implements the logic that allows service descriptors to be matched by service queries. This is a generic component, which can have multiple implementations for different deployments of SLIM. This way SLIM is not forced to use a single service description and query language, but can adapt to multiple languages, from the simplest ones, in which services are described by a small set of attributes as in nanoSLP (Jardak, Meshkova, Riihijrvi, Rerkrai, & Mahonen, 2008), to the most complex ones, like those using XML descriptors and queries. Since SLIM performs service discovery using a query message that reaches every node in the network (more on this later), the matching is solved at each node separately. This reduces the effort that each node has to spend (minimizing power consumption) and simplifies the issue of coding the *Service Matcher*, which has only to check if the incoming query matches the services exported by the local node.

On top of this is the *Service Location & Invocation Layer*, which offers a simple API to register a service, discover the services of interest, and invoke them. The main operations provided by this API are shown in Figure 3 using the C syntax, as most of the WSAN platforms use this language (or a dialect, like NesC). The first two operations allow applications to register a service into the SLIM runtime and to locate the services they need. The sendMsg operation is used to send messages to single services, chosen among those returned by the lookup operation. Finally, the mcastMsg operation sends the same message, in multicast, to all those services that satisfy the query q.

Figure 3. The SLIM API

```
registerSvc(SvcDescr s, void (*callback)(void *msg,
               int msize, void *reply, int *rsize))
SvcDescr *lookup(SvcQuery q)
void sendMsg(SvcDescr s, void *msg, int msize)
void mcastMsg(SvcQuery q, void *msg, int msize)
```

THE SLIM PROTOCOL

Previous experience with routing protocols (Cugola & Migliavacca, 2009; Baldoni, Beraldi, Cugola, Migliavacca, & Querzoni, 2005; Cugola, Murphy, & Picco, 2006) convinced us that using traditional routing tables, holding the next hop for each destination, together with unicast link-layer transmission to forward packets, hop-by-hop, from the source to the destination, is a bad idea in mobile, pervasive scenarios, like those we target. Accordingly, SLIM uses a radically different approach based on link-layer broadcast transmission, opportunistic forwarding, and soft state.

Brief summary of the protocol. More specifically, each SLIM node keeps a *distance table* storing the distance from those nodes it heard about recently (information in this table expires after a short period of time). During the service location phase the distance table of each node is filled with the distance (in hops) from the service consumer who originates the query. This information is subsequently used to bring service advertisements back. At the same time, during this "return" phase, the distance table of the nodes along the route toward the service consumer is filled with information about the service providers who replied. Finally, both information (i.e., the distance from service consumers and service providers) are used to route request and reply messages back and forth. During this message exchange, tables of nodes along the route are renewed, while new nodes have the opportunity of discovering their distance from the communicating peers.

Service location phase. The traditional approach to service location, which uses a single node acting as a registry to store information about all the services available in a given domain, is hardly applicable to a dynamic, pervasive scenario like the one we address in this paper. Starting from this consideration, the SLIM location protocol adopts a different, peer-to-peer approach, which does not rest upon the existence of a specific registry node.

In particular, when a service consumer *C* invokes the lookup primitive, its MAC address and the query describing the services it is interested in are encapsulated into a SVCREQ packet (see Figure 4), which is sent out in broadcast (link-layer), while a timeout is started to determine the maximum time to wait for collecting results.

Upon receiving a SVCREQ packet, each node:

1. Uses the *service matcher* component to check if the query matches one of the services it registered;
2. Adds a new record to its distance table (see Figure 4) with the address of *C* as the first field, the number of hops to reach *C* (i.e., the hops traveled by the SVCREQ packet so far) as field *DH*, and the RSSI (Received Signal Strength Indication) of the packet as field *SS*;
3. Starts a timer that is inversely proportional to the RSSI of the packet and the remaining battery of the node.

If, during this waiting period, the same packet (each packet is identified through the address of the source and a progressive number) is heard again, the node cancels retransmission; otherwise, when the timer expires, it forwards the packet in broadcast. This algorithm is repeated at each hop with two results:

1. The SVCREQ packet floods the network being retransmitted only by those nodes that are farther by the previous forwarder (thus having the opportunity of covering the largest new area, minimizing the number of retransmissions) and whose battery is more charged;
2. The distance table of each node in the network is filled with the distance *DH* (in hops)

Figure 4. SLIM distance table and packets

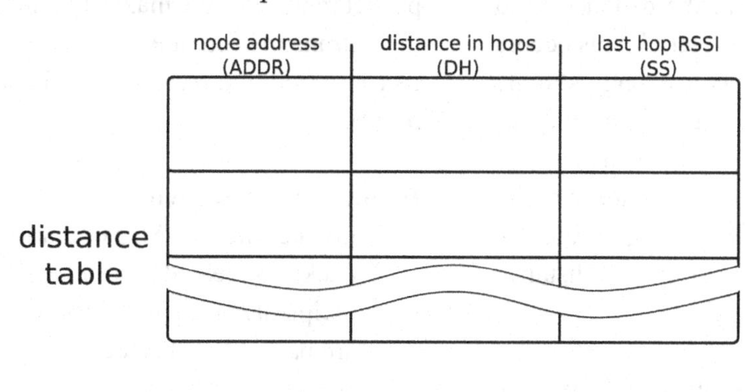

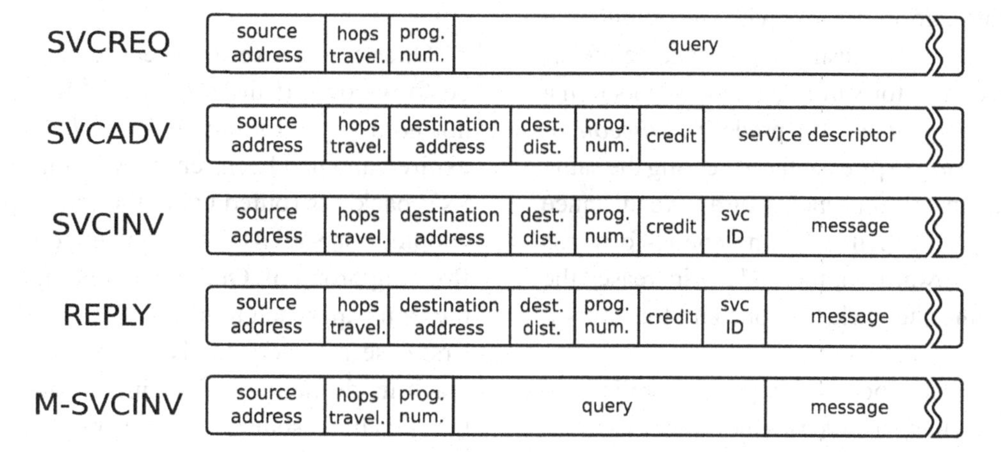

from the service requester *C*, plus the signal strength *SS* of the last hop traveled.

When the SVCREQ packet reaches the provider *P* of a service that matches the request, a SVCADV packet is created, which holds *P*'s MAC address, the address of the requester *C* (i.e., the source of the SVCREQ packet), the distance of *C* from *P*, and the descriptor of the matching service.

Like SVCREQs, SVCADVs are also sent out in broadcast, but routing is different. Indeed, when a node *N* hears a SVCADV packet:

1. It updates its distance table from *P*;
2. If its distance from the destination *C* (field *DH* of the distance table) is greater or equal to the distance included into the packet itself, *N* drops the packet;
3. Otherwise it updates the distance from the destination recorded into the packet and starts a timer inversely proportional to a combination of *DH*, *SS*, and the remaining battery.

As for SVCREQs, if during this waiting period the same packet is heard again, the node cancels retransmission, otherwise it forwards the packet in broadcast.

Under ideal conditions, this protocol results in an efficient, greedy forwarding of SVCADV packets, which chooses opportunistically, as forwarders, those nodes that are closer to the destination *C* (small *DH*) and whose link from the probable next forwarder are stronger (large *SS*).

Mobility and local minima. On the other hand, mobility may break this protocol, by introducing local minima in the distance field toward the service requester *C*. This occurs whenever a node *N* has a wrong estimate of its distance from *C*, e.g., because it was once closer to *C* but now moved in a region where the real distance is higher. Nearby nodes will not forward packets generated by *N* because of the (wrong) smaller distance it puts in those packets.

To solve this issue we complemented the basic forwarding algorithm above with a *retransmission mechanism*. After transmitting a packet (either as a source or as a forwarder), a node *N* puts it in a *retransmission queue*. If a predefined *timeout of retransmission* expires without hearing the same packet again, this time including a lower distance from *C* (this happens when no one re-forwards the packet toward *C*), the node *N* increases the distance into the packet by one and transmits it again.

The consequence of this mechanism is two-fold: on one hand it increases the protocol's resistance to collisions, which is good since link-layer broadcast is particularly subject to collisions. On the other hand, increasing the distance for packets that were not forwarded by neighbors, it also allows overcoming local minima, by increasing the set of potential forwarders for the retransmitted packet.

Unfortunately, asymmetric links may trigger the retransmission mechanism even when it was not required, thus increasing the network traffic without any positive effect on delivery. To limit this problem we allow each node to retransmit each packet at most once. Moreover, we also introduce in SLIM a mechanism of *credits*, which further reduces retransmission. When a SVCADV packet is created, it is assigned a predefined credit (an integer), which is decremented each time the retransmission timeout expires at a node and the packet is retransmitted. When the credit value goes to zero the retransmission mechanism is not used anymore, i.e., the initial credit of a SVCADV packet represents the maximum number of times the retransmission mechanism may fire along its route from the service provider to the service requester.

Invocation phase. When a service consumer *C* invokes the sendMsg operation, a SVCINV packet is created holding the address of the recipient *P*, the identifier of the service (both are part of the service description), and the message to be sent.

Once this packet has been created it must be forwarded. If the distance table of *C* does not hold any information about *P* (because it expired and had been removed) then the SVC-INV packet is routed using the same approach adopted for SVCREQs, i.e., using our "smart" flooding protocol. Otherwise it is routed using the same approach used for SVDADVs. In the first case it is very likely that the packet will reach its destination: flooding is robust and it rebuilds the distance field toward *C* that is easily followed back by replies. As a consequence, in this case we do not adopt any mechanism to further increase delivery. Conversely, if the packet is sent using the SVCADV approach, it may happen that it gets lost. To cope with this risk, *C* starts a timer just after sending the packet. If the timer expires without receiving a reply, the SVCINV packet is sent again, this time using the flooding approach.

Finally, replies are encoded into REPLY packets, which are routed using the same approach adopted for SVCADVs.

The last case we have to consider is that of a node *C* invoking the mcastMsg operation. In this case we build an M-SVCINV packet, similar to SVCREQs, but including the service query provided by *C* as an implicit destination address. This packet is routed as SVCREQs, flooding the network in search of matching services. A timeout is used to set the maximum time to wait for collecting results.

EVALUATION

To provide an accurate and replicable analysis of SLIM under different conditions we used the OMNeT++ (Varga, 2001) network simulator. To simulate a WSAN with mobile nodes we used the Mobility Framework (http://mobility-fw.sourceforge.net) in its most recent incarnation, which includes a fairly accurate model of a CC2420 card and the 802.15.4 MAC. For the channel, we used a path loss model taking into account interferences from parallel transmissions to compute (at runtime) the SNR of each frame.

To put the SLIM results in context we compared it against a solution based on DYMO (Chakeres & Perkins, 2008), a well known protocol (the successor of AODV) for unicast routing in MANETs. Since SOP usually adopts unicast invocation (using a unicast transport on top of an IP network) and being interested in a solution for mobile scenarios, the choice of DYMO was the most natural one. This also accounts for the fact that the IETF, while proposing IPv6 for WSAN (Hui & Culler, 2008), has not yet chosen which protocol to use for routing but is considering unicast protocols only.

DYMO is an *on-demand* routing protocol. It creates routes between nodes only when they are required for communication. Routes are generated by flooding the network with *route request* packets that create entries in the routing tables of receiving nodes. Entries are temporary: they are trusted and used for a limited period of time and then expire; after expiration a new route creation process is needed for further communication. After routes are created they are followed by forwarding packets in unicast, hop-by-hop. Notice that to offer a fair comparison with SLIM, we adapted DYMO to WSAN and to the specific application service we are interested in, by reducing the number and length of fields in packet headers.

Reference scenario. We considered a reference scenario in which a single node behaves as a client, periodically sending service requests and invocations, while all other nodes act as service providers. We chose this simplified scenario after observing how the results in scenarios including several clients (not included here for space's sake) did not exhibit remarkable differences w.r.t. the scenarios involving a single client issuing a large number of invocations (which we discuss below).

In particular, our reference scenario includes 75 service providers moving (with the client) in an area of 0.25 Km^2 ($500m$ x $500m$) at a maximum speed of 3 m/s (minimum is 0 m/s). Each service provider exposes a single service and client queries are matched by 10% of services on average. The exact behavior of the client is the following: it periodically sends a lookup request, chooses one service provider among all responding ones and sends 10 invocation messages to it, with an average delay among an invocation and the following one of 10 seconds. Starting from this reference scenario, we analyzes the performance of SLIM under different conditions, changing, one by one, the main parameters of the scenario: the density of the network (number of nodes per Km^2), the area in which nodes are located, the maximum speed of nodes, the frequency of service invocation, and the number of credits used by the SLIM routing protocol.

Service invocation. Our first analysis focuses on service invocation, forgetting about the performance and cost of service location. In particular, we measured *delivery* as the percentage of service invocations actually receiving a reply from the service provider, and *traffic* as the average traffic (in KB/sec) generated to obtain this result. For SLIM, the latter includes all the traffic generated by SVCINV (either when sent in unicast or in flooding) and REPLY packets, while for DYMO it includes the traffic generated to build routes and to transport messages forth and back.

Actual results are shown in Figure 5: we run each simulation 20 times, varying the seeds of the random number generators used in our models and plotting the sample mean we measured and its 95% confidence interval. Generally speaking, we observe that our protocol delivers much more messages that DYMO, especially when using at least one credit, while generating significantly less traffic.

In particular, first row of graphs in Figure 5 shows how SLIM and DYMO behave while changing the density of nodes. We observe how both protocols present higher delivery at higher densities, which is reasonable since a lower density increases the probability of partitioning the network. At our reference density of 300 nodes per Km^2, the delivery of SLIM with one or two credits is very close to 100% while DYMO still misses some packets. As for traffic, we notice how the SLIM forwarding mechanism, which opportunistically privileges long links when available, results in using a more or less constant number of hops to reach the interested service provider, and hence generates the same traffic, independently from the density of the network. On the contrary, DYMO does not include any mechanism to filter hops based on the distance between nodes: as a consequence its traffic grows linearly with the number of nodes.

A similar behavior can be observed when increasing the area of simulation (with a fixed density), as shown by the second row of graphs. Both protocols decrease their overall delivery while the area grows, as packets have to travel a higher number of hops to reach their destination; but SLIM limits this trend due to its robust forwarding mechanism, which does not choose forwarders deterministically as in DYMO, while also adopting ad-hoc mechanisms like packet retransmission and renewal of routes through overhearing, which reduce packet loss and help keeping distance tables in sync even in presence of mobility. As for the traffic, also in this case SLIM behaves better than DYMO thanks to its opportunistic selection of retransmitting nodes during packet forwarding.

If we focus on the impact of credits on performance, we may observe that credits increase delivery with a minimal impact on traffic. Actually, increasing the number of credits from 0 to 1 reduces the overall traffic since the presence of credits increases the chance for a packet to reach its destination, thus reducing the need of reissuing the same packet again. Further increasing credits (e.g., from 1 to 2) slightly increases traffic as the retransmission mechanism may wrongly fire due to the presence of asymmetric links. In conclusion, we may observe that the ideal number of credits to use depends on the area of the network (see second row of graphs), the greater the area the more beneficial to delivery is increasing the number of credits, but in common scenarios like those we considered (a maximum area of 0.75 Km^2 with 225 service providers), two is the maximum number of credits worth using.

As our default scenario takes into account node mobility, we also studied the impact of the maximum speed of nodes (third row in Figure 5). Both protocols reduce delivery when maximum speed grows, as fast mobility contributes to a faster invalidation of routing information. On the other hand, DYMO suffers this problem much more than SLIM, as it chooses routes deterministically and it does not include any mechanism for retransmission or routing information renewal (apart rebuilding the route entirely). Notice that DYMO cannot reach a 100% delivery even in a fixed scenario with no mobile nodes. While this appeared strange to us at first, a more in-depth study has revealed that the packet loss is once again due to the route creation mechanism: in particular, DYMO does not include any mechanism to filter out long hops while flooding route requests to build the routing tables to be used for packet forwarding. As a result, especially in presence of asymmetric links (our channel model includes them) it may create routes that are hard to follow back. As for the traffic, SLIM uses more bandwidth when the speed increases, as packet loss causes retransmission of packets

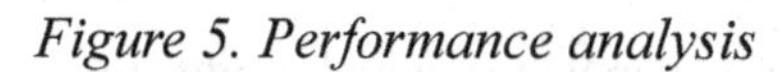

Figure 5. Performance analysis

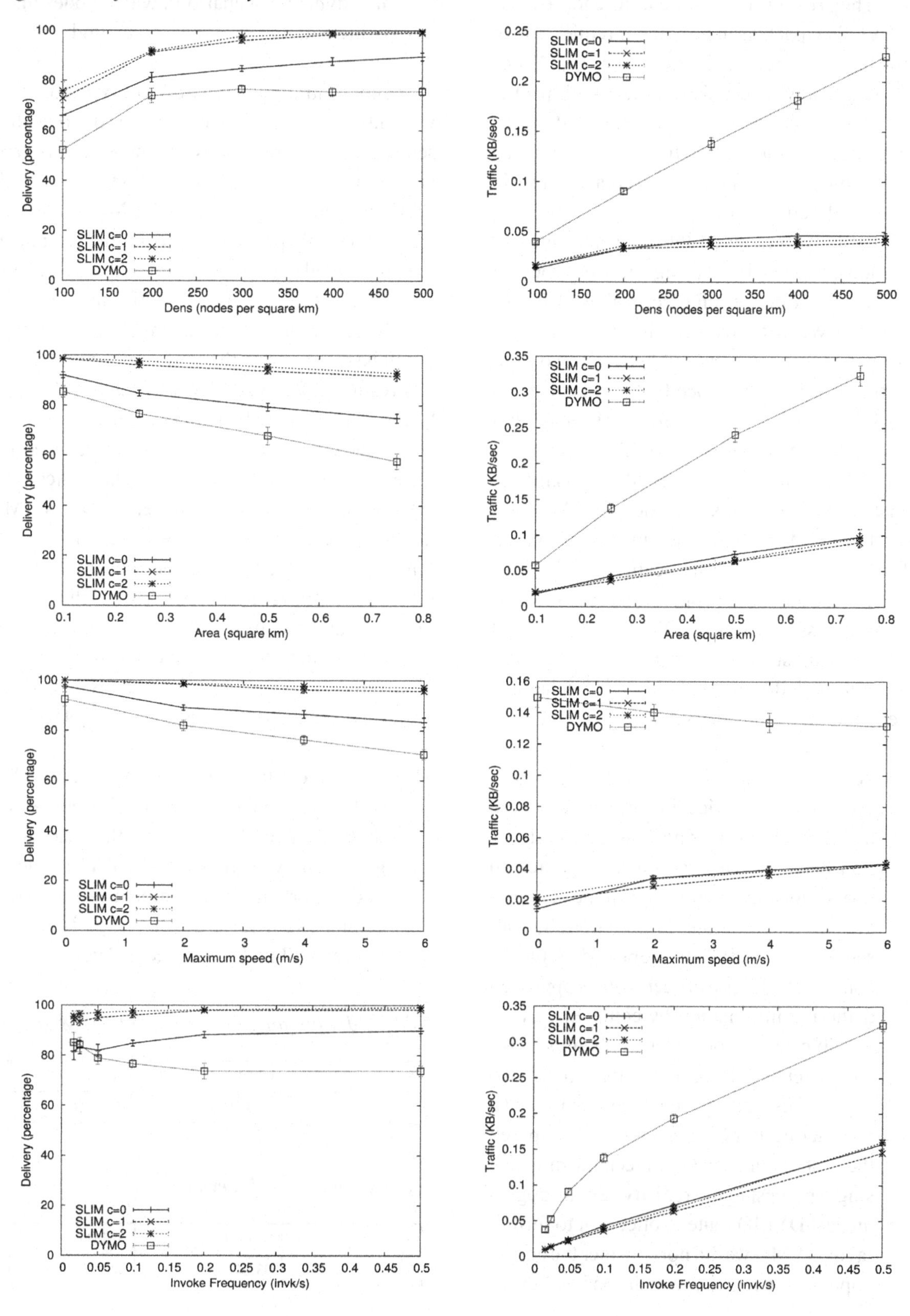

at each hop (using credits) or end-to-end (resending SVCINV packets in flooding if replies do not come back in time). On the contrary DYMO does not include any mechanism to deal with packet loss: as a consequence, once a packet is dropped it does not generate further traffic. This explains why we measured a moderate decrease in traffic as speed of nodes increases.

Finally, we measured the impact of the application behavior by varying the frequency of invocations from the reference value of 0.1 (1 every 10s). When the frequency is very low, routing information becomes invalid between two consecutive invocations and need to be updated. Both DYMO and SLIM use flooding in this phase, but our mechanism produces less traffic, minimizing the number of nodes that have to retransmit the packet, while keeping a similar delivery. At higher rates both protocols may use the same routing information for multiple invocations; however, a failure in the unicast transmission directly brings to packet loss when using DYMO while SLIM may opportunistically leverage different routes, while also including a retransmission mechanism when credits are greater than zero.

Service discovery. After measuring the performance of service invocation we were interested in measuring how service location behaves. The traditional approach to service location uses a single node acting as a *registry*, which stores information about all services exposed by service providers. On the contrary, SLIM uses a peer-to-peer approach gathering information by contacting every possible service provider to obtain service advertisements back. To compare these two approaches we supposed the availability of a node acting as the registry and we compared the cost of our discovery mechanism with a single request to the registry sent through a unicast DYMO route. Notice that to further favor DYMO we did not consider the cost to populate the registry, which can be relevant in a dynamic scenario in which nodes may come and go while moving around.

To concentrate on service discovery, in this case we built our simulations with the client periodically sending out service requests without performing any service invocation; Table 1 shows the overall traffic generated by SLIM w.r.t. DYMO. For space reasons we only provide results for the reference scenario, as all other ones reflect the trends already presented in Figure 5. Notice that, even if DYMO uses a unicast route to invoke the registry, it still has to undergo the route creation process: this results in an overall cost that is about twice SLIM's one, even when using credits.

Besides the costs, we were also interested in the effectiveness of the discovery phase. Accordingly, we measured the overall delivery for SLIM, i.e., the percentage of services the client receive information about w.r.t. to the total number of services matching the query. Table 2 shows the results obtained in the standard scenario: with zero credits less than 80% of matching services are discovered; however, only one credit is enough to bring this percentage to a very good value of 93.2%.

Multicast invocation. As already mentioned, the SLIM API includes a mcastMsg operation to contact all the services that satisfy a given query. Unfortunately, DYMO does not support multicast communication so we cannot use it for a direct comparison, but we can still make some considerations.

Table 1. Traffic for discovery (w.r.t. DYMO)

SLIM c=0	SLIM c=1	SLIM c=2
33.27%	49.78%	56.16%

Table 2. Discovered services

SLIM c=0	SLIM c=1	SLIM c=2
78.39%	93.24%	94.19%

In particular, we notice that multicast service invocation in SLIM behaves exactly as service discovery, the only difference being a slight increase in the size of used packets (to account for the size of the message to be sent, as shown in Figure 4). With the payloads we used in our simulation, this results in multicast service invocation generating less than 5% greater more traffic than service discovery.

Starting from this number and remembering the results reported in Table 1, we can say that the cost of multicast invocation is still significantly lower than that of a single, unicast invocation in DYMO, as deduced by looking at the cost for DYMO to interact (in unicast) with the registry.

Along the same line, Table 2 demonstrates that a flooded request receives the vast majority of expected replies. The same result still holds when dealing with multicast invocation (we have done all the tests in the different scenarios, including several repliers for each request). More than 93% of expected replies are received using a single credit.

RELATED WORK

Different works have proposed the service oriented approach as a suitable abstraction for application developers to access the resources of WSANs. Most of them, however, consider an entire network of sensors as a single service provider and focus on the interaction between external (possibly remote) applications and a gateway server, which acts as a bridge between the sensor network and the outside, service oriented world. One of the first proposals in this direction is (Golatowski et al., 2003), which presents, at a very high level, a simple model in which an external server is adopted to communicate with the WSAN. A similar approach is described in (Avilés-López & García-Macías, 2009), which focuses on the abstraction layer offered to programmers. None of these works, however, provides details on how communication takes place between nodes.

As opposed to these approaches, SLIM does not introduce a full-fledged macro-programming style for a WSAN (Picco & Mottola, in press); instead, it moves services directly into the WSAN, thus making the task of programming distributed applications simpler. Some works take the same approach.

In (Leguay, Lopez-Ramos, Jean-Marie, & Conan, 2008) authors propose a protocol stack, WSN-SOA which reproduces the architectural concepts and information exchange of a regular SOA implementation. A topic based publish-subscribe communication paradigm is used to receive desired information from the network. Each service is registered as a topic on a gateway and applications choose which topics they are interested in. A unicast routing protocol is used to forward data from sensors to the gateway. Authors observe its lack of efficiency and scalability and promise further investigations on this aspect. The same communication paradigm is adopted in (Delicato, Pires, Pirmez, & Carmo, 2003), using Directed Diffusion (Intanagonwiwat, Govindan, & Estrin, 2000) to exchange data with the sensor nodes. OASiS (Kushwaha, Amundson, Koutsoukos, Neema, & Sztipanovits, 2007) proposes a programming paradigm in which developers define *logical objects* which are mapped to observed properties in the sensor network. Each sensor cooperates in the management of a logical object according to its capabilities, exposed as services. Authors propose requests flooding to locate services but don't provide implementation details. SYLPH (Tapia, Fraile, Rodríguez, Paz, & Bajo, 2009) proposes a layered architecture which can be adopted on top of heterogeneous devices. Gateways are used to connect devices that belong to different networks and use different communication protocols. None of the works above takes into account mobility, which instead is typical in various application scenarios for WSANs.

A Service Location Protocol for pervasive embedded devices, nanoSLP, has been proposed as part of the nanoIP protocol stack (Jardak et al.,

2008). As SLIM, nanoSLP uses query flooding to locate matching services. The same approach is adopted by the Simple Service Discovery Protocol (SSDP), used in the UPnP protocol (Jeronimo & Weast, 2003). DEAPspace (Hermann et al., 2001) investigates completely decentralized discovery solutions, by continuously pushing information from node to node, so that all devices hold a list of all known services; as a consequence, discovery is performed locally. Similarly, Konark (Helal, Desai, Verma, & Lee, 2003) and PDP (Campo, Munoz, Perea, Marin, & Garcia-Rubio, 2005) use a hybrid push/pull communication style to exchange information about known services. Such information is cached in devices, so that a limited number of hops have to be travelled for discovery. In (Sailhan & Issarny, 2005), the authors propose a service discovery protocol aimed at large scale mobile ad-hoc networks, in which multiple directories are used to store known services. The nodes where directories have to be located are chosen dynamically at run-time, according to given parameters, involving resources of nodes and environmental variables. In (Bromberg & Issarny, 2005), the authors focused on the integration of heterogeneous discovery protocols, to provide interoperability and flexibility.

The main benefit of SLIM w.r.t. these proposals is the full integration of service discovery and invocation with the routing protocol. Discovery, in fact, has the effect of populating the routing tables that are exploited to propagate service invocations and replies.

The SLIM protocol shares many of its core mechanisms with CCBR (Cugola & Migliavacca, 2009), but applies them to a radically different communication paradigm. Indeed, CCBR is a data-aware routing protocol that allows sinks to express their interest in data, which is subsequently pushed by sensors when they read it. This enables the traditional "publish-subscribe like" communication paradigm found in many protocols for WSNs. Conversely, SLIM adopts a pull style of interaction toward sensors, while also supporting commands to be sent to actuators. Moreover, some of the mechanisms that were originally developed for CCBR, have been modified in SLIM (e.g., by making larger use of RSSI) to benefit from the experience we gained in testing CCBR both in simulation and in real scenarios.

Finally, several protocols have been proposed so far to opportunistically find the next hop forwarder in a wireless network (Biswas & Morris, 2005; Zorzi & Rao, 2003; Jain & Das, 2008; Choudhury & Vaidya, 2004; Li, Sun, Ma, & Chen, 2008). They differ in the way this choice is made and how potential forwarders coordinate to limit multiple routes forwarding. With respect to these issues, the SLIM protocol adopts a very simple but efficient mechanism based on overhearing to let nodes coordinate, using a combination of hop distance from the destination (when available), link length and quality (through the RSSI), and remaining energy to elect the next forwarder. It also couples these mechanisms with a unique retransmission protocol to increase delivery while keeping overhead under control. Moreover, differently from the protocols above, which use opportunistic forwarding to provide unicast multi-hop routing, SLIM uses its own opportunistic mechanism within a complex protocol, which jointly supports service location and invocation on a mobile WSAN.

CONCLUSION

In this paper we presented SLIM, a middleware to support service oriented programming (SOP) in mobile Wireless Sensor and Actuator Networks (WSANs). SLIM brings SOP directly inside the sensor network, allowing each node to act as a service provider, a service consumer, or both. While this holds the promise of simplifying the issue of programming large and complex WSANs, the kind of services each node may offer are those tailored to a sensing and actuating scenario, like closing a valve, getting the temperature and humidity sensed by a specific node, or collecting and averaging the values measured in an area. These services will seamlessly integrate with more complex, application-level services offered by PCs located in a standard network.

SLIM exposes a simple yet powerful API to applications, allowing service registration, discovery, and invocation. To better address the needs of the typical applications for WSANs, SLIM supports both unicast and multicast invocations. Moreover, the layered architecture of SLIM enables the usage of virtually every language for service descriptions and queries.

SLIM includes an advanced routing protocol explicitly designed for a mobile, multi-hop WSAN. The results we measured comparing this protocol with a more traditional one, DYMO, which defines and manages unicast routes between communicating nodes, shows that SLIM outperforms DYMO, by providing higher message delivery at a significantly lower cost.

Our plan for the future is to implement SLIM in TinyOS in order to apply it to real scenarios, like those we are considering in the WASP (http://www.wasp-project.org/) project.

ACKNOWLEDGMENT

This work was partially supported by the European Commission, Programme IDEAS-ERC, Project 227977-SMScom; and by the Italian Government under the projects FIRB INSYEME and PRIN D-ASAP.

REFERENCES

Akyildiz, I. F., & Kasimoglu, I. H. (2004). Wireless sensor and actor networks: Research challenges. *Ad Hoc Networks*, *2*(4). doi:10.1016/j.adhoc.2004.04.003

Andonovic, I., Michie, C., Gilroy, M., Goh, H., Kwong, K., & Sasloglou, K. (2009). *Wireless sensor networks for cattle health monitoring*. New York: Springer.

Avilés-López, E., & Garcìa-Macìas, J. A. (2009). Tinysoa: a service-oriented architecture for wireless sensor networks. *Service Oriented Computing and Applications*, *3*(2), 99–108. doi:10.1007/s11761-009-0043-x

Baldoni, R., Beraldi, R., Cugola, G., Migliavacca, M., & Querzoni, L. (2005, December). Content-based routing in highly dynamic mobile ad hoc networks. *Int. Journal of Pervasive Computing and Communications*, *1*(4), 277–288. doi:10.1108/17427370580000131

Biswas, S., & Morris, R. (2005, October). Exor: opportunistic multi-hop routing for wireless networks. *ACM SIGCOMM Computer Communication Review, 35*(4).

Bromberg, Y. D., & Issarny, V. (2005). Indiss: interoperable discovery system for networked services. In *Proceedings of the acm/ifip/usenix 2005 international conference on middleware (Middleware '05)* (pp. 164-183). New York: Springer Verlag.

Campo, C., Munoz, M., Perea, J. C., Marin, A., & Garcia-Rubio, C. (2005). Pdp and gsdl: A new service discovery middleware to support spontaneous interactions in pervasive systems. In *Proceedings of the third ieee international conference on pervasive computing and communications workshops (Percomw '05)* (pp. 178-182). Washington, DC: IEEE Computer Society.

Chakeres, I. D., & Perkins, C. E. (2008, February). *Dynamic MANET On-demand Routing Protocol* (IETF Internet Draft, draft-ietf-manet-dymo-12.txt).

Choudhury, R., & Vaidya, N. (2004, January). Mac-layer anycasting in ad hoc networks. *ACM SIGCOMM Computer Communication Review*, *34*(1), 75–80. doi:10.1145/972374.972388

Cugola, G., & Migliavacca, M. (2009). A context and content-based routing protocol for mobile sensor networks. In *Proceedings of the 6th european conference on wireless sensor networks (Ewsn '09)* (pp. 69-85). Berlin: Springer Verlag.

Cugola, G., Murphy, A., & Picco, G. (2006). The handbook of mobile middleware. In *Content-Based Publish-Subscribe in a Mobile Environment*. New York: CRC Press.

Delicato, F. C. P. F., Pires, F. P., Pirmez, L., & Carmo, L. F. (2003). A flexible web service based architecture for wireless sensor networks. In *Proceedings of the 23rd international conference on distributed computing systems (Icdcsw '03)* (p. 730). Washington, DC: IEEE Computer Society.

Evers, L., & Havinga, P. (2007, Oct). Supply chain management automation using wireless sensor networks. In *Proceedings of the ieee int. conf. on mobile adhoc and sensor systems*, Pisa, Italy (pp. 1-3). Washington, DC: IEEE.

Golatowski, F., Blumenthal, J. H. M., Haase, M., Burchardt, H., & Timmermann, D. (2003). Service-Oriented Software Architecture for Sensor Networks. In *Proceedings of the int. workshop on mobile computing (imcâ™03)* (pp. 93-98).

Helal, S., Desai, N., Verma, V., & Lee, C. (2003, march). *Konark - a service discovery and delivery protocol for ad-hoc networks* (Vol. 3, pp. 2107-2113).

Hermann, R., Husemann, D., Moser, M., Nidd, M., Rohner, C., & Schade, A. (2001). Deapspace: transient ad hoc networking of prevasive devices. *Computer Networks*, *35*(4), 411–428. doi:10.1016/S1389-1286(00)00184-5

Hui, J. W., & Culler, D. E. (2008, July-August). Extending IP to Low-Power, Wireless Personal Area Networks. *Internet Computing*, *12*(4), 37–45. doi:10.1109/MIC.2008.79

Intanagonwiwat, C., Govindan, R., & Estrin, D. (2000). Directed diffusion: a scalable and robust communication paradigm for sensor networks. In *Proceedings of the 6th annual international conference on mobile computing and networking (Mobicom '00)* (pp. 56-67). New York: ACM.

Jain, S., & Das, S. (2008). Exploiting path diversity in the link layer in wireless ad hoc networks. *Ad Hoc Networks*, *6*(5), 805–825. doi:10.1016/j.adhoc.2007.07.002

Jardak, C., Meshkova, E., & Riihij, Ã. ¤rvi, J., Rerkrai, K., & Mahonen, P. (2008, March 31-April 3). Implementation and performance evaluation of nanoip protocols: Simplified versions of tcp, udp, http and slp for wireless sensor networks. In *Proceedings of the IEEE Wireless Communications & Networking Conference (Wcnc 2008),* Las Vegas, NV (pp. 2474-2479). Washington, DC: IEEE.

Jeronimo, M., & Weast, J. (2003). *Upnp design by example: A software developer's guide to universal plug and play*. Intel Press.

Kushwaha, M., Amundson, I., Koutsoukos, X., Neema, S., & Sztipanovits, J. (2007, January). Oasis: A programming framework for service-oriented sensor networks. In *Proceedings of the Communication systems software and middleware the 2nd international conference (comsware 2007)* (pp. 1-8).

Leguay, J., Lopez-Ramos, M., Jean-Marie, K., & Conan, V. (2008, October). An efficient service oriented architecture for heterogeneous and dynamic wireless sensor networks. In Proceedings of the *33rd IEEE Conference on Local Computer Networks (LCN 2008)* (pp. 740-747).

Li, L., Sun, L., Ma, J., & Chen, C. (2008, June). A receiver-based opportunistic forwarding protocol for mobile sensor networks. In *the 28th International Conference on Distributed Computing Systems Workshops.*

Picco, G., & Mottola, L. (in press). Programming wireless sensor networks: Fundamental concepts and state-of-the-art. *Comp. Surv. Proceedings of the 3rd Int. Conf. on Pervasive Computing Technologies for Healthcare.* (2009, April). London. Washington, DC: IEEE.

Sailhan, F., & Issarny, V. (2005). Scalable service discovery for manet. In *Proceedings of the Third IEEE International Conference on Pervasive Computing and Communications (Percom '05)* (pp. 235-244). Washington, DC: IEEE Computer Society.

Sikka, P., Corke, P., Valencia, P., Crossman, C., Swain, D., & Bishop-Hurley, G. (2006). Wireless ad-hoc sensor and actuator networks on the farm. In *Proceedings of the 5th Int. Conf. on Information Processing in Sensor Networks* (pp. 492-499). New York: ACM.

Sohraby, K., Minoli, D., & Znati, T. (2007). *Wireless sensor networks: Technology, protocols, and applications*. New York: John Wiley. doi:10.1002/047011276X

Tapia, D. I., Fraile, J. A., Rodrìguez, S., de Paz, J. F., & Bajo, J. (2009). Wireless sensor networks in home care. In *Proceedings of the 10th International Work-Conference on Artificial Neural Networks (IWANN '09)* (pp. 1106-1112). Berlin: Springer Verlag.

Varga, A. (2001, June). The OMNeT++ Discrete Event Simulation System. In *Proceedings of the European Simulation Multiconference (ESM'2001)*.

Zorzi, M., & Rao, R. (2003). Geographic random forwarding (geraf) for ad hoc and sensor networks: multihop performance. *IEEE Transactions on Mobile Computing, 2*(4), 337–348. doi:10.1109/TMC.2003.1255648

This work was previously published in International Journal of Systems and Service-Oriented Engineering, Volume 1, Issue 1, edited by Dickson K.W. Chiu, pp. 62-78, copyright 2010 by IGI Publishing (an imprint of IGI Global).

Chapter 19
A Method to Support Fault Tolerance Design in Service Oriented Computing Systems

Domenico Cotroneo
Università degli Studi di Napoli Federico II and Complesso Universitario Monte Sant'Angelo, Italy

Antonio Pecchia
Università degli Studi di Napoli Federico II, Italy

Roberto Pietrantuono
Università degli Studi di Napoli Federico II, Italy

Stefano Russo
Università degli Studi di Napoli Federico II and Complesso Universitario Monte Sant'Angelo, Italy

ABSTRACT

Service Oriented Computing relies on the integration of heterogeneous software technologies and infrastructures that provide developers with a common ground for composing services and producing applications flexibly. However, this approach eases software development but makes dependability a big challenge. Integrating such diverse software items raise issues that traditional testing is not able to exhaustively cope with. In this context, tolerating faults, rather than attempt to detect them solely by testing, is a more suitable solution. This paper proposes a method to support a tailored design of fault tolerance actions for the system being developed. This paper describes system failure behavior through an extensive fault injection campaign to figure out its criticalities and adopt the most appropriate countermeasures to tolerate operational faults. The proposed method is applied to two distinct SOC-enabling technologies. Results show how the achieved findings allow designers to understand the system failure behavior and plan fault tolerance.

INTRODUCTION

Service Oriented Computing (SOC) is emerging as a leading paradigm in the context of scalable distributed software development. As a matter of fact, it enables complex applications to be seamlessly developed by integrating services and components rather then building them entirely *from scratch*. This significantly reduces development cost and time to market.

To achieve flexible, interoperable, and massively distributed applications, the SOC paradigm relies on a number of support technologies, middleware platforms and ad-hoc protocols, which allow

DOI: 10.4018/978-1-4666-1767-4.ch019

developers to focus mainly on service development and composition. However, the integration of these heterogeneous software items raises significant dependability challenges that need to be addressed for achieving trusted SOC applications. In fact, such items are usually conceived to be integrated with other systems, hence developed without any specific *operational context* in mind. Consequently, testing activities carried out during their development may be not enough to guarantee a proper service during operations (Weyuker, 1998; Moraes, Durães, Barbosa, Martins, & Madeira, 2007), because of unforeseeable interactions with the rest of the system integrating them.

A viable solution to cope with this issue is *tolerating* residual faults rather than trying to avoid or to remove them before the operational phase. *N-version programming* (Avizienis & Chen, 1977; Avizienis, 1985) and *recovery blocks* (Randell, 1987; Kim &Welch, 1989) are well known software fault tolerance strategies; however, for many industrial purposes, these techniques have been neglected due to their high cost (Saha, 2006) and to the lack of data about their effectiveness (Anderson, Barrett, Halliwell, & Moulding, 1995). Rather than developing additional versions of the target program, focusing on the improvement of its *single-version* (e.g., by adding extra code to handle exceptional bad situations (Bishop & Fríncke, 2004)) is a more suitable solution in the context of SOC technologies. This strategy can allow system designers to achieve a good trade-off among cost, time to market and a proper dependability level.

Designing fault tolerance actions to improve the single-version of the program presents several challenging issues. First, tracking and handling all potential conditions to be tolerated may be not feasible in practice, especially when dealing with large size software systems. Indeed, this would lead to programming cost comparable with (or even greater than) the implementation of additional versions. Moreover, a mere *brute-force* addition of fault-handling code may inadvertently lead to (i) the introduction of new faults and to (ii) heavy performance depletion. As a consequence, a crucial task, which cannot be driven only by the developer's experience or attitude, is the choice of the most appropriate fault tolerance actions to apply where actually needed.

This requires a deep knowledge about the system features and its behavior in presence of faults, which probably is the real challenge to be addressed. Without such knowledge, the single-version fault tolerance approach, even if potentially more suitable than multiple-version solutions (especially for business-critical contexts), would be not applicable. This paper proposes a method to support the design of fault tolerance actions specifically tailored for the system being developed. Through an extensive software fault injection campaign we identify the most critical software modules, as well as related fault types, which heavily impact the correct behavior of a system during the operational time. This knowledge allows designers to figure out (i) *where* to place additional code, by exploiting the information about critical modules and (ii) *what* to place, in terms of the specific action to implement, by exploiting the information about the fault types. Our ultimate aim is to drive improvements by applying the proper fault tolerance action only where actually needed. Developers would thus characterize, in terms of failure behavior, the underlying *context-independent* software items before their deployment in a SOA, with the possibility to effectively intervene by enriching their fault tolerance ability.

The proposed method is applied to two distinct SOC enabling technologies (i) the popular Apache Web Server, whose availability level is crucial for many web-based service-oriented applications, and (ii) the TAO Open Data Distribution Service (DDS). Experimental campaigns show how these systems have been characterized through an in-depth injection of software faults, and how the achieved findings can be used to tailor the design of fault tolerance strategies to the target system.

The rest of this paper is organized as follows. We present related work in the area of dependability assurance of service-oriented applications and fault tolerance design solutions. We subsequently discuss the basic steps of the proposed approach, and we describe how we conducted experimental campaigns. Finally, findings coming out from the two reference case-studies are reported, followed by the conclusion and lessons learned from the experience.

RELATED WORK

The increasing adoption of Service Oriented Architectures (SOAs) calls for effective approaches to ensure high dependability. However, current approaches for ensuring service and system dependability are still far from meeting the needs of trustworthy computing. Mostly, research on SOC focuses on protocols, functionality, transactions, ontology, composition, the semantic Web, and interoperability, while little research has explored dependability issues (Tsai, Zhou, Chen, & Bai, 2008). To pursue dependability goals and to increase confidence in SOC, different strategies can be considered: so far, the most explored solution in the literature has been *testing*.

Several studies addressed the problems of testing service-oriented applications. Canfora and Di Penta (2006) dealt with integration testing, functional testing, non-functional testing, and regression testing, discussing the roles and the needs of both service providers and service consumers. Bertolino et al. (2009) presented several testing methods for SOAs embedded inside the PLASTIC validation framework. The latter aims to combine different verification techniques for both functional and extra-functional properties, spanning over off-line and on-line testing stages. In Bartolini, Bertolino, Elbaum, and Marchetti (2009), authors propose an approach, named *Service Oriented Coverage Testing* (SOCT), to provide a tester with feedback about how a service, called a *Testable Service*, is exercised. In other words, they provide dedicated testing means to make services more transparent to an external tester, enabling a white-box testing approach for SOA, contrarily to the typical black-box approach.

Tsai, Zhou, Chen, and Bai (2008) propose an open framework, implementing a *group testing* strategy, which attempts to increase the test efficiency by identifying and eliminating test cases with overlapping coverage. Baresi and Di Nitto (2007) discussed various aspects of testing and validation of service-oriented architectures. In particular they presented a number of SOA verification techniques, including static analysis, testing, monitoring, modeling, and reliability analysis.

Recently, Canfora, and Di Penta (2009) surveyed the existing SOC testing approaches, outlining a set of issues that limit the testability of SOC systems. Among others, they emphasized as key points (i) the *lack of control*, determined by the services physical implementation on an independent infrastructure under the control of the provider, (ii) the *dynamicity* of services, which are concretely retrieved from a registry and instantiated only when needed, thus hindering testing, and (iii) the *cost* of a service-level testing, where a repeated invocation of a service for testing may be expensive or even not applicable whenever the service produces side effects.

In a broader view, the problem is even more complex than what described by Canfora and Di Penta (2009). Indeed, even if a service-level testing would suffice to guarantee a correct implementation of services and a right services composition/aggregation, the final dependability is mainly determined by how much the underlying infrastructures, platforms, technologies and support systems are able to ensure the desired quality, i.e., *by how much the core of SOC is dependable*. In literature, this crucial issue has been quite neglected, since efforts mainly focused on service-level quality assurance (as in the cited papers).

At infrastructural level, testing turns out to be inadequate, since the mentioned support software items are typically designed, implemented and tested without a specific context in mind. They are conceived to be integrated into other software systems; hence the final perceived dependability depends upon the interaction patterns between them and the rest of the system, *which are often unpredictable at testing time*. This causes unexpected behaviors at runtime, which may easily lead to serious failures. Finally, the changing and evolving contexts for which SOC is designed makes testing activities even harder.

In this work, we rather intend to provide a support for achieving high availability level through *fault tolerance*. Fault tolerance design can rely on various code and time redundancy techniques proposed in the literature. The conventional fault tolerance approaches, e.g., Recovery Blocks (RB) (Randell, 1987; Kim &Welch, 1989), N-Version Programming (NVP) (Avizienis & Chen, 1977; Avizienis, 1985), rely on multiple versions of the software application, to be developed independently using different languages, tools, compiler and so on.

For this reason, these techniques are very expensive, hence not applicable for many ordinary medium/low cost applications. They found wider room for very complex applications in safety-critical contexts.

For SOC based systems, solutions based on the improvement of the single version are more suitable. In this case, redundancy is not given by extra versions of the application, but by extra code added to the single version to implement single fault tolerance *actions*. In particular, the concept is to introduce code with the aim of handling exceptional (bad) situations in a reasonable way (Bishop & Frincke, 2004), e.g., by intercepting an error and warning the user, or terminating the application gracefully. This form of redundancy is sometimes referred to as *robust programming*, or *fault-tolerant programming*.

The approach can be implemented in various ways, ranging from assertions placement (Zenha Rela, Madeira, & Silva, 1996; Saha, 1997) and control flow checking (Yau & Chen, 1980; Oh, Shirvani, & McCluskey, 2002), to procedure duplication (PD) (Pradhan, 1996) and triplication solutions (Saha, 2006). There are several fault tolerance actions like the mentioned ones that a designer could apply in response to faults. However, all of them have the same common problems, exacerbated in large size systems that preclude their practicability, i.e., (i) where it is more convenient to place such an action, and (ii) what type of action is more suitable for the system in hand. The objective of this work is to provide developers with a method to design a *system-specific* fault tolerance strategy, rather than subjectively applying generic and expensive one-fits-all actions.

PROPOSED METHOD

In order to properly plan a tailored fault tolerance strategy, engineers should be able to quantitatively evaluate the system behavior in presence of faults. They should gain a preliminary knowledge of the system's ability to react to the activation of various kinds of fault, and exploit it to select the best fault tolerance mechanisms.

More specifically, designers have to figure out *how* the system fails (i.e., its *failure modes*) and what kinds of faults are more likely responsible for its failure. In this way, they can adopt punctual countermeasures in the most critical parts of the system itself against the most influential classes of faults, as well as to manage specific failure modes by which the system is typically affected. To obtain such a characterization, the approach proposed in this paper includes the following steps:

1. *Failure Modes Definition.* The first step is the definition of potential modes by which the system could fail, such as *crash*, *passive/ active hang*, *value*, and so on. Let us denote these failure modes for the system S with $(F_1, F_2, \ldots, F_K)$.

2. *Fault Injection.* In this step the fault injection campaign is carried out, in order to create a faulty version of the system and evaluate its behavior against the faults activation. Faults of several types have to be injected in the software, since we are interested in figuring out how a given fault type impacts on the system. To this aim, we classified faults according to a well-known classification scheme, namely the *Orthogonal Defect Classification* (ODC) (Chillarege, Bhandari, Chaar, Halliday, Moebus, Ray, & Wong, 1992), extended according to Durães and Madeira (2006). The adopted classification is a well-established scheme and a reference for the state of the art in fault injection studies. During these experiments, all the observed failure data are collected, along with the faults that caused them (type and location).
3. *Failure Analysis.* The following two steps allow obtaining the system characterization to be used for figuring out *where* greater fault tolerance abilities are required, *and what kinds* of fault need more to be tolerated. In this step, designers evaluate how the system failed during experiments, and what modules are more critical:
    - *Failure Modes Distribution*: collected failure data are classified according to their modes (F_1, F_2, ..., F_k) and to the responsible modules. Hence the output is a set of indexes $\#F_{j,mi}$ denoting the number of failures of type j (among the identified failure modes) occurred due to the module m_i.

Failure Proneness Computation: per each module m_i[1], a failure proneness value indicating its criticality is computed as follows:

$$FP_{mi} = \frac{1}{K} \cdot \sum_{j=1}^{K} \frac{\# F_{j,mi}}{\# F_j} \quad (1)$$

where K denotes the number of considered failure modes, and F_j is the number of failures of type j. This index is a weighted average of the m_i 's failures caused by the identified failure modes. It represents how much, in the average, a module caused the system failures, i.e., its *criticality*. The output of this step is therefore the list of modules criticality.

Fault Analysis. Per each module, it is important to figure out what types of fault caused, more likely, the observed failures, in order to adopt appropriate fault tolerance actions. In this step, the most relevant fault types are identified, i.e., the ones mainly responsible for the failures: a further index is then computed, *FC* (i.e., Fault Criticality):

$$FC_{t,i} = \frac{1}{K} \cdot \sum_{j=1}^{K} \frac{\# Fault_{t,i,j}}{\# SeededFault_{i,j}} \quad (2)$$

where $FC_{t,i}$ is the average criticality of faults of type t in the module i. It is computed by considering $\#FC_{t,i,j}$, which denotes the number of faults of type t in the module m_i that caused a failure of type j, $\#SeededFault_{i,j}$, which denotes the number of total faults injected in m_i that caused a failure of type j, with K denoting the number of failure modes. The index represents how much the fault type t impacted, in the average, the observed number of failures due to m_i.

At the end of these steps, designers will have a complete picture of the system behavior. They know: the average criticality index FP_{mi} per module, the failure modes that more affect a module (the $\#F_{j,mi}$ values), and the fault types more responsible for the observed failures. This characterization allows designers to focus their attention on the most critical modules and on what types of fault the system has to tolerate. They would first identify the modules that in the average (with respect to all the considered failure modes) exhibited more failures, by examining the FP_{mi} values. Starting from these modules, they consider the $FC_{t,i}$ values to figure out the most critical fault types for that module. Inside each module, designers also know the portion of the code in which the injected faults have been more or less tolerated. In other words, they can infer the modules that need more interventions, what types of fault have to be more tolerated and where, inside a module, it is more convenient to place a *fault tolerance* action.

Given a *Fault Type / Failure Mode* pair for a module m_i, there is no single solution or criterion to decide what action is better to apply. This choice is bound to several factors:

- How much a given action costs, considering the number of potential portion of the code to which it would be applied (and what is the available budget for the purpose of making the system fault tolerant);
- The programmers' expertise and ability in dealing with the different fault types and in placing the corresponding fault tolerance actions;
- How much the considered module m_i, is critical for the system. More resources should be devoted to more critical modules, hence assigning more expensive and safer actions to them;
- How much the considered fault type impacts the overall dependability; for critical faults it is worth to spend more;
- The structure and programming style of the program, that can make the final system more prone to be modified in one way instead of another;
- The size of the program; this impacts the abstraction level at which one can apply the action (e.g., for large systems it could be hard to apply an action at each method);

For instance, suppose that a module m_i, turned out to be one of the most critical modules and particularly prone to cause crash failures. Moreover, suppose that from the analysis of the $FC_{t,i}$ index results that crash failures are in the majority caused by faults on the parameters list (i.e., the ODC *Interface* fault type). One can decide (i) to add a *checking* code portion at the entry of each m_i's methods as fault tolerance action, in order to check the argument values and make the module more robust to these faults; (ii) to act at higher level, e.g., by inserting checking code where values are exchanged with other modules; (iii) to be more focused on fewer places (e.g., some methods) where the injected *Interface* faults shown, during the experiments, to be less tolerated (and caused relatively more failures than other places). The listed criteria represent guidelines to make such choice judiciously.

EXPERIMENTS

Once identified failure modes potentially affecting the system, we inject software faults in the target platform by means of *changes* in the source code of the program. Changes are introduced according to specific *fault-operators* based upon faults actually uncovered in several open-source projects. In particular, fault operators are based on a large field data study encompassing 668 faults over 12 real-world software systems (Durães & Madeira, 2006).

It is recognized that this approach needs the source code of the program to be applied; nevertheless its accuracy is greater than other fault injection techniques (e.g., G-SWIFT (Durães & Madeira, 2006)). Figure 1 summarizes fault operators.

Exactly one software fault is introduced in the target source code for each fault injection experiment. The target platform is compiled and its resulting *faulty* version is stored for the subsequent experimental campaign. We use a software fault injection technique acting at source code level (Natella & Cotroneo, 2010), and we adopt a support tool[2] to automate the injection process.

Once completed the injection process, we execute the experimental campaign, which consists of 4 distinct logic phases (Figure 2), detailed in the following. In particular, for each faulty version of the target platform, a **campaign manager** program

Figure 1. Fault operators (adapted from Durães & Madeira, 2006)

Acronym	Explanation
OMFC	Missing function call
OMVIV	Missing variable initialization using a value
OMVAV	Missing variable assignment using a value
OMVAE	Missing variable assignment with an expression
OMIA	Missing IF construct around statements
OMIFS	Missing IF construct plus statements
OMIEB	Missing IF construct plus statements plus ELSE before statements
OMLC	Missing clause in expression used as branch condition
OMLPA	Missing small and localized part of the algorithm
OWVAV	Wrong value assigned to variable
OWPFV	Wrong variable used in parameter of function call
OWAEP	Wrong arithmetic expression in parameter of a function call

1. Initializes the target platform;
2. Starts a **tester** program, which exploits commonly-used platform capabilities by running a specific workload;
3. Identifies the *experiment outcome*. In particular, the campaign manager figures out the specific failure (among the failure modes chosen by the analyst (step 1 of the method), if any, resulting from the injected fault. To this aim it exploits both (i) *operating system* level information (e.g., memory dumps generation) and (ii) *tester* level information (e.g., a timeout expiration, computation results provided by the platform in hand to the tester program);
4. Produces a **report** containing information intended to be used during the analysis phase (e.g., injected fault type, target source file, experienced failure mode) and cleans up the system for the successive fault injection experiment.

Figure 2. Experimental campaign

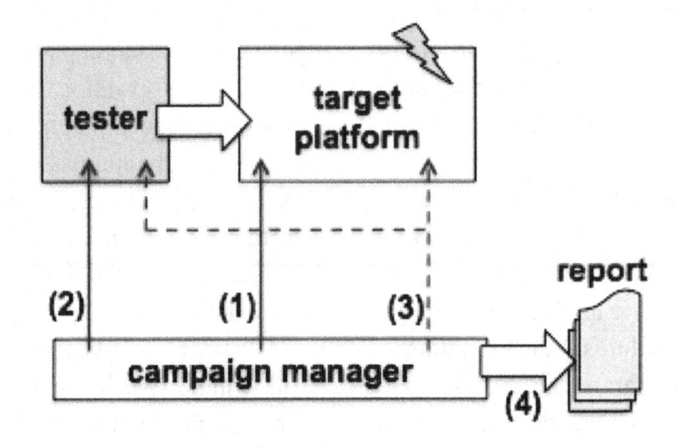

Machines composing the experimental testbed (Intel Pentium 4 3.2 GHz, 4GB RAM, 1,000 Mb/s Network Interface equipped) run a RedHat Linux Enterprise 4. An Ethernet LAN interconnects the machines. Platform-specific details (e.g., failure modes, and the used tester program), when needed, will be clarified in the context of the reference case-studies.

CASE STUDY 1. APACHE WEB SERVER

In this section we provide a detailed failure analysis for a widely used software platform in SOC environments, i.e., the Apache Web Server[3] (version 1.3.41). To this aim we tailor the experimental setup to the proposed case study. In particular, the tester program is the httperf[4] tool, which has been configured to exploit most of the features offered by the Web Server (e.g., virtual-hosts, multiple methods, cookies) by means of a specific stress workload.

We identify three significant failure modes for the reference platform:

- *Crash*: unexpected termination of the Web Server. A memory dump is generated by the operating system;
- *Hang*: the Web Server process is up but, (i) one or more HTTP requests supplied by the httperf tool or (ii) the Web Server *start* or *stop* phases, are not acknowledged within a proper timeout (i.e., tuned before the campaign by means of several fault-free runs of the Web Server);
- *Content*: all error conditions that are not the result of a hang or crash (e.g., wrong value delivered to the client).

We execute 8,433 fault injection experiments involving 11 source modules under the /src/main folder, i.e., the core of the Web Server. Figure 3 reports the experiments breakup by fault operator.

During the campaign 1,395 experiments result in a failure outcome (i.e., 743, 101, and 551, crashes, hangs, and contents, respectively). Figure 4 depicts the relative failures distribution. Crash and content failures are the most likely to occur. We only experience 7% of faulty runs to result in a hang outcome, mainly due to the triggering of infinite loops in the code.

Reports produced by the campaign manager are used to estimate the *failure proneness* of each of the source modules involved in the fault injection experiment (i.e., FP_{mi}). To this aim we use equation 1 described in the third step of the proposed method. Figure 5 reports the achieved results. For each of the considered failure modes (i.e., columns 1,2,3), we identify the most contributing source modules, by assessing $\#F_{j,mi}$. The last column reports the most *failure prone* source modules, in descending order.

Figure 3. Experiments breakup by fault operator (Apache Web Server)

Operator	#	Operator	#
OMFC	819	OMIA	791
OMIEB	282	OMIFS	812
OMLC	325	OMVAE	1,149
OMLPA	2,183	OMVIV	65
OMVAV	221	OWPFV	1,148
OWAEP	361		=====
OWVAV	277	**Total**	**8,433**

Figure 4. Failure modes distribution (Apache Web Server)

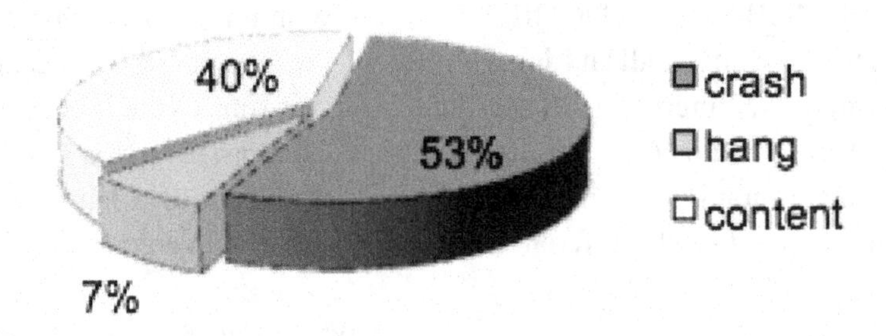

The alloc and buff modules, which mainly provide support capabilities to the Web Server, result to be the most critical ones, according to the proposed method. As a remark, it should be noted that alloc is ranked in the first three positions for each of the considered failure modes. On the other hand, the buff module is the most significant contributor to hang failures (about 33% of cases). This feature makes the buff module result in high failure proneness. The http_main and http_protocol modules are the most critical among the ones encapsulating the Web Server HTTP requests handling/processing code. The analysis also reveals that the http_log module, even if not the most critical one, is responsible for a significant amount of content failures (i.e., 40%). Even if http_log does not provide critical capabilities, i.e., logging, it is very likely that a latent fault in this module is triggered. As a matter of fact, the Apache Web Server makes an extended use of logs as, for example, to keep trace of resource accesses.

Findings related to critical software modules within the Apache Web Server clearly suggest *where* it may be useful to take fault tolerance actions in a strategic way. Nevertheless, this information alone is not enough to address *what* should be placed within the source code, in terms of specific fault tolerance actions.

This information is enabled by the fourth step of the proposed method. In particular, we carry out the fault analysis, by applying equation 2, in order to figure out critical fault types for the alloc, buff, and http_main modules, respectively. Figure 6 summarizes the results of the analysis. In particular, the OMLPA (i.e., *missing small and*

Figure 5. Failure proneness (Apache Web Server)

Crash	#	Hang	#	Content	#	FP_{m_i}	value
alloc	183	buff	34	http_log	220	**alloc**	**0.19293**
http_config	155	http_protocol	21	alloc	85	**buff**	**0.14236**
http_main	105	alloc	18	http_main	61	**http_main**	**0.14011**
util	99	http_main	17	http_core	57	**http_log**	**0.13623**
http_core	73	http_request	5	http_config	51	**http_protocol**	**0.10736**
http_protocol	70	util	3	buff	35	**http_config**	**0.10369**
buff	20	http_core	2	util	25	**http_core**	**0.07383**
http_request	16	http_config	1	http_protocol	11	**util**	**0.06944**
http_vhost	11	-	-	http_request	6	**http_request**	**0.02730**
http_log	7	-	-	-	-	**http_vhost**	**0.00493**
util_uri	4	-	-	-	-	**util_uri**	**0.00179**
total	743	*total*	101	*total*	551	*total*	1.00000

Figure 6. Fault analysis (Apache Web Server)

alloc	$FC_{t,i}$	buff	$FC_{t,i}$	http_main	$FC_{t,i}$
OMLPA	0.486	OMLPA	0.284	OMLPA	0.342
OMVAE	0.260	OMVAE	0.152	OMIA	0.139
OMFC	0.093	OWAEP	0.108	OMVAE	0.138
OWPFV	0.051	OWVAV	0.096	OMIFS	0.087
OMIFS	0.028	OWPFV	0.089	OMFC	0.084

localized part of the algorithm) is a critical fault for each of the identified module. In other words, it is very likely that *algorithmic* errors, even if small or localized, result in a failure outcome. Additionally, missing *variable assignments* and *if constructs,* i.e., OMVAE and OMIA, respectively, turn out to be the second most critical fault categories in the considered modules.

Driven by these results, we can provide guidelines concerning the most suitable fault tolerance mechanisms in case of the Apache Web Server. The high occurrences number experienced for the OMLPA fault type suggests the need for a replication-based approach at least for the most critical functions provided by the identified modules. The OMVAE faults may be detected by checking the value of a variable before its usage by means of assertions or acceptance tests placed in the code.

CASE STUDY 2. TAO OPEN DATA DISTRIBUTION SYSTEM

Case study 2 consists of a failure analysis of TAO Open Data Distribution Service (DDS)[5], i.e., an open source C++ implementation of the OMG's v1.0 DDS specification[6]. The tester program is a DDS-based application. It is composed by 2 processes, i.e., a *publisher*, which sends data bound to a specific DDS topic, and a *subscriber* process, which receives them. DDS middleware assures communication services between the described processes.

We identify three possible failure modes:

- *Crash*: unexpected termination of at least one of the DDS processes. A memory dump is generated by the operating system;
- *No messages* (no_msg, in the following): none of the messages sent by the publisher is delivered by the DDS middleware to the subscriber process within a proper timeout (i.e., tuned before the campaign by means of several fault-free runs of the DDS-based application);
- *Value*: messages delivered to the subscriber process are different from the ones sent by the publisher.

We execute 5,892 fault injection experiments involving 60 source modules under the /dds/DCPS and /dds/DCPS/transport folders, i.e., the core of the DDS library. Figure 7 reports the experiments breakup by fault operator.

During the campaign 1,685 experiments result in a failure outcome (i.e., 880, 712, and 93, crashes, no_msgs, and values, respectively). Figure 8 depicts the relative failures distribution. *Crash* and *no_msg* failures are the most likely to occur. *Value* failures are almost rare (only 6%). It should be noted that this behavior is somewhat different compared to the Apache Web Server (Figure 3). As a matter of fact, *hang* failures (i.e., the complete lack of service with the platform still up), resulted only in few experiments.

Reports produced by the campaign manager program are used to estimate the *failure proneness*,

Figure 7. Experiments breakup by fault operator (TAO Open DDS)

Operator	#	Source	#
OMFC	1022	OMIA	440
OMIEB	306	OMIFS	324
OMLC	86	OMVAE	321
OMLPA	1921	OMVIV	254
OMVAV	614	OWPFV	396
OWAEP	24		=====
OWVAV	184	**Total**	**5,892**

i.e., FP_{mi}, for each of the source modules involved in the fault injection experiment.

Figure 9 reports achieved results. For each of the considered failure modes (columns 1,2,3), we identify the most contributing source files, by assessing $\#F_{j,mi}$ (for the sake of clarity we only report the 10 most occurring modules out of the 60 involved in the experiments). The last column reports the most *failure prone* source modules, in descending order.

The Service_Participant module is the most critical. As a matter of fact it is always ranked in the first two positions for each of the considered failure modes. This module provides support

Figure 8. Failure modes distribution (TAO Open DDS)

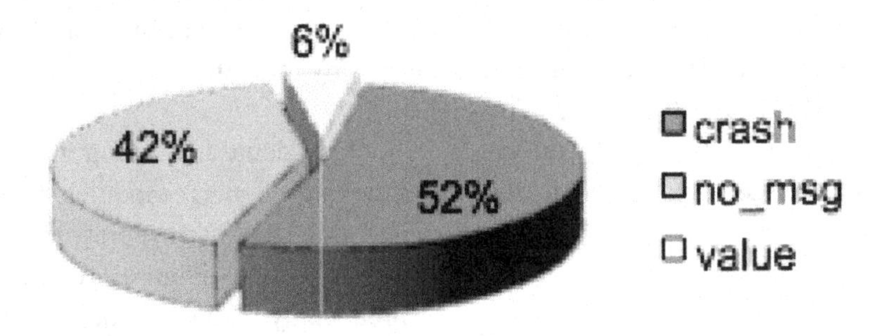

Figure 9. Failure proneness (TAO Open DDS)

Crash	#	No_msg	#	Value	#	FP_{m_i}	value
DataWriterImpl	106	Service_Participant	118	DataWriterImpl	12	**Service_Participant**	**0.11515**
Service_Participant	73	DataWriterImpl	65	Service_Participant	9	**DataWriterImpl**	**0.11359**
DataReaderImpl	64	DataReaderImpl	63	DataReaderImpl	8	**DataReaderImpl**	**0.08241**
TransportImpl	32	DataSampleHeader	38	TransportImpl	5	**DataLink**	**0.03587**
TransportFactory	29	DataLink	31	WriteDataContainer	4	**WriteDataContainer**	**0.03252**
DataLink	28	SimpleTCPRecStr	28	SimpleTcpTransport	4	**SimpleTcpTransport**	**0.03177**
SimpleTCPRecStr	28	WriteDataContainer	27	TopicDescriptionImpl	3	**TransportImplFact.**	**0.03155**
DomainPartic.Impl	27	SubscriberImpl	27	SubscriberImpl	3	**DataSampleHeader**	**0.03063**
PublisherImpl	27	PublisherImpl	27	QueueRemoveVisitor	3	**DomainPartic.Impl**	**0.03012**
SubscriberImpl	27	DomainPartic.Impl	27	DomainPartic.Impl	3	**TransportRecStr**	**0.02637**
...	...	...	...	...	...	...	...
total	880	*total*	712	*total*	93	*total*	1.00000

capabilities (e.g., QoS setup) to a DDS-based application during the startup phase. As in the case of the Apache Web Server, we experience that a module providing *support* capabilities to the platform, is the most critical one.

The DataWriterImpl and DataReaderImpl modules follow Service_Participant in the FP_{mi} ranking. These modules are invoked by a DDS-based application to send and receive a message, respectively. We experience that faults injected in these modules are very likely to result in an application failure. We hypothesize that this behavior is the result of the lack of an inter-lying, i.e., between the DDS application and the DataWriter(Reader) Impl module, which may mitigate, if not tolerate, a fault once triggered.

The analysis also reveals that the most critical modules involving the communication layer are DataLink and SimpleTCPTransport. The subsequent fault analysis is carried out focusing on the three most critical modules, i.e., Service_Participant, DataWriterImpl, and DataReaderImpl, respectively. Figure 10 summarizes the results. The OMLPA is a critical fault type for all the identified source files. This finding, according to what experienced with Apache Web Server, confirms that the *algorthmic* errors usually result in a failure. On the contrary, the DDS middleware is more sensitive to the OMFC and OMVAV faults.

Obtained results allow us to provide guidelines concerning the most suitable fault tolerance mechanisms for the most critical modules of TAO Open DDS. The number of high occurrences experienced for the OMLPA fault type suggests the need for a replication-based approach at least for the most critical function provided by the identified modules. Additionally, while OMVAV faults may still be detected by checking the value of a variable before its usage, control flow checking strategies are more effective to deal with the OMFC class.

CONCLUSION AND LESSONS LEARNED

This paper presented a method to drive source code improvements in a software system by a tailored design of code-level fault tolerance actions. The method is based on a characterization of the software item by means of an extensive fault injection campaign and a subsequent failure analysis. Such a characterization is conceived to understand the failure behavior of a system whose target employment context is not known in advance. This is the case of a number of SOC enabling technologies and core software infrastructures, which constitute the base of SOC applications. Testing such software items does not suffice, due to context unawareness; therefore, we focused on a system-specific fault tolerance strategy to be planned from the provided characterization. Developers would thus characterize, in terms of failure behavior, the software items before their deployment in a SOA, with the possibility to intervene where actually needed for enriching them with fault tolerance ability. The method has been

Figure 10. Fault analysis (TAO Open DDS)

Srv_Part	$FC_{t,i}$	DataWriI	$FC_{t,i}$	DataReaI	$FC_{t,i}$
OMLPA	0.623	OMLPA	0.503	OMLPA	0.356
OMFC	0.125	OMFC	0.111	OMVAV	0.167
OMVAE	0.095	OMVAE	0.089	OWPFV	0.131
OMIEB	0.076	OWPFV	0.072	OMFC	0.084
OMIA	0.019	OMVAV	0.068	OMVIV	0.073

applied to two distinct SOC-enabling technologies (i) the popular Apache Web Server, and (ii) the TAO Open Data Distribution Service (DDS), by injecting several thousands of faults in both cases. Results allowed identifying the software modules that in the average greatly affected the system failure behavior and the most relevant fault types for each of them, in both cases. Additionally, from the analyzed case studies, we noted that:

- *Software systems may exhibit different failure behaviors.* Reported failures distributions (Figures 2 and 3) shown that each platform is more prone to experience a specific class of failures. This highlights the need for a platform-specific approach, in order to plan specific mechanisms via a detailed analysis of the failure behavior of the system in hand. On the contrary, providing a *general-enough* fault tolerance strategy may be quite hard and ineffective.
- *Modules providing support facilities are critical.* Experimental campaigns revealed that the alloc and Service_Participant modules, for the Apache Web Server and TAO Open DDS, respectively, strongly impact on the failure behavior of the system. This confirms that often the most of failures are due to few critical modules; thus, being able to quantitatively identify such modules is crucial.
- *Algorithmic errors usually result in a failure of the system.* This finding has been experienced in both the Apache Web Server and the TAO Open DDS. No general rules can be figured out for the other fault type, which mainly remain platform-dependent. This enforces the need for a system-specific fault tolerance strategy.

Achieved findings allow improving considered platforms. As a matter of fact, guidelines provided by the presented approach can be exploited to start up the design of *fault-tolerant* versions of both Apache Web Server and TAO Open DDS. Moreover, experiments can be extended to other *context-independent* software technologies (addressing the class of *Open Source Software* (OSS) systems), such as the MySQL Server, or CORBA-compliant ORBs, in order to provide a characterization of their failure behavior, potentially useful to the developers' community and to quality manager teams.

ACKNOWLEDGMENT

This work has been partially supported by the Consorzio Interuniversitario Nazionale per l'Informatica (CINI) and by the Italian Ministry for Education, University, and Research (MIUR) within the frameworks of the "Dependable Off-The-Shelf based middleware systems for Large-scale Complex Critical Infrastructures" (DOTS-LCCI) Project of National Research Interest (PRIN), and the "CRITICAL Software Technology for an Evolutionary Partnership" (CRITICAL-STEP) Project (http://www.critical-step.eu), Marie Curie Industry-Academia Partnerships and Pathways (IAPP) number 230672, in the context of the Seventh Framework Programme (FP7).

REFERENCES

Anderson, T., Barrett, P. A., Halliwell, D. N., & Moulding, M. R. (1995). Software Fault Tolerance: An Evaluation. *IEEE Transactions on Software Engineering, 11*(12), 1502–1510. doi:10.1109/TSE.1985.231894

Avizienis, A. (1985). The N-Version Approach to Fault-Tolerant Systems. *IEEE Transactions on Software Engineering, 11*(12), 1491–1501. doi:10.1109/TSE.1985.231893

Avizienis, A., & Chen, L. (1977). On the implementation of N-version programming for software fault-tolerance during program execution. *In Proceedings of the 1st IEEE International Computer Software and Applications Conference* (pp. 149-155). Washington, DC: IEEE Computer Society.

Baresi, L., & Di Nitto, E. (Eds.). (2007). *Test and Analysis of Web Services*. New York: Springer Verlag. doi:10.1007/978-3-540-72912-9

Bartolini, C., Bertolino, A., Elbaum, S., & Marchetti, E. (2009). Whitening SOA Testing. *In Proceedings of the 7th Joint Meeting of the European Software Engineering Conference and the ACM SIGSOFT Symposium on the Foundations of Software Engineering* (pp. 161-170). New York: ACM.

Bertolino, A., De Angelis, G., Frantzen, L., & Polini, A. (2009). The PLASTIC Framework and Tools for Testing Service-Oriented Applications. In A. Lucia & F. Ferrucci (Eds.), *Proceedings of Software Engineering: international Summer Schools, ISSSE 2006-2008,* Salerno, Italy (LNCS 5413, pp. 106-139). Berlin: Springer Verlag.

Bishop, M., & Frincke, D. (2004). Teaching Robust Programming. *IEEE Security and Privacy, 2*(2), 54–57. doi:10.1109/MSECP.2004.1281247

Canfora, G., & Di Penta, M. (2006). Testing services and service-centric systems: Challenges and opportunities. *IT Professional, 8*, 10–17. doi:10.1109/MITP.2006.51

Canfora, G., & Di Penta, M. (2009). Service Oriented Architecture Testing: A Survey. In A. Lucia & F. Ferrucci (Eds.), *Proceedings of Software Engineering: international Summer Schools, ISSSE 2006-2008,* Salerno, Italy (LNCS 5413, pp. 78-105). Berlin: Springer Verlag.

Chillarege, R., Bhandari, I. S., Chaar, J. K., Halliday, M. J., Moebus, D. S., Ray, B. K., & Wong, M. Y. (1992). Orthogonal Defect Classification-A Concept for In-Process Measurements. *IEEE Transactions on Software Engineering, 18*(11), 943–956. doi:10.1109/32.177364

Durães, J., & Madeira, H. (2006). Emulation of Software Faults: A Field Data Study and a Practical Approach. *IEEE Transactions on Software Engineering, 32*(11), 849–867. doi:10.1109/TSE.2006.113

Goseva-Popstojanova, K., Mathur, A. P., & Trivedi, K. S. (2001). Comparison of architecture-based software reliability models. *In Proceedings of the 12th International Symposium on Software Reliability Engineering* (pp. 22-31). Washington, DC: IEEE Computer Society.

Kim, K. H., & Welch, H. O. (1989). Distributed Execution of Recovery Blocks: An Approach for uniform Treatment of Hardware and Software Faults in Real-Time Applications. *IEEE Transactions on Computers, 38*(5), 626–636. doi:10.1109/12.24266

Liu, W. (2005). Trustworthy service selection and composition - reducing the entropy of service-oriented Web. In *Proceedings of the 3rd IEEE International Conference on Industrial Informatics* (pp. 104-109). Washington, DC: IEEE Computer Society.

Moraes, R. L. O., Durães, J., Barbosa, R., Martins, E., & Madeira, H. (2007). Experimental Risk Assessment and Comparison Using Software Fault Injection. In *Proceedings of the 37th IEEE/IFIP International Conference on Dependable Systems and Networks* (pp. 512-521). Washington, DC: IEEE Computer Society.

Natella, R., & Cotroneo, D. (2010). Emulation of Transient Software Faults for Dependability Assessment: A Case Study. In *Proceedings of the eight European Dependable Computing Conference*. Washington, DC: IEEE Computer Society.

Oh, N., Shirvani, P. P., & McCluskey, E. J. (2002). Control-flow checking by software signatures. *IEEE Transactions on Reliability, 51*(2), 111–122. doi:10.1109/24.994926

Pradhan, D. K. (1996). *Fault-Tolerant Computer System Design*. Upper Saddle River, NJ: Prentice Hall, Inc.

Randell, B. (1987). Design-Fault Tolerance. In Avizienis, A., Kopertz, H., & Laprie, J.-C. (Eds.), *The Evolution of Fault-Tolerant Computing* (pp. 251–270). Vienna, Italy: Springer Verlag.

Saha, G. K. (1997). EMP- Fault Tolerant Computing: A New Approach. *International Journal of Microelectronic Systems Integration*, *5*(3), 183–193.

Saha, G. K. (2006). A Single-Version Scheme of Fault Tolerant Computing. *Journal of Computer Science & Technology*, *6*(1), 22–27.

Tsai, W. T., Zhou, X., Chen, Y., & Bai, X. (2008). On Testing and Evaluating Service-Oriented Software. *Computer*, *41*(8), 40–46. doi:10.1109/MC.2008.304

Weyuker, E. J. (1998). Testing Component-Based Software: A Cautionary Tale. *IEEE Software*, *15*(5), 54–59. doi:10.1109/52.714817

Yau, S., & Chen, F. (1980). An Approach in Concurrent Control Flow Checking. *IEEE Transactions on Software Engineering*, *6*(2), 126–137. doi:10.1109/TSE.1980.234478

Zenha Rela, M., Madeira, H., & Silva, J. G. (1996). Experimental Evaluation of the Fail-Silent Behaviour in Programs with Consistency Checks. In *Proceedings of the Twenty-Sixth Annual international Symposium on Fault-Tolerant Computing* (pp. 394-403). Washington, DC: IEEE Computer Society.

ENDNOTES

1. A module is intended as a logically independent unit performing a well-defined function (Goseva-Popstojanova, Mathur, & Trivedi, 2001). The granularity of a module is an analysis choice, addressed by a tradeoff between a large number of small units and a small number of large units; in our experiments a module is a *file*
2. http://www.mobilab.unina.it/SFI.htm
3. http://httpd.apache.org/
4. http://www.hpl.hp.com/research/linux/httperf/
5. http://download.ociweb.com/OpenDDS/
6. http://www.omg.org

This work was previously published in International Journal of Systems and Service-Oriented Engineering, Volume 1, Issue 3, edited by Dickson K.W. Chiu, pp. 75-89, copyright 2010 by IGI Publishing (an imprint of IGI Global).

Chapter 20
Demand-Driven Development of Service Compositions in Organizational Networks

Ralph Feenstra
Delft University of Technology, The Netherlands

Marijn Janssen
Delft University of Technology, The Netherlands

Sietse Overbeek
Delft University of Technology, The Netherlands

ABSTRACT

Organizational collaborate more and more in organizational networks to remain competitive. New systems can be created by assembling a set of elementary services provided by various organizations. Several composition methods are available, yet these methods are not adopted in practice as they are primarily supply-driven and cannot deal the complex characteristics of organizational networks. In this paper, the authors present a service composition development method and a quasi-experiment to evaluate this method by comparing it with existing ones. The development method is able to deal with incomplete information, to take the demand as a starting point, to deal with news services that do not exist yet, to include and to evaluate non-functional requirements, to show various stakeholder views, and to help to create a shared vision. Visualization and evaluation of alternative compositions and negotiation about the desired results are important functions of any composition method in organizational networks.

INTRODUCTION

Organizations are more and more forced to collaborate with each other (Chiu, Cheung, & Zhuang, 2010), share information and reuse system components to reduce cost. The creation in organizational networks results in the creation of flows that are no longer self-contained within a single organization. The performance depends more and more on the performance of external partners that are often unknown. Whereas in the past each organization has developed applications independently of other organizations, the current development is to develop components

DOI: 10.4018/978-1-4666-1767-4.ch020

only once and reuse them as services in the organizational network (Janssen & Joha, 2008). This has resulted into the trend of creating new systems as composites of web services (Liang, Huang, & Chuang, 2007). Web services technologies promise to create new business applications by composing existing services and to publish these applications as services for further composition (Liu et al., 2010). This is known as web service composition or composition for short. A *composition* combines a set of web services following a certain composition pattern to achieve a certain objective (Curbera, Khalaf, Mukhi, Tai, & Weerawarana, 2003). Web services are self-contained, self-describing software modules that can be published, and remotely invoked and in this way reused (Fremantle, Weerawarana, & Khalaf, 2002).An example of a composition is the buying or selling of travel insurances. There can be a service for checking the identity, for adding the required information, calculating the costs of an insurance policy, and for financial settlement and payment. Such compositions are synthesized by the reuse of elementary web services. It is not hard to imagine that many of these services can be re-used for all kind of other (insurance) products. Yet, the benefits of reusing information and functionality have not been attained as creating service composition is a difficult endeavor (Feenstra, Janssen, & Wagenaar, 2007). Services are often not developed for reuse. There are many obstacles blocking the easy composition resulting in limited use of service compositions in practice.

The creation of a composition within a organizational network can easily be viewed as a simple and straightforward process, but in fact is a very complex endeavor and can be characterized by organizations having an already available set of services, having many interdependencies, and different aims (Feenstra et al., 2007). For example the comparison of travel insurance policies can potentially be reused for comparing other types of insurances and even other types of products like books and cars. Yet this would requires that characteristics can be changed, the interface used matches the interface of single items that are compared and so on. Furthermore whereas customers want to compare products, the product providers (insurance companies) might not want this or are afraid that comparison will take place based on a limited amount of characteristics which might not be in favor of their products. Existing composition methods are often supply-driven as they assume the presence of a set of interoperable services an consensus about the objectives (Milanovic & Malek, 2004). What might be viewed as a process of selecting components can easily result in the need to develop new components, struggling about the interpretation of functionality and performance and finally resulting in expensive and time-consuming implementation projects. The complexity and heterogeneity of the organizations might serious hamper development. Components might simply not be available or not fit the desired purpose. Further, Liang, Huang and Chuang (2007) found that there were the constraints of non-functional service properties are hardly taken into account. Already existing services can complicate the reuse process, as the modules are not determined using a conscious design process and the effectiveness of a "modularization" is dependent upon the criteria used in dividing the system into modules (Parnas, 1972). As a result, existing composition methods are not adopted by the field.

The goal of this paper is to present a demand-driven service composition method and to evaluate the merits and disadvantages of this method. This method should be able to deal with the distinguished characteristics of organizational networks. In the following section we discuss the theoretical background and the requirements on a composition approach for organizational networks. This is followed by our research approach. In section three the composition method is presented and subsequently evaluated in section five. Finally, we discuss the results and draw conclusions.

BACKGROUND

Organizational Networks

Organizations collaborate in networks to benefit from each other's resources and competition is shifted towards competition between networks instead of single organizations. Organizational networks consist of many organizations that are independent and make their decisions autonomously. Their decisions and behavior might affect other network partners. Designing a service composition in a network of agencies involves addressing many challenges outside the control of a single agency (Milward & Provan, 2003). Organizational networks can be viewed as large and complex systems consisting of a large number of independent entities having many interdependencies that seek to accomplish their own goals, but operate according to some rules and norms in the context of relationships with the other entities. An organization can be viewed as "a set of multiple, goals-shifting political coalitions" (March & Simon, 1958).

Network partners have mostly different interests and seek their own goals, but are nevertheless interdependent of each other. Due to these interdependencies no single actor can solve the problem autonomously. Further, none of the actors can impose a particular solution on the others. The creation of solutions depends on mutual agreement and understanding (Bruijn & Heuvelhof, 2002). The creation of a solution has to take into account contradictory goals and interests. This is an important starting point in organizational network development efforts, as no one actor can tell others what to do or what to think. Consequently, it is important that all actors involved are able to express their own point of view and no solution is chosen in advance. Furthermore actors might favor different type of solutions and need to negotiate with each other to come to a final decision.

The decision makers involved in designing compositions face the difficult task to create service composition. A large set of alternatives is available and there is no support for the evaluation and comparison of service composition. Human decision making capacity is limited, persons can only deal with a limited set of details (Simon, 1960), in this case a limited set of alternative compositions. The limits to human information handling capacity introduce a non-rational element to the decision making. This element grows larger if the decision makers are overwhelmed with detailed information. The limits to human rationality are often described using the term bounded rationality (Simon, 1960). In complex organizational networks there is often no single 'best solution'. Networks are characterized by interdependence between actors, differences between actors that interfere with cooperation, and reticence of actors that don't see their interests met by choosing a particular solution and dynamics regarding the group of actors involved in the decision making process (Bruin & Heuvelhof, 1995). As a result, the problem addressed cannot be only viewed as a 'hard science' that can be based on a multidimensional optimization process that can be solved using mathematic algorithms. One stakeholder might favor efficiency over service levels, whereas another stakeholder might have opposing preferences. Still the solution should satisfy both stakeholders. It is impossible to derive a formal specification of the preferences of each actor in every situation, as these might not be necessarily based on rational reasons.

The organization in the network will need to cooperate to a certain extent in order to realize a solution that at the same time helps realizing common goal and pays sufficient attention to their specific interests. The decision making process is effective only if it leads to commonly supported decisions (Bruin, Heuvelhof, & Veld, 1998). The realization and success of service composition in organizational networks strongly depends on the participation. From a technology perspective, collaboration is complicated as different stakeholders will be used to different names and terms (syntax and semantics) to describe services. From an organizational perspective collaboration is difficult as stakeholders might have different and even opposing requirements and the selection of a service composition might need extensive negotiation. This all results in the need for a service composition development process in which stakeholders are involved and actively participate to express their interest and requirements.

Existing Service Composition Methods

Service composition is the constructing a new system (or service) by invoking a number of elementary services that are often provided by organizations. Functional decomposition is a common approach found in systems theory (Sage & Armstrong, 2000). The idea is that a system can be decomposed into parts, each part can be implemented and in turn integrated. Functional decomposition refers broadly to the process of resolving a functional relationship into its constituent parts in such a way that the original function can be reconstructed (i.e., recomposed) from those parts by composition. This assumes that first an analysis is made and then the functional components are determined.

Several evaluations of the composition approaches can be found in literature (Beek, Bucchiarone, & Gnesi, 2006; Feenstra et al., 2007; Liang et al., 2007; Milanovic & Malek, 2004). Milanovic and Malek (2004) and Beek et al. (2006) compare composition approaches. In their evaluation Beek et al. and Milanovic et al. assume that a set of interoperable services exists. In other words they take the available supply as a starting point. This is not a realistic starting point as in organizational networks services are often not developed for the purpose of reuse and are not interoperable. Furthermore that existing set of services might not be sufficient to satisfy the need and new services might need to be developed.

Milanovi and Malek (2004) and Beek (2006) found that most composition approaches neglect specification of non-functional properties. Only the Semantic Markup for Web Services (OWL-S) lets users define non-functional properties, but that capability has yet to be fully specified (Milanovic & Malek, 2004). Liang, Huang and Chuang found that there were no publications that take into account the constraints of non-functional service properties (Liang et al., 2007). These kinds of properties include cost, security, dependability, response time, reliability and scalability. In decision-making concerning the selection of services these non-functional properties have a large influence. A typical concern of stakeholders has to do with a trade-off in non-functional requirements and as such it is essential that they are addressed.

Requirements on Service Composition

The analysis and selection of services for web services composition is, in general, much more complex than the analysis and selection of parts for product design because selection criteria are difficult to define and composition components are diverse (Liang et al., 2007). Different stakeholders might have different selection criteria and look at other types of characteristics. Components can be small or large grained, might be configurable and interfaces might be completely different.

Existing composition methods assume that all services are well-defined, that there are unambiguous selection requirements and that there is a clear, undisputed goal regarding the performance. In fact, the above approaches can be better denoted using the term *representational formalism* than using the term composition approach (Feenstra et al., 2007). The existing methods are not able to deal with the interdependencies and various aims that characterize organizational networks. The following list of requirements on a demand-driven service composition development method for organizational networks is derived based on a large number of interviews at two organizations. A composition method should be able to

1. *to deal with incomplete information*. Stakeholders are not able to provide, or might even not want to provide, an overview of all their services in advance. Some information might be available whereas other information might be omitted. For example there might be limited information available about the scalability and availability of a service or about the security when used in an interorganizational setting;
2. *take the demand as a starting point*: Current composition methods use existing services and in this way take the supply as a starting point. Not the problem but the solution is central in these methods. Instead, the need to solve a certain problem should be the starting point and the service composition solution should solve this need;
3. *deal with news services that do not exist yet*: Some services exist, others are under development or the development has been planned for. For example the rise of the number of customers might result into the need to develop a service that is able to deal with large volumes. Furthermore, management might wonder which services need to be developed in the future. In the travel insurance example a comparison services to compare various insurance options might not exist yet, but might become desirable in the future to show trade-offs to customers. Existing methods cannot support the planning process;
4. *to include non-functional requirements and evaluate compositions based on these type of requirements*: In the situation of designing a composition many organizations already have developed services and there is no greenfield situation. Instead a composition needs to be created from services that might not be designed for that purpose. Most composition methods neglect non-functional requirements and only some methods use quality metrics (Liang et al., 2007). Non-functional requirements like security, scalability, availability and so on should often be satisfied. In many cases no hard data about quality metrics are available. This tacit knowledge needs to be mobilized;
5. *show various stakeholder views*: There might be different stakeholder requirements and different stakeholders might prefer different solutions. For example, one party might focus on the cost, whereas others might be more interested in reliability and ensuring high service levels. There is a need for negotiation and supporting the trade-off during a service composition process.
6. *create a shared vision*: Requirements, functionality and performance might be ambiguous and interpreted differently by stakeholders. Therefore, the creation of a shared view is crucial. For example, identity management might be interpreted as a service used for acquiring basic data about persons or as a facility that can support authentication. Both interpretations might be correct, but might also be incorrect in a specific case.

These six main requirements will serve as a starting point for the development of the composition method described in this paper. In the following section we will discuss the research approach to develop and test a composition method that is better equipped to deal with the specific aspects characterizing or organizational networks.

1. DEMAND-DRIVEN SERVICE COMPOSITION DEVELOPMENT METHOD

System development is often viewed as a highly complex process which is difficult to structure and has many interactions. In such situations a variety of different mental templates or cognitive filters can be used to influence managerial decision making (Walsh, 1995). Methods can be used to guide actions and decisions. The aim of this composition method is to improve the development process. This should help to simplify complicated composition processes and provide knowledge at an appropriate level of abstraction for decision-makers. We opted for an approach consisting of steps and support for each of these steps, aimed at stimulating both the interactions among participants and the group processes. The basic idea is that in this way the majority of stakeholders are represented and tacit (expert) knowledge is mobilized. The main differences with existing composition method are that the need will be taken as a starting point and that the method deals with the requirement originating from the characteristics of organizational networks. The service composition method consists of a number of steps and support for these steps that will be discussed in the following subsections.

Service Composition Steps

Our method is based on the idea that services are not well-described, and if they are described the description might be ambiguous. The need (demand) to solve a certain problem is the starting point instead of the supply of services. Therefore, the involvement of stakeholders is necessary to 1) represent their goals and purposes and 2) to mobilize their knowledge. The system development process is viewed as an interactive process where many parties with different interests can negotiate on the outcomes. Nevertheless, this process needs to be guided. Burstein et al. (2005) used three main steps, which are: service discovery, engagement and enactment. Our efforts are concentrated on service discovery and engagement and not on enactment as new services might not to be developed before execution. Our method is based on the requirements of the compositions as a starting point and therefore a requirement specification step is added. In addition, we wanted to have the evaluation of possible compositions as an explicit step, which might result in increasing understanding and the identification of new compositions. As a result, our method consists of five steps that are always part of a service composition development process. The steps provide guidance at a high level of abstraction and could help decision-makers to deal with a conscious development process. The steps are interactive and there is no predefined sequence. The only condition is that all steps need to be completed before a service composition is created.

Step 1: Requirements Specification

The first stage focuses on identifying and understanding the need for a new system or service. This stage starts with rough ideas concerning the desired functionality and finishes with a structured list of desired properties of the design. During this stage the stakeholder needs are indentified, general ideas for the use of the composition are listed and a list of desirable properties of the composition is made.

Step 2: Description of Existing Services

Stakeholders need to communicate with each other and discuss the use of possible services. To support this, a kind of service register is collaboratively created or modified at this stage. Not all data need to be included in the register, as they might not be available. Moreover, services that are under development can be included. Decisions on the attributes that will be included in the register have to be made. Probably someone is held responsible for the correctness of the descriptions. The maintaining of this service register can be a continuous effort.

Step 3: Design of Sets of Service Compositions

In this phase the participants discuss and negotiate on possible service compositions. The compositions are designed and each stakeholder designs his or her individual composition.

Step 4: Evaluation of the Composition Designs

The designs created in the previous step need to be assessed. Evaluation can be used to select a single composition or to select a subset of compositions for implementation or for further investigation and refinement. Various evaluation methods can be supported and evaluations might come from various stakeholder perspectives. The evaluation consists of a dependency checking part (which checks whether the composition is feasible) and a part that checks stakeholder specific criteria.

Step 5: Planning and Reporting

During the last step the decisions that were made are listed including their motivation. Furthermore, a list of changes to existing services that are needed and/or a list of new (to be developed) services is made.

The five steps of the service composition process are visualized as a UML activity diagram (Rumbaugh, Jacobson, & Booch, 2004) in Figure 1. Arrows in an activity diagram indicate transitions coming out of each activity, connecting it to the next activity. A transition may *fork* into two or more parallel activities. The fork and the subsequent *join* of the threads coming out of the fork appear in the diagram as solid bars.

A starting point of our method is that all participants are able to express their own point of view and no solution direction is chosen in advance. Consequently, these steps should stimulate dialogues among stakeholders and clarify the assumptions on underlying views of other stakeholders. In this way judgmental pitfalls and advocacy of ones own viewpoint without support by any facts or explicit mentioning of assumptions can be avoided. For example, the steps should help to make explicit that a participant assumes that a service is reliable while there is no information to support this. This might result in the need for more research or the creation of a set of shared starting points that are plausible. The focus of the method is on exploiting rather than eliminating each others' viewpoints. Schein (1996) argues that experiencing and acknowledging differences of interpretations is a critical step for further inquiry of underlying assumptions. These rules of the composition design process will be communicated in advance.

The steps described above need to be taken, but our method is not aimed at prescribing the sequence of steps or iterations. Our starting point is the knowledge in the organization that needs to be used in the process and consequently stakeholder interaction and exchange of knowledge to come up with better and more feasible service compositions is key. In the next subsection we describe the support for these steps.

Service Composition Support

The service composition steps indicate the essential steps that need to be taken. A prototype was developed that supports each of these steps. The central idea of the support for these steps is the use of the service portfolios concept, analogous to the concept of application portfolios (e.g., Jefferey & Leliveld, 2004). The service portfolio is intended to support decision makers involved in a service composition process. The prototype consists of the following elements:

1. A centralized database containing all existing services described using codified service data. This database contains both functional and non-functional properties of services. Stakeholders can add properties and in this

Figure 1. UML activity diagram of the service composition process

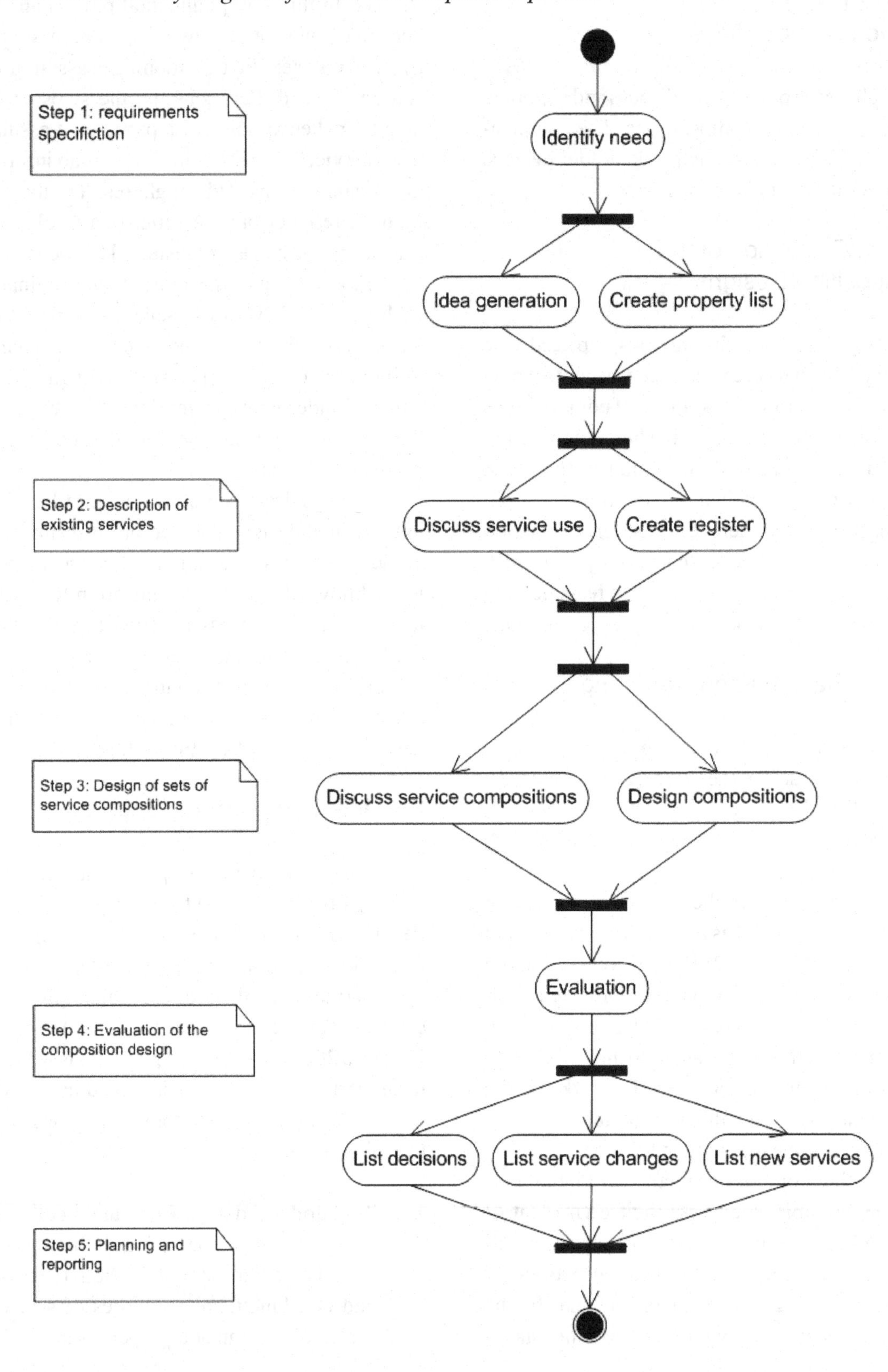

way include their own evaluation criteria. For example, if a stakeholder thinks it is necessary, the stakeholder can add reliability as a non-functional property.
2. The database includes current services and services that will come available in short time and identifies services that need to be developed in order to support new developments;
3. The service portfolio describes the desired composite service that needs to be developed. In this portfolio not only the supply of elementary services is described, but also the needs and requirements on the new services from the various stakeholders' perspectives is included.

An UDDI can be used as the basis for a service directory, making it possible to publish and locate service (www.uddi.org). Yet, UDDI is not suitable for registering in-depth functional requirements and non-functional requirement (Tsai, Xiao, Paul, & Chen, 2006). As such a database was developed which was able to store thies data. This database provided the opportunity to add new requirements by the participants. The service portfolio can be extended during the service composition process and stakeholders can express their own view on the desired composition.

Various commenting, sorting and ranking methods were included, as actors might favor their own way of evaluation or might have different standards for evaluation. Aspects like costs, reliability, continuity, availability, reliability and performance of services might be decisive factors and should be included in the evaluation of compositions. On the other hand Jeffery and Leliveld (Jefferey & Leliveld, 2004) found that a common pitfall for application portfolio is to include too many metrics. It is essential to have distinctive sets of metrics for different types of investments. Instead of taking a predefined set of metrics as a starting point, support was created for participants to add their own performance evaluation criteria.

The tool that was developed for supporting the service composition provides support during the five steps discussed in the previous subsection. Each stakeholder can add services, and can provide functional and non-functional descriptions for the services. The ability to select services and create a new composition and the function to evaluate the resulting compositions based on metrics like cost, reliability, etc. Often information will not be complete and the tool will only show what is known. Next, the participants can add their own expectations about the performance of services or use this to discuss the assumptions with others. In this way, the tool is aimed at facilitating the steps and at supporting the identification of differences between starting points and objectives among stakeholders. In addition they can discuss their own preferred service compositions with each other. In this way it supports the creation of a dialogue among stakeholders, which should ultimately help to identify service compositions that are acceptable to all stakeholders.

EVALUATION

Evaluation Approach

We conducted the quasi-experimentation by comparing the service composition method as currently used by the participants, labeled as 'best practice method' and the demand-driven composition method proposed in this paper by employing a quasi-experiment. A *quasi-experiment* is a study in which interventions, in our case two composition methods, are deliberately introduced to observe their effects, by non-randomly assigning the units (Cook & Campbell). To assess the impact of the newly designed demand-driven composition method, different methods have to be used to solve exactly the same case. In this way the conditions remain the same and the difference in outcomes can be contributed to the different composition methods. A workshop approach was chosen because it enables

Figure 2. The composition evaluation approach

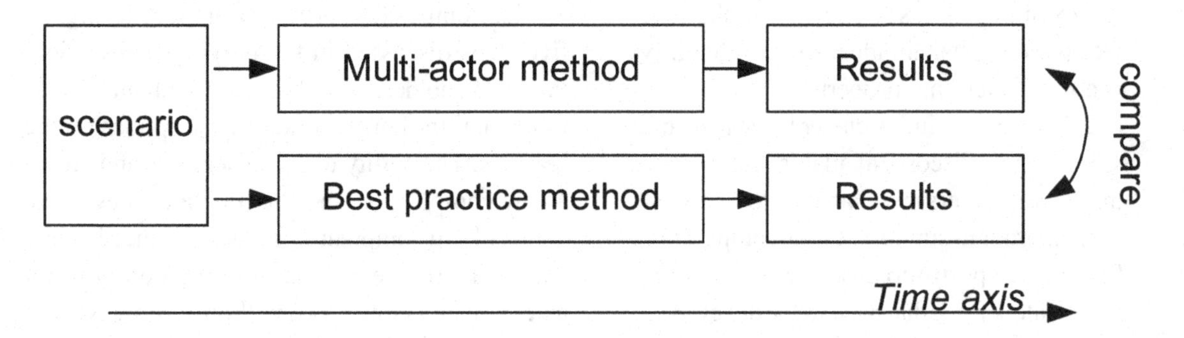

the comparison of different composition approaches within a controlled environment and did not require long term efforts from busy professionals. During the workshops, the participants designed compositions for several given scenarios within a relatively short time. In total, there were 21 participants during a 4-hour workshop. All participants were experienced in the field and were able to compare our composition method with their own practices. The workshop resembled the real situation closely, as multiple organizations having various objectives had to design one composition. This means that they had to search for possible solutions, negotiated on the use of certain services, and the development of new services.

Figure 2 depicts the structure of the quasi-experiment. Two types of workshops were held, using exactly the *same* scenario as input to ensure that the methods are evaluated using the same circumstances. In one workshop the participants could use their own method as used by their organizations. We label their own method as 'best practice method', as they are using the methods used by their own organization which is based on aspects like their organizational standards and past experiences. In the other workshop the demand-driven service composition method presented was used by the participants. We did not opt for comparing our method with existing composition methods, as they are based on other starting points which do not match the scenarios which are derived from practice.

At the end of both workshops, the opinions of the participants regarding the method and the provided support were asked. For this purpose, a survey was developed based on the requirements discussed in section two and aimed at evaluating the main characteristics of our composition method. The results of the survey were analyzed using SPSS. Next to the quantitative comparison of the method, the behavior of the workshop participants was observed and interviews were held. This resulted in a qualitative evaluation of the strengths and weaknesses of the best practice and demand-driven service composition method.

Composition Scenario

In the workshop a use-case scenarios for developing a composition was created based on the travel insurance example briefly discussed at the beginning of this paper. When a customer indicates he likes to order insurance the following steps will be taken by the customer:

1. Login to reuse existing data
2. Add data. The client adds the desired start date and end date of the insurance. An overview of insurance offers is shown to the client. The possibility to buy a permanent travel insurance policy has to be part of the offers.
3. View offer. Based on a start and end date the price for each type of insurance can be calculated.

4. Select offer. When the client decides to buy one of the insurances, he indicates which offer it concerns. In the next step the type of payment is selected.
5. Payment. The payment is settled by means of various payment options. The bank or credit card company provides a set of interfaces to directly complete the payment procedure.
6. Confirmation. In the last step of the process the customer requires a confirmation which includes the policy number and all details.

This use-case scenarios was used in a sessions resulting in the following results.

Quantitative Results

Our composition method relies heavily on group interactions and supporting the interactions among the stakeholders. In this light, the questions shown in Table 1 were formulated to perform the quantitative evaluation. The best practice method and our method were compared. The column named 'sig.' indicated if there is a significant difference in the scores between these methods. In total, 10 of the 18 questions show significant differences in favor of our method (given a 95% level of confidence). Three questions were even significantly different given a 99% confidence interval.

The participants were especially positive by the ability to manage personal requirements, to add and evaluate non-functional properties and to evaluate compositions. These aspects are related to the involvement of different stakeholders who might have different preferences. By including non-functional requirements, they were able to evaluate the effect on metrics like cost, quality, reliability, scalability and so on. They were able to add the type of performance indicators they preferred.

The evaluation shows that the participants found that the ability to add their own perspective was a benefit. They were able to make a set of requirements on the composition from their own point of view and share this with the other participants in the workshop. In this way support for requirement negotiation was created.

Also the adding of non-functional requirements was viewed as benefit of the composition method over the methods they used in their daily practice. In this way the cost, reliability of resulting composition and response time of a service can be evaluated. Response time was viewed as an important factor, as the subsequent invocation of web services causes overhead resulting in longer response time for the composition.

Further, the participants considered they way in which a shared view was created as better. The method and supporting tool show the services used in the compositions, the performance assumption underlying each of the services and this makes the desires of participants explicit. In this way the various stakeholder views were made visible to the participants.

Not surprisingly, the participants did not find that the method supported better functional requirements and the evaluation of functional requirements, as in their own practice the functions are also taken as a starting point. The participants indicated that the use of multiple evaluation methods was not viewed as a benefit of the supporting tool. In their own practices they were used to use various evaluation methods. Surprisingly, the question results indicated that the shared view provided no more insight in the requirements and services available. The motivation of the participants was that an overview of services was always a necessary condition and that in one way or another a requirement would be discussed in the process. Also the ability to add services that are not developed yet was not considered as a difference with their own practice. They argued that they often discussed the development of new services, as the set of existing services is hardly never been able to create the desired new system or at least needs to be modified.

Table 1. Qualitative comparison of composition methods

	Best practice method	Demand-driven Composition method
Strengths	• Proven method in other situations (for example based on RUP, well known to the system developers) • Participants are familiar with the method in other situations • Not formal and easy to work around certain aspect • Focus on creating working products • No detection of infeasible combinations of services	• Takes the problem as a starting point and each stakeholder can define their own requirements and evaluation criteria for the compositions • Enables comparison of multiple alternatives • Facilitates comparisons in a group of stakeholders and makes requirements and desires on the composition explicit • Focuses on dependencies among services and feasibility of compositions • Enables the use of non-functional (QoS) data regarding services • Process followed is easy to understand to non-technical people ('this service register is comparable to a shopping site where people compare different articles and make recommendations') • Is able to deal with incomplete requirements
Weaknesses	• Method is not designed to work with a service register, no focus on reuse of available services • Not easy to follow by outsiders and non-IT experts • No guidelines for the descriptions of services • Dominated by technical considerations • No one is responsible for the data collected on services • The quality and correctness of data is not checked or discussed • Differences in skills of actors are clearly reflected in the process • Results and steps are not clearly defined in case services are composed • Much effort is needed to reach agreement among multiple stakeholders, as there is no overview • Lack of support for evaluation of multiple designs • No detection of infeasible combinations of services	• No executable result, no formal correctness checks, instead reliance of stakeholder judgment • No guaranteed outcome. Further, provides no single outcome and participants are sometimes left puzzled, there are many alterative having their own strengths and weaknesses • Demands involvement and time of all stakeholders • Depends on facilitator to keep to the rules of the game. The starting point of the method that the focus of the method is on exploiting rather than eliminating each others viewpoints might not hold during a workshop. Instead of having an open discussion, the focus might shift on defending and attacking each other viewpoints

Figure 3. Quantitative comparison of composition methods

	N=21	Best Practice		Multi Actor			
	Related questions:	Avg.	Std.	Avg.	Std.	Sig.	Z score
1	The support of functional requirements	4.95	1.60	4.76	1.64		-0.595
2	Managing a personal list of requirements	2.90	1.67	4.57	2.01	**	-2.812
3	Making a requirements inventory with a group of involved parties	4.05	1.66	5.24	1.30	*	-2.115
4	Support for requirement negotiation	3.90	1.18	4.81	1.86	*	-2.156
5	Functional attributes are supported during evaluation	3.81	1.54	4.57	1.69		-1.867
6	Non-functional attributes are supported during evaluation	3.05	1.60	4.19	2.32	*	-2.246
7	The composition method supports a set of evaluation methods	3.14	1.49	3.76	2.32		-1.179
8	The evaluation method supports attributes beyond functional fit	3.38	1.60	4.90	1.67	*	-2.258
9	Costs, reliability and speed are evaluated	4.67	1.28	5.57	1.25	*	-2.406
10	The composition method supports the creation of a shared view	4.29	1.85	5.24	1.55	*	-2.088
11	The shared view provides insight into the requirements	4.05	1.69	4.52	1.86		-1.100
12	The shared view provides insight into the services available	4.24	1.55	4.67	1.88		-0.915
13	The shared view enables communication between stakeholders	4.00	1.92	5.19	1.47	*	-2.242
14	The service register also contains new services in short time	2.90	2.10	2.81	2.29		-0.248
15	It is possible to add new services to the register	2.90	2.21	2.62	2.11		-0.402
16	The register contains non-functional attributes of services	2.24	1.51	4.38	2.38	**	-3.074
17	The register provides evaluations of services by users	2.52	1.75	3.86	2.24	**	-2.629
18	The register provides functionality to add/edit service attributes	3.52	2.18	3.05	2.13		-0.984
	Legend: ** = p<0,01; * = p<0,05						

Qualitative Results

The composition methods were also evaluated qualitatively during the session by the participants. Based on observations and the comments of the participants during the session and interviews Table 1 containing strengths and weaknesses was created.

The qualitative results confirm the outcomes of the quantitative evaluation presented in Table 1. The participants liked that the problem was taken as a starting point and that there were various ways to compare and discuss composition. Although the focus of the composition method is on dialogue and understanding each other interests and assumptions and avoiding judgmental pitfalls and advocacy of their own viewpoint, they felt that this cannot be avoided. They indicated that several times they felt that people were defending their position and judging others instead of looking for new solutions. They indicated that the role of the facilitator is crucial, as the facilitator should intervene in such situations and ensure that the rules of the game in the workshop are not violated.

Additional comments are related to the fact that the participants were familiar with the method used in their organizations, whereas they had no experience with the demand-driven service composition development method. The idea of bringing stakeholders together requires that they should be available at the same time, which was viewed as a weakness. The scheduling of such a session needs to be planned a long time in advance, as most stakeholders have a very busy schedule.

Our composition method does not generate a single outcome that is optimal in a certain way. It is aimed at supporting the creation of various service compositions and evaluating them. In this way trade-offs are made transparent and understandable to others. The participants indicated that his was also a weakness, as it is sometimes hard to understand which outcome would be the best.

Both the current methods and the demand-driven service composition development method are able to deal well with a set of services that are not interoperable and need to be modified to create a composition. The participants indicate that the reliance on experts was a strength but also a weakness at the same time. After understanding that formal methods can be used to prove the correctness, automatically generate result and so on, they would prefer such elements included in the method. In short, they would like to have the best of both worlds. As the starting points of the composition method and the demand-driven composition service method are different, it might be difficult to combine both worlds and this needs further research.

DISCUSSION AND CONCLUSION

Service composition might be viewed as a relatively straightforward process, but in networks of organizations composition is a complicated endeavor. The gap between the characteristics of organizational networks and existing service compositions methods is bridged in this research. To overcome this gap a demand-driven service composition method was developed which fulfill six basic requirements as determined by the characteristics of organizational networks. The development method is able to deal with incomplete information, to take the demand as a starting point, to deal with news services that do not exist yet, to include and to evaluate non-functional requirements, to show various stakeholder views and to help to create a shared vision. This resulted in a method consisting of five steps (requirements specification, description of existing services, design of sets of service compositions, evaluation of the composition designs and as last stage planning and reporting) and providing support during these steps, which puts dialogue and conversational forms to the centre rather than optimization logic. The method is aimed at encouraging participants to think about the assumptions underlying their decisions and creating a shared view. In this way deeper inquiry into possible service compositions was created.

The method was evaluated using quasi-experimentation and conducting both quantitative and qualitative analyses. This evaluation shows that the participants found that the ability to add their own perspective, manage their personal requirements and discuss the various perspectives with others are viewed as a strength of the composition method. Also the inclusion of non-functional properties and the way compositions are evaluated are viewed as a strength. Further, the participants stressed the need for taken the demand as a starting point, as at the end the organizational problem needs to be solved. The evaluation showed that the focus on dialogues instead of judgment was sometimes difficult and depends on the skills of the facilitator.

At the heart of the service portfolio is a centralized database containing descriptions of existing services described and prescriptions of new services were used. This database contains both functional and non-functional properties of services and goes beyond the data stored in a UDDI directory. The service portfolio helps to create a systematic overview of the characteristics and limitations of services, which data is used to evaluate the feasibility of service composition. More research is necessary which type of data should be stored about the services and how the dependencies among services can be analyzed to determine which compositions are feasible and results in which outcomes.

The evaluation shows that ways of handling the detection of infeasible combinations of services has not been given attention. Methods might come up with combinations of services that might not be feasible to implement and are as such useless. Therefore we recommend that composition methods include systematic analysis of the dependencies among services in compositions. The method and evaluation was focused on a relatively small group of persons. The scaling of this approach might be more cumbersome as more people get involved. Our final recommendation is to research the scalability of the proposed approach.

REFERENCES

Beek, M. H. T., Bucchiarone, A., & Gnesi, S. (2006). *A Survey on Service Composition Approaches: From Industrial Standards to Formal Methods* (Tech. Rep. No. 2006-TR-15). Consiglio Nazionale delle Ricerche: ISTI.

Bruijn, J. A., & Heuvelhof, E. F. (2002). *Process management: why project management fails in complex decision making processes*. Dordrecht, The Netherlands: Kluwer Academic.

Bruin, J. A. D., & Heuvelhof, E. F. T. (1995). *Netwerkmanagement, Strategieën, Instrumenten en Normen*. Utrecht, The Netherlands: Lemma.

Bruin, J. A. D., Heuvelhof, E. F. T., & Veld, R. J. I. T. (1998). *Procesmanagement, over procesontwerp en besluitvorming*. Amsterdam, The Netherlands: Academic Service.

Burstein, M., Bussler, C., Finin, T., Huhns, M. N., Paolucci, M., & Sheth, A. P. (2005). A Semantic Web Services Architecture. *IEEE Internet Computing, 9*(5), 72–81. doi:10.1109/MIC.2005.96

Chiu, D., Cheung, S. C., & Zhuang, H.-F. L. P. H. E. K. H. H. M. W. H. H. Y. (2010). Engineering e-Collaboration Services with a Multi-Agent System Approach. *International Journal of Systems and Service-Oriented Engineering, 1*(1), 1–25.

Cook, T. D., & Campbell, D. T. (n.d.). *Quasi-Experimentation: Design and Analysis Issues for Field Settings*. Houghton Mifflin Company.

Curbera, F., Khalaf, R., Mukhi, N., Tai, S., & Weerawarana, S. (2003). The next step in Web Services. *Communications of the ACM, 46*(10), 29–34. doi:10.1145/944217.944234

Feenstra, R. W., Janssen, M., & Wagenaar, R. W. (2007). Evaluating Web Service Composition Methods: The need for including Multi-Actor Elements. *The Electronic. Journal of E-Government, 5*(2), 153–164.

Fremantle, P., Weerawarana, S., & Khalaf, R. (2002). Enterprise services. Examine the emerging files of web services and how it is intergraded into existing enterprise infrastructures. *Communications of the ACM, 45*(20), 77–82.

Janssen, M., & Joha, A. (2008). Emerging shared service organizations and the service-oriented enterprise: Critical management issues. *Strategic Outsourcing: An International Journal, 1*(1), 35–49. doi:10.1108/17538290810857466

Jefferey, M., & Leliveld, I. (2004). Best Practices in IT Portfolio Management. *MIT Sloan Management Review, 45*(3), 41–49.

Liang, W.-Y., Huang, C.-C., & Chuang, H.-F. (2007). The design with object (DwO) approach to Web services composition. *Computer Standards & Interfaces, 29*(1), 54–68. doi:10.1016/j.csi.2005.11.001

Liu, A., Liu, H., Lin, B., Huang, L., Gu, N., & Li, Q. (2010). A Survey of Web Services Provision. *International Journal of Systems and Service-Oriented Engineering, 1*(1), 26–45.

March, J. G., & Simon, H. A. (1958). *Organizations*. New York: John Wiley & Sons Inc.

Milanovic, N., & Malek, M. (2004). Current solutions for Web service composition. *IEEE Internet Computing, 8*(6), 51–59. doi:10.1109/MIC.2004.58

Milward, H. B., & Provan, K. G. (2003, October 9-11). *Managing Networks Effectively.* Paper presented at the the 7th National Public Management Research Conference, Georgetown University, Washington, DC.

Parnas, D. L. (1972). On the criteria to be used in decomposing systems into modules. *15*(12), 1053-1058.

Rumbaugh, J., Jacobson, I., & Booch, G. (2004). *The Unified Modeling Language Reference Manual* (2nd ed.). Boston: The Addison-Wesley Object Technology Series.

Sage, A. P., & Armstrong, J. E. (2000). *Introduction to Systems Engineering*. New York: Wiley.

Schein, E. H. (1996). Kurt Lewin's Change Theory in the Field and in the Classroom: Notes Toward a Model of Managed Learning. *Reflections: The SoL Journal, 1*(1), 59–74. doi:10.1162/152417399570287

Simon, H. A. (1960). *The new science of Management Decisions*. New York: Harper & Row.

Tsai, W. T., Xiao, B., Paul, R. A., & Chen, Y. (2006). *Consumer-centric service-oriented architecture: A new approach.* Paper presented at the Software Technologies for Future Embedded and Ubiquitous Systems, 2006 and the 2006 Second International Workshop on Collaborative Computing, Integration, and Assurance.

Walsh, J. P. (1995). Managerial and organizational cognition: notes from at rip down memory lane. *Organization Science, 6*(3), 280–321. doi:10.1287/orsc.6.3.280

This work was previously published in International Journal of Systems and Service-Oriented Engineering, Volume 1, Issue 4, edited by Dickson K.W. Chiu, pp. 27-41, copyright 2010 by IGI Publishing (an imprint of IGI Global).

Chapter 21
Capacity-Driven Web Services: Concepts, Definitions, Issues, and Solutions

Samir Tata
TELECOM SudParis and CNRS UMR Samovar, France

Zakaria Maamar
Zayed University, UAE

Djamel Belaïd
TELECOM SudParis and CNRS UMR Samovar, France

Khouloud Boukadi
SFAX, Tunisia

ABSTRACT

This paper presents the concepts, definitions, issues, and solutions that revolve around the adoption of capacity-driven Web services. Because of the intrinsic characteristics of these Web services compared to regular, mono-capacity Web services, they are examined in a different way and across four steps denoted by description, discovery, composition, and enactment. Implemented as operations to execute at run-time, the capacities that empower a Web service are selected with respect to requirements put on this Web service such as data quality and network bandwidth. In addition, this paper reports on first the experiments that were conducted to demonstrate the feasibility of capacity-driven Web services, and also the research opportunities that will be pursued in the future.

INTRODUCTION

It is widely known that Web services standards/specifications (e.g., Web Services Definition Language (WSDL) and Universal Description Discovery and Integration (UDDI)) are not keeping the pace with the increasing number of challenges that dynamic environments (e.g., ubiquitous) and modern enterprises pose on Web services (Ardissono, Goy, & Petrone, 2003; Langdon, 2003; Papazoglou, Traverso, Dustdar, & Leymann 2007). Sudden drop in network bandwidth, limited power of mobile platforms, and ad-hoc joint ventures are

DOI: 10.4018/978-1-4666-1767-4.ch021

examples of these challenges. As a result, these standards/specifications are regularly subject to enhancements and extensions as the literature indicates (Most′efaoui & Younas, 2007; Luo, Montrose, Kim, Khashnobish, & Kang, 2006; Beek, Gnesi, Mazzanti, & Moiso, 2006). Since Web services are here to fulfill the promise of developing loosely-coupled, cross-enterprise business applications, empowering them with other mechanisms on top of these extensions and enhancements would be highly appreciated by those who advocate for Web services' benefits. In this paper we denote these mechanisms by *capacities*.

We embrace capacities to address the particular issue of Web services limited-ability (i.e., in terms of willingness and responsiveness) to cater to different levels of service offers. Chatterjee et al. mention that "*most Web services platforms are based on a best-effort model, which treats all requests uniformly, without any type of service differentiation or prioritization*" (Chatterjee, Chaudhari, Das, Dias, & Erradi, 2005). Today's Web services are designed and deployed without taking into account if the requests they receive originate from regular or new customers (Tao & Yang 2007) or from users who are at work or on the roads (Vukovic & Robinson, 2004). Some users are strict with the minimum network throughput to use while others are flexible with the freshness of the data to receive. Web services simply ignore the requirements and constraints of the environments in which users reside!

The objective of this paper is to discuss the design and development of capacity-driven Web services across the steps of description, discovery, composition, and enactment[1]. In the description step we show how to structure a capacity. In the discovery step we show how to look for relevant Web services based on their capacities and the environment's requirements and constraints. In the composition step we show how the discovered Web services take part in composition scenarios. Last but not least, in the enactment step we show how to invoke capacities of Web services that are now engaged in these scenarios.

Our contributions are as follows: (1) provide a structured way to define and load capacities into Web services, (2) develop mechanisms to select and activate a capacity out of several ones in a Web service, and (3) extend current standards/specifications if necessary to support the characteristics of capacity-driven Web services. The rest of this paper is organized as follows: we provide a running example while discussing the foundations upon which the capacity-driven Web Services are built. Then we present how to describe, discover, compose, and enact capacity-driven Web Services. Next we review the implementation work that was conducted. Then we report on related work, Followed by a conclusion and discussion in context to future.

RUNNING EXAMPLE

Our running example concerns a real-state office that runs and sells different types of properties such as villas and flats. The office is equipped with a set of PCs, while the staff in charge of conducting the visits is equipped with various heterogeneous handheld devices. The staff is usually in contact with customers as per the following description. The customers contact the office to request an estimate, purchase, or rent a property. The office can also contact customers as per their initial requests.

"Get the map of properties in the vicinity" is among the services that the real-state staffs use in their day-to-day business. This service returns all the properties that the office manages and is within a walking distance from a staff. This distance and other criteria like price range and number of bedrooms are set by the staff. "Get the map of properties in the vicinity" is treated as a composite Web service that relies on internal and external Web services as per the following description:

1. "LocateAgent" returns the current position of a staff in terms of latitude and longitude.
2. "LocatetProperties" retrieves the list of properties (identifier, location, etc.) that are in the vicinity of the staff (at a maximum distance from her) and whose characteristics match her criteria.
3. "GenerateMap" produces a navigation map that shows some properties in response to the staff's request.
4. "Display" displays the map on the device of the staff.

It is worth mentioning that "LocateAgent" and "GenerateMap" services are context aware. Their performance depends on the characteristics of the staff's device: network interfaces type (GSM, WiFi, GPS, etc.), quality of the signal strength of the connexion, screen size, etc.

CAPACITY-DRIVEN WEB SERVICES

This section consists of four parts. The first part contains definitions related this time to capacity-driven Web services. The second part discusses the foundations upon which these Web services are built. The third part presents the structure of a capacity. And the last part defines the types of requirements.

Definitions

Capacity is an aggregation of a set of actions that implement the functionality of a Web service. Functionality (e.g., currencyConversion) is usually used to differentiate a Web service from other peers, though it is common that independent bodies develop Web services with similar functionalities but different non-functional properties. Concretely speaking, the actions in a capacity correspond to operations in a WSDL document and vary according to the business-application domain. As per our proposed running example, the actions in LocateProperties Web service could be:

- Retrieve a list of properties' identifiers in the vicinity of the staff using the coordinates (latitude, longitude) of the staff and the maximum distance.
- Select from the list of identifiers the properties that satisfy the staff's requirements.
- Locate (latitude, longitude) the identified properties.

Several capacities along with their respective actions satisfy the unique functionality of a Web service in different ways. Which capacity to select and make active out of the available capacities in a Web service requires assessing the environment so that appropriate details are collected and prepared for this selection. Depending on the characteristics of the staff's device, "LocateAgent" can be carried out with one of the following capacities:

- "GPSGeoLocation" capacity that is an internal function of the device if the device is equipped with a Global Positioning System (GPS).
- "GSMGeoLocation" capacity that provides the position of the staff's device if the device is equipped with a Global System for Mobile (GSM).
- "AddressGeoLocation" capacity that provides the position (latitude, longitude) of the staff from the standardized address.

In term of assessing the environment, the following needs to be carried out:

- identifying the requirements that a Web service has to satisfy before it accepts to participate in a new composition scenario (on top of its participations in other ongoing compositions) and then,
- selecting a capacity to trigger following the participation acceptance.

This assessment concerns the requirements for composition needs and the requirements for

Figure 1. Capacity-driven Web services representation

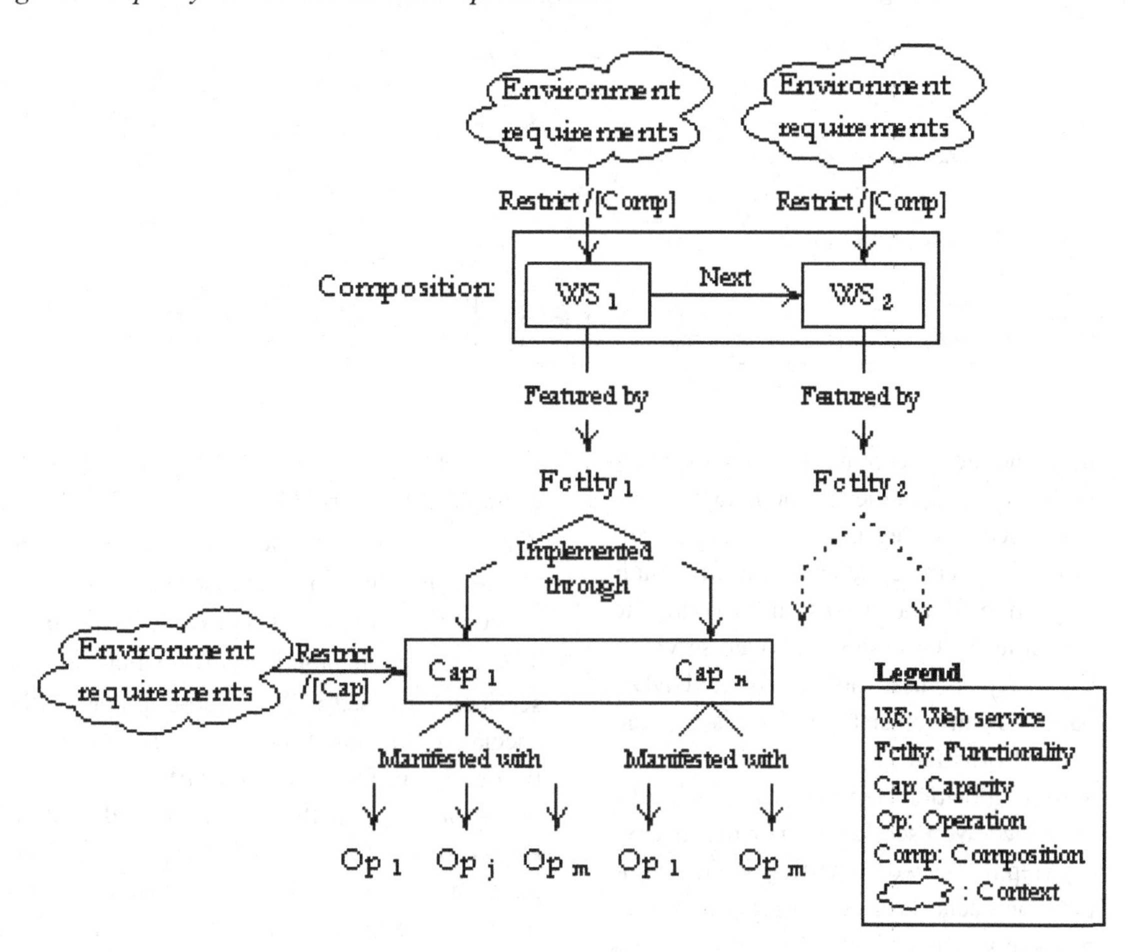

capacity-activation needs. An example of these latter requirements is the minimum network bandwidth to maintain so that a Web service guarantees critical content delivery to users of handheld devices.

Foundations

Figure 1 shows how capacity-driven Web services are put into action. To this end a composition scenario like the real-state office service is used through two Web services: WS_1 ("LocateAgent") returning the geo-location of the staff as functionality and *WS*2 ("LocateProperties") returning a selection of properties in the vicinity of the staff as functionality. WS1 and WS2 need to satisfy some composition requirements (*Restrict*/[*Comp*]) before they participate in this scenario. At run-time both Web services are enacted so their respective functionalities are executed. This requires selecting and activating the appropriate capacities with respect to the current execution requirements (*Restrict*/[*Cap*]).

The way capacity-driven Web services make composition scenarios progress through the steps of description, discovery, composition, and enactment is illustrated as follows.

- The description step is to define the functionalities of Web services, the different capacities that implement these functional-

Figure 2. Capacity categorization

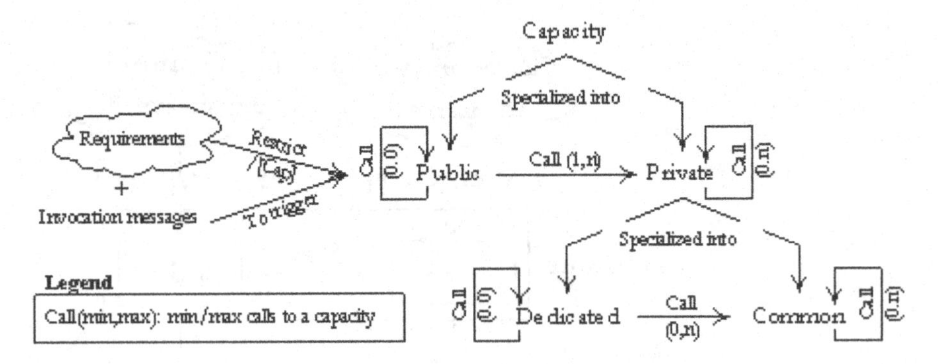

ities, and the execution requirements under which these capacities are activated.

- The discovery step is to look for the relevant Web services according to first, their respective functionalities and second, the capacities of the discovered Web services. This step is user needs-driven, UDDI-based (or any other type of registry), and context-oriented.
- The composition step is to check with the identified Web services as per the discovery step if they would participate in a composition scenario. Prior to replying either positively or negatively, these Web services assess their ability to satisfy the current composition requirements. Example of requirement is not to exceed the maximum number of compositions that a Web service can take part in at a time.
- The enactment step is to prepare the Web services as per the composition step for execution. This includes deploying these Web services on top of computing platforms and routing invocation requests to their respective capacities with respect to the assessed execution requirements.

When these four steps are completed, the composition scenario that is represented in Figure 1 could be implemented as follows where → stands for next: **WS** (Cap_1: Op_1, Op_j ,Op_m) → WS_2 (Cap_2 :Op_1). Another configuration for example $\mathbf{WS}_1$ (Cap_2 :Op_1 ,Op_m) → WS_2 (Cap_2 :Op_1) could have been adopted to implement this scenario, but this is not the case due to the current execution requirements that excluded $\mathbf{Cap}_2$ in $\mathbf{WS}_2$. Common practices in the field of Web services, *e.g.* based on WSDL descriptions, neither specialize the operations to execute nor confine the operations into dedicated parts. Instead, these operations are usually woven into and scattered across the business logic of a Web service, which makes any change in this business logic or these operations extensive, expensive, and error-prone (Ortiz, Hernández, & Clemente, 2006; Kouadri Mostéfaoui, Maamar, & Narendra, 2009).

Capacity analysis

A capacity in a Web service manifests itself with operations (or actions) that implement the functionality of this Web service. Since a Web service is multi-capacity, the capacity to activate at run-time is subject to satisfying requirements whose some are given and we discuss how a capacity is structured using Figure 2.

- Separate capacities in a Web service could have common operations (e.g., verification of username/password). Cap_1 and Cap_n in WS_1 have Op_1 and Op_m in common. Contrarily, Op_j exists in Cap_1, only.

- Common operations in separate capacities could be overloaded if needed. Op_1 takes one input argument in WS_1's Cap_1 (e.g., username only because of the use of a default password) and two input arguments in WS_1's Cap_n (e.g., username and password are both required).
- A default capacity Cap_{def} could be planned on top of other capacities in a Web service[2]. This default capacity is free of any execution requirement and is automatically activated when none of the other capacities in a Web service is able to satisfy the current execution requirements. By having a default capacity, the execution of a Web service is guaranteed independently of the status of the environment.
- In a Web service, capacities could be either enabled or disabled, if necessary. Only the enabled capacities are eligible for possible activation. Enabling/Disabling capacities goes along with the way a default capacity would be managed, i.e., considered for possible activation upon enabling decision.

Like access modifiers in object-oriented programming, we categorize capacity into public and private (Figure 2). Public capacities are exposed to the external environment that consists of users and Web services. Submitting invocation messages to public capacities is subject to satisfying execution requirements. Private capacities are hidden as their names hint and just public capacities can call (or invoke) them. By doing so, the privacy of a Web service in terms of offered and existing capacities is maintained. Contrarily to private capacities that might call each other, public capacities do not since each public capacity implements the functionality of a Web service in a different way.

In addition to public and private types, the private type of a capacity is refined into dedicated and common. A dedicated capacity is exactly called by one public capacity through a private capacity. Dedicated capacities do not call each other as shown in Figure 2. Finally, a common capacity could be called from several private capacities either dedicated or common.

Types of Requirements

A Web service implements the functionality it offers through different capacities. Which capacity to activate at run-time depends on the execution requirements that this Web service can satisfy. The requirements could be of different types with focus on data, network, and resource in this paper.

1. Data requirements concern the quality of data that a Web service manipulates. This could be about freshness (when were the data obtained), source (who was the sender of data), and validity (when do the data expire). Different thresholds (or termination conditions) could maintain data-requirements satisfaction over-time, e.g., data obtained 12 hours ago are still considered as valid. Thresholds are set by future consumers of Web services.
2. Network requirements concern the communication channels that a Web service uses during interaction with users and peers. This could be about bandwidth, throughput, and reliability. Like with data requirements, thresholds could maintain network-requirements satisfaction over-time, e.g., 10% drop in network bandwidth is still considered as acceptable. Thresholds are set by developers of Web services.
3. Resource requirements concern the computing facilities in terms of availability and reliability upon which a Web service is scheduled to run. Thresholds are set by developers of Web services and could maintain resource-requirements satisfaction over-time, e.g., 5% reduction in resource time-availability is still considered as acceptable.

Let's consider the "LocateAgent" Web service that provides the position of a staff using longitude and latitude. We have informally presented in Section 3[REMOVED REF FIELD] the capacities of this Web service. The different requirements in the context of this Web service could be as follows:

- the data requirements: Authorization tracking staff (no or yes).
- the network requirements: Strength of the GPS or WiFi signal.
- the resource requirements: type of device connectivity (GSM+GPS, GSM+WiFi, or GSM).

In the following we detail the different requirements of the capacities of the "LocateAgent" Web service:

- the "GPSGeoLocation" capacity: this capacity requires a GPS-enabled device and a minimum power of the signal to operate the GPS.
- the "GSMGeoLocation" capacity: this capacity requires a WiFi-enabled terminal with sufficient signal strength to operate and a authorization of staff to be located. This capacity can be considered as default capacity because it only requires a GSM connection, which is available in most mobile phones.
- the "AddressGeoLocation" capacity: this capacity requires only the WiFi to provide the service with sufficient signal strength to operate.

STEP-BY-STEP EMPOWERMENT OF WEB SERVICES WITH CAPACITIES

This section presents how to describe, discover, compose, and enact capacity-driven Web Services. We discuss for each step the changes that need to be made in Web services and their respective standards/specifications so that this empowerment can happen.

Description Step - Weaving Capacities into Web Services' Descriptions

To define capacities in a Web service, we thought of extending the description of this Web service. We started by examining the *de-facto* standard to describe Web services, namely WSDL, for possible extensions. These extensions are about (i) associating capacities with WSDL interfaces and (ii) using unique string literals to name these capacities in a WSDL document. As a result, the description of a Web service now includes multiple interfaces, one interface per capacity. Capacities are defined with two new WSDL interface attributes: **capacityName** and **capacityType**. **capacityName** uniquely identifies a capacity in a WSDL document, and **capacityType** specifies the access modifier of a capacity as per the types in Figure 2. This way of describing capacities is in line with the current practices in the semantic Web through the semantic annotation of WSDL (Kopecký, Vitvar, Bournez, & Farrell, 2007) and could enable as well to enhance capacity description with semantic details if needed.

The extensions we propose have an impact on the latest version of WSDL specification (*2.0*), precisely on the section of the WSDL service description. Quoted from this specification, a Web service *"specifies a single interface that the WSDL service will support, and a list of endpoint locations where that WSDL service can be accessed. Each endpoint must also refer to a previously defined binding to indicate what protocols and transmission formats are to be used at that endpoint. A WSDL service is only permitted to have one interface"*. Going back to our work, different capacities through their respective interfaces satisfy the unique functionality of a Web service in various ways. Building on top of the WSDL specification (*2.0*), we propose *C*-WSDL standing

Figure 3. C-WSDL document in XML

```
01:<description targetNamespace="xs:anyURI">
02:  <documentation />*
03:  [<import />|<include />]*  <types />?
04:  <interface  name="xs:NCName"
05:    cwsdl:capacityName="xs:NCName"
06:    cwsdl:capacityType="Public|Private|Dedicated|Common">*
07:    <cwsdl:requirement
08:          type="Data|Network|Resource" value="xs:int"/>*
09:    <operation>+
10:  </interface>
11:  <binding />*
12:  <service name="xs:NCName" interface="xs:QName">
13:    <endpoint />+
14:  </service>
15:</description>
```

for *Capacity-driven Web Services Description Language* as a new way to describe first, the multiple interfaces that a Web service supports and second, the list of endpoint locations, one per interface binding (implementation), that give access to this Web service.

Figure 3 suggests a BNF-based pseudo schema of a *C*-WSDL document that structures a capacity-driven Web service. This schema handles the WSDL-document of regular (i.e., mono-capacity) Web service as well.

WSDL document. Here a description has a possible empty set of **documentation** (Figure 3, line: 02), **import** and/or **include** elements, and a possible empty set of data **types** (line: 03). In addition, a WSDL interface has a single **name** attribute (line: 04). A WSDL service has a single **name** and a single **interface** attributes (line: 12), and a set of **endpoint** elements (line: 13).

***C*-WSDL document.** *C*-WSDL proposes two extended attributes to describe capacity names and types (Figure 3, lines: 05/06) of a given *C*-WSDL interface (line: 04). In addition *C*-WSDL proposes an extended element to describe the requirements that could be put on a capacity (lines: 07/08). A *C*-WSDL service has a single **interface** attribute (line: 12) that corresponds to the interface of the default capacity (Cap_{def}). Finally, a *C*-WSDL service has a set of **endpoint** elements (line: 13).

In conclusion, the major difference between WSDL and *C*-WSDL resides in the endpoint elements. Initially, the endpoints refer to a single WSDL interface but they refer now to a set of different *C*-WSDL interfaces that provide the different capacities. Note, that this extension has no impact on the existing Web services framework and tools.

Discovery Step: Tuning Web Services Discovery Using Capacities

Current practices in Web services discovery rely on registries like UDDI and ebXML to match users' needs (or service requests) with Web services' functionalities. Recently these practices have been extended in response to different concerns such as semantics (Benbernou & Hacid, 2005) and security (Shehab, Bhattacharya, & Ghafoor, 2007). Unfortunately, these practices and extensions do not take into account the status of the environment during the discovery of Web services. This status would mean for example level of user (novice or expert), type of device (desktop or handheld), delivery time (am or pm period), etc. To address this shortcoming, the context comes into play by fine tuning the initial list of discovered Web services in order to retain only the Web services that can respond to changes (through their capacities) in the environment (Blefari-Melazzi, Casalicchio, & Salsano, 2007; Hesselman, Tokmakoff, Pawar, & Iacob, 2006).

As part of our research on capacity-driven Web services, we considered that the context-based discovery needs to be fine tuned further if new requirements (like the ones we suggest in Section 3.4[REMOVED REF FIELD]) are posed on Web services at run-time. Indeed we do not restrict the discovery step to the Web services that satisfy a user's needs[3], but "drill" into these same Web services to discover the capacities they will need to satisfy these requirements at run-time. This extra discovery action is subject to the following assumption: the context of the requirements associated with capacities remains unchanged between the discovery and enactment moments. This assumption is not difficult to maintain if one considers situations where the time elapsed to execute the steps of discovery, composition, and enactment is negligible compared to the time that sees changes in the context (frequency of change). If the context is subject to continuous changes, it would be appropriate to adopt a dynamic binding of capacities, which would permit to delay their selection to the last moment prior to invocation. Though this aspect does not fall into the scope of this paper, this could be achieved by provisioning an agile framework for supporting service level agreements and a monitoring system of requirements (Skene, Lamanna, & Emmerich, 2004).

Following the discovery of the candidate Web services that can satisfy the user's need, a matrix denoted by **suitability** $\mathbf{M}_s$ (Table 1) is generated for each Web service using its respective *C*-WSDL document (Figure 3, lines: 07/08). In M_s, the number of rows corresponds to the number of capacities (excluding the default capacity) in a Web service, and the number of columns corresponds to the number of requirement types that are considered (three types as per Section 3.4[REMOVED REF FIELD]). While the number of columns is similar to all Web services, the number of rows changes as this depends on the number of capacities that are available in each Web service. Each cell (*i,j*) in M_s reflects the suitability of a capacity for the satisfaction of a certain requirement. Moreover, each row *i* in M_s reflects the suitability of a capacity for the satisfaction of all the requirement types at a time. The following notation is used to fill in M_s's cells: 1: poor, 2: medium, and 3: high. For example, $M_s(cap_{il}, datarequirement / [cap_{il}]) = 1$ stands that Cap_{il} in WS_i is suitable for an assessment of the data level to poor. Completing M_s's cells is done by the suppliers of Web services based on their respective *C*-WSDL documents.

Concurrently to establishing the suitability matrices, an assessment of the current level of data, network, and resource requirements happens.

Table 1. Suitability matrix per candidate Web service

Web service $_i$	Requirement types		
	Data	Network	Resource
Cap $_{il}$	1	2	1
M	M	M	M
Cap $_{ij}$	1	1	2

To this end, an assessment structure denoted by S_a is computed on the fly. S_a consists of three couples where each is structured as follows:

- (dr, Val_{rr}): dr stands for data requirement and Val_{rr} is the value that corresponds to the current level of data using either 0: unknown, 1: poor, 2: medium, or 3: high. This value is obtained after running Val_{rr} function shown in Figure 7.
- (nr, Val_{rr}): nr stands for network requirement and Val_{rr} is the value that corresponds to the current level of network using either 0: unknown, 1: poor, 2: medium, or 3: high. This value is obtained after running Val_{rr} function shown in Figure 7.
- (rr, Val_{rr}): rr stands for resource requirement and Val_{rr} is the value that corresponds to the current level of resource using either 0: unknown, 1: poor, 2: medium, or 3: high. This value is obtained after running Val_{rr} function shown in Figure 7.

For the sake of simplicity, the values of data, network, and recourse requirements are considered in this paper as integer. Note that they can be considered as concepts from an ontology that models the semantics of data, network, and resource requirements. All the value comparisons in this paper will be then replaced by concept matchmaking.

Table 2. Suitability matrix for AgentLocation Web service

AgentLocation Service	Requirement types		
	Data	Network	Resource
GPSGeoLocation	0	1	3
GSMGeoLocation	2	3	2
AddressGeoLocation	0	3	2

Example: S_a={(dr,0), (nr,1), (rr,2)} means that the current assessment levels of data, network, and resource are unknown, poor, and medium, respectively.

In the following we detail the values of the different requirements of the capacities of the "LocateAgent" Web service:

- the "GPSGeoLocation" capacity: GSM is available (Resource=1) and there is a strong connection available (Network=3).
- the "GSMGeoLocation" capacity: GSM and WiFi are available (Resource=2), authorization tracking staff is needed (Data=3) and there is a strong network connection available (Network=3).
- the "AddressGeoLocation" capacity: GSM and WiFi are available (Resource=2) and there is a strong network connection available (Network=3).

Table 2 gives the suitability matrix for "Agent-Location" Web Ssrvice.

The details of the algorithm implementing the Web services discovery step are given in Figure 4 and Figure 7. This algorithm uses four functions.

1. ServiceDiscovery function takes four input arguments namely S_a, M_s, a service request R, and the registry of Web services to screen Reg, and returns the identifier of Web services ServList $_R$ that (1) matches the functional properties of R based on the positive outcome of SemanticMatch function and

Figure 4. Web services discovery algorithm (1/4)

```
Funct ServiceDiscovery (Sa,Ms,R,Reg)
Input: Sa - assessment structure
Input: R - service request
Input: Reg - Web services registry to search
Input: Ms - suitability matrix
Output: ServListR- list of discovered services
Auxiliary: I - integer
Begin
    for each Web Service WS in Reg do
        Mws ← suitability matrix of WS
        if SemanticMatch (R,WS) then
            ID ← CapacityToActivate (Sa,MWS)
            ▷ MWS is the suitability matrix of WS
            if ID is not null then
                ServListR ← ServListR.add(WS, ID)
            end if
        end if
    end for
    return ServListR
End
```

Figure 5. Web services discovery algorithm (2/4)

```
Funct SemanticMatch (R,WS)
Input: WS - Web service WS
Input: R - service request
Output: match- Boolean
Begin
    for each interface i of WS do
        match ← true
        for each operation opR in R do
            if thers is no opi operation of i s.t. subsume(opR, opi)
        then
                match ← false
                break
            end if
        end for
        if match = true then
            break
        end if
    end for
    return match
End
```

Figure 6. Web services discovery algorithm (3/4)

```
Funct CapacityToActivate (Sa,Ms)
Input: Sa- assessment structure
Input: Ms- suitability matrix
Output: Capid- identifier of capacity including default capacity to activate
Auxiliary: I - integer
Auxiliary: n - number of rows in Ms
Begin
    for i=0 to n do
        if AssessmentMatch (Sa,Ms.Rowi) = true then
            return ID.Capi
            ▷ identifier of the capacity to activate
        end if
    end for
    return ID.DefCap
    ▷ identifier of the default capacity to activate
    ▷ if there is no suitable capacity
End
```

Figure 7. Web services discovery algorithm (4/4)

```
Funct AssessmentMatch (Sa,Row)
Input: Sa - assessment structure
Input: Row - complete row in the suitability matrix
Output: cmp - Boolean
Begin
    ▷ situation assessment by calling appropriate functions
    Sa.Valdr ← Compute(Valdr)
    Sa.Valnr ← Compute(Valnr)
    Sa.Valrr ← Compute(Valrr)
    if (Sa.Valdr|nr|rr = Rowi,0|1|2) then
        cmp ← true
        ▷ detection of a perfect match
    else cmp ← false
    end if
    return cmp
End
```

(2) has at least one specific row in M_s that is satisfied by the values contained in S_a based on the outcome of CapacityToActivate function. In case of no perfect match, the identifier of the default capacity is returned.

2. SemanticMatch function takes two input arguments namely a service request R and a Web service WS, and returns a boolean about either the success or the failure of the semantic match between R and WS. This function is based on our previous work (DBLP:conf/wetice/MBareckTM07).
3. CapacityToActivate function identifies a capacity to activate in a Web service. It takes S_a and M_s as input arguments and returns the identifier of the capacity. The match locates a specific row in the suitability matrix based on the values that the assessment structure carries. If there is no perfect match, the identifier of the default capacity is returned.
4. AssessmentMatch function takes two input arguments namely S_a and a complete row Row in M_s, and returns a boolean about either the success or the failure of the comparison between the current level of data, network, and resource requirements and the available capacities reported in the suitability matrix.

Composition Step: Inviting Web Services for Composition

Composition puts the Web services that are identified in the discovery step together according to the business logic of a certain case study like the one in Section 2[REMOVED REF FIELD]. In compliance with Figure 1 (*Restrict*/[*Comp*]), we give these Web services the opportunity to either accept or reject participating in a composition scenario. This is based on a set of requirements that suppliers put on their Web services including number of participations, period of unavailability, reputation of peers, etc.

- *Maximum number of participations* limits the number of concurrent compositions over time in which a Web service can participate at a time. This limitation is strictly motivated by the non-functional properties that a Web service announces to potential users and thus, needs to guarantee.
- *Next period of unavailability* informs when a Web service is not available for composition due to various reasons such as maintenance.
- *Reputation of peers* limits the list of Web services that a Web service accepts interacting with during compositions. This list could be established using some recommender techniques (Adomavicius & Tuzhilin, 2005).

In (Maamar, Kouadri Mostéfaoui, & Yahyaoui, 2005) we discuss how Web services either accept or reject participating in composition scenarios. That was done through the Web services instantiation principle that stresses the conditional participation of a Web service in any composition. Concretely, a participation happens through what we called Web service instance. Figure 8 shows the organization of a Web service with respect to the instantiation principle: compositions already deployed, compositions currently deployed, and compositions to be deployed. Interesting to highlight the requirements (*Restrict*/[*Comp*]) that suppliers put on Web services with respect to future participations in compositions.

In Figure 9 we illustrate a simple specification of the real-state office service using a finite state machine (Harel & Naamad 1996). The component Web services that agreed to participate in this composition are “LocateAgent”, “LocateProperties”, “GenerateMap” and “Display”. If one of

Figure 8. Organization of a Web service

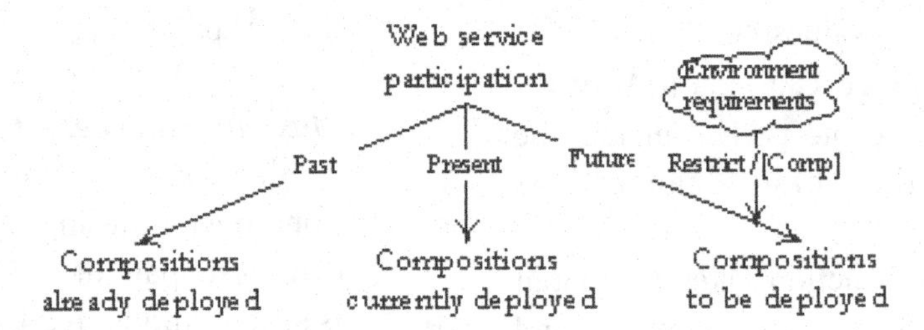

these Web services rejects the participation, the discovery step is reactivated again.

Enactment Step: Invoking Capacities of Web Services

This step concludes the process of putting capacity-driven Web services into action. After discovering the needed Web services along with their respective capacities and confirming their acceptance of participating in a composition scenario, it is now time to invoke these capacities. As reported in the discovery step, the assessment level of the requirements on capacities continues to be valid in the enactment step. As a result, there is no need to start the search for other capacities.

The enactment step is to deploy capacity-driven Web services, receive and route requested to them, collect results from them, and submit these results back to requesters. During this step, the different capacities in the same Web service could be involved according to Figure 2. Public capacities are the first to be invoked. Afterwards, additional private capacities (whether dedicated or common) are invoked according to the current requirements to satisfy and the business logic that implements the functionality of this Web service. The description we provide on this step illustrates an intra-capacity collaboration in the same Web service. An inter-capacity collaboration in separate Web services exists as well and illustrates the implementation of a composition scenario.

One of the aspects that need to be looked into in the enactment step is the consistency of the capacity-related requirements at the composition level. Indeed it is unlikely that all suppliers would agree on the requirements that could be put on their Web services. Conflicting requirements could arise, which call for corrective actions prior to executing any composition. These actions start by labelling the requirements either *local* for the component level or *global* for the composition level. In the following we briefly illustrate how global data-requirements could be set using data freshness as an example. We assume crisp values for local requirements and a composition scenario of type $WS_1 \rightarrow WS_2$. Local data-requirements are WS_1:$dr_{data.freshness}$ 2 day(s) and WS_2 :$dr_{data.freshness}$ 1 day(s) (e.g., data should not have been collected more than 2 days ago). In case there are common data between WS_1 and WS_2 (i.e., WS_1(Data$_{output}$ WS_2(Data$_{input}$), the global data-requirements for data freshness should be

Figure 9. Specification of the real-state office example

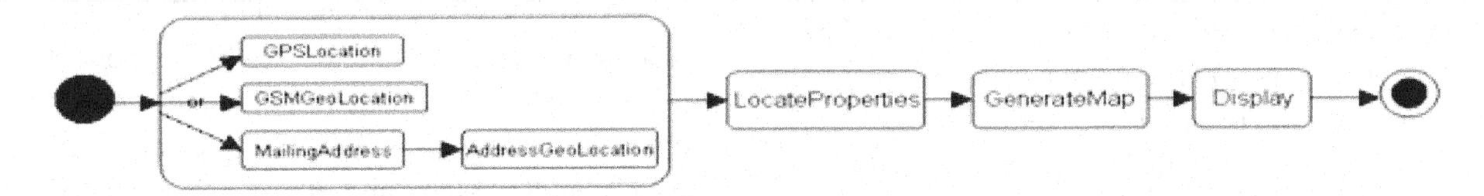

the minimum value among these two local data-requirements, i.e., WS_1 WS_2 :$dr_{data.freshness}$ 1 day(s). As a result of this agreement on the global data-requirement, WS_1 changes its local data-requirement value to 1 day(s).

IMPLEMENTATION

This section consists of two parts. The first part describes the architecture of the prototype we developed and summarizes the technologies we used. The second part explains how a capacity-driven Web service is enacted.

Prototype Architecture

Figure 10 focuses on the discovery and composition steps and illustrates the types of managers in the prototype that implements capacity-driven Web services. The composition manager assists designers define composition scenarios. The discovery manager identifies the Web services that satisfy users' needs by searching the *C*-WSDL service registry. Finally, the requirement manager assesses the environment in terms of data, resource, and network prior it sends the details collected back to either the composition or the discovery manager. To manipulate *C*-WSDL documents we use the Eclipse Modeling Framework (EMF)[4]. From the XML schema of *C*-WSDL, the EMF provides tools and runtime support to produce a set of Java classes for *C*-WSDL manipulation.

When the composition manager receives a user's request (Figure 10, action 1), it asks the discovery manager to look for the component Web services that could take part in the composition scenario that will satisfy this request (action 2). These components Web services are established after the discovery manager screens the *C*-WSDL service registry (actions 3 and 4). Afterwards, the discovery manager asks the requirement manager to assess the current environment so that the data-, resource-, and network-related requirements are identified (actions 5, 6, 7). This would permit to identify the capacities to use per component Web service and per type of requirement. When the composition manager receives the list of component Web services and capacities (action 8) from the discovery manager, it selects the component Web services that will participate in the composition using the composition requirements that the requirement manager sends (actions 9, 10).

In addition to the manipulation of *C*-WSDL documents, we implemented a semantically-enhanced *C*-WSDL service registry that builds upon the UDDI specification along with a discovery manager that deals with capacity-driven Web services discovery. The *C*-WSDL service registry and discovery manager implement the algorithms presented in Figure 4, Figure 5, Figure 6, and Figure 7. The requirement manager is based on COSMOS, which is a framework for managing context information (Rouvoy, Conan & Seinturier 2008).

Figure 11 focuses on the enactment step and illustrates the types of managers in charge of this step. Given a selected Web service, the invocation manager identifies the capacity to trigger. The requirement manager assesses the environment in terms of data, resource, and network as described above. The execution manager executes the selected capacity. In the next section, we present how we have used Aspect-Oriented Programming (AOP) to implement the execution manager.

Capacity-Driven Web Services Enactment

Existing mechanisms that support Web services' containers such as Apache Axis are mainly developed to meet the needs of regular (i.e., mono-capacity) Web services in terms of hosting them, receiving and routing SOAP requests to them, collecting results from them, and submitting these results back to requestors. Unfortunately these mechanisms are not enough to support the execution of capacity-driven Web services. To

Figure 10. Prototype architecture with focus on discovery and composition steps

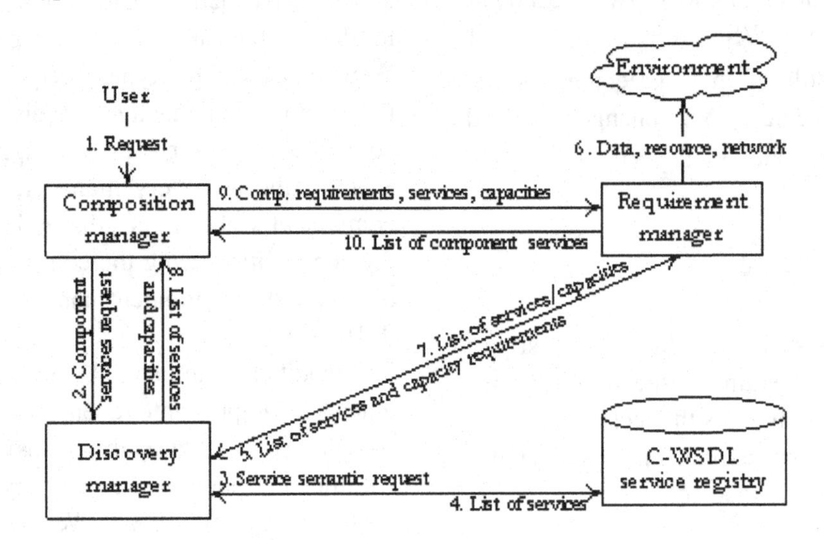

address these shortcomings, we implement capacities of Web services using AOP. AOP provides mechanisms to separate crosscutting concerns into single modules called aspects. Aspects can be dynamically integrated into a system thanks to the dynamic weaving principle (Bockisch, Haupt, & Ostermann, 2004). AOP introduces three key concepts: join points, pointcuts, and advices (AOP 2007). Join points are points in the execution of a program that indicate where new behaviors could be included. Pointcuts designate a set of join points. Whenever a program execution reaches a join point that is described in a pointcut, a piece of code associated with this pointcut (called advice) is executed. Finally, advices specify behavioral effect at joint points.

Figure 12 depicts how we used AOP concepts to implement capacity-driven Web services. In this case, an aspect includes a pointcut that matches a given Web service and one or more advices. These advices refer to the capacities of this Web service. Advices are defined as Java methods and pointcuts are specified in XML format.

Our implementation approach for capacity-driven Web services deployment and enactment

Figure 11. Prototype architecture with focus on the enactment step

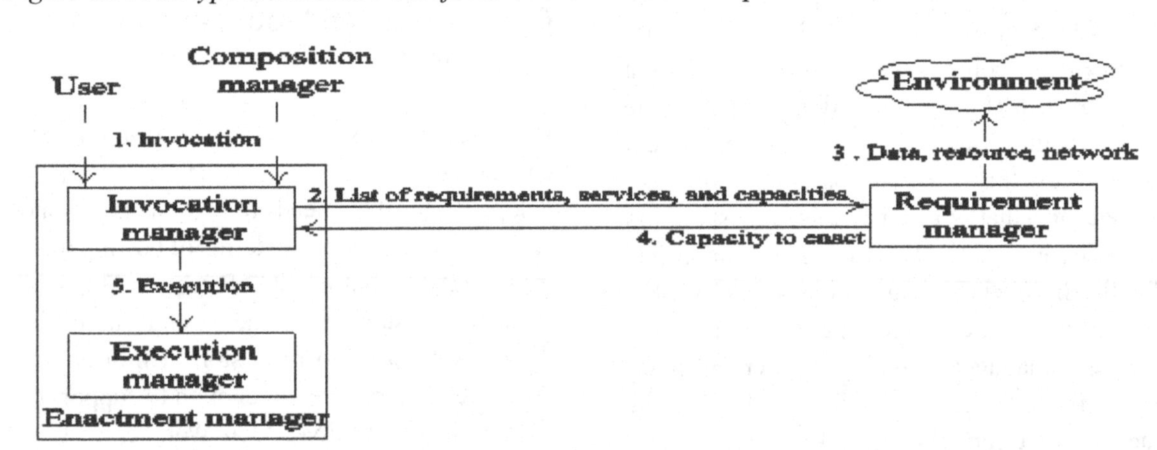

Figure 12. AOP to support capacity-driven Web services implementation

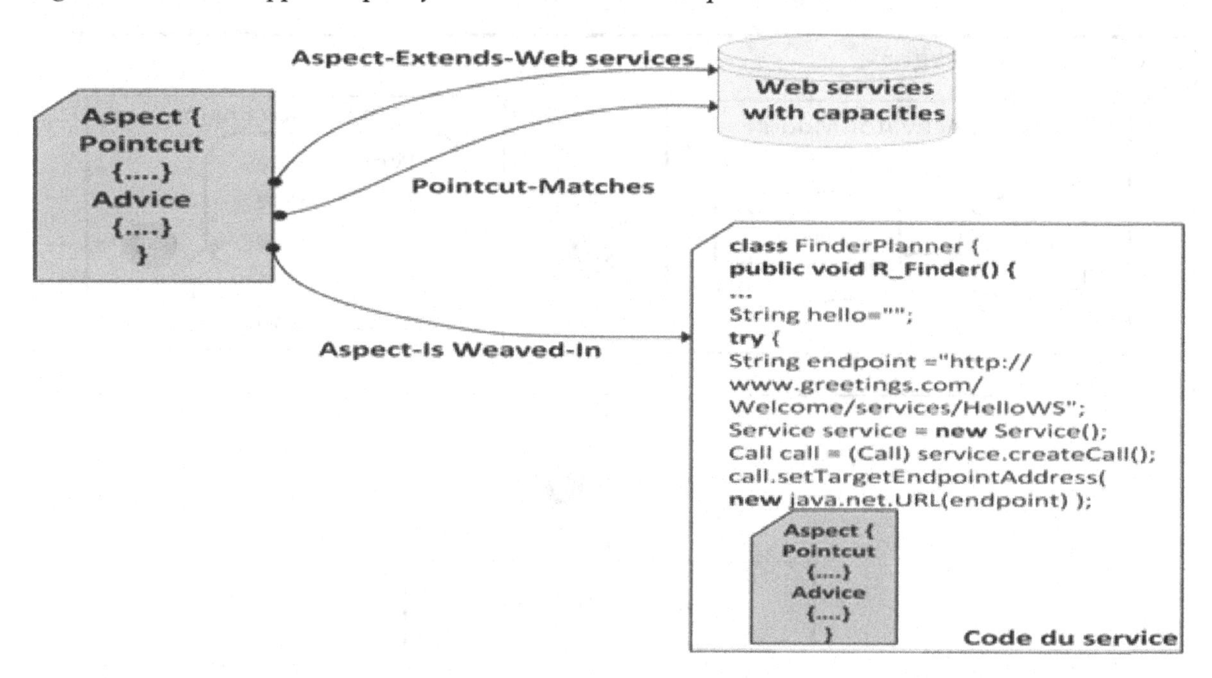

is presented in Figure 13. In this figure, the execution manager previously presented in Figure 11 is detailed. Figure 13 includes the Aspect Manager Module (AMM), the Aspect Activator Module (AAM), and the Weaver Module (WM).

- The AMM is responsible for adding new aspects to a corresponding aspect registry. In addition, the AMM can deal with a new advice implementation, which could be added to this registry. The aspect registry contains the method names of the different advices related to a given Web service. For example, a developer can send a register command in order to add a new advice implementation with a given service to the corresponding aspect registry.
- The AAM is responsible for storing aspects, advices, and capacities, and providing advice binding to the Web service engine. This binding consists of identifying and specifying the advice implementation (the Java method that implements a capacity) to apply to a Web service.
- The WM is based on an AOP mechanism known as weaving. The WM performs a run time weaving, which consists of injecting an advice implementation into the core logic of a Web service. The WM works closely with the Web service engine. In the proposed prototype, we adopt Apache Axix2/Java as a Web service implementation engine.

The invocation of a capacity-driven Web service is as follows. The composition manager sends the AAM a Web service's ID and the capacity that should be used (Figure 13, action 1). This AAM identifies the advice implementation namely the Java class to be executed to the Web service based on the information sent by the composition manager (i.e., Web service's ID and Capacity's ID). The advice implementation and pointcut related to the Web service to which the advice will be executed constitute the aspect binding (action 2). When the Web service engine receives the aspect binding, it activates the WM that integrates the advice implementation into the core logic of the Web service. By doing so, the

Figure 13. Step-by-step execution of a capacity-driven Web service

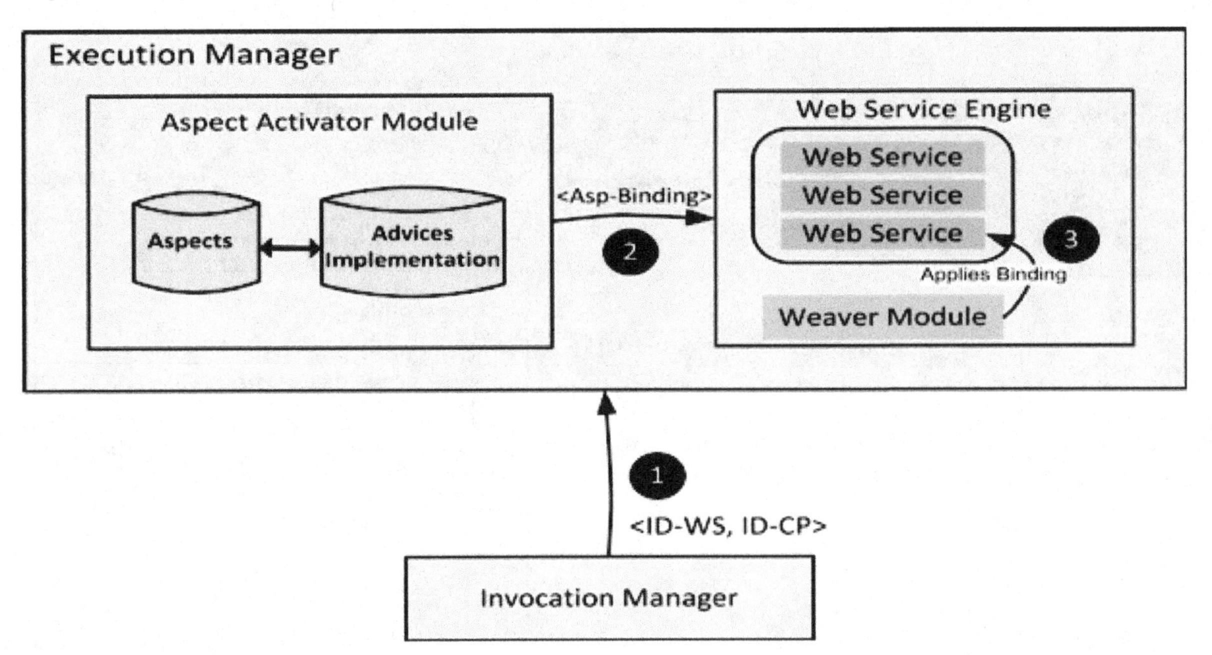

Web service will execute the appropriate capacity in response to the current execution requirements in the environment.

Validation

Few studies on multi-level services are reported in the literature, which results in the scarcity of experimental results. To highlight the importance of the adoption of capacity-driven Web services, we compare them to regular Web services. This comparison relies on the efforts to put into accommodating the context changes in both cases (capacity-driven Web service and regular Web service).

Capacities are implemented as aspects. An important advantage of AOP is the ability to weave capacities into service code, in an automated way. Automation is important for maintaining the modularity of the capacities. Without automation, changes in a capacity would require manual weaving, a tedious process. Besides, reacting to context evolution, as it arises, can be supported by the non-invasive capacities compositions, which is provided by AOP. For example, if a new capacity feature is required, developers add only the corresponding aspect code without changing the base system.

However in a regular Web service, managing changes requires the intervention of the programmer who is responsible for adapting the service code and re-deploying it. Web services providers have no means to dynamically adapt an existing Web service to context requirements changes. From the other side, Web services' customers have no way to dynamically adapt themselves to the service changing in order to avoid execution failures. Another important issue concerns the adaptation cost. Capacities implemented as aspects uses decomposition principal at design and development-time to separate basic core service from aspects. However the decomposition is resolved during compilation, when capacity code is woven into the service code. The effort for service adaptation, that is for integrating common and variant functionality is reduced to the minimum because the aspect weaver does this automatically. So, significant

Algorithm 1.

1 // Aspect for Geolocat ionWebService
2 **public class** Aspect Aspect4 LocateAgentWebService {
3 **public** String AddressGeoLocation () {
4 **return** coordinates ;
5 }
6 **pubic** Object GSMGeoLocation (Invocation invocation) throws **Throwable** {
7 **public** String getLocation () {
8 **return** Locat ionCoor ;
9 }
10 **return** invocation.invokeNext();
11 }
12 }

efforts are found only at design-time. However in a regular service scenario, the adaptation cost involves both the engineering cost and the implementation cost.

Example

In the following example, Aspect4 LocateAgentWebService class (line 2) is declared along with two advices namely AddressGeoLocation (line 3) and GSMGeoLocation (line 6). Both advices represent two capacities in "Geolocation" WS defined at design-time.

Figure 14 depicts the selection and execution of a capacity that satisfies the current level of data, network and resource requirements reported when running the LocateAgentWebService Web service. We assume that current context information is: (DataLevel= 0), (NetworkLevel=1), and (ResourceLevel=3). Taking into account this information, the composition manager selects the "GPSGeoLocation" capacity for execution (step 1). This capacity finds a staff's geographic coordinates from his GPS position. Once, the capacity is selected, the next step consists of binding the advice implementing this capacity to the Aspect4 LocateAgentWebService (already defined for the LocateAgentWebService Web service) (step2).

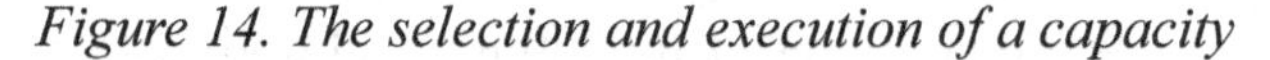

Figure 14. The selection and execution of a capacity

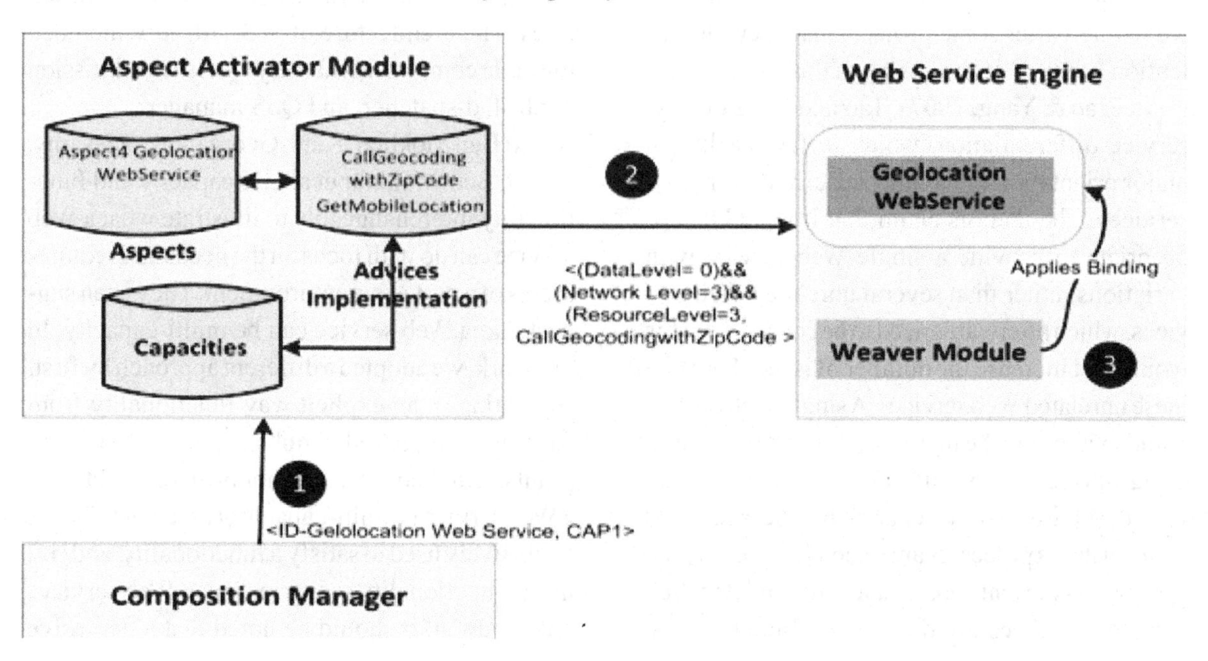

Technically, step 2 consists of declaring the Aspect4 LocateAgentWebService class and then deploying it using the aspect factory and the dynamic binding of the advices. These concepts have recently been defined by the JBoss community as the only available way of using dynamic AOP in standalone applications (JBoss AOP, 2009). First, an aspect factory is created. The constructor of GenericAspectFactory receives the class of the aspect (line 3). Next, an aspect definition (line 4) using the factory, defining also the name of the aspect and the scope of the aspect, is created. Finally, an advice factory is defined. This factory receives the aspect definition (line 10, 19 or 25) and the name of the advice. This advice factory is added to a new advice binding that is registered at AspectManager by a call to AspectManager.addAdviceBinding (line 15, 22 or 28). This binding is done dynamically depending on the data level, network, and resource information.

RELATED WORK

Our literature review indicates a lack of appropriate models that would make Web services cater to different levels of "service" offers. Nevertheless, we list hereafter some projects that drew our attention because of the concepts they use.

In (Tao & Yang, 2007), Tao and Yang discuss service differentiation, which in fact backs the major orientation of having capacity-driven Web services. The authors claim that it would be appropriate to provide a single Web service with variations rather than several unrelated Web services, which afterwards need to be composed. This would just increase the number of issues between these unrelated Web services. A single Web service could exist along the line of service differentiation design-principle. A differentiated Web service would require first, to separate business policies from business processes and second, to identify usage contexts so that appropriate differentiated Web service interfaces are developed. Tao and Yang use an online pharmacy service to motivate and illustrate their approach. For example, different tasks are performed according to customer profiles (e.g., new, returning) and requested medicine (e.g., prescribed, unprescribed). Overall, the approach of Tao and Yang revolves around the definition of four concepts namely abstract business process, business policies, context configured business process, and service interfaces. These concepts could to a certain extent be mapped onto our work's concepts. For instance, an abstract business process could map onto either the operations that are common to all capacities or the operations that constitute the default capacity, and the business policies could map onto the algorithms we suggested for capacity selection.

In (Erradi, Padmanabhuni, & Varadharaja, 2006), Erradi et al. discuss the rationale of reviewing Web services management mechanism from a differential QoS perspective. To this end, they developed WS-DiffServ that uses admission controls and request scheduling to enable Web services remain responsive even when their capacity loads are exceeded. Erradi et al. discuss the various configurations that a Web service could comply with depending on various factors like data quality, result formatting, and notification rates. The architecture of WS-DiffServ included multiple components namely classifier, admission control, dispatcher, and QoS manager.

In (Ben Mokhtar, Kaul, Georgantas, & Issarny, 2006), Ben-Mokhtar et al. use capacity and functionality interchangeably to illustrate what a Web service can do with focus on the needs and requirements of a pervasive environment. They even suggest that a Web service can be multi-capacity. In our work, we adopted a different approach by first, separating in an explicit way functionality from capacity and second, emphasizing the following points: (i) a Web service is mono-functionality, (ii) a Web service is multi-capacity, (iii) capacities are exclusively used to satisfy a functionality, and (iv) multi-functionality scenarios mean Web services composition. It should be noted that a pervasive

Algorithm 2.

```
1. public static void main(String [] args) {
2. // aspect factory declaration
3. AspectFactory aspectFactory = new GenericAspectFactory(
. Aspect4GeolocationWebService.class, null);
4. AspectDefinition aspectDefinition = new AspectDefinition(
. Aspect4GeolocationWebService, Scope.PER_INSTANCE, aspectFactory);
5. AspectManager manager = AspectManager. instance();
6. // capacity aspect deployment
7. manager. addAspectDefinition(aspectDefinition);
8. if (GetDataLevel() == 0)&& (GetNetwork Level()==1)&&(GetResourceLevel()==3)
{
9. // advice factory creation
10. AdviceFactory adviceFactory = new AdviceFactory(aspectDefinition,
. GPSGeoLocation);
11. // dynamic binding creation
12 AdviceBinding binding = new AdviceBinding(execution (public ! static void
. GPSGeoLocation->*, null);
13. binding. addInterceptorFactory(adviceFactory);
14. // binding addition
15. manager.addAdviceBinding(binding);
16. }
17. // GSMGeoLocation advice should be activated
18. if (GetDataLevel() == 2)&& (GetNetwork Level()==3)&&(GetResourceLevel()==2)
{
19. AdviceFactory adviceFactory = new AdviceFactory(aspectDefinition,
. GSMGeoLocation);
20. AdviceBinding binding = new AdviceBinding(execution (public ! static void
. GSMGeoLocation->*, null);
21. binding. addInterceptorFactory(adviceFactory);
22. manager.addAdviceBinding(binding);
23. }
24. if (GetDataLevel() == 0)&& (GetNetwork Level()==3)&&(GetResourceLevel()==2)
{
25. AdviceFactory adviceFactory = new AdviceFactory(aspectDefinition,
. GetweatherCondition);
26. AdviceBinding binding = new AdviceBinding(execution(public ! static void
. AddressGeoLocation->*, null);
27. binding. addInterceptorFactory(adviceFactory);
28. manager. addAdviceBinding(binding);
29. }
30 }
```

environment is a perfect case-study for deploying capacity-based Web services. Indeed, this environment's needs and requirements constantly change, which requires a continuous adaptation of the actions to perform. For example, mobile devices appear and disappear without prior notice. As a result, adaptation could be handled by activating the appropriate capacities in a Web service.

Another use of capacity can be found in the field of service-level agreements with some works reported in (Heiko, Toshiyuki, Philipp, Oliver, & Wolfgang, 2006) and (Oldham, Verma, Sheth, & Hakimpour, 2006). These agreements contain clauses on the way services are delivered according to different factors like computing resource availabilities, users' preferences, network bandwidth, etc. This type of delivery requires mechanisms that permit guaranteeing the satisfaction of these clauses at run time. These mechanisms could be implemented through capacities that, as we proposed, are activated after constraint satisfaction. Each time a clause in an agreement needs to be guaranteed, a dedicated capacity is selected. As a result, capacity perfectly fits into the main idea of maintaining a certain level of satisfaction between what users request and what software applications offer.

Although capacity is not explicitly referred to in the project that Vukovic and Robinson report in (Vukovic & Robinson, 2004), the adaptive planning approach to compose Web services of this project highlights the importance of selecting the appropriate actions, means, and resources based on the current requirements of the environment. The mail replication system that was used as a running scenario indicates the suitable procedure to dynamically adopt based on user location, activity, computing device, and network bandwidth. Examples of procedures consist of displaying mail headers when the user's activity and location is walking and street, respectively, displaying full mails when the user' activity and location is working and at desk, respectively, etc. Each procedure calls for a specific set of operations, which are in line with the way we use capacities.

FUTURE WORK

This section illustrates some of the research directions that we plan to pursue in the near future. These directions stress the additional efforts that need to be put into capacity-driven Web services prior to making those "viable" solutions to the issues that still hinder regular Web services operation. Among others, we plan to study requirement consistency, capacity types, and capacity-driven Web services engineering.

Firstly, one of the concerns in the enactment step is the consistency between capacity-related requirements at the composition level. Indeed it is unlikely that all suppliers would agree on the same set of requirements to put on Web services. Conflicting requirements in terms of minimum and maximum values for example could arise, which calls for reconciliating actions prior to any composition execution.

Secondly, the classification of capacities into public and private types can be taken one step further by having more types such as dynamic and static. On the one hand, a dynamic capacity can be for example consumable or perishable. A consumable capacity is bound to a maximum number of use (e.g., number of login attempts must be limited due to some security concerns), and a perishable capacity is bound to an expiry date (e.g., in a university student registration must be stopped after a certain date). In either case, a capacity becomes unavailable for further uses after either reaching this maximum number or going beyond this date. On the other hand, a static capacity is neither consumable nor perishable. To deal with dynamic capacities one should revisit how capacity-driven Web services are now described, discovered, composed, and enacted.

Last but not least, we plan to work on a goal-based approach for capacity-driven Web services engineering. The added-value of goals to engineering systems is reported in different works (Damas, Lambeau, & Lamsweerde, 2006; Donzelli, 2004). We expect that goals would define the roles that these Web services play in business applications, frame the requirements on these Web services, and identify the processes in term of business logic that these Web services implement. Because of the specificities of capacity-driven Web services compared to "regular" Web services, their engineering in terms of design, development, and deployment needs to be conducted in a complete different way. It is expected that goals should be geared towards the following related aspects: business logic (does the business logic that a capacity-driven Web service implements need to be adapted?), requirement (what are the requirements that would trigger the adaptation of this business logic?), and capacity (what are the operations per capacity that need to be developed in response to these requirements?).

CONCLUSION

In this paper we presented the concepts, definitions, issues, and solutions that underpin the design, development, and deployment of capacity-driven Web services. We introduced the concept of capacity as a new way of making Web services take appropriate actions in response to some environmental requirements. To describe capacities we suggested the *C*-WSDL as an extension to the WSDL. A prototype was discussed and built upon aspects. In addition we identified the issues related to and proposed solutions for capacity-driven Web services over the description, discovery, composition, and enactment steps. The description step is about weaving capacities into Web services' descriptions. The discovery step is about fine tuning Web services discovery using capacities. The composition step is about checking Web services' acceptances in composition scenarios. And the enactment step is about invoking capacities of Web services, which concludes the life-cycle of putting capacity-driven Web services into action. In term of future we identified different research issues that need to be addressed such as requirement consistency, capacity types, and capacity-driven Web services engineering.

REFERENCES

Adomavicius, G., & Tuzhilin, A. (2005). Toward the Next Generation of Recommender Systems: A Survey of the State-of-the-Art and Possible Extensions. *IEEE Transactions on Knowledge and Data Engineering*, *17*(6). doi:10.1109/TKDE.2005.99

AOP. (2007). *AspectOriented Software Development.* Retrieved from http://www.aosd.net

Ardissono, L., Goy, A., & Petrone, G. (2003). Enabling Conversations with Web Services. In *Proceedings of the international joint conference on autonomous agents & multi-agent systems,* Melbourne, Australia.

Ben Mokhtar, S., Kaul, A., Georgantas, N., & Issarny, V. (2006). Efficient Semantic Service Discovery in Pervasive Computing Environments. In *Proceedings of the ACM/IFIP/USENIX 7th international middleware conference,* Melbourne, Australia.

Benbernou, S., & Hacid, M. S. (2005). Resolution and Constraint Propagation for Semantic Web Services Discovery. *Distributed and Parallel Databases*, *18*(1). doi:10.1007/s10619-005-1074-8

Blefari-Melazzi, N., Casalicchio, E., & Salsano, S. (2007). Context-aware Service Discovery in Mobile Heterogeneous Environments. In *Proceedings of the 16th IST mobile & wireless communications summit,* Budapest, Hungary.

Bockisch, C., Haupt, M., & Ostermann, K. (2004). Virtual Machine Support for Dynamic Join Points. In *Proceedings of the 3rd international conference on aspect-oriented software development,* Lancaster, UK.

Chatterjee, A. M., Chaudhari, A. P., Das, A. S., Dias, T., & Erradi, A. (2005). Differential QoS Support in Web Services Management. *SOA World Magazine, 5*(8).

Damas, C., Lambeau, B., & van Lamsweerde, A. (2006). Scenarios, Goals, and State Machines: a Win-Win Partnership for Model Synthesis. In *Proceedings of the 14th ACM SIGSOFT international symposium on foundations of software engineering,* Portland, OR.

Donzelli, P. (2004). A Goal-driven and Agent-based Requirements Engineering Framework. *Requirement Engineering, 9*(1).

Erradi, A., Padmanabhuni, S., & Varadharaja, N. (2006). Differential QoS Support in Web Services Management. In *Proceedings of the IEEE international conference on web services,* Chicago.

Harel, D., & Naamad, A. (1996). The STATEMATE Semantics of Statecharts. *ACM Transactions on Software Engineering and Methodology, 5*(4). doi:10.1145/235321.235322

Heiko, L., Toshiyuki, N., Philipp, W., Oliver, W., & Wolfgang, Z. (2006). *Reliable Orchestration of Resources using WS-Agreement* (Tech. Rep. No. TR-0050). Institute on Grid Systems, Tools, and Environments.

Hesselman, C., Tokmakoff, A., Pawar, P., & Iacob, S. (2006). Discovery and Composition of Services for Context-Aware Systems. In *Proceedings of the European conference on smart sensing and context,* Enschede, The Netherlands.

JBoss AOP. (2009). *JBoss AOP Reference Documentation.* Retrieved from http://docs.jboss.org/aop/1.3/aspectframework/reference/en/html/dynamic.html

Kopecký, J., Vitvar, T., Bournez, C., & Farrell, J. (2007). SAWSDL: Semantic Annotations for WSDL and XML Schema. *IEEE Internet Computing, 11*(6). doi:10.1109/MIC.2007.134

Kouadri Mostéfaoui, G., Maamar, Z., & Narendra, N. C. (2009). Managing Web Service Quality: Measuring Outcomes and Effectiveness. In Khan, K. (Ed.), *Aspect-oriented Framework for Web Services (AoF4WS): Introduction and Two Example Case Studies*. Hershey, PA: IGI Global Publishing.

Langdon, C. S. (2003, July). The State of Web Services. *IEEE Computer, 36*(7).

Luo, J., Montrose, B., Kim, A., Khashnobish, A., & Kang, M. (2006). Adding OWL-S Support to the Existing UDDI Infrastructure. *Web Services, IEEE International Conference on*, 153-162.

Maamar, Z., Kouadri Mostéfaoui, S., & Yahyaoui, H. (2005). Towards an Agent-based and Context-oriented Approach for Web Services Composition. *IEEE Transactions on Knowledge and Data Engineering, 17*(5). doi:10.1109/TKDE.2005.82

Mostéfaoui, S. K., & Younas, M. (2007). Context-oriented and transaction-based service provisioning. *International Journal of Web and Grid Services, 3*(2), 194–218. doi:10.1504/IJWGS.2007.014074

Oldham, N., Verma, K., Sheth, A., & Hakimpour, F. (2006). Semantic WS-Agreement Partner Selection. In *Proceedings of the 15th international conference on world wide web conference,* Edinburgh, Scotland.

Ortiz, G., Hernández, J., & Clemente, P. J. (2006). Web Services Orchestration and Interaction Patterns: An Aspect-Oriented Approach. In *Proceedings of the international conference on service oriented computing,* New York.

Ould Ahmed M'Bareck, N., Tata, S., & Maamar, Z. (2007). Towards An Approach for Enhancing Web Services Discovery. In *Proceedings of the 16th IEEE international workshops on enabling technologies: Infrastructure for collaborative enterprises,* Paris.

Papazoglou, M. P., Traverso, P., Dustdar, S., & Leymann, F. (2007). Service-Oriented Computing: State of the Art and Research Challenges. *IEEE Computer, 40*(11).

Rouvoy, R., Conan, D., & Seinturier, L. (2008). Software Architecture Patterns for a Context Processing Middleware Framework. *IEEE Distributed Systems Online, 9*(6).

Shehab, M., Bhattacharya, K., & Ghafoor, A. (2007). Web Services Discovery in Secure Collaboration Environments. *ACM Transactions on Internet Technology, 8*(1). doi:10.1145/1294148.1294153

Skene, J., Lamanna, D. D., & Emmerich, W. (2004). Precise Service Level Agreements. In *Proceedings of the 26th international conference on software engineering,* Edinburgh, UK.

Tao, A., & Yang, J. (2007). Context Aware Differentiated Services Development With Configurable Business Processes. In *Proceedings of the IEEE international enterprise computing conference,* Annapolis, MD.

ter Beek, M. H., Gnesi, S., Mazzanti, F., & Moiso, C. (2006). Formal Modelling and Verification of an Asynchronous Extension of SOAP. In *Proceedings of the European Conference on Web Services* (pp. 287-296). Washington, DC: IEEE Computer Society.

Vukovic, M., & Robinson, P. (2004). Adaptive, Planning-based, Web Service Composition for Context Awareness. In *Proceedings of the European conference on web services,* Erfurt, Germany.

ENDNOTES

1. These steps form the life cycle of a Web service.
2. A default capacity illustrates the current way of developing Web services, which we refer to them as mono-capacity.
3. For the sake of simplicity, non-functional properties are excluded from Web services selection.
4. http://www.eclipse.org/modeling/emf

This work was previously published in International Journal of Systems and Service-Oriented Engineering, Volume 1, Issue 4, edited by Dickson K.W. Chiu, pp. 65-88, copyright 2010 by IGI Publishing (an imprint of IGI Global).

Chapter 22
Investigating the Role of Service Encounter in Enhancing Customer Satisfaction

Irene Y. L. Chen
National Changhua University of Education, Taiwan

ABSTRACT

Recently, it is found that several pure e-tailers set up a customer service center where on-line shoppers can access a real person over the phone to answer their questions. However, there has been little systematic research examining how service encounter help to enhance customer satisfaction when a pure e-retail company set up a call center to provide additional services. This study conducted a questionnaire survey and collected data from persons who shopped on-line and had experiences in requesting help from customer service centers. 116 responses were collected and the data were then analyzed to examine the four relationships posited in the research model. The proposed research model suggests that service encounter significantly influences service quality and information quality, which can jointly predict customer satisfaction. Findings of this study help to advance the understanding of the role that service encounters play in enhancing customer satisfaction.

INTRODUCTION

Virtual stores exist in the cyberspace and offer merchandise and services through an electronic channel to their customers with a fraction of the overhead required in a bricks-and-mortar store (Chen & Tan, 2004; Hoffman, Novak, & Chatterjee, 1996; Yesil, 1997). Given the lower setup costs, lower cost per customer contact, and lower maintenance cost of virtual stores, individuals can now easily own a virtual store (i.e., to become an e-tailer). In recent years, several pure e-retail companies have created wealth and successfully built up reputations of their business and brands

DOI: 10.4018/978-1-4666-1767-4.ch022

of product. It is found that some of these e-tailers set up a customer service center where on-line shoppers can access a real person over the phone to answer their questions.

Levary and Mathieu (2000) indicated that market share is determined by a retailer's ability to attract new customers and retain existing customers; customer total satisfaction with the purchasing experience affects retailer ability to attract new customers and retain existing customers. Krampf (2003) suggested that customer satisfaction is a key to a firm's survival in today's marketplace, it has been embraced by practitioners and academics alike as "the highest order goal of a company" (Peterson & Wilson, 1992). Even though the online retailing is becoming a common business model, and some of the e-tailers have set up a customer service center, there has been a lack of research examining the role of such service encounter in enhancing customer satisfaction.

There is growing consensus that service quality is an antecedent of satisfaction with services (Birgelen, Ruyter, Jong, & Wetzels, 2002). Prior studies have suggested that service quality and information quality can predict customer satisfaction (Wang, 2008; Schaupp, Belanger, & Fan, 2009), and service encounter can predict service quality and information quality. In the context of online purchase, Zeithaml et al. (Zeithaml, Parasuraman, & Malhorta, 2002, p. 362) argued that companies need to shift the focus of e-business from e-commerce (the transaction) to e-service (all cues and encounters that occur before, during and after the transactions). Buckley (2003) also suggested that understanding, measuring and managing service quality on the web has become a topical issue for ensuring customer satisfaction.

Gounaris et al. (2005) indicated that in the context of e-shopping the replacement of human-to-human interactions with human-to-machine interactions may cause a constraint on the growth of online purchase. Although firms gained efficiencies from selling online (e-commerce), their failure to focus on customers' needs and wants resulted in poor online service performance (Loonam & O'Loughlin, 2008). Therefore, according to extant literature of Internet marketing (e.g., Buckley, 2003; Gounaris, Dimitriadis, & Stathakopoulos, 2005), providing high quality human-to-human service over the phone in virtual stores is as important as that in physical stores.

A service encounter is defined as the period of time during which a customer interacts with a service (Shostack, 1985). Groth et al. (Groth, Butek, & Douma, 2001) suggested that there are different service modes of the service delivery: face-to-face, telephone, and Internet. For customers who shop on-line, they can contact a 'real' customer service representative over the phone. Mallalieu (2006) stated that we typically trust the advice of experts to help us make effective decisions; salespeople often spend time trying to establish some type of rapport with a consumer prior to attempting to complete a sale. Customers' perceived quality of the service and information they receive based on the service encounter interactions will impact their overall feeling of online purchase experiences.

This study hence attempts to advance the understanding of the roles of service encounter interactions in enhancing overall customer satisfaction in the online retail companies. The two research questions of interest to this study are: (1) do the pre-validated quality–satisfaction relationship in the marketing and consumer behavior literature sustains in the online retail model? (2) To what extent do service encounter interactions influence levels of customers' satisfaction in online retail companies? Empirical data were collected from online shoppers. Survey questionnaires were administered to online shoppers who had experiences of contacting a customer service representative. This study then analyzed the data and discussed the theoretical and practical implications of the findings.

Figure 1. The research model

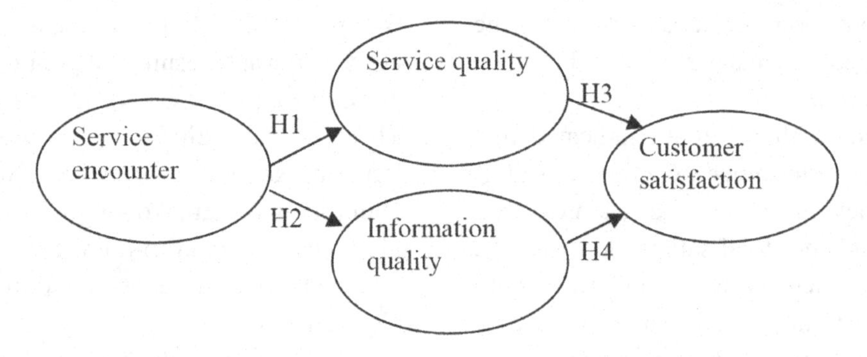

THEORETICAL BACKGROUND AND RESEARCH MODEL

Information technology acceptance literature and marketing literature have extensively discussed the topic of customer satisfaction. Yet, little is known concerning the enhancement of customer satisfaction in the case of pure e-tailers. Based on an extensive review of extant literature, this study proposes a customer satisfaction model for the case of online retailing. The proposed research model is depicted in Figure 1.

Service Encounter, Service Quality and Information Quality

Laing and McKee (2001) indicated that the concept of the service encounter represents the process of interaction between the consumer and the service provider, which, results in the actual delivery of the service. Carlzon (1987) graphically describes the service encounter as "the moment of truth" where the service is actually delivered. More specifically Suprenant and Solomon (1987) defined the service encounter as being "the dyadic interaction between the customer and the service provider firm". The service encounter can consequently be viewed as the juncture at which the consumer can evaluate the service offering and where the service supplier can attempt to manage the consumer perception of the service (John, 1996).

In this study, information quality (IQ) is defined as the degree to which product information delivered by the virtual store is accurate, relevant, complete, and in the format required by the consumers. According to Massad et al. (Massad, Heckman, & Crowston, 2006), on the Web, customers engage in service encounters with a business by visiting its Web site, navigating through it, searching for product and service information, communicating with customer service representatives, and perhaps purchasing a product or service. According to delivery ways, service encounters were classified into remote encounter, indirect personal encounter and direct personal encounter based on interactions. Consumers can make deals via one or more of these service encounters when they are purchasing (Gabbott & Hogg, 1998; Lin, 2007; Shostack, 1985). Groth et al. (Groth, Gutek, & Douma, 2001) suggested that there are three types of service delivery-face to face, telephone and Internet.

Mallalieu (2006) indicated that consumers regularly interact with retail salespeople in order to make purchases across a wide range of products and services; retail sales encounters occur on a daily basis yet we know little about these complex interactions, especially from the consumer's perspective in terms of his or her perceptions with regard to these types of persuasive attempts.

Service encounter does not merely explain the interaction between customers and the entire service system, no matter what the meaning, attribute, or type of the service encounter is; in other words, service encounter occurring directly between the customer and service provider is helpful for mutual interaction and communication (Lin, 2007). Interpersonal service encounters not only reflect the emphasis of consumers on a specific product or service, but also provide a foundation for maintaining high level of customer satisfaction. Prior studies have found that service encounter interaction is an important antecedent of customers' evaluation of the qualities of the service and the information they receive. For example, Lin (2007) provided a customer satisfaction model and found that interpersonal–based service encounter has a positive and significant effect on service quality. Schaupp et al. (Schaupp, Belanger, & Fan, 2009) suggested that the opinions provided by referent others will influence information quality of a website. Accordingly, this study suggests that the opinions provided by a customer service representative will influence consumers' perceived information quality of a virtual store. Therefore, the following hypotheses are proposed:

H1: Service encounter is positively associated with service quality.
H2: Service encounter is positively associated with information quality.

Service Quality, Information Quality and Customer Satisfaction

Parasuraman et al. (Parasuraman, Zeithaml, & Berry, 1985) suggested that service quality is derived from the comparison between what the customer feels should be offered and what is provided. Past research indicated that the service quality attributes of a website include effectiveness of product search and comparison, interactivity, responsiveness, clarity on security and privacy policies, assurance, empathy and product tracking (Kuan, Bock, & Vathanophas, 2008). Several studies have incorporated service quality dimension in measuring the performance of IS systems or the IS department (Ferguson & Zawacki, 1993; Kettinger & Lee, 1994). Results indicated that the effectiveness of an IS unit can be partially assessed by its capability to provide quality service to its users. Cronin and Taylor (1992) evaluated service quality based on performance and found that service quality was the cause of customer satisfaction.

Building and enhancing customer satisfaction has long been perceived as an enabler of electronic commerce and thus of electronic retailing. Wang (2008) indicated that the updated D&M model (DeLone & McLean, 2003) can be adapted to an e-commerce context. The findings by D&M model (2003), Wang (2008) and Schaupp et al. (Schaupp, Belanger, & Fan, 2009) all demonstrated that service quality and information quality are helpful for predicting customer satisfaction. Therefore, the following hypotheses are proposed.

H3: Service quality is positively associated with customer satisfaction.
H4: Information quality is positively associated with customer satisfaction.

RESEARCH METHODOLOGY

Structure of the Survey and Measurement Scales

In order to test the hypotheses, research aimed at online shoppers was developed. Scales that measure the constructs were generated from previous research and modified to fit the channel context. New items were developed through a thorough literature review on the topics. The items were written in the form of statements with which the respondent was to agree or disagree on a 5-point Likert scale.

Construct of service encounter was measured by the items adapted from Lin (2007), Massad et al. (Massad, Heckman, & Crowston, 2006) and Noone et al. (Noone, Kimes, Mattila, & Wirtz, 2009). Service quality was measured by items adapted from Lin (2007) and Myers et al. (Myers, Kappelman, & Prybutok, 1997). The scale for customer satisfaction was adapted from Lin (2007) and Noone et al. (Noone, Kimes, Mattila, & Wirtz, 2009).

Data Collection

Five EMBA students with working or research experience within the framework of Internet users' behavior collaborated in the final survey definition. Thus, all the necessary changes were made according to the experts' comments starting from an initial draft survey. Also, a preliminary survey was conducted with another five graduate students majoring in information management and business administration in order to check adequate understanding of all questions.

The research subjects of this study are the online shoppers. The survey questionnaires were administered to 296 full-time senior college students majoring in different area. Each of the subjects was asked to complete one questionnaire. Only 116 of them had experiences of contacting a customer service representative in a virtual store. All of the 116 questionnaires were completed and valid, and were analyzed in the present study.

DATA ANALYSIS

Measurement Model

This study used the Cronbach's alpha values to examine the reliability. Table 1 lists all these values. All of them were greater than 0.76, well above the recommended threshold value of 0.6. Convergent validity was evaluated for the measurement scales using two criteria suggested by Fornell and Larcker (1981): (1) all indicator factor loadings should be significant and exceed 0.70, and (2) average variance extracted (AVE) by each construct should exceed the variance due to a measurement error for that construct (i.e., should exceed 0.50). As summarized in Table 1, AVEs ranged from 0.88 to 0.91, greater than variance due to a measurement error. All items exhibited loadings higher than 0.7 on their respective constructs. These figures signify desirable convergent validity of the measurement.

Discriminant validity was evaluated through comparison of the AVE of construct pairs to the squared correlation between construct pairs. Fornell and Larcker (1981) recommended a test of discriminant validity, where the AVE should exceed the squared correction between that and any other construct. All of the items in Table 1 exhibited AVE greater than 0.56, which is the highest squared correlation between construct pairs. The overall results provide support for acceptable discriminant validity of constructs.

Model Testing Results

The structural equation modeling (SEM) approach is applied using LISREL 8.80 to examine the overall fit of the model, the explanatory power of research model and the relative strengths of the individual causal path. Several model-fit measures were used to express the model's overall goodness of fit: normed fit index (NFI), non-normed fit index (NNFI), comparative fit index (CFI), goodness of fit index (GFI), and standardized root mean square residual (SRMSR). For models with good fit, GFI, NFI, NNFI and CFI should exceed

Table 1. Composite reliability and discriminant validity

Construct	CR	1	2	3	4
1. Service encounter	.86	.88			
2. Service quality	.76	.42	.91		
3. Information quality	.90	.45	.48	.90	
4. Customer satisfaction	.88	.55	.56	.53	.89

1. CR: Composite reliability.
2. Correlations were all significant at the 0.01 level (2-tailed). Diagonal elements represent the Average Variance Extracted, while off diagonal elements are represented by the correlation among constructs. For discriminant validity, diagonal elements should be larger than off-diagonal elements.

Figure 2. The SEM Analysis Result

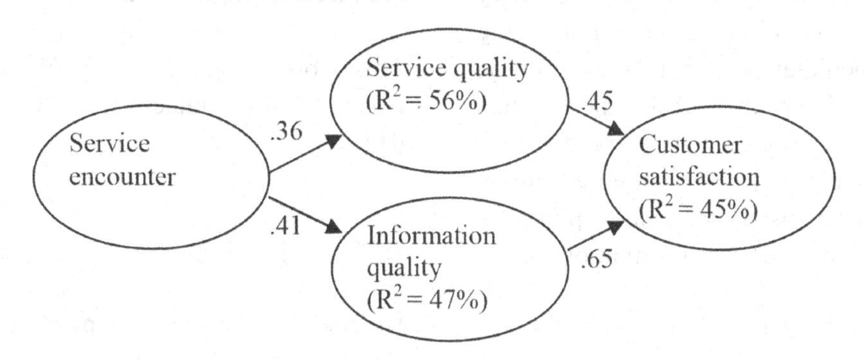

0.9, and SRMSR should be less than 0.1. Most indices met these conditions (GFI=0.90; NFI=0.91; CFI=0.93; NNFI=0.91; SRMSR=0.051). The overall results suggested that the research model provided a reasonably good fit to the data.

The significance of individual paths was examined and summarized in Figure 2. All of the hypothesized paths exhibited a P-value of < 0.05. The effects of service encounter on system quality and information quality perceived usefulness were both acceptable with path coefficients of 0.36 ($p < 0.001$) and 0.41 ($p < 0.01$), respectively. Service encounter accounted for approximately 56% of the observed variance for service quality, and explained 47% of the observed variance for information quality. Service quality and information quality both had strong effects on customer satisfaction with path coefficients of 0.45 ($p < 0.001$) and 0.65 ($p < 0.001$), respectively. Service quality and information quality jointly explained 45% of the variance for customer satisfaction. Thus, all hypotheses were supported.

RESEARCH LIMITATION AND FUTURE WORK

This study suffers from two limitations. The effects of the determinants in the research model may change over time as consumers gain experiences in using the website of a store and selecting merchandise from previous experiences. However, the user responses in our research were cross-sectional data and did not present an opportunity to examine the long-term trend of these hypothesized relationships. Further longitudinal studies are recommended to validate the research findings.

Moreover, this study demonstrated the importance of service encounter in enhancing customer satisfaction. Consumers buying different types of merchandise may require different types and levels of services and information. For example, consumers buying books may not request as much information from customer service center as those buying clothes and generally need to know more about the actual size, color and material of the clothes. Hence, further studies are encouraged to find out whether the research findings can be generalized in all types of virtual store.

CONCLUSION

The goal of the present study was to further the understanding of service encounter's role in enhancing customer satisfaction. Several findings are derived from the results. Firstly, consistent with prior studies related to the IS success (Wang, 2008; Schaupp, Belanger, & Fan, 2009), service quality and information quality are both important for predicting customer satisfaction. Information quality has a relatively stronger influence on customer satisfaction than service quality does. This leads to some implications.

Online shoppers usually search information by themselves and would spend time in browsing web page. When they need further information that is not provided by the website, they contact a customer service representative to request for help. In relatively less cases would an online shopper contact a customer service representative without first reads the information on the web page.

Secondly, consistent with prior studies based on Internet marketing, service encounter has a positive and significant impact on both service quality and information quality. This implies that no matter how well a website is designed and structured, it is always important to have a customer service center to provide prompt help to customers. Notwithstanding, in designing the website of a virtual store, the manager can provide functions such as online chat, forums, video conference or voice over IP to have multiple channels in place for responding consumer requests to avoid problems like busy-line waiting.

Using a virtual store as a merchandise selling channel has become a very common business model. Increasing profit by maintaining a high customer satisfaction level is a principle goal that most business firms are striving for. One of the strategies to enhance customer satisfaction is to provide prompt responses to customers' requests. Following this line of thought, making customers satisfied with the experience of contacting a customer service representative will receive increasing attention in firms managing online business. The findings of this study provide an avenue for the research of virtual store and practical implications for managers of virtual stores.

ACKNOWLEDGMENT

This work is supported by National Science Council, Taiwan under grants NSC 99-2410-H-018 -021.

REFERENCES

Birgelen, M., Ruyter, K., Jong, A., & Wetzels, M. (2002). Customer evaluations of after-sales service contact modes: An empirical analysis of national culture's consequences. *International Journal of Research in Marketing*, *19*, 43–64. doi:10.1016/S0167-8116(02)00047-2

Buckley, J. (2003). E-service quality and the public sector. *Managing Service Quality*, *13*(6), 453–462. doi:10.1108/09604520310506513

Carlzon, J. (1987). *Moments of Truth*. Cambridge, MA: Ballinger Books.

Chen, L. D., & Tan, J. (2004). Technology adaptation in e-commerce: key determinants of virtual stores acceptance. *European Management Journal*, *22*(1), 74–86. doi:10.1016/j.emj.2003.11.014

Cronin, J. J., & Taylor, S. A. (1992). Measuring service quality: a reexamination and extension. *Journal of Marketing*, *56*, 55–68. doi:10.2307/1252296

DeLone, W. H., & McLean, E. R. (2003). The DeLone and McLean model of information systems success: a ten-year update. *Journal of Management Information Systems*, *19*, 9–30.

Ferguson, J. M., & Zawacki, R. A. (1993). Service quality: A critical success factor for IS organizations. *Information Strategy: The Executive's Journal*, *9*(2), 24–30.

Fornell, C., & Larcker, D. F. (1981). Evaluating structural equation models with unobservable variables and measurement error. *JMR, Journal of Marketing Research*, *18*(1), 39–50. doi:10.2307/3151312

Gabbott, M., & Hogg, G. (1998). *Consumers and services*. New York: John Wiley & Sons Ltd.

Gounaris, S., Dimitriadis, S., & Stathakopoulos, V. (2005). Antecedents of perceived quality in the context of Internet retail stores. *Journal of Marketing Management, 21*, 669–700. doi:10.1362/026725705774538390

Groth, M., Gutek, B. A., & Douma, B. (2001). Effects of service mechanisms and modes on customers' attributions about service delivery. *Journal of Quality Management, 6*, 331–348. doi:10.1016/S1084-8568(01)00043-8

Hoffman, D. L., Novak, T. P., & Chatterjee, P. (1996). Commercial scenarios for the web: opportunities and challenges. *Journal of Computer-Mediated Communication, 1*, 3.

John, J. (1996). A dramaturgical view of the health care service encounter: Cultural value-based impression management guidelines for medical professional behaviour. *European Journal of Marketing, 30*(9), 60–75. doi:10.1108/03090569610130043

Kettinger, W. J., & Lee, C. C. (1994). Perceived service quality and user satisfaction with the information services function. *Decision Sciences, 25*(6), 737–766. doi:10.1111/j.1540-5915.1994.tb01868.x

Krampf, R., Ueltschy, L., & d'Amico, M. (2003). The contribution of emotion to consumer satisfaction in the service setting. *Marketing Management Journal, 13*(1), 32–52.

Kuan, H. H., Bock, G. W., & Vathanophas, V. (2008). Comparing the effects of website quality on customer initial purchase and continued purchase at e-commerce websites. *Behaviour & Information Technology, 27*(1), 3–16. doi:10.1080/01449290600801959

Laing, A., & McKee, L. (2001). Willing volunteers or unwilling conscripts? Professionals and marketing in service organisations. *Journal of Marketing Management, 17*, 559–575. doi:10.1362/026725701323366935

Levary, R., & Mathieu, R. G. (2000). Hybrid retail: integrating e-commerce and physical stores. *Industrial Management (Des Plaines), 42*(5), 6–13.

Lin, W. B. (2007). The exploration of customer satisfaction model from a comprehensive perspectives. *Expert Systems with Applications, 33*, 110–121. doi:10.1016/j.eswa.2006.04.021

Loonam, M., & O'Loughlin, D. (2008). Exploring e-service quality: a study of Irish online banking. *Marketing Intelligence & Planning, 26*(7), 759–780. doi:10.1108/02634500810916708

Mallalieu, L. (2006). Consumer perception of salesperson influence strategies: an examination of the influence of consumer goals. *Journal of Consumer Behaviour, 5*, 257–268. doi:10.1002/cb.177

Massad, N., & Heckman, R., & Crowston. (2006). Customer satisfaction with electronic service encounters. *International Journal of Electronic Commerce, 10*(4), 73–104. doi:10.2753/JEC1086-4415100403

Myers, B. L., Kappelman, L. A., & Prybutok, V. R. (1997). A comprehensive model for assessing the quality and productivity of the information systems function: toward a theory for information systems assessment. *Information Resources Management Journal, 10*(1), 6–25.

Noone, B. M., Kimes, S. E., Mattila, A. S., & Wirtz, J. (2009). Perceived service encounter pace and customer satisfaction. *Journal of Service Management, 20*(4), 380–403. doi:10.1108/09564230910978494

Parasuraman, A., Zeithaml, V. A., & Berry, L. L. (1985). A conceptual model of service quality and its implications for future research. *Journal of Marketing, 49*, 41–50. doi:10.2307/1251430

Peterson, R. A., & Wilson, W. R. (1992). Measuring customer satisfaction: Fact and artifact. *Journal of the Academy of Marketing Science, 20*(1), 61–71. doi:10.1007/BF02723476

Schaupp, L. C., Belanger, F., & Fan, W. (2009). Examining the success of websites beyond e-commerce: an extension of the IS success model. *Journal of Computer Information Systems*, *49*(4), 42–52.

Shostack, L. (1985). Planning the service encounter. In Czepiel, J., Solomon, M., & Surprenant, C. (Eds.), *The Service Encounter* (pp. 243–254). Lexington, MA: Lexington Books.

Surprenant, C., & Solomon, M. R. (1987). Predictability and personalisation in the service encounter. *Journal of Marketing*, *51*, 73–80. doi:10.2307/1251131

Wang, Y. S. (2008). Assessing e-commerce systems success: a respecification and validation of the DeLone and McLean model of IS success. *Information Systems Journal*, *18*, 529–557. doi:10.1111/j.1365-2575.2007.00268.x

Yesil, M. (1997). *Creating the Virtual Store: Taking Your Web Site from Browsing to Buying*. New York: John Wiley.

Zeithaml, V. A., Parasuraman, A., & Malhorta, A. (2002). Service Quality Delivery Through Web Sites: A Critical Review of Extant Knowledge. *Journal of the Academy of Marketing Science*, *30*(4), 362–375. doi:10.1177/009207002236911

This work was previously published in International Journal of Systems and Service-Oriented Engineering, Volume 1, Issue 4, edited by Dickson K.W. Chiu, pp. 19-26, copyright 2010 by IGI Publishing (an imprint of IGI Global).

Compilation of References

Aalst, W. M. P., Dumas, M., Ouyang, C., Rozinat, A., & Verbeek, E. (2008). Conformance checking of service behavior. *ACM Transactions on Internet Technology*, *8*(3). doi:10.1145/1361186.1361189

Aalst, W. M. P., Hofstede, A. H. M., Kiepuszewshi, B., & Barros, A. P. (2003). Workflow patterns. *Distributed and Parallel Databases*, *14*(3), 5–51. doi:10.1023/A:1022883727209

Abdelzaher, T., Thaker, G., & Lardieri, P. (2004). A feasible region for meeting aperiodic end-to-end deadlines in resource pipelines. In *Proceedings 24th International Conference on Distributed Computing Systems* (pp. 436-445).

Abeni, L., & Buttazzo, G. (2004). Resource reservation in dynamic real-time systems. *Real-Time Systems*, *27*(2), 123–167. doi:10.1023/B:TIME.0000027934.77900.22

Access to Knowledge through the Grid in a Mobile World (Akogrimo). (2007). *Web Site home page.* Retrieved from www.akogrimo.org

Adams, C., & Lloyd, S. (1999). *Understanding public-key infrastructure: concepts, standards, and deployment considerations.* Macmillan Technical Publishing.

Adomavicius, G., & Tuzhilin, A. (2005). Toward the Next Generation of Recommender Systems: A Survey of the State-of-the-Art and Possible Extensions. *IEEE Transactions on Knowledge and Data Engineering*, *17*(6). doi:10.1109/TKDE.2005.99

Agarwal, S., & Studer, R. (2006). Automatic matchmaking of web services. In *Proceedings of IEEE International Conference on Web Services (ICWS)* (pp. 45-54).

Agarwal, V., Dasgupta, K., Karnik, N. M., Kumar, A., Kundu, A., Mittal, S., & Srivastava, B. (2005). A service creation environment based on end to end composition of web services. In *Proceedings of International Conference on World Wide Web (WWW)*, (pp. 128-137).

Aggarwal, R., Verma, K., Miller, J. A., & Milnor, W. (2004). Constraint driven web service composition in METEOR-S. In *Proceedings of IEEE International Conference on Service Computing (SCC)* (pp. 23-30).

Aggestam, L. (2006, May 21-24). Wanted: A Framework for IT-supported KM. In *Proceedings of the 17th Information Resources Management Association (IRMA'06)*, Washington (pp. 46-49).

Aggestam, L., & Backlund, P. (2007). Strategic knowledge management issues when designing knowledge repositories. In H. Österle, J. Schelp, & R. Winter (Eds.), *Proceedings of the Fifteenth European Conference on Information Systems*, University of St. Gallen, St. Gallen, Switzerland (pp.528-539).

Aggestam, L., & Söderström, E. (2010). Seven types of knowledge loss in the knowledge capture process. In *Proceedings of the 18th European Conference on Information Systems (ECIS 2010).*

Aggestam, L., Backlund, P., & Persson, A. (in press). Supporting Knowledge Evaluation to Increase Quality in Electronic Knowledge Repositories. *International Journal of Knowledge Management.*

Akaike, H. (1974). A new look at the statistical model identification. *IEEE Transactions on Automatic Control*, *19*(6), 716–723. doi:10.1109/TAC.1974.1100705

Akkiraju, R., Farrell, J., Miller, J. A., Nagarajan, M., Schmidt, M.-T., Sheth, A., & Verma, K. (2005). *Web service semantics - WSDL-S.* Retrieved May 31, 2009, from http://www.w3.org/Submission/WSDL-S/.

Akogrimo Deliverable D4. 1.4. (2007). *Consolidated Network Service Provisioning Concept*. Retrieved from http://www.akogrimo.org/modules02d2.pdf?name=UpDownload&req=getit&lid=124

Akyildiz, I. F., & Kasimoglu, I. H. (2004). Wireless sensor and actor networks: Research challenges. *Ad Hoc Networks*, *2*(4). doi:10.1016/j.adhoc.2004.04.003

Alavi, M., & Leidner, D. E. (2001). Knowledge management and knowledge management systems: conceptual foundations and research issues. *Management Information Systems Quarterly*, *25*(1), 107–136. doi:10.2307/3250961

Albinsson, L., Forsgren, K., & Lind, M. (2007). *Towards a Co-Design Approach for Open Innovation. Designed for Co-designers workshop.* Paper presented at the meeting of the Participatory Design Conference (PDC 2008), School of Informatics, Indiana University, Bloomington, IN.

Allan, J., Wade, C., & Bolivar, A. (2003). Retrieval and Novelty Detection at the Sentence Level. In. *Proceedings of SIGIR*, *2003*, 314–321.

Allcock, B. (2002). Data Management and Transfer in High Performance Computational Grid Environments. *Parallel Computing Journal*, *28*(5), 749–771. doi:.doi:10.1016/S0167-8191(02)00094-7

Alrifai, M., & Risse, T. (2009). Combining global optimization with local selection for efficient QoS-aware service composition. In *Proceedings of International Conference on World Wide Web (WWW)* (pp. 881-890).

Altenhofen, M., Borger, E., & Lemcke, J. (2005). An abstract model for process mediation. In *Proceedings of International Conference on Formal Engineering Methods (ICFEM)* (pp. 81-95).

Alvesson, M., & Kaarreman, D. (2001). Odd couple: Making sense of the curious concept of knowledge management. *Journal of Management Studies*, *38*(7), 995–1018. doi:10.1111/1467-6486.00269

Anantatmula, V., & Kanungo, S. (2005). Establishing and structuring criteria for measuring knowledge management efforts. In *Proceedings of the 38th Hawaii International Conference on System Sciences (HICSS-38)*. Washington, DC: IEEE Press.

Andersen, V. K., & Medaglia, R. (2008) eGovernment Front-End Services: Administrative and Citizen Cost Benefits. In M. A. Wimmer & E. Ferro (Eds.), *Proceedings of EGOV 2008.* Berlin: Springer.

Anderson, T., Barrett, P. A., Halliwell, D. N., & Moulding, M. R. (1995). Software Fault Tolerance: An Evaluation. *IEEE Transactions on Software Engineering*, *11*(12), 1502–1510. doi:10.1109/TSE.1985.231894

Andonovic, I., Michie, C., Gilroy, M., Goh, H., Kwong, K., & Sasloglou, K. (2009). *Wireless sensor networks for cattle health monitoring*. New York: Springer.

Ando, R., & Takefuji, Y. (2003, January). Location-driven watermark extraction using supervised learning on frequency domain. *WSEAS Transactions On Computers*, *2*(1), 163–169.

Andreeva, J., Campana, S., Fanzago, F., & Herrala, J. (2008). High-Energy Physics on the Grid: the ATLAS and CMS Experience. *Journal of Grid Computing*, *6*(1), 3–13. doi:10.1007/s10723-007-9087-3

Andrieux, A., Czajkowski, K., Dan, A., Keahey, K., Ludwig, H., Nakata, T., Pruyne, J., Rofrano, J., Tuecke, S., & Xu, M. (2007). *Web Services Agreement Specification (WS-Agreement)* (No. GFD-R-P. 107). Global Grid Forum.

Andrieux, A., Czajkowski, K., Dan, A., Keahey, K., Ludwig, H., & Pruyne, J. (2008). *Web services agreement negotiation specification (WSAgreementNegotiation) (draft) (Tech. Rep.)*. Open Grid Forum.

Anthopoulos, L. G., Siozos, P. S., & Tsoukalas, I. A. (2007). Applying participatory design and collaboration in digital public services for discovering and re-designing e-Government services. *Government Information Quarterly*, *2*(24), 353–376. doi:10.1016/j.giq.2006.07.018

AOP. (2007). *AspectOriented Software Development.* Retrieved from http://www.aosd.net

Ardagna, D., & Pernici, B. (2007). Adaptive service composition in flexible processes. *IEEE Transactions on Software Engineering*, *33*(6), 369–384. doi:10.1109/TSE.2007.1011

Ardissono, L., Goy, A., & Petrone, G. (2003). Enabling Conversations with Web Services. In *Proceedings of the international joint conference on autonomous agents & multi-agent systems,* Melbourne, Australia.

Armenio, F., Barthel, H., Dietrich, P., et al. (2009). *The EPCglobal Architecture Framework*. Retrieved from http://www.epcglobalinc.org/

Arnold-Moore, A., Fuller, M., Kent, A., Sacks-Davis, R., & Sharman, N. (2000). Architecture of a content management server for XML document applications. In *Proceedings of 1st International Conference on Web Information Systems Engineering (WISE 2002)*, Hong Kong, China (Vol. 1, pp. 97-108).

Artz, D. (2001, May/June). Digital steganography: hiding data within data. *IEEE Internet Computing*, *5*(3), 75–80. doi:10.1109/4236.935180

Asgarkhani, M. (2005). The effectiveness of e-Service in Local Government: A Case Study. *Electric journal of e-government, 3*(4), 157-166.

Associated Press. (2005). *Body ID: Barcodes for Cadavers.* Retrieved from http://www.wired.com/medtech/health/news/2005/02/66519

AUTOSAR (2008). Specification of the virtual functional bus, v1.0.2, r3.1, rev 0001. *Automotive Open System Architecture*.

Aversano, L., Canfora, G., & Ciampi, A. (2004). An algorithm for web service discovery through their composition. In *Proceedings of IEEE International Conference on Web Services (ICWS)* (pp. 332-339).

Aversano, L., de Canfora, G., Lucia, A., & Gallucci, P. (2002). Integrating document and workflow management tools using XML and web technologies: a case study. In *Proceedings of Sixth European Conference on Software Maintenance and Reengineering* (pp. 24-33).

Avilés-López, E., & García-Macías, J. A. (2009). Tinysoa: a service-oriented architecture for wireless sensor networks. *Service Oriented Computing and Applications*, *3*(2), 99–108. doi:10.1007/s11761-009-0043-x

Avizienis, A., & Chen, L. (1977). On the implementation of N-version programming for software fault-tolerance during program execution. *In Proceedings of the 1st IEEE International Computer Software and Applications Conference* (pp. 149-155). Washington, DC: IEEE Computer Society.

Avizienis, A. (1985). The N-Version Approach to Fault-Tolerant Systems. *IEEE Transactions on Software Engineering, 11*(12), 1491–1501. doi:10.1109/TSE.1985.231893

Axelsson, K., & Melin, U. (2008). Citizen Participation and Involvement in eGovernment Projects – An emergent framework. In M. A. Wimmer & E. Ferro (Eds.), *Proceedings of EGOV 2008* (pp. 207-218). Berlin: Springer.

Axelsson, K., & Melin, U. (2007). Talking to, not about, citizens – Experiences of focus groups in public e-service development. In M. A. Wimmer, H. J. Scholl, & A. Grönlund (EdS.) *Proceedings of EGOV, 2007*, 179–190.

Baader, F., Calvanese, D., McGuinness, D. L., Nardi, D., & Patel-Schneider, P. F. (2003). *The description logic handbook: theory, implementation, and applications*. Cambridge University Press.

Bafoutsou, G., & Mentzas, G. (2001). A Comparative Analysis of Web-based Collaborative Systems. In *Proceeding 12th International. Workshop Database and Expert Systems Applications* (pp. 496-500).

Bagchi, A., Caruso, F., Mayer, A., Roman, R., Kumar, P., & Kowtha, S. (2009). *Framework to achieve multi-domain service management*. Paper presented at the IFIP/IEEE International Symposium on Integrated Network Management (IM '09) (pp 287-290).

Baldoni, R., Beraldi, R., Cugola, G., Migliavacca, M., & Querzoni, L. (2005, December). Content-based routing in highly dynamic mobile ad hoc networks. *Int. Journal of Pervasive Computing and Communications, 1*(4), 277–288. doi:10.1108/17427370580000131

Banerjee, S., & Pedersen, T. (2003). Lesk: Extended gloss overlaps as a measure of semantic relatedness. In *Proceedings of the 18th International Joint Conference on Artificial Intelligence*, Acapulco, Mexico (pp. 805-810). Menlo Park, CA: AAAI Press.

Baresi, L., Guinea, S., & Plebani, P. (2007). Policies and aspects for the supervision of BPEL processes. In J. Krogstie, A.L. Opdahl, & G. Sindre (Eds.), *Proceedings of CAiSE 2007, LNCS, Vol. 4495* (pp. 340-395). Berlin, Germany: Springer.

Baresi, L., & Di Nitto, E. (Eds.). (2007). *Test and Analysis of Web Services*. New York: Springer Verlag. doi:10.1007/978-3-540-72912-9

Barry, D. K. (2003). *Web Services and Service-Oriented Architectures: The savvy manager's guide*. San Francisco, CA: Morgan Kaufmann Publishers.

Bartlett, C. (1998). *McKinsey & Company: Managing knowledge and learning* (Case 9-396-357). Cambridge, MA: Harvard Business School Press.

Bartolini, C., Bertolino, A., Elbaum, S., & Marchetti, E. (2009). Whitening SOA Testing. *In Proceedings of the 7th Joint Meeting of the European Software Engineering Conference and the ACM SIGSOFT Symposium on the Foundations of Software Engineering* (pp. 161-170). New York: ACM.

Bartolini, C., Sahai, A., & Sauve, J. P. (Eds.). (2006). *Proceedings of the First IEEE/IFIP Workshop on Business-Driven IT Management, BDIM'06*. Piscataway, USA: IEEE Press.

Bartolini, C., Sahai, A., & Sauve, J. P. (Eds.). (2007). *Proceedings of the Second IEEE/IFIP Workshop on Business-Driven IT Management, BDIM'07*. Piscataway, USA: IEEE Press.

Bartolini, C., Salle, M., & Trastour, D. (2006). IT service management driven by business objectives: An application to incident management. In J.L. Hellerstein & B. Stiller (Eds.), *Proceedings of NOMS 2006* (pp. 45-55). Piscataway, USA: IEEE Press.

Beek, M. H. T., Bucchiarone, A., & Gnesi, S. (2006). *A Survey on Service Composition Approaches: From Industrial Standards to Formal Methods* (Tech. Rep. No. 2006-TR-15). Consiglio Nazionale delle Ricerche: ISTI.

Bellavista, P., Corradi, A., Montanari, R., & Stefanelli, C. (2003). Context-Aware Middleware for Resource Management in the Wireless Internet. *IEEE Transactions on Software Engineering, 29*(12), 1086–1099. doi:10.1109/TSE.2003.1265523

Ben Mokhtar, S., Kaul, A., Georgantas, N., & Issarny, V. (2006). Efficient Semantic Service Discovery in Pervasive Computing Environments. In *Proceedings of the ACM/IFIP/USENIX 7th international middleware conference*, Melbourne, Australia.

Benatallah, B., Casati, F., Grigori, D., Nezhad, H. R., & Toumani, F. (2005). Developing adapters for web services integration. In *Proceedings of International Conference on Advanced Information System Engineering (CAiSE)* (pp. 415–429).

Benatallah, B., Hacid, M. S., Leger, A., Rey, C., & Toumani, F. (2005). On automating web services discovery. *The VLDB Journal, 14*(1), 84–96. doi:10.1007/s00778-003-0117-x

Benbernou, S., & Hacid, M. S. (2005). Resolution and Constraint Propagation for Semantic Web Services Discovery. *Distributed and Parallel Databases, 18*(1). doi:10.1007/s10619-005-1074-8

Berbner, R., Spahn, M., Repp, N., Heckmann, O., & Steinmetz, R. (2006). Heuristics for QoS-aware web service composition. In *Proceedings of International Conference on Web Services (ICWS)* (pp. 72-82).

Berhe, G., Brunie, L., & Pierson, J. M. (2004). Modeling Service-Based Multimedia Content Adaptation in Pervasive Computing. *the First Conference on Computing Frontiers on Computing Frontiers*, ACM Press, 60-69.

Bernardo, J., & Smith, A. (2001). Bayesian theory. *Measurement Science & Technology, 12*, 221–222.

Berners-Lee, T., Hendler, J., & Lassila, O. (2001). The Semantic Web. *Scientific American, 284*(5), 34–43. doi:10.1038/scientificamerican0501-34

Bertino, E., Castano, S., Ferrari, E., & Mesiti, M. (2002). Protection and administration of XML data sources. *Data & Knowledge Engineering, 43*(3), 237–260. doi:10.1016/S0169-023X(02)00127-1

Bertolino, A., De Angelis, G., Frantzen, L., & Polini, A. (2009). The PLASTIC Framework and Tools for Testing Service-Oriented Applications. In A. Lucia & F. Ferrucci (Eds.), *Proceedings of Software Engineering: international Summer Schools, ISSSE 2006-2008*, Salerno, Italy (LNCS 5413, pp. 106-139). Berlin: Springer Verlag.

Betin-Can, A., Bultan, T., & Fu, X. (2005). Design for verification for asynchronously communicating web services. In *Proceedings of International Conference on World Wide Web (WWW)* (pp. 750-759).

Bettman, J. R. (1970). Information processing models of consumer behavior. *JMR, Journal of Marketing Research, 7*(3), 370–376. doi:10.2307/3150297

Bichler, M., & Lin, K.-J. (2006, March). Service-oriented computing. *IEEE Computer, 39*(3), 99–101.

Bieberstein, N., Bose, S., Fiammante, M., Jones, K., & Shah, R. (2005). *Service-Oriented Architecture (SOA) Compass: Business Value, Planning, and Enterprise Roadmap (Developerworks)*. Indianapolis, IN: IBM Press.

Bierly, P., & Chakrabarti, A. (1996). Generic knowledge strategies in the U.S. pharmaceutical industry. *Strategic Management Journal, 7*(4), 123–135.

Biffl, S., & Aurum, A. Boehm, B., Erdogmus, H., & Gruenbacher, P. (Eds.). (2006). *Value-based software engineering*. Berlin, Germany: Springer.

Binney, D. (2001). The knowledge management spectrum – understanding the KM landscape. *Journal of Knowledge Management, 1*(5), 33–42. doi:10.1108/13673270110384383

Birgelen, M., Ruyter, K., Jong, A., & Wetzels, M. (2002). Customer evaluations of after-sales service contact modes: An empirical analysis of national culture's consequences. *International Journal of Research in Marketing, 19*, 43–64. doi:10.1016/S0167-8116(02)00047-2

Bishop, C. M. (1995). *Neural networks for pattern recognition*. Oxford University Press, USA.

Bishop, C. M. (2006). *Pattern recognition and machine learning (information science and statistics)* (New edition ed.). Springer-Verlag.

Bishop, M., & Frincke, D. (2004). Teaching Robust Programming. *IEEE Security and Privacy, 2*(2), 54–57. doi:10.1109/MSECP.2004.1281247

Bistarelli, S., & Santini, F. (2008). Propagating Multitrust within Trust Networks. In *Proceedings of the Special Interest Group on Applied Computing 2008, Track on Trust, Recommendations, Evidence and other Collaboration Know-how*, Fortaleza, Ceara, Brazil. New York: ACM Publications.

Biswas, S., & Morris, R. (2005, October). Exor: opportunistic multi-hop routing for wireless networks. *ACM SIGCOMM Computer Communication Review, 35*(4).

Blake, S., Black, D., Carlson, M., Davies, M., Wang, Z., & Weiss, W. (1998). *An architecture for differentiated services, RFC 2475*.

Blefari-Melazzi, N., Casalicchio, E., & Salsano, S. (2007). Context-aware Service Discovery in Mobile Heterogeneous Environments. In *Proceedings of the 16th IST mobile & wireless communications summit*, Budapest, Hungary.

Blodgett, J. G., & Anderson, R. D. (2000). A Bayesian network model of the consumer complaint process. *Journal of Service Research, 2*(4), 321–338. doi:10.1177/109467050024002

Blodgood, J. M., & Salisbury, W. D. (2001). Understanding the influence of organizational change strategies on information technology and knowledge management strategies. *Decision Support Systems, 1*(31), 55–69. doi:10.1016/S0167-9236(00)00119-6

Bockisch, C., Haupt, M., & Ostermann, K. (2004). Virtual Machine Support for Dynamic Join Points. In *Proceedings of the 3rd international conference on aspect-oriented software development*, Lancaster, UK.

Bohn, R. (1994). Measuring and managing technological knowledge. *Sloan Management Review, 36*(1), 61–73.

Boisot, M. (1999). *Knowledge Assets - The Paradox of Value*. Oxford, UK: Oxford University Press.

Bollella, G., Brosgol, D., Furr, H., et al. (2000). *The real-time specification for java*. Addison-Wesley. Bou-Ghannam, A., & Faulkner, P. (2008, December). Enable the real-time enterprise with business event processing. *IBM Business Process Management Journal*(1.1).

Bonatti, P. A., & Festa, P. (2005). On optimal service selection. In *Proceedings of International Conference on World Wide Web (WWW)* (pp. 530-538).

Bordeaux, L., Salaun, G., Berardi, D., & Mecella, M. (2004). When are two web services compatible. In *Proceedings of International Workshop on Technologies for E-Services (TES)* (pp. 15-28).

Bose, R., & Frew, J. (2005). Lineage retrieval for scientific data processing: a survey. *ACM Computing Surveys, 37*(1), 1–28. doi:10.1145/1057977.1057978

Bouras, C., Campanella, M., & Sevasti, A. (2002). *SLA definition for the provision of an EF-based service*. Paper presented at the 16th International Workshop on Communications Quality & Reliability (CQR 2002), Okinawa, Japan (pp. 17-21).

Bracciali, A., Brogi, A., & Canal, C. (2005). A formal approach to component adaptation. *Journal of Systems and Software, 74*(1), 45–54. doi:10.1016/j.jss.2003.05.007

Braden, R., Clark, D., & Shenker, S. (1998). *Integrated services in the internet architecture: an overview, RFC 1633*. Common Object Request Broker Architecture: Core Specification. (2004, March).

Briscombe, N., et al. (2006). *Enabling Integrated Emergency Management: Reaping the Akogrimo Benefits.* Retrieved from http://www.akogrimo.org/download/White_Papers_and_Publications/ Akogrimo_whitePaper_Disas terCrisisMgmt_v1-1.pdf

Brogi, A., Canal, C., Pimentel, E., & Vallecillo, A. (2004). Formalizing web service choreographies. *Electronic Notes in Theoretical Computer Science, 105*, 73–94. doi:10.1016/j.entcs.2004.05.007

Brogi, A., Corfini, S., & Popescu, R. (2008). Semantics-based composition-oriented discovery of web services. *ACM Transactions on Internet Technology, 8*(4). doi:10.1145/1391949.1391953

Bromberg, Y. D., & Issarny, V. (2005). Indiss: interoperable discovery system for networked services. In *Proceedings of the acm/ifip/usenix 2005 international conference on middleware (Middleware '05)* (pp. 164-183). New York: Springer Verlag.

Broomhead, D. S., & Lowe, D. (1988). Multivariable functional interpolation and adaptive networks. *Complex Systems, 2*, 321–355.

Bruijn, J. A., & Heuvelhof, E. F. (2002). *Process management: why project management fails in complex decision making processes*. Dordrecht, The Netherlands: Kluwer Academic.

Bruin, J. A. D., & Heuvelhof, E. F. T. (1995). *Netwerk-management, Strategieën, Instrumenten en Normen.* Utrecht, The Netherlands: Lemma.

Bruin, J. A. D., Heuvelhof, E. F. T., & Veld, R. J. I. T. (1998). *Procesmanagement, over procesontwerp en besluitvorming*. Amsterdam, The Netherlands: Academic Service.

Buckley, J. (2003). E-service quality and the public sector. *Managing Service Quality, 13*(6), 453–462. doi:10.1108/09604520310506513

Bui, T. X. (1987). *Co-oP: a Group Decision Support System for Cooperative Multiple Criteria Group Decision Making*, Springer, LNCS 290.

Bui, T. X., Bodart, F., & Ma, P.-C. (1998). ARBAS: A Formal Language to Support Argumentation in Network-Based Organization. *Journal of Management Information Systems, 14*(3), 223–240.

Bultan, T., Fu, X., Hull, R., & Su, J. (2003). Conversation specification: a new approach to design and analysis of e-service composition. In *Proceedings of International Conference on World Wide Web (WWW)* (pp. 403–410).

Burges, C. (1998). A tutorial on support vector machines for pattern recognition. *Data Mining and Knowledge Discovery, 2*(2), 121–167. doi:10.1023/A:1009715923555

Burstein, M., Bussler, C., Finin, T., Huhns, M. N., Paolucci, M., & Sheth, A. P. (2005). A Semantic Web Services Architecture. *IEEE Internet Computing, 9*(5), 72–81. doi:10.1109/MIC.2005.96

Bustos, J., & Watson, K. (2002). *Beginning. NET Web Services using C*. New York: Wrox Press Ltd.

Cabri, G., Leonardi, L., Mamei, M., & Zambonelli, F. (2003). Location-Dependent Services for Mobile Users. *IEEE Transactions on Systems*, Man and Cybernetics. *Part A., 33*(6), 667–681.

Cahill, V., Gray, E., Seigneur, J.M., Jensen, C.D., Chen, Y., Shand, B., Dimmock, N., Twigg, A., Bacon, J., English, C., Wagealla, W., Terzis, S., Nixon, P., Serugendo, G., Bryce, C., Carbone, M., Krukow, K., & Nielsen, M. (2006). *Pervasive Computing*, 52-61.

Cahill, V., Gray, E., Seigneur, J. M., Jensen, C. D., Chen, Y., & Shand, B. (2007). A Survey of Trust and Reputation Systems for on-line Service Provision. *Decision Support Systems, 43*(2), 618–644. doi:10.1016/j.dss.2005.05.019

Camarillo, G., & Garcia-Martin, M. *The 3G IP Multimedia Subsystem*. New York: John Wiley and Sons.

Campanella, M., Przybylski, M., Roth, R., Sevasti, A., & Simar, N. (2003). *Multidomain End to End IP QoS and SLA* (LNCS 2601, pp. 171-184). New York: Springer.

Campo, C., Munoz, M., Perea, J. C., Marin, A., & Garcia-Rubio, C. (2005). Pdp and gsdl: A new service discovery middleware to support spontaneous interactions in pervasive systems. In *Proceedings of the third ieee international conference on pervasive computing and communications workshops (Percomw '05)* (pp. 178-182). Washington, DC: IEEE Computer Society.

Canfora, G., & Di Penta, M. (2009). Service Oriented Architecture Testing: A Survey. In A. Lucia & F. Ferrucci (Eds.), *Proceedings of Software Engineering: international Summer Schools, ISSSE 2006-2008,* Salerno, Italy (LNCS 5413, pp. 78-105). Berlin: Springer Verlag.

Canfora, G., Penta, M. D., Esposito, R., & Villani, M. L. (2006). An approach for QoS-aware service composition based on genetic algorithms. In *Proceedings of International Conference on Genetic and Evolutionary Computation (GECCO)* (pp. 1069-1075).

Canfora, G., Penta, M. D., Esposito, R., Perfetto, F., & Villani, M. L. (2006). Service composition (re)binding driven by application-specific QoS. In *Proceedings of International Conference on Service Oriented Computing (ICSOC)* (pp. 141-152).

Canfora, G., & Di Penta, M. (2006). Testing services and service-centric systems: Challenges and opportunities. *IT Professional*, *8*, 10–17. doi:10.1109/MITP.2006.51

Cao, J., et al. (2003, May 12-15). GridFlow: Workflow Management for Grid Computing. In *Proceedings of the 3rd International Symposium on Cluster Computing and the Grid (CCGrid),* Tokyo, Japan. Los Alamitos, CA: IEEECS Press.

Cardoso, J., Sheth, A., Miller, J., Arnold, J., & Kochut, K. (2004). Quality of service for workflows and web service processes. *Journal of Web Semantics*, *1*(3), 281–308. doi:10.1016/j.websem.2004.03.001

Carlson, D. (2001). *Modeling XML Applications with UML*. Addison-Wesley.

Carlsson, S. A., & Kalling, T. (2006). Why is it that a knowledge management initiative works or fails. In *Proceedings of the Fourteenth European Conference on Information Systems*, Gothenburg, Sweden.

Carlzon, J. (1987). *Moments of Truth*. Cambridge, MA: Ballinger Books.

Carroll, J. M., & Rosson, M. B. (2007). Participatory design in community informatics. *Design Studies*, *3*(28), 243–261. doi:10.1016/j.destud.2007.02.007

Casati, F., Shan, E., Dayal, U., & Shan, M.-C. (2003). Business-oriented management of Web services. *Communications of the ACM*, *46*(10), 55–60. doi:10.1145/944217.944238

Castelfranchi, C., & Falcone, R. (2000). Trust is much more than subjective probability: Mental components and sources of trust. In *Proceedings of the 33rd Annual Hawaii International Conference on System Sciences* (Vol. 1, pp. 4-7).

Chakeres, I. D., & Perkins, C. E. (2008, February). *Dynamic MANET On-demand Routing Protocol* (IETF Internet Draft, draft-ietf-manet-dymo-12.txt).

Charalabidis, Y., Askounis, D., Gionis, G., Lampathaki, F., & Metaxiotis, K. (2006). Organising Municipal e-Government Systems: A Multi-facet Taxonomy of e-Services for Citizens and Businesses. In W. E. Al (Ed.), *Proceedings of EGOV 2006* (pp. 195-206). Berlin: Springer.

Chatterjee, A. M., Chaudhari, A. P., Das, A. S., Dias, T., & Erradi, A. (2005). Differential QoS Support in Web Services Management. *SOA World Magazine, 5*(8).

Chavez, A., & Maes, P. (1996). Kasbah: An Agent Marketplace for Buying and Selling Goods. In *Proceedings of the 1st International Conference on the Practical Application of Intelligent Agents and Multi-Agent Technology*, 75-90.

Chebotko, A. (Manuscript submitted for publication). S, Lu., & Fotouhi, F. (2007). Semantics preserving SPARQL-to-SQL query translation. *Data & Knowledge Engineering*.

Chen, L. Q., Xie, X., Fan, X., Ma, W. Y., Zhang, H. J., & Zhou, H. Q. (2002b). *A Visual Attention Model for Adapting Images on Small Displays*. Technical report MSR-TR-2002-125, Microsoft Research.

Chen, L. Q., Xie, X., Ma, W. Y., Zhang, H. J., Zhou, H. Q., & Feng, H. Q. (2002a). *DRESS: A Slicing Tree Based Web Representation for Various Display Sizes*. Technical report MSR-TR-2002-126, Microsoft Research.

Chen, Y., Tsai, F. S., & Chan, K. L. (2007). Blog search and mining in the business domain. In *Proceedings of the 2007 International Workshop on Domain Driven Data Mining (DDDM '07)* (pp. 55-60).

Chen, L. D., & Tan, J. (2004). Technology adaptation in e-commerce: key determinants of virtual stores acceptance. *European Management Journal*, *22*(1), 74–86. doi:10.1016/j.emj.2003.11.014

Cheung, S.-C., Chiu, D. K. W., & Ho, C. (2008). The use of digital watermarking for intelligence multimedia document distribution. *Journal of Theoretical and Applied Electronic Commerce Research*, *3*(3), 103–118. doi:10.4067/S0718-18762008000200008

Chillarege, R., Bhandari, I. S., Chaar, J. K., Halliday, M. J., Moebus, D. S., Ray, B. K., & Wong, M. Y. (1992). Orthogonal Defect Classification-A Concept for In-Process Measurements. *IEEE Transactions on Software Engineering*, *18*(11), 943–956. doi:10.1109/32.177364

Chiu, D. K. W., & Hung, P. C. K. (2005). Privacy and Access Control in Financial Enterprise Content Management. In *Proceedings of the 38th Hawaii International Conference on System Sciences*, Big Island, HI. Washington, DC: IEEE Computer Society Press.

Chiu, D. K. W., Chan, W. C. W., Lam, G. K. W., Cheung, S. C., & Luk, F. T. (2003). An Event Driven Approach to Customer Relationship Management in an e-Brokerage Environment. In *Proceedings of the 36th Hawaii International Conference on System Sciences*, Big Island, HI. Washington, DC: IEEE Computer Society Press.

Chiu, D. K. W., Cheung, S. C., & Hung, P. C. K. (2002b). A Contract Template Driven Approach to e-Negotiation Processes. In *Proceedings of the 6th Pacific Asia Conf. on Information Systems*, Tokyo, Japan, CDROM.

Chiu, D. K. W., Cheung, S. C., & Leung, H.-F. (2006). Mobile Workforce Management in a Service-Oriented Enterprise: Capturing Concepts and Requirements in a Multi-Agent Infrastructure. In R. Qiu (Ed.), *Enterprise Service Computing: From Concept to Deployment. Enterprise Service Computing: From Concept to Deployment*, Idea Group Publishing.

Chiu, D. K. W., Cheung, S. C., Hung, P. C. K., & Leung, H.-F. (2004b). Constraint-based Negotiation in a Multi-Agent Information System with Multiple Platform Support. In *Proceedings of the 37th Hawaii International Conference on System Sciences* (HICSS37), Big Island, Hawaii, CDROM, IEEE Computer Society Press.

Chiu, D., & Leung, H. (2005). Towards ubiquitous tourist service coordination and integration: a multi-agent and semantic web approach. In *Proceedings of the 7th international conference on Electronic commerce* (pp. 574–581).

Chiu, D. K. W., Cheung, S. C., Hung, P. C. K., Chiu, S. Y. Y., & Chung, K. K. (2005). Developing e-Negotiation Process Support with a Meta-modeling Approach in a Web Services Environment. *Decision Support Systems*, *40*(1), 51–69. doi:10.1016/j.dss.2004.04.004

Chiu, D. K. W., Cheung, S. C., Kafeza, E., & Leung, H.-F. (2004). A Three-Tier View Methodology for adapting M-services. *IEEE Transactions on Systems, Man and Cybernetics. Part A*, *33*(6), 725–741.

Chiu, D. K. W., Cheung, S. C., Karlapalem, K., Li, Q., Till, S., & Kafeza, E. (2004). Workflow View Driven Cross-Organizational Interoperability in a Web Service Environment. *Information Technology and Management*, *5*(3-4), 221–250. doi:10.1023/B:ITEM.0000031580.57966.d4

Chiu, D. K. W., Kafeza, M., Cheung, S. C., Kafeza, E., & Hung, P. C. K. (2009). Alerts in Healthcare Applications: Process and Data Integration. *International Journal of Healthcare Information Systems and Informatics*, *4*(2), 36–56.

Chiu, D. K. W., Karlapalem, K., Li, Q., & Kafeza, E. (2002). Workflow Views Based E-Contracts in a Cross-Organization E-Service Environment. *Distributed and Parallel Databases*, *12*(2-3), 193–216. doi:10.1023/A:1016503218569

Chiu, D. K. W., Leung, H., & Lam, K. (2009). On the making of service recommendations: An action theory based on utility, reputation, and risk attitude. *Expert Systems with Applications*, *36*(2), 3293–3301. doi:10.1016/j.eswa.2008.01.055

Chiu, D. K. W., & Li, Q. (1997). A Three-Dimensional Perspective on Integrated Management of Rules and Objects. *International Journal of Information Technology*, *3*(2), 98–118.

Chiu, D. K. W., Li, Q., & Karlapalem, K. (2001). Web Interface-Driven Cooperative Exception Handling in ADOME Workflow Management System. *Information Systems*, *26*(2), 93–120. doi:10.1016/S0306-4379(01)00012-6

Chiu, D. K. W., Yueh, Y. T. F., Leung, H.-f., & Hung, P. C. K. (2009). Towards Ubiquitous Tourist Service Coordination and Process Integration: a Collaborative Travel Agent System with Semantic Web Services. *Information Systems Frontiers*, *11*(3), 241–256. doi:10.1007/s10796-008-9087-2

Chiu, D., Cheung, S. C., & Zhuang, H.-F. L. P. H. E. K. H. H. M. W. H. H. Y. (2010). Engineering e-Collaboration Services with a Multi-Agent System Approach. *International Journal of Systems and Service-Oriented Engineering*, *1*(1), 1–25.

Choudhury, R., & Vaidya, N. (2004, January). Mac-layer anycasting in ad hoc networks. *ACM SIGCOMM Computer Communication Review*, *34*(1), 75–80. doi:10.1145/972374.972388

Choy, M. C., Srinivasan, D., & Cheu, R. L. (2003). Cooperative, Hybrid Agent Architecture for Real-time Traffic Signal Control. *IEEE Transactions on Systems, Man and Cybernetics. Part A*, *33*(5), 597–607.

Christensen, E., Curbera, F., Meredith, G., & Weerawarna, S. (2001). *Web Services Description Language (WSDL) 1.1., W3C, Note 15*. Retrieved from http://www.w3.org/TR/wsdl

Chung, J. C. S., Chiu, D. K. W., & Kafeza, E. (2007). An Alert Management System for Concrete Batching Plant. In *Proceedings of the 12th International Conference on Emerging Technologies and Factory Automation* (pp. 591-598), Patras, Greece.

Cimpian, E., & Mocan, A. (2005). WSMX process mediation based on choreographies. In *Proceedings of Business Process Management Workshops* (pp. 130-143).

Cimpian, E., Mocan, A., & Stollberg, M. (2006) Mediation enabled semantic web services usage. In *Proceedings of Asian Semantic Web Conference (ASWC)* (pp. 459-473).

Clement, L., Hately, A., Riegen, C., & Rogers, T. (2004). *UDDI Version 3.0.2*. Retrieved May 31, 2009, from http://www.uddi.org/pubs/uddi_v3.htm.

Coalition, O. W. L.-S. (2004). OWL-S: Semantic markup for web services. *W3C Member Submission*. Retrieved April 05, 2010, from http://www.w3.org/Submission/OWL-S/

Colgrave, J., Akkiraju, R., & Goodwin, R. (2004). External matching in UDDI. In *Proceedings of IEEE International Conference on Web Services (ICWS)* (pp. 226-233).

Conrad, M., French, T., Huang, W., & Maple, C. (2006). A Lightweight Model of Trust Propagation in a Multi-Client Network Environment: To What Extent Does Experience Matter? In *Proceedings of the 1st International Conference on Availability, Reliability and Security* (pp. 482-487). Washington, DC: IEEE Publications.

Cook, T. D., & Campbell, D. T. (n.d.). *Quasi-Experimentation: Design and Analysis Issues for Field Settings*. Houghton Mifflin Company.

Corritore, C., & Krasher, B., & Wiedenbeck. (2003). On-line trust: concepts, evolving themes, a model. *International Journal of Human-Computer Studies*, *58*, 737–758. doi:10.1016/S1071-5819(03)00041-7

Cortes, C., & Vapnik, V. (1995). Support-vector networks. *Machine Learning*, *20*(3), 273–297.

Courcoubetis, C., & Siris, V. (1999). *Managing and Pricing Service Level Agreements for Differentiated Services. Paper presented in the IEEE/IFIP (IWQoS'99)*. London: UCL.

Cox, I., Kilian, J., Leighton, T., & Shamoon, T. (1996). Secure spread spectrum watermarking for images, audio and video. *Image Processing, 1996. Proceedings., International Conference on, 3*.

Cox, I., Miller, M., Bloom, J., Fridrich, J., & Kalker, T. (2007). *Digital watermarking and steganography* (2nd ed.). Morgan Kaufmann.

Cox, I., Doerr, G., & Furon, T. (2006). Watermarking is not cryptography. *Lecture Notes in Computer Science*, *4283*, 1–15. doi:10.1007/11922841_1

Croll, M., Lee, A., & Parnall, S. (1997). Content Management - the Users Requirements. In *Proceedings of the International Broadcasting Convention*, Amsterdam, The Netherlands (pp. 12-16).

Cronin, J. J., & Taylor, S. A. (1992). Measuring service quality: a reexamination and extension. *Journal of Marketing*, *56*, 55–68. doi:10.2307/1252296

Cross, R., Parker, A., Prusak, L., & Borgatti, S. (2001). Knowing What We Know: Supporting Knowledge Creation and Sharing in Social Networks. *Organizational Dynamics*, *30*(2), 100–120. doi:10.1016/S0090-2616(01)00046-8

Cugola, G., & Migliavacca, M. (2009). A context and content-based routing protocol for mobile sensor networks. In *Proceedings of the 6th european conference on wireless sensor networks (Ewsn '09)* (pp. 69-85). Berlin: Springer Verlag.

Cugola, G., Murphy, A., & Picco, G. (2006). The handbook of mobile middleware. In *Content-Based Publish-Subscribe in a Mobile Environment*. New York: CRC Press.

Curbera, F., Duftler, M. J., Khalaf, R., Nagy, W., Mukhi, N., & Weerawarana, S. (2002). Unraveling the web services web: an introduction to SOAP, WSDL, and UDDI. *IEEE Internet Computing*, *6*(2), 86–93. doi:10.1109/4236.991449

Curbera, F., Khalaf, R., Mukhi, N., Tai, S., & Weerawarana, S. (2003). The next step in Web Services. *Communications of the ACM*, *46*(10), 29–34. doi:10.1145/944217.944234

Czajkowski, K., et al. (2004, March 5). *The WS-Resource Framework Version 1.0*. Retrieved from. http://www.106.ibm.com/developerworks/library/

Czajkowski, K., et al. (2005). *From Open Grid Services Infrastructure to WS-Resource Framework Refactoring & Evolution*. Retrieved from http://www.globus.org/wsrf/specs/ogsi_to_wsrf_1.0.pdf

Damas, C., Lambeau, B., & van Lamsweerde, A. (2006). Scenarios, Goals, and State Machines: a Win-Win Partnership for Model Synthesis. In *Proceedings of the 14th ACM SIGSOFT international symposium on foundations of software engineering,* Portland, OR.

Davenport, T. H., Jarvenpaa, S. L., & Beers, M. C. (1996). Improving knowledge work processes. *Sloan Management Review*, *4*(37), 53–65.

Davenport, T. H., Long, D., & Beers, M. C. (1998). Successful knowledge management projects. *Sloan Management Review*, *30*(2), 43–57.

Davenport, T. H., & Prusak, L. (1998). *Working Knowledge*. Boston: Harvard Business School Press.

Davis, J. C. (2000). Protecting privacy in the cyber era. *IEEE Technology and Society Magazine*, *19*(2), 10–22. doi:10.1109/44.846270

Davulcu, H., Kifer, M., & Ramakrishnan, I. V. (2004). CTR-S: a logic for specifying contracts in semantic web services. In *Proceedings of International Conference on World Wide Web on Alternate Track Papers & Posters (WWW)* (pp. 144-153).

Dayal, U. (1989). Active Database Management Systems. In *Proceedings of the 3rd International Conference on Data and Knowledge Bases* (pp. 150-169).

Dayal, S., Landesberg, H., & Zeisser, M. (1999). How to build trust online. *Marketing Management*, *8*(3), 64–69.

Delaigle, J. F., Vleeschouwer, C. D., & Macq, B. (1998). Watermarking algorithm based on a human visual model. *Signal Processing*, *66*(3), 319–335. doi:10.1016/S0165-1684(98)00013-9

Delicato, F. C. P. F., Pires, F. P., Pirmez, L., & Carmo, L. F. (2003). A flexible web service based architecture for wireless sensor networks. In *Proceedings of the 23rd international conference on distributed computing systems (Icdcsw '03)* (p. 730). Washington, DC: IEEE Computer Society.

DeLone, W. H., & McLean, E. R. (2003). The DeLone and McLean model of information systems success: a ten-year update. *Journal of Management Information Systems*, *19*, 9–30.

Deutsch, A., Sui, L., Vianu, V., & Zhou, D. (2006). Verification of communicating data-driven web services. In *Proceedings of ACM SIGACT-SIGMOD-SIGART Symposium on Principles of Database Systems (PODS)* (pp. 90-99).

Deutsch, M. (1958). Trust and suspicion. *The Journal of Conflict Resolution*, *2*(3), 265–279. doi:10.1177/002200275800200401

Dey, A. K., & Abowd, G. D. (1999). *Toward A Better Understanding of Context and Context-Awareness*. GVU Technical Report GIT-GVU-99-22, College of Computing, Georgia Institute of Technology.

Diaz, O. G. F., Salgado, R. S., & Moreno, I. S. (2006). Using case-based reasoning for improving precision and recall in web services selection. *International Journal of Web and Grid Services*, *2*, 306–330. doi:10.1504/IJWGS.2006.011358

DiffServ Provisioning. In *Proceedings of the International Conference on Software Engineering Research, Management & Applications (SERA2005)*, Mt. Pleasant, USA (pp. 325-330).

Dimitrakos, T., Golby, D., & Kearney, P. (2004, October). Towards a Trust and Contract Management Framework for Dynamic Virtual Organisations. In *Proceedings of the eChallenges Conference (eChallenges 2004),* Vienna, Austria.

Dimitrakos, T., Martrat, J., & Wesner, S. (Eds.). (2010). *A selection of common capabilities validated in real-life business trials by the BEinGRID consortium XV* (p. 210). New York: Springer.

Dincbas, M., Simons, H., & van Hentenryck, H. (1998). Solving the Car-Sequencing Problem in Constraint Logic Programming. In *Proceeding of ECAI-88* (pp. 290-295).

Do, H. H., Melnik, S., & Rahm, E. (2003). Comparison of Schema Matching Evaluations. In A. B. Chaudhri, M. Jeckle, E. Rahm, & R. Unland, Eds. *Revised Papers from the Node 2002 Web and Database-Related Workshops on Web, Web-Services, and Database Systems* (LNCS 2593, pp. 221-237). London: Springer Verlag.

Doan, A., Madhavan, J., Domingos, P., & Halevy, A. (2004). Ontology matching: a machine learning approach. In Staaband, S., & Studer, R. (Eds.), *Handbook of ontologies: International handbookon informations systems* (pp. 385–404). Berlin: Springer Verlag.

Dobson, G., Lock, R., & Sommerville, I. (2005). QoSOnt: a QoS Ontology for Service-Centric Systems. In *Proceedings of the 31st EUROMICRO Conference on Software Engineering and Advanced Applications*. Washington, DC: IEEE Computer Society

Domencich, T., & McFadden, D. (1975). *Urban Travel Demand: A Behavioural Analysis*. Amsterdam: North-Holland.

Dong-Jin, S., & Young-Tak, K. (2003). Design and Implementation of Performance Management for the DiffServ-aware-MPLS Network. In *Proceedings of the Conference on APNOMS 2003*, Fukuoka, Japan.

Donzelli, P. (2004). A Goal-driven and Agent-based Requirements Engineering Framework. *Requirement Engineering, 9*(1).

Drew, S. (1999). Building Knowledge Management into Strategy: Making Sense of a New Perspective. *Long Range Planning, 32*(1), 130–136. doi:10.1016/S0024-6301(98)00142-3

Dubois, D., & Prade, H. (1994, August). Possibility theory, belief revision and nonmonotonic logic. In A. L. Ralescu (Ed.), *Proceedings on the IJCAI-93 Workshop on Fuzzy Logic in Artificial Intelligence* (LNCS 847, pp. 51-61). New York: Springer.

Dubois, D., Lang, J., & Prade, H. (1987, August). Theorem proving under uncertainty - a possibility theory based approach. In J. McDermott (Ed.), *Proceedings of the Tenth International Joint Conference on Artificial Intelligence* (pp. 984-986). San Francisco, CA: Morgan Kaufmann Publishers Inc.

Dubois, D., Lang, J., & Prade, H. (1991, October). A brief overview of possibilistic logic. In R. Krause & P. Siegel (Eds.), *Proceedings of Symbolic and Quantitative Approaches to Uncertainty (ECSQAU '91)* (LNC 548, pp. 53-57). New York: Springer.

Dubois, D., Lang, J., & Prade, H. (1993). Possibilistic Logic. In Gabbay, D. M., Hogger, C. J., Robinson, J. A., & Nute, D. (Eds.), *Handbook of Logic in Artificial Intelligence and Logic Programming* (*Vol. 3*, pp. 439–513). Oxford, UK: Oxford University Press.

Dubois, D., Lang, J., & Prade, H. (1994). Automated Reasoning using Possibilistic Logic: Semantics, Belief Revision, and Variable Certainty Weights. *IEEE Transactions on Knowledge and Data Engineering, 6*(1), 64–71. doi:10.1109/69.273026

Dubois, D., & Prade, H. (1990). Resolution principles in possibilistic logic. *International Journal of Approximate Reasoning, 4*(1), 1–21. doi:10.1016/0888-613X(90)90006-N

Dumas, M., Sport, M., & Wang, K. (2006). Adapt or perish: algebra and visual notation for service interface adaptation. In *Proceedings of International Conference on Business Process Management (BPM)* (pp. 65-80).

Duquennoy, S., Grimaud, J. J., & Vandewalle, G. (2009). Smart and Mobile Embedded Web Server. In *Proceedings of the International Conference on Comlex, Intelligent and Software Intensive Systems* (pp. 571- 576).

Durães, J., & Madeira, H. (2006). Emulation of Software Faults: A Field Data Study and a Practical Approach. *IEEE Transactions on Software Engineering, 32*(11), 849–867. doi:10.1109/TSE.2006.113

Duraisamy, S. (2008). *SOAP*. Retrieved from http://searchsoa.techtarget.com/sDefinition/0,sid26gci214295,00.html

Dustdar, S., & Kramer, B. J. (2008): Introduction to Special Issue on Service-Oriented Computing (SOC). *ACM Transactions on the Web, 2*(2).

Edvardsson, B., & Larsson, P. (2004). *Service guarantees*. Lund: Studentlitteratur.

Edwards, J., Coutts, I., & McLeod, S. (2000). Support for system evolution through separating business and technology issues in a banking system. In *Proceedings of the International Conference on Software Maintenance* (pp. 271-276).

Edwards, W. (1992). *Utility Theories: Measurements and Applications*. Norwell, MA: Kluwer Academic Publishers.

Egger, F. (2003). *From Interactions to Transactions: designing the Trust Experience for Business-to-Consumer Electronic Commerce*. Unpublished doctoral dissertation, Technical University Eindhoven, The Netherlands.

Elman, J. (1990). Finding structure in time. *Cognitive Science*, *14*(2), 179–211.

Elmasri, R. A., & Navathe, S. B. (2006) *Fundamentals of Database Systems* (5th ed.). Addison-Wesley.

Erl, T. (2006). *Service-Oriented Architecture: Concepts, Technology, and Design*. Upper Saddle River, NJ: Prentice-Hall.

Erradi, A., Padmanabhuni, S., & Varadharaja, N. (2006). Differential QoS Support in Web Services Management. In *Proceedings of the IEEE international conference on web services,* Chicago.

Erradi, A., Tosic, V., & Maheshwari, P. (2007). MASC -. NET-based middleware for adaptive composite Web services. In K. Birman, L.-J. Zhang, & J. Zhang (Eds.), *Proceedings of ICWS 2007* (pp. 727-734). Los Alamitos, USA: IEEE-CS Press.

Eschenauer, L., & Gligor, V. (2002). A key-management scheme for distributed sensor networks. In *Proceedings of the 9th ACM conference on computer and communications security* (pp. 41–47).

Evers, L., & Havinga, P. (2007, Oct). Supply chain management automation using wireless sensor networks. In *Proceedings of the ieee int. conf. on mobile adhoc and sensor systems*, Pisa, Italy (pp. 1-3). Washington, DC: IEEE.

Fakhfakh, K., Chaari, T., Tazi, S., Drira, K., & Jmaiel, M. (2009). Semantic Enabled Framework for SLA Monitoring. *International Journal on Advances in Software*, *2*(1), 36–46.

Farrell, J., & Lausen, H. (2007). *Semantic annotations for WSDL and XML schema*. Retrieved May 31, 2009, from http://www.w3.org/TR/sawsdl/.

Feenstra, R. W., Janssen, M., & Wagenaar, R. W. (2007). Evaluating Web Service Composition Methods: The need for including Multi-Actor Elements. *The Electronic. Journal of E-Government*, *5*(2), 153–164.

Fei, Y., et al. (2009). RFID Middleware Event Processing Based on CEP. In *Proceedings of the 2009 IEEE International Conference on e-Business Engineering.*

Fensel, D. (2001). Challenges in Content Management for B2B Electronic Commerce. In *Proceedings of the 2nd International Workshop on User Interfaces to Data Intensive Systems (UIDIS 2001)*, Zurich, Switzerland (pp. 2-4).

Ferguson, J. M., & Zawacki, R. A. (1993). Service quality: A critical success factor for IS organizations. *Information Strategy: The Executive's Journal*, *9*(2), 24–30.

Ferraiolo, D. F., Kuhn, D. R., & Chandramouli, R. (2003). *Role-based access control*. Norwood, MA: Artech House Publishers.

Ferraiolo, D. F., Sandhu, R., Gavrila, S., Kuhn, D. R., & Chandramouli, R. (2001). Proposed NIST Standard for Role-Based Access Control. *ACM Transactions on Information and System Security*, *4*(3), 224–274. doi:10.1145/501978.501980

Fischer-Hübner, S. (2001). *IT-Security and Privacy – Design and Use of Privacy-Enhancing Security Mechanism* (LNCS 1958). New York: Springer.

Flood, M. (1952). *Some Experimental Games. Research Memorandum RM0789*. Santa Monica, CA: RAND Corporation.

Foo, S., Sharma, R., & Chua, A. (2007). *Knowledge Management Tools and Techniques* (2nd ed.). Upper Saddle River, NJ: Prentice Hall.

Forman, G. H., & Zahorjan, J. (1994). The challenges of mobile computing. *IEEE Computer*, *27*(4), 38–47.

Fornell, C., & Larcker, D. F. (1981). Evaluating structural equation models with unobservable variables and measurement error. *JMR, Journal of Marketing Research*, *18*(1), 39–50. doi:10.2307/3151312

Foster, I. (2005). Globus Toolkit Version 4: Software for Service-Oriented Systems. In *Proceedings of the* IFIP *International Conference on Network and Parallel Computing* (LNCS 3779, pp. 2-13). New York: Springer Verlag.

Foster, I. T., Fidler, M., Roy, A., Sander, V., & Winkler, L. (2004). End-to-end quality of service for high-end applications. *Computer Communications*, *27*(14), 1375–1388. doi:10.1016/j.comcom.2004.02.014

Foster, I., & Kesselman, C. (1998). *The Grid: Blueprint for a new computing infrastructure*. San Francisco, CA: Morgan Kaufmann Publishers Inc.

Foster, I., Kesselman, C., & Tuecke, S. (2001). The anatomy of the grid: Enabling scalable virtual organizations. *International Journal of High Performance Computing Applications, 15*(3), 200–222. doi:.doi:10.1177/109434200101500302

Fraile, J.-C., Paredis, C. J. J., Wang, C.-H., & Khosla, P. K. (1999). Agent-based Planning & Control of a Multi-manipulator Assembly System. In. *Proceedings of the IEEE International Conference on Robotics & Automation, 2*, 1219–1225.

Francis, W. (2008). *The NextGRID Final Report (Project Final Report)*. Edinburgh, UK: The University of Edinburgh.

Fremantle, P., Weerawarana, S., & Khalaf, R. (2002). Enterprise services. Examine the emerging files of web services and how it is intergraded into existing enterprise infrastructures. *Communications of the ACM, 45*(20), 77–82.

Friedman, N., Geiger, D., & Goldszmid, M. (1997). Bayesian Network Classifiers. *Machine Learning, 29*(2-3), 131–163. doi:10.1023/A:1007465528199

Fukuyama, F. (1995). *Trust: The Social Virtues and the Creation of Prosperity*. London: Hamish Hamilton.

Gabbott, M., & Hogg, G. (1998). *Consumers and services*. New York: John Wiley & Sons Ltd.

Gambetta, D. (2000). *Trust: Making and Breaking Cooperative Relations*. Retrieved from www.sociology.ox.ac.uk/papers/trustbook.html

Gambetta, D. (1988). Can we trust trust? In Gambetta, D. (Ed.), *Trust: Making and Breaking Cooperative Relations* (pp. 213–237). London: Basil Blackwell.

Gamma, E., Helm, R., Johnson, R., & Vlissides, J. (1995). *Design Patterns: Elements of Reusable Object-Oriented Software*. Addison Wesley, Boston, MA, USA.

Garcia, J. M., Toma, I., Ruiz, D., & Ruiz-Cortes, A. (2008, November). A service ranker based on logic rules evaluation and constraint programming. In *Proceedings of the 2nd ECOWS Non-Functional Properties and Service Level Agreements in Service Oriented Computing Workshop*, Dublin, Ireland.

Garg, M., Saran, R., Randhawa, H., & Singh, R. S. (2002). *A SLA framework for QoS provisioning and dynamic capacity allocation*. Paper presented at the 10th IEEE International Workshop on Quality of Service.

Garmany, J. (2003). *Oracle Replication: Snapshot, Multi-master & Materialized Views Scripts*. Kittrell, NC: Rampant TechPress.

Garrido, L., Brena, R., & Sycara, K. (1996). Cognitive Modeling & Group Adaptation in Intelligent Multi-Agent Meeting Scheduling. In *Proceedings of the 1st Iberoamerican Workshop on Distributed Artificial Intelligence & Multi-Agent Systems* (pp. 55–72).

Gentzsch, W. (2002). Grid computing: a new technology for the advanced web. In *Proceedings of the NATO Advanced Research Workshop on Advanced Environments, Tools, and Applications for Cluster Computing* (LNCS 2326, pp. 1-15). New York: Springer.

Gerst, M. H. (2003). The Role of Standardisation in the Context of E-collaboration: A SNAP shot. In *Proceedings of the 3rd Conferernce on Standardization & Innovation in Information Technology* (pp. 113-119).

Gibbons, A. (1985). *Algorithmic Graph Theory*. Cambridge, UK: Cambridge University Press.

Gibelin, N., & Makpangou, M. (2005). Efficient and transparent web-services selection. In *Proceedings of International Conference on Service Oriented Computing (ICSOC)* (pp. 527-532).

Giunchiglia, F., & Shvaiko, P. (2003). Semantic matching. *The Knowledge Engineering Review, 18*(3), 305–332. doi:10.1017/S0269888904000074

Globus. (2006). *Information Services (MDS): Key Concepts*. Retrieved from http://www.globus.org/toolkit/docs/4.0/info/key-index.html

Goderis, D., et al. (2002, January). *Service Level Specification Semantics and Parameters* (Internet Draft, <draft-tequila-sls-02.txt>).

Golatowski, F., Blumenthal, J. H. M., Haase, M., Burchardt, H., & Timmermann, D. (2003). Service-Oriented Software Architecture for Sensor Networks. In *Proceedings of the int. workshop on mobile computing (imcâ™03)* (pp. 93-98).

Golbeck, J., & Hendler, J. (2006). FilmTrust: Movie Recommendations using Trust in Web-based Social Networks. In *Proceedings of the IEEE Consumer Communications and Networking Conference*.

Golbek, J. A. (2005). *Computing and applying Trust in Web-based Social Networks*. Unpublished doctoral dissertation, University of Maryland, College Park, MD.

Gore, C., & Gore, E. (1999). Knowledge management: The way forward. *Total Quality Management, 4/5*(10), 554–560.

Goseva-Popstojanova, K., Mathur, A. P., & Trivedi, K. S. (2001). Comparison of architecture-based software reliability models. *In Proceedings of the 12th International Symposium on Software Reliability Engineering* (pp. 22-31). Washington, DC: IEEE Computer Society.

Gounaris, S., Dimitriadis, S., & Stathakopoulos, V. (2005). Antecedents of perceived quality in the context of Internet retail stores. *Journal of Marketing Management, 21*, 669–700. doi:10.1362/026725705774538390

Governatori, G., Milosevic, Z., & Sadiq, S. W. (2006). Compliance checking between business process and business contracts. In *Proceedings of IEEE International Enterprise Distributed Object Computing Conference (EDOC)* (pp. 221-232).

Grabner-Krauter, S., Kaluschia, E. A., & Fladnitzer, M. (2005). Perspectives of Online Trust and Similar Constructs–A Conceptual Clarification. In *Proceedings of the 8th International Conference on Electronic Commerce*, Fredericton, Canada (pp. 235-243).

Grabner-Krauter, S., & Kaluschia, E. A. (2003). Empirical research in on-line trust: a review and critical assessment. *International Journal of Human-Computer Studies, 58*, 783–812. doi:10.1016/S1071-5819(03)00043-0

Groth, M., Gutek, B. A., & Douma, B. (2001). Effects of service mechanisms and modes on customers' attributions about service delivery. *Journal of Quality Management, 6*, 331–348. doi:10.1016/S1084-8568(01)00043-8

Gulliver, P. H. (1979). *Disputes and Negotiation: A Cross-Cultural Perspective*. New York: Academic.

Guo, W.-Y. (2008). Reasoning with semantic web technologies in ubiquitous computing environment. *Journal of Software, 3*(8), 27–33. doi:10.4304/jsw.3.8.27-33

Gupta, K., & Govindarajan, V. (2000). Knowledge Flows Within Multinational Corporations. *Strategic Management Journal, 21*(4), 473–496. doi:10.1002/(SICI)1097-0266(200004)21:4<473::AID-SMJ84>3.0.CO;2-I

Guttman, R., & Maes, P. (1998). Cooperative vs. Competitive Multi-Agent Negotiations in Retail Electronic Commerce. In *Proceedings of the Second International Workshop on Cooperative Information Agents (CIA'98)*, Paris (LNAI 1435, pp. 135-147). New York: Springer Verlag.

Haddad, J. E., Manouvrier, M., Ramirez, G., & Rukoz, M. (2008). QoS-driven selection of web services for transactional composition. In *Proceedings of International Conference on Web Services (ICWS)* (pp. 653-660).

Hardin, R. (Ed.). (2002). *Trust and Trustworthiness*. New York: Russell Sage Foundation.

Harel, D., & Naamad, A. (1996). The STATEMATE Semantics of Statecharts. *ACM Transactions on Software Engineering and Methodology, 5*(4). doi:10.1145/235321.235322

Hari, S., Egbu, C., & Kumar, B. (2005). A knowledge capture awareness tool: An empirical study on small and medium enterprises in the construction industry. *Engineering, Construction, and Architectural Management, 6*(12), 533–567. doi:10.1108/09699980510634128

Hasselmeyer, P., Mersch, H., Koller, B., Quyen, H. N., Schubert, L., & Wieder, P. (2007). *Implementing an SLA Negotiation Framework*. Paper presented at eChallenges.

Hasselmeyer, P., Qu, C., Schubert, L., Koller, B., & Wieder, P. (2006). Towards Autonomous Brokered SLA Negotiation. In Cunningham, P. M. (Ed.), *Exploiting the Knowledge Economy - Issues, Applications, Case Studies (Vol. 3)*. Amsterdam: IOS Press.

Hau, J., Lee, W., & Newhouse, S. (2003). Autonomic service adaptation in ICENI using ontological annotation. In *Proceedings of International Workshop on Grid Computing (GRID)* (pp. 10-17).

Heiko, L., Toshiyuki, N., Philipp, W., Oliver, W., & Wolfgang, Z. (2006). *Reliable Orchestration of Resources using WS-Agreement* (Tech. Rep. No. TR-0050). Institute on Grid Systems, Tools, and Environments.

He, J., Gao, T., Hao, W., Yen, I.-L., & Bastani, F. (2007). A Flexible Content Adaptation System Using a Rule-Based Approach. *IEEE Transactions on Knowledge and Data Engineering*, *19*(1), 127–140. doi:10.1109/TKDE.2007.250590

Helal, S., Desai, N., Verma, V., & Lee, C. (2003, march). *Konark - a service discovery and delivery protocol for ad-hoc networks* (Vol. 3, pp. 2107-2113).

He, M., Jennings, N. R., & Leung, H. (2003). On agent-mediated electronic commerce. *IEEE Transactions on Knowledge and Data Engineering*, *15*(4), 985–1003. doi:10.1109/TKDE.2003.1209014

Henriksen, Z. H. (2004). The diffusion of e-services in Danish municipalities. In Traunmüller (Ed.), *Proceedings of EGOV 2004* (pp.164-171). Berlin: Springer.

Hermann, R., Husemann, D., Moser, M., Nidd, M., Rohner, C., & Schade, A. (2001). Deapspace: transient ad hoc networking of prevasive devices. *Computer Networks*, *35*(4), 411–428. doi:10.1016/S1389-1286(00)00184-5

Hesselman, C., Tokmakoff, A., Pawar, P., & Iacob, S. (2006). Discovery and Composition of Services for Context-Aware Systems. In *Proceedings of the European conference on smart sensing and context,* Enschede, The Netherlands.

Hinde, S. (2002). The perils of privacy. *Computers & Security*, *21*(3), 424–432. doi:10.1016/S0167-4048(02)00508-4

Hirst, G., & St-Onge, D. (1998). Hso: Lexical chains as representations of context for the detection and correction of malapropisms. In Fellbaum, C. (Ed.), *WordNet: An Electronic Lexical Database* (pp. 305–332). Cambridge, MA: MIT Press.

Hoffman, D. L., Novak, T. P., & Chatterjee, P. (1996). Commercial scenarios for the web: opportunities and challenges. *Journal of Computer-Mediated Communication*, *1*, 3.

Hoffman, D., Novak, T., & Peralta, M. (1999). Building consumer trust online. *Communications of the ACM*, *42*(4), 80–85. doi:10.1145/299157.299175

Hoffner, Y., Field, S., Grefen, P., & Ludwig, H. (2001). Contract driven creation and operation of virtual enterprises. *Computer Networks*, *37*, 111–136. doi:10.1016/S1389-1286(01)00210-9

Holbrook, M. B., & Hirschman, E. C. (1982). The Experiential Aspects of Consumption: Consumer Fantasies. *Feelings and Fun*, *9*(2), 132–140.

Holtzman, S. (1989). *Intelligent Decision Systems*. Reading, MA: Addison-Wesley Publishing Co.

Hong, D., Chiu, D. K. W., Cheung, S. C., Shen, V. Y., & Kafeza, E. (2007). Ubiquitous Enterprise Service Adaptations Based on Contextual User Behavior. *Information Systems Frontiers*, *9*(4), 343–358. doi:10.1007/s10796-007-9039-2

Horrocks, I., Patel-Schneider, P. F., Boley, H., Tabet, S., Grosof, B., & Dean, M. (2004). *SWRL: A semantic web rule language combining OWL and RuleML*. W3C Member Submission.

Howard, J. A. (1989). *Consumer behavior in marketing strategy*. Upper Saddle River, NJ: Prentice Hall.

Howard, J. A., & Sheth, J. N. (1969). *The theory of buyer behavior*. New York: John Wiley & Sons.

Hua, Z., Xie, X., Liu, H., Lu, H., & Ma, W.-Y. (2006). Design and Performance Studies of an Adaptive Scheme for Serving Dynamic Web Content in a Mobile Computing Environment. *IEEE Transactions on Mobile Computing*, *5*(12), 1650–1662. doi:10.1109/TMC.2006.182

Hudert, S., Ludwig, H., & Wirtz, G. (2009). Negotiating SLAs-An Approach for a Generic Negotiation Framework for WS-Agreement. *Journal of Grid Computing*, *7*(2), 225–246. doi:10.1007/s10723-009-9118-3

Hugo, H., & Allan, B. (2004). *Web Services Glossary*. Retrieved from http://www.w3.org/TR/wsgloss/

Huhns, M. N., & Singh, M. P. (2005, January-February). *Service-oriented computing: Key concepts and principles.* IEEE Internet Computing.

Hui, J. W., & Culler, D. E. (2008, July-August). Extending IP to Low-Power, Wireless Personal Area Networks. *Internet Computing*, *12*(4), 37–45. doi:10.1109/MIC.2008.79

Hultgren, G. (2007). *E-services as social interaction via the use of IT systems: a practical theory.* Unpublished doctoral dissertation, University of Linköping, Sweden.

Hung, P. C. K. (2001). *Secure Workflow Model. Unpublished doctoral disseration.* Hong Kong, China: Department of Computer Science, The Hong Kong University of Science and Technology.

Hung, P. C., Chiu, D. K., Fung, W. W., Cheung, W. K., Wong, R., & Choi, S. P. (2007). End-to-end privacy control in service outsourcing of human intensive processes: A multi-layered web service integration approach. *Information Systems Frontiers*, *9*(1), 85–101. doi:10.1007/s10796-006-9019-y

Hung, S., Chang, Y., & Yu, T. J. (2006). A classification of business-to-business buying decisions: Risk importance and probability as a framework for e-business benefits. *Government Information Quarterly*, *1*(23), 97–122. doi:10.1016/j.giq.2005.11.005

Hwang, S.-Y., Wang, H., Srivastava, J., & Paul, R. A. (2004). A probabilistic QoS model and computation framework for web services-based workflows. In *Proceedings of International Conference on Conceptual Modeling (ER)* (pp. 596-609).

Hwang, S.-Y., Wang, H., Tang, J., & Srivastava, J. (2007). A probabilistic approach to modeling and estimating the QoS of web-services-based workflows. *Information Sciences*, *177*(23), 5484–5503. doi:10.1016/j.ins.2007.07.011

IBM. (2003, May). *Business Process Execution Language for Web Services (BPEL), Version 1.1.* Retrieved from http://www.ibm.com/developerworks/library/ws-bpel

IBM. (2004). *WS-ResourceLifetime specification.* Retrieved from http://www.ibm.com/developerworks/library/ws-resource/ws-resourcelifetime.pdf

Imai, H., Izawa, D., Yoshida, K., & Sato, Y. (2004). On Detecting Interactions in Hayashi's Second Method of Quantification. *Modeling Decisions for Artificial Intelligence, 3131*, 1–22.

Intanagonwiwat, C., Govindan, R., & Estrin, D. (2000). Directed diffusion: a scalable and robust communication paradigm for sensor networks. In *Proceedings of the 6th annual international conference on mobile computing and networking (Mobicom '00)* (pp. 56-67). New York: ACM.

International Business Machines. (2003). *An architectural blueprint for autonomic computing.* IBM.

Ion, M., Telesca, L., Botto, F., & Koshutanski, H. (2007). An Open Distributed Identity and Trust Management Approach for Digital Community Ecosystems. In *Proceedings of the International Workshop on ICT for Business Clusters in Emerging Markets.* Retrieved from http://www.create-net.org/osco/publications/ion-tele-bott-kosh-eBusiness-07.pdf

Jaeger, M. C., Rojec-Goldmann, G., & Muhl, G. (2004). QoS aggregation for service composition using workflow patterns. In *Proceedings of International Conference on Enterprise Distributed Object Computing (EDOC)* (pp. 149-159).

Jain, S., & Das, S. (2008). Exploiting path diversity in the link layer in wireless ad hoc networks. *Ad Hoc Networks*, *6*(5), 805–825. doi:10.1016/j.adhoc.2007.07.002

James, K. (2001). *Overview of WSDL.* Retrieved from http://developers.sun.com/appserver/reference/techart/overviewwsdl.html

Janssen, M., & Joha, A. (2008). Emerging shared service organizations and the service-oriented enterprise: Critical management issues. *Strategic Outsourcing: An International Journal*, *1*(1), 35–49. doi:10.1108/17538290810857466

Japanese Ministry of Economy. Trade and Industry. (2007). *Towards Innovation and Productivity, Improvement in Service Industries.* Retrieved March 27, 2009, from http://www.meti.go.jp/english/report/downloadfiles/0707ServiceIndustries.pdf

Jardak, C., Meshkova, E., & Riihij, Ã. ¤rvi, J., Rerkrai, K., & Mahonen, P. (2008, March 31-April 3). Implementation and performance evaluation of nanoip protocols: Simplified versions of tcp, udp, http and slp for wireless sensor networks. In *Proceedings of the IEEE Wireless Communications & Networking Conference (Wcnc 2008)*, Las Vegas, NV (pp. 2474-2479). Washington, DC: IEEE.

Jarvenpaa, S. L., Tractinsky, J., & Vitale, M. (2000). Consumer Trust in an Internet Store. *Journal of Information Technology Management*, *1*(1/2), 45–71. doi:10.1023/A:1019104520776

Järvinen, R., & Lehtinen, U. (2004). Services, e-Services and e-Service innovations – combination of theoretical and practical knowledge. In Hannula, M., Järvelin, A.-M., & Seppä, M. (Eds.), *Frontiers of e-Business research* (pp. 78–89).

Javalgi, R., Martin, C., & Todd, P. (2004). The export of e-services in the age of technology transformation: challenges and implications for international service providers. *Journal of Services Marketing*, *7*(18), 560–573. doi:10.1108/08876040410561884

JBoss AOP. (2009). *JBoss AOP Reference Documentation.* Retrieved from http://docs.jboss.org/aop/1.3/aspectframework/reference/en/html/dynamic.html

Jefferey, M., & Leliveld, I. (2004). Best Practices in IT Portfolio Management. *MIT Sloan Management Review*, *45*(3), 41–49.

Jennex, M. E., Smolnik, S., & Croasdell, D. (2007). Towards Defining Knowledge Management Success. In *Proceedings of the 40th Hawaii International Conference on the System Sciences,* Waikoloa, HI (p. 193).

Jensen, F. (1996). *Introduction to Bayesian networks.* Springer-Verlag New York, Inc. Secaucus, NJ, USA.

Jensen, P. E. (2005). A Contextual Theory of Learning and the Learning Organization. *Knowledge and Process Management*, *1*(12), 53–64. doi:10.1002/kpm.217

Jeronimo, M., & Weast, J. (2003). *Upnp design by example: A software developer's guide to universal plug and play*. Intel Press.

JESS. "Jess Expert System Shell." from http://www.jessrules.com/.

Johannesson, P., Andersson, B., Bergholtz, M., & Weigand, H. (2008). Enterprise Modeling for Value Based Service Analysis. In Stirna, J., & Persson, A. (Eds.), *Proceedings of PoEM 2008*. Berlin: Springer.

John, J. (1996). A dramaturgical view of the health care service encounter: Cultural value-based impression management guidelines for medical professional behaviour. *European Journal of Marketing*, *30*(9), 60–75. doi:10.1108/03090569610130043

Johnson, J. (1994). Pulse-coupled neural nets: translation, rotation, scale, distortion, and intensity signal invariance for images. *Applied Optics*, *33*, 6239–6253. doi:10.1364/AO.33.006239

Jones, V. E., Ching, N., & Winslett, M. (1995). Credentials for privacy and interoperation. In *Proceedings of the New Security Paradigms Workshop* (pp. 92-100).

Jordan, D., & Evdemon, J. (2007). *Web services business process execution language version 2.0*. Retrieved May 31, 2009 from http://docs.oasis-open.org/wsbpel/2.0/OS/wsbpel-v2.0-OS.html.

Jordan, M. I. (1998). *Learning in graphical models*. Cambridge, MA: MIT Press.

Julien, C., & Roman, G.-C. (2006). EgoSpaces: Facilitating Rapid Development of Context-Aware Mobile Applications. *IEEE Transactions on Software Engineering*, *32*(5), 281–298. doi:10.1109/TSE.2006.47

Jupp, V. (2003). Realizing the vision of e-government. In Curtin, G., Sommer, M., & Sommer-Vis, V. (Eds.), *The world of E-government*. New York: Haworth Press.

Jurca, R., Faltings, B., & Binder, W. (2007). Reliable QoS monitoring based on client feedback. In *Proceedings of International Conference on World Wide Web (WWW)* (pp. 1003-1012).

Kafeza, E., Chiu, D., Cheung, S., & Kafeza, M. (2004). Alerts in mobile healthcare applications: requirements and pilot study. *IEEE Transactions on Information Technology in Biomedicine*, *8*(2), 173–181. doi:10.1109/TITB.2004.828888

Kahn, D. (1996). The history of steganography. In *Proceedings of the First International Workshop on Information Hiding* (pp. 1–5). London, UK: Springer-Verlag.

Kahneman, D., & Tversky, A. (1979). Prospect theory: An analysis of decision under risk. *Journal of the Econometric Society*, *47*(2), 263–291. doi:10.2307/1914185

Kanellos, M. (2005). *RFID chips used to track dead after Katrina.* Retrieved from http://news.cnet.com/RFID-chips-used-to-track-dead-after-Katrina /2100-11390_3-5869708.html

Kankanhalli, A., Tan, B. C. Y., & Wei, K.-K. (2005). Contributing knowledge to electronic knowledge repositories: an empirical investigation. *Management Information Systems Quarterly*, *1*(29), 113–143.

Kanso, H., Soulé-Dupuy, C., & Tazi, S. (2007, June). Representing Author's Intentions of Scientific Documents. In *Proceedings of the International Conference on Enterprise Information Systems*, *3*, 489–492.

Kaplan, R. S., & Norton, D. P. (1996). Using the balanced scorecard as a strategic management system. *Harvard Business Review*, *74*(1), 75–85.

Kari, L., & Rozenberg, G. (2008). The many facets of natural computing. *Communications of the ACM*, *51*(10), 72–83. doi:10.1145/1400181.1400200

Karvonen, K. (2007). Enabling Trust between Humans and Machines. In *Proceedings of the European e-Identity Conference, Eema's 20th Annual Conference,* Paris, France. Retrieved from http://www.tml.tkk.fi/~kk/publications.html

Katzenbeisser, S., & Fabien, A. P. (Eds.). (2000). *Information hiding techniques for steganography and digital watermarking*. Artech House Publishers.

Kavantzas, N., Burdett, D., Ritzinger, G., Fletcher, T., & Lafon, Y. (2004, December). Web Services Choreography Description Language Version 1.0. In *Proceedings of the World Wide Web Consortium*.

Keeney, R., & Raiffa, H. (1993). *Decisions with Multiple Objectives: Preferences and Value Tradeoffs*. Cambridge, UK: Cambridge University Press.

Keller, A., & Ludwig, H. (2003). The WSLA Framework: Specifying and Monitoring Service Level Agreements for Web Services. *Journal of Network and Systems Management*, *11*(1), 57–81. doi:10.1023/A:1022445108617

Kephart, J. O., & Walsh, W. E. (2004). An artificial intelligence perspective on autonomic computing policies. In M. Lupu & M. Kohli (Eds.), *Proceedings of Policy 2004* (pp. 3-12). Los Alamitos, USA: IEEE-CS Press.

Kephart, J. O., & Chess, D. M. (2003). The vision of autonomic computing. *Computer*, *36*(1), 41–50. doi:10.1109/MC.2003.1160055

Kettinger, W. J., & Lee, C. C. (1994). Perceived service quality and user satisfaction with the information services function. *Decision Sciences*, *25*(6), 737–766. doi:10.1111/j.1540-5915.1994.tb01868.x

Khedr, M., & Karmouch, A. (2004). Negotiating Context Information in Context-Aware Systems. *IEEE Intelligent Systems*, *19*(6), 21–29. doi:10.1109/MIS.2004.70

Kim, J., & Moon, J. Y. (1998). Designing Towards Emotional Usability in Customer Interfaces: Trustworthiness of cyber-banking system interfaces. *Interacting with Computers*, *10*, 1-29. doi:10.1016/S0953-5438(97)00037-4

Kim, K. H., & Welch, H. O. (1989). Distributed Execution of Recovery Blocks: An Approach for uniform Treatment of Hardware and Software Faults in Real-Time Applications. *IEEE Transactions on Computers*, *38*(5), 626–636. doi:10.1109/12.24266

Kim, L. (1995). Crisis Construction and Organisational Learning: Capability Building in Catching-up at Hyundai Motor. *Organization Science*, *9*(4), 506–521. doi:10.1287/orsc.9.4.506

Kinno, A., Yukitomo, H., & Nakayama, T. (2004). An Efficient Caching Mechanism for XML Content Adaptation. *the 10th International Multimedia Modeling Conference* (MMM'04), Jan., IEEE Press, 308-315.

Kitayama, F., Hitose, S., Kondoh, G., & Kuse, K. (1999). Design of a framework for dynamic content adaptation to Web-enabled terminals and enterprise applications. In *Proceedings of the Sixth Asia Pacific Software Engineering Conference* (pp. 72-79).

Klir, G. J., & Yuan, B. (Eds.). (1996). Fuzzy Sets, Fuzzy Logic, and Fuzzy Systems: Selected Papers by Lotfi A. Zadeh. World Scientific Publishing.

Klir, G., & Yuan, B. (1995). *Fuzzy sets and fuzzy logic: theory and applications*. Prentice Hall Upper Saddle River, NJ.

Kohonen, T. (1988). An introduction to neural computing. *Neural Networks*, *1*(1), 3–16. doi:10.1016/0893-6080(88)90020-2

Kona, S., Bansal, A., & Gupta, G. (2007). Automatic composition of semantic web services. In *Proceedings of International Conference on Web Services (ICWS)* (pp. 150-158).

Konig, D., Lohmann, N., Moser, S., Stahl, C., & Wolf, K. (2008). Extending the compatibility notion for abstract WS-BPEL processes. In *Proceedings of International Conference on World Wide Web (WWW)* (pp. 785-794).

Kopecký, J., Vitvar, T., Bournez, C., & Farrell, J. (2007). SAWSDL: Semantic Annotations for WSDL and XML Schema. *IEEE Internet Computing*, *11*(6). doi:10.1109/MIC.2007.134

Kortuem, G., & Kawsar, F. (2010). Smart Objects as Building Blocks for the Internet of Things. *IEEE Internet Computing*, *1/2*, 44–51. doi:10.1109/MIC.2009.143

Kouadri Mostéfaoui, G., Maamar, Z., & Narendra, N. C. (2009). Managing Web Service Quality: Measuring Outcomes and Effectiveness. In Khan, K. (Ed.), *Aspect-oriented Framework for Web Services (AoF4WS): Introduction and Two Example Case Studies*. Hershey, PA: IGI Global Publishing.

Kraetzschmar, G. (1997). *Distributed Reason Maintenance for Multiagent Systems*. Unpublished doctoral dissertation, School of Engineering, University of Erlangen-Nürnberg, Erlangen, Germany.

Krampf, R., Ueltschy, L., & d'Amico, M. (2003). The contribution of emotion to consumer satisfaction in the service setting. *Marketing Management Journal*, *13*(1), 32–52.

Krause, A., Smailagic, A., & Siewiorek, D. P. (2006). Context-Aware Mobile Computing: Learning Context-Dependent Personal Preferences from A Wearable Sensor Array. *IEEE Transactions on Mobile Computing*, *5*(2), 113–127. doi:10.1109/TMC.2006.18

Krukow, K., & Nielsen, M. (2003). *Pervasive Computing*, 52-61.

Kuan, H. H., Bock, G. W., & Vathanophas, V. (2008). Comparing the effects of website quality on customer initial purchase and continued purchase at e-commerce websites. *Behaviour & Information Technology*, *27*(1), 3–16. doi:10.1080/01449290600801959

Kumar, V. (1998). Algorithms for Constraint-Satisfaction Problems: A Survey. *AI Magazine*, *13*(1), 32–44.

Küng, J., Luckeneder, T., Steiner, K., Wagner, R. R., & Woss, W. (2001). Persistent topic maps for knowledge and Web content management. In *Proceedings of the 2nd International Conference on Web Information Systems Engineering*, Kyoto, Japan (Vol. 2, pp. 151-158).

Kurz, B., Popescu, I., & Gallacher, S. (2004). FACADE - A Framework for Context-Aware Content Adaptation and Delivery. *Second Annual Conference on Communication Networks and Services Research*, IEEE CS Press, 46-55.

Kushwaha, M., Amundson, I., Koutsoukos, X., Neema, S., & Sztipanovits, J. (2007, January). Oasis: A programming framework for service-oriented sensor networks. In *Proceedings of the Communication systems software and middleware the 2nd international conference (comsware 2007)* (pp. 1-8).

Kuter, U., & Golbeck, J. (2007). SUNNY: A New Algorithm for Trust Inference in Social Networks, using Probabilistic Confidence Models. In *Proceedings of the Twenty-Second National Conference on Artificial Intelligence (AAAI-07)*, Vancouver, BC.

Kwee, A. T., Tsai, F. S., & Tang, W. (2009). *Sentence-level Novelty Detection in English and Malay* (LNCS 5476, pp. 40-51).

Kwee, A. T., & Tsai, F. S. (2009). Mobile Novelty Mining. *International Journal of Advanced Pervasive and Ubiquitous Computing*, *1*(4), 43–68.

Kwok, K. H. S., & Chiu, D. K. W. (2004). An Integrated Web Services Architecture for Financial Content Management. In *Proceedings of the 37th Hawaii International Conference on System Sciences*, Big Island, HI. Washington, DC: IEEE Computer Society Press.

Kwok, S., Cheung, S., Wong, K., Tsang, K., Lui, S., & Tam, K. (2003). Integration of digital rights management into the Internet Open Trading Protocol. *Decision Support Systems*, *34*(4), 413–425. doi:10.1016/S0167-9236(02)00067-2

Laing, A., & McKee, L. (2001). Willing volunteers or unwilling conscripts? Professionals and marketing in service organisations. *Journal of Marketing Management*, *17*, 559–575. doi:10.1362/026725701323366935

Lamparter, S., Ankolekar, A., Studer, R., & Grimm, S. (2007). Preference-based selection of highly configurable web services. In *Proceedings of the 16th international Conference on World Wide Web (WWW '07)*, Banff, Canada (pp. 1013-1022). New York: ACM.

Langdon, C. S. (2003, July). The State of Web Services. *IEEE Computer*, *36*(7).

Lau, R. Y. K., ter Hofstede, A. H. M., & Bruza, P. D. (2000, April). Adaptive Profiling Agents for Electronic Commerce. In J. Cooper (Ed.), *Proceedings of the 4th CollECTeR Conference on Electronic Commerce*, Breckenridge, CO.

Lau, R. Y. K. (2007). Towards A Web Services and Intelligent Agents Based Negotiation System for B2B eCommerce. *Electronic Commerce Research and Applications*, *6*(3), 260–273. doi:10.1016/j.elerap.2006.06.007

Lau, R. Y. K., Tang, M., Wong, H., Milliner, S., & Chen, Y. (2006). An Evolutionary Learning Approach for Adaptive Negotiation Agents. *International Journal of Intelligent Systems*, *21*(1), 41–72. doi:10.1002/int.20120

Lau, R. Y. K., Wong, O., Li, Y., & Ma, L. C. K. (2008). Mining Trading Partners' Preferences for Efficient Multi-Issue Bargaining in e-Business. *Journal of Management Information Systems*, *25*(1), 81–106. doi:10.2753/MIS0742-1222250104

Law, L., Menezes, A., Qu, M., Solinas, J., & Vanstone, S. (2003). An efficient protocol for authenticated key agreement. *Designs, Codes and Cryptography*, *28*(2), 119–134. doi:10.1023/A:1022595222606

Leacock, C., & Chodorow, M. (1998). Lch: Combining local context and WordNet similarity for word sense identification. In Fellbaum, C. (Ed.), *WordNet: An Electronic Lexical Database* (pp. 265–283). Cambridge, MA: MIT Press.

Lecue, F., Delteil, A., & Leger, A. (2007). Applying abduction in semantic web services composition. In *Proceedings of IEEE International Conference on Web Services (ICWS)* (pp. 94-101).

Lee, R. C. M., Mark, K. P., & Chiu, D. K. W. (2007). Enhancing Workflow Automation in Insurance Underwriting Processes with Web Services and Alerts. In *Proceedings of the. 40th Hawaii International Conference on System Sciences*, Big Island, Hawaii, CDROM, IEEE Computer Society Press.

Lee, Y. W., Chandranmenon, G., & Miller, S. C. (2003). *GAMMA: A Content Adaptation Server for Wireless Multimedia Applications*. Bell-Labs, Technical Report.

Legout, A., Urvoy-Keller, G., & Michiardi, P. (2006). Rare first and choke algorithm are enough. In *Proceedings of the IMC'06*, Rio de Jameiro, Brazil.

Leguay, J., Lopez-Ramos, M., Jean-Marie, K., & Conan, V. (2008, October). An efficient service oriented architecture for heterogeneous and dynamic wireless sensor networks. In Proceedings of the *33rd IEEE Conference on Local Computer Networks (LCN 2008)* (pp. 740-747).

Lei, Z., & Georganas, N. D. (2001). Context-based Media Adaptation in Pervasive Computing. *Canadian Conference on Electrical and Computer Engineering*, May

Leino-Kilpi, H., Valimaki, M., Dassen, T., Gasull, M., Lemonidou, C., Scott, A., & Arndt, M. (2001). Privacy: A review of the literature. *International Journal of Nursing Studies*, *38*(6), 663–671. doi:10.1016/S0020-7489(00)00111-5

Lemlouma, T., & Layaida, N. (2001). The Negotiation of Multimedia Content Services in Heterogeneous Environments. *The 8th International Conference on Multimedia Modeling (MMM)*, Amsterdam, The Netherlands, Nov. 5-7, 187-206.

Lemlouma, T., & Layaida, N. (2004). Context-Aware Adaptation for Mobile Devices. *2004 IEEE International Conference on Mobile Data Management*, IEEE CS Press, 106-111.

Levary, R., & Mathieu, R. G. (2000). Hybrid retail: integrating e-commerce and physical stores. *Industrial Management (Des Plaines)*, *42*(5), 6–13.

Li, F., Yang, F., Shuang, K., & Su, S. (2007). Q-Peer: A decentralized QoS registry architecture for web services. In *Proceedings of International Conference on Service Oriented Computing (ICSOC)* (pp. 145-156).

Li, L., Liu, C., & Wang, J. (2007). Deriving transactional properties of composite web services. In *Proceedings of IEEE International Conference on Web Services (ICWS)* (pp. 631-638).

Li, L., Sun, L., Ma, J., & Chen, C. (2008, June). A receiver-based opportunistic forwarding protocol for mobile sensor networks. In *the 28th International Conference on Distributed Computing Systems Workshops.*

Li, L., Wei, J., & Huang, T. (2007). High performance approach for Multi-QoS constrained web services selection. In *Proceedings of International Conference on Service Oriented Computing (ICSOC)* (pp. 283-294).

Liang, H., Tsai, F. S., & Kwee, A. T. (2009). Detecting Novel Business Blogs. In *Proceedings of the Seventh International Conference on Information, Communications, and Signal Processing (ICICS).*

Liang, W.-Y., Huang, C.-C., & Chuang, H.-F. (2007). The design with object (DwO) approach to Web services composition. *Computer Standards & Interfaces*, *29*(1), 54–68. doi:10.1016/j.csi.2005.11.001

Liao, C., Chen, J.-L., & Yen, D. (2007). Theory of planning behavior (TPB) and customer satisfaction in the continued use of e-service: An integrated model. *Computers in Human Behavior*, *6*(23), 2804–2822. doi:10.1016/j.chb.2006.05.006

Lin, B., Li, Q., & Gu, N. (2007). A semantic specification framework for analyzing functional composability of autonomous web services. In *Proceedings of IEEE International Conference on Web Services (ICWS)* (pp. 695-702).

Lin, F.-R., Tan, G. W., & Shaw, M. J. (1998). Modeling Supply-chain Networks by a Multi-agent System. In *Proceedings of the 31st Hawaii International Conference on System Sciences* (HICSS31), *5*, 105-114.

Lin, Y.-B., & Chlamtac, I. (2000). *Wireless & Mobile Network Architectures*. John Wiley & Sons.

Lindblad-Gidlund, K. (2008). Driver or Passenger? An analysis of Citizen-Driven eGovernment. In M.A. Wimmer & E. Fererro (Eds.), *Proceedings of EGOV 2008* (pp.267-278). Berlin: Springer.

Lin, F.-R., & Pai, Y.-H. (2000). Using Multi-agent Simulation & Learning to Design New Business Processes. *IEEE Transactions on Systems, Man and Cybernetics. Part A*, *30*(3), 380–384.

Lin, K.-J., Panahi, M., Zhang, Y., Zhang, J., & Chang, S.-H. (2009). Building accountability middleware to support dependable SOA. *IEEE Internet Computing*, *13*(2), 16–25. doi:10.1109/MIC.2009.28

Linthicum, D. S. (2003). *Next Generation Application Integration: From Simple Information to Web Services*. Reading, MA: Addison-Wesley.

Lin, W. B. (2007). The exploration of customer satisfaction model from a comprehensive perspectives. *Expert Systems with Applications*, *33*, 110–121. doi:10.1016/j.eswa.2006.04.021

Lin, Y.-B., & Chlamtac, I. (2000). *Wireless and Mobile Network Architectures*. New York: John Wiley & Sons.

Litke, A., Konstanteli, K., Andronikou, V., Chatzis, S., & Varvarigou, T. (2008). Managing service level agreement contracts in OGSA-based Grids. *Future Generation Computer Systems*, *24*(4), 245–258. doi:.doi:10.1016/j.future.2007.06.004

Liu, A., Li, Q., Huang, L., & Liu, H. (2008). Building profit-aware service-oriented business applications. In *Proceedings of IEEE International Conference on Web Services (ICWS)* (pp. 489-496).

Liu, H., Li, Q., Gu, N., & Liu, A. (2008, April 21-25). A logical framework for modeling and reasoning about semantic web services contract. In J. Huai, R. Chen, H.-W. Hon, Y. Liu, W.-Y. Ma, A. Tomkins, & X. Zhang (Eds.), *Proceedings of the 17th International Conference on World Wide Web (WWW 2008)*, Beijing, China (pp. 1057-1058). New York: ACM.

Liu, L., Song, H., & Liu, Y. (2001). HDBIS Supporting E-collaboration in E-business. In *Proceedings of the 6th International Conference on Computer Supported Cooperative Work in Design* (pp. 157-160).

Liu, W. (2005). Trustworthy service selection and composition - reducing the entropy of service-oriented Web. In *Proceedings of the 3rd IEEE International Conference on Industrial Informatics* (pp. 104-109). Washington, DC: IEEE Computer Society.

Liu, Y. H. A., Ngu, H., & Zeng, L. (2004). QoS computation and polcing in dynamic web service selection. In *Proceedings of International Conference on World Wide Web (WWW)* (pp. 66-73).

Liu, A., Liu, H., Lin, B., Huang, L., Gu, N., & Li, Q. (2010). A Survey of Web Services Provision. *International Journal of Systems and Service-Oriented Engineering*, *1*(1), 26–45.

Lo, G., & Kersten, G. K. (1999). Negotiation in Electronic Commerce: Integrating Negotiation Support & Software Agent Technologies. In *Proceedings of the 29th Atlantic Schools of Business Conference*.

Loermans, J. (2002). Synergizing the learning organization. *Journal of Knowledge Management*, *3*(6), 285–294. doi:10.1108/13673270210434386

Loonam, M., & O'Loughlin, D. (2008). Exploring e-service quality: a study of Irish online banking. *Marketing Intelligence & Planning*, *26*(7), 759–780. doi:10.1108/02634500810916708

Lorraine, S., Lee, K., Fiedler, D., & Smith, J. S. (2008). Radio frequency identification (RFID) implementation in the service sector: A customer - facing diffusion model. *International Journal of Production Economics*, *112*(2), 587–600. doi:10.1016/j.ijpe.2007.05.008

Luce, R. D. (1959). *Individual choice behaviour: a theoretical analysis*. New Yokr: Wiley.

Ludwig, H., Dan, A., & Kearney, R. (2004). Cremona: An architecture and library for creation and monitoring of WS-Agreements. In B.J. Kraemer, K.-J. Lin, & P. Narasimhan (Eds.), *Proceedings of ICSOC 2007* (pp. 65-74). New York, USA: ACM.

Ludwig, H., Gimpel, H., & Kearney, R. D. (2005). Template-Based Automated Service Provisioning - Supporting the Agreement-Driven Service Life-Cycle. In *Proceedings of the International Conf. Service Oriented Computing (ICSOC'05)* (pp. 283-295).

Lum, W. Y., & Lau, F. C. M. (2002). A Context-Aware Decision Engine for Content Adaptation. *IEEE Pervasive Computing / IEEE Computer Society and IEEE Communications Society*, *1*(3), 41–49. doi:10.1109/MPRV.2002.1037721

Luo, J., Montrose, B., Kim, A., Khashnobish, A., & Kang, M. (2006). Adding OWL-S Support to the Existing UDDI Infrastructure. *Web Services, IEEE International Conference on*, 153-162.

Luo, Z., et al. (2010). A Coordinated P2P Message Delivery Mechanism in Local Association Networks for the Internet of Things. In *Proceedings of the IEEE Intl Conference on e-Business Engineering*, Shanghai, China.

Luo, Z., Lai, M., et al. (2010). Adopting RFID in Body Tagging: a Local Association Network Approach. In *Proceedings of the IEEE Intl Conference on RFID TA*, Guangzhou, China.

Luo, Y., Liu, K., & Davis, D. N. (2002). A Multi-agent Decision Support System for Stock Trading. *IEEE Network*, *16*(1), 20–27. doi:10.1109/65.980541

Luo, Z. (2008). Value Analysis Framework for Technology Adoption with Case Study on China Retailers. *Communications of the Association for Information Systems*, *23*, 295–318.

Lupu, E., & Sloman, M. (1999). Conflicts in policy-based distributed systems management. *IEEE Transactions on Software Engineering*, *25*(6), 852–869. doi:10.1109/32.824414

Lyytinen, K., & Yoo, Y. (2002). Research Commentary: The Next Wave of Nomadic Computing. *Information Systems Research*, *13*(4), 377–388. doi:10.1287/isre.13.4.377.75

Maamar, Z., Kouadri Mostéfaoui, S., & Yahyaoui, H. (2005). Towards an Agent-based and Context-oriented Approach for Web Services Composition. *IEEE Transactions on Knowledge and Data Engineering*, *17*(5). doi:10.1109/TKDE.2005.82

MacInnis, D. J., & Jaworski, B. J. (1989). Information processing from advertisements: Toward and integrative framework. *Journal of Marketing*, *53*, 1–23. doi:10.2307/1251376

Maes, P., Guttman, R., & Moukas, A. (1999). Agents that buy and sell. *Communications of the ACM*, *42*(3), 81–91. doi:10.1145/295685.295716

Mallalieu, L. (2006). Consumer perception of salesperson influence strategies: an examination of the influence of consumer goals. *Journal of Consumer Behaviour*, *5*, 257–268. doi:10.1002/cb.177

Mamat, A., Lu, Y., Deogun, J., & Goddard, S. (2008, July). Real-time divisible load scheduling with advance reservation. In *Euromicro conference on real-time systems, ecrts '08* (pp. 37-46).

March, J. G., & Simon, H. A. (1958). *Organizations*. New York: John Wiley & Sons Inc.

Margaritis, D., & Thrun, S. (2001). A Bayesian Multiresolution independence test for continuous variables. In *Proceedings Seventeenth Conference on Uncertainty in Artificial Intelligence*.

Marsh, S. P. (1994). *Formalising Trust as a Computational Concept*. Unpublished doctoral dissertation, Stirling University, UK.

Martens, A. (2005). Process oriented discovery of business partners. In *Proceedings of International Conference on Enterprise Information Systems (ICEIS)* (pp. 57-64).

Martens, A., Moser, S., Gerhardt, A., & Funk, K. (2006). Analyzing compatibility of BPEL processes. In *Proceedings of Advanced International Conference on Telecommunications and International Conference on Internet and Web Applications and Services (AICT/ICIW)*.

Martens, A. (2003). On compatibility of web services. *Petri Net Newsletter*, *65*, 12–20.

Martin, R. (2007, May). Data latency playing an ever increasing role in effective trading. *InformationWeek*.

Massad, N., & Heckman, R., & Crowston. (2006). Customer satisfaction with electronic service encounters. *International Journal of Electronic Commerce*, *10*(4), 73–104. doi:10.2753/JEC1086-4415100403

Massy, W. F., Montgomery, D., & Morrison, D. G. (1970). *Stochastic Models of Buyer Behavior*. Cambridge, MA: MIT Press.

Matsumoto, I. T., Stapleton, J., Glass, J., & Thorpe, T. (2005). A knowledge capture report for multidisciplinary design environments. *Journal of Knowledge Management*, *3*(9), 83–92. doi:10.1108/13673270510602782

Matthews, B., Bicagregui, C., & Dimitrakos, T. (2003). *Building Trust on the Grid: Trust Issues Underpinning Scalable Virtual Organizations*. Retrieved from http://epubs.cclrc.ac.uk/bitstream/643/trustedgridERCIM.pdf

Maurer, U. (1996). Modelling a public-key infrastructure. *Lecture Notes in Computer Science*, 325–350.

Maximilien, E. M., & Singh, M. P. (2004). A framework and ontology for dynamic web services selection. *IEEE Internet Computing*, *8*(5), 84–93. doi:10.1109/MIC.2004.27

Ma, Y., & Zhang, C. (2008). Quick convergence of genetic algorithm for QoS-Driven web service selection. *Computer Networks*, *52*(5), 1093–1104. doi:10.1016/j.comnet.2007.12.003

Mayer, R. C., Davis, J. H., & Schoorman, F. D. (1995). An Integrative model of organisational trust. *Academy of Management Review*, *20*(3), 709–734. doi:10.2307/258792

Mc Guiness, D., & van Harmelen, F. (2004). *OWL Web Ontology Language Overview, W3C Recommendation*. Retrieved April 05, 2010, from http://www.w3.org/TR/owl-features/

McIlraith, S. A., Son, T. C., & Zeng, H. (2001). Semantic web services. *IEEE Intelligent Systems*, *16*(2), 46–53. doi:10.1109/5254.920599

McKnight, D. H., Choudhury, V., & Kacmar, C. (2002). Developing and validating trust measures for e-commerce: an integrative typology. *Information Systems Research*, *13*(3), 334–359. doi:10.1287/isre.13.3.334.81

McNay, H. E. (2002). Enterprise Content Management: an Overview. In *Proceedings of the IEEE International Professional Communication Conference* (pp. 396-402).

Melin, U., Axelsson, K., & Lundsten, M. (2008). Talking to, not about, Entrepreneurs – Experiences of Public e-service Development in a Business Start Up Case. In Cunningham, P., & Cunningham, M. (Eds.), *eChallenges*. Amsterdam: IOS Press.

Melnik, S., Garcia-Molina, H., & Rahm, E. (2002). Similarity Flooding: A Versatile Graph Matching Algorithm and Its Application to Schema Matching. In *Proceedings of the 18th international Conference on Data Engineering (ICDE)* (p. 117). Washington, DC: IEEE Computer Society.

Menasce, D. A. (2002). QoS issues in web services. *IEEE Internet Computing*, *6*(6), 72–75. doi:10.1109/MIC.2002.1067740

Microsoft. "*Microsoft Reader*." from http://www.microsoft.com.

Mikic-Rakic, M., Malek, S., & Medvidovic, N. (2005). Improving availability in large, distributed component-based systems via redeployment. In *Proceedings of International Working Conference on Component Deployment (CD)* (pp. 83-98).

Milanovic, N., & Malek, M. (2004). Current solutions for Web service composition. *IEEE Internet Computing*, *8*(6), 51–59. doi:10.1109/MIC.2004.58

Milward, H. B., & Provan, K. G. (2003, October 9-11). *Managing Networks Effectively*. Paper presented at the the 7th National Public Management Research Conference, Georgetown University, Washington, DC.

Mitchell, A. A. (1981). The dimensions of advertising involvement. *Advances in Consumer Research. Association for Consumer Research (U. S.)*, *8*, 25–30.

Mocan, A., & Cimpian, E. *WSMX data mediation*. Retrieved May 31, 2009, from http://www.wsmo.org/TR/d13/d13.3/v0.2/20051011/d13.3v0.2_20051011.pdf.

Mohan, R., Smith, J. R., & Li, C. S. (1999). Adapting Multimedia Internet Content for Universal Access. *IEEE Transactions on Multimedia*, *1*(1), 104–114. doi:10.1109/6046.748175

Montagnat, J., Glatard, T., Plasencia, I. C., Castejón, F., Pennec, X., & Taffoni, G. (2008). Workflow-Based Data Parallel Applications on the EGEE Production Grid Infrastructure. *Journal of Grid Computing*, *6*(4), 369–383. doi:10.1007/s10723-008-9108-x

Moraes, R. L. O., Durães, J., Barbosa, R., Martins, E., & Madeira, H. (2007). Experimental Risk Assessment and Comparison Using Software Fault Injection. In *Proceedings of the 37th IEEE/IFIP International Conference on Dependable Systems and Networks* (pp. 512-521). Washington, DC: IEEE Computer Society.

Moses, S. A., Gruenwald, L., & Dadachanji, K. (2008). A scalable data structure for real-time estimation of resource availability in build-to-order environments. *Journal of Intelligent Manufacturing, 19*(5), 611-622. Object Management Group. (2003, November). *Realtime-corba specification, v 2.0.*

Mostefaoui, S. K., & Hirsbrunner, B. (2004). Context Aware Service Provisioning. *the IEEE International Conference on Pervasive Services (ICPS)*, Beirut, Lebanon, Jul. 19-23, 71-80.

Mostefaoui, S. K., Tafat-Bouzid, A., & Hirsbrunner, B. (2003). Using Context Information for Service Discovery and Composition. *5th International Conference on Information Integration and Web-based Applications and Services (iiWAS)*, Jakarta. *Indonesia*, (Sep): 15–17, 129–138.

Mostéfaoui, S. K., & Younas, M. (2007). Context-oriented and transaction-based service provisioning. *International Journal of Web and Grid Services, 3*(2), 194–218. doi:10.1504/IJWGS.2007.014074

Motomura, Y. (2001). BAYONET: Bayesian Network on Neural Network. *Foundations of Real-World Intelligence*, 28-37.

Motomura, Y., & Kanade, T. (2005). Probabilistic human modeling based on personal construct theory. *Journal of Robot and Mechatronics, 17*(6), 689–696.

Mukherjee, D., Jalote, P., & Nanda, M. G. (2008).Determining QoS of WS-BPEL compositions. In *Proceedings of International Conference on Service Oriented Computing(ICSOC)* (pp. 378-393).

Mukherjee, D., Delfosse, E., Kim, J.-G., & Wang, Y. (2005). Optimal Adaptation Decision-Taking for Terminal and Network Quality-of-Service. *IEEE Transactions on Multimedia, 7*(3), 454–462. doi:10.1109/TMM.2005.846798

Muller, K. R., Kika, S., Ratsch, G., Tsuda, K., & Scholkopf, B. (2001). An introduction to kernel-based learning algorithms. *IEEE Transactions on Neural Networks, 12*(2), 181–202. doi:10.1109/72.914517

Murphy, K., Weiss, Y., & Jordan, M. I. (1999, July 30-August 1). *Loopy belief propagation for approximate inference: An empirical study.* Paper presented at Uncertainty in Artificial Intelligence, London.

Myers, B. L., Kappelman, L. A., & Prybutok, V. R. (1997). A comprehensive model for assessing the quality and productivity of the information systems function: toward a theory for information systems assessment. *Information Resources Management Journal, 10*(1), 6–25.

Nagarajan, M., Verma, K., Sheth, A., & Miller, J. (2007). Ontology driven data mediation in web services. *International Journal of Web Services Research, 4*(4), 104–126.

Nahapiet, J., & Ghoshal, S. (1998). Social capital, intellectual capital and organizational advantage. *Academy of Management Review, 23*(2), 242–266. doi:10.2307/259373

Naiburg, E. J., & Maksimchuk, R. A. (2001) *UML for Database Design*. Addison-Wesley.

Nam, J., Ro, Y. M., Huh, Y., & Kim, M. (2005). Visual Content Adaptation According to User Perception Characteristics. *IEEE Transactions on Multimedia, 7*(3), 435–445. doi:10.1109/TMM.2005.846801

Naoe, K., & Takefuji, Y. (2008, September). Damageless Information Hiding using Neural Network on YCbCr Domain. *International Journal of Computer Sciences and Network Security, 8*(9), 81–86.

Natella, R., & Cotroneo, D. (2010). Emulation ofTransient Software Faults for Dependability Assessment: A Case Study. In *Proceedings of the eight European Dependable Computing Conference*. Washington, DC: IEEE Computer Society.

Nezhad, H., Benatallah, B., Martens, A., Curbera, F., & Casati, F. (2007). Semi-automated adaptation of service interactions. In *Proceedings of International Conference on World Wide Web (WWW)* (pp. 993-1002).

Ng, K. W., Tsai, F. S., & Goh, K. C. (2007). Novelty Detection for Text Documents Using Named Entity Recognition. In *Proceedings of the 2007 Sixth International Conference on Information, Communications and Signal Processing* (pp. 1-5).

Nichols, K., & Carpenter, B. (2000). *Definition of Differentiated Services Behavior Aggregates and Rules for their Specification* (draft-ietf-diffserv-ba-def-00.txt).

Nikander, P., & Karvonen, K. (2000). Users and Trust in Cyberspace. In *Proceedings of the Cambridge Security Protocols Workshop 2000.*

Nonaka, I., & Konno, N. (1998). The Concept of "Ba": Building a Foundation for Knowledge Creation. *California Management Review*, *40*(3), 40–54.

Nonaka, I., & Takeuchi, H. (1995). *The Knowledge-Creating Company*. New York: Oxford University Press.

Nonaka, I., Toyama, R., & Konno, N. (2000). SECI, Ba and Leadership: A unified model of dynamic knowledge creation. *Long Range Planning*, *33*(1), 5–34. doi:10.1016/S0024-6301(99)00115-6

Noone, B. M., Kimes, S. E., Mattila, A. S., & Wirtz, J. (2009). Perceived service encounter pace and customer satisfaction. *Journal of Service Management*, *20*(4), 380–403. doi:10.1108/09564230910978494

O'Reilly, T. (2005). What is Web 2.0? Design Patterns and Business Models for the Next Generation of Software. *O'Reilly Media, Inc*. Retrieved June 15, 2008, from http://oreilly.com/pub/a/oreilly/tim/news/2005/09/30/what-is-web-20.html

O'Sullivan, J., Edmond, D., & Hofstede, A. T. (2002). What's in a service? Towards accurate description of non-functional service properties. *Distributed and Parallel Databases*, *12*, 117–133. doi:10.1023/A:1016547000822

OASIS. (2005). *Web Services Base Notification 1.3 (WS-BaseNotification)*. Retrieved from http://docs.oasis-open.org/wsn/wsn-ws_base_notification-1.3 -spec-os.pdf

Oh, N., Shirvani, P. P., & McCluskey, E. J. (2002). Control-flow checking by software signatures. *IEEE Transactions on Reliability*, *51*(2), 111–122. doi:10.1109/24.994926

Oldham, N., Verma, K., Sheth, A., & Hakimpour, F. (2006). Semantic WS-Agreement Partner Selection. In *Proceedings of the 15th international conference on world wide web conference,* Edinburgh, Scotland.

Olmedilla, D., Rana, O., Matthews, B., & Nejdi, W. (2006). Security and Trust Issues in Semantic Grids. In *Proceedings Semantic Grid: The Convergence of Technologies*. Retrieved from http://drops.dagstuhl.de/popus/volltexte/2006/408

Olmedo, V., Villagrá, V., Konstanteli, K., Burgos, J., & Berrocal, J. (2009). Network mobility support for Web Service-based Grids through the Session Initiation Protocol. *Future Generation Computer Systems*, *25*(7), 758–767. doi:.doi:10.1016/j.future.2008.11.007

Omelayenko, B. (2001) Preliminary Ontology Modeling for B2B Content Integration. In *Proceedings of the 12th International Workshop on Database and Expert Systems Applications*, Munich, Germany (pp. 7-13).

Ong, C. L., Kwee, A. T., & Tsai, F. S. (2009). Database Optimization for Novelty Detection. In *Proceedings of the Seventh International Conference on Information, Communications, and Signal Processing (ICICS).*

Ono, C., Kurokawa, M., Motomura, Y., & Aso, H. (2007). A context-aware movie preference model using a bayesian network for recommendation and promotion. In *Proceedings of User modeling 2007, 4511*, 257-266.

Opitz, A., König, H., & Szamlewska, S. (2008). What Does Grid Computing Cost? *Journal of Grid Computing*, *6*(4), 385–397. doi:10.1007/s10723-008-9098-8

Ortiz, G., Hernández, J., & Clemente, P. J. (2006). Web Services Orchestration and Interaction Patterns: An Aspect-Oriented Approach. In *Proceedings of the international conference on service oriented computing,* New York.

Ould Ahmed M'Bareck, N., Tata, S., & Maamar, Z. (2007). Towards An Approach for Enhancing Web Services Discovery. In *Proceedings of the 16th IEEE international workshops on enabling technologies: Infrastructure for collaborative enterprises,* Paris.

Panahi, M., Lin, K.-J., Zhang, Y., Chang, S.-H., Zhang, J., & Varela, L. (2008). The llama middleware support for accountable service-oriented architecture. *In Icsoc '08: Proceedings of the 6th international conference on service-oriented computing* (pp. 180–194). Berlin, Heidelberg: Springer-Verlag.

Pan, L., & Scarbrough, H. (1999). Knowledge Management in Practice: an Exploratory case study. *Technology Analysis and Strategic Management*, *11*(3), 370–387.

Paolucci, M., Kawamura, T., Payne, T. R., & Sycara, K. P. (2002). Importing the semantic web in UDDI. In *Proceedings of International Workshop on Web Services, E-Business, and the Semantic Web (WES)* (pp. 225-236).

Paolucci, M., Kawamura, T., Payne, T. R., & Sycara, K. P. (2002). Semantic matching of web services capabilities. In *Proceedings of International Semantic Web Conference (ISWC)* (pp. 333-347).

Papazoglou, M. P., Traverso, P., Dustdar, S., & Leymann, F. (2007). Service-Oriented Computing: State of the Art and Research Challenges. *IEEE Computer, 40*(11).

Papazoglou, M. P., & Georgakopulos, D. (2003). Service-oriented computing. *Communications of the ACM, 46*(10), 25–28. doi:10.1145/944217.944233

Papazoglou, M. P., & Van den Heuvel, W. J. (2005). Web services management: a survey. *IEEE Internet Computing, 9*, 58–64. doi:10.1109/MIC.2005.137

Parasuraman, A., Zeithaml, V. A., & Berry, L. L. (1985). A conceptual model of service quality and its implications for future research. *Journal of Marketing, 49*, 41–50. doi:10.2307/1251430

Park, S., Ko, Y., & Jai-Hoon. (2003). Disconnected Operation Service in Mobile Grid Computing. In *Proceedings of ICSOC 2003.* Retrieved from www.unitn.it/convegni/download/icsoc03/papers/Jai.pdf

Parkin, M., Kuo, D., & Brooke, J. (2006). A framework & negotiation protocol for service contracts. In Proceedings of the IEEE SCC (pp. 253-256). Washington, DC: IEEE Computer Society.

Parnas, D. L. (1972). On the criteria to be used in decomposing systems into modules. *15*(12), 1053-1058.

Pashtan, A., Kollipara, S., & Pearce, M. (2003). Adapting Content for Wireless Web Service. *IEEE Internet Computing, 7*(5), 79–85. doi:10.1109/MIC.2003.1232522

Patwardhan, S. (2003). *Vector: Incorporating dictionary and corpus information into a context vector measure of semantic relatedness*. Unpublished master's thesis, University of Minnesota, Duluth, MN.

Pautasso, C., Heinis, T., & Alonso, G. (2007). Autonomic execution of Web service compositions. In C.K. Chang & L.-J. Zhang (Eds.), *Proceedings of ICWS 2005* (pp. 435-442). Los Alamitos, USA: IEEE-CS Press.

Pearl, J. (2000). *Causality: Models, Reasoning, and Inference*. Cambridge, UK: Cambridge University Press.

Pedersen, T., Patwardhan, S., & Michelizzi, J. (2004) WordNet:Similarity - Measuring the Relatedness of Concepts. In *Proceedings of Fifth Annual Meeting of the North American Chapter of the Association for Computational Linguistics* (pp. 38-41). Boston: Association for Computational Linguistics.

Peer, H., Changtao, Q., Bastian, K., Lutz, S., & Philipp, W. (2006, October). *Towards Autonomous Brokered SLA Negotiation Exploiting the Knowledge Economy: Issues, Applications, Case Studies (eChallenges 2006)* (pp. 44-51). Barcelona, Spain: IOS Press.

Peng, P., et al. (2008). A P2P Based Collaborative RFID Data Cleaning Model. In *Proceedings of the 2008 International Symposium on Advances in Grid and Pervasive Systems (AGPS 2008).*

Peterson, R. A., & Wilson, W. R. (1992). Measuring customer satisfaction: Fact and artifact. *Journal of the Academy of Marketing Science, 20*(1), 61–71. doi:10.1007/BF02723476

Petty, R. E., & Cacioppo, J. T. (1986). *Communication and persuasion: central and peripheral routes to attitude change*. Berlin: Springer Verlag.

Phan, T., Zorpas, G., & Bagrodia, R. (2002). An Extensible and Scalable Content Adaptation Pipeline Architecture to Support Heterogeneous Clients. *the 22nd International Conference on Distributed Computing Systems*, IEEE CS Press, 507-516.

Picco, G., & Mottola, L. (in press). Programming wireless sensor networks: Fundamental concepts and state-of-the-art. *Comp. Surv. Proceedings of the 3rd Int. Conf. on Pervasive Computing Technologies for Healthcare.* (2009, April). London. Washington, DC: IEEE.

Ponnekanti, S. R., & Fox, A. (2002). SWORD: A developer toolkit for web service composition. In *Proceedings of International Conference on World Wide Web on Alternate Track Papers (WWW).*

Powers, C. S., Ashley, P., & Schunter, M. (2002) Privacy promises, access control, and privacy management - Enforcing privacy throughout an enterprise by extending access control. In *Proceedings of the 3rd International Symposium on Electronic Commerce* (pp. 13- 21).

Pradhan, D. K. (1996). *Fault-Tolerant Computer System Design*. Upper Saddle River, NJ: Prentice Hall, Inc.

Prasithsangaree, P., & Krishnamurthy, P. (2003). Analysis of energy consumption of RC4 and AES algorithms in wireless LANs. In *Proceeding of IEEE Global Telecommunications Conference* (pp. 1445-1449).

Preist, C. (2004). A Conceptual Architecture for Semantic Web Services. In F. Van Harmelen, S. A. McIlraith, & D. Plexousakis (Eds.), *Proceedings of ISWC 2004*. Berlin: Springer Verlag.

Pujolie, G. (2006). An autonomic-oriented architecture for the Internet of Things. In *Proceedings of the IEEE 2006 International Symposium on Modern Computing* (pp. 163-168).

Putnam, R. D. (2000). *Bowling Alone: The Collapse and revival of American Community*. New York: Touchstone Press.

Ramaswamy, L., Iyengar, A., Liu, L., & Douglis, F. (2005). Automatic Fragment Detection in Dynamic Web Pages and Its Impact on Caching. *IEEE Transactions on Knowledge and Data Engineering*, *17*(6), 859–874. doi:10.1109/TKDE.2005.89

Randell, B. (1987). Design-Fault Tolerance. In Avizienis, A., Kopertz, H., & Laprie, J.-C. (Eds.), *The Evolution of Fault-Tolerant Computing* (pp. 251–270). Vienna, Italy: Springer Verlag.

Ran, S. (2003). A model for web services discovery with QoS. *ACM SIGecom Exchanges*, *4*(1), 1–10. doi:10.1145/844357.844360

Rao, A. S., & Georgeff, M. P. (1995). BDI Agents: from theory to practice. In *Proceedings 1st International Conference on Multiagent Systems* (pp. 312-319).

Ratnasingam, P. (2002). The Importance of Technology Trust in Web Services Security. *Information Management & Computer Security*, *10*(5), 255–260. doi:10.1108/09685220210447514

Richard, P., Devinney, T., Yip, G., & Johnson, G. (2009). Measuring Organisational Performance: Towards Methodological Best Practice. *Journal of Management*, *35*(3), 718–804. doi:10.1177/0149206308330560

Richard, T., & Nory, E. (2005). Knowledge Management and Business Intelligence: The importance of integration. *Journal of Knowledge Management*, *9*(4), 45–55. doi:10.1108/13673270510610323

Robertson, S., & Soboroff, I. (2002). *The TREC 2002 Filtering Track Report*. Paper presented at TREC 2002 - the 11th Text Retrieval Conference.

Roman, D., & Kifer, M. (2007, September 23-27). Reasoning about the behavior of semantic web services with concurrent transaction logic. In C. Koch, J. Gehrke, M. N. Garofalakis, D. Srivastava, K. Aberer, A. Deshpande, D. Florescu, C. Y. Chan, V. Ganti, C.-C. Kanne, W. Klas, & E. J. Neuhold (Eds.), *Proceedings of the 33rd International Conference on Very Large Data Bases,* University of Vienna, Austria (pp. 627-638). New York: ACM.

Roman, M., Hess, C. K., Cerqueira, R., Ranganathan, A., Campbell, R. H., & Nahrstedt, K. (2002). Gaia: A Middleware Infrastructure to Enable Active Spaces. *IEEE Pervasive Computing / IEEE Computer Society and IEEE Communications Society*, *1*(4), 74–83. doi:10.1109/MPRV.2002.1158281

Romer, T., & Teece, D. (1998). Capturing Value from Knowledge Assets: The New Economy, Markets for Know-How, and Intangible Assets. *California Management Review*, *40*(3), 55–79.

Rong-xiao, G., Jing-bo, X., Shu-fu, D., & Kai, W. (2009). Research on Multi-domain Policy-Based SLA Management Model. In. *Proceedings of the International Conference on Networking and Digital Society*, *1*, 209–212.

Rosario, S., Benveniste, A., Haar, S., & Jard, C. (2008). Probabilistic QoS and soft contracts for transaction-based web services orchestrations. *IEEE Transactions on Service Computing*, *1*(4), 187–200. doi:10.1109/TSC.2008.17

Rosenberg, F., Enzi, C., Michlmayr, A., Platzer, C., & Dustdar, S. (2007) Integrating quality of service aspects in top-down business process development using WS-CDL and WS-BPEL. In M. Spies & M.B. Blake (Eds.), *Proceedings of EDOC 2007* (pp. 15-26). Los Alamitos, USA: IEEE-CS Press.

Rosenberg, F., Platzer, C., & Dustdar, S. (2006). Bootstrapping performance and dependability attributes of web services. In *Proceedings of International Conference on Web Services (ICWS)* (pp. 205-212).

Rosenberg, J., Schulzrinne, H., Camarillo, G., Johnston, R., Peterson, J., Sparks, R., Handley, M., & Schooler, E. (2002, June). SIP: session initiation protocol. *RFC 3261, Internet Engineering Task Force.*

Rosenberg, J., Schulzrinne, H., et al. (2002). *SIP: Session Initiation Protocol" IETF RFC 3261*. Retrieved from http://www.ietf.org/rfc/rfc3261.txt"http://www.ietf.org/rfc/rfc3261.txt

Rosenberg, J., & Remy, D. (2004). *Securing Web Services with WS-Security: Demystifying WS-Security, WS-Policy, SAML, XML Signature, and XML Encryption*. Sams.

Rosenblatt, F. (1958). The perceptron: A probabilistic model for information storage and organization in the brain. *Psychological Review*, *65*(6), 386–408. doi:10.1037/h0042519

Rosis, F., Novielli, N., Carofiglio, V., & Cavalluzzi, A. (2006). User modeling and adaptation in health promotion dialogs with an animated character. *Journal of Biomedical Informatics*, *39*(5), 514–531. doi:10.1016/j.jbi.2006.01.001

Rossi, P. E., Allenby, G., & McCulloch, R. (2005). *Bayesian statistics and marketing*. New York: John Wiley and Sons.

Rothe, J. (2002). Some facets of complexity theory and cryptography: A five-lecture tutorial. *ACM Computing Surveys*, *34*(4), 504–549. doi:10.1145/592642.592646

Rouvoy, R., Conan, D., & Seinturier, L. (2008). Software Architecture Patterns for a Context Processing Middleware Framework. *IEEE Distributed Systems Online*, *9*(6).

Rowley, J. (2006). An analysis of the e-service literature: towards a research agenda. *Internet Research: Electronic Networking Applications and Policy*, *6*(16), 879–897.

Rumbaugh, J., Jacobson, I., & Booch, G. (2004). *The Unified Modeling Language Reference Manual* (2nd ed.). Boston: The Addison-Wesley Object Technology Series.

Rumelhart, D., & McClelland, J. (1986). *Parallel distributed processing: explorations in the microstructure of ognition, vol. 1: foundations*. MIT Press Cambridge, MA, USA.

Rutkowski, A. F., Vogel, D. R., van Genuchten, M., Bemelmans, T. M. A., & Favier, M. (2002). E-collaboration: the Reality of Virtuality. *IEEE Transactions on Professional Communication*, *45*(4), 219–230. doi:10.1109/TPC.2002.805147

Sabater-Mir, J., Pinyol, I., Villatoro, D., & Cuní, G. (2007). Towards Hybrid Experiments on reputation mechanisms: BDI Agents and Humans in Electronic Institutions. In *Proceedings of the 12th Conference of the Spanish Association for Artificial Intelligence (CAEPIA-07)* (Vol. 2, pp. 299-308).

Sage, A. P., & Armstrong, J. E. (2000). *Introduction to Systems Engineering*. New York: Wiley.

Saha, G. K. (1997). EMP- Fault Tolerant Computing: A New Approach. *International Journal of Microelectronic Systems Integration*, *5*(3), 183–193.

Saha, G. K. (2006). A Single-Version Scheme of Fault Tolerant Computing. *Journal of Computer Science & Technology*, *6*(1), 22–27.

Sailhan, F., & Issarny, V. (2005). Scalable service discovery for manet. In *Proceedings of the Third IEEE International Conference on Pervasive Computing and Communications (Percom '05)* (pp. 235-244). Washington, DC: IEEE Computer Society.

Salle, M. (2004). *IT service management and IT governance: Review, comparative analysis and their impact on utility computing*. (Technical Report HPL-2004-98). Palo Alto, USA: HP Laboratories. Retrieved June 1, 2009, from http://www.hpl.hp.com/techreports/2004/HPL-2004-98.pdf

Salle, M., & Bartolini, C. (2004). Management by contract. In R. Boutaba & S.-B. Kim (Eds.), *Proceedings of NOMS 2004* (pp. 787-800). Piscataway, USA: IEEE Press.

Saltelli, A., Chan, K., & Scotte, E. M. (2008). *Sensitivity Analysis*. New York: Wiley.

Samarati, P., & Vimercati, S. (2001) *Access Control: Policies, models, mechanisms* (LNCS 2171). New York: Springer.

Sandhu, R. S., Bhamidipati, V., & Munawer, Q. (1999). The ARBAC97 model for role-based administration of roles. *ACM Transactions on Information and System Security*, *1*(2), 105–135. doi:10.1145/300830.300839

Sandip, S. (1997). Developing an Automated Distributed Meeting Scheduler. *IEEE Expert*, *12*(4), 41–45. doi:10.1109/64.608189

Sasaki, H. (Ed.). (2007). *Intellectual property protection for multimedia information technology*. IGI Global.

Satyanarayanan, M. (2004). The Many Faces of Adaptation. *IEEE Pervasive Computing / IEEE Computer Society and IEEE Communications Society*, *3*(3), 4–5. doi:10.1109/MPRV.2004.1321017

Schaupp, L. C., Belanger, F., & Fan, W. (2009). Examining the success of websites beyond e-commerce: an extension of the IS success model. *Journal of Computer Information Systems*, *49*(4), 42–52.

Schein, E. H. (1996). Kurt Lewin's Change Theory in the Field and in the Classroom: Notes Toward a Model of Managed Learning. *Reflections: The SoL Journal*, *1*(1), 59–74. doi:10.1162/152417399570287

Schilit, B. N., Adams, N. I., & Want, R. (1994). Context-Aware Computing Applications. *IEEE Workshop on Mobile Computing Systems and Applications*, Santa Cruz, CA, USA, 85-90.

Schoeman, E. D. (1984). *Philosophical dimensions of privacy: an anthology*. New York: Cambridge University Press. doi:10.1017/CBO9780511625138

Schreiber, G., Akkermans, H., Anjewierden, A., de Hoog, R., Shadbolt, N., Van de Velde, W., & Wielinga, B. (2000). *Knowledge Engineering and Management: The CommonKADS Methodology*. Cambridge, MA: MIT Press.

Sebastiani, P., Ramoni, N., & Crea, A. (2000). Profiling your customers using Bayesian networks. *ACM SIGKDD Explorations Newsletter*, *1*(2), 91–96. doi:10.1145/846183.846205

SECURE. *Secure Environments for Collaboration among Ubiquitous Roaming Entities*. (2001). Retrieved from http://www.dsg.cs.tcd.ie/dynamic/?category_id=-30

Serhani, M. A., Dssouli, R., Hafid, A., & Sahraoui, H. A. (2005). A QoS broker based architecture for efficient web services selection. In *Proceedings of International Conference on Web Services (ICWS)* (pp. 113-120).

Shakshuki, E., Ghenniwa, H., & Kamel, M. (2000). A Multi-agent System Architecture for Information Gathering. In *Proceedings of the 11th International Workshop on Database & Expert Systems Applications* (pp. 732-736), Los Alamitos, CA, USA.

Shao, L., Zhang, J., Wei, Y., Zhao, J., Xie, B., & Mei, H. (2007). Personalized QoS prediction for web services via collaborative filtering. In *Proceedings of International Conference on Web Services (ICWS)* (pp. 439-446).

Shao, L., Zhang, L., Xie, T., Zhao, J., Xie, B., & Mei, H. (2008). Dynamic availability estimation for service selection based on status identification. In *Proceedings of IEEE International Conference on Web Services (ICWS)* (pp. 645-652).

Sharma, S., & Bock, G.-W. (2005). Factor's Influencing Individual's Knowledge Seeking Behaviour in Electronic Knowledge Repository. In D. Bartmann, F. Rajola, J. Kallinikos, D. Avison, R. Winter, P. Ein-Dor, Jr. Becker, F. Bodendorf, & C. Weinhardt (Eds.), *Proceedings of the Thirteenth European Conference on Information System*, Regensburg, Germany (pp.390-403). ISBN 3-937195-09-2

Sharma, R., & Chowdhury, N. (2007). On the Use of a Diagnostic Tool for Knowledge Audit. *Journal of Knowledge Management Practice*, *8*(4).

Shehab, M., Bhattacharya, K., & Ghafoor, A. (2007). Web Services Discovery in Secure Collaboration Environments. *ACM Transactions on Internet Technology*, *8*(1). doi:10.1145/1294148.1294153

Shen, D., Yu, G., Yin, N., & Nie, T. (2004). Heterogeneity resolution based on ontology in web services composition. In *Proceedings of IEEE International Conference on E-Commerce Technology for Dynamic E-Business* (pp. 274-277).

Sheng, M. (2009). Ubiquitous RFID: Where are We? In *Information Systems Frontiers*. New York: Springer.

Shimonski, R. J., Schmied, W., Chang, V., & Shinder, T. W. (2002). *Building DMZs for Enterprise Networks*. Syngress.

Shitani, T., Ito, T., & Sycara, K. (2000) Multiple Negotiations among Agents for a Distributed Meeting Scheduler. In *Proceedings of the 4th International Conference on MultiAgent Systems* (pp. 435-436).

Shi, X. (2006). Sharing service semantics using SOAP-based and REST Web services. *IT Professional*, *8*(2), 18–24. doi:10.1109/MITP.2006.48

Shostack, L. (1985). Planning the service encounter. In Czepiel, J., Solomon, M., & Surprenant, C. (Eds.), *The Service Encounter* (pp. 243–254). Lexington, MA: Lexington Books.

Sikka, P., Corke, P., Valencia, P., Crossman, C., Swain, D., & Bishop-Hurley, G. (2006). Wireless ad-hoc sensor and actuator networks on the farm. In *Proceedings of the 5th Int. Conf. on Information Processing in Sensor Networks* (pp. 492-499). New York: ACM.

Sim, K. M., & Wang, S. Y. (2004). Flexible negotiation agent with relaxed decision rules. *IEEE Transactions on Systems, Man, and Cybernetics. Part B, Cybernetics*, *34*(3), 1602–1608. doi:10.1109/TSMCB.2004.825935

Simon, H. A. (1960). *The new science of Management Decisions*. New York: Harper & Row.

Sirin, E., Parsia, B., & Hendler, J. A. (2004). Filtering and selecting semantic web services with interactive composition techniques. *IEEE Intelligent Systems*, *19*(4), 42–49. doi:10.1109/MIS.2004.27

Sirin, E., Parsia, B., Wu, D., Hendler, J. A., & Nau, D. S. (2004). HTN planning for web service composition using SHOP2. *Journal of Web Semantics*, *1*(4), 377–396. doi:10.1016/j.websem.2004.06.005

Skene, J., Lamanna, D. D., & Emmerich, W. (2004). Precise Service Level Agreements. In *Proceedings of the 26th international conference on software engineering*, Edinburgh, UK.

Skyrme, D. (2000). *Knowledge Management Assessment: A Practical Tool*. London: David Skyrme Associates Limited.

Sloman, M. (1995). Management issues for distributed services. In N. Davies (Ed.), *Proceedings of SDNE'95* (pp. 52-59). Los Alamitos, USA: IEEE-CS Press.

Sloman, M. (1994). Policy driven management for distributed systems. *Journal of Network and Systems Management*, *2*(4), 333–360. doi:10.1007/BF02283186

Soboroff, I. (2004). *Overview of the TREC 2004 Novelty Track*. Paper presented at TREC 2004 - the 13th Text Retrieval Conference.

Sohraby, K., Minoli, D., & Znati, T. (2007). *Wireless sensor networks: Technology, protocols, and applications*. New York: John Wiley. doi:10.1002/047011276X

Song, S., Hwang, K., & Kwok, Y. (2005). Trusted Grid Computing with Security Binding and Trust Integration. *Journal of Grid Computing*, *3*(1/2), 53–73. doi:10.1007/s10723-005-5465-x

Song, W., Chen, D., & Chung, J.-Y. (2006). Web services: an approach to business integration models for micro-payment. *International Journal of Electronic Business*, *4*, 265–280. doi:10.1504/IJEB.2006.010866

Steffen, B., Thomas, W., & Kurt, G. (2006). An Ontology for Quality-Aware Service Discovery. *International Journal of Computer Systems Science & Engineering*, *21*(5).

Steller, L., & Krishnaswamy, S. (2009). Efficient mobile reasoning for pervasive discovery. In S. Y. Shin & S. Ossowski (Eds.), *Proceedings of the 2009 ACM Symposium on Applied Computing (SAC)*, Honolulu, HI (pp. 1247-1251). New York: ACM.

Stroulia, E., & Hatch, M. P. (2003). An Intelligent-agent Architecture for Flexible Service Integration on the Web. *IEEE Transactions on Systems, Man and Cybernetics. Part C*, *33*(4), 468–479.

Studer, R., Benjamins, V. R., & Fensel, D. (1998). Knowledge engineering: principles and Methods. *IEEE Transactions on Data and Knowledge Engineering*, *25*(1/2), 161–197.

Stufflebeam, W. H., Antón, A. I., He, Q., & Jain, N. (2004, October 28). Specifying privacy policies with P3P and EPAL: lessons learned. In *Proceedings of the 2004 ACM Workshop on Privacy in the Electronic Society*, Washinton, DC (pp. 35-35).

Super Quant Monte-Carlo Challenge - V Grid Plugtests. (2008). Retrieved October 1, 2009, from http://www-sop.inria.fr/oasis/plugtests2008/ProActiveMonteCarloPricingContest.html

Surjanto, B., Ritter, N., & Loeser, H. (2000). XML Content Management Based on Object-Relational Database Technology. In *Proceedings of the 1st International Conference on Web Information Systems Engineering*, Hong Kong, China (Vol. 1, pp. 70-79).

Surprenant, C., & Solomon, M. R. (1987). Predictability and personalisation in the service encounter. *Journal of Marketing*, *51*, 73–80. doi:10.2307/1251131

Sveiby, K.-E. (1997). *The New Organizational Wealth*. San Francisco, CA: Berrett-Koehler.

Sycara, K., & Zeng, D. (1996). Coordination of Multiple Intelligent Software Agents. *International Journal of Cooperative Information Systems*, *5*(2&3), 181–212. doi:10.1142/S0218843096000087

Tang, W., & Tsai, F. S. (2009). Threshold Setting and Performance Monitoring for Novel Text Mining. In *Proceedings in Applied Mathematics 3 Society for Industrial and Applied Mathematics - 9th SIAM International Conference on Data Mining 2009* (pp. 1310-1319).

Tang, W., Kwee, A. T., & Tsai, F. S. (2009). Accessing Contextual Information for Interactive Novelty Detection. In *Proceedings of the European Conference on Information Retrieval (ECIR) Workshop on Contextual Information Access, Seeking, and Retrieval Evaluation.*

Tang, W., Tsai, F. S., & Chen, L. (2010). *Blended Metrics for Novel Sentence Mining*. Expert Syst. Appl.

Tao, A., & Yang, J. (2007). Context Aware Differentiated Services Development With Configurable Business Processes. In *Proceedings of the IEEE international enterprise computing conference,* Annapolis, MD.

Tapia, D. I., Fraile, J. A., Rodrìguez, S., de Paz, J. F., & Bajo, J. (2009). Wireless sensor networks in home care. In *Proceedings of the 10th International Work-Conference on Artificial Neural Networks (IWANN '09)* (pp. 1106-1112). Berlin: Springer Verlag.

Teege, G. (1994). Making the difference: a subtraction operation for description logic. In *Proceedings of International Conference on Principles of Knowledge Representation and Reasoning (KR)* (pp. 540-550).

Teicher, J., Hughes, O., & Dow, N. (2002). E-government: a new route to public sector quality. *Managing Service Quality, 6*(12), 384–393. doi:10.1108/09604520210451867

Teich, J., Wallenius, H., & Wallenius, J. (1999). Multiple-Issue Auction and Market Algorithms for the World Wide Web. *Decision Support Systems, 26*(1), 49–66. doi:10.1016/S0167-9236(99)00016-0

TeleManagement Forum. (2004). *SLA Management Handbook, TeleManagement* (Tech. Rep. No. GB917, Version 2, Vol. 4). Berkshire, UK: The Open Group.

ter Beek, M. H., Gnesi, S., Mazzanti, F., & Moiso, C. (2006). Formal Modelling and Verification of an Asynchronous Extension of SOAP. In *Proceedings of the European Conference on Web Services* (pp. 287-296). Washington, DC: IEEE Computer Society.

The Open Grid Forum (OGF). (2006). *The Open Grid Services Architecture, Version 1.5.* Retrieved from http://www.ogf.org/documents/GFD.80.pdf

Tierney, K. J. (2007). Testimony on Needed Emergency Management Reforms. *Journal of Homeland Security and Emergency Management, 4*(3). Retrieved from http://www.bepress.com/jhsem/vol4/iss3/15. doi:10.2202/1547-7355.1388.

Tiitinen, P. (2003). User Roles in Document Analysis. In *Proceedings of CAISE'03 Forum Short Paper Proceedings* (pp. 205-208). University of Maribor Press.

Tiwana, A. (2001). *The Essential Guide to Knowledge Management: E-Business and CRM Applications.* New York, NY: Prentice Hall.

Toivonen, S., Kolari, J., & Laakko, T., USA. (2003). Facilitating Mobile Users with Contextualized Content. *Artificial Intelligence in Mobile System Workshop*, Seattle, WA, USA, October,

Toma, I., Foxvog, D., & Jaeger, M. C. (2006). Modeling QoS characteristics in WSMO. In *Proceedings of the 1st Workshop on Middleware for Service Oriented Computing MW4SOC 2006*, Melbourne, Australia (Vol. 184, pp. 42-47). New York: ACM.

Tosic, V. (2008). On modeling and maximizing business value for autonomic service-oriented systems. In D. Ardagna, M. Mecella, & J. Yang (Eds.), Proceedings of *BPM 2008 Workshops, LNBIP, Vol. 17* (pp. 407-418). Berlin, Germany: Springer.

Tosic, V., Erradi, A., & Maheshwari, P. (2007). WS-Policy4MASC - A WS-Policy extension used in the Manageable and Adaptable Service Compositions (MASC) middleware. In van der Aalst, W.M.P., L.-J. Zhang, & P.C.K. Hung (Eds.), *Proceedings of SCC 2007* (pp. 458-465). Los Alamitos, USA: IEEE-CS Press.

Tosic, V., Kruti, P., & Pagurek, B. (2002). WSOL – Web Service Offerings Language. In *Proceedings of WES 2002* (LNCS 2512, pp. 57-67). Berlin: Springer Verlag.

Tosic, V., Lutfiyya, H., Yazhe, T., Sherdil, K., & Dimitrijevic, A. (2005, September 28-30). On requirements for management of mobile XML Web services and a corresponding management system. In *Proceedings of the 7th International Conference on Telecommunications in Modern Satellite, Cable and Broadcasting Services* (Vol.1, pp. 57- 60).

Tosic, V., Suleiman, B., & Lutfiyya, H. (2007). UML profiles for WS-Policy4MASC as support for business value driven engineering and management of Web services and their compositions. In M. Spies & M.B. Blake (Eds.), *Proceedings of EDOC 2007* (pp. 157-168). Los Alamitos, USA: IEEE-CS Press.

Tosic, V., Ma, W., Pagurek, B., & Esfandiari, B. (2004). Web Service Offerings Infrastructure (WSOI) - A Management Infrastructure for XML Web Services. In *Proceedomgs of NOMS 2004, Seoul, Korea* (pp. 817–830). Washington, DC: IEEE.

Tosic, V., Pagurek, B., Patel, K., Esfandiari, B., & Ma, W. (2005). Management applications of the Web Service Offerings Language (WSOL). *Information Systems, 30*(7), 564–586. doi:10.1016/j.is.2004.11.005

Tsai, W. T., Xiao, B., Paul, R. A., & Chen, Y. (2006). *Consumer-centric service-oriented architecture: A new approach.* Paper presented at the Software Technologies for Future Embedded and Ubiquitous Systems, 2006 and the 2006 Second International Workshop on Collaborative Computing, Integration, and Assurance.

Tsai, F. S. (2009). Network intrusion detection using association rules. *International Journal of Recent Trends in Engineering, 2*(1), 202–204.

Tsai, F. S., & Chan, K. L. (2007). Detecting cyber security threats in weblogs using probabilistic models. *Intelligence and Security Informatics, 4430*, 46–57. doi:10.1007/978-3-540-71549-8_4

Tsai, F. S., & Chan, K. L. (2010). *Redundancy and novelty mining in the business blogosphere*. The Learning Organization.

Tsai, F. S., Chen, Y., & Chan, K. L. (2007). *Probabilistic techniques for corporate blog mining* (*. LNCS, 4819*, 35–44.

Tsai, F. S., Etoh, M., Xie, X., Lee, W.-C., & Yang, Q. (2010). Introduction to Mobile Information Retrieval. *IEEE Intelligent Systems, 25*(1), 11–15. doi:10.1109/MIS.2010.22

Tsai, F. S., Han, W., Xu, J., & Chua, H. C. (2009). Design and Development of a Mobile Peer-to-peer Social Networking Application. *Expert Systems with Applications, 36*(8), 11077–11087. doi:10.1016/j.eswa.2009.02.093

Tsai, F. S., Tang, W., & Chan, K. L. (2010). *Evaluation of Metrics for Sentence-level Novelty Mining*. Information Sciences.

Tsai, W. T., Zhou, X., Chen, Y., & Bai, X. (2008). On Testing and Evaluating Service-Oriented Software. *Computer, 41*(8), 40–46. doi:10.1109/MC.2008.304

Tsang, E. (1993). *Foundations of Constraint Satisfaction.* Academic Press.

Tucker, A. (1950). A two-person dilemma. Lecture at Stanford University, Palo Alto, California. Stanford University Press. In Poundstone, W. (Ed.), *Prisoner's Dilemma* (2nd ed.). New York: Anchor Books.

Tyrväinen, P., Salminen, A., & Päivärinta, T. (2003) Introduction to the Enterprise Content Management Minitrack. In *Proceedings of the 36th Hawaii International Conference on System Sciences*, Big Island, HI. Washington, DC: IEEE Computer Society Press.

UK Cabinet Office. (2004). *Civil Contingencies Act 2004: Emergency Preparedness.* Retrieved from http://www.ukresilience.info/preparedness/ccact/eppdfs.aspx

UK National Grid Service (NGS). (2010). *Homepage.* Retrieved from http://www.grid-support.ac.uk

University of Virginia. (2005). *WSRF.net*. Retrieved from http://www.cs.virginia.edu/~gsw2c/wsrf.net.html

Vaculín, R., & Sycara, K. (2007). Towards automatic mediation of OWL-S process models. In *Proceedings of IEEE International Conference on Web Services (ICWS)* (pp. 1032-1039).

Van der Mei, R. D., & Meeuwissen, H. B. (2006). *Modelling End-to-end Quality-of-Service for Transaction-Based Services in Multi-Domain Environments.* Paper presented at the IEEE International Conference on Web Services (ICWS'06), Hyatt Regency at 0' Hare Airport, Chicago.

van Lamsweerde, A., Darimont, R., & Massonet, P. (1995). Goal-Directed Elaboration of Requirements for a Meeting Scheduler: Problems & Lessons Learnt. In *Proceedings of the 2nd IEEE International Symposium on Requirements Engineering* (pp. 194-203).

Van Velsen, L., Van der Geest, T., Ter Hedde, M., & Derks, W. (2008). Engineering User Requirements for e-Government Services: A Dutch Case Study. In M. A. Wimmer, H. J. Scholl, & E. Fererro (Eds.), *Proceedings of EGOV 2008* (pp.243-254). Berlin: Springer Verlag.

Van Velsen, L., Van der Geest, T., Ter Hedde, M., & Derks, W. (2009). Requirements engineering for e-Government services: A citizen-centric approach and case study. *Government Information Quarterly, 3*(26), 477–486. doi:10.1016/j.giq.2009.02.007

Varga, A. (2001, June). The OMNeT++ Discrete Event Simulation System. In *Proceedings of the European Simulation Multiconference (ESM'2001)*.

Vetro, A., & Timmerer, C. (2005). Digital Item Adaptation: Overview of Standardization and Research Activities. *IEEE Transactions on Multimedia*, *7*(3), 418–426. doi:10.1109/TMM.2005.846795

Vukovic, M., & Robinson, P. (2004). Adaptive, Planning-based, Web Service Composition for Context Awareness. In *Proceedings of the European conference on web services*, Erfurt, Germany.

W3C Web Services Policy Working Group. (2007). *Web Services Policy 1.5 – Framework*. (W3C Recommendation 04 September 2007). Retrieved June 1, 2009, from http://www.w3.org/TR/ws-policy/

W3C. (2002). *Web Service Choreography Interface (WSCI) 1.0*. Retrieved from www.w3.org/TR/wsci/

W3C. (2004). "*OWL-S: Semantic Markup for Web Services*." from http://www.w3.org/Submission/OWL-S/.

W3C. "*Document Object Model (DOM)*." from http://www.w3.org/DOM/.

Waldburger, S., & Stiller, B. (2005). *Toward the Mobile Grid: Service Provisioning in a Mobile Dynamic Virtual Organization* (Tech. Rep. No. 2005.07). Zurich, Switzerland: University of Zurich

Waldburger, M. (2007). Grids in a Mobile World: Akogrimo's Network and Business Views. *Praxis der Informationsverarbeitung und Kommunikation*, *30*(1), 32–43. doi:.doi:10.1515/PIKO.2007.32

Walsh, J. P. (1995). Managerial and organizational cognition: notes from at rip down memory lane. *Organization Science*, *6*(3), 280–321. doi:10.1287/orsc.6.3.280

Wang, H., Xu, J., Li, P., & Hung, P. (2008). Incomplete preference-driven web services selection. In *Proceedings of IEEE International Conference on Service Computing (SCC)* (pp. 75-82).

Wang, J., et al. (2007). RFID Assisted Object Tracking for Automating Manufacturing Assembly Lines. In *Proceedings of the 2009 IEEE International Conference on e-Business Engineering (ICEBE 2007)*.

Wang, X., Lao, G., DeMartini, T., Reddy, H., Nguyen, M., & Valenzuela, E. (2002, November 22). XrML - eXtensible rights Markup Language. In *Proceedings of the 2002 ACM Workshop on XML Security*, Fairfax, VA (pp. 71-79).

Wang, X., Vitvar, T., Kerrigan, M., & Toma, I. (2006). A QoS-aware selection model for semantic web services. In *Proceedings of Service Oriented Computing (ICSOC 2006)* (LNCS 4294, pp. 390-401). New York: Springer.

Wang, M., Wang, H., Vogel, D., Kumar, K., & Chiu, D. K. W. (2009). Agent-Based Negotiation and Decision Making for Dynamic Supply Chain Formation. *Engineering Applications of Artificial Intelligence*, *4*(2), 36–56.

Wang, Y. S. (2008). Assessing e-commerce systems success: a respecification and validation of the DeLone and McLean model of IS success. *Information Systems Journal*, *18*, 529–557. doi:10.1111/j.1365-2575.2007.00268.x

Wang, Y., Kim, J.-G., Chang, S.-F., & Kim, H.-M. (2007). Utility-Based Video Adaptation for Universal Multimedia Access (UMA) & Content-Based Utility Function Prediction for Real-Time Video Transcoding. *IEEE Transactions on Multimedia*, *9*(2), 213–220. doi:10.1109/TMM.2006.886253

Web Services Agreement Specification. (2007, March). *The Open Grid Forum*. Retrieved from http://www.ogf.org/documents/GFD.107.pdf

Weerawarana, S., Curbera, F., Leymann, F., Storey, T., & Ferguson, D. F. (2005). *Web Services Platform Architecture: SOAP, WSDL, WS-Policy, WS-Addressing, WS-BPEL, WS-Reliable Messaging, and More*, Prentice Hall.

Wegner, L., Paul, M., Thamm, J., & Thelemann, S. (1996). A Visual Interface for Synchronous Collaboration and Negotiated Transactions. In *Proceeding of Workshop on Advanced Visual Interfaces* (pp. 156-165).

Weissman, J. B., Kim, S., & England, D. (2005). Supporting the Dynamic Grid Service Lifecycle. In *Proceedings of the IEEE/ACM CCGrid International Symposium on Cluster Computing and the Grid* (Vol. 2, pp. 808-815).

Weitzman, L., Dean, S. E., Meliksetian, D., Gupta, K., Zhou, N., & Wu, J. (2002). Transforming the content management process at IBM.com. In *Proceedings of the Conference on Human Factors and Computing Systems*, Minneapolis, MN (pp. 1-15).

Welch, V., et al. (2003, June). Security for Grid Services. In *Proceedings of the Twelfth International Symposium on High Performance Distributed Computing (HPDC-12)*. Washington, DC: IEEE Press.

Wenger, E., McDermott, R., & Snyder, W. (2002). *Cultivating communities of practice: a guide to managing knowledge*. Boston, MA: Harvard Business School Press.

Wesner, S., Dimitrakos, T., & Jeffery, K. (2004). Akogrimo - The Grid goes Mobile. *ERCIM news* (No. 59).

Weverka, P. (2001). *Mastering ICQ: The Official Guide*. Hungry Minds.

Weyuker, E. J. (1998). Testing Component-Based Software: A Cautionary Tale. *IEEE Software*, *15*(5), 54–59. doi:10.1109/52.714817

WhatIs.com. (2003). *UDDI*. Retrieved from http://searchsoa.techtarget.com/sDefinition/0,sid26gci508228,00.html

Wiig, K. M. (1993). *Knowledge Management Foundations – Thinking About Thinking – How People and Organizations Create, Represent, and use Knowledge*. Arlington, TX: Schema Press LTD.

Wiig, K. M. (1994). *Knowledge Management The Central Management Focus for Intelligent-Acting Organizations*. Arlington, TX: Schema Press LTD.

Wikipedia. (2009). *Web Service*. Retrieved from http://en.wikipedia.org/wiki/Webservice#Stylesofuse

Wilson, M., Arenas, A., & Schubert, L. (2007). *TrustCOM Framework Version 4*. Retrieved from http://www.eu-trustcom.com

Wolf, P., Steinebach, M., & Diener, K. (2007). Complementing DRM with digital watermarking: mark, search, retrieve. *Online Information Review*, *31*(1), 10–21. doi:10.1108/14684520710731001

Wombacher, A., Fankhauser, P., Neuhold, E., & Mahleko, B. (2004). Matchmaking for business processes based on choreographies. In *Proceedings of IEEE International Conference one-Technology, e-Commerce, and e-Science (EEE)* (pp. 359-368).

Wong, J. Y. Y., Chiu, D. K. W., & Mark, K. P. (2007). Effective e-Government Process Monitoring and Interoperation: A Case Study on the Removal of Unauthorized Building Works in Hong Kong. In *Proceedings of the 40th Hawaii International Conference on System Sciences*, Big Island, Hawaii, CDROM, IEEE Press.

World Wide Web Consortium. (W3C). (2002, April 16). *The Platform for Privacy Preferences 1.0 (P3P1.0) Specification*. Retrieved February 21, 2010, from http://www.w3.org/TR/P3P/

Wu, Z., & Palmer, M. (1994). Verbs semantics and lexical selection. In *Proceedings of the 32nd Annual Meeting on Association For Computational Linguistics* (pp 133-138). Las Cruces, NM: Association for Computational Linguistics

Wu, Z., Ranabahu, A., Gomadam, K., Sheth, A., & Miller, J. (2007). *Automatic composition of semantic web services using process and data mediation*. Technical Report, LSDIS Lab, University of Georgia.

Xarchos, C., & Charland, B. (2008). Innovapost uses Web 2.0 tools to engage its employees. *Strategic HR Review*, *7*(3), 13–18. doi:10.1108/14754390810865766

Xie, X., Liu, H., Ma, W.-Y., & Zhang, H.-J. (2006). Browsing Large Pictures Under Limited Display Sizes. *IEEE Transactions on Multimedia*, *8*(4), 707–715. doi:10.1109/TMM.2006.876294

Xu, J., & Reiff-Marganiec, S. (2008). Towards heuristic web services composition using immune algorithm. In *Proceedings of IEEE International Conference on Web Services (ICWS)* (pp. 238-245).

Yang, S. J. H., Tsai, J. J. P., & Chen, C. C. (2003). Fuzzy Rule Base Systems Verification Using High Level Petri Nets. *IEEE Transactions on Knowledge and Data Engineering*, *15*(2), 457–473. doi:10.1109/TKDE.2003.1185845

Yang, S. J. H., Zhang, J., & Chen, I. Y. L. (2008). A JESS enabled context elicitation system for providing context-aware Web services. *Expert Systems with Applications*, *34*(4), 2254–2266. doi:10.1016/j.eswa.2007.03.008

Yang, S. J. H., Zhang, J., Chen, R. C. S., & Shao, N. W. Y. (2007). A Unit of Information-Based Content Adaptation Method for Improving Web Content Accessibility in the Mobile Internet. *Electronics and Telecommunications Research Institute (ETRI). Journal*, *29*(6), 794–807.

Yang, S. J. H., Zhang, J., Tsai, J. J. P., & Huang, A. F. M. (2009). SOA-based Content Delivery Model for Mobile Internet Navigation. *International Journal of Artificial Intelligence Tools, 18*(1), 141–161. doi:10.1142/S0218213009000081

Yau, S., & Chen, F. (1980). An Approach in Concurrent Control Flow Checking. *IEEE Transactions on Software Engineering, 6*(2), 126–137. doi:10.1109/TSE.1980.234478

Ye, X., & Mounla, R. (2008). A hybrid approach to QoS-aware service composition. In *Proceedings of IEEE International Conference on Web Services (ICWS)* (pp. 62-69).

Yee, K. Y., Tiong, A. W., Tsai, F. S., & Kanagasabai, R. (2009). OntoMobiLe: A Generic Ontology-centric Service-Oriented Architecture for Mobile Learning. In *Proceedings of the Tenth International Conference on Mobile Data Management (MDM) Workshop on Mobile Media Retrieval (MMR)* (pp. 631-636).

Yellin, D., & Strom, R. (1997). Protocol specifications and component adaptors. *ACM Transactions on Programming Languages and Systems, 19*(2), 292–333. doi:10.1145/244795.244801

Yesil, M. (1997). *Creating the Virtual Store: Taking Your Web Site from Browsing to Buying*. New York: John Wiley.

Yoon, K. P., & Hwang, C.-L. (1995). *Multiple attribute decision making: an introduction*. Sage University Paper series on Quantitative Applications in the Social Sciences, 07-104, Thousand Oaks, CA: Sage.

Young-Tak, K. (2005). Inter-AS Session & Connection Management for QoS-guaranteed

Yu, T., & Lin, K.-J. (2005, December). Service selection algorithms for composing complex services with end-to-end QoS constraints. In *Proc. 3rd Int. Conference on Service Oriented Computing* (ICSOC2005), Amsterdam, The Netherlands.

Yu, T., Zhang, Y., & Lin, K.-J. (2007, May). Efficient algorithms for web services selection with end-to-end QoS constraints. *ACM Transactions on the Web.*

Yueh, Y. T. F., Chiu, D. K. W., Leung, H. F., & Hung, P. C. K. (2007). A Virtual Travel Agent System for M-Tourism with a Web Service Based Design and Implementation. In *Proceedings of the IEEE 21st International Conference on Advanced Information Networking and Applications*, Niagara Falls, Canada.

Yu, T., Zhang, Y., & Lin, K.-J. (2007). Efficient algorithms for web services selection with end-to-end QoS constraints. *ACM Transactions on Web, 1*(1), 1–26. doi:10.1145/1232722.1232723

Zack, M. (1999). Developing a Knowledge Strategy. *California Management Review, 41*(3), 125–145.

Zadeh, L. A. (1978). Fuzzy sets as a basis for a theory of possibility. *Fuzzy Sets and Systems, 1*, 3–28. doi:10.1016/0165-0114(78)90029-5

Zeithaml, V. A., Parasuraman, A., & Malhorta, A. (2002). Service Quality Delivery Through Web Sites: A Critical Review of Extant Knowledge. *Journal of the Academy of Marketing Science, 30*(4), 362–375. doi:10.1177/009207002236911

Zeleny, M. (1982). *Multiple Criteria Decision Making*. New York: McGraw-Hill.

Zeng, L., Benatallah, B., Dumas, M., Kalagnanam, J., & Sheng, Q. Z. (2003). Quality driven web services composition. In *Proceedings of International Conference on World Wide Web (WWW)* (pp. 411-421).

Zeng, L., Benatallah, B., Xie, G., & Lei, H. (2006). Semantic service mediation. In *Proceedings of International Conference on Service Oriented Computing (ICSOC)* (pp. 490-495).

Zeng, L., Benatallah, B., Ngu, A. H. H., Dumas, M., Kalagnanam, J., & Chang, H. (2004). QoS-Aware middleware for web services composition. *IEEE Transactions on Software Engineering, 30*(5), 311–327. doi:10.1109/TSE.2004.11

Zenha Rela, M., Madeira, H., & Silva, J. G. (1996). Experimental Evaluation of the Fail-Silent Behaviour in Programs with Consistency Checks. In *Proceedings of the Twenty-Sixth Annual international Symposium on Fault-Tolerant Computing* (pp. 394-403). Washington, DC: IEEE Computer Society.

Zhai, C., Jansen, P., Bai, S., Stoica, E., Grot, N., & Evans, D. A. (1999). Threshold calibration in CLARIT adaptive filtering. In *Proceedings of Seventh Text Retrieval Conference TREC-7* (pp. 149-156).

Zhang, H.-P., Xu, H.-B., Bai, S., Wang, B., & Cheng, X.-Q. (2004). Experiments in TREC 2004 Novelty Track at CAS-ICT. In *Proceedings of TREC 2004 - the 13th Text Retrieval Conference.*

Zhang, T., et al. (2008). Developing a Trusted System for Tracking Asset on the Move. In *Proceedings of the 9th International Conference for Young Computer Scientists* (pp. 2008-2013).

Zhang, T., Luo, Z., Wong, E. C., Tan, C. J., & Zhou, F. (2008). Mobile Intelligence for Delay Tolerant Logistics and Supply Chain Management. In *Proceedings of the 2008 IEEE International Conference on Sensor Networks, Ubiquitous, and Trustworthy Computing* (pp. 280-284).

Zhang, Y., & Tsai, F. S. (2009a). Chinese Novelty Mining. In *Proceedings of the Conference on Empirical Methods in Natural Language Processing (EMNLP '09)* (pp. 1561-1570).

Zhang, Y., & Tsai, F. S. (2009b). Combining Named Entities and Tags for Novel Sentence Detection. In *Proceeding of the WSDM '09 Workshop on Exploiting Semantic Annotations in Information Retrieval (ESAIR '09)* (pp. 30-34).

Zhang, Y., Gill, C., & Lu, C. (2008). Reconfigurable real-time middleware for distributed cyber-physical systems with aperiodic events. *International Conference on Distributed Computing Systems* (pp.581-588).

Zhang, Y., Kwee, A. T., & Tsai, F. S. (2010). Multilingual Sentence Categorization and Novelty Mining. *Information Processing and Management: an International Journal.*

Zhang, J., Zhang, L.-J., Quek, F., & Chung, J.-Y. (2005). A Service-Oriented Multimedia Componentization Model. *International Journal of Web Services Research, 2*(1), 54–76.

Zhang, Y., & Callan, J. (2001). Maximum Likelihood Estimation for Filtering Thresholds. In. *Proceedings of SIGIR, 2001*, 294–302.

Zhang, Y., Callan, J., & Minka, T. (2002). Novelty and Redundancy Detection in Adaptive Filtering. In. *Proceedings of SIGIR, 2003*, 81–88.

Zhang, Y., Lin, K.-J., & Hsu, J. Y. (2007). Accountability monitoring and reasoning in service-oriented architectures. *Journal of Service-Oriented Computing and Applications, 1*(1).

Zhang, Y., & Tsai, F. S. (2010). *D2S: Document-to-Sentence Framework for Novelty Detection.* Tech. Rep.

Zhang, Z., Tan, Y., & Dey, D. (2009). Price competition with service level guarantee in web services. *Decision Support Systems, 47*(2), 93–104. doi:10.1016/j.dss.2009.01.004

Zhao, L., Zheng, M., & Ma, S. (2006). The Nature of Novelty Detection. *Information Retrieval, 9*, 537–541. doi:10.1007/s10791-006-9000-x

Zheng, G., & Bouguettaya, A. (2009). Service Mining on the Web. *IEEE Transactions on Service Computing, 2*(1), 65–78. doi:10.1109/TSC.2009.2

Zhou, Z., et al. (2007). Interconnected RFID Reader Collision Model and its Application in Reader Anti-collision. In *Proceedings of the IEEE RFID Conference*, TX.

Zhou, C., Chia, L.-T., & Lee, B.-S. (2004). QoS-aware and federated enhancement for UDDI. *International Journal of Web Services Research, 1*(2), 58–85.

Zhou, C., Chia, L.-T., & Lee, B.-S. (2005). Web services discovery with DAML-QoS ontology. *International Journal of Web Services Research, 2*(2), 43–66.

Ziegler, C. N., & Lausen, G. (2005). Propagation models for trust and distrust in social networks. *Information Systems Frontiers, 7*(4/5), 337–358. doi:10.1007/s10796-005-4807-3

Zorzi, M., & Rao, R. (2003). Geographic random forwarding (geraf) for ad hoc and sensor networks: multihop performance. *IEEE Transactions on Mobile Computing, 2*(4), 337–348. doi:10.1109/TMC.2003.1255648

About the Contributors

Dickson K.W. Chiu received the B.Sc. (Hons.) degree in Computer Studies from the University of Hong Kong in 1987. He received the M.Sc. (1994) and the Ph.D. (2000) degrees in Computer Science from the Hong Kong University of Science and Technology (HKUST). He started his own computer company while studying part-time. He has also taught at several universities in Hong Kong. His research interest is in service computing with a cross-disciplinary approach, involving workflows, software engineering, information technologies, agents, information system management, security, and databases. The results have been widely published in over 150 papers in international journals and conference proceedings (most of them have been indexed by SCI, SCI-E, EI, and SSCI), including many practical master and undergraduate project results. He received a best paper award in the 37th Hawaii International Conference on System Sciences in 2004. He is the founding Editor-in-Chief of the *International Journal on Systems and Service-Oriented Engineering* and serves in the editorial boards of several international journals. He co-founded several international workshops and co-edited several journal special issues. He also served in the program committees for over 100 international conferences and workshops. He is a Senior Member of both the ACM and the IEEE, and a life member of the Hong Kong Computer Society.

* * *

Djamel Belaid is currently Professor in the Computer Science Department at Institut Telecom. He is working in the area of distributed computing systems. His interests include middleware for pervasive environments, Service Oriented Architecture (SOA), service composition, functional and non-functional service monitoring, and adaptation of context aware service based applications. He has worked in a number of French and European collaborative researches funded projects involving collaboration with other academic institutions, research laboratories, large companies and SMEs.

Nik Bessis is currently a principal lecturer (Associate Professor) in the Department of Computer Science and Technology at the University of Bedfordshire (UK). His research interest is the analysis, research, and delivery of user-led developments with regard to data integration, annotation, and data push in distributed environments. These have a particular focus on the use of next generation and grid technologies for the benefit of various virtual organisational settings. He is involved in and leading a number of funded projects in these areas. Dr. Bessis has published numerous papers and articles in international conferences and journals, served as a reviewer, program committee member, session chair, associate editor, conference chair, editor of three books and the Editor-in-Chief of the International Journal of Distributed Systems and Technologies.

Khouloud Boukadi is an associate professor in Computer Science in the Multimedia, InfoRmation systems & Advanced Computing Laboratory –Miracl (Faculty of Economics and Management of Sfax - Tunisia). Her research interests include service computing, Web services, context-aware computing and agility of Information systems. She got a Ph.D. in computer science from Ecole des Mines, Saint Etienne, France.

Antony Brown has been a member of staff at the University of Bedfordshire for 8 years, lecturing in computer science with a speciality in system modelling, databases and internet technologies. He was awarded his PhD by the University of Bedfordshire in 2008 for his thesis on 'Multilevel modelling and prediction of melanoma incidence'. Most recently his research interest has been the focused on the mathematics of trust within Grid structures and VOs. He also runs a web development company specialising in advising and developing academic projects.

Tarak Chaari is a computer science Associate Professor at the Higher Institute of Electronics and Communication of Sfax in Tunisia where he teaches distributed information systems modeling and development. He obtained a PHD in 2007 and a M.S. (DEA) degree in 2003 from the National Institute of Applied Sciences of Lyon (France). He received his Engineering diploma in 2002 from the National Engineering School of Sfax (Tunisia). His research interests focus on semantic integration in distributed systems and adaptation in context–aware architectures. Dr Tarak Chaari is an IEEE and an ACM professional member since 2008. He has more than 20 regular papers in international conferences and journals. More details are available on his home page: http://www.redcad.org/members/tarak.chaari/.

Kap Luk Chan obtained his PhD degree in Robot Vision from Imperial College of Science, Technology and Medicine, University of London, U.K. in 1991. He is now an associate professor in the School of Electrical and Electronic Engineering, Nanyang Technological University, Singapore. His research interests are in image analysis and computer vision, particularly in texture analysis, statistical image analysis, perceptual grouping, image and video retrieval, application of machine learning in computer vision, computer vision for human computer interaction, and biomedical signal and image analysis. He is also a member of the IEEE and IET.

Irene Y.L. Chen received her PhD degree in management of information systems from National Kaohsiung First University of Science and Technology, Taiwan. She is now an assistant professor of the Department of Accounting at National Changhua University of Education, Taiwan. Her current research interests include e-commerce, online retailing, online shopping behavior and virtual community. She has published journal articles in Journal of Information Science, Educational Technology & Society, International Journal of Human-Computer Studies, Computers & Education.

Winston Cheng is a software engineer by training and has been in this profession since 1995. His key research interests are in knowledge management, enterprise resource planning and information technology project management. He received his MSc in Knowledge Management from the National Technological University in 2009 and his BCS from the National University of Singapore in 1995. Winston is currently a user project manager with the Singapore government managing the civil service-wide Human Resource Management System Project.

Mary Cheung graduated from University of Washington, Seattle, WA, majoring in Business Management Systems, Marketing and International Business. Ms CHEUNG has been working in sales and marketing positions in various industries including I.T, electronics, retail, market research, etc. Ms CHEUNG is currently working in E-Business Solutions Limited, a wholly owned subsidiary of The University of Hong Kong, as Senior Manager of Strategic Sales Group, focusing in RFID business. She contributed to several major RFID commercial projects in Hong Kong, including RFID Library Management System for Main Library of the University of Hong Kong, RFID Baggage Monitoring System for Airport Authority of Hong Kong, and RFID Mortuary Management System for Hospital Authority.

Shing-Chi (S.C.) Cheung is an Associate Professor of Computer Science and Engineering at the Hong Kong University of Science and Technology. Professor Cheung has participated actively in the program and organizing committees of major international conferences on software engineering and services computing. He is serving on the editorial board of the IEEE Transactions on Software Engineering (TSE), the Journal of Computer Science and Technology (JCST), the International Journal of RF Technologies: Research and Applications, and the International Journal on Systems and Service-Oriented Engineering (IJSSOE). He is an executive committee member of the ACM SIGSOFT. His research interests include software engineering, services computing, ubiquitous computing, and embedded software engineering. His work has been reported by more than 100 publications at international journals and conferences, which include TOSEM, TSE, ASE, DSS, TR, ICSE, FSE, ESEC and ICDCS. He is a Chartered Fellow of the British Computer Society, a senior member of the IEEE, and a registered professional engineer of Hong Kong. He co-found the first International Workshop on Services Engineering (SEIW) in 2005, the first International Workshop on Automation of Software Testing (AST) in 2006, and was the tutorials chair of ICSE 2006. He has co-edited special issues for the Computer Journal, the Journal of Service-Oriented Computing and Applications, and the International Journal of Web Services Research (JWSR). He is the HKUST representative co-founding the RFID Benchmarking Test Consortium in China and the Global RF Lab Alliance.

Domenico Cotroneo is Assistant Professor at DIS, the Department of Computer and Systems Engineering of Federico II University of Naples, from 2004. His current scientific interests are in the following areas: i) dependability aspects of distributed and middleware-based systems; ii) middleware technologies; and iii) dependability issues of complex software systems. Domenico Cotroneo has been involved with University of Naples in several international and national projects. Furthermore, he has served as Program Committee member or as Local Chair in a number of scientific conferences, related to his research topics.

Gianpaolo Cugola received his Dr.Eng. degree in Electronic Engineering from Politecnico di Milano. In 1998 he received the Prize for Engineering and Technology from the Dimitri N. Chorafas Foundation for his Ph.D. thesis on Software Development Environments. He is currently Associate Professor at Politecnico di Milano where he teaches several courses in the area of Computer Science. He has been involved in several projects financed by the EU commission (IDEAS-ERC-227977 SMSCom, IST-034963 WASP, IST-511556 POMPEI, IST-11400 MOTION, ESPRIT-34840 PIE, ESSI-21244 MIDAS), by Microsoft Research ("Network Aware Programming" and "Virtual Campus"), and by the Italian governor (PRIN D-ASAP, FIRB Insyeme, CNR IS_MANET). He is co-author of tens of scientific papers published in international journals and conference proceedings. His research interests are in the area of Software Engineering and Distributed Systems. In particular, his current research focuses on middleware technology for largely distributed and highly reconfigurable distributed applications.

Masako Dohi is an Associate Professor of the Otsuma Women's University and an Invited Research Scientist of the Digital Human Research Center at the National Institute of Advanced Industrial Science and Technology of Japan. She received her Ph.D degree from the Sinshu University in 2007. Her research interests are: relation between 3D human body shape and industrial products design (clothes, shoes etc.); physiological and psychological measurement and evaluation concerning shoe comfort. Dr. Dohi has received a presentation award from the Japan Society of Kansei Engineering in 2008. She is a member the Japan Ergonomic Society, the Japan Society of Kansei Engineering, the Japan Society of Home Economics, International Federation for Home Economics, and the Japan Research Association for Textile End-Uses.

Khalil Drira received the Engineering and M.S. (DEA) degrees in Computer Science from ENSEEIHT (INP Toulouse), in June and September 1988 respectively. He obtained the Ph.D. and HDR degrees in Computer Science from UPS, University Paul Sabatier Toulouse, in October 1992, and January 2005 respectively. Since 1992, he is "Chargé de Recherche"a full-time research position at the French National Center for Scientific Research (CNRS). His research interests include formal design, implementation, testing and provisioning of distributed communicating systems and cooperative networked services. His research activity addresses different topics in this field focusing on model-based analysis and design of correctness properties including testability, robustness, adaptability and reconfiguration. He is involved in several national and international projects in the field of distributed and concurrent communicating systems. He is author of more than 150 regular and invited papers in international conferences and journals. More details are available on his wiki: http://www.laas.fr/~khalil/wiki

Kaouthar Fakhfakh is a PHD student attached to the Tools and Software for Communication (OLC) research team in the LAAS Laboratory and the Research Unit on Development and Control of Distributed Applications (ReDCAD). Her research interests focus on automating service level agreements semantic management. She holds a master degree in information systems from INSA Lyon on Web service composition using business protocols. She teaches Web information systems programming in the National School of Engineers (ENIS) in the University of Sfax (Tunisia). She is a student member in the Association for Computing Machinery (ACM) and the Institute of Electric and Electronic Engineering (IEEE) since 2008. More information can be found at: http://www.redcad.org/members/kaouthar.fakhfakh

Kwok Fai So is currently Chair Professor and Head of the Department of Anatomy, Jessie Ho Professor in Neuroscience at *The University of Hong Kong* (http://www.hku.hk/anatomy/staff/KFso.html), a member of the *Chinese Academy of Sciences*, Co-Chairman of the Board of Director of the *China Spinal Cord Injury Network (*www.chinascinet.org,*)*, and Co-Director of the *State Key laboratory of Brain and Cognitive Sciences at HKU (*www.hku.hk/fmri/). He received his Ph.D. degree from MIT. He is one of the pioneers in the field of axonal regeneration in visual system. He was the first to show lengthy regeneration of retinal ganglion cells in adult mammals with peripheral nerve graft. He is currently using multiple approaches to promote axonal regeneration in the optic nerve and spinal cord. His team identifies neuroprotective and regenerative factors including: peptide nanofiber scaffold, trophic factors, herbal extracts, other small molecules, immune responses and environmental manipulation. He is the author and co-author of over 230+ publications.

Ralph Feenstra is a PhD candidate at the Faculty of Technology, Policy and Management of Delft University of Technology. He holds an MSc in systems engineering, policy analysis and management. During his studies he obtained experience in Group Support Systems at the Council for Scientific and Industrial Research (CSIR), in Pretoria, South Africa. During his masters project he was involved in the design of a quick scan method for the rapid evaluation of road traffic networks at TNO (Netherlands Organization for Applied Scientific Research). His PhD research is focused on service composition methods for multi-actor networks. During his time as a PhD candidate he was involved in diverse Service Oriented Architecture related research and design projects for the Dutch police force, ABN Amro, Ernst & Young and Capgemini. He also assisted four master students with the completion of their thesis. He has a strong interest in the area of service compositions, collaborative service design and service reuse. He has his own company and provides consultancy services for online applications. This company is also involved in research activities supported by NL Agency, a department of the Dutch Ministry of Economic Affairs.

Tim French PhD MA FCollP MBCS is a Senior Lecturer in Computing at Bedfordshire University. He has recently completed a part-time PhD in E-Trust at Reading University, School of Systems Engineering. He has an active professional and personal interest in cross-cultural localisation. He has worked freelance for usability consultancies for "blue chip".

Julian Gallop has an MA in Mathematics and a Diploma in Computer Science from Cambridge University. He recently retired from the Rutherford Appleton Laboratory (RAL), the largest of the Laboratories of the UK STFC (Science and Facilities Research Council), where he was a Principal Scientist and where he continues as a Visitor. He was leader of groups and teams at RAL, coordinator of national programmes in the UK and team leader of EU project work packages. He has publications in computer graphics, visualization, computer animation, product data standards, data mining, distributed computer graphics architecture and service-oriented grids.

Maria Grammatikou (Ph.D.) is a Senior Researcher and Coordinator of the Network Management and Optimal Design Laboratory (NETMODE) of NTUA of Greece. She works as a teaching assistant in NTUA. Her research interests include topics on management and optimal design Quality of Service of computer communication networks, Security and Intrusion Detection Systems, Electronic commerce and web-based technologies, parallel/distributed systems and communication systems evaluation. She has a PhD in Computer Science from the NTUA of Greece. She has an extensive experience in software systems design & development and she has participated in many European projects. She is a member of the WG7/TC48 Working Group.

Naijie Gu is currently a professor and Ph.D. supervisor at the University of Science and Technology of China. His research interests include parallel algorithm, parallel architecture, and distributed computing.

Shuihua Han is a professor from school of management, Xiamen University. Prof Han is specialised in information system and risk/security management, esp in supply chain management field. His recent interest is related to the security/privacy issues using RFID track and trace in the healthcare.

Haiyang Hu received the BS, MS, and PhD degree from Nanjing University, Nanjing, China, in 2000, 2003 and 2006, respectively. He is currently an Associate Professor in the College of Computer Science and Information Engineering, Zhejiang Gongshang University, Hangzhou, China. His current research interests include mobile computing, and software engineering. He has published about 20 research papers in international journals and conferences. Dr. Hu has served as the PC members in conferences and workshops (NASAC'08, MBC'09, and WCMT'09). He is a senior member of CCF. He serves in the editorial board of the International Journal on Systems and Service-Oriented Engineering.

Hua Hu received the BS, MS, and PhD degrees from Zhejiang University, Hangzhou, China, in 1987, 1990, and 1998, respectively. He is currently a Vice President of the Hangzhou Dianzi University, Zhejiang, China. He was also a visiting professor at the State University of New York, USA. His current research interests include distributed computing, software agents, and workflow technology. He has published more than 60 research papers in international journals and conferences. Prof. Hu has served as the general chair in the MBC'09 and WCMT'09 international workshops. He is an Editorial Advisory Board member of the International Journal on Systems and Service-Oriented Engineering.

Liusheng Huang received the M.S. degree from the University of Science and Technology of China in 1988. He is currently a professor and Ph.D. supervisor at the University of Science and Technology of China. He has published 6 books and more than 80 papers. His research interests are in the areas of information security, distributed computing and high performance algorithms.

Patrick C. K. Hung is an Assistant Professor at the Faculty of Business and Information Technology in UOIT and an Adjunct Assistant Professor at the Department of Electrical and Computer Engineering in University of Waterloo. Patrick is currently collaborating with Boeing Phantom Works (Seattle, USA) and Bell Canada on security- and privacy-related research projects, and he has filed two US patent applications on "Mobile Network Dynamic Workflow Exception Handling System." In addition, Patrick is also cooperating on Web services composition research projects with Southeast University in China. Patrick has been serving as a panelist of the Small Business Innovation Research and Small Business Technology Transfer programs of the National Science Foundation (NSF) in the States since 2000. He is an executive committee member of the IEEE Computer Society's Technical Steering Committee for Services Computing, a steering member of EDOC "Enterprise Computing," and an associate editor/editorial board member/guest editor in several international journals such as the IEEE Transactions on Services Computing, International Journal of Web Services Research (JWSR), International journal of Business Process and Integration Management (IJBPIM), and the International Journal on Systems and Service-Oriented Engineering.

Tsukasa Ishigaki is a postdoctoral fellow of the National Institute of Advanced Industrial Science and Technology (AIST), Tokyo, Japan. He received B.E. and M.E. degrees from Hosei University, Tokyo, Japan in 2003 and 2005, respectively and his Ph.D. degree from the Graduate University for

Advanced Studies at the Institute of Statistical Mathematics (ISM) in 2007. From 2007 to 2008 he served as a postdoctoral fellow for Core Research for Evolutional Science and Technology, Japan Science and Technology Agency in ISM. His current interests are service engineering, machine learning, and information fusion methods, with particular emphasis on their application to real systems. He is a member the IEEE, the Society of Instruments and Control Engineers and the Institute of Electronics, Information and Communication Engineers.

Marijn Janssen is an Associate Professor within the Information and Communication Technology section and Director of the interdisciplinary SEPAM Master programme of the Faculty of Technology, Policy and Management at Delft University of Technology. He is also in charge of the "IT and business architecture" of the Toptech executive master. He has been a consultant for the Ministry of Justice and received a Ph.D. in information systems (2001). He serves on several editorial boards (including International Journal of E-Government Research, International Journal of E-Business Research Government Information Quarterly and Information Systems Frontiers), has many research projects and is involved in the organization of a number of conferences. His research interests are in the field of e-government, design science, orchestration, composition and shared services. He was ranked as one of the leading e-government researchers in 2009 and published over 180 refereed publications. More information: www.tbm.tudelft.nl/marijnj.

Mohamed Jmaiel obtained his diploma of engineer in Computer Science from Kiel (Germany) University in 1992 and his Ph.D. from the Technical University of Berlin in 1996. He joined the National School of Engineers of Sfax (Tunisia) as Assistant Professor of Computer Science in 1995. He became an Associate Professor in 1997 and full Professor in January 2009. He participated to the initiation of many graduate courses at the University of Sfax. His current research areas include software engineering of distributed systems, formal methods in model-driven architecture, component oriented development, self-adaptive and pervasive systems, autonomic middleware. He published more than 100 regular and invited papers in international conferences and journals, and has co-edited four conferences proceedings and three journals special issues on these subjects. More details are available on his home page: http://www.redcad.org/members/jmaiel/.

Ian Johnson is a researcher at the e-Science Centre of the Science and Technology Facilities Council in the UK. His research interests are Grid security and Virtual organisations.

Eleanna Kafeza is a lecturer at Athens University of Economics and Business. She received the PhD from Hong Kong University of Science and Technology, where she also held a visiting assistant lecturer position. Her research interests are in workflow management systems, legal issues in web contracting, web services and grid computing. She co-founded several international workshops, co-edited several special issues in journals, and has served in the program committee of many international conferences. She is an Associate Editor of International Journal on Systems and Service-Oriented Engineering and the International Journal of Organizational and Collective Intelligence.

Magdalini Kardara has obtained a Diploma in Electrical and Computer Engineering in 2004 from the National Technical University of Athens and an MSc in Advanced Computing from Imperial College

London in 2006. She is a PhD candidate in the department of Electrical and Computer Engineering of NTUA and is currently working as Research Associate and Software engineer in the Telecommunications Laboratory. She has a significant expertise on Web services and Grid technologies through participation in related EU and national projects. She is currently participating in the BEinGRID and GRIA projects.

Tom Kirkham is a researcher at Nottingham University on the EU sponsored TAS3 project. In 2007 after a few years of research in the application and development Virtual Organisations in eScience and SME environments he gained his PhD from the University of Wales Bangor. His specialist areas are in Grid Computing, security, Virtual Organisations and the integration of business legacy systems in SOA applications.

Kleopatra Konstanteli was born in Athens, Greece in 1981. She received her diploma in Electrical and Computer Engineering in 2004 by the National Technical University of Athens (NTUA). In 2007 she received an MBA by the NTUA in corporation with University of Athens and University of Piraeus. She is currently pursuing her doctoral-level research in computer science while at the same working as a research associate in the Telecommunications Laboratory of Electrical and Computer Engineering of NTUA and participating in EU funded projects. Her research interests are mainly focused in the field of service-oriented and real-time computing.

Makiko Kouchi is a Prime Senior Researcher in the Digital Human Research Center at the National Institute of Advanced Industrial Science and Technology of Japan. She studied physical anthropology and received her Doctoral degree in Science from the University Tokyo. Her research interests include variations in human body due to aging, secular change and movement, as well as modeling and analysis of 3D human body shape. Dr. Kouchi has received such awards as the Promotion prize from the Anthropological Science for a Significant Paper (1992), the AIST Director-General's Award (2000), and a Basic Research Award from the International Society of Biomechanics, 5th Footwear Symposium (2001). She is a member of the Anthropological Society of Nippon, the Japan Ergonomic Society, and the Japan Society of Biomechanisms.

Agus T. Kwee is a project officer staff in School of Computer Engineering, Nanyang Technological University, Singapore. He received his bachelor's degree in computing science from Simon Fraser University, Canada, in 2005. His current research interests include software engineering, text mining, and social networking.

Kevin K.S. Kwok received the M.Sc. degree in Computer Science from the Chinese University of Hong Kong in 2003. The case study of this paper is based on his degree project supervised by Dr. Dickson K.W. Chiu.

Chun Wai Martin Lai graduated from the University of Hong Kong. He is a software developer and project manager of system software and commercial application systems. His technical speciality includes RFID, software internationalization, operating systems, banking systems, Internet e-business solutions, network applications, databases and Java technologies. He is currently the Manager of Technology Support in the Hong Kong R&D Centre for Logistics and Supply Chain Management Enabling Technologies, where he actively working on RFID based research and development works.

Raymond Y. K. Lau is an assistant professor in the Department of Information Systems at the City University of Hong Kong. His research interests include information retrieval, data mining, agent-mediated e-commerce, and service-oriented computing. He holds a Ph.D. in Information Technology from Queensland University of Technology, a M.S. in Business Systems Analysis and Design from the City University (U.K.), and an M.AppSc. in Information Studies from Charles Sturt University. He has worked in the ICT industry for ten years before joining the academic field in Australia in 1998. He is the author of over 70 refereed international journals and conference papers. His research work has been published in renowned journals such as ACM Transactions on Information Systems, IEEE Transactions on Knowledge and Data Engineering, IEEE Internet Computing, Journal of Management Information Systems, Decision Support Systems, International Journal of Pattern Recognition and Artificial Intelligence, Electronic Commerce Research and Applications, etc. He is the technical editor of the Journal of Asian Information Management and Associate Editor of the International Journal on Systems and Service-Oriented Engineering. He also served as a program committee member for over 30 international conferences and workshops. He is a Senior Member of both the ACM and the IEEE, and a life member of the Hong Kong Computer Society.

Ho-fung Leung is currently a full Professor in Computer Science and Engineering, The Chinese University of Hong Kong. He has been active in research on intelligent agents, multi-agent systems, game theory, and semantic web, and has published close to 150 papers in these areas. Professor Leung has served on the program committee of many conferences. He is a member of the planning committee of PRIMA and a senior PC member of AAMAS'08. He is currently serving on the program committees of CEC'08 & EEE'08, EDOC 2008, ICMLC 2008, IDPT 2008, ISEAT 2008, M2AS'08, SAC 2008 (PL Track) and SLAECE 2008. He was the chairperson of ACM (Hong Kong Chapter) in 1998. Professor Leung is a Senior Member of ACM, a Senior Member of the IEEE, and a Chartered Fellow of the BCS. He is a Chartered Engineer registered by the ECUK and is awarded the designation of Chartered Scientist by the Science Council, UK. Professor Leung received his B.Sc. and M.Phil. degrees in Computer Science from The Chinese University of Hong Kong, and his Ph.D. degree in Computing from Imperial College of Science, Technology and Medicine, University of London. He is an Associate Editor of the International Journal on Systems and Service-Oriented Engineering.

Qing Li received the B.Eng. degree from Hunan University, Changsha, China, and the M.Sc. and Ph.D. degrees from the University of Southern California, Los Angeles, all in computer science. He is currently a Professor at the City University of Hong Kong, Kowloon, a Visiting Professor at Zhejiang University, Hangzhou, China, a Guest Professor at the University of Science and Technology of China, Hefei, and an Adjunct Professor at Hunan University. His research interests include database modeling, web services, multimedia retrieval and management, and e-learning systems. He has authored over 200 papers in technical journals and international conferences. He is actively involved in the research community by serving as a Journal Reviewer, Program Committed Chair/Co-Chair, and Organizer/Co-Organizer of several international conferences. Prof. Li serves as the Chairman of the Hong Kong Web Society, a Councilor of the Database Society of Chinese Computer Federation, and a Steering Committee Member of the International WISE Society. He serves in the International Advisory Board of the International Journal of Systems and Service-Oriented Engineering.

Baoping Lin is currently a PhD candidate both in the Computer Science Department of the University of Science and Technology of China and in the Computer Science Department of the City University of Hong Kong. His research interests are in the area of service-oriented computing.

Kwei-Jay Lin is a professor in the Department of Electrical Engineering and Computer Science at the University of California, Irvine. His research interests include service- oriented architectures, e-commerce technology, and real- time systems. Lin has a PhD in computer science from the University of Maryland at College Park. He is a senior member of the IEEE. Contact him at klin@uci.edu.

An Liu received the B.Eng. degree from Anhui Normal University, Wuhu, China, and the Ph.D. degree from the University of Science and Technology of China, Hefei, China, all in computer science. He is currently working at the University of Science and Technology of China. His research interests are in the area of service-oriented computing. He serves in the editorial board of the International Journal of Systems and Service-Oriented Engineering.

Hai Liu is currently a PhD candidate both in the Computer Science Department of the University of Science and Technology of China and in the Computer Science Department of the City University of Hong Kong. His research interests are in the area of service-oriented computing.

Zhongjun Luo had trainings in both computer science (M.Sc.) and life sciences including medicine (M.D.), biochemistry (Ph.D.), and genetics (Postdoctoral). Dr. Luo started his career in bioinformatics and information technology in 2001 when he was a research assistant professor in Vanderbilt University Medical Center. Late on, Dr. Luo has worked at National Cancer Institute Center for Bioinformatics on caBIG projects and at Department of Commerce on IT projects, SNAPR and IMSR. Dr. Luo, has recently accomplished FDA (the Food and Drug Administration) commissioner's Fellowship program and is now working as Regulatory Information Specialist at FDA of the United States.

Zongwei Luo is a senior researcher at the E-business Technology Institute, The University of Hong Kong (China). His research has been supported by various funding sources, including China NSF, HKU seed funding, HK RGC, and HK ITF. His research results have appeared in major international journals and leading conferences. He is the founding Editor-in-Chief of the International Journal of Applied Logistics and serves as an associate editor and editorial advisory board member in many international journals. Dr. Luo's recent interests include applied research and development in the area of service science and computing, innovation management and sustainable development, technology adoption and risk management, and e-business model and practices, especially for logistics and supply chain management.

Zakaria Maamar is a full professor in the College of Information Technology at Zayed University, Dubai, United Arab Emirates. His research interests include Web services, social networks, and context-aware computing. He has a PhD in computer science from Laval University, Quebec City, Canada.

Vasilis Maglaris was born in Athens in 1952 and is the founder and director of the Network Management and Optimal Design Laboratory at NTUA since 1989. He received the Diploma in Mechanical & Electrical Engineering from NTUA, Greece in 1974, the M.Sc. in Electrical Engineering from the

Polytechnic Institute of Brooklyn, New York in 1975 and the Ph.D. degree in Electrical Engineering & Computer Science from Columbia University, New York in 1979. In 1989 he joined the faculty of the Electrical & Computer Engineering Department of the NTUA, where he is Professor of Computer Science. He served as the Chairman of the Board of the Greek Research & Technology Network – GRNET, the National Research & Education Network (NREN) of Greece from its inception in September 1998 until June 2004. Since October 2004 he is the Chairman of the National Research & Education Networks Policy Committee (NREN PC).

Carsten Maple is Head of the Computer Science and Technology Department at the University of Bedfordshire, UK and a Professor of Applicable Computing. His research interests are in the area of applicable computing, information security, trust and authentication distributed systems, graph theory and omptimisation techniques. He is involved in and leading a number of funded research and commercial projects in these areas. Professor Maple has published numerous papers and articles in international conferences and journals, and he is the co-Editor of the International Journal of Applied Informatics. In addition, Professor Maple is a regular reviewer of several journals and conferences and has served as a guest speaker, an associate editor, a conference chair, a scientific program committee member, and a session chair in numerous international conferences.

Alessandro Margara graduated at Politecnico di Milano in 2008, obtaining a Laurea Specialistica (M.Sc. equivalent) in Computer Science Engineering with a final grade of 110/110. He is currently a 2nd year PhD student at DeepSE group, advised by prof. Gianpaolo Cugola. His research focuses on event-based middleware for large scale distributed systems.

Constantinos Marinos received the diploma in electrical engineering from the National Technical University of Athens, Greece, in 2002. He is currently a PhD candidate in the Department of Electrical and Computer Engineering at National Technical University of Athens (NTUA). He is also a member of the Network Management and Optimal Design Laboratory at NTUA where he has participated in a number of projects both national and European. . He has been involved in numerous European research projects (e.g. SEEGRID, HELLASGRID, ARGUGRID, GN2 / 3, EGEE, EFIPSANS) trying to resolve issues concerning the design of architectures, network management and security of systems. His main research interests lie in the areas of network management using middleware technologies.

Brian Matthews is leader of the Information Management Group of the e-Science Centre of the Science and Technology Facilities Council in the UK. He has been working in the area of using web systems to support scientific data management and in digital libraries for some eight years. His interests include Semantic Web and Metadata.

Masaaki Mochimaru is the Deputy Director of the Digital Human Research Center at the National Institute of Advanced Industrial Science and Technology of Japan. He received his Doctoral degree in Engineering from Keio University. His research interests are: human body shape measurement technologies, deformation and motion; footwear biomechanics (shoe design based on foot factors); digital human modeling for product assessment; and the application of ergonomics to product design (clothes, eyeglasses, gloves, computer mouse etc). Dr. Mochimaru has received many awards for his work. He is a member the IEEE Computer Society, SAE International, the International Ergonomics Association, the Japan Ergonomic Society, the Society of Biomechanisms Japan, the Society of Instruments and Control Engineers.

Yoichi Motomura is the team leader of Large Scale Data Based Modeling Research Team, Center for Service Research (CfSR) and the senior research scientist of Digital Human Research Center (DHRC) at National Institute of Advanced Industrial Science and Technology (AIST) in Japan. He received B.E, M.E and Ph.D from the University of Electro-Communications. He joined to the Real World Computing Project in the Electrotechnical Laboratory, AIST, MITI (the ministry of international trade and industry, Japan) in 1993. His current research works are statistical learning theory, probabilistic inference algorithms on Bayesian networks, and their applications to user modeling and human behavior understanding and prediction from sensory data. He also conducts the venture company, Modellize Inc., as the chief technology officer. He received a best presentation award, research promotive award from Japanese Society of Artificial Intelligence, Docomo Mobile Science award and IPA super creator award.

Weiran Nie is a PhD candidate in Electrical Engineering and Computer Science Department of University of California at Irvine. He is interested in the fields of distributed systems and QoS management in the context of service-oriented computing. Email him at wnie@uci.edu.

Kensuke Naoe, a Ph.D. candidate, graduated from the Faculty of Environment and Information Studies of Keio University in 2002. He received the Master degree from the Graduate School of Media and Governance at Keio University in 2004. His major is artificial neural network and information security. His interested research areas are digital information hiding, machine learning, network intrusion detection and malware detection.

Sietse Overbeek is an Assistant Professor in the field of conceptual modelling of (web) information systems and service-oriented computing within the Information and Communication Technology section of the Faculty of Technology, Policy and Management at Delft University of Technology. Sietse has (co-)authored several journal papers, conference publications, and book chapters and is a member of multiple international program committees related to conferences on (service-oriented) systems analysis and design. His main research interests include (web) information systems, conceptual modelling (including meta modelling and ontology modelling), service-oriented and system architecting, formal methods, and matchmaking mechanisms. Sietse received his Master's degree from the Radboud University Nijmegen, The Netherlands in April 2005, and received his Ph.D. from the same university in April 2009. The subject of his Ph.D. thesis was the bridging of supply and demand for knowledge intensive tasks based on knowledge, cognition, and quality. For more information, see: www.sietseoverbeek.nl.

Mark Panahi is a PhD candidate in the department of Electrical Engineering and Computer Science at the University of California, Irvine. His research interests include real-time performance for service-oriented infrastructures. Contact him at mpanahi@gmail.com.

Symeon Papavassiliou (Ph.D.) is Associate Professor of Electrical and Computer Engineering at NTUA. Before joining NTUA he was an Associate Professor at the New Jersey Institute of Technology

(NJIT), USA, while from 1995 till 1999 he was a Senior Technical Staff Member at AT&T Laboratories in New Jersey. Dr. Papavassiliou was awarded the Best Paper Award in INFOCOM'94, the AT&T Division Recognition and Achievement Award in 1997, and the National Science Foundation (NSF) Career Award in 2003. He was the Director of the Broadband, Mobile and Wireless Networking Laboratory (2000-2004), and the Associate Director of New Jersey Center for Wireless Networking and Internet Security (2002-2004). From 08/2006 to 08/2009 he served on the Board of the Greek National Regulatory Authority on Telecommunications and Posts. His main research interests lie in the areas of mobile wireless networks, sensor networks, ad-hoc networks, autonomic networks, and stochastic system and protocol performance evaluation.

Antonio Pecchia is a PhD student at the Department of Computer and Systems Engineering (DIS) at the Federico II University of Naples, where he received the BS and MS degrees in computer engineering in 2005 and 2008, respectively. His main research interests are in the area of dependability evaluation of complex distributed systems. His recent work focuses on system modelling and the design of novel logging infrastructures suitable for critical applications.

Roberto Pietrantuono received the BS and the MS degrees in computer engineering in 2003 and 2006, respectively, from the Federico II University of Naples, Italy. He received the PhD degree in computer and automation engineering from the same university, at the Department of Computer and Systems Engineering (DIS). His main research interests are in the area of software engineering, particularly in the software verification of critical systems and software reliability modelling and evaluation. He is a member of the IEEE.

Vassiliki Pouli is a research assistant at the Computer Science Division of the Electrical & Computer Engineering Department of NTUA (NETMODE laboratory). She received the diploma in electrical and computer engineering at the National Technical University of Athens in 2005. Since 2005, she has been a teaching assistant in the course "Network Management and Intelligent Networks" of the last semester of the Department of Electrical and Computer Engineering, National Technical University of Athens. She has been involved in numerous European research projects (e.g. GRIDCC, HELLASGRID, ARGUGRID, GN2 / 3, EGEE) conducting research to resolve various issues concerning the design and development of architectures, ensuring quality of service applications and security of systems. Her main research interests are in networking, security, service level agreements, trust and negotiation of agents.

Stefano Russo, PhD, IEEE Member, was born in Naples, Italy, in 1963. He is Professor and Deputy Head at the Department of Computer and Systems Engineering (DIS) of the Federico II University of Naples, Italy, where he leads the Distributed and Mobile Computing Group. He graduated in Electronic Engineering in 1988, and became Ph.D. in 1993; he was Assistant Professor at DIS from 1994 to 1998, and Associate Professor from 1998 to 2002. He is Chairman of the Curriculum in Computer Engineering, and Director of the National Laboratory of CINI (National Inter-universities Consortium for Informatics) in Naples. He teaches the courses of Advanced Programming and Distributed Systems. His current scientific interests are in the following areas: distributed software engineering; dependable systems; middleware technologies; mobile computing.

Paul Sant completed a BSc (Hons) in Computing and Information Systems from the University of Liverpool in 1999, and obtained his PhD in Computer Science from Kings College London in 2003. His main research interests are in application of graph theory to practical problems, and in the general area of algorithm design. Since 2006 he has been a Senior Lecturer in Computer Science, and his most recent research has been to develop graph theoretic models for trust within the computer security arena.

Hideyasu Sasaki, Ph.D. & Esq. is the Editor-in-Chief of the International Journal of Organizational and Collective Intelligence (IJOCI). Professor Sasaki is a graduate of the University of Tokyo, B.A. and LL.B., in 1992, 1994, received an LL.M., from the University of Chicago Law School in 1999, an M.S. and a Ph.D. in Cybernetic Knowledge Engineering with honors from Keio University in 2001, 2003, respectively. He is an associate professor at Department of Information Science and Engineering, Ritsumeikan University, Kyoto, Japan. He was an assistant professor at Keio University from 2003 to 2005. His research interests include decision science and intelligence computing, especially mathematical modeling of decision making under time constraint. Dr. Sasaki has also experienced lawyering and litigations as Attorney-at-Law in N.Y., U.S.A. He is an associate editor at the International Journal of Systems and Service-Oriented Engineering (IJSSOE) and a reviewer for the Journal of Information Sciences, Elsevier, in 2008, 2009 and ACM Transactions on Knowledge Discovery from Data (KDD) in 2008. He is active in program committees of the ACM International Conference on Management of Emergent Digital EcoSystems (MEDES), ICIW/SLAECE, SOMK, SoCPaR, ICADL, LAoIS, etc.

Ravi S. Sharma is Associate Professor at the Wee Kim Wee School of Communication and Information at the Nanyang Technological University. Ravi had spent the previous 10 years in industry as Asean Communications Industry Principal at IBM Global Services and Director of the Multimedia Competency Centre of Deutsche Telekom Asia. Ravi's teaching, consulting and research interests are in strategies for the knowledge and digital economy. His work has appeared in leading journals, conferences, trade publications and the broadcast media. He also sits on the boards of several professional journals and non-profit organizations. Ravi received his PhD in engineering from the University of Waterloo.

Yoshiyasu Takefuji (Ph.D.) is a tenured professor at the Faculty of Environment and Information Studies of Keio University, since April 1992 and was on tenured faculty of Electrical Engineering at Case Western Reserve University, since 1988. Before joining Case, he taught at the University of South Florida and the University of South Carolina. He received his BS (1978), MS (1980), and Ph.D. (1983) from Electrical Engineering from Keio University. His research interests focus on neural computing, security, internet gadgets, and nonlinear behaviors. He received the National Science Foundation/ Research Initiation Award in 1989, the distinct service award from IEEE Trans. on Neural Networks in 1992, and has been an NSF advisory panelist. He has received the best paper award from AIRTC in 1998 and a special research award from the US air force office of scientific research in 2003. He is currently an associate editor of International Journal of Multimedia Tools and Applications, editor of International Journal on Computational Intelligence and Applications, and editor of International Journal of Knowledge-based intelligent engineering systems. He was an editor of the Journal of Neural Network Computing, an associate editor of IEEE Trans. on Neural Networks, Neural/parallel/scientific computations, and Neurocomputing. He has published more than 120 journal papers and more than 100 conference papers.

Dwight Tan is an Engineer by training and has been in this profession since 2003. His key research interests are in knowledge management strategies, business intelligence and organizational learning. He received his MSc in Knowledge Management from the National Technological University in 2009 and his BEng from the University of Glasgow in 2002. Dwight is a Quality Engineer in an international manufacturing firm and has also spent a year and a half as a General Education Officer.

Wenyin Tang is a research fellow in School of Electrical & Electronic Engineering, Nanyang Technological University, Singapore. She received B.E. degree from School of Electrical Engineering, Shanghai Jiao Tong University, China, in 2002, and received the PhD degree in computer science from Nanyang Technological University, Singapore, in 2008. Her research interests cover machine learning, text mining, and bioinformatics.

Samir Tata is currently a Professor in TELECOM SudParis, France. His current research area includes business process management, service oriented computing and their applications in virtual enterprises. He is the responsible of the SIMBAD research group (http://www-inf.it-sudparis.eu/SIMBAD). He was chair of several international workshops. He was/is member of the steering or the program committee of several international conferences.

Saïd Tazi is an Associate Professor at the 'Université Toulouse1 Capitole', France, and permanent Research Senior at LAAS-CNRS laboratory. His scientific project is about the comprehension of cognitive processes of creation, communication, co-operation and interaction in order to develop models and software to sustain and improve these processes. Among these cognitive processes, Dr S. Tazi has been interested particularly by the externalization of such processes as a mean to represent knowledge of self describing and autonomous artifacts. The study of the adaptability and modeling of the adaptation of content as well as its processing constitute a challenge he wants to defeat. The developed models are investigated in the framework of projects that aim at help people while they are designing and learning artifacts. His current technological interests are related to Semantic Web services, E-services (E-Learning, E-Health, E-Commerce...), Distributed Systems and Ubiquitous computing.

George L. Tipoe's first degree is a Bachelor of Arts (BA). He proceeded to take a postgraduate degree in Medicine (Doctor of Medicine). After his MD degree, he continued his training in Cardiovascular Radiology specializing in Interventional Radiology. He went to clinical practice for 3 years and joined the Department of Anatomy on July 1, 1988. He took his academic post as a Lecturer in this department on Feb. 1, 1991. Dr. Tipoe finished his PhD in Anatomy at this Department (HKU) in 1995 and was promoted to Associate Professor in 1996. In the first ten years of his career, his research focused on Cancer and Cell Biology. He has published about 50 papers in this field. He shifted his research into antioxidant, inflammation, oxidative stress and animal models in liver diseases in the next ten years. He has published about 45 papers in this field. Overall, Dr. Tipoe has published about 95 papers and 5 book chapters. His teaching portfolio is centered on topographic human anatomy and has been teaching in this area for almost 22 years. He is actively involved in medical education research since 1998 and has published 5 papers in this field. He is currently the Associate Editor of the Anatomical Science Education, Mayo Clinic, USA, and the Assistant Dean (Academic Networking and Student Affairs, LKS Faculty of Medicine, HKU).

Vladimir Tosic (vladat@computer.org; http://www.nicta.com.au/people/tosicv) is a Researcher at NICTA in Sydney, Australia; a Visiting Fellow at the University of New South Wales, Sydney, Australia; and an Adjunct Research Professor at the University of Western Ontario, London, Canada. He previously held several positions in industry and academia in Europe, Canada, and Australia. He completed his Ph.D. degree at Carleton University, Ottawa, Canada. Dr. Tosic received many academic awards, including the 2001 Upsilon Pi Epsilon / IEEE Computer Society Award for Academic Excellence and several Natural Sciences and Engineering Research Council of Canada (NSERC) awards and grants. Most of his peer-reviewed papers were in the area of management of service-oriented systems and his current research is focused on autonomic business-driven management. Additionally, he presented several conference tutorials about these topics, co-organized several international workshops, and serves as an editorial board or program committee member of many international journals, conferences, and workshops.

Flora S. Tsai is currently with the School of Electrical & Electronic Engineering, Nanyang Technological University, Singapore. She is a graduate of MIT and Columbia University with degrees in Electrical Engineering and Computer Science. Her current research focuses on data and text mining, information retrieval, service-oriented computing, and software engineering.

Theodora A. Varvarigou received the B. Tech degree from the National Technical University of Athens, Athens, Greece in 1988, the MS degrees in Electrical Engineering (1989) and in Computer Science (1991) from Stanford University, Stanford, California in 1989 and the Ph.D. degree from Stanford University as well in 1991. She worked at AT&T Bell Labs, Holmdel, New Jersey between 1991 and 1995. Between 1995 and 1997 she worked as an Assistant Professor at the Technical University of Crete, Chania, Greece. Since 1997 she was elected as an Assistant Professor while since 2007 she is a Professor at the National Technical University of Athens, and Director of the Postgraduate Course "Engineering Economics Systems".

Minhong Wang is an Assistant Professor of Information &Technology Studies at The University of Hong Kong. Her current research interests include business process management, workflow systems, knowledge management, information systems, and e-learning. She has published papers in *Information & Management, Expert Systems with Applications, Knowledge-based Systems, Engineering Applications of Artificial Intelligence, Journal of Educational Technology and Society, International Journal of Intelligent Information Technologies, International Journal of Internet & Enterprise Management, International Journal of Technology and Human Interaction*, and presented papers at international conferences, including CAiSE, BPM, HICSS, AMCIS, ICEIS, PRICAI, CEC/EEE, ICELW among others. She is the Editor-in-Chief of *Knowledge Management & E-Learning: an International Journal (KM&EL)*, and *The Handbook of Research on Complex Dynamic Process Management: Techniques for Adaptability in Turbulent Environments*. She is the Guest Editor of *International Journal of Internet & Enterprise Management*. She is an Associate Editor of the International Journal on Systems and Service-Oriented Engineering. She also serves on the Editorial Board for a number of international journals.

Patrick Wong received his B.Sc. training in Applied Medical Science at The University of Hong Kong / Napier University, Edinburgh. Then, he registered as a Medical Laboratory Technologist Part I in Supplementary Medical Professions Council of Hong Kong. Mr. Wong joined Department of Orthopaedics and Traumatology, HKU in 1986 as Technician. He received a Continuing Professional Grants by HKU to pursue his training in undecalcified tissue at AO Institute at Davos, Switzerland. In 2001, he received the first prize of the student Best Poster Presentation Award at European Cells & Materials Conference at Davos, Switzerland. He was the person-in-charge of histology, animal surgery laboratories as well as Bone Bank at Queen Mary Hospital. He has more than twenty years research experience in animals and bone substitutes. In 2008, Mr. Wong joined Department of Anatomy as Senior Technical Officer. He mainly focuses in database management, embalming and teaching materials.

Tianle Zhang's recent interests include applied research and development in the area of service science and computing, wireless network, network modeling, intelligent transportation system. He is now work in Services Sciences and Intelligent Transportation Research Center, School of Computer Science and Technology, Beijing University of Posts and Telecommunications, Beijing, China.

Wenping Zhang holds a Bachelor of Engineering degree from University of Science and Technology of China. He is now a research assistant at the Department of Information Systems of the City University of Hong Kong. He also worked as a research assistant for various computing projects at the Hong Kong Polytechnic University before. His research interests include information retrieval, opinion mining, and service-oriented computing.

Yi Zhuang received the CCF Doctoral Dissertation Award conferred by Chinese Computer Federation in 2008 and IBM Ph.D Fellowship 2007-2008, participating in the study of an optimal hybrid storage model based on DB2 as a research intern in IBM China Research Lab. His research interests include database systems, index techniques, parallel computing and multimedia retrieval and indexing. He is currently an Associate Professor at the College of Computer & Information Engineering in Zhejiang Gongshang University. He obtained his PhD degree in computer science from Zhejiang University in 2008. Dr. Zhuang is currently a member of ACM, a member of IEEE, and senior member of CCF. Dr. Zhuang has served as PC co-chairs in the MBC'09 and WCMT'09 international workshops as well as PC members and reviewers for some top technical journals and leading international conferences such as TMM, TKDE, ACM MM'08, ICDE'08,'09, MMM'05,'06, WAIM'05,'08. Dr. Zhuang has published more than 20 papers including IEEE International Conference on Distributed Computing Systems (ICDCS'08), ACM Transactions on Asian Language Information Processing (TALIP), Journal of Computer Science and Technology (JCST), Science in China (E: Information Science), etc. Dr. Zhuang has also co-authored two books and received 3 patents. He is an Associate Editor of the International Journal on Systems and Service-Oriented Engineering.

Index

E

F

G

H

I

K

L

M

N

O

P

Q

R

S

T

U

V

W

X